Name	Symbol	Atomic Number		
Neodymium	Nd	60		
Neon	Ne	10		
Neptunium	Np	93		
Nickel	Ni	28		
Niobium	Nb	41		
Nitrogen	N	7	14.00674(7)	c,d
Nobelium	No	102	(259)	a
Osmium	Os	76	190.2(1)	c
Oxygen	O	8	15.9994(3)	c,d
Palladium	Pd	46	106.42(1)	c
Phosphorus	P	15	30.973762(4)	
Platinum	Pt	78	195.08(3)	
Plutonium	Pu	94	(244)	a
Polonium	Po	84	(209)	a
Potassium	K	19	39.0983(1)	
Praseodymium	Pr	59	140.90765(3)	
Promethium	Pm	61	(145)	a
Protactinium	Pa	91	231.03588(2)	a
Radium	Ra	88	226.0254	a,b
Radon	Rn	86	(222)	a
Rhenium	Re	75	186.207(1)	
Rhodium	Rh	45	102.90550(3)	
Rubidium	Rb	37	85.4678(3)	c
Ruthenium	Ru	44	101.07(2)	c
Samarium	Sm	62	150.36(3)	c
Scandium	Sc	21	44.955910(9)	
Selenium	Se	34	78.96(3)	
Silicon	Si	14	28.0855(3)	d
Silver	Ag	47	107.8682(2)	c
Sodium	Na	11	22.989768(6)	
Strontium	Sr	38	87.62(1)	c,d
Sulfur	S	16	32.066(6)	c,d
Tantalum	Ta	73	180.9479(1)	
Technetium	Tc	43	(98)	a
Tellurium	Te	52	127.60(3)	c
Terbium	Tb	65	158.92534(3)	
Thallium	Tl	81	204.3833(2)	
Thorium	Th	90	232.0381(1)	a,c
Thulium	Tm	69	168.93421(3)	
Tin	Sn	50	118.710(7)	c
Titanium	Ti	22	47.88(3)	
Tungsten	W	74	183.85(3)	
Unnilhexium	Unh	106	(263)	a
Unnilpentium	Unp	105	(262)	a
Unnilquadium	Unq	104	(261)	a
Unnilseptium	Uns	107	(262)	a
Uranium	U	92	238.0289(1)	a,c,e
Vanadium	V	23	50.9415(1)	
Xenon	Xe	54	131.29(2)	c,e
Ytterbium	Yb	70	173.04(3)	c
Yttrium	Y	39	88.90585(2)	
Zinc	Zn	30	65.39(2)	
Zirconium	Zr	40	91.224(2)	c

[a] Element has no stable nuclides. However, three such elements (Th, Pa, and U) do have a characteristic terrestrial isotopic composition, and for these an atomic weight is tabulated.

[b] Element for which the value A_r is that of the radioisotope of longest half-life.

[c] Geological specimens are known in which the element has an isotopic composition outside the limits for normal material. The difference between the atomic weight of the element in such specimens and that given in the table may exceed the implied uncertainty.

[d] Range in isotopic composition of normal terrestrial material prevents a more precise $A_r(E)$ being given; the tabulated $A_r(E)$ value should be applicable to any normal material.

[e] Modified isotopic compositions may be found in commercially available material because it has been subjected to an undisclosed or inadvertent isotopic separation. Substantial deviations in atomic weight of the element from that given in the table can occur.

Reagent Chemicals

Reagent Chemicals

NINTH EDITION

American Chemical Society Specifications

Official from January 1, 2000

American Chemical Society
New York   Oxford
OXFORD UNIVERSITY PRESS
2000

Oxford University Press

Oxford New York
Athens Auckland Bangkok Bogotá Buenos Aires Calcutta
Cape Town Chennai Dar es Salaam Delhi Florence Hong Kong Istanbul
Karachi Kuala Lumpur Madrid Melbourne Mexico City Mumbai
Nairobi Paris São Paulo Singapore Taipei Tokyo Toronto Warsaw

and associated companies in
Berlin Ibadan

Copyright © 2000 by the American Chemical Society

Developed and distributed in partnership by the
American Chemical Society and Oxford University Press

Published by Oxford University Press, Inc.
198 Madison Avenue, New York, New York 10016

Oxford is a registered trademark of Oxford University Press

Library of Congress Cataloging-in-Publication Data

American Chemical Society.
 Reagent chemicals: American Chemical Society specifications, official
from January 1, 2000. —9th ed.
  p.   cm.
 Includes index.
 ISBN 0–8412–3671–2
 1. Chemical tests and reagents. I. Title.
QD77.A54
543′.01—dc21   99-38870
                                                        CIP

1 3 5 7 9 8 6 4 2

Printed in the United States of America
on acid–free paper

Contents

DEFINITIONS, PROCEDURES, STANDARDS, AND SPECIFICATIONS

SPECIFICATIONS FOR
REAGENT CHEMICALS

SPECIFICATIONS FOR
STANDARD-GRADE REFERENCE MATERIALS

INDEX

Committee on Analytical Reagents
1993–1999

Paul A. Bouis, 1987–
Chair, 1992–

Stephen J. Arpie	1997–	Lyle H. Phifer	1997–
Michael Bolgar	1997–	Paul H. Piscia	1999–
Kishor D. Desai	1981–	Joseph M. Rao	1991–
Christine M. Foster	1998–	Michael A. Re	1997–
Benjamin D. Halpern	1988–1993	Nancy S. Simon	1991–
Norman C. Jamieson	1982–1998	Michael Spratt	1993–
Richard S. Juvet	1986–1996	Vernon A. Stenger	1962–
Richard A. Link	1991–1996	*Chair*	1967–1973
Clarence Lowery	1974–	*Consultant*	1992–
Chair	1985–1991	William A. Telliard	1998–
Robert W. Marshman	1997–	Samuel M. Tuthill	1958–
Edward F. Martz	1997–	*Chair*	1974–1980
Loren C. McBride	1983–	*Consultant*	1992–
Rajendra V. Mehta	1992–	Thomas Tyner	1998–
John R. Moody	1986–	Ruth H. Wantz	1991–1996
Thyagaraja Parasaran	1996–	Charles M. Wilson	1988–

William E. Schmidt, 1967–
Secretary, 1967–

vii

Notice to Readers

As new information becomes available, the Committee on Analytical Reagents updates the specifications in this book and occasionally adds specifications for new reagents. All updates are available to the reader free of charge on the World Wide Web at the following URL:

http://pubs.acs.org/reagents

Preface

The American Chemical Society (ACS) Committee on Analytical Reagents sets the specifications for most chemicals used in analytical testing. Currently, ACS is the only organization in the world that sets requirements and develops validated methods for determining the purity of reagent chemicals. These specifications have also become the de facto standards for chemicals used in many high-purity applications. Publications and organizations that set specifications or promulgate analytical testing methods—such as the *United States Pharmacopeia* and the U.S. Environmental Protection Agency—specify that ACS reagent-grade purity be used in their test procedures.

The ACS Committee on Analytical Reagents evolved from the Committee on the Purity of Chemical Reagents, which was established in 1903. Analysts at that time were disturbed by the quality of reagents available and by the discrepancies between labels and the actual purity of the materials. The Committee's role in resolving these issues expanded rapidly after its 1921 publication of specifications for ammonium hydroxide and for hydrochloric, nitric, and sulfuric acids. Specifications appeared initially in *Industrial and Engineering Chemistry* and later in its *Analytical Edition*. In 1941, the existing specifications were reprinted in a single pamphlet. Revisions and new specifications were later gathered into a book, the 1950 edition of *Reagent Chemicals*. Two editions per decade followed, except during the 1970s, when only one edition was published.

The word *specification*, as used in this book, means the individual entry for each reagent chemical. The specification has two components, the *requirements* (usually expressed in numerical terms) and the *tests* (instructions for the analytical methods to be used in determining conformity to the requirements). The choice of nomenclature, abbreviations, units, and editing policies has been generally guided by current ACS practices. Structural formulas are shown for most organic chemicals. Atomic weights have been calculated using the 1989 IUPAC Table of Standard Atomic Weights.

The commonplace introduction of instrumentation into analytical laboratories, beginning in the late 1950s, resulted in dramatic improvements in the sensitivity and accuracy of analytical measurements. In line with these improved instruments, the requirements for reagent chemicals and the tests used to measure their purity needed to be improved. Under the capable leadership of Stanley Clabaugh and, later, Vernon A. Stenger, Samuel M. Tuthill, and Wallace Rohrbough, these improvements were made during the 1960s and 1970s. The first instrumental technique, flame emission spectroscopy, was introduced into the compendium in 1961. It was followed by other appropriate techniques in an effort to develop test methods that would be as accurate and cost-effective as possible.

The Eighth Edition, which became official in 1993, substantially changed and updated the general procedures and attempted to make the book easier to read. Format changes included a new layout for requirements and a continuation of the trend toward detailed general methods in the front of the book, with only reagent-specific test conditions under each individual chemical. Updated methods included gas chromatography (in which capillary columns were first used), water determination (in which coulometric methods were added), and the replacement of flame emission techniques by atomic absorption for metal determinations. Sections on determining method detection limits and on the preparation and standardization of volumetric solutions were added. In addition, the Eighth Edition saw the elimination of boiling point and density requirements and the replacement of the test for substances not precipitated by ammonium sulfide by specific metals determination using atomic absorption.

The changes in this volume, the Ninth Edition, continue the trend toward eliminating or simplifying some of the tedious classical procedures for trace analysis and adding instrumental methods where possible. Classical procedures such as gravimetry and titrimetry, however, continue to be the cornerstone of the assay procedures for inorganic reagents because of their inherent higher precision and accuracy for the determination of major components. New analytical techniques introduced in this edition include mass, infrared, and plasma emission spectroscopy, as well as ion and thin-layer chromatography. For the first time, reagents suitable for use in ultratrace analysis, such as sub-boiled acids packaged in Teflon containers, are specified; this has required that "clean room" analytical practices be covered and accounts for the inclusion of more sophisticated analytical techniques, such as plasma emission spectroscopy. The most significant change in this edition is the addition of standard-grade reference materials, which was done at the request of the Chemical Reference Materials Manufacturers Association, with the support and encouragement of the National Institute of Standards and Technology and the U.S. Environmental Protection Agency. Approximately 400 standards are specified in a separate section following the traditional reagent chemicals.

As new information becomes available, the Committee updates the specifications in this book and occasionally adds specifications for new reagents. All updates are available to the reader free of charge on the World Wide Web at the following URL: http://pubs.acs.org/reagents.

The membership of the Committee was increased substantially during the preparation of this edition in order to provide the specific expertise needed for the new section on standard-grade reference materials. Michael Re serves as subcommittee chair for the Environmental Standards group. Charles Wilson leads the new reagents subcommittee, William Schmidt steers the trace metals subcommittee, and Norman Jamieson and Michael Spratt direct the organics subcommittee. Special acknowledgment goes to Barbara K. Barr, Lawrence Becker, Thomas J. Bzik, Nick Csikai, Joseph Gammell, William Hahn, William Hehemanic, Ron Lacock, Edward F. Martz, Scott May, Belvard (Skip) Prichard, Thomas D. Schware, and Preston D. Wright for their valuable contributions in developing and validating some of the analytical procedures that have greatly improved the quality of the test methods. Special thanks also go to Bennie L. Jones, Betsy Kulamer, and Susan Kurasz for their diligence and attention to detail in the preparation of the manuscript.

The Committee has formal written operating procedures that describe and govern its operations. Interested parties may obtain a copy of these procedures by addressing their requests to:

Secretary, ACS Committee on Analytical Reagents
c/o Books Department
American Chemical Society
1155 16th Street, NW
Washington, DC 20036

The Committee urges that any errors observed be reported, invites constructive criticism, and welcomes suggestions, particularly for new reagents and improved test methods. Organizations wishing to adopt the ACS specifications for their own purposes are encouraged to do so by requesting permission. In this way, it is hoped that worldwide harmonization of reagent chemical specifications might occur. Communications on these subjects should be sent to the secretary at this address. Anyone interested in serving on the committee should also contact the secretary.

PAUL A. BOUIS
Chair, ACS Committee on Analytical Reagents

Definitions, Procedures, Standards, and Specifications

Introduction and Definitions

The specifications prepared by the Committee on Analytical Reagents of the American Chemical Society are intended to serve for reagents and standard-grade reference materials to be used in precise analytical work of a general nature. Standard-grade reference materials are suitable for preparation of analytical standards used for a variety of applications, including instrument calibration, quality control, analyte identification, method performance, and other applications requiring high-purity materials. (For the sake of brevity, standard-grade reference materials may be referred to as *standards* throughout this book.) It is recognized that there may be special uses for which reagents and standard-grade reference materials conforming to other, or more rigorous, specifications may be needed. Therefore, where known and where feasible, some of the specifications herein include requirements and tests for certain specialized uses. However, it is impossible to include specifications for all such uses, and thus there may be occasions when it will be necessary for the analyst to further purify reagents known to have special purity requirements for certain uses.

The American Chemical Society has adopted the practice of expressing the concentration of chemical solutions in terms of mol/L (moles per liter). However, the concentrations of volumetric solutions used in analytical chemistry usually are expressed as *normality*, N, (the number of gram equivalent weights in each liter of solution) or sometimes as *molarity*, M, (the number of gram molecular weights in each liter of solution). In this book, it has been the practice to express such concentrations as either normality or molarity. Two examples showing equivalent concentrations by these two practices are: "…1 N sodium hydroxide (1 mol/L)" and "…1 N sulfuric acid (0.5 mol/L)". However, as can be seen, using mol/L destroys the concept of equivalents, as they are applied in volumetric analysis. Further, it is very unwieldy to write a calculation formula where the concentrations of the volumetric solutions are expressed in mol/L. Therefore, the use of normality and molarity are retained in this edition of *Reagent Chemicals,* which makes its practice consistent with that of other chemical testing societies, such as the American Society for Testing and Materials (ASTM); the *United States Pharmacopeia*; and the Institute of Medicine's *Food Chemicals Codex.*

The requirements and the details of tests are based on published work, on the experience of members of the Committee in the examination of reagent chemicals and standards on the market, and on studies of the tests made by members of the Committee. The limits and procedures are designed for application to reagents and standards in freshly opened containers. Reagents and standards in containers of extended age, in containers subject to constant changes in humidity or headspace gas content (as by repetitive opening and closing of the container), or in containers subjected to potential inadvertent contamination by repeated opening of the container may not conform to the designated requirements. Where the possibility of change due to age, humidity, light, or headspace contamination is recognized, the specification usually contains a warning; nonetheless, the analyst is cautioned to take appropriate steps to ensure the continued purity of the reagents and standards, especially after opening the container.

In determining quality levels to be defined by new or revised specifications, the Committee is guided by the following general principles. When a specification is first prepared, it will usually be based on the highest level of purity (of the reagent or standard to which it applies) that is competitively available in the United States. Generally, the term "competitively available" is understood to mean that the material is available from two or more suppliers. If a significantly higher level of purity subsequently becomes available on the same competitive basis, the specification generally will be revised accordingly. There may be cases where a material is available from only one producer. This does not preclude it from becoming an ACS reagent or standard, if a suitable specification can be prepared. If the reagent or standard later becomes available on a competitive basis, it will be appropriate to review the specification for possible revision.

Because the requirements of a specification relating to the content of designated impurities must necessarily be expressed in terms of maximum permissible limits, products conforming to the specification will normally contain less than the maximum permissible proportion of some or all of these impurities. A given preparation of a reagent chemical or standard that has less than the maximum content of one or more impurities permitted by the specification, therefore, is not considered as of higher quality than that defined by the specification.

A lower permissible limit for a given impurity will be adopted only if it is significantly different from the one it is intended to supersede. In general, a new requirement for an impurity whose content is not greater than 0.01% will not be considered significantly different unless it decreases the maximum permissible content of the impurity by at least 50%. This principle will also be approximated in the revision of those requirements defined by the term, "Passes test".

Tests as written are considered to be applicable only to the accompanying requirements. Modification of a requirement, especially if the change is toward a higher level of purity, will necessitate reconsideration, and often revision, of the test to ensure its validity.

The assays and tests described herein constitute the methods upon which the ACS specifications for reagent chemicals and standards are based. The analyst is not prevented, however, from applying alternative methods of analysis. Such methods shall be validated to assure that they produce results of at least equal reliability. The validation protocols used are left to the discretion of the testing laboratory, so long as it has a documented standard operating procedure in place. In the absence of such a procedure, the reader is referred to a well-established protocol that would clearly meet the needs of any ACS reagent requirements and tests (such as "Validation of Compendial Methods" in *United States Pharmacopeia*, 1999, pp 2149–2152). In the event of doubt or disagreement concerning a substance purported to comply with the ACS specifications, only the methods described herein are applicable.

REAGENTS FOR SPECIAL PURPOSES

For some reagents, the 6th and 7th editions of *Reagent Chemicals* had separate specifications defining them, such as "suitable for use in ultraviolet spectrophotometry", "suitable for use in determining pesticide residues", or "suitable for use in high-performance liquid chromatography". Beginning with the 8th edition, these special-use reagent chemicals have been treated in a single integrated presentation. The seller shall designate in product labeling the suitability for one or more of these special uses on the basis of the relevant requirements and tests.

DISCLAIMER

The reagent chemicals and standards included herein may be hazardous substances, and the use of such reagent chemicals and the application of the various test methods may involve hazardous substances, operations, and equipment. The American Chemical Society (ACS) and the ACS Committee on Analytical Reagents do not purport in this book or in any other publication to specify minimum legal standards or to address all of the risks and safety problems associated with reagent chemicals and standards, their use, or the methods prescribed for testing them.

No warranty, guarantee, or representation is made by ACS or the ACS Committee on Analytical Reagents as to the accuracy or sufficiency of the information contained herein, and ACS and the ACS Committee on Analytical Reagents assume no liability or responsibility in connection therewith. It is the responsibility of whoever uses the reagent chemicals and/or the testing methods set forth in this book to establish appropriate safety and health practices and to determine the applicability of any regulatory standards and/or limitations. Users of this book should consult and comply with pertinent local, state, and federal laws and should consult legal counsel if there are any questions or concerns about the applicable laws, safety issues, and reagent chemicals or the testing methods set forth herein.

CONTAINERS

The *container* is the device that holds the reagent or standard; it is, or may be, in direct contact with the reagent or standard. The *closure* is part of the container.

The container in which a reagent or standard is sold and/or stored must be suitable for its intended purpose and should not interact physically or chemically with the contained reagent or standard so as to alter its quality (within a reasonable period of time or when mandated by an expiration date) beyond the requirements of the specification.

Containers for solids normally have wide mouths to facilitate both the filling of the container and the removal of the contents. Containers for liquids normally have narrow mouths so the contents may be easily poured into other, frequently smaller, containers.

Prior to being filled, the container should be free from extraneous particulate matter, otherwise clean, and dry as necessary.

INTERPRETATION OF REQUIREMENTS

The requirements of reagent chemicals and standards can be divided into two main classes: an assay or quantitative determination of the principal or active constituent and the determination of the impurities or minor constituents. The requirements of standard-grade reference materials are divided into identity and assay sections. In some cases, physical properties are specified.

Method Detection Limits

The purity of reagent chemicals and standards continues to improve, driven by customer demand and the evolution of manufacturing processes and analytical technology. A clear and practical definition of analytical detection limits is vital to the accurate and precise determination of purity. An excellent review and definition of detection limits has been published. The following subsections, "Introduction" and "Determining Detection Limits", are adapted from APHA, 1998, pp 1–20. This reference provides other details, including the use of control charts to measure and monitor the accuracy and precision of any analytical procedure.

In case the limit of detection of trace metal impurities in ultratrace reagents needs to be determined, a more relevant method has been published and is recommended. This method is covered in the section on trace and ultratrace elemental analysis on page 35.

Introduction. Detection limits are controversial, principally because of inadequate definition and confusion of terms. Frequently, the instrumental detection limit is used for the method detection limit and vice versa. Whatever term is used, most analysts agree that the smallest amount that can be detected above noise in a procedure and within a stated confidence limit is the detection limit. The confidence limits are set so that probabilities of both Type I and Type II errors (see below) are acceptably small.

Current practice identifies several detection limits, each of which has a defined purpose. These are instrument detection limit (IDL), lower limit of detection (LLD), method detection limit (MDL), and limit of quantitation (LOQ). Occasionally the instrument detection limit is used as a guide for determining the MDL. The relationship among these limits is approximately IDL:LLD:MDL:LOQ = 1:2:4:10.

Determining Detection Limits. An operating analytical instrument usually produces a signal (noise) even when no sample is present or when a blank is being analyzed. Because any quality assurance (QA) program requires frequent analysis of blanks, the mean and standard deviation become well known; the blank signal becomes very precise, i.e., the Gaussian curve of the blank distribution becomes very narrow. The IDL is the constituent concentration that produces a signal noise greater than three standard deviations of the mean noise level or that can be determined by injecting a standard to produce a signal that is five times the signal-to-noise ratio. The IDL is useful for estimating the constituent concentration or amount in an extract needed to produce a signal to permit calculating an estimated method detection limit.

The LLD is the amount of constituent that produces a signal sufficiently large that 99% of the trials with that amount will produce a detectable signal. Determine the LLD by multiple injections of a standard at near-zero concentration (concentration no greater than five times the IDL). Determine the standard deviation by the usual method. To reduce the probability of a Type I error (false detection) to 5%, multiply s by 1.645 from a cumulative normal probability table. Also, to reduce the probability of a Type II error (false nondetection) to 5%, double this amount to 3.290. As an example, if 20 determinations of a low-level standard yielded a standard deviation of 6 µg/L, the LLD is $3.29 \times 6 = 20$ µg/L.

The MDL differs from the LLD in that samples containing the constituent of interest are processed through the complete analytical method. The method detection limit is greater than the LLD because of extraction efficiency and extract concentration factors. The MDL can be achieved by experienced analysts operating well-calibrated instruments on a nonroutine basis. For example, to determine the MDL, add a constituent to reagent water, or to the matrix of interest, to make the concentration near the estimated MDL. Analyze seven portions of this solution and calculate the standard deviation(s). From a table of the one-sided t distribution, select the value of t for $7 - 1 = 6$ degrees of freedom and at the 99% level; this value is 3.14. The product $3.14 \times s$ is the desired MDL.

Although the LOQ is useful within a laboratory, the practical quantitation limit (PQL) has been proposed as the lowest level achievable among laboratories within specified limits during routine laboratory operations. The PQL is significant because different laboratories will produce different MDLs even though using the same analytical procedures, instruments, and sample matrices. The PQL is about five times the MDL and represents a practical and routinely achievable detection limit with a relatively good certainty that any reported value is reliable.

Rounding Procedures

For comparison of analytical results with requirements for assays and impurities, the observed or calculated values are rounded to the number of digits carried in the requirement. The method and the rounding procedure are in accord with those in the *United States Pharmacopeia* (1999, p 4). When rounding is required, consider only one digit in the place to the right of the last digit in the limit expression. If this digit is smaller than 5, it is eliminated and the preceding digit is unchanged. If this digit is greater than 5, it is eliminated and the preceding digit is increased by one. If this digit equals 5, the 5 is eliminated and the preceding digit is increased by one. This rounding procedure is illustrated in Table I.

The foregoing procedure conforms to the common electronic calculator and computer procedure of rounding up when the digit to be dropped is 5 or 5 followed by zeros. The rounded value is obtained in a single step by direct rounding of the most precise value available and not in two or more steps of successive rounding. For example, 97.5487 rounds to 97.5 against a requirement of 97.6 and not in two possible steps of 97.55 and then 97.6.

The formula weights and factors for computing results are based on the 1989 International Atomic Weights shown on the inside of the front cover of this book. The formula weights are rounded to two decimal places.

Table I. Rounding Procedures

Requirement	Observed Value	Rounded Value	Pass/Fail
Not less than 98%	97.6	98	pass
	97.5	98	pass
	97.4	97	fail
Not less than 98.0%	97.95	98.0	pass
	97.94	97.9	fail
Not more than 0.01%	0.014	0.01	pass
	0.015	0.02	fail
	0.016	0.02	fail
Not more than 0.02%	0.015	0.02	pass
	0.025	0.03	fail
	0.026	0.03	fail

Assay Requirements

Assay requirements are included for most of the reagent chemicals and all of the standard-grade reference materials in this book. An *assay value*, in the sense used herein, is the content or concentration of a stated major component in the reagent. Unless otherwise specified, assay requirements are on an as-is basis (i.e., without drying, ignition, or other pretreatment of the sample).

Unless described in great detail and carried out with exceptional skill, available assay methods seldom are accurate enough to permit using a weighed quantity of a reagent so assayed in an exacting stoichiometric operation. This use of reagent chemicals should be limited to those designated as standards (for example, acidimetric or reductometric standards) because especially exacting assay methods are provided for such reagents.

Except in the case of standards, assays, through their minimum and maximum limits, mainly serve to assure acceptable consistency of the strength of reagents offered in the marketplace. They are particularly useful, for example, in the requirements for acid–water systems to control strength; for alkalies to limit the content of water and carbonate; for oxidizing and reducing substances that may change strength during storage; and for hydrates to control, within reasonable limits, deviations in the amount of water from that indicated in the formula. If, however, it should be necessary to use such reagents in stoichiometric operations, the user should ascertain that the assay values produced are of sufficient accuracy and precision for such use.

Assayed values for standard-grade reference materials are sufficient for the intended use described in this book. Assay requirements for all standard-grade reference materials include a minimum of two assay methods. Additional assay methods (such as thin layer chromatography) may include a requirement of "passes test." These secondary methods are intended to support the primary numerical assay methods.

Impurity Requirements

Requirements for impurities are expressed in the following ways: (1) as numerical limits; (2) in terms of the expression "passes test" with an accompanying approximate numerical limit; or (3) in terms of the expression "passes test" without an approximate numerical limit. The distinction among these forms of expression is based on the Committee's opinion as to the relative quantitative significance of the prescribed test methods. The methods given for determining conformity to requirements of the first type are considered to yield, in competent hands, what are usually thought of as "quantitative" results, whereas those of the second type can be expected to yield only approximate values. Those in class 3 give definitions that cannot be expressed in numbers. It is obvious, however, that these distinctions as to quantitative significance cannot be sharp and that even the numerically expressed requirements are not all defined with equal accuracy. The final and essential definition of any requirement, therefore, must reside in the described test method rather than in its numerical expression.

If a test method yields results that are adequately reproducible on repeated trials in different laboratories, it offers a satisfactory definition of the content of an impurity whether or not the result can be expressed by a number. Although the Committee has endeavored to base requirements, so far as possible, on test methods that meet this criterion, a considerable number are based on essentially undefined statements such as "no turbidity", "no color", or "the color shall not be completely discharged in … minutes". Although some of the requirements of this kind could be replaced by others based on quantitative comparisons or measurements, to do so would require more costly or time-consuming procedures than appear at this time to be justified. The approach to an ultimate goal of replacing in every instance the word "none" or its equivalent by the expression "maximum allowable…", therefore, is limited both by deficiencies of knowledge and by practical considerations of expediency.

Unlisted Impurities

The primary objective of the Committee in preparing reagent or standard specifications is to assure the user of the strength, quality, and purity of the reagents. It is, however, manifestly impossible to include in each specification a test for every impurity and contaminant that may be present. The Committee recognizes that for certain uses more stringent or additional requirements may be appropriate, and for such uses additional testing beyond that described in the specifications should be employed by the user. The Committee's intent in establishing the specifications is to recognize the common uses for which the reagent or standard is employed and to establish requirements that are consistent both with these uses and with the manufacturing processes and quality of the available reagents. The presence of moisture, either as water of crystallization or as an impurity, falls within the purview of contamination unless permitted by the requirements in the applicable specification.

While tests for foreign particulate matter are not usually included in the specifications for solid reagents, such matter constitutes contamination. Similarly, although tests for clarity are not usually included in the specifications for liquid reagents or for solutions of solid reagents, the presence of haze, turbidity, or foreign particulate matter also constitutes contamination.

In some instances, residual amounts of substances that have been added as aids in the process of purification may be present. An example is the use of complexing agents to keep certain metal ions in solution during recrystallization. These substances not only are impurities but may interfere with the tests. Certain reagents, such as desiccants and indicators, have requirements that assure suitability for their intended use. Such reagents may contain impurities that do not interfere with their intended use, but that may make these reagents unsuitable for other uses.

When the Committee becomes aware of an unlisted impurity that affects adversely the known or specified uses of reagents or standards, a new requirement is added to the specifications, provided a suitable method of test is available. Users of reagents can protect themselves against the effects of unlisted impurities on a specific analytical procedure by applying, ad hoc, an appropriate suitability test.

Added Substances

Unless otherwise specified for an individual reagent chemical, the reagents described in this volume may contain suitable preservatives or stabilizers, intentionally added to retard or inhibit natural processes of deterioration. Such preservatives or stabilizers may be regarded as suitable only if the following conditions are met:

1. They do not exceed the minimum quantity required to achieve the desired effect.
2. They do not interfere with the tests and assays for the individual reagents unless specifically waived by this book.
3. The presence of any added substance must be declared on the label of the individual package. Unless of a proprietary nature, the name and concentration of any added substance should be stated on the label.

Identity Requirements

Identity requirements and tests are not included in the specifications for reagent chemicals. If there is any question as to the identity of a chemical, identity can be ascertained by appropriate analytical methods. Identity requirements are included for standard-grade reference materials. For each material, a minimum of one identity method is required. Requirements are designed to unequivocally prove identity. Many of the methods rely on comparisons of data to standardized special libraries, such as the National Institute of Standards and Technology (NIST) infrared and mass spectral libraries. In most instances, a combination of complementary techniques is used.

PRECAUTIONS FOR TESTS

The descriptions of the individual tests are intended to give all essential details without repetition of considerations that should be obvious to an experienced analyst. A few suggestions are given for precautions and procedures that are particularly applicable to the routine testing of reagent chemicals.

Samples for Testing

To eliminate accidental contamination or possible change in composition, samples for testing must be taken from freshly opened containers.

ing drying, igniting, cooling in a suitable desiccator, and weighing. In general, desiccators should be charged with indicating-type silica gel. Stronger desiccants may be required for certain applications. The length of the drying or ignition and the temperature employed must be the same as specified in the procedure in which the tared equipment is to be used. Where it is directed to dry or ignite to constant weight, two successive weighings may not differ by more than ±0.0002 g, the second weighing following a second drying or ignition period.

In those operations wherein a filtering crucible is specified, a fritted-glass crucible, a porous porcelain crucible, or a crucible having a sponge platinum mat may be used. Certain solutions may attack the filtering vessel; for example, both fritted glass and porcelain are affected by strongly alkaline solutions. Even platinum sponge mats are attacked by hydrochloric acid unless the crucible is first washed with boiling water to remove oxygen.

Collection of Precipitates and Residues

In many of the tests, the amount of residue or precipitate may be so small as to escape easy detection. Therefore, the absence of a weighable residue or precipitate must never be assumed. However small, it must be properly collected, washed, and dried or ignited. The size and type of the filter paper to be used is selected according to the amount of precipitate to be collected, not the volume of solution to be filtered.

Often when barium chloride is added in slight excess to precipitate barium sulfate, the amount of precipitate produced is so small that it is hard to see and collect. If a small amount of a suspension of ashless filter paper pulp is added toward the end of the period of digestion on the steam bath or hot plate, the flocculation and collection of the sulfate are facilitated.

Some precipitates have a strong tendency to "creep", while others, like magnesium ammonium phosphate, stick fast to the walls of the container. The use of the rubber-tip "policeman" is recommended, sometimes supplemented by a small piece of ashless filter paper. The analyst should guard against the possibility of any of the precipitate escaping beyond the upper rim of the filter paper.

Precipitates, especially if recently produced, may be partially dissolved unless proper precautions are taken with both their formation and subsequent washing. In general, precipitates should not be washed with water alone but with water containing a small amount of common ion to decrease the solubility of the precipitate. It is much better to wash with several small portions of washing solution than with fewer and larger portions. Calcium oxalate precipitates should be washed with dilute (about 0.1%) ammonium oxalate solution; therefore, mixed precipitates that may contain calcium oxalate and magnesium ammonium phosphate should be washed with a 1% ammonia solution containing also 0.1% ammonium oxalate.

Ignition of Precipitates

The proper conditions for obtaining the precipitate are given in the individual test directions. The precipitate is collected on an ashless filter paper of suitable size and porosity (for example, a paper of 2.5-μm retention for a precipitate made up of fine particles), the residue and paper are thoroughly washed with the proper solution, and the moist filter paper is folded about the residue. The filter paper is then placed in a suitable tared, preconditioned crucible, and the paper is dried. The drying may be done by using a 105 °C oven, by heating on a hot plate, by using an infrared lamp, or by carefully using a gas microburner in a covered crucible. When dry, the paper is charred at the lowest possible temperature, the charred paper is destroyed by gentle ignition, and the crucible is finally ignited at 800 ± 25 °C for 15 min. The ignited crucible is cooled in a suitable desiccator and weighed.

Insoluble Matter

The intent of these tests is to determine the amount of insoluble foreign matter (for example, filter fibers and dust particles) present under test conditions designed to dissolve completely the substance being tested. After dissolution and prior to filtration, it is a good practice to check visually (using a white background) for insoluble foreign matter.

▶**Procedure.** Prepare a solution of the sample as specified in the individual test directions. Unless otherwise specified, heat to boiling in a covered beaker, and digest on a low-temperature (≈ 100 °C) hot plate (or a steam bath) for 1 h. Filter the hot solution through a suitable tared, medium-porosity (10–15-μm) filtering crucible. Unless otherwise specified, wash the beaker and filter thoroughly with hot water, dry at 105 °C, cool in a desiccator, and weigh.

Residue after Evaporation

This test is designed to determine the amount of any higher-boiling impurities or nonvolatile dissolved material that may be present in a reagent chemical. It is used chiefly in testing organic solvents and some acids.

The preferred container for the evaporation of almost all reagents is a platinum dish. However, in many cases other containers (for example, dishes of porcelain, silica, or aluminum) may be found suitable. In a few cases, such as when strong oxidants are evaporated, platinum should not be used.

A low-temperature (≈ 100 °C) hot plate is usually specified in the tests in this book for evaporations. A steam bath may be used, if the steam is generated from deionized or distilled water in a properly maintained apparatus.

The residue is not dried to constant weight in this test because continued heating may slowly volatilize some of the high-boiling impurities. The drying

MEASUREMENT OF PHYSICAL PROPERTIES

Boiling Range

The boiling-range procedure is essentially an empirical method. Hence adherence to the specified procedural details is required to obtain consistent and reproducible results. The recommended method is that developed by ASTM, and the analyst is referred to ASTM (1997c) for information concerning the description of the apparatus and the proper performance of the test. If the liquid to be tested is extremely volatile or produces noxious fumes during distillation, suitable cooling and venting facilities must be provided.

Color (APHA)*

The color intensity of liquids may be estimated rapidly by using platinum–cobalt standards, as described in ASTM (1997b). This method is particularly applicable to those materials in which the color-producing substances have light-absorption characteristics nearly identical with those in the standards. Colors having hues other than light yellow or reddish yellow cannot be determined with these standards.

The platinum–cobalt color standards contain carefully controlled amounts of potassium chloroplatinate and cobaltous chloride. Each platinum–cobalt color unit is equivalent to 1 mg of platinum per liter of solution (1 ppm), and the standards are named accordingly. For example, the No. 20 platinum–cobalt standard contains 20 ppm of platinum. These platinum–cobalt standards are also called APHA (for American Public Health Association) and Hazen standards.

Apparatus. The apparatus for this measurement consists of a series of Nessler tubes, a color comparator, and a light source. These tubes must match each other with respect to the color of the glass and height of the graduation mark. They must be fitted with suitable closures to prevent loss of liquid by evaporation and contamination of the standards by dust or dirt. The most commonly used type of Nessler tube is the 100-mL tall-form tube. However, for some samples, particularly those having darker colors, a better color match may be obtained by using 100-mL short-form Nessler tubes.

For the most accurate estimation of color, it is desirable to use a comparator constructed so that white light is reflected from a white glass plate with equal intensity through the longitudinal axes of the tubes being compared. The tubes are shielded so that no light enters the tubes from the side. In most cases, satisfactory estimates of the color can be obtained by holding the sample and standards close to each other over a white plate.

---

*See APHA, 1998.

The best source of light is generally considered to be diffuse daylight. However, for general routine analyses, the use of a titrating lamp equipped with a daylight-type fluorescent tube is satisfactory.

Preparation of APHA No. 500 Platinum–Cobalt Standard. (This standard is also commercially available.) Add approximately 500 mL of reagent water to a 1000-mL volumetric flask, add 100 mL of hydrochloric acid, and mix well. Weigh 1.245 g of potassium chloroplatinate (K_2PtCl_6) and 1.000 g of cobalt chloride hexahydrate ($CoCl_2 \cdot 6H_2O$) to the nearest milligram, and transfer to the flask. Swirl until complete dissolution is effected, dilute to the mark with reagent water, and mix thoroughly. The spectral absorbance of this APHA No. 500 standard must fall within the limits given below when measured in a suitable spectrophotometer, with a 1-cm light path and reagent water as the reference liquid in a matched cell.

Wavelength (nm)	Absorbance
430	0.110 to 0.120
455	0.130 to 0.145
480	0.105 to 0.120
510	0.055 to 0.065

Preparation of APHA Platinum–Cobalt Standards. Prepare the required color standards by diluting the volume of APHA No. 500 platinum–cobalt standard listed in Table II with reagent water to a total volume of 100 mL. The use of a buret is recommended in measuring the No. 500 standard. This series of standards is usually sufficient to permit an experienced analyst to make color comparisons with the necessary precision and accuracy. If a more exact estimate of color is desired, additional standards may be prepared to supplement those given by using proportional amounts of the No. 500 platinum–cobalt standard.

▶**Procedure.** Transfer 100 mL of the sample to a matched 100-mL tall-form Nessler tube. (If the sample is turbid, filter or centrifuge before filling the tube to remove visible turbidity.) Compare the color of the sample with the colors of the series of platinum–cobalt standards in matching Nessler tubes. View vertically down through the tubes against a white background. Report as the color the number of the APHA standard that most nearly matches the sample. In the event that the color lies midway between two standards, report the darker of the two.

Conductivity (Specific Conductance)

Conductivity is a measure of the ability of a material to conduct electricity. For liquids, this ability depends on the presence of ionic species (electrolytes) in a

Table II. APHA Pt–Co Color Standard Preparation

APHA Pt–Co Color Standard Number	APHA No. 500 Pt–Co Standard (mL to dilute to 100 mL)
0	0.00
1	0.20
3	0.60
5	1.00
10	2.00
15	3.00
18	3.60
20	4.00
25	5.00
30	6.00
35	7.00
40	8.00
50	10.00
60	12.00
70	14.00
80	16.00
90	18.00
100	20.00
120	24.00
140	28.00
160	32.00
180	36.00
200	40.00
250	50.00
300	60.00
350	70.00
400	80.00
450	90.00
500	100.00

solution to carry an electric current. *Conductance* is defined as the reciprocal of resistance. The unit of conductance is the mho, which is the conductance of a body through which 1 A of current flows when the potential difference is 1 V. This unit is also referred to as the reciprocal ohm (ohm^{-1}) or the siemens (S). *Conductivity*, also called *specific conductance*, is defined as the reciprocal of the resistance of a one meter cube of liquid at a specified temperature ($\Omega^{-1} \cdot m^{-1}$).

Laboratory conductivity measurements are used to assess the degree of mineralization of distilled or deionized water. The most convenient unit for reporting water conductivity is the μmho/cm, or the numerically equal $\mu S \cdot cm^{-1}$. Numerous conductivity instruments are available with various means of reporting units of conductance that can be converted to the previously mentioned units of measure. For the purposes described in this book, the cell should be standardized by using known low concentration solutions of potassium chloride (KCl) in ion-free water. Conductivity reference data can be obtained in Lind et al., 1959.

Density at 25 °C

Density (g/mL) may be determined by any reliable method in use in the testing laboratory, such as by a pycnometer or a density meter. A satisfactory method uses a 5-mL U-shaped bicapillary pycnometer, as described by Smith and others, (1950) and later adopted by ASTM.

▶**Procedure.** Tare a pycnometer, the volume of which has been accurately determined at 25 °C. Fill the pycnometer with the liquid to be tested and immerse in a constant-temperature (25 ± 0.1 °C) water bath for at least 30 min. Observe the volume of liquid at this temperature. Remove the pycnometer from the bath, wipe it dry, and weigh.

$$D \text{ at } 25\,^{\circ}\text{C} = \frac{W}{V} + 0.0010$$

where D is density in g/mL; W is weight, in grams, of liquid in pycnometer at 25 °C; V is volume, in mL, of liquid in pycnometer at 25 °C; and 0.0010 is the correction factor to compensate for buoyancy.

The correction of 0.0010 for air buoyancy is accurate to ±0.0001 g/mL if the density varies from 0.7079 g/mL (anhydrous ethyl ether) to 1.585 g/mL (carbon tetrachloride) and the pycnometer is weighed at a temperature of from 20 to 30 °C at an atmospheric pressure of from 720 to 775 mm of mercury.

Freezing Point

Place 15 mL of sample in a test tube (20 × 150 mm) in which is centered a thermometer traceable to NIST. The sample tube is centered by corks in an outer tube about 38 × 200 mm. Cool the whole apparatus, without stirring, in a bath of shaved or crushed ice with water enough to wet the outer tube. When the temperature is about 3 °C below the normal freezing point of the reagent, stir to induce freezing, and read the thermometer every 30 s. The temperature that remains constant for 1 to 2 min is the freezing point.

Melting Point

Where required, the melting point may be determined by the capillary tube method. Calibration of the method by the use of reference standards is required. A satisfactory capillary tube method is as follows.

Apparatus. A suitable melting-point apparatus consists of a glass container for a bath of transparent fluid, an appropriate stirring device, a thermometer traceable to NIST, and a controlled source of heat. The requisite temperature is considered in selecting the bath fluid, but light paraffin is generally suitable and certain liquid silicones are well adapted to the higher temperature ranges. The fluid is deep enough to permit immersion of the thermometer to its specified immersion depth such that the bulb is still about 2 cm above the bottom of the bath. The capillary tube is about 10 cm long and 1.0 ± 0.2 mm internal diameter, with walls 0.2–0.3 mm thick. Any commercially available apparatus meeting the above requirements may be used.

▶**Procedure.** Grind the material to be tested to a very fine powder, and render it anhydrous if it contains water of hydration by drying it as directed. If the substance contains no water of hydration, dry it over a suitable desiccant.

Charge a capillary glass tube, one end of which is heat-sealed, with sufficient quantity of the dry powder to form a column 2.5–3.5 mm high in the bottom of the tube when packed down as closely as possible by moderate tapping on a solid surface.

Heat the bath until the temperature is about 10 °C below the expected melting point and is rising at a rate of 1 ± 0.5 °C/min. Remove the thermometer, quickly attach the capillary tube to the thermometer by wetting both with a drop of the bath liquid, and adjust the height so that the material in the capillary tube is level with the thermometer bulb.

Replace the thermometer and attached capillary tube when the temperature is 5 °C below the expected melting point, and continue heating until the solid coalesces and is completely melted. Record the temperature.

Particle Size for Granular Materials

When a mesh or a mesh range is stated on the label for a reagent chemical, the label shall include reference to a coarse sieve and to a finer sieve. The sieve number (mesh) is related to the sieve opening as indicated in Table III.

When tested according to the following procedures, at least 95% of the material shall pass through the coarse sieve and at least 70% shall be retained on the finer sieve. This requirement applies only to materials that are 60-mesh or coarser.

▶**Procedure.** The sieves used in this procedure shall be those known as the U.S. Standard Sieve Series. Details of the standardization of such sieves can be found in ASTM (1995).

Place 25 to 100 g of the material to be tested upon the appropriate coarser standard sieve, which is mounted above the finer sieve, to which a close-fitting pan is attached. Place a cover on the coarser sieve, and shake the stack in a rotary horizontal direction and vertically by tapping on a hard surface for not less than

Table III. Sieve Numbers and Sieve Openings

Sieve No. (mesh)	Sieve Opening (mm)
2	9.52
4	4.76
8	2.38
10	2.00
20	0.84
30	0.59
40	0.42
50	0.297
60	0.250
80	0.177

20 min or until sifting is practically complete. Weigh accurately the amount of material remaining on each sieve.

An alternate procedure may be used in which the screening through the standard sieves is carried out in a mechanical sieve shaker. This shaker reproduces the circular and tapping motion given to the testing sieves in hand sifting, but with a mechanical action. Follow the directions provided by the manufacturer of the shaker.

Specific Rotation

To measure optical rotation, any suitable commercially available instrument may be used. Manufacturer's supplied standards or certified standards must be used to verify accuracy of the instrument at the time of optical rotation determination. Specific rotation is to be calculated from the observed optical rotations in the prepared test solution, as directed in the test. Unless otherwise specified, measurements of optical rotation are made using the sodium D-line (a doublet at 589.0 nm and 589.6 nm) at 25 °C. For optical rotation measurements using a polarimeter, a minimum of five readings is required. Blank correction consists of measuring the optical rotation of the solvent used to dissolve the sample. For a photoelectric instrument, a single, stable reading of the sample solution is taken and corrected for the solvent blank. Specified temperature of the test solution is to be maintained within 0.2 °C during the determination.

Calculation:

$$[\alpha]_D^{25°} = \frac{A \times 100}{L \times \text{Sample wt (g)}}$$

Where $[\alpha]$ is specific rotation at 25 °C, A is corrected specific rotation, and L is the length of the polarimeter cell in decimeters.

COLORIMETRY AND TURBIDIMETRY

Conditions and quantities for colorimetric or turbidimetric tests have been chosen to produce colors or turbidities that can be observed easily. These conditions and quantities approach but do not reach the minimum that can be observed. Five minutes, unless some other time is specified, must be allowed for the development of the colors or turbidities before the comparisons are made. If solutions of samples contain any turbidity or insoluble matter that might interfere with the later observation of colors or turbidities, they must be filtered before the addition of the reagent used to produce the color or turbidity. Conditions of the tests will vary from one reagent to another, but most of the tests fall close to the limits indicated in Table IV.

Table IV. Usual Limits for Colorimetric or Turbidimetric Standards

Sample	Quantity Used for Comparison (mg)	Approximate Volume (mL)
Ammonium (NH_4)	0.01	50
Arsenic (As)	0.002–0.003	3
Barium (Ba) chromate test	0.05–0.10	50
Chloride (Cl)	0.01	20
Copper(Cu)		
Dithizone extraction	0.006	10–20
Hydrogen sulfide	0.02	50
Pyridine thiocyanate	0.02	10–20
Heavy metals (as Pb)	0.02	50
Iron (Fe)		
1,10-Phenanthroline	0.01	25
Thiocyanate	0.01	50
Lead (Pb)		
Chromate	0.05	25
Dithizone	0.002	10–20
Sodium sulfide	0.01–0.02	50
Manganese (Mn)	0.01	50
Nitrate (NO_3)		
Brucine sulfate	0.005–0.02	50
Diphenylamine	0.005	25
Phosphate (PO_4)		
Direct molybdenum blue	0.02	25
Molybdenum blue ether extraction	0.01	25
Sulfate (SO_4)	0.04–0.06	12
Zinc (Zn)		
Dithizone extraction	0.006–0.01	10–20

Ammonium (Test for Ammonia and Amines)

▶**Procedure.** Dissolve the specified quantity of sample in 60 mL of ammonia-free water in a Kjeldahl flask connected through a spray trap to a condenser, the end of which dips below the surface of 10 mL of 0.1 N hydrochloric acid. Add 20 mL of freshly boiled sodium hydroxide solution (10%), distill 35 mL, and dilute the distillate with water as directed. Add 2 mL of freshly boiled sodium hydroxide solution (10%), mix, add 2 mL of Nessler reagent, and again mix. Any color should not be darker than that produced in a standard containing the amount of ammonium ion (NH_4) specified in the individual test description in an equal volume of solution containing 2 mL of the sodium hydroxide solution and 2 mL of Nessler reagent.

Arsenic

This colorimetric comparative procedure is based upon the reaction between silver diethyldithiocarbamate and arsine.

Interferences. Metals or salts of metals—such as chromium, cobalt, copper, mercury, molybdenum, nickel, palladium, and silver—may interfere with the evolution of arsine.

Antimony, which forms stibine, may produce a positive interference in color development with silver diethyldithiocarbamate solution. Although potassium iodide and stannous chloride in the generator tend to repress the evolution of stibine at low levels of antimony, at higher levels repression is incomplete, a situation that can lead to erroneously high results for arsenic.

Interference from antimony, however, can be essentially eliminated by adding ferric ion to the generator. An arsenic test employing this addition is included in the standard for phosphoric acid (page 466).

Antimony can be determined, if desired, by atomic absorption or by differential pulse polarography. An atomic absorption procedure for antimony is included in the aforementioned standard for phosphoric acid.

Apparatus. The apparatus (*see* Figure 1) consists of an arsine generator (a) fitted with a scrubber unit (c) and an absorber tube (e), with standard-taper or ground-glass ball-and-socket joints (b and d) between the units. Alternatively, any apparatus embodying the principle of the assembly described and illustrated may be used.

Sample Solution. Dissolve the specified weight of sample in water and dilute with water to 35 mL, or use the solution prepared as directed in the individual reagent standard.

▶**Procedure.** Unless otherwise specified in the individual test description, proceed as follows: To the sample, add 20 mL of dilute sulfuric acid (1+4), 2 mL of

are expressed "as lead" by comparing the color developed in a test solution with that developed in a lead standard/control solution. The optimum conditions for proper performance of the test are: pH of the solution between 3 and 4, total volume of 50 mL, use of freshly prepared hydrogen sulfide water, and use of a standard/control containing 0.02 mg of lead ion (Pb). These conditions are rarely modified in the individual procedures, and it would only be after a method validation that the modifications would be suitable for the individual reagent. The color comparisons should be made, using matched 50-mL color-comparison tubes, by viewing vertically over a white background. In some instances, an appropriate amount of the sample is added to the standard, thereafter called the control (*see* page 91), because the color developed in the test may be affected by the sample being tested. In other instances, no such effort occurs and the standard may be prepared without any of the sample. In general, the individual standard procedures provide only for the preparation of a sample solution. The remainder of the test directions, being the same for all reagents, are provided in this chapter.

Two general test procedures are provided. Method 1 is to be used unless otherwise specified in the individual standard. This method is used, generally, for those substances that yield clear, colorless solutions under the specified test conditions. Method 2 is used, generally, only for certain organic compounds.

Unless otherwise directed in the individual specification, the test shall be carried out as follows.

▶**Procedure.**

Method 1

Test Solution. Using the solution prepared as directed in the individual standard, adjust the pH to between 3 and 4 (by using a pH meter) with 1 N acetic acid or ammonium hydroxide (10% NH_3). Dilute with water to 40 mL, if necessary, and mix.

Standard/Control Solution. To 20 mL of water, or to the portion of that solution specified in the individual standard, add 0.02 mg of lead (Pb).* Dilute with water to 25 mL and mix. Adjust the pH (within 0.1 unit) to the value established for the test solution (using a pH meter) with 1 N acetic acid or ammonium hydroxide (10% NH_3), dilute with water to 40 mL, and mix.

To each of the tubes containing the test solution and the standard/control solution, add 10 mL of freshly prepared hydrogen sulfide water, and mix. Any brown color produced within 5 min in the test solution should not be darker than that produced in the standard/control solution.

---

*When a control solution is specified in the individual specification and a direction is given therein to add a weight of lead ion (Pb), no further lead shall be added. Proceed to the dilution or pH adjustment, as necessary.

Method 2

Test Solution. Transfer the quantity of substance specified in the individual specification to a suitable crucible, add sufficient sulfuric acid to wet the sample, and carefully heat at a low temperature until thoroughly charred. (The crucible may be loosely covered with a suitable lid during the charring.) Add to the carbonized mass 2 mL of nitric acid and 0.25 mL of sulfuric acid, and heat cautiously until white fumes of sulfur trioxide are no longer evolved. Ignite, preferably in a muffle furnace, at 500 to 600 °C until the carbon is burned off completely. Cool, add 4 mL of 6 N hydrochloric acid, and cover. Digest on a hot plate ($\approx$100 °C) for 15 min, uncover, and slowly evaporate on the hot plate to dryness. Moisten the residue with 0.05 mL of hydrochloric acid, add 10 mL of hot water, and digest for 2 min. Add ammonium hydroxide (10%) dropwise, until the solution is just alkaline, and dilute with water to 25 mL. Adjust the pH of the solution to between 3 and 4 (by using a pH meter) with 1 N acetic acid, dilute with water to 40 mL, and mix.

Standard/Control Solution. To 20 mL of water, add 0.02 mg of lead ion (Pb), and dilute with water to 25 mL. Adjust the pH (within 0.1 unit) to the value established for the test solution (by using a pH meter) with 1 N acetic acid or ammonium hydroxide (10% NH_3), dilute with water to 40 mL, and mix.

To each of the tubes containing the test solution and the standard/control solution, add 10 mL of freshly prepared hydrogen sulfide water, and mix. Any brown color produced in the test solution within 5 min should not be darker than that produced in the standard/control solution.

Iron

Two procedures are provided for the determination of iron. Method 1, which employs thiocyanate, is usually performed in 50 mL of solution containing 2 mL of hydrochloric acid, to which 30 to 50 mg of ammonium peroxydisulfate crystals and 3 mL of ammonium thiocyanate reagent solution (about 30%) are added. These quantities of acid and thiocyanate were selected so that the production of the red color would not be affected significantly by anything except the iron. The use of peroxydisulfate to oxidize the iron prevents fading of the color and eliminates the need for immediate comparison with the standard. For certain alkaline earth salts, permanganate is used to oxidize the iron.

Method 2, which employs 1,10-phenanthroline, is used generally for phosphate reagents. It is usually performed by adjusting the pH (when necessary) to between 4 and 6, adding 6 mL of hydroxylamine hydrochloride reagent solution and 4 mL of 1,10-phenanthroline reagent solution, and diluting with water to 25 mL. Because development of the color may be slow, the comparison of the color produced is not made until 60 min after addition of the 1,10-phenanthro-

tected from loss of and exposure to ammonia. Distill 35 mL, and dilute the distillate with water to 50 mL. Add 2 mL of freshly boiled sodium hydroxide solution (10%), mix, add 2 mL of Nessler reagent, and again mix. Any color should not be darker than that produced in a standard containing the amount of nitrogen (N) specified in the individual test description and treated exactly like the sample. Commercially available apparatus may be used.

Phosphate

Three general procedures are provided to test for phosphate: direct molybdenum blue (method 1), extracted molybdenum blue (method 2), and precipitation (method 3). In method 2, conditions are established that allow only the phosphomolybdate to be extracted into ether, thus eliminating any interference from arsenate and silicate.

▶Procedure.

Method 1
Prepare the sample solution as directed in the individual test description. Add 1 mL of ammonium molybdate reagent solution and 1 mL of 4-(methylamino)phenol sulfate reagent solution, and allow to stand at room temperature for 2 h. Any blue color should not exceed that produced by 0.02 mg of phosphate ion (PO_4) in an equal volume of solution containing the quantities of reagents used in the test.

Method 2
Prepare the sample solution as directed in the individual test description. Transfer to a separatory funnel, add 35 mL of ether, shake vigorously, and allow the layers to separate. Draw off and discard the aqueous layer unless it is indicated in the individual specification that the aqueous layer should be saved for the silica determination. Wash the ether layer twice with 10 mL of dilute hydrochloric acid (1 + 9), drawing off and discarding the aqueous layer each time. Add 0.2 mL of a freshly prepared 2% solution of stannous chloride in hydrochloric acid. If the solution is cloudy, shake with a small amount of dilute hydrochloric acid (1 + 9) to clear. Any blue color should not exceed that produced by a standard containing the concentration of phosphate ion (PO_4) cited in the individual test description and treated exactly as is the sample.

Method 3
Dissolve 20 g of sample in 100 mL of dilute ammonium hydroxide (1 + 4). Prepare a standard containing 0.1 mg of phosphate ion (PO_4) in 100 mL of dilute ammonium hydroxide (1+ 4). To each solution, add dropwise, with rapid stirring, 3.5 mL of ferric nitrate reagent solution, and allow to stand

about 15 min. If necessary, warm gently to coagulate the precipitate, but do not boil. Filter and wash several times with dilute ammonium hydroxide (1 + 9). Dissolve the precipitate by pouring 60 mL of warm dilute nitric acid (1 + 3) through the filter and catching the solution in a 250-mL glass-stoppered conical flask. Add 13 mL of ammonium hydroxide, warm the solution to 40 °C, and add 50 mL of ammonium molybdate–nitric acid reagent solution. Shake vigorously for 5 min, and allow to stand for 2 h at 40 °C. Any yellow precipitate obtained from the sample should not exceed that obtained from the standard.

Silicate

Small amounts of silica or silicate are determined by a molybdenum blue method. Conditions have been established that allow the extraction of the silicomolybdate and thus eliminate any interference from arsenate or phosphate.

Sulfate

▶Procedure.

Method 1

Unless otherwise stated in the specification, dissolve (0.005 ÷ % maximum allowable) g of sample in a minimum volume of water. This gives a sample size that provides 0.05 mg of SO_4 at limit. The standard also contains 0.05 mg of SO_4. To determine the sample weight in grams, divide 0.005 by the maximum allowable (in actual percent). For example, if the maximum allowable is 0.01%, 0.005 is divided by 0.01 to arrive at 0.5 g of sample. Add 1 mL of dilute hydrochloric acid (1 + 19). Filter through a prewashed small filter paper, and wash with two 2-mL portions of water. Dilute to 10 mL, and add 1 mL of barium chloride reagent solution. Any turbidity should not exceed that produced by 0.05 mg of sulfate ion (SO_4) in an equal volume of solution containing the quantities of reagents used in the test. Unless otherwise specified, compare turbidity 10 min after adding the barium chloride to the sample and standard solutions.

Method 2

Unless otherwise stated in the specification, dissolve (0.005 ÷ % maximum allowable) g of sample in water, and evaporate to dryness on a hot plate ($\approx$100 °C). Proceed as in method 1.

Method 3

Unless otherwise stated in the specification, add 10 mg of sodium carbonate to (0.005 ÷ % maximum allowable) g of sample, and evaporate to dryness on a hot plate ($\approx$100 °C). Proceed as in method 1.

Method 4

Unless otherwise stated in the specification, dissolve (0.005 ÷ % maximum allowable) g of sample in water, evaporate to 5 mL, add 1 mL of bromine water, and evaporate to dryness on a hot plate (≈100 °C). Proceed as in method 1.

TRACE AND ULTRATRACE ELEMENTAL ANALYSIS

Analysis of individual elements can be performed by a variety of techniques, which differ in their sensitivity, selectivity, and cost. For the purpose of determining the purity of reagent chemicals in this book, plasma emission and several modes of atomic absorption spectroscopy (AAS) are used. AAS techniques are recommended for most reagent chemicals, whereas plasma emission spectroscopy is used only in the analysis of ultratrace reagents.

General Background

Trace metal analysis is routinely done by emission or absorption spectroscopy techniques. Flame atomic absorption spectroscopy (FAAS), cold vapor atomic absorption spectroscopy (CVAAS), hydride-generation atomic absorption spectroscopy (HGAAS), and inductively coupled plasma optical emission spectroscopy (ICP–OES) are the principal instrumental techniques used for trace metal analysis. In this book, AAS is used when accurate determination of only a few elements in solution is required. FAAS is the most cost-effective analytical technique for analysis of common metallic impurities for most reagent chemicals. CVAAS is used for the determination of mercury, and HGAAS is used for selenium analysis. AAS analysis has few interferences, especially when compared to plasma emission analysis.

ICP–OES is usually the technique of choice when rapid quantitative, multielement analysis of solutions is required. The ultratrace reagents in this book are analyzed by ICP–OES for approximately 25 of the most laboratory-significant metallic impurities. Ultratrace reagents generally are preconcentrated prior to analysis.

The detailed principles of atomic absorption and plasma emission spectroscopy are available from many fine references, and only those details that describe how the technique should be used to assure that the specifications for the required elements are met will be covered in this book. A specific instrumental procedure for each technique is not given, since operating details will vary with instrument design.

Reagents and Standards

Precautions should be taken to avoid contamination. Sample containers and glassware should be carefully cleaned and stored. The highest-purity water should be used, obtained preferably from a mixed-bed strong acid, strong-base ion-exchange cartridge capable of producing water with an electrical resistivity of 18 megohm-cm. Teflon or plasticware can be substituted when leaching of metals from glassware is possible.

Accurate standard solutions of elements to be determined are needed for the preparation of working standards. Suitable standard solutions are commercially available. These solutions should be traceable to NIST standards. Concentrations of 1–10 g/L are recommended for storage. More dilute solutions and working standards may be adsorbed on the walls of glassware and should not be used after more than 24 h unless experience has shown that they are stable for a longer period.

The Ultratrace Laboratory Environment

The analysis of reagent chemicals intended for use in ultratrace applications must be done under "clean air" conditions, since the limit of detection is very often dependent on the procedural "blank" obtained (Moody, 1982). The specifications used to determine the quality of "clean air" in a laboratory are the class standards of the U.S. Federal Standard 209e.* A "Class 100" air quality, where the airborne particle count is less than 100 particles of 0.6 μm diameter per cc of air, is considered sufficient in this book for analysis of ultratrace reagents. Class 100 conditions are usually achieved by the use of high-efficiency particulate filters in conjunction with proper laboratory design. Clean air conditions can be used in either an entire laboratory or only in a laminar flow hood. Very often, in the absence of such facilities, clean air is supplied only to the specific area where contamination of a sample might occur. Extensive texts on these subjects exist, and the reader is encouraged to examine these in detail prior to undertaking any ultratrace analysis.

Limit of Detection

The determination of the limit of detection (LOD) for ultratrace analysis should be done using SEMI (1999). This regression-based method, which uses the calibration data to determine the LOD, is available as a download at www.airproducts.com/gases/mol-estimator. This guide, although applicable to many analytical techniques (for instance, gas and liquid chromatography), has been specifically adopted for trace metal analysis of semiconductor-grade chemicals of similar quality to the ultratrace reagents found in this monograph.

Atomic Absorption Spectroscopy

AAS is used when accurate determinations of only a few elements in solution are required. AAS has been chosen as the most cost-effective analytical technique for

---

*The U.S. Federal Standard 209e will be superseded by a new ISO standard that is in the process of approval in 1999.

trace metal analysis of many reagents contained in this book. AAS is a very sensitive technique capable of analyses measured in parts per million (ppm) with flame atomization. When furnace atomization is employed, analyses can be measured in parts per billion (ppb).

General Background. In FAAS, the sample is aspirated into a high-temperature flame, usually air–acetylene or nitrous oxide–acetylene. After the solvent evaporates, the flame produces metal atoms that absorb light at wavelengths characteristic of the specific atom. (The wavelength at which each element absorbs light is usually unique to that element.) The amount of light absorbed is proportional to the concentration of the element present; this relationship is referred to as Beer's law. For Beer's law to be valid, the bandwidth of the radiation to be absorbed by the atom must be narrower than the absorption line for the absorbing species. The hollow cathode lamp, which is used in AAS is an emission source, has line widths that are typically one-tenth the width of the absorption lines used in AAS. The optical system in AAS isolates the desired emission line from the lamp, focuses it through the sample to maximize the signal, separates it from background radiation, and finally passes it into the detector.

Conventional AAS spectrometers are equipped with a turret to accommodate multiple hollow cathode lamp sources. The burner is slotted and will vary in design and path length according to the gases to be used. The monochromator will have a resolution of at least 0.1 nm. The signal from the photo-multiplier tube detector is processed by a microprocessor and displayed on a digital readout or a terminal screen. A deuterium background corrector or equivalent is provided.

In this book, specific analytical instrument conditions, such as the flame gases to be used, the wavelengths, and the slit width, are given in the procedures for individual reagents. The lamp and burner position and the flame characteristics should be optimized for a given determination.

Sensitivity and Detection Limits. Sensitivity in AAS is defined as the concentration of a test element in an aqueous solution that will produce absorption of 1% or 0.0044 absorbance units. It is normally expressed in µg/mL or µg/g per 1% absorption. Note that this expression is related to the slope of the calibration curve.

The sensitivity of the instrument is normally optimized during the method setup. A tabulation of expected absorbance vs. standard concentration is shown in Table V. Excellent detailed definitions and procedures for determining detection limits exist; in particular, *see* APHA (1998). Several instrumental variables can be manipulated to improve the detection limit.

Instrumentation. The manufacturer's instruction manual should be followed for detailed operating procedures and for matters of care, cleaning, and calibration. It is assumed that the analyst is familiar with the texts (such as Kenkel, 1994) available on AAS and is aware of the various interferences and sources of error.

Table V. Expected Absorbance of Standards

Element	Wave-Length (nm)	Sens. Check[a] (mg/L)	Linear Range[b] (mg/L)	Min.[c] (mg/L)	Min. (mg/25 mL)	Max.[d] (mg/25 mL)	Recommended Standard (mg/25 mL)	Expected Absorbance Units
Antimony	217.6	15.0	100.0	1.5	0.038	2.50	0.10–0.20	0.07–0.14
Barium	553.6	10.0	50.0	1.0	0.025	1.25	0.10–0.20	0.06–0.12
Bismuth	223.1	10.0	50.0	1.0	0.025	1.25	0.10–0.20	0.12–0.24
Cadmium	228.8	0.5	3.0	0.05	0.0013	0.075	0.01–0.02	0.18–0.36
Calcium	422.7	0.5	3.0	0.05	0.0013	0.075	0.02–0.04	0.25–0.50
Cobalt	240.7	2.5	15.0	0.25	0.0063	0.375	0.02–0.04	0.08–0.16
Copper	324.8	1.5	10.0	0.15	0.0038	0.250	0.02–0.04	0.11–0.22
Iron	248.3	2.5	15.0	0.25	0.0063	0.375	0.05–0.10	0.20–0.40
Lead	217.0	5.0	30.0	0.50	0.013	0.750	0.10–0.20	0.16–0.32
Lithium	670.8	1.0	5.0	0.10	0.0025	0.125	0.01–0.02	0.11–0.22
Magnesium	285.2	0.15	1.0	0.015	0.0004	0.025	0.005–0.01	0.25–0.50
Manganese	279.5	1.0	5.0	0.10	0.0025	0.125	0.02–0.04	0.22–0.44
Molybdenum	313.3	15.0	100.0	1.50	0.038	2.50	0.25–0.50	0.13–0.26
Nickel	232.0	4.0	20.0	0.40	0.01	0.50	0.05–0.10	0.18–0.36
Potassium	766.5	0.4	2.0	0.04	0.001	0.050	0.01–0.02	0.20–0.40
Silver	328.1	1.5	10.0	0.15	0.0038	0.250	0.02–0.04	0.11–0.22
Sodium	589.0	0.15	1.0	0.015	0.0004	0.025	0.005–0.01	0.25–0.50
Strontium	460.7	2.0	10.0	0.20	0.005	0.250	0.05–0.10	0.35–0.70
Zinc	213.9	0.4	2.0	0.04	0.001	0.050	0.01–0.02	0.20–0.40

NOTE: This table is based on instrument-specific conditions. In some instances, the recommended standard addition is lower than cited in this table. Dilution may be modified to meet maximum allowable limits.

[a]Sens. Check is the concentration giving approximately 0.2 AU.   [b]Linear Range is the upper concentration of linear range.   [c]Min. mg/L is the concentration giving 0.02 AU (Sens. Check divided by 10).   [d]Max. is the upper limit in mg per 25 mL (linear range divided by 40).

The procedures for individual reagents take the following interferences into account insofar as possible. However, because of differences among instruments, the analyst should be aware of these effects and check the particular instrument.

Interferences. Surface tension, viscosity, acid content, and solute content may affect the nebulization rate and influence the sensitivity of an instrument by altering the concentration of absorbing species in the flame volume under observation. Oxide formation within the flame may make it desirable to add complexing or releasing agents. Reduction of ionization in the presence of alkali elements can cause serious errors when these elements are present in some samples and not in others. Flame gas absorption becomes appreciable at 250–300 nm and becomes increasingly severe at lower wavelengths. Scattering of radiant energy from particles in the flame is usually negligible above about 300 nm or at sample concentrations on the order of 10 g/L or less. Spurious absorbance by flame gas or scattering will bias results obtained by standard additions.

Where spurious absorption or other background correction is required, details are given in the procedure for an individual chemical. For an element with the stated resonance line in the ultraviolet region (<350 nm), the correction can be performed with a deuterium discharge lamp. For an element with the stated resonance line in the visible region (>350 nm), the adjacent-line technique can be employed. In this method, the background absorption is determined at a wavelength adjacent, ± nm, to the line being used for element of interest; that is, background absorption is measured at a wavelength where significant atomic absorption by that element is absent.

Analysis. Quantitative analysis by AAS depends on the applicability of Beer's law. In this book, the method of standard additions is used for trace-metal determination of reagents prepared as solutions. The method of standard additions is applicable at low concentrations where a linear relationship exists between signal and concentration. Most instruments can indicate the signal as transmission (%T) or absorbance (A). Absorbance is directly proportional to concentration but %T is a logarithmic relationship.

The use of standard additions places the standards in a matrix similar to that of the sample. It is essential that the sample solution, along with the additions, have absorbencies in the range of which linearity is expected, routinely 0.01–0.5 A. Standard addition does not eliminate interferences, but it does assure that the element of interest behaves similarly in the sample and the standard additions. A calibration line obtained for aqueous standards can be used if its slope and the slope of the standard addition line are identical. In that event, chemical (matrix) effects are negligible.

As indicated above, reagent-specific conditions for AAS analysis are listed under the individual reagents. This information includes but is not limited to sample preparation, element wavelength, use of background correction, fuel-oxidant mixture, use of additives, and the recommended standard additions.

▶**Procedure.** Prepare the sample for analysis by dissolving in water or by following a reagent-specific procedure. Prepare four solutions for standard additions: the reagent blank (which may be only water), the sample solution, and two standard additions. The recommended additions usually correspond to half and equal to the specification limit. The signal for the reagent blank preparation can be subtracted from the signals of the three other solutions, or the auto-zero feature of the instrument can be employed. The three results can be treated graphically or mathematically to determine the analyte concentration in the sample. When several elements are to be determined, they can be added simultaneously to the standard addition solutions. A separate set of standard additions may be required if different sample weights must be used to stay in the linear absorbance (A) range.

Mercury Determination by CVAAS

The sensitive determination of mercury is based upon its reduction in aqueous solution to the free element, vaporization of the latter into a stream of air or inert gas, and passage of the vapors through an optical cell. The cell is substituted for the burner, and a high-intensity, mercury, electrodeless discharge lamp can be used to increase sensitivity. The apparatus for the reduction and vaporization is shown in Figure 2. Commercially available fluorescence or gold film mercury analyzers may be substituted if the test method is validated. Care must be taken to guard against contamination of samples and solutions by mercury. Glassware should be washed with nitric acid (1 + 3), followed by thorough rinsing with water prior to analysis. Dedicated glassware is a good practice.

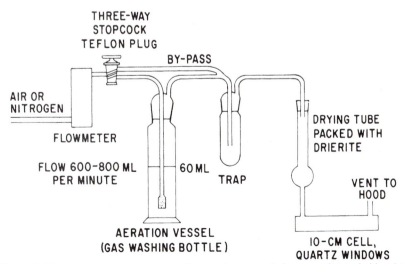

Figure 2. Mercury aeration apparatus. Connections are of glass or poly(vinyl chloride).

Ultraviolet radiation from the lamp is directed through the cell and into the detector. The absorbance of mercury in the gas stream is measured and compared to a calibration curve from known solutions. Standard additions can also be used.

▶**Procedure.** Sample preparation is described in the test for each reagent. Reagent blanks should also be run as directed in the test. The solution of the sample, or aliquot thereof, is transferred to the aeration vessel before addition of any reducing agent. If the sample has been treated with permanganate, the container is rinsed with the quantity of hydroxylamine hydrochloride solution specified, and the rinsings are also added to the aeration vessel with enough water to bring the level to a 60-mL calibration line. The instrument is adjusted to give a smooth baseline with carrier gas flowing through the bypass and cell and with the radiation of the 253.7-nm mercury resonance line focused through the cell onto the detector. Stannous chloride solution as specified is introduced into the aeration vessel, and the valve is set to direct the carrier gas through the vessel and the cell. Bubbling is continued until the instrument returns to its zero point. Prior to running the next sample, the aeration vessel should be drained and rinsed. Periodically, any deposit of stannic oxide should be removed. Generally, 2.5% stannous chloride solution is used; however, the analyst should follow the manufacturer's operating manual requirements.

Selenium Determination by HGAAS

Hydride-generation is commonly used for trace-level analysis of hydride-forming elements such as antimony, arsenic, and selenium. Hydride-generation gives better sensitivity for these elements than FAAS. Selenium analysis using conventional FAAS is insensitive because its analytical wavelength (196.0 nm) is too close to the vacuum ultraviolet region where absorption of source light by air and flame gases occurs. In this book, HGAAS is used for selenium analysis of ultratrace reagents. The determination of selenium is based on its conversion to a volatile hydride (SeH_2) by sodium borohydride reagent and aspiration into an atomic absorption spectrometer. Quartz atomization cells are required for ultratrace analysis.

Equipment. AAS systems equipped with a hydride reaction cell and an externally mounted, heated quartz cell or a quartz cell with an internal fuel-rich oxygen–hydrogen or air–hydrogen flame are commercially available and recommended.

Sensitivity and Interferences. Quartz atomization cells minimize background noise caused by the flame and are required for ultratrace analysis. The formation or presence of chlorine during the hydride generation will result in low values due to the reoxidation of Se IV to Se VI. Only Se IV readily forms a hydride using sodium borohydride.

Analysis. Selenium determination is done using standard additions. The reagent-specific conditions can be found under the individual reagent. (Selenium is tested at 196.0 nm wavelength, depending on the instrument setup; 1.0 µg of selenium in 100-mL volume gives approximately 0.12 absorbance units.)

▶**Procedure.** Prepare the sample for analysis by following the reagent-specific procedure. The standard additions can also be found under the individual reagent.

Plasma Emission Spectroscopy

Optical emission spectroscopy (OES) using inductively coupled plasma (ICP) is a rapid and convenient method to determine trace metals in various matrices. In this book, ultratrace reagents are analyzed by ICP–OES. In order to achieve the requirements specified for ultratrace reagents, preconcentration of samples by evaporation under clean air conditions and reconstitution to a 1% nitric acid matrix is required prior to analysis by ICP–OES

Coupling an ICP with a mass spectrometer (ICP–MS) as a detector has enabled direct detection levels of concentrations below parts-per-billion. While individual reagents do not specify the use of ICP–MS, it would be acceptable to substitute this technique for the ultratrace reagents as long as method validation has been done. It must be noted that this ICP–MS method must be augmented by the use of graphite furnace atomic absorption spectroscopy (GFAAS), since interferences limit the capability of most ICP–MS. Most notably calcium, iron, potassium, and sodium are better suited to analysis by GFAAS.

General Background. ICP–OES is based on the ionization or excitation of atoms by a source consisting of a stream of argon gas being ionized by an applied radio frequency field. This field is inductively coupled to the ionized argon by a water-cooled coil surrounding a quartz torch. The torch supports and confines the plasma. A sample aerosol is generated in a nebulizer and spray chamber and subsequently introduced into the plasma through an injector located within the torch. The atoms are instantly dissociated at temperatures in excess of 6000 K. The high temperature of the plasma facilitates rapid and complete dissociation. This minimizes the effects of interferences and substantially expands the linear dynamic range of analysis.

The ionization generated by the torch results in emission spectra that are mainly in the optical (visible) part of the spectrum (160–900 nm) and that enable multielemental detection using one of three spectrometric detection methods. The emitted light can be focused on the entrance slit of either a monochromator or a dispersing polychromator, thus enabling the instrument to function as a sequential or simultaneous analyzer, respectively. The sequential system provides greater wavelength selection and resolution, while the simultaneous system allows greater sample throughput. In both cases, very sensitive photo-multiplier tubes are used as the detector. Recently, spectrometers that use solid state detectors have become available. These instruments are usually equipped with high-resolution Echelle gratings that disperse the spectrum into two dimensions. This results in a spectrometer that is affordable, has high resolution and sensitivity, and provides flexibility similar to a sequential instrument.

Interferences, Detection Limits, and Sensitivity. Interference effects, limits of detection, sensitivity, and linear dynamic range must be investigated and established for each individual analyte line on the particular instrument being used in analysis of the individual reagent. Types of interferences other than the simple overlap of a concomitant line on an analyte line must be considered. Special attention must be paid to the possibility of external contamination during sample preparation and analysis.

Instrumentation. A grating instrument with sufficient resolution to separate the relevant spectral lines of the elements specified in the requirements for an individual reagent is required. No specific procedure is given since operating details will vary with instrument design.

Analysis. Working standard solutions should be prepared on the day of use by dilution of a stock standard solution. Standards should be matrix-matched to the sample solution, typically 1% nitric acid. Multielement standards may be prepared. The components should be selected to provide stable standards within the required concentration range. Analyze the samples versus matrix-matched calibration standards. If a new or unusual matrix is encountered, the method of standard additions should be used. The standard addition will not detect coincident spectral overlap; an alternate wavelength or method is recommended for verification. The suggested analytical lines for determination of the various metals in ultratrace reagents are shown in Table VI. Other analytical lines may be used if a different sensitivity is desired.

▶**Procedure.** The achievable detection limits can be improved by preconcentration of a sample by evaporation under clean air conditions, and reliability can be assured by use of a blank or control sample determination. The sample size will depend on the detection limits required and the capability of the instrument used. A background correction technique is required. The procedure for the individual reagent will provide instructions for any required additives in the sample and blank solutions. The manufacturer's manual should be consulted for proper use of the instrument. The user should check the performance of the instrument at regular intervals to assure that adequate sensitivity is being attained to meet the specified requirements.

The following sample preparation has given satisfactory results for the ultratrace-grade reagents in this book. Special conditions can be found under the trace metal test for the individual reagent.

Table VI. Suggested Plasma Emission Analytical Lines

Element	Line
Aluminum (Al)	396.15
Barium (Ba)	455.40
Boron (B)	249.77
Cadmium (Cd)	214.44
Calcium (Ca)	393.27
Chromium (Cr)	205.55
Cobalt (Co)	228.63
Copper (Cu)	324.75
Iron (Fe)	238.20
Lead (Pb)	220.35
Lithium (Li)	670.78
Magnesium (Mg)	279.55
Manganese (Mn)	257.60
Molybdenum (Mo)	202.02
Potassium (K)	766.49
Silicon (Si)	212.41
Sodium (Na)	589.59
Strontium (Sr)	407.77
Tin (Sn)	189.99
Titanium (Ti)	334.94
Vanadium (V)	292.40
Zinc (Zn)	213.86

Sample Preparation and Analysis. In a clean air environment, place 100 g of sample in one 100-mL PTFE evaporating dish and a reagent blank in another. Slowly evaporate on a hot plate, avoiding loss of sample by effervescence or spattering, until approximately one drop of liquid remains. Do not allow the sample to evaporate to complete dryness. Cool. Transfer quantitatively to individual volumetric flasks using a 1% solution of ultratrace-grade nitric acid for rinsing and dilution to volume. Analyze the reagent sample by ICP–OES. Correct for the reagent blank.

DIRECT ELECTROMETRIC METHODS

pH Potentiometry

pH Range. This pH requirement is intended to limit the amount of free acid or alkali that is allowed in a reagent. Sodium chloride, for example, has a requirement that the pH of a 5% solution should be from 5.0 to 9.0 at 25 °C. This

Table VII. Effect of Excess Acid or Base on the pH of 5% Solution of Salts of Strong Acids and Bases

Excess Acid/Base	0.01%, pH	0.001%, pH	0.0001%, pH
HCl	3.9	4.9	5.9
H_2SO_4	4.0	5.0	6.0
HNO_3	4.1	5.1	6.1
HBr	4.2	5.2	6.2
$HClO_4$	4.3	5.3	6.3
KOH	9.9	8.9	7.9
NaOH	10.1	9.1	8.1

requirement limits the free hydrochloric acid or sodium hydroxide to about 0.001%. The effect of excess acid or base on the pH of a 5% solutions of salts of strong acids and bases is given in Table VII.

The pH of the solution is determined by means of any suitable pH meter equipped with a glass electrode and a reference half cell, usually a saturated calomel electrode (SCE), while the solution is protected from absorption of atmospheric gases. The meter and electrodes are standardized at pH 4.00 at 25 °C with 0.05 M NIST potassium hydrogen phthalate ($KHC_8H_4O_4$) and at pH 9.18 at 25 °C with 0.01 M NIST sodium borate decahydrate ($Na_2B_4O_7 \cdot 10H_2O$). Because the pH values are reported to only 0.1 pH unit, the meter and electrodes are considered to be in satisfactory working order if they show an error of less than 0.05 pH unit in the pH range of 4.00 to 9.18. (Use exact pH from NIST certificate.)

For more accurate measurement of a test solution, the meter and electrodes should be standardized against a buffer standard solution whose pH is close to the pH of the test solution.

▶**pH of a 5% Solution at 25 °C.** The solution to be tested is prepared by dissolving 5.0 g of the sample in 100 mL of carbon dioxide-free water* while protecting the solution from absorption of carbon dioxide from the atmosphere. The pH measurement is made as previously described.

For information, Table VIII reports the pH of a 5% solution of the pure salts. This value was determined experimentally for each salt by dissolving 10.0 g of salt in approximately 200 mL of water, then making the solution slightly acid and titrating with standard alkali. Another similar solution was prepared, made

---

*Carbon dioxide-free water can be prepared by purging reagent water with carbon dioxide-free air (using a gas dispersion tube) or nitrogen for at least 15 min or by boiling reagent water vigorously for at least 5 min and allowing it to cool while protected from absorption of carbon dioxide from the atmosphere; alternatively, fresh 18-mΩ deionized water can be used.

Table VIII. pH of a 5% Solution of Pure Salts at 25 °C

Salt	pH
Ammonium acetate	7.0
Ammonium bromide	4.9
Ammonium chloride	4.7
Ammonium nitrate	4.8
Ammonium phosphate, monobasic	4.1
Ammonium sulfate	5.2
Ammonium thiocyanate	4.9
Barium chloride, dihydrate	6.9
Barium nitrate	6.9
Calcium chloride, dihydrate	6.6
Calcium nitrate, tetrahydrate	6.6
Lithium perchlorate	7.0
Magnesium nitrate, hexahydrate	6.6
Magnesium sulfate, heptahydrate	6.8
Manganese chloride, tetrahydrate	5.4
Potassium acetate	9.0
Potassium bromate	7.0
Potassium bromide	7.0
Potassium chloride	7.0
Potassium chromate	9.3
Potassium iodate	7.0
Potassium iodide	7.0
Potassium nitrate	7.0
Potassium phosphate, monobasic	4.2
Potassium sodium tartrate, tetrahydrate	8.4
Potassium sulfate	7.3
Potassium thiocyanate	7.0
Sodium acetate, anhydrous	8.9
Sodium acetate trihydrate	8.9
Sodium bromide	7.0
Sodium chloride	7.0
Sodium molybdate dihydrate	8.1
Sodium nitrate	7.0
Sodium phosphate, dibasic, heptahydrate	9.0
Sodium phosphate, monobasic, monohydrate	4.2
Sodium pyrophosphate, decahydrate	10.4
Sodium sulfate	7.2
Sodium tartrate dihydrate	8.4
Sodium thiosulfate, pentahydrate	7.9
Zinc sulfate heptahydrate	5.4

slightly alkaline, and titrated with standard acid. Graphs were constructed for each titration by plotting pH vs. milliliters of titrating solution. The average of the two end points so determined is reported as the pH of a 5% solution containing no free acid or alkali.

pH of Other Concentrations. Directions for the preparation of solutions at concentrations of other than 5% are given in individual specifications. The pH measurements are made as previously described.

pH of Buffer Standard Solutions. For reagents that are suitable for use as pH standards, concentrations are stated in the individual specifications. They are expressed in the same units (molal) as the NIST standard reference materials to which they are compared. The meter and electrodes must be capable of a precision of 0.005 pH unit at the specified pH for satisfactory results.

Polarographic Analysis

Polarography offers a rapid and sensitive method of determining a number of ions or compounds that are electrolytically reducible, usually at a dropping mercury cathode, in a solution with appropriate electrical conductivity. Various commercial instruments are available. The measurements depend upon obtaining the curve of current transported vs. applied potential as the latter is scanned in a negative direction. Typical apparatus consists of a reference half cell (usually a saturated calomel electrode), an electrolyte container or cell that can be de-aerated by bubbling with nitrogen or argon, the dropping mercury electrode, a source of uniformly increasing potential, means of converting the resulting current to a measurable signal, and a recorder with suitable ranges of sensitivity down to 1 μA full scale.

As the voltage applied to the cell rises, the residual current of the supporting electrolyte increases very gradually until the potential at the dropping electrode reaches the reduction potential of the most easily reduced species, ordinarily the substance under determination. At this point the current rises rapidly above the residual value to a limiting value dependent on the concentration of the electroactive species. The difference between the residual current and the limiting current comprises the diffusion current of the electroactive species. The resultant current–voltage curve, which is overall S-shaped, is called a *polarographic curve* or *polarogram*. The midpoint in the rise of the curve, one-half the distance between the residual current and the limiting current, is called the *half-wave potential* and is characteristic of the electroactive substance. Ultimately, as the applied voltage continues to rise, the current increases again as a result of the reduction of the supporting electrolyte.

Because the observed current is also a function of the size of the growing mercury drop at the cathode, the curve will show pronounced oscillations as the

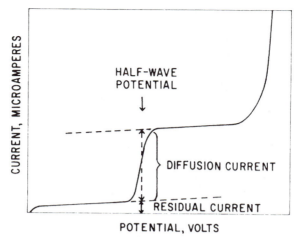

Figure 3. Polarographic curve.

drops grow and fall. Most instruments incorporate means of damping out such oscillations. Reading of the current is simplified with the damped curve, but on some early-model instruments the potential readings may be shifted by up to 0.1 V from those observed without damping. A typical damped polarographic curve is shown in Figure 3. Calibration is performed with known solutions of the material to be determined.

Directions for individual reagent chemicals normally include appropriate means of treating the sample to convert it to a suitable electrolyte, free from oxygen and containing any needed pH buffer and maximum suppressor. *Maxima* are abnormally high current readings that may distort a polarogram—generally at the top of the step in the wave. Most maxima can be suppressed by gelatin, some indicators, or a surface-active agent. Use of too much suppressor, however, may lower the diffusion current of the material being determined. Maxima can be avoided if the method is sensitive and the concentration of the substance being determined is low.

Typical substances that may be determined polarographically in this monograph include many metals (especially lead, zinc, and cadmium), some anions (such as bromate, iodate, and nitrate), and several types of organic materials (such as aldehydes, ketones, and peroxides).

A number of extensions of the original polarographic method have advanced the capabilities of polarography. For example, differential pulse polarography and square wave polarography, for which commercial instruments are widely available, permit both higher sensitivity and greater resolution of constituents. With the increased sensitivity, however, greater precautions against contamination and interfering materials are necessary.

Differential Pulse Polarography. The detection limits in conventional dc polarography, typically 10^{-5} M, are set by the charging current resulting from the continuous growth of the mercury-drop electrode. One approach to improving the sensitivity of polarographic methods has been the development of pulse methods that reduce the contribution of charging current to the overall measured current. In pulse polarography, instead of applying a continuously increasing potential (as in dc polarography), a voltage pulse is applied to the mercury drop near the end of its drop time. The reason for this can be seen in Figure 4. The faradaic current, i_f (or that due to the reduction of the electroactive species), increases continuously during the time of the drop and is greatest at that moment just before the drop falls. The charging current (i_c), however, decreases steadily over the time of the drop. The ratio of faradaic to charging current (i.e., the sensitivity) can be optimized by sampling the current at the end of the drop time.

In differential pulse polarography, a voltage pulse, typically 50 ms, is applied to the electrode during the last portion of the drop's time (which is controlled by a mechanical drop dislodger and is commonly 0.5, 1.0, or 2.0 s). When the voltage is first applied, the charging current is very large, but it decays exponentially. The pulses have a constant amplitude of between 5 and 100 mV, and they are superimposed on a slowly increasing voltage ramp, as shown in Figure 5. The current is measured twice: immediately preceding the pulse and near the end of the drop time. The overall response output to the recorder is the difference in the two currents (Δi) sampled. The plot of Δi as a function of voltage is peak-shaped, as shown in Figure 6. By measuring the difference in current before the pulse and toward the end of the pulse, the contribution of the charging current to the overall measured current is significantly reduced. In addition, the backgrounds are flat, as opposed to the sloping backgrounds found in dc and some other polarographic techniques. Because the polarograms are peak-shaped, the limiting currents are easy to determine in differential pulse polarography. The reduction in charging current results in detection limits as low as 10^{-8} M.

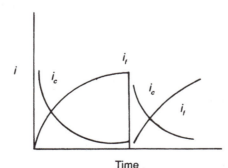

Time

Figure 4. Comparison of faradaic current (i_f) and charging current (i_c) vs. time.

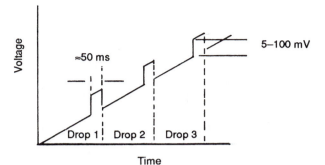

Figure 5. Voltage pulse excitation for differential pulse polarography.

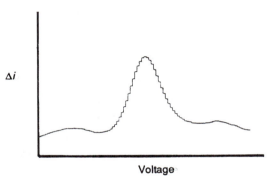

Figure 6. Differential pulse polarogram.

Instruments suitable for differential pulse polarography and other voltametric techniques are commercially available from several manufacturers. The instrument settings and conditions should be set in accordance with the manufacturer's recommendations, and these may not necessarily agree with those described in the next section.

Specific Procedures

▶**Ammonium.** Transfer the specified amount of sample to each of two 25-mL volumetric flasks containing 5 mL of water. Neutralize, if necessary, with the appropriate quantity of ammonia-free 6 N sodium hydroxide (prepared by boiling and cooling 6 N sodium hydroxide solution). To one of the flasks, add the indicated quantity of ammonium ion. To each, add 5 mL of acetate buffer (to establish pH 4) and 5 mL of ammonia-free formaldehyde (see below). Heat on a hot plate (≈100 °C) for 5 min with occasional shaking, cool, and dilute to volume with water. Transfer a suitable portion to the cell, deoxygenate for 5 min with inert gases such as nitrogen or argon, and record the polarogram from −0.60 to −1.10 V vs. SCE. The following instrumental settings have proved to be satisfactory: drop time, 1 s; scan rate, 5 mV/s; sensitivity, 0.5 μA full scale; modulation amplitude, 50 mV. The peak for the ammonium–formaldehyde derivative occurs at about −0.85 V, but its position is dependent on pH.

The peak for the sample should not exceed one-half of the peak for the sample plus standard.

Reagents

1. Acetate Buffer. Dissolve 10.4 g of sodium acetate trihydrate, CH_3COONa · $3H_2O$, in 200 mL of water, and add 32 mL of glacial acetic acid.

2. Ammonia-Free Formaldehyde. Add 50 g of a cation-exchange resin in hydrogen form (e.g., Dowex HCR-W2, 50–100 mesh) to 250-mL bottle of 37% formaldehyde solution. Place the bottle in a shaker for 1 h, or let it stand for several hours and mix occasionally.

►**Bromate and Iodate.** Add to the polarographic cell an appropriate volume of solution prepared as described in the specification. Deoxygenate for 5 min with inert gases such as nitrogen or argon, and record the polarogram from −0.6 V to −1.8 V vs. SCE. The following conditions are satisfactory: drop time, 1 s; scan rate, 5 mV/s; sensitivity, 0.5 μA full scale; modulation amplitude, 25 mV. The peak occurs at about −1.25 V for iodate and −1.6 V for bromate. If the ionic strength is not sufficiently high, the bromate wave becomes more negative and the cathodic wave may interfere. In this case, calcium chloride may be added, and a reagent blank correction may be applied.

The peak for the sample should not exceed one-half of the peak for the sample plus standard.

►**Carbonyl Compounds.** To the specified quantity of sample in each of two 25-mL volumetric flasks, add 10 mL of phosphate buffer (see below) and 2 mL of 2% hydrazine sulfate solution. To one of the flasks, add the indicated quantities of standard solutions. Dilute each to volume with water and mix. Promptly transfer a suitable portion to the cell, deoxygenate for 5 min with nitrogen, and record the polarogram from −0.6 V to −1.6 V vs. SCE. The following conditions are satisfactory: drop time, 1 s; scan rate, 5 mV/s; sensitivity, 0.5 μA full scale; modulation amplitude, 50 mV.

The peak heights for the sample should not be greater than one-half of the peak heights for the sample plus standards. Approximate peak potentials for the hydrazones of known carbonyl compounds are as follows: acetaldehyde, −1.1 V; acetone, −1.3 V; 2-butanone, −1.30 V; butyraldehyde, −1.20 V; formaldehyde, −1.0 V; propionaldhyde, −1.1 V. When formaldehyde alone is specified as the standard, the sum of the carbonyl peak heights in the sample should not be greater than the height due to added formaldeyde in the sample plus standard.

Reagent

Phosphate Buffer. Dissolve 10.0 g each of monobasic sodium phosphate monohydrate ($NaH_2PO_4 \cdot H_2O$) and anhydrous dibasic sodium phosphate (Na_2HPO_4) in water, and dilute with water to 500 mL.

▶**Lead (and Cadmium) in Zinc Compounds.** Prepare a solution with the prescribed quantities of sample and hydrochloric acid in water to make 25.0 mL. Prepare a standard with the same volume of hydrochloric acid and the specified quantities of lead (and cadmium) in 25.0 mL. Place 10 mL of either solution in the polarographic cell, deoxygenate with nitrogen for 5 min, and record the polarogram from –0.25 to –0.80 V vs. SCE. The following conditions are satisfactory: drop time, 1 s; scan rate, 5 mV/s; sensitivity, 0.5 μA full scale; modulation amplitude, 50 mV. Read the peak heights for lead at approximately –0.44 V and those for cadmium at approximately –0.63 V.

The peak(s) for the sample should not be greater than the peak(s) for the standard.

▶**Nitrite.** Dissolve the prescribed weight of sample in 20 mL of mixed reagent solution (see below), and dilute with water to 25 mL. Prepare a standard with the specified quantity of nitrite and 20 mL of reagent solution, also diluted to 25 mL. Place a suitable volume of either solution in the polarographic cell, deoxygenate for 5 min with nitrogen, and record the polarogram from –0.30 to –0.80 V vs. Ag/AgCl electrode. The following conditions are satisfactory: drop time, 0.5 s; scan rate, 10 mV/s; sensitivity, 0.2 μA full scale; modulation amplitude, 50 mV. Read the peak height at approximately –0.52 V. The peak for the sample should not be greater than that for the standard.

Mixed Reagent Solution. Prepare by adding 5 mL of solution A, 10 mL of solution B, and 10 mL of solution C to 100 mL of water in an amber bottle. A fresh mixture should be prepared each week.

Solution A. Dissolve 0.044 g of diphenylamine in 40 mL of methanol, and dilute to 100 mL with water.

Solution B. Potassium thiocyanate, 0.1 N.

Solution C. Dilute 7.0 mL of 60% perchloric acid to 250 mL with water.

TITRIMETRY: ELECTROMETRIC END-POINT DETECTION

Potentiometric Titrations

Potentiometric titrations are titrations in which the equivalence point is determined from the rate of change of the potential difference between two electrodes.

The potential difference can be measured by any reliable millivoltmeter, including a pH meter that can be read either in pH units or millivolts. (These titrations can also be made by using commercially available automatic titrators, which either plot the complete titration or act to close an electrically operated buret valve exactly at the equivalence point.) The electrode pair, which consists of an indicating electrode and a reference electrode, depends on the type of titration and is indicated in the individual specifications.

Relatively large increments of the titrant may be added until the equivalence point is approached, usually within 1 mL. In the vicinity of the end point, small equal increments (0.1 mL, for example) are added; after allowing sufficient time for the indicator electrode to reach a constant potential, the voltage and volume of titrant are recorded. The titration is continued for several increments of titrant beyond the end point. For acid–base titrations, the pH is recorded instead of the potential difference but is treated the same as potential difference for the determination of the end point.

▶**First Derivative Method for the Determination of the Equivalence Point.** When the titration is symmetrical about the end point (same rate of change of potential before and after the end point), the position of the end point may be found by a graphical plot of $\Delta E/\Delta V$ vs. V, where E is the potential and V is the volume of titrant. The end point corresponds to the maximum value of $\Delta E/\Delta V$. (This is a plot of the first derivative, dE/dV, of $E = f(V)$ vs. the volume of titrant and is referred to as the first derivative method.)

▶**Second Derivative Method for the Determination of the Equivalence Point.** The end point can be determined by the use of the second derivative, d^2E/dV^2, of $E = f(V)$ and is the point where d^2E/dV^2 becomes zero. This point can be determined mathematically, assuming there is no significant difference between the average slope $\Delta E/\Delta V$ and the true slope dE/dV. The end point lies in the increment where the ΔE is the greatest. The amount of titrant to be added to the buret reading at the beginning of the interval is found by multiplying the volume of the increment by the factor in which the numerator is the last $+\Delta^2E$ value and the denominator is the sum of the last $+\Delta^2E$ value and the first $-\Delta^2E$ value, disregarding the sign. When the volume increments are identical, the method can be reduced to the use of the first and second differences. This simplification is employed in a typical example shown in Table IX.

Water by the Karl Fischer Method

The water content of most of the organic reagents discussed in *Reagent Chemicals* is determined by the Karl Fischer method (for more on Karl Fischer titration, see Scholz, 1984; Schilt, 1991). This method involves the titration of the sample in methanol or any suitable solvent—for example, pyridine, formamide, or petroleum ether—with the Karl Fischer reagent. This reagent consists of iodine, sulfur

Table IX. Typical Example of the Second Derivative Method for Determination of the Equivalence Point

V (mL)	E (mV)	ΔE	Δ²E
16.00	504		
		6	
16.10	510		+4
		10	
16.20	520		+20
		30	
16.30	550		+60
		90	
16.40	640		−50
		40	
16.50	680		−30
		10	
16.60	690		

$$V(\text{endpoint}) = 16.30 + 0.10 \times \frac{60}{60 + 50} = 16.35 \text{ mL}$$

dioxide, an amine, and a solvent in which the iodine and the sulfur dioxide are rapidly and quantitatively consumed by the water in the sample. The end point is detected either visually from the color change caused by free iodine or electrometrically. The latter method is preferred in most cases and necessary in the case of colored solutions.

Two different techniques can be used to add the iodine needed in the reaction. In the volumetric Karl Fischer titration, the Karl Fischer reagent is added by means of a volumetric buret. With this technique, it is necessary to standardize the reagent often, and the amount of reagent added must be measured accurately. This accuracy is especially important in determining small amounts of water. A small error in delivering this required amount of reagent can lead to a large error in the determined water content.

The development of commercial coulometric instrumentation has enabled the alternative coulometric titration method to be introduced into the analytical laboratory. Coulometry is an electrochemical process involving the generation of various materials in direct proportion to their equivalent weights. As applied to the determination of water by the Karl Fischer method, the iodine needed in the reaction is generated in situ (inside the actual titration vessel) from an iodide-containing solution. As in the volumetric Karl Fischer method, the excess iodine beyond the end point is usually determined by an electrometric technique. The amount of iodine generated for the reaction is determined accurately and related to the amount of water present in the sample. The coulometric Karl Fischer titration is a more sensitive procedure (i.e., it can determine smaller amounts of water). Because of this, exclusion of extraneous sources of moisture is absolutely necessary. No standardization of reagent is required because this is an absolute method depending only on electrochemical laws and the accurate measurement of electrical current. The method is generally recommended when the samples are liquid and have a low water content.

Volumetric Procedure

The apparatus described here, and the Karl Fischer reagent described under Reagent Solutions, page 87, are suitable for the determination of water by this method. There are, however, a number of commercial titration instruments and Karl Fischer reagents available. These instruments and reagents may be used in accord with their manufacturers' instructions.

Apparatus. *Enclosed Titration System.* The Karl Fischer reagent is highly sensitive to water, and exposure to atmospheric moisture must be avoided. The reagent is best stored in a reservoir connected to an automatic buret with all exits to the atmosphere protected with a suitable desiccant—for example, silica gel or anhydrous calcium chloride. The closed titration flask, with provision for insertion of platinum–platinum electrodes and insertion of the sample through a septum or removable closure, must be connected to the buret in such a way that atmospheric moisture cannot enter through the joint. Because of the necessity for a closed titration system, stirring is best accomplished magnetically.

Polarizing Unit for Use with the Electrometric End Point. A simple polarizing unit consists of a 1.5-V dry battery connected across a radio-type potentiometer of about 100 ohms. (One platinum electrode is connected to one end of the potentiometer; the other platinum electrode is connected through a microammeter to the movable control of the potentiometer.) By adjustment of the movable contact, the initial polarizing current can be set to the desired value of 100 microamperes through a 2000-ohm precision resistor. This current passes through the platinum–platinum electrodes, which are immersed in the solution to be titrated.

Reagents. *Preparation of the Reagent.* Preparation should proceed according to Mitchell and Smith (1980). *See* Reagent Solutions, page 82.

Commercial Reagents. Stabilized Karl Fischer reagents for volumetric water determinations are available from several suppliers. Pyridine and pyridine-free reagents can be used for water determination. For ketones and aldehydes, specially formulated reagents are available and should be used. Pyridine and pyridine-free reagents have defined water capacities, which should be considered when these reagents are used.

Standardization of Karl Fischer Reagents. The Karl Fischer reagent, pre-pared as described under Reagent Solutions, gradually deteriorates. It should be standardized within an hour before use, or daily if it is in continuous use. The pyridine-free reagents are more stable, and titrant–solvent systems exist that are free from deterioration.

Method A

Place 25 to 50 mL of methanol in the titration flask, and titrate with Karl Fis-cher reagent to the electrometric end point. The deflection should be main-tained for at least 30 s. This is a blank titration on the water contained in the methanol and the titration flask. Record the volume. Add 0.05 mL of water, accurately weighed, and again titrate with Karl Fischer reagent to the end point. Calculate the strength of the reagent in milligrams of water per millili-ter of Karl Fischer reagent from the weight of water added and the net vol-ume of Karl Fischer reagent used.

Method B

Use the same general procedure but add a weighed amount of reagent sodium tartrate dihydrate instead of the water, and calculate the Karl Fischer titration factor (expressed in milligrams of water per milliliter) by the following:

$$\text{Karl Fischer titration factor} = \frac{\text{wt. of sodium tartrate (mg)} \times 0.1566}{\text{Karl Fischer reagent (mL)}}$$

Sodium tartrate does not dissolve completely in methanol and thus can cause premature, transient end points near the true end point. Hence the titration must be performed slowly, and the suspension must be well stirred when near the final end point.

▶Procedure for Samples.

Method 1

Titrate 25 to 50 mL of methanol or other suitable solvent, as specified, to an end point with the Karl Fischer reagent, as in the standardization. Unstopper the titration flask and rapidly add an accurately weighed portion of the solid or liquid sample to be tested. (For liquids, a weight buret or syringe is conve-nient or, when the density is known, a measured volume of sample may be introduced with a syringe or pipet.) Quickly restopper the flask, and titrate again with Karl Fischer reagent to an end point. The percentage of water in the sample is calculated from the known strength of the Karl Fischer reagent, the net volume used, and the sample weight.

Coulometric Procedure

Apparatus. Follow the instrument manufacturer's instructions for specific operation procedures.

Reagents. Pyridine and pyridine-free coulometric reagents have defined water capacities. Several suppliers offer solutions that are used for assuring the performance of coulometric instruments. These usually are nonhygroscopic mixtures of solvents with a known water content. Reagent water may be used if the small amount required is weighed accurately.

▶**Procedure for Samples.**

Method 2
Titrate the sample cell to dryness. Activate the sample titration procedure, and add the specified amount of sample with a syringe (if liquid) or as a weighed solid. Determine the weight of the liquid sample by weighing the syringe before and after injection or by calculating from the volume injected and the known sample density. The instrument will indicate the end of the titration and the amount of water in the sample is automatically calculated.

Ion-Exchange Column Assays

A 100-mL capacity class B buret with Teflon TFE or PTFE stopcock or a commercially available column having 1.5–2 cm bore may be used as a column. For fluoride assays, acrylic–polyethylene buret having Teflon stopcocks is preferred. All other assays may be carried out in a glass column of the same dimensions.

NOTE: All water used in the assay must be free of carbon dioxide and ammonia.

Column Preparation
Cation-exchange resin, Dowex HCR-W2 H^+ form 16–20 mesh or Amberlite IR-120 or equivalent is suitable for assays. About 75 mL of cation-exchange resin is placed in a liter-size container and filled with water, stirred, and let settle for a minute or two. After settling, water is gently decanted, and the washing repeated at least 4–5 times. In a suitable column, a cotton ball is inserted above the stopcock to prevent resin from plugging the stopcock and to achieve an even flow of the eluate. Following the last washing of the resin and decanting of the wash, the slurry is poured into the column with the stopcock open. The water level is always kept above the resin. Ion-exchange resin in the H^+ form should be washed with water until about 50 mL of eluate requires about 0.05 mL of 0.1 N sodium hydroxide to neutralize, using phenolphthalein as an indicator. An ion-exchange resin, which is in Na^+ form, should be regenerated by adding approximately 100 mL of 4 N hydrochloric acid through the column and then being washed until 0.05 mL of 0.1 N sodium hydroxide is required to neutralize about 50 mL of eluate, using phenolphthalein as an indicator solution. About 3–4 assays may be carried out, but drastically lower assays indicate that a new ion-exchange resin is required.

Method

Prepare the sample as directed in the individual tests of the monographs. Pass the sample solution through a cation-exchange column with water at a rate of about 5 mL/min., and collect about 250 mL of the eluate in a 500-mL titration flask. Wash the resin in the column at a rate of about 10 mL/min into the same titration flask, and titrate with 0.1 N sodium hydroxide. Continue the elution until 50 mL of the eluate requires no further titration.

CHROMATOGRAPHY

Chromatography is an analytical technique used in quantitative determination of purity of most organic and an increasing number of inorganic reagent chemicals and standard-grade reference materials. The broad scope of chromatography allows it to be used in the separation, identification, and assay of diverse chemical species, ranging from simple metal ions to compounds of complex molecular structure, such as proteins.

In chromatography, the separation of individual components in a mixture is achieved when a mobile phase is passed over a stationary phase. Differences in affinities of various substances for these phases result in their separation.

Chromatography can be divided into two main branches, depending on whether the mobile phase is a gas or a liquid. Gas chromatography is principally used for analysis of volatile, thermally stable materials. Liquid chromatography is particularly useful for analysis of nonvolatile or thermally unstable organic substances. Ion chromatography, a technique in which anions and cations can be determined by using the principles of ion exchange, is a form of liquid chromatography. Thin-layer chromatography, often called planar chromatography, is also a form of liquid chromatography.

Gas Chromatography

Gas chromatography (GC) is used in this monograph to determine the assay and/or the trace impurities in both organic reagents and standards. Gas chromatography may be subdivided into gas–liquid and gas–solid chromatography. Gas–liquid chromatography is by far the most widely used form of gas chromatography.

The heart of a GC system is the column, which is contained in an oven operated in either the isothermal or the temperature-programmed mode. Columns packed with a nonvolatile liquid phase coated on a porous solid support were used extensively in early editions of this monograph. In this edition, only capillary columns, constructed of fused silica onto which is bonded the liquid phase, are used because of their greater resolving power and chemical inertness.

Conventional capillary gas chromatography uses long, narrow-bore columns with an inside diameter (i.d.) from 0.22 to 0.32 mm, coated or bonded with a thin film of the liquid phase. This arrangement results in high resolution but low sample capacity. Special injection techniques and hardware are employed. The intro-duction of wide-bore capillary columns, which have an i.d. of typically 0.53 mm and a relatively thick film of the liquid phase (1 μm to 5 μm), can allow a labora-tory to use a standard packed-column instrument and conditions while gaining the advantages of capillary technology.

Direct flash vaporization or on-column injection of the sample with stan-dard-gauge needles can be used with capillary columns. Two types of flash vapor-ization injection techniques can be employed: a split mode of operation, in which part of the sample is vented from the injector, or a splitless mode, in which a smaller volume is injected and no portion of the sample is vented from the injec-tor. The splitless mode, using a 0.1-μL sample size, is recommended for the assay of most reagent solvents, while the split mode often is recommended for analysis of most of the standard-grade reference materials.

The conventional split injector is a flash vaporization device. The liquid plug, introduced with a syringe, is immediately volatilized, and a small fraction of the resultant vapor enters the column while the major portion is vented to waste. To perform a GC analysis using a split injection technique, the GC instrument should be configured such that a preheated carrier gas, controlled by a pressure regulator or a combination of a flow controller and a back pressure regulator, enters the injector. The flow is divided into two streams. One stream of carrier gas flows upward and purges the septum. The septum purge flow is controlled by a needle valve. Septum purge flow rates are usually between 3 and 5 mL/min. A high flow of carrier gas enters the vaporization chamber, which is a glass or quartz liner, where the vaporized sample is mixed with the carrier gas. The mixed stream is split at the column inlet, and only a small fraction enters the column. A needle valve or flow controller regulates the split ratio.

Split ratios (measured column flow/measured inlet flow) typically range from 1:50 to 1:500 for conventional capillary columns (0.22 mm to 0.32 mm i.d.). For high sample capacity columns, such as wide bore columns and/or thick film columns, low split ratios (1:5 to 1:50) are commonly used.

The method most commonly recommended for analysis of pure chemical components (for example, matrices with a narrow constant boiling range) is the hot needle, fast sample introduction. In this method, the sample is taken into the syringe barrel (typically 2–5 μL in a 10-μL syringe) without leaving an air plug between the sample and plunger. After insertion into the injection zone, the nee-dle is allowed to heat up for 3 to 5 s. This period of time is sufficient for the needle to be heated to the injector temperature. Then, the sample is injected by rapidly pushing the plunger down (fast injection), after which the needle is withdrawn from the injector within one second. Either manual or automatic sample intro-ductions can be used. The measurement reproducibility will be enhanced by not varying the injected volume, which typically should be 0.5 μL to 2.0 μL. The use of an automatic injection system can significantly enhance measurement preci-sion. Also, loosely packing the injection liner with deactivated glass wool or glass beads can provide thorough mixing between sample and carrier gas, yielding less sample discrimination and better measurement precision. However, analysts should be aware of adsorption and decomposition.

In conjunction with wide-bore columns, on-column injection can minimize sample degradation while increasing the reproducibility of results. On-column injection allows the injection of a liquid sample directly into the inlet of the column. Excellent quantitative precision and accuracy for thermally labile compounds and wide volatility range samples have been reported. Sample sizes range from 0.5 μL to 2 μL. The injection should be performed as fast as possible, with the column oven temperature below or equal to the boiling point of the solvent.

After injection, the liquid is allowed to form a stable film (flooded zone). This takes several seconds. If the solutes to be analyzed differ much in boiling points, in comparison to the solvent, ballistic heating to high temperature is allowed. On the other hand, if the solute boiling points do not differ too much compared to the solvent, temperature programming is applied to fully exploit the solvent effect. When peak splitting and/or peak distortion is observed, the connection of a retention gap can provide the solution.

If the composition of the sample is not that complex, the use of megabore columns is recommended, particularly since automated injection becomes easier. If high resolution with automated injection is needed, a deactivated but uncoated wide bore precolumn (20 cm to 50 cm in length) should be connected to a conventional analytical narrow bore column. Hydrogen is the carrier gas of choice. If H_2 cannot be used for safety reasons, helium may be substituted. High carrier-gas velocities (50–80 cm/sec H_2, 30–50 cm/sec He) ensure negligible band broadening.

In comparison to vaporizing injectors (that is, split and splitless), there are some disadvantages to on-column injections. The two primary disadvantages pertain to sample pretreatment. First, because the sample is introduced directly onto the column, relatively "clean" samples must be prepared. Nonvolatile and less-volatile materials collect at the head of the column, causing a loss of separation efficiency; therefore, sample cleanup is a prerequisite. Second, many samples may be too concentrated for on-column injection and will need to be diluted.

Wide-bore columns can be used either in a high-resolution mode (carrier-gas flow rates less than 10 mL/min) to obtain optimum resolution of sample components or at higher flow rates (10–30 mL/min), which will generate packed column-quality separations in a shorter time. The increased length of capillary columns allows for better separation of components, with the result that as few as three columns of high, moderate, and low polarity can handle the majority of analytical requirements.

A wide variety of detectors are used to quantify and/or identify the components in the eluent from the column. Thermal conductivity detectors (TCD) and flame ionization detectors (FID) are examples of general detectors that provide a linear response to most organic compounds. Electron capture detectors (ECD) and photoionization are examples of class-specific detectors often used in trace environmental analysis. Mass spectrometric-type detectors are used to provide positive component identification. The detector output after amplification is used to produce the *chromatogram*, a plot of component response vs. elution time. Modern systems convert the analog output to digital form, which allows for further manipulation and interpretation of the data.

Results from a chromatogram, when used to determine the assay of a reagent or standard, are often expressed as area percent. A response factor correction for each component is required for the most accurate results, especially when the sample components differ markedly in their detector response. An internal standard reduces error due to variations in injection quantities, column conditions, and detector conditions. Use of a TCD or FID minimizes the need to correct for response and is usually sufficient for determining reagent or standard assay.

The use of control charts to aid the analyst in visualizing chromatographic variability is suggested. To certify that assay results are valid, use of a system suitability test as described in a later section is recommended.

Gas Chromatography–Mass Spectrometry

Gas chromatography–mass spectrometry (GC–MS) is one of the techniques used in the identity requirement for standard-grade reference materials. Mass spectrometers are used for many kinds of chemical analysis, especially those where identity or proof of chemical structure is critical. Examples range from environmental analysis to the analysis of petroleum products and biological materials, including the products of genetic engineering. Mass spectrometers use the difference in mass-to-charge ratio (m/e) of ionized atoms or molecules to separate them from each other. Mass spectrometry is therefore useful for quantitation of atoms or molecules and also for determining chemical and structural information about molecules. Molecules have distinctive fragmentation patterns that provide structural information to identify structural components. The largest peak in a mass spectrum is called the *base peak* and is assigned an arbitrary height of 100. The remaining peaks are then normalized to the base peak.

The general operation of a mass spectrometer is:

1. creating gas-phase ions;
2. separating the ions in space or time based on their mass-to-charge ratio; and
3. measuring the quantity of ions of each mass-to-charge ratio.

The ion separation power of a mass spectrometer is described by the resolution, which is defined as $R = m/\Delta m$, where m is the ion mass and Δm is the difference in mass between two resolvable peaks in a mass spectrum with similar mass values. For example, a mass specrometer with a resolution of 1000 can resolve an ion with a m/e of 100.0 from an ion with an m/e of 100.1.

In general, a mass spectrometer consists of an ion source, a mass-selective analyzer, and an ion detector. Since mass spectrometers create and manipulate gas-phase ions, they operate in a high-vacuum system. The magnetic-sector, quadrupole, and time-of-flight designs also require extraction and acceleration ion optics to transfer ions from the source region into the mass analyzer.

Liquid Chromatography

Many reagent chemicals and standard-grade reference materials described in this monograph can be assayed by and/or tested for suitability for use in liquid chromatography (LC). Liquid chromatography is a technique in which the sample interacts with both the liquid mobile phase and the stationary phase to effect a separation. An LC system consists of a pump that delivers a liquid, usually a solvent for the sample, at a constant flow rate through a sample injector, a column, and a detector. The solvent is called the mobile phase. Typically, the flow rate is 1 to 10 mL/min for conventional liquid chromatography. In isocratic operation, the composition of the mobile phase is kept constant, while in gradient elution, it is varied during the analysis. Gradient elution is required when the sample mixture contains components with a wide range of affinity for the stationary phase. In the isocratic mode, the purity of the solvents is less critical, as the impurities are adsorbed–desorbed at a constant rate, whereas in gradient elution impurities may result in extraneous peaks and/or shifts in the baseline. This characteristic necessitates the incorporation of a gradient elution test in this monograph to verify the quality of a solvent when used in this most stringent LC system mode of operation.

In liquid chromatography, a dilute solution of the sample to be analyzed is introduced into the mobile phase via the sample injector and enters the column, where it is separated into its individual components. The columns are usually steel or glass, densely packed with semirigid organic gels or rigid inorganic silica microspheres, to which a variety of substrates can be chemically bonded and whose typical particle size is 3 μm to 10 μm.

Several mechanisms of separation are possible in liquid chromatography. In *adsorption chromatography,* separation is based on adsorption–desorption kinetics, whereas in *partition chromatography,* the separation is based on partitioning of the components between the mobile and stationary phases. Ion exchange is the dominant mechanism in ion chromatography. In practice, a successful separation may involve a combination of separation mechanisms.

A commonly used term, coined by early chromatographers for describing separations dominated by adsorption–desorption, is *normal phase.* In normal phase chromatography, the stationary phase is strongly polar (for instance, silica or aminopropyl), and the mobile phase is less polar (for instance, hexane). Polar components are thus retained on the column longer than less-polar materials.

A second term of historical origin is *reverse phase.* Reverse phase generally applies to separations dominated by partition chromatography, and the elution of components in a reverse phase separation are more or less reversed from the order that would be obtained in a normal phase separation. Reverse phase, the more widely used mode of chromatography, uses a nonpolar (hydrophobic) stationary phase, such as C-18 (octadecyl) chemically bonded to silica, while the mobile phase is a polar liquid such as acetonitrile–water. Hydrophobic (non-polar) components are retained longer than hydrophilic (polar) components. The elution properties of the mobile phase can be adjusted to modify a separation by addition of appropriate ionic modifiers. These modifiers can be chosen for their ability to either suppress or enhance ion formation in the sample. When enhancement is chosen, the separation mechanism may involve both ion exchange and partition.

The wide range of available stationary phases in combination with changes in mobile-phase composition makes liquid chromatography a very flexible separation-assay technique. The back pressure on the pump is several hundred to several thousand psi, depending upon the particle size of the packing material, mobile-phase flow rate, and viscosity. The column eluent is continuously monitored by a sensitive detector chosen to respond either to the sample component alone (for instance, an ultraviolet photometer) or to a change in some physical property of the mobile phase due to the presence of the solute (for example, a differential refractometer). Other detectors such as electrochemical, fluorescence, etc., are also used for more specialized applications. When photometric detectors are used, solvents are often specified in absorbance units at a specific wavelength. The response of the detector is related to the concentration or weight of the solute and is displayed on a recorder. A computer is usually interfaced to the LC system to control the method and to collect and analyze data.

Ion Chromatography

Ion chromatography (IC) is a subset of liquid chromatography. Whereas conventional liquid chromatography is mainly used for the analysis of nonionic organic compounds, ion chromatography separates and determines ionic species or ionizable compounds, both organic and inorganic. Under ideal conditions, quantitation to the sub-part-per-billion level is attainable. For trace analysis, water and reagents of the highest quality need to be used. Deionized water of >18 MΩ cm resistivity is recommended.

For the testing of reagent chemicals, ion chromatography is ordinarily applied to the determination of anions present as minor impurities. Generally, the major component should be eliminated or greatly reduced in concentration to avoid overloading the anion-exchange column used for separation.

IC columns are usually 5–25 cm in length and 2–10 mm in diameter. Columns may be either metallic or plastic. Packing materials can be either anionic or cationic resins, depending on the separation desired, and there are several variations within each type. Typical particle sizes range from 5 to 20 μm.

In many anion applications of ion chromatography, buffered, aqueous, mobile phases (eluents) of approximately millimole strength are used. In most cases, IC analyses are carried out isocratically, and gradient elutions are used only for the more complex separations. Typical eluent flow rates for ion chromatography are 1–2 mL/min, with system operating pressures at 1000 psi or less. Microbore columns are operated at lower flow rates of 0.2–0.5 mL/min.

Standard and sample injections are normally made by using a loop injection system. The most widely used detection method for ion chromatography is conductivity. However, other detection methods, including ultraviolet-visible spectroscopy, electrochemistry, and fluorescence can be used. Conductivity is a universal detection mode for ions, whereas the other detectors provide selective, analyte-specific detection.

In conductimetric IC analysis, the impact of the significant background conductivity of the eluent on analyte determinations is minimized by chemical or electronic suppression. Chemical suppression involves notably reducing the eluent conductivity, as well as enhancing analyte response by means of a chemical reaction. Several innovative techniques have been developed to achieve these changes, including packed-column and membrane-based devices. Electronic suppression minimizes eluent background noise via the design of the electronic circuitry in the conductivity detector, although the actual background eluent conductivity is not reduced, as it is with chemical suppression.

Because of the two different modes of suppression, the columns and eluents used also fall into two categories. Columns for chemical suppression typically have higher ion-exchange capacities than those used for electronic suppression. Eluents used in chemical suppression also differ from those used in electronic suppression in that they are defined by the chemical reactions that occur in the suppression of the eluent.

Thin-Layer Chromatography

Thin-layer chromatography (TLC) is a very simple form of solid–liquid adsorption chromatography. It is probably the quickest, easiest, and most frequently applied technique for determining purity of organic compounds. As in other forms of liquid chromatography, the sample interacts with a liquid mobile phase and a stationary phase to effect partitioning of components based on their affinity to the solid and liquid phase. Thin-layer chromatography serves many purposes in the laboratory because of its simplicity. It is commonly used in organic synthesis to monitor chemical reactions. Starting materials, intermediates, and products often elute differently, and product formation or starting-material disappearance may be observed.

Another very common use is in purity determinations of organic compounds. Typically, it is not used as a quantitative technique, but it is useful for looking for impurities. In many cases, thin-layer chromatography is very sensitive and can be used to detect impurities of less than 1% in the sample. A single spot on a TLC plate is a good indication of purity. Thin-layer chromatography is also applied in selected requirements for standard-grade reference materials as a technique for purity confirmation. The technique is used with other complementary purity assays to screen standard-grade reference materials for impurities. Standard-grade reference materials have passed the purity test when, having used the specific conditions in the method, an analyst can observe a single spot.

In thin-layer chromatography, the stationary phase is spread as a thin layer over glass or plastic. Calcium sulfate or an organic polymer (such as starch) is added to the solid phase to bind the solid to the glass or plastic. Often, fluorescent material is added to the solid phase to aid in detection of analytes. TLC plates are commercially available or can be prepared in the laboratory. Glass or liquid plates are available commercially as sheets that are cut into strips for use. Plates can be prepared in the laboratory by preparing a slurry of the solid phase with a solvent, such as chloroform or methanol, and applying a thin coat over a glass plate. Plates are most effective when dried in an oven before use. The most common stationary phases used in thin-layer chromatography are silica gel and alumina. Reverse-phase TLC plates are also available for elution of polar compounds. It is common to perform TLC analysis with both silica and alumina to maximize the effectiveness of the technique.

Performing TLC analysis requires minimal equipment. The plates are cut into strips approximately 10 cm in height. The sample is dissolved in a volatile solvent and applied to the bottom of the plate. This is accomplished by applying or "spotting" the solution about 0.5 cm from the bottom of the plate using a thin capillary tube. The plate is placed into a development chamber that contains enough elution solvent to come just below the sample spot. The solvent then travels up the plate and moves the components of the sample at different rates according to their affinity. When the solvent is about 1 cm from the upper end of the plate, the plate is removed, the solvent line is marked and the plate is allowed to dry.

Detection of spots on the TLC plate is accomplished in a number of ways. The most common method is to view the plate under ultraviolet (UV) light. Compounds that fluoresce will be detected. Alternatively, the plates can contain a fluorescent material within the solid phase. In this case, the plate will fluoresce and compounds that do not fluoresce will appear as black spots on the plate. Another common method is to use iodine as a complexing agent. Many organic compounds form charge-transfer complexes with iodine, resulting in a dark-colored species. The TLC plate is placed in a chamber containing iodine and "developed" for several minutes. Upon removal from the chamber, the plates will contain dark spots; these spots should be immediately circled with a pencil, since the complex is reversible and the spots will fade away. In other detection methods, the plate is sprayed with a solution to "stain" the analytes. Common reagents are sulfuric acid solution, which will char many organic compounds, and potassium permanganate solution, which will oxidize many organic compounds. Combinations of two or more methods may be used to detect a broader range of analytes.

It is often useful to determine the distance that a particular compound has moved up the plate in relation to the solvent front. This value is called the retention factor and designated R_f. The R_f value is calculated by measuring the distance that the analyte moved from the origin and dividing that by the distance the solvent moved from the origin. For purity assays, it is desirable to have an R_f value in the range of 0.3–0.5.

RECOMMENDED CHROMATOGRAPHY PROCEDURES

Assay Methods

Procedures published in this monograph list the parameters and values established to give satisfactory results. These procedures should be considered adequate only for determining the constituents specified in the individual tests, and they may not necessarily separate other components in a given sample. An attempt is made to keep each procedure for reagents as simple as possible, while still retaining the desired sensitivity, because simplicity may lead to wider usage. The procedures used for standard-grade reference materials are, in general, more specific, since it is necessary to separate similar eluting impurities.

The list of parameters for gas chromatography includes such variables as column type, diameter and length of column, carrier-gas flow rate, detector type, and operating parameters. The list for liquid chromatography includes such variables as diameter and length of column, packing type and particle size, mobile-phase composition and flow rate, detector type, and operating parameters. For reagents, the parameters represent one set of conditions that result in reproducible analyses. Because of differences among instruments, one or more of the stated parameters may require adjustment for any given instrument. Therefore these parameters, except for the column and type of detector, should be regarded as guides to aid the analyst in establishing the optimum conditions for a particular instrument. Relative retention times are given as an aid in peak identification. Exact reproduction of those times is not essential. Parameters given for the standard-grade reference materials section, however, must be adhered to exactly as stated in the test.

Gas Chromatography

The volatile organic reagents in this monograph can be assayed by gas chromatography. A single set of instrument conditions shown here has been chosen and found to give satisfactory results for the assay of these reagents. Three columns of varying polarity can be used to achieve the desired separation: type I, low polarity, methyl silicone, 5 μm; type II, moderate polarity, mixed cyano, phenylmethyl silicone, 1.5 μm; and type III, high polarity, polyethylene glycol, 1 μm. The recommended column type and reagent-specific conditions can be found listed alphabetically under the individual reagents. The retention times and relative retention times vs. methanol of the organic reagents are listed alphabetically (Table X) and in increasing order (Table XI).

Table X. GC Retention Time of Reagents in Alphabetical Order

Compounds	Column of Choice	Type I, Nonpolar		Type II, Med. Polar		Type III, High Polar	
		RT	RR	RT	RR	RT	RR
Acetaldehyde	I	2.9	0.97	2.2	0.89	2.8	0.4
Acetic acid	I	8.8	2.9	N/A	N/A	N/A	N/A
Acetic anhydride	I	11.4	3.8	N/A	N/A	N/A	N/A
Acetone	I	4.9	1.6	3.4	1.4	4.1	0.6
Acetonitrile	I	4.7	1.6	4.4	1.8	8.7	1.4
Acetyl chloride	I	5.2	1.7	N/A	N/A	N/A	N/A
2-Aminoethanol	I	10.6	3.5	9.9	4.0	N/A	N/A
Aniline	I	19.5	6.5	17.3	7.0	21.9	3.4
Benzene	I	11.2	3.7	6.7	2.7	6.9	1.1
Benzyl alcohol	I	20.6	6.8	18.6	7.5	23.5	3.7
2-Butanone	I	8.5	2.8	6.0	2.4	6.0	1.0
Butyl acetate	I	15.3	5.1	11.7	4.7	10.6	1.7
Butyl alcohol	I	11.1	3.7	9.0	3.6	12.3	1.9
t-Butyl alcohol	I	6.3	2.1	4.1	1.7	6.5	1.0
Carbon disulfide	I	7.4	2.5	3.0	1.2	3.0	0.5
Carbon tetrachloride	I	11.2	3.7	6.0	2.4	5.3	0.8
Chlorobenzene	I	16.6	5.5	12.5	5.0	16.5	2.6
Chloroform	I	9.2	3.1	6.0	2.4	8.9	1.4
Cyclohexane	I	11.4	3.8	5.4	2.2	3.0	0.5
Cyclohexanone	I	17.5	5.8	14.9	6.0	15.4	2.4
1,2-Dichloroethane	I	10.3	3.4	7.4	3.0	10.0	1.6
Dichloromethane	I	6.2	2.0	3.5	1.4	6.4	1.0
Diethanolamine	I	20.6	6.9	20.0	8.0	N/A	N/A
Diethylamine	I	11.2	3.0	5.0	2.5	N/A	N/A
N,N-Dimethylformamide	I	14.2	4.7	13.6	5.5	16.2	2.5
Dimethylsulfoxide	III	15.7	5.2	16.0	6.4	20.5	3.2
Dioxane	I	12.2	4.1	9.1	3.6	10.3	1.6
Ethyl acetate	I	9.2	3.1	5.8	2.3	5.6	0.9
Ethyl alcohol	I	4.3	1.4	3.2	1.3	7.4	1.2
Ethyl ether	I	5.6	1.9	2.7	1.1	2.4	0.4
Ethylbenzene	I	17.0	5.6	12.5	5.0	11.8	1.8
Formamide	III	N/A	N/A	N/A	N/A	23.0	3.7
2-Furancarboxyaldehyde	I	15.8	5.2	14.2	5.7	17.9	2.8
Glycerol	II	19.8	6.6	19.7	7.9	N/A	N/A

Continued in next page

Table X—*Continued.*

Compounds	Column of Choice	Type I, Nonpolar		Type II, Med. Polar		Type III, High Polar	
		RT	RR	RT	RR	RT	RR
2-Hexane	I	8.6	2.9	3.2	1.3	2.3	0.4
n-Hexane	I	9.2	3.1	3.5	1.4	2.3	0.4
Isobutyl alcohol	I	10.0	3.3	7.9	3.2	11.2	1.7
Isopentyl alcohol	I	13.2	4.4	11.0	4.4	13.5	2.1
Isopropyl alcohol	I	5.4	1.8	3.8	1.5	7.3	1.1
Isopropyl ether	I	9.1	3.0	2.6	1.0	2.6	0.4
Methanol	I	3.0	1.0	2.5	1.0	6.4	1.0
2-Methoxyethanol	III	10.2	3.4	8.4	3.4	13.2	2.1
Methyl-*tert*-butyl ether	I	11.2	3.0	5.9	2.0	2.8	0.4
4-Methyl-2-pentanone	I	13.3	4.4	10.2	4.1	8.9	1.4
1-Methyl-2-pyrrolidone	I	20.6	6.8	19.3	7.7	21.8	3.4
Nitrobenzene	I	22.0	7.3	19.4	7.8	22.1	3.5
Nitromethane	III	7.2	2.4	7.0	2.8	12.2	1.9
1-Octanol	I	21.2	7.0	17.8	7.2	19.2	3.0
2-Octanol	I	20.2	6.7	16.0	6.4	18.2	2.9
1-Pentanol	I	14.2	4.7	11.8	4.7	14.3	2.2
Perchloroethylene	I	15.8	5.2	10.7	4.3	9.4	1.5
1,2-Propanediol	III	13.4	4.5	13.4	5.4	19.8	3.1
Perchloroethylene	I	15.8	5.2	10.7	4.3	9.4	1.5
Propionic acid	III	14.0	4.6	12.5	5.0	18.5	2.9
η-Propyl alcohol	III	7.6	2.5	5.9	2.4	10.1	1.6
Pyridine	I	13.4	4.5	10.6	4.2	13.2	2.1
Quinoline	II	N/A	N/A	121.6	8.7	25.0	3.9
Tetrahydrofuran	I	9.8	3.3	5.8	2.3	5.1	0.8
Toluene	I	14.4	4.8	10.0	4.0	9.7	1.5
1,1,1-Trichloroethane	I	10.6	3.5	6.0	2.4	5.4	0.9
Trichloroethylene	I	12.3	4.1	7.8	3.1	8.4	1.3
1,1,2-Trichlorotrifluoro-ethane	I	6.7	2.2	2.7	1.1	2.4	0.4
2,2,4-Trimethylpentane	I	12.4	4.1	5.9	2.4	2.6	0.4
1,2-Xylene	I	17.8	5.9	13.3	5.4	12.8	2.0
1,3-Xylene	I	17.2	5.7	12.6	5.1	12.0	1.9
1,4-Xylene	I	16.9	5.6	12.4	5.0	11.6	1.8

Note: RT is retention time; RR is relative retention (methanol 1.0); and N/A is not available.

Table XI. GC Retention Time of Reagents by Chronological Order on Type I Column

Compounds	Column of Choice	Type I, Nonpolar		Type II, Med. Polar		Type III, High Polar	
		RT	RR	RT	RR	RT	RR
Acetaldehyde	I	2.9	0.97	2.2	0.89	2.8	0.4
Methanol	I	3.0	1.0	2.5	1.0	6.4	1.0
Ethyl alcohol	I	4.3	1.4	3.2	1.3	7.4	1.2
Acetonitrile	I	4.7	1.6	4.4	1.8	8.7	1.4
Acetone	I	4.9	1.6	3.4	1.4	4.1	0.6
Acetyl chloride	I	5.2	1.7	N/A	N/A	N/A	N/A
Isopropyl alcohol	I	5.4	1.8	3.8	1.5	7.3	1.1
Ethyl ether	I	5.6	1.9	2.7	1.1	2.4	0.4
Dichloromethane	I	6.2	2.0	3.5	1.4	6.4	1.0
t-Butyl alcohol	I	6.3	2.1	4.1	1.7	6.5	1.0
1,1,2-Trichlorotrifluoro-ethane	I	6.7	2.2	2.7	1.1	2.4	0.4
Nitromethane	III	7.2	2.4	7.0	2.8	12.2	1.9
Carbon disulfide	I	7.4	2.5	3.0	1.2	3.0	0.5
η-Propyl alcohol	III	7.6	2.5	5.9	2.4	10.1	1.6
2-Butanone	I	8.5	2.8	6.0	2.4	6.0	1.0
2-Hexane	I	8.6	2.9	3.2	1.3	2.3	0.4
Acetic acid	I	8.8	2.9	N/A	N/A	N/A	N/A
Isopropyl ether	I	9.1	3.0	2.6	1.0	2.6	0.4
Chloroform	I	9.2	3.1	6.0	2.4	8.9	1.4
Ethyl acetate	I	9.2	3.1	5.8	2.3	5.6	0.9
n-Hexane	I	9.2	3.1	3.5	1.4	2.3	0.4
Tetrahydrofuran	I	9.8	3.3	5.8	2.3	5.1	0.8
Isobutyl alcohol	I	10.0	3.3	7.9	3.2	11.2	1.7
2-Methoxyethanol	III	10.2	3.4	8.4	3.4	13.2	2.1
1,2-Dichloroethane	I	10.3	3.4	7.4	3.0	10.0	1.6
2-Aminoethanol	I	10.6	3.5	9.9	4.0	N/A	N/A
1,1,1-Trichloroethane	I	10.6	3.5	6.0	2.4	5.4	0.9
Butyl alcohol	I	11.1	3.7	9.0	3.6	12.3	1.9
Benzene	I	11.2	3.7	6.7	2.7	6.9	1.1
Carbon tetrachloride	I	11.2	3.7	6.0	2.4	5.3	0.8
Diethylamine	I	11.2	3.0	5.0	2.5	N/A	N/A
Methyl-tert-butyl ether	I	11.2	3.0	5.9	2.0	2.8	0.4

Continued on next page

Table XI—*Continued.*

Compounds	Column of Choice	Type I, Nonpolar		Type II, Med. Polar		Type III, High Polar	
		RT	RR	RT	RR	RT	RR
Acetic anhydride	I	11.4	3.8	N/A	N/A	N/A	N/A
Cyclohexane	I	11.4	3.8	5.4	2.2	3.0	0.5
Dioxane	I	12.2	4.1	9.1	3.6	10.3	1.6
Trichloroethylene	I	12.3	4.1	7.8	3.1	8.4	1.3
2,2,4-Trimethylpentane	I	12.4	4.1	5.9	2.4	2.6	0.4
Isopentyl alcohol	I	13.2	4.4	11.0	4.4	13.5	2.1
4-Methyl-2-pentanone	I	13.3	4.4	10.2	4.1	8.9	1.4
1,2-Propanediol	III	13.4	4.5	13.4	5.4	19.8	3.1
Pyridine	I	13.4	4.5	10.6	4.2	13.2	2.1
Propionic acid	III	14.0	4.6	12.5	5.0	18.5	2.9
1-Pentanol	I	14.2	4.7	11.8	4.7	14.3	2.2
N,N-Dimethylformamide	I	14.2	4.7	13.6	5.5	16.2	2.5
Toluene	I	14.4	4.8	10.0	4.0	9.7	1.5
Butyl acetate	I	15.3	5.1	11.7	4.7	10.6	1.7
Dimethylsulfoxide	III	15.7	5.2	16.0	6.4	20.5	3.2
2-Furancarboxyaldehyde	I	15.8	5.2	14.2	5.7	17.9	2.8
Perchloroethylene	I	15.8	5.2	10.7	4.3	9.4	1.5
Chlorobenzene	I	16.6	6.6	12.5	5.0	16.5	2.6
1,4-Xylene	I	16.9	5.6	12.4	5.0	11.6	1.8
Ethylbenzene	I	17.0	5.6	12.5	5.0	11.8	1.8
1,3-Xylene	I	17.2	5.7	12.6	5.1	12.0	1.9
Cyclohexanone	I	17.5	5.8	14.9	6.0	15.4	2.4
1,2-Xylene	I	17.8	5.9	13.3	5.4	12.8	2.0
Aniline	I	19.5	6.5	17.3	7.0	21.9	3.4
Glycerol	II	19.8	6.6	19.7	7.9	N/A	N/A
2-Octanol	I	20.2	6.7	16.0	6.4	18.2	2.9
Benzyl alcohol	I	20.6	6.8	18.6	7.5	23.5	3.7
Diethanolamine	I	20.6	6.9	20.0	8.0	N/A	N/A
1-Methyl-2-pyrrolidone	I	20.6	6.8	19.3	7.7	21.8	3.4
1-Octanol	I	21.2	7.0	17.8	7.2	19.2	3.0
Nitrobenzene	I	22.0	7.3	19.4	7.8	22.1	3.5
Quinoline	II	N/A	N/A	21.6	8.7	25.0	3.9
Formamide	III	N/A	N/A	N/A	N/A	23.0	3.7

Note: RT is retention time; RR is relative retention (methanol 1.0); and N/A is not available.

►Procedure.

Column: 30-m × 0.53-mm i.d. fused silica capillary coated with a film that has been surface-bonded and cross-linked

Column Temperature: 40 °C isothermal for 5 min, then programmed to 220 °C at 10 °C/min., hold 2 min

Injector Temperature: 150–220 °C

Detector Temperature: 250 °C

Sample Size: 0.2 µL splitless

Carrier gas: helium at 3 mL/min

Detector: Thermal conductivity or flame ionization

Standard-grade reference materials must be assayed by multiple methods. Specific multiple GC methods are given under most classes of standard-grade reference materials. Methods differ mainly in the type of stationary phase employed and the physical dimensions of the capillary column.

Identity by Gas Chromatography–Mass Spectrometry

Standard-grade reference materials must be identified by multiple methods. GC–MS is one of the primary identity methods used. Specific conditions are found under each class of standard-grade reference materials.

Liquid Chromatography

Reagents and standard-grade reference materials can be analyzed by various modes of liquid chromatography. In this book, reagents not suitable for assay by gas chromatography are analyzed by liquid chromatography. Some standard-grade reference materials are used for LC applications. Specific tests for LC suitability for these materials are described in the class specifications. Thin-layer chromatography is used to confirm the purity of standard-grade reference materials. Ion chromatography is used to determine the anion level of some reagents.

Solvent Suitability for Liquid Chromatography. Solvents used in liquid chromatography are tested for suitability in gradient elution analysis. The objective is to assess the performance suitability of the solvents under routine LC operating conditions. The operating parameters found under the specific reagent were selected to simulate a typical gradient run. The reverse gradient portion of the chromatogram is of little practical interest in actual sample analysis. Thus, only the peaks eluted in the "up-gradient" portion of the program are recorded (e.g., water to acetonitrile or water to methanol). Commercially available reverse-phase columns come in a variety of supports and stationary phases. Thus, parameters such as monomeric vs. polymeric phase, endcapped vs. nonendcapped, particle size, and %C loading are specified.

►**Procedure.**

Column: Octadecyl (C-18 polymeric phase), 250 × 4.6 mm i.d., 5μm, 12% C loading, endcapped

Mobile Phase: A. Solvent to be tested

B. LC reagent-grade water

Conditions: Detector: Ultraviolet at 254 nm

Sensitivity: 0.02 AUFS (absorbance units full scale)

Gradient Elution:

1. Program the LC system to develop a solvent–water gradient as follows:

Time (min)	Flow (mL/min)	% A (Solvent)	% B (Water)
0	2.0	20	80
30	2.0	20	80
50	2.0	100	0
60	2.0	100	0

2. Start the program after achieving a stable baseline.

3. Disregard the results of the first program. Repeat the program and use these results in the Step 4 evaluation.

4. No peak should be greater than 25% of full scale (0.005 absorbance units).

Reagent-specific conditions can be found under the individual reagents.

Extraction–Concentration

Solvent Suitability for Extraction–Concentration. Trace-level analysis, such as for pollutants in the environment and toxic chemicals in the workplace, frequently requires the use of solvent-extraction procedures to isolate the analyte. In some cases, the solvent is then removed to concentrate the analyte for detection and quantification. It is critical that the solvent be of the highest purity, so that potential interferences during subsequent chromatographic analysis are minimized. For GC analysis, a pure solvent is usually characterized by a small or narrow solvent "front" and the absence of extraneous peaks under the temperature-programmed conditions commonly used for these analyses.

In this monograph, the purity of the solvent is specified as suitable for use with FID (universal) and ECD (class-specific) detectors. For analysis of solvents that are not compatible with a given detector, as in the case of halogenated solvents and ECD, a solvent-exchange step is used prior to GC analysis. In LC analysis using gradient elution and UV detection, a pure solvent would not show extraneous peaks or cause unusual baseline drift when analyzed as a sample. In this monograph, this suitability is specified by measuring the absorbance through a particular spectral region for each solvent. The solvent should give a smooth curve and contain no extraneous absorptions throughout the specified region.

►Procedure: ECD and FID Suitability.

Preconcentration. Preconcentration of the solvent is required to reach the specification levels set for the FID and ECD detectors. Preconcentration can be accomplished with a conventional Kuderna–Danish evaporator-concentrator or a vacuum rotary evaporator. The solvent should never be evaporated to dryness.

Procedure. Accurately transfer 500 mL of test solvent to a 1-L evaporator. Concentrate (100×) to approximately 5 mL. If solvent exchange is necessary, add two separate 50-mL portions, and evaporate to 5 mL each time.

Analysis. Analyze the concentrated solvent and an external standard by using a gas chromatograph equipped with wide-bore capillary columns and FID and ECD detectors. The parameters cited here have given satisfactory results.

Instrument: GC–ECD.

Column: 30-m × 530-μm i.d. fused silica capillary, coated with 1.5-μm film of 5% diphenyl, 94% dimethyl, 1% vinyl polysiloxane that has been surface-bonded and cross-linked

Column Temperature: 40 to 250 °C at 15 °C/min; hold at 250 °C for 15 min

Injector Temperature: 250 °C

Detector Temperature: 320 °C

Carrier Gas: helium at 2–5 mL/min

Sample Size: 5 μL

Standard: 1 μg/L of heptachlor epoxide in hexane

Standard Size: 5 μL

Instrument: GC–FID.

Column: 30-m × 530-μm i.d. fused silica capillary, coated with 1.5-μm film of 5% diphenyl, 94% dimethyl, 1% vinyl that has been surface-bonded and cross-linked

Column Temperature: 40 to 250 °C at 15 °C/min; hold at 250 °C for 15 min

Injector Temperature: 250 °C

Detector Temperature: 250 °C

Carrier Gas: helium at 2–5 mL/min

Sample Size: 5 μL

Standard: 1 mg/L of 2-octanol in hexane

Standard Size: 5 µL

Results. The purity of the solvents is expressed in micrograms per liter for FID and nanograms per liter for ECD, measured against an external standard appropriate for the analysis being performed. In this monograph, 2-octanol is used for FID and heptachlor epoxide is used for ECD. Measure the peak area-height for all peaks inside a window of 0.5 to 2 times the retention time of 2-octanol for GC–FID suitability. The sum of the peaks must be no greater than 10 µg/L, with no single peak greater than 5.0 µg/L. Measure the peaks for all peaks inside a window of 0.2 to 2 times the retention time of heptachlor epoxide for GC–ECD suitability. The sum of the peaks must be no greater than 10 ng/L, with no single peak greater than 5.0 ng/L. For pesticide residue analysis, the standards might also be a series of pesticides, or, for drinking water, a series of halomethanes. Detailed procedures for such analysis can be found in many excellent references.

LC Suitability

▶**Absorbance.** Measure the absorbance of the sample in a 1-cm cell against water (as the reference liquid) in a matched cell from 400 nm to the cutoff of the reagent solvent. The absorbance should be less than 1.00 at the cutoff wavelength and not exceed the absorbances at each wavelength specified in the requirements of the reagent.

Thin-Layer Chromatography

A solvent system is prepared as specified in the method. A developing chamber that can be sealed is prepared. The chamber is filled to a depth of approximately 3 mm with the solvent system. A piece of filter paper cut to the appropriate size is placed inside the chamber, and the chamber is closed. The chamber can then be shaken gently to ensure saturation of the filter paper and chamber. A solution of the sample is prepared in a volatile solvent such as methanol or dichloromethane at a concentration of about 1–10%. A TLC plate approximately 10 cm in height should be dried in an oven at 110 °C and stored in a desiccator prior to use. Place a spot of the solution onto the plate approximately 0.5 cm from the bottom using a capillary tube. The diameter of the spot should be 1–2 mm. Allow the solvent to evaporate and repeat the spotting procedure one or two more times, each time allowing the solvent to dry (the amount of sample applied will vary depending on sample concentration). Excess sample will result in loss of resolution due to the sample "streaking" up the plate or to a broad and undefined spot. Carefully place the plate into the developing chamber on the opposite side from the filter paper. Make sure that the sample spot is above the solvent level. Allow the solvent to climb to about 1 cm from the top of the plate. Remove the plate and mark the solvent front. Detect the analytes as specified in the method.

Ion Chromatography

Ion chromatography is used in this monograph to determine the anion content of various reagents. The specific procedure is given under the individual reagent. The methods are based on the initial removal of any anionic matrix, when one is present. The use of a chemical suppression method is recommended. The water used in the preparation of the standards and as a diluent should be on the order of 0.06 micromho cm^{-1}. The use of a preconcentration column can dramatically lower the concentration at which anions can be determined.

Calibration for GC and LC Systems

Because the response of a detector is not always the same for equal concentrations of all components, an instrument must be calibrated to obtain concentrations in terms of weight percentages. Correction of area values to weight values is accomplished by the use of calibration factors (also referred to as response factors) that may be obtained in one of the following ways.

Standard Addition Technique. The sample is analyzed by the prescribed procedure before and after the addition of a measured amount of the component of interest. The ratio of the area obtained for that component, before and after the addition, combined with the weight of added component, can be used to calculate the calibration factor.

Calibration Mixture. A mixture is prepared to correspond as closely as possible in composition to the sample to be analyzed. For the preparation of a calibration mixture, it is desirable, but not always possible, to start with components of negligible impurity. When the impurity content cannot be ignored, it may be established by an independent means. The mixture is analyzed by the prescribed procedure, and the area of each peak is measured. The calibration factor for each component is the ratio of its weight or concentration to its area. For an unidentified component, the calibration factor cannot be determined and is usually assumed to be the same as that of a known component in the calibration mixture. The frequency of calibration is left up to the analyst and should follow normal good laboratory practices.

System Suitability Tests for GC and LC Systems

To determine whether a chromatographic system can provide acceptable, repeatable results, it can be subjected to a suitability test at the time of use. Underlying such a test is the concept that electronics, equipment, column, and standard or samples, coupled with sample handling, constitute a single system. Specific data are collected from repeated injections of the assay preparation or standard preparation. The results are compared with limiting values for key parameters, such as peak symmetry, efficiency, precision, and resolution.

The most useful test parameter is the precision of replicate injections of the analytical reference solution, prepared as directed under the individual reagent. The precision of replicate injections is expressed as the relative standard deviation as follows:

$$S_R(\%) = (100/\overline{X}) \left[\sum_{i=1}^{N} (X_i - \overline{X})^2/(N-1) \right]^{1/2}$$

where S_R is the relative standard deviation in percent, $\overline{X}$ is the mean of the set of N measurements, and X_i is an individual measurement. When an internal standard is used, the measurement X_i usually refers to the measurement of the relative area, A_S:

$$X_I = A_S = a_R/a_i$$

where a_R is the peak area of the substance being analyzed and a_i is the peak area of the internal standard.

The peak asymmetry value is useful and can be limited by specifying an asymmetry factor. The asymmetry factor, T, is defined as the ratio of the distance from the leading edge to the trailing edge divided by twice the distance from the leading edge to the peak perpendicular measured at 5% of the peak height (*see* Figure 7). Therefore,

$$T = AB/2AC$$

This asymmetry factor is unity for a symmetrical peak and increases in value as the asymmetry becomes more pronounced.

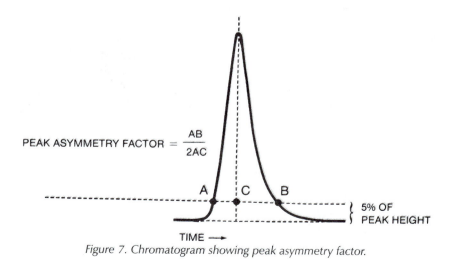

$$\text{PEAK ASYMMETRY FACTOR} = \frac{AB}{2AC}$$

Figure 7. Chromatogram showing peak asymmetry factor.

Resolution, R, can be specified to ensure separation of closely eluting components or to establish the general separation efficiency of the system. R can be calculated as follows:

$$R = \frac{1.178(t_2 - t_1)}{W_{(h/2)_1} + W_{(h/2)_2}}$$

where t_2 and t_1 are the retention times of the two component peaks, and $W_{(h/2)_1}$ and $W_{(h/2)_2}$ are the half-height widths of the two peaks in units of time. An R value of 1.0 means that the resolution is 98% complete. This condition is sufficient, in most cases, for peak area calculations. An R value of 1.5 or greater represents baseline, or complete, separation of the peaks.

INFRARED SPECTROSCOPY

Infrared (IR) spectroscopy is an absorption method widely used in both qualitative and quantitative analysis. The infrared region of the spectrum includes wave numbers ranging from about 12,800 to 10 cm^{-1}. The IR range can be separated into three regions: the near-infrared (12,800–4,000 cm^{-1}), the mid-infrared (4,000–200 cm^{-1}), and the far-infrared (200–10 cm^{-1}). Most analytical applications fall in the mid-infrared region of the spectrum. IR spectroscopy has primarily been used to assist in identification of organic compounds. The IR spectrum of an organic compound is a unique physical property and can be used to identify unknowns by interpretation of characteristic absorbances and comparison to spectral libraries. IR spectroscopy is also employed in quantitative techniques. Because of its sensitivity and selectivity, IR spectroscopy can be used to quantitate analytes in complex mixtures. Quantitative analysis is used extensively in detection of industrial pollutants in the environment. The near-infrared region has been used for quantitative applications, such as determination of water in glycerol and determination of aromatic amines in complex mixtures. The IR technique is discussed here primarily for application in identification of organic compounds and will focus on the mid-infrared region. Instrumental operating procedures are not given because they will vary depending on instrument design. A brief discussion of the theory will be followed by a discussion of instrumentation, sample handling techniques, and qualitative analysis.

General Background

Unlike UV and visible spectroscopy, which uses larger energy absorbances from electronic transitions, IR spectroscopy relies on the much smaller energy absorbances that occur between various vibrational and rotational states. When molecular vibrational or rotational events occur that cause a net dipole moment, IR absorption can occur. Molecular vibrations can be classified as either *stretching* or *bending.* Stretching is a result of continuous changing distances in a bond between two atoms. Bending refers to a change in the angle between two bonds. Bending motions include scissoring, rocking, wagging, and twisting. The various types of vibrations and rotations absorb at different frequencies within the infrared region, thus resulting in unique spectral properties for different molecular species.

Instrumentation

Basic instrumentation for IR spectroscopy includes a radiation source, wavelength selector, sample container, detector, and signal processor. Continuous sources are used in the mid-infrared region and include the incandescent wire source, the Nernst glower, and the Globar. The Nernst glower is hotter and brighter than the incandescent wire. The Globar is a rod of silicon carbide that is electrically heated and that typically requires a water cooling system. The Globar provides greater output than the Nernst glower in the region below 5 μm. Otherwise, their spectral energies are similar. Newer technologies have also employed various lasers as sources for infrared applications.

The three general types of infrared detectors are thermal, pyroelectric, and photoconducting detectors. Thermal detectors measure very minute temperature changes as a method of detection. They operate over a wide range of wavelengths and can be operated at room temperature. The main disadvantages of thermal detectors are slow response and less sensitivity when compared to other detectors. Pyroelectric detectors are specialized thermal detectors that provide faster response and more sensitivity. Photoconducting detectors rely on interactions between incident photons and a semiconductor material. These detectors also provide increased sensitivity and faster response.

The two main types of instruments used for qualitative analysis are dispersive-grating spectrophotometers and multiplex instruments that employ Fourier transform. Wavelength selection in dispersive-grating instruments is accomplished using filters, prisms, or reflection gratings. Traditional instruments have consisted of a filter-grating or prism-grating system that covers the range between 4000–650 cm^{-1}. The most common is a double-beam instrument that uses reflection grating for dispersing radiation.

Fourier transform has been applied to infrared spectroscopy and is now a very common technique. Fourier transform infrared (FTIR) spectroscopy offers enhanced sensitivity compared to dispersive IR spectroscopy. Signal-to-noise ratios are often improved by an order of magnitude. Most commercial FTIR instruments are based on the Michelson interferometer and require a computer interface to perform the Fourier transform and process data. Using an interferometer provides an increase in resolution and results in accurate and reproducible frequency determinations. This offers an advantage when using background subtraction techniques. Most Fourier transform instruments are now benchtop size and relatively easy to maintain. These instruments are largely replacing traditional dispersive instruments.

Sample Handling

Sample containers and handling can present a challenge in the infrared region. Materials used to produce cuvettes are not transparent and cannot be used. Cells prepared from alkali halides such as sodium chloride are widely used due to their transparent properties in the infrared region. A common problem with sodium chloride cells is that they absorb moisture and become fogged. Polishing is required to restore the cells to a more transparent state.

Liquids may be analyzed in their neat form by placing a small amount of sample on a sodium chloride plate and then placing a second plate on top to form a sample film. The plates are then placed in an appropriate holder in the sample compartment of the instrument. This technique provides adequate spectra for qualitative use.

Solutions of liquid or solid materials can also be analyzed by IR spectroscopy. Solvents should be chosen that do not have absorbances in the region of interest. Unfortunately, no solvent is completely transparent in the mid-infrared region. With double-beam instruments, a reference cell containing blank solvent can be employed. Common moderate absorbances will not be observed. Solvent transmission should always be above 10% when using a solvent reference cell. Influences of solvent on the absorbance of the solute should be considered. For example, hydrogen bonding of alcohols or amines with the solvent may affect characteristic vibrational frequency of the functional group. When practical, it is desirable to analyze neat materials for qualitative analysis.

A method commonly used for analysis of neat solid samples is the *mull* technique. The technique consists of grinding the material into a fine powder and then dispersing it into a liquid or solid matrix to form a mull. Liquid mulls have been formed by combining the powdered analyte with Nujol (a heavy hydrocarbon oil). The liquid mull is analyzed between salt plates as described above. The disadvantage of Nujol is that hydrocarbon bands may interfere with analyte absorbances. A second method of forming a mull involves grinding the powdered analyte with dry potassium bromide and forming a disk. The ratio of analyte to potassium bromide is usually about 1:100. The materials are ground together using a mortar and pestle or a small ball mill. The mixture is then pressed in a die at 10,000–15,000 psi to form a small transparent disk and analyzed. Care must be taken when preparing the disk to protect it from moisture. It is very common to see absorbances for moisture when using potassium bromide disks.

Another technique for handling samples is the use of disposable sample cards. These commercially available cards contain an IR-transparent material on which a neat liquid or solution can be placed for analysis. An analyte can be dissolved in a volatile solvent, placed on the card, and the solvent evaporated to form a thin coating of sample, which can then be analyzed. This technique may be used when a limited amount of analyte material is available.

Qualitative Analysis

The most widely used application of IR spectroscopy is for qualitative analysis of organic compounds. Compounds have unique spectra that depend on molecular attributes. A common method of interpreting IR spectra is to consider two regions: the functional group frequency region (3600–1200 cm^{-1}) and the "fingerprint" region (1200–600 cm^{-1}). A combination of interpreting the functional group region and comparing the fingerprint region with those in spectral libraries provides, in many cases, sufficient evidence to positively identify a compound.

The functional group region provides evidence of functional groups in a molecule based on the absorbance frequency. Tables are available in standard spectra manuals that provide ranges of absorbances for specific functional groups. Common groups with characteristic absorbances include aldehydes, ketones, esters, alkenes, alkynes, alcohols, amines, amides, carboxylic acids, nitro groups, and nitriles. While most functional groups fall in the 3600–1200 cm^{-1} range, some can also fall in the fingerprint region. For example, C–O bonds can be around 1000 cm^{-1}, and C–Cl absorbances are typically found in the range of 600–800 cm^{-1}.

The fingerprint region is often unique to the analyte. In this region, very small differences in structure can lead to differences in absorbances. Most single bonds absorb in this area, and differences in skeletal structure of molecules will result in frequency and intensity differences. The fingerprint region is most effectively used by comparison to existing spectra. Many commercial suppliers if IR instrumentation also offer searchable electronic spectral libraries. Computer-based search systems offer a very rapid method of comparing unknown samples to known spectra. The systems usually yield a list of possible compounds ranked in order of best fit of the unknown spectrum to the library spectrum.

▶Procedure

IR spectroscopy is used in selected standard-grade reference materials specifications as a technique for identity confirmation. The criteria under which standard-grade reference materials pass identity tests are based on observation of characteristic absorbances and by comparison to a spectral library, if one is available. Each entry to be analyzed by IR spectroscopy will contain a minimum of three absorbances that must be present to pass the test. In addition, NIST should be used as a source to obtain comparison spectra. A standard-grade reference material will pass the comparison test if all absorbances found in the NIST spectrum are present in the test spectrum. The NIST spectral library contains many of the standard-grade reference materials listed in this book, and spectral comparisons should be made if NIST spectra are available. NIST spectra are available by contacting NIST or by accessing NIST's Web site, http://webbook.nist.gov/chemistry/.

Reagent Solutions: Preparation

Throughout the Specifications section of this book, the term *reagent solution* is used to designate solutions described in this section. The tests and limits in the specifications are based on the use of the solutions in the strength indicated below.

Wherever the use of ammonium hydroxide or an acid is prescribed with no indication of strength or dilution, the reagent is to be used at full strength as described in its specification. Dilutions are indicated either by the percentage of some constituent or by the volumes of reagents and water mixed to prepare a dilute reagent. Dilute acid or ammonium hydroxide $(1 + X)$ means a dilute solution prepared by mixing 1 volume of the strong acid or ammonium hydroxide with X volumes of reagent water.

Unless otherwise indicated, the reagent solutions are prepared and diluted with reagent water by using standard class A volumetric pipets and volumetric flasks.

REAGENT WATER

Throughout the specifications for reagent chemicals, the term "water" means distilled water or deionized water that meets the requirements of Water, Reagent, page 686. However, for specific applications—such as UV determinations or liquid and ion chromatography—ASTM Type I reagent water (ASTM, 1999) or water for which the suitability has been determined should be used. In tests for nitrogen compounds, water should be "ammonia-free" or "nitrogen-free". Water for use in analysis of ultratrace metals must meet the requirements on page 688. For some tests, freshly boiled water must be used to ensure freedom from material absorbed from the air, such as ammonia, carbon dioxide, or oxygen.

WATER (Suitable for LC Gradient Elution Analysis)

Analyze the sample by using the following gradient elution conditions. Use ACS reagent-grade or better acetonitrile.

Column: Octadecyl (C-18 polymeric phase), 250 × 4.6 mm i.d., 5 μm, 12% C loading, endcapped

Mobile Phase: A. Water to be tested

B. ACS reagent-grade acetonitrile

Conditions: Detector: Ultraviolet at 254 nm

Sensitivity: 0.02 AUFS

Gradient Elution:

1. Program the liquid chromatograph to develop a water–acetonitrile gradient as follows:

Time (min)	Flow (mL/min)	% A (Water)	% B (Acetonitrile)
0	2.0	100	0
20	2.0	100	0
40	2.0	0	100
50	2.0	0	100

2. Equilibrate the system at 100% water. Start the program after achieving a stable baseline. Repeat the gradient program, and use these results to evaluate the water. No peak should be greater than 10% of full scale (0.002 absorbance units).

REAGENT SOLUTIONS

NOTE. Smaller quantities may be made having the same concentrations as described below.

Acetic Acid, 1 N. Dilute 57.5 mL of glacial acetic acid with water to 1 L.

Alcohol. "Alcohol" refers to the specification of ethyl alcohol, page 291. This is distinguished from the specification of reagent alcohol, page 544, which is a denatured form of ethyl alcohol.

Ammoniacal Buffer, pH 10. Dissolve 180 g of ammonium chloride in 750 mL of freshly opened ammonium hydroxide, and dilute to 1 L with water. (The pH of a 1 + 10 dilution with water should be about 10.)

Ammonia–Cyanide Mixture, Lead-Free. Dissolve 20 g of potassium cyanide in 150 mL of ammonium hydroxide, and dilute with water to 1 L. Remove lead by shaking the solution with small portions of dithizone extraction solution in chloroform until the dithizone retains its original green color; discard the extraction solution.

Ammonium Acetate–Acetic Acid Buffer Solution. Dissolve 77.1 g of ammonium acetate in water, add 57 mL of glacial acetic acid, and dilute with water to 1 L. (The pH of a 1 + 10 dilution with water should be about 4.5.)

Ammonium Citrate, Lead-Free. Dissolve 40 g of citric acid in 100 mL of water, and make alkaline to phenol red with ammonium hydroxide. Remove lead by shaking the solution with small portions of dithizone extraction solution in chloroform until the dithizone solution retains its original green color; discard the extraction solution.

Ammonium Hydroxide, 10% NH_3. Dilute 400 mL of ammonium hydroxide with water to 1 L.

Ammonium Hydroxide, 2.5% NH_3. Dilute 100 mL of ammonium hydroxide with water to 1 L.

Ammonium Metavanadate. Dissolve 2.5 g of ammonium metavanadate, NH_4VO_3, in 500 mL of boiling water, cool, and add 20 mL of nitric acid. Dilute with water to 1 L, and store in a polyethylene bottle.

Ammonium Molybdate. Dissolve 5 g of ammonium molybdate tetrahydrate, $(NH_4)_6Mo_7O_{24} \cdot 4H_2O$, in water, and dilute with water to 100 mL. Store in a polyethylene bottle. Discard if a precipitate forms.

Ammonium Molybdate–Nitric Acid Solution. Mix thoroughly 100 g of molybdic acid, 85%, with 240 mL of water, and add 140 mL of ammonium hydroxide. Filter and add 60 mL of nitric acid. Cool and pour, with constant stirring, into a cool mixture of 400 mL of nitric acid and 960 mL of water. Add 0.1 g of ammonium phosphate dissolved in 10 mL of water, allow to stand 24 h, and filter through glass wool.

Ammonium Molybdate–Sulfuric Acid Solution. Dissolve 50 g of ammonium molybdate tetrahydrate, $(NH_4)_6Mo_7O_{24} \cdot 4H_2O$, in about 500 mL of water, add 27.5 mL of sulfuric acid, and dilute with water to 1 L.

Ammonium Oxalate. Dissolve 35 g of ammonium oxalate monohydrate, $(COONH_4)_2 \cdot H_2O$, in water, and dilute with water to 1 L.

Ammonium Phosphate. Dissolve 130 g of dibasic ammonium phosphate, $(NH_4)_2HPO_4$, in water, and dilute with water to 1 L.

Ammonium Thiocyanate. Dissolve 300 g of ammonium thiocyanate, NH_4SCN, in water, and dilute with water to 1 L.

Aqua Regia. In a fume hood, cautiously mix 5 mL of nitric acid in 15 mL of hydrochloric acid.

Barium Chloride, 12%. Dissolve 120 g of barium chloride dihydrate, $BaCl_2 \cdot 2H_2O$, in water, filter, and dilute with water to 1 L.

Barium Chloride, 30%. Dissolve 30 g of barium chloride dihydrate, $BaCl_2 \cdot 2H_2O$, in 80 mL of water, filter, and dilute with water to 100 mL.

Benedict's Solution. Dissolve 173 g of sodium citrate dihydrate and 100 g of anhydrous sodium carbonate, Na_2CO_3, in 800 mL of water. Heat to aid dissolution, filter if necessary, and dilute with water to 850 mL. Dissolve 17.3 g of copper sulfate pentahydrate, $CuSO_4 \cdot 5H_2O$, in 100 mL of water. Add this solution, with constant stirring, to the alkaline citrate solution, and dilute with water to 1L.

Bromine Water. A saturated aqueous solution of bromine. Add about 10 mL of liquid bromine to 75 ml of water in a 100-mL bottle so that when the mixture is shaken, undissolved bromine remains in a separate phase.

Bromphenol Blue Indicator. Dissolve 0.10 g of the salt form of bromphenol blue in water, and dilute with water to 100 mL (pH 3.0–4.6).

Bromthymol Blue Indicator. Dissolve 0.10 g of bromthymol blue in 100 mL of dilute alcohol (1 + 1), and filter if necessary (pH 6.0–7.6).

Brucine Sulfate. Dissolve 0.6 g of brucine sulfate in dilute sulfuric acid (2 + 1), previously cooled to room temperature, and dilute to 1 L with the dilute acid. The sulfuric acid should be nitrate-free acid prepared as follows: dilute the concentrated sulfuric acid (about 96% H_2SO_4) to about 80% H_2SO_4 by adding it to water, heat to dense fumes of sulfur trioxide, and cool. Repeat the dilution and fuming 3 or 4 times.

Chromotropic Acid. Dissolve 1 g of chromotropic acid, 4,5-dihydroxy-2,7-naphthalenedisulfonic acid, in water, and dilute with water to 100 mL.

Crystal Violet Indicator. Dissolve 100 mg of crystal violet in 10 mL of glacial acetic acid.

5,5-Dimethyl-1,3-cyclohexanedione, 5% in Alcohol. Dissolve 5 g of 5,5-dimethyl-1,3-cyclohexanedione in alcohol, and dilute with alcohol to 100 mL.

Dimethylglyoxime. Dissolve 1 g of dimethylglyoxime in alcohol, and dilute with alcohol to 100 mL.

N,N-Dimethyl-p-phenylenediamine. Add 200 mL of sulfuric acid to approximately 700 mL of water. Cool to room temperature, add 1 g of N,N-dimethyl-p-phenylenediamine (p-aminodimethylaniline) monohydrochloride, and dilute with water to 1 L.

Diphenylamine. Dissolve 10 mg of colorless diphenylamine in 100 mL of sulfuric acid. In a separate beaker, dissolve 2 g of ammonium chloride in 200 mL of water. Cool both solutions in an ice bath, and cautiously add the sulfuric acid solution to the water solution, taking care to keep the resulting solution cold. The solution should be nearly colorless.

Dithizone Extraction Solution. Dissolve 30 mg of dithizone in 1 L of chloroform plus 5 mL of alcohol. Store the solution in a refrigerator.

Dithizone Indicator Solution. Dissolve 25.6 mg of dithizone in 100 mL of alcohol. Store in a cold place, and use within 2 months.

Dithizone Test Solution. Dissolve 10 mg of dithizone in 1 L of chloroform. Keep the solution in a glass-stoppered, lead-free bottle, protected from light and stored in a refrigerator.

Eosin Y. Dissolve 50 mg of eosin Y in 10 mL of water.

(Ethylenedinitrilo)tetraacetic Acid, Disodium Salt, 0.2 M. Dissolve 74.4 g of (ethylenedinitrilo)tetraacetic acid, disodium salt, dihydrate in 950 mL of water, and dilute with water to 1 L.

Ferric Ammonium Sulfate Indicator. Dissolve 80 g of clear crystals of ferric ammonium sulfate dodecahydrate, $FeNH_4(SO_4)_2 \cdot 12H_2O$, in water, and dilute with water to 1 L. A few drops of sulfuric acid may be added, if necessary, to clear the solution.

Ferric Chloride. Dissolve 100 g of ferric chloride hexahydrate $FeCl_3 \cdot 6H_2O$, in dilute hydrochloric acid (1 + 99), and dilute with the dilute acid to 1 L.

Ferric Nitrate. Dissolve 17 g of ferric nitrate nonahydrate, $Fe(NO_3)_3 \cdot 9H_2O$, in 100 mL of water.

Ferric Sulfate. Add 25 mL of sulfuric acid to 125 g of ferric sulfate n-hydrate, $Fe_2(SO_4)_3 \cdot nH_2O$. Add about 700 mL of water, digest on a steam bath to dissolve, cool to room temperature, and dilute with water to 1 L.

Ferroin Indicator, 0.025 M. Dissolve 0.70 g of ferrous sulfate heptahydrate, $FeSO_4 \cdot 7H_2O$ and 1.5 g of 1,10-phenanthroline in 100 mL of water.

Hexamethylenetetramine Solution, Saturated. Dissolve about 20 g of hexamethylenetetramine in 100 mL of water.

Hydrazine Sulfate, 2%. Dissolve 2 g of hydrazine sulfate, $(NH_2)_2 \cdot H_2SO_4$, in water, and dilute with water to 100 mL.

Hydrochloric Acid, 20% HCl. Dilute 470 mL of hydrochloric acid with water to 1 L.

Hydrochloric Acid, 10% HCl. Dilute 235 mL of hydrochloric acid with water to 1 L.

Hydrochloric Acid, 7 M. Dilute hydrochloric acid (1 + 1) with water.

Hydrogen Sulfide Water. Saturate 100 mL of water by bubbling hydrogen sulfide gas for 1 min. This solution must be freshly prepared.

Hydroxylamine Hydrochloride. Dissolve 10 g of hydroxylamine hydrochloride, $NH_2OH \cdot HCl$, in water, and dilute with water to 100 mL.

Hydroxylamine Hydrochloride for Dithizone Test. Dissolve 20 g of hydroxylamine hydrochloride, $NH_2OH \cdot HCl$, in about 65 mL of water and add 0.10 to 0.15 mL of thymol blue indicator solution. Add ammonium hydroxide until a yellow color appears. Add 5 mL of a 4% solution of sodium diethyldithiocarbamate. Mix thoroughly, and allow to stand for 5 min. Extract with successive portions of chloroform until no yellow color is developed in the chloroform layer when the extract is shaken with a dilute solution of a copper salt. Add hydrochloric acid until the indicator turns pink, and dilute with water to 100 mL.

Indigo Carmine. Dissolve 1 g of FD&C Blue No. 2, dried at 105 °C, in a mixture of 800 mL of water and 100 mL of sulfuric acid, and dilute with water to 1 L.

Jones Reductor. The zinc is amalgamated by immersing it in a solution of mercuric chloride in hydrochloric acid. A 250-g portion of 20-mesh zinc is covered with reagent water in a 1-L suction flask, a solution containing 11 g of mercuric chloride ($HgCl_2$) in 100 mL of hydrochloric acid is poured into the flask, and the system is slowly mixed and shaken for about 2 min. The solution is poured off and the amalgam is washed thoroughly with hot tap water, then finally with reagent water. The column is charged with six 250-g portions.

Apparatus. A dispensing buret, about 22 in. long and 2 in. in diameter, equipped with a glass stopcock and delivery tube, 6 mm wide × 3.5 in. long.

The reductor is charged with an 8-in. column of 20-mesh amalgamated zinc (1500 g) and, on top of this, a 6-in. column of larger (1–2 cm) amalgamated zinc (about 750 g). The delivery tube is connected to a 1-L flask through a two-hole rubber stopper. One hole is used as an inlet; the other functions as an outlet for carbon dioxide gas.

Karl Fischer Reagent. Dissolve 762 g of iodine in 2420 mL of pyridine in a 9-L glass-stoppered bottle, and add 6 L of methanol. To prepare the active reagent, add 3 L of the foregoing stock to a 4-L bottle, and cool by placing the bottle in a slurry of chopped ice. Add carefully about 135 mL of liquid sulfur dioxide, collected in a calibrated cold trap, and stopper the bottle. Shake the mixture until it is homogeneous, and set it aside for one or two days before use.

Lead Acetate. Dissolve 100 g of lead acetate trihydrate, $(CH_3COO)_2Pb \cdot 3H_2O$, in water. If necessary, add a few drops of acetic acid to clear the solution, and dilute with water to 1 L.

Metalphthalein-Screened Indicator. Dissolve 0.18 g of metal-phthalein and 0.02 g of Naphthol Green B (i.e., Acid Green 1; C.I. 10020) in water containing 0.5 mL of ammonium hydroxide, and dilute to 100 mL with water.

4-(Methylamino)phenol Sulfate. Dissolve 2 g of 4-(methyl-amino)phenol sulfate in 100 mL of water. To 10 mL of this solution add 90 mL of water and 20 g of sodium bisulfite. Confirm the suitability of the reagent solution by the following test: Add 1 mL of this reagent solution to each of four solutions containing 25 mL of 0.5 N sulfuric acid and 1 mL of ammonium molybdate 5% reagent solution. Add 0.005 mg of phosphate ion (PO_4) to one of the solutions, 0.01 mg to a second, and 0.02 mg to a third.

Allow to stand at room temperature for 2 h. The solutions in the three tubes should show readily perceptible differences in blue color corresponding to the relative amounts of phosphate added, and the one to which 0.005 mg of phosphate was added should be perceptibly bluer than the blank.

Methyl Orange Indicator. Dissolve 0.10 g of methyl orange in 100 mL of water (pH 3.2–4.4).

Methyl Red Indicator. Dissolve 0.10 g of methyl red in 100 mL of alcohol. See page 426 for a description of the three forms of methyl red (pH 4.2–6.2).

Nessler Reagent. Dissolve 143 g of sodium hydroxide in 700 mL of water. Dissolve 50 g of red mercuric iodide and 40 g of potassium iodide in 200 mL of water. Pour the iodide solution into the hydroxide solution, and dilute with water to 1 L. Allow to settle, and use the clear supernatant liquid. Nessler reagent prepared by any of the recognized methods may be used, but the reagent described reacts promptly and consistently.

Nitric Acid, 10% HNO₃. Dilute 105 mL of nitric acid with water to 1 L.

Nitric Acid, 1% HNO₃. Dilute 10.5 mL of nitric acid with water to 1 L.

Oxalic Acid. Dissolve 40 g of oxalic acid dihydrate, $(COOH)_2 \cdot 2H_2O$, in water, and dilute with water to 1 L.

PAN Indicator. Dissolve 0.05 g of 1-(2-pyridylazo)-2-naphthol in 50 mL of reagent alcohol.

PAR Indicator. Dissolve 0.10 g of 4-(2-pyridylazo)resorcinol in 100 mL of alcohol.

1,10-Phenanthroline. Dissolve 0.1 g of 1,10-phenanthroline monohydrate in 100 mL of water containing 0.1 mL of 10% hydrochloric acid reagent solution.

Phenoldisulfonic Acid. Dissolve 50 g of phenol, C_6H_5OH, in 300 mL of sulfuric acid, add 150 mL of fuming sulfuric acid (15% SO_3), and heat at 100 °C for 2 h.

Phenolphthalein Indicator. Dissolve 1 g of phenolphthalein in 100 mL of alcohol (pH 8.0–10.0).

Phenol Red Indicator. Dissolve 0.10 g of phenol red in 100 mL of alcohol, and filter if necessary (pH 6.8–8.2).

Potassium Chloride, 2.5% KCl. Dissolve 2.5 g of potassium chloride, KCl, in water, and dilute with water to 100 ml.

Potassium Chromate. Dissolve 10 g of potassium chromate, K_2CrO_4, in water and dilute with water to 100 mL.

Potassium Cyanide, Lead-Free. Dissolve 10 g of potassium cyanide, KCN, in sufficient water to make 20 mL. Remove lead by shaking with portions of dithizone extraction solution. Part of the dithizone remains in the aqueous phase but can be removed, if desired, by shaking with chloroform. Dilute the potassium cyanide solution with water to 100 mL.

Potassium Dichromate. Dissolve 10 g of potassium dichromate, $K_2Cr_2O_7$, in water, and dilute with water to 100 mL.

Potassium Ferricyanide. Dissolve 5 g of potassium ferricyanide, $K_3Fe(CN)_6$, in 100 mL of water. Prepare the solution at the time of use.

Potassium Ferrocyanide. Dissolve 10 g of potassium ferrocyanide trihydrate, $K_4Fe(CN)_6 \cdot 3H_2O$, in 100 mL of water. Prepare the solution at the time of use.

Potassium Hydroxide, 0.5 N in Methanol. Dissolve 34 to 36 g of potassium hydroxide, KOH, in 20 mL of water, and dilute with methanol to 1 L. Allow the solution to stand in a stoppered bottle for 24 h. Decant the clear supernatant solution into a bottle provided with a tight-fitting stopper.

Potassium Iodide, 16.5% (for Arsenic Determination). Dissolve 16.5 g of potassium iodide, KI, in water, and dilute with water to 100 mL.

Potassium Iodide, 10% (for Free Chlorine). Dissolve 10 g of potassium iodide, KI, in water, and dilute with water to 100 mL.

Silver Diethyldithiocarbamate. Dissolve 1 g of silver diethyldithiocarbamate, $(C_2H_5)_2NCS_2Ag$, in 200 mL of freshly distilled pyridine.

Silver Nitrate. Dissolve 17 g of silver nitrate, $AgNO_3$, in 1 L of water. Store in an amber bottle.

Sodium Acetate, 10%. Dissolve 10 g of sodium acetate trihydrate in water, and dilute with water to 100mL.

Sodium Borohydride. Dissolve 0.6 g sodium borohydride and 0.5 g sodium hydroxide, and dilute with stirring to 100 mL with water.

Sodium Carbonate, 1%. Dissolve 5.0 g of sodium carbonate, anhydrous, Na_2CO_3, in 500 mL of water. Store in a plastic bottle (10 mg/mL).

Sodium Citrate, 1 M. Dissolve 294 g of sodium citrate dihydrate, $Na_3C_6H_5O_7 \cdot 2H_2O$, in 950 mL of water, and dilute with water to 1 L.

Sodium Cyanide. Dissolve 10 g of sodium cyanide, NaCN, in water, and dilute with water to 100 mL.

Sodium Diethyldithiocarbamate. Dissolve 1.0 g of sodium diethyl-dithiocarbamate in water, and dilute with water to 1 L.

Sodium Hydroxide. Dissolve 100 g of sodium hydroxide, NaOH, in water, and dilute with water to 1 L.

Sodium Sulfite. Dissolve 12.5 g of sodium sulfite, Na_2SO_3, in water, and dilute with water to 100 mL. Keep in a tightly closed bottle. For best results, freshly prepared solution should be used.

Stannous Chloride, 40%. Dissolve 40 g of stannous chloride dihydrate, $SnCl_2 \cdot 2H_2O$, in 100 mL of hydrochloric acid. Store this solution in polyethylene or Teflon containers and use within 3 months.

Stannous Chloride, 2%. Dissolve 2 g of stannous chloride dihydrate, $SnCl_2 \cdot 2H_2O$, in hydrochloric acid, and dilute with hydrochloric acid to 100 mL.

Starch Indicator. Mix 1 g of soluble starch with 10 mg of red mercuric iodide and enough cold water to make a thin paste, add 200 mL of boiling water, and boil for 1 min while stirring. Cool before use.

Sulfuric Acid, Chloride- and Nitrate-Free. In a hood, gently fume sulfuric acid in a crucible or dish for at least 30 min. Allow to cool and transfer to a tightly capped glass bottle for storage.

Sulfuric Acid 0.5 N. Slowly add 15 mL of sulfuric acid to 750 mL of water, and dilute to 1 L.

Sulfuric Acid, 10%. Slowly add 60 mL of sulfuric acid to 750 mL of water, cool, and dilute with water to 1 L.

Sulfuric Acid, 25%. Slowly add 170 mL of sulfuric acid to 750 mL of water, cool, and dilute with water to 1 L.

Thymol Blue Indicator. Dissolve 0.10 g of thymol blue in 100 mL of alcohol (acid range, pH 1.2–2.8; alkaline range, pH 8.0–9.2).

Thymolphthalein Indicator. Dissolve 0.10 g of thymolphthalein in 100 mL of alcohol (pH 8.8–10.5).

Titanium Tetrachloride. Cool separately in small beakers surrounded by crushed ice 10 mL of 20% hydrochloric acid and 10 mL of clear, colorless titanium tetrachloride. Add the tetrachloride dropwise to the chilled acid. Allow the mixture to stand at ice temperature until all of the solid dissolves, then dilute the solution with 20% hydrochloric acid to 1 L.

Triton X-100, 0.20%. Dissolve 0.20 g of Triton X-100 (polyethylene glycol ether of isooctylphenol) in water, and dilute with water to 100 mL.

Variamine Blue B Indicator. Dissolve 0.20 g of Variamine Blue B in 20 mL of water, and stir for 5 min.

SOLID REAGENT MIXTURES

Eriochrome Black T Indicator. Grind 0.2 g of eriochrome black T to a fine powder with 20 g of potassium chloride.

Hydroxy Naphthol Blue Indicator. Use the mixture with sodium chloride commercially supplied (*see* specifications for this indicator, page 345).

Methylthymol Blue Indicator. Grind 0.2 g of methylthymol blue to a fine powder with 20 g of potassium nitrate.

Murexide Indicator. Grind 0.2 g of murexide to a fine powder with 20 g of potassium nitrate.

Xylenol Orange Indicator. Grind 0.2 g of xylenol orange (either free acid or sodium salt form) to a fine powder with 20 g of potassium nitrate. Alternatively, use a solution of 0.1 g of xylenol orange in 100 mL of alcohol.

CONTROL AND STANDARD SOLUTIONS

Control Preparation

Many of the tests for impurities require the comparison of the color or the turbidity produced under specified conditions by an impurity ion with the color or the turbidity produced under similar conditions by a known amount of the impurity. Three types of solutions are used to determine amounts of impurities in reagent chemicals: (1) the blank, (2) the standard, and (3) the control.

(1) Blank. A solution containing the quantities of solvents and reagents used in the test.

(2) Standard. A solution containing the quantities of solvents and reagents used in the test plus an added known amount of the impurity to be determined.

(3) Control. A solution containing the quantities of solvents and reagents used in the test plus an added known amount of the impurity to be determined and a known amount of the reagent chemical being tested. Some of the chemical being tested must be added to the solution because it is known that the chemical interferes with the test.

Throughout this book, the expressions "for the control" or "for the standard" add *X* mg of *Y* ion mean to add the proper aliquot of one of the following solutions. These solutions must be prepared at the time of use or checked often enough to make certain that the tests will not be affected by changes in strength of the standard solution during storage. Calibrated volumetric glassware must be used in preparing these solutions.

Standard Preparation

NOTE. In place of standard solutions prepared as described below, commercially available standards may be used after proper dilution.

Acetaldehyde (0.1 mg of CH₃CHO in 1 mL). Prepare in water by appropriate dilution of an assayed acetaldehyde solution.

Acetone (0.1 mg of (CH₃)₂CO in 1 mL). Dilute 6.4 mL of acetone with water to 1 L. Dilute 2.0 mL of this solution with water to 100 mL. Prepare the solution at the time of use.

Aluminum (0.01 mg of Al in 1 mL). Dissolve 0.100 g of metallic aluminum, Al, in 10 mL of dilute hydrochloric acid (1 + 1), and dilute with water to 100 mL. To 10 mL of this solution, add 25 mL of dilute hydrochloric acid (1 + 1), and dilute with water to 1 L.

Ammonium (0.01 mg of NH₄ in 1 mL). Dissolve 0.296 g of ammonium chloride, NH₄Cl, in water and dilute with water to 100 mL. Dilute 10 mL of this solution with water to 1 L.

Antimony (0.1 mg of Sb in 1 mL). Dissolve 0.2743 g of antimony potassium tartrate hemihydrate, KSbOC₄H₄O₆ · ½H₂O, in 100 mL of water. Add 100 mL of hydrochloric acid, and dilute with water to 1 L.

Arsenic (0.001 mg of As in 1 mL). Dissolve 0.132 g of arsenic trioxide, As₂O₃, in 10 mL of 10% sodium hydroxide reagent solution, neutralize with 10% sulfuric acid reagent solution, add an excess of 10 mL, and dilute with water to 1 L. To 10 mL of this solution, add 10 mL of 10% sulfuric acid reagent solution, and dilute with water to 1 L.

Barium (0.1 mg of Ba in 1 mL). Dissolve 0.178 g of barium chloride dihydrate, BaCl₂ · 2H₂O, in water, and dilute with water to 1 L.

Bismuth (0.01 mg of Bi in 1 mL). Dissolve 0.232 g of bismuth trinitrate pentahydrate, Bi(NO₃)₃ · 5H₂O, in 10 mL of dilute nitric acid (1 + 9), and dilute with water to 100 mL. Dilute 10 mL of this solution plus 10 mL of nitric acid with water to 1 L.

Bromate (0.1 mg of BrO₃ in 1 mL). Dissolve 0.131 g of potassium bromate, KBrO₃, in water, and dilute with water to 1 L.

Bromide (0.1 mg of Br in 1 mL). Dissolve 1.49 g of potassium bromide, KBr, in water, and dilute with water to 100 mL. Dilute 10 mL of this solution with water to 1 L.

Cadmium (0.025 mg of Cd in 1 mL). Dissolve 0.10 g of cadmium chloride 2.5-hydrate, $CdCl_2 \cdot 2\frac{1}{2}H_2O$, in water, add 1 mL of hydrochloric acid, and dilute with water to 500 mL. To 25 mL of this solution, add 1 mL of hydrochloric acid, and dilute with water to 100 mL.

Calcium (0.1 mg of Ca in 1 mL). Dissolve 0.250 g of calcium carbonate, $CaCO_3$, in 20 mL of water plus 5 mL of 10% hydrochloric acid reagent solution, and dilute with water to 1 L.

Calcium (0.01 mg of Ca in 1 mL). Dilute 100 mL of the preceding solution with water to 1 L.

Carbon (4 mg of C in 1 mL). Dissolve 8.5 g of potassium hydrogen phthalate, $HOCOC_6H_4COOK$, in water, and dilute to 1 L.

Carbonate (0.3 mg of CO_3, 0.2 mg of CO_2, or 0.06 mg of C in 1 mL). Dissolve 0.53 g of sodium carbonate, Na_2CO_3, in carbon dioxide-free water, and dilute with carbon dioxide-free water to 1 L. Prepare the solution at the time of use.

Chloride (0.01 mg of Cl in 1 mL). Dissolve 0.165 g of sodium chloride, NaCl, in water, and dilute with water to 100 mL. Dilute 10 mL of this solution with water to 1 L.

Chlorine (0.01 mg of Cl_2 in 1 mL). Dilute 1.0 mL of a 2.5% sodium hypochlorite, NaOCl, solution with water to 1 L. Correct for any assay not equal to 2.5%. Standardize the sodium hypochlorite solution before use by the following method: Pipet 3 mL into a tared, glass-stoppered flask. Weigh accurately, and add 50 mL of water, 2 g of potassium iodide, and 10 mL of acetic acid. Titrate the liberated iodine with 0.1 N sodium thiosulfate, adding 3 mL of starch indicator solution near the end point. One milliliter of 0.1 N sodium thiosulfate corresponds to 3.723 mg of NaOCl.

Chlorobenzene (0.1 mg of C_6H_5Cl in 1 mL). Dilute 1.0 mL of chlorobenzene to 100 mL with 2,2,4-trimethylpentane. Dilute 9.0 mL of this solution to 100 mL with 2,2,4-trimethylpentane.

Chromate (0.01 mg of CrO_4 in 1 mL). Dissolve 1.268 g of potassium dichromate, $K_2Cr_2O_7$, in water, and dilute with water to 1 L. Dilute 1.0 mL of this solution to 100 mL.

Copper Stock Solution (0.1 mg of Cu in 1 mL). Dissolve 0.393 g of cupric sulfate pentahydrate, $CuSO_4 \cdot 5H_2O$, in water, and dilute with water to 1 L.

Copper Standard for Dithizone Test (0.001 mg of Cu in 1 mL). Dilute 1 mL of the copper stock solution with water to 100 mL. This dilute solution must be prepared immediately before use.

Ferric Iron (0.1 mg of Fe^{3+} in 1 mL). Dissolve 0.863 g of ferric ammonium sulfate dodecahydrate, $FeNH_4(SO_4)_2 \cdot 12H_2O$, in water, and dilute with water to 1 L.

Fluoride Stock Solution (1 mg of F in 1 mL). Dissolve 2.21 g of sodium fluoride, NaF, previously dried at 110 °C for 2 h, and dissolve with 200 mL of water in a 400-mL plastic beaker. Transfer to a 1-L volumetric flask, dilute with water to the mark, and store the solution in a plastic bottle.

Fluoride (0.005 mg in 1 mL). Transfer 5.0 mL of the fluoride stock solution to a 1-L plastic volumetric flask, dilute with water to the mark, and store in a plastic bottle. Prepare the solution at the time of use.

Formaldehyde (0.1 mg of HCHO in 1 mL). Dilute 2.7 g (2.7 mL) of 37% formaldehyde solution to 1 L with water (prepare freshly as needed). Dilute 10 mL of this solution to 100 mL.

Formaldehyde (0.01 mg of HCHO in 1 mL). Dilute 10 mL of the preceding solution with water to 100 mL. Prepare the solution at the time of use.

Hydrogen Peroxide (0.1 mg of H$_2$O$_2$ in 1 mL). Dilute 3.0 mL of 30% hydrogen peroxide with water to 100 mL. Dilute 1.0 mL of this solution with water to 100 mL. Prepare the solution at the time of use.

Iodate (0.10 mg of IO$_3$ in 1 mL). Dissolve 0.123 g of potassium iodate, KIO_3, in water, and dilute to 1 L.

Iodate (0.01 mg of IO$_3$ in 1 mL). Dilute 10 mL of the preceding iodate solution (0.10 mg of IO$_3$ in 1 mL) to 100 mL.

Iodide (0.01 mg of I in 1 mL). Dissolve 0.131 g of potassum iodide, KI, in water, and dilute with water to 100 mL. Dilute 10 mL of this solution to 1 L.

Iron (0.01 mg of Fe in 1 mL). Dissolve 0.702 g of ferrous ammonium sulfate hexahydrate, $Fe(NH_4)_2(SO_4)_2 \cdot 6H_2O$, in 10 mL of 10% sulfuric acid reagent solution, and dilute with water to 100 mL. To 10 mL of this solution, add 10 mL of 10% sulfuric acid reagent solution, and dilute with water to 1 L.

Lead Stock Solution (0.1 mg of Pb in 1 mL). Dissolve 0.160 g of lead nitrate, $Pb(NO_3)_2$, in 100 mL of dilute nitric acid (1 + 99), and dilute with water to 1 L. The solution should be prepared and stored in containers free from lead. Its strength should be checked every few months to determine whether the lead content has changed by reaction with the container.

Lead Standard for Dithizone Test (0.001 mg of Pb in 1 mL). Dilute 1 mL of the lead stock solution to 100 mL with dilute nitric acid (1 + 99). This solution must be prepared immediately before use.

Lead Standard for Heavy Metals Test (0.01 mg of Pb in 1 mL). Dilute 10 mL of the lead stock solution to 100 mL with water. This dilute solution must be prepared at the time of use.

Magnesium (0.01 mg of Mg in 1 mL). Dissolve 1.014 g of clear crystals of magnesium sulfate heptahydrate, $MgSO_4 \cdot 7H_2O$, in water, and dilute with water to 100 mL. Dilute 10 mL of this solution with water to 1 L.

Magnesium (0.002 mg of Mg in 1 mL). Dilute 100 mL of the preceding solution with water to 500 mL. Prepare the solution at the time of use.

Manganese (0.01 mg of Mn in 1 mL). Dissolve 3.08 g of manganese sulfate monohydrate, $MnSO_4 \cdot H_2O$, in water, and dilute with water to 1 L. Dilute 10 mL of this solution with water to 1 L.

Mercury (0.05 mg of Hg in 1 mL). Dissolve 1.35 g of mercuric chloride, $HgCl_2$, in water, add 8 mL of hydrochloric acid, and dilute with water to 1 L. To 50 mL of this solution, add 8 mL of hydrochloric acid, and dilute with water to 1 L.

Methanol (0.1 mg of CH_3OH in 1 mL). Dilute 5.1 mL of methanol, CH_3OH, with water to 200 mL. Dilute 5.0 mL of this solution with water to 1 L. Prepare the solution at the time of use.

Molybdenum (0.01 mg of Mo in 1 mL). Dissolve 0.150 g of molybdenum trioxide, MoO_3, in 10 mL of dilute ammonium hydroxide (1 + 9), and dilute with water to 100 mL. Dilute 10 mL of this solution with water to 1 L.

Nickel (0.01 mg of Ni in 1 mL). Dissolve 0.448 g of nickel sulfate hexahydrate, $NiSO_4 \cdot 6H_2O$, in water, and dilute with water to 100 mL. Dilute 10 mL of this solution with water to 1 L.

Nitrate (0.01 mg of NO_3 in 1 mL). Dissolve 0.163 g of potassium nitrate, KNO_3, in water, and dilute with water to 100 mL. Dilute 10 mL of this solution with water to 1 L.

Nitrilotriacetic Acid (10 mg of $(HOCOCH_2)_3N$ in 1 mL). Transfer 1.0 g of nitrilotriacetic acid, $(HOCOCH_2)_3N$, to a 100-mL volumetric flask. Dissolve in 10 mL of 10% potassium hydroxide solution, and dilute with water to volume.

Nitrite (0.01 mg of NO$_2$ in 1 mL). Dissolve 0.150 g of sodium nitrite (NaNO$_2$) in water, and dilute with water to 100 mL. Dilute 1.0 mL of this solution with water to 100 mL. This solution should be freshly prepared.

Nitrite (0.001 mg of NO$_2$ in 1 mL). Dissolve 0.150 g of sodium nitrite, NaNO$_2$, in water, and dilute with water to 100 mL. Dilute 0.1 mL of this solution with water to 100 mL. This solution must be prepared at the time of use.

Nitrogen (0.01 mg of N in 1 mL). Dissolve 0.382 g of ammonium chloride, NH$_4$Cl, in water, and dilute with water to 100 mL. Dilute 10 mL of this solution with water to 1 L.

Phosphate (0.01 mg of PO$_4$ in 1 mL). Dissolve 0.143 g of monobasic potassium phosphate, KH$_2$PO$_4$, in water, and dilute with water to 100 mL. Dilute 10 mL of this solution with water to 1 L.

Platinum–Cobalt Standard (APHA No. 500). *See* page 19.

Potassium (0.01 mg of K in 1 mL). Dissolve 0.191 g of potassium chloride, KCl, in water, and dilute with water to 1 L. Dilute 100 mL of this solution with water to 1 L.

Selenium (1 μg of Se in 1 mL). Transfer 0.10 mL of standard (1000 μg/g Se, commercially available standard) to a 100-mL volumetric flask, and dilute to the mark with water. Prepare fresh before use.

Silica (0.01 mg of SiO$_2$ in 1 mL). Dissolve 0.473 g of sodium silicate, Na$_2$SiO$_3$ · 9H$_2$O, in 100 mL of water in a platinum or polyethylene dish. Dilute 10 mL of this solution (polyethylene graduate) with water to 1 L in a polyethylene bottle.

Silver (0.1 mg of Ag in 1 mL). Dissolve 0.157 g of silver nitrate, AgNO$_3$, in water, and dilute with water to 1 L.

Sodium (0.01 mg of Na in 1 mL). Dissolve 0.254 g of sodium chloride, NaCl, in water, and dilute with water to 1 L. Dilute 100 mL of this solution with water to 1 L.

Strontium Stock Solution (1 mg of Sr in 1 mL). Dissolve 0.242 g of strontium nitrate, Sr(NO$_3$)$_2$, in water, and dilute with water to 100 mL.

Strontium (0.1 mg of Sr in 1 mL). Dilute 10 mL of the strontium stock solution with water to 100 mL.

Strontium (0.01 mg of Sr in 1 mL). Dilute 10 mL of the strontium stock solution with water to 1 L.

Sulfate (0.01 mg of SO$_4$ in 1 mL). Dissolve 0.148 g of anhydrous sodium sulfate, Na$_2$SO$_4$, in water, and dilute with water to 100 mL. Dilute 10 mL of this solution with water to 1 L.

Sulfide (0.01 mg of S in 1 mL). Dissolve 0.75 g of sodium sulfide nonahydrate, Na$_2$S $\cdot$ 9H$_2$O, in water, and dilute with water to 100 mL. Dilute 10 mL of this solution with water to 1 L. This solution must be prepared immediately before use.

Tin (0.05 mg of Sn in 1 mL). Dissolve 0.100 g of metallic tin (Sn) in 10 mL of dilute hydrochloric acid (1 + 1), and dilute with water to 100 mL. Dilute 5 mL of this solution with dilute hydrochloric acid (1 + 9) to 100 mL. This solution must be prepared immediately before use.

Titanium (0.1 mg of Ti in 1 mL). Dissolve 0.739 g of potassium titanyl oxalate dihydrate, K$_2$TiO(C$_2$O$_4$)$_2$ $\cdot$ 2H$_2$O, in water, and dilute with water to 1 L.

Zinc Stock Solution (0.1 mg of Zn in 1 mL). Dissolve 0.124 g of zinc oxide, ZnO, in 10 mL of dilute sulfuric acid (1 + 9), and dilute with water to 1 L.

Zinc Standard for Dithizone Tests (0.001 mg of Zn in 1 mL). Dilute 1 mL of the zinc stock solution with water to 100 mL. This dilute solution must be prepared immediately before use.

Volumetric Solutions: Preparation and Standardization

GENERAL INFORMATION

Throughout the Specifications section of this book, the term *volumetric solution* is used to designate a solution prepared and standardized as described in this section.

Two background references that the analyst may find helpful are ASTM (1997a) and the section on Volumetric Solutions in *United States Pharmacopeia* (1999). Various textbooks on chemical analysis also deal with volumetric solutions and volumetric analysis, on both a theoretical and a practical basis.

The concentrations of volumetric solutions usually are expressed as N, normality (the number of gram equivalent weights in each liter of solution), although concentration sometimes is expressed as M, molarity (the number of gram molecular weights in each liter of solution).

The accuracy of many analytical procedures is dependent on the manner in which such solutions are prepared, standardized, and stored. The directions that follow describe the preparation of these solutions and their standardization against primary standards or against already-standardized solutions. Alternatively, in cases where suitable primary standard chemicals are available, volumetric solutions may be prepared by dissolving accurately weighed portions of these substances to make an accurately known volume of solution. All volumetric solutions, of course, must be thoroughly mixed as part of their preparation. Stronger or weaker solutions than those described are prepared and standardized in the same general manner, using proportional amounts of reagents and standards. Lower strengths frequently may be prepared by accurately diluting a stronger solution. However, when necessary because of the accuracy requirements of the analysis being performed, volumetric solutions prepared in this manner should be standardized before use against a primary standard or by comparison with an appropriate volumetric solution of known strength.

Certain reagents, such as arsenic trioxide, Reductometric Standard, are designated as *standards* in this book. Certified standards of a number of chemical compounds are available as standard reference materials (SRMs)* from NIST.

It is desirable to standardize a solution in such a way that the end point is observed at the same pH value (or oxidation–reduction potential) anticipated during subsequent use. For example, a sodium hydroxide solution found to be 0.1000 N when standardized against potassium hydrogen phthalate to phenolphthalein (pH 8.5) may behave as though 0.1002 N when used for titrating a strong acid to methyl red (pH 5). Only a small part of this difference is due to the NaOH required to bring pure water from pH 5 to 8.5. In this case, the error can be avoided by titrating the strong acid to phenolphthalein (pH 8.5), the same indicator that was used in standardizing the sodium hydroxide.

Storage

Glass containers are suitable for the storage of most volumetric solutions. However, polyolefin containers are recommended for alkaline solutions. Volumetric solutions are stable for varying lengths of time, depending on their chemical nature and their concentration. Dilute solutions are likely to be less stable than those that are 0.1 N or stronger. If there is any doubt about the reliability of a solution, it should be restandardized at the time of use.

Temperature Considerations

If possible, volumetric solutions should be prepared, standardized, and used at 25 °C. If a titration is carried out at a temperature different from that at which the solution was standardized, a temperature correction may be needed, depending on the accuracy requirement of the analysis being performed. If the temperature of use is higher than the temperature of standardization, the correction is to be subtracted from the standardization value; if the temperature of use is lower than the temperature of standardization, the correction is to be added. For guidance with respect to temperature corrections, consider that a 5 °C temperature change will alter the normalities of solutions approximately as follows: For aqueous solutions 0.1 N or less, by 1 part in 1000; 0.5 N, by 1.2 parts in 1000; and 1 N, by 1.4 parts in 1000. For nonaqueous solutions, based on the expansion coefficients of the solvents, by 5.4 parts in 1000 for glacial acetic acid; 5.5 parts in 1000 for dioxane; and 6.2 parts in 1000 for methanol.

Errors caused by temperature differences, as well as by improper drainage of burets or pipets, can be avoided if measurements are made on a weight basis. With a sensitive direct-reading balance and the use of squeeze bottles, titrations by weight have become very practical and precise.

---

*Available from the Office of Standard Reference Materials, National Institute of Standards and Technology, Gaithersburg, MD 20899.

Correction Factors

It is not necessary that volumetric solutions have the identical normalities (or molarities) indicated in this section, or as specified in many of the standardization and assay calculations in this book. It is necessary, however, that the exact value be known, because if it is different from the nominal value specified, a correction factor will have to be applied. Thus, if a calculation is based on an exactly 0.1 N volumetric solution, and the solution actually used is 0.1008 N, its volume must be multiplied by a factor of 1.008 to obtain the corrected volume of 0.1 N solution to use in the calculation. Similarly, if the actual normality is 0.0984 N, the factor is 0.984.

Replication

Because the results obtained by titrations depend on the reliability of the volumetric solutions used, it is imperative that the latter be standardized at least in duplicate, with particular care, and preferably by experienced analysts. Duplicate standardizations for solutions from 0.1 N to 1 N should agree at least to within 2 parts in 1000. Duplicates for 0.05 N solutions may differ by as much as 10 parts in 1000. If deviations are greater than indicated, the determinations should be repeated until the criteria are met.

VOLUMETRIC SOLUTIONS

NOTE. When "measure accurately" is specified in this section, the volumetric ware used shall conform to the tolerance accepted by NIST, as referenced in NIST (1974).

Acetic Acid, 1 N. Add 61 g (58.1 mL) of glacial acetic acid to sufficient water to make 1 L, and mix thoroughly. Standardize as follows: Measure accurately 40 mL of the solution into a 250-mL conical flask, add 0.10 mL of phenolphthalein indicator solution, and titrate with 1 N sodium hydroxide volumetric solution to a permanent pink color.

$$N = \frac{mL\ NaOH \times N\ NaOH}{mL\ CH_3COOH\ solution}$$

Arsenious Acid, 0.1 N. Weigh accurately 4.946 g of arsenic trioxide, Reductometric Standard, previously dried for 1 h at 105 °C, and dissolve in 60 mL of 1 N sodium hydroxide in a 250-mL beaker, warming on a hot plate ($\approx$100 °C), if necessary. Neutralize with 59–60 mL of 1 N sulfuric acid, cool, transfer to a 1-L volumetric flask, dilute to volume, and mix thoroughly. The solution should have a pH between 5 and 9.

Bromine, 0.1 N, Solution I. (Prepared from bromine.) Pipet 2.6 mL of bromine into a 1-L glass-stoppered volumetric flask containing 25 g of potassium bromide and 5 mL of hydrochloric acid dissolved in 500 mL of water. Stopper and swirl until the bromine is dissolved, dilute to volume, and mix thoroughly. Standardize as follows: Measure accurately 40 mL of the solution into a 500-mL iodine flask containing 100 mL of water and 2 g of potassium iodide, and titrate the liberated iodine with 0.1 N sodium thiosulfate volumetric solution. Add 3 mL of starch indicator solution near the end of the titration, and continue to a colorless end point.

$$N = \frac{mL\ Na_2S_2O_3 \times N\ Na_2S_2O_3}{mL\ Br_2\ solution}$$

Bromine, 0.1 N, Solution II. (Prepared as a potassium bromate–potassium bromide mixture.) Transfer 2.8 g of potassium bromate, plus 15 g of potassium bromide, to a 1-L volumetric flask, dilute to volume, and mix thoroughly. Standardize as follows: Measure accurately 40 mL of the solution into a 500-mL iodine flask, and dilute with 100 mL of water. Add 2 g of potassium iodide, stopper the flask, and swirl the contents carefully; then add 5 mL of hydrochloric acid, stopper the flask again, swirl vigorously, and allow to stand for 5 min. Titrate the liberated iodine with 0.1 N sodium thiosulfate volumetric solution. Add 3 mL of starch indicator solution near the end of the titration, and continue to a colorless end point.

$$N = \frac{mL\ Na_2S_2O_3 \times N\ Na_2S_2O_3}{mL\ Br_2\ solution}$$

Ceric Ammonium Sulfate, 0.1 N. Place 65 g of ceric ammonium sulfate dihydrate in a 1500-mL beaker, and slowly add 30 mL of sulfuric acid. Stir to a smooth paste, and cautiously add water in portions of 20 mL or less, with stirring, until the salt is dissolved. Dilute, if necessary, to approximately 500 mL, and mix thoroughly by stirring. Allow to stand overnight. If a residue has formed, or the solution is turbid, filter through a fine-porosity sintered-glass filter (do not filter through paper or similar material). Transfer the filtrate to a 1-L volumetric flask, dilute to volume, and mix thoroughly. Standardize as follows: Measure accurately 40 mL of 0.1 N arsenious acid volumetric solution into a 500-mL conical flask. Add 100 mL of water and 10 mL of 25% sulfuric acid solution, followed by 0.1 mL of 1,10-phenanthroline reagent solution and 0.2 mL of 0.1 N ferrous ammonium sulfate solution. Add 0.1 mL of osmium tetroxide solution,* and titrate slowly with the ceric solution to a change from pink to pale blue. Perform a blank titra-

---

*To prepare oxmium tetroxide solution, in a well-ventilated hood dissolve 0.1 g of osmium tetroxide in 40 mL of 0.1 N H_2SO_4. *CAUTION:* Osmium tetroxide vapor is extremely hazardous.

tion on all of the reagents except the arsenious acid solution, and subtract the volume of ceric sulfate solution consumed in the blank titration from that consumed in the first titration.

$$N = \frac{mL\ As_2O_3 \times N\ As_2O_3}{mL\ Ce^{IV}\ solution\ (corrected\ for\ blank)}$$

Alternatively, the solution may be standardized against arsenic trioxide, Reductometric Standard: Weigh accurately into a 500-mL Erlenmeyer flask, 0.21 ± 0.01 g of arsenic trioxide, Reductometric Standard, previously dried for 1 h at 105 °C. Add 15 mL of 1 N sodium hydroxide, warm gently, and swirl to hasten dissolution, being certain that no particles of arsenic trioxide remain on the sides of the flask, and then neutralize with 15 mL of 1 N sulfuric acid. Add 110 mL of water, and proceed as directed above, beginning with "add 10 mL of 25% sulfuric acid...", except substitute "arsenic trioxide" for "arsenious acid solution" in the sentence that calls for a blank titration.

$$N = \frac{g\ As_2O_3}{0.04946 \times mL\ Ce^{IV}\ solution\ (corrected\ for\ the\ blank)}$$

EDTA, 0.1 M. Dissolve 37.22 g of $Na_2EDTA \cdot 2H_2O$ in 500 mL of water in a 1-L volumetric flask, dilute to mark, and mix thoroughly. Standardize as follows: Weigh accurately 0.40 g of calcium carbonate, chelometric standard, transfer to a 400-mL beaker, and add water. Cover the beaker with a watch glass, and introduce 4 mL of hydrochloric acid (1 + 1) from a pipet inserted between the lip of the beaker and the edge of the watch glass. Swirl the contents of the beaker to dissolve the calcium carbonate. Wash down the sides of the beaker and the watch glass with water, and dilute to 200-mL. While stirring, add from a 50-mL buret about 30 mL of EDTA standard solution. Adjust the solution to pH 12–13 with 10% sodium hydroxide and 300 mg of hydroxy naphthol blue indicator, and continue titration with the EDTA standard solution to a blue endpoint.

$$M = \frac{(g\ CaCO_3)(Assay/100)(1000)}{100.09 \times mL\ EDTA}$$

Ferrous Ammonium Sulfate, 0.1 N. Dissolve 40 g of ferrous ammonium sulfate hexahydrate in 200 mL of 25% sulfuric acid, dilute to 1 L with water, and mix thoroughly. On each day of use, standardize as follows: Measure accurately 40 mL of the solution into a 250-mL conical flask, add 0.1 mL of 1,10-phenanthroline reagent solution, and titrate with 0.1 N ceric ammonium sulfate volumetric solution to the change from pink to pale blue.

$$N = \frac{mL\ Ce^{IV} \times N\ Ce^{IV}}{mL\ Fe^{II}\ solution}$$

Hydrochloric Acid, 1 N. Dilute 85 mL of hydrochloric acid to 1 L, and mix thoroughly. Standardize as follows: Weigh accurately about 2.2 g of sodium carbonate, alkalimetric standard, that previously has been heated at 285 °C for 2 h. Dissolve in 100 mL of water, and add 0.1 mL of methyl red indicator solution. Add the acid slowly from a buret, with stirring, until the solution becomes faintly pink. Heat the solution to boiling, cool, and again titrate until the solution becomes faintly pink. Repeat this procedure until the faint color is no longer discharged on further boiling.

$$N = \frac{g\ Na_2CO_3}{0.05300 \times mL\ HCl\ solution}$$

Iodine, 0.1 N. Place approximately 40 g of potassium iodide and 10 mL of water in a large, glass-stoppered weighing bottle. Mix the contents by swirling; allow to come to room temperature, and then weigh accurately. Add approximately 12.7 g of assayed iodine (page 348), previously weighed on a rough balance, and reweigh the stoppered bottle and contents to obtain the exact weight of the iodine added. Mix and transfer quantitatively to a 1-L volumetric flask, dilute to the mark, and mix thoroughly. For approximate work, calculate the normality from the weight and assay value of the iodine taken (atomic weight = 126.90). For more exact work, standardize as follows: Measure accurately 40 mL of 0.1 N arsenious acid volumetric solution into a 250-mL conical flask. Add 2 g of sodium bicarbonate dissolved in 20 mL of water, and titrate with the iodine solution until within about 2 mL of the anticipated equivalence point. Then add 3 mL of starch indicator solution, and continue the titration to the first permanent blue tinge.

$$N = \frac{mL\ As_2O_3 \times N\ As_2O_3}{mL\ I_2\ solution}$$

Lead Nitrate, 0.1 M. Dissolve 33.1 g of lead nitrate in 300 mL of water. Transfer to a 1-L volumetric flask, and dilute to the mark with water. Mix thoroughly. Standardize as follows: Measure accurately 40 mL of the solution into a 400-mL beaker. Add 0.15 mL of acetic acid and 15 mL of a saturated aqueous solution of hexamethylenetetramine. Add a few milligrams of xylenol orange indicator mixture, and titrate with 0.1 M EDTA volumetric solution to a yellow color.

$$M = \frac{mL\ EDTA \times M\ EDTA}{mL\ Pb(NO_3)_2\ solution}$$

Oxalic Acid 0.1 N. Dissolve 6.5 g of oxalic acid dihydrate in 200 mL of water, and dilute to 1000 mL in a volumetric flask. Mix well. Measure accurately 40 mL of the solution in a 250-mL conical flask, add 50 mL of water and 7 mL of sulfuric acid and slowly with stirring add 30 mL of 0.1 N potassium permanganate volumetric solution. Heat the solution to 70 °C and continue the titration until the pink color persists for 30 s.

$$N = \frac{\text{mL KMnO}_4 \times \text{N KMnO}_4}{\text{mL OxalicAcid}}$$

Perchloric Acid in Glacial Acetic Acid, 0.1 N. Add slowly, with stirring, 8.5 mL of 70% perchloric acid* to a mixture of 500 mL of glacial acetic acid and 21 mL of acetic anhydride. Cool, dilute to 1 L with glacial acetic acid, and mix thoroughly. Determine the water content of the solution by titrating 30 g with Karl Fischer reagent (page 87). If the water content is greater than 0.05%, add a small amount of acetic anhydride, mix thoroughly, and again determine the water. Repeat until the water content is between 0.02% and 0.05%. In a similar manner, if the original water content is below 0.02%, add water to bring the water content into the 0.02–0.05% range. Allow the solution to stand for 24 h, and check the water content to be certain that it remains in this range. Standardize as follows: Weigh accurately about 0.7 g of potassium hydrogen phthalate, Acidimetric Standard, previously lightly crushed and dried for 2 h at 120 °C. Dissolve in 50 mL of glacial acetic acid, add 0.1 mL of crystal violet indicator solution, and titrate with the perchloric acid solution until the violet color changes to blue-green. Perform a blank titration on 50 mL of glacial acetic acid in the same manner, and subtract the volume of perchloric acid solution consumed in the blank titration from that consumed in the first titration.

$$N = \frac{\text{g KHC}_8\text{H}_4\text{O}_4}{0.20423 \times \text{mL HClO}_4 \text{ solution (corrected for the blank)}}$$

Potassium Bromate, 0.1 N. Weigh accurately 2.784 g of potassium bromate, transfer to a 1-L volumetric flask, dilute to volume, and mix thoroughly. Standardize as follows: Measure accurately 40 mL of the solution into a 250-mL glass-stoppered conical flask, add 3 g of potassium iodide and 3 mL of hydrochloric acid, stopper the flask, and allow to stand for 5 min. Titrate the liberated iodine with the 0.1 N sodium thiosulfate volumetric solution. Add 3 mL of starch indicator solution near the end of the titration, and continue to a colorless end point.

$$N = \frac{\text{mL Na}_2\text{S}_2\text{O}_3 \times \text{N Na}_2\text{S}_2\text{O}_3}{\text{mL KBrO}_3 \text{ solution}}$$

Potassium Dichromate, 0.1 N. Weigh exactly 4.903 g of potassium dichromate, NIST Standard Reference Material, previously dried for 4 h at 120 °C, transfer to a 1-L volumetric flask, dilute to volume, and mix thoroughly.

CAUTION: Perchloric acid in contact with some organic materials can form explosive mixtures. Use safety goggles and rubber gloves when preparing the solution. Rinse any glassware that has been in contact with perchloric acid before setting it aside.

Potassium Hydroxide, Methanolic, 0.1 N. Dissolve 6.8 g of potassium hydroxide in a minimum amount of water (about 5 mL), dilute to 1 L with methanol, and mix thoroughly. Allow to stand in a container protected from carbon dioxide for 24 h. Carefully decant the clear solution, leaving any residue behind, into a suitable container, mix well, and store protected from carbon dioxide. Standardize as follows: Weigh accurately about 0.85 g of potassium hydrogen phthalate, Acidimetric Standard, previously lightly crushed and dried for 2 h at 120 °C. Dissolve in 100 mL of carbon dioxide-free water, add 0.15 mL of phenolphthalein indicator solution, and titrate with the methanolic potassium hydroxide solution to a permanent faint pink color.

$$N = \frac{g\ KHC_8H_4O_4}{0.20423 \times mL\ of\ methanolic\ KOH\ solution}$$

Potassium Iodate, 0.05 M. Weigh exactly 10.700 g of potassium iodate, previously dried at 110 °C to constant weight, transfer to a 1-L volumetric flask, dilute to volume, and mix thoroughly.

Potassium Permanganate, 0.1 N. Dissolve 3.3 g of potassium permanganate in 1 L of water in a 1500-mL conical flask, heat to boiling, and boil gently for 15 min. Allow to cool, stopper, and keep in the dark for 48 h. Then filter through a fine-porosity sintered-glass filter, and place in a glass-stoppered amber bottle. Standardize as follows: Weigh accurately 0.24–0.28 g of sodium oxalate, NIST Standard Reference Material, prepared for use as directed in the applicable NIST certificate, transfer to a 500-mL conical flask containing 250 mL of water and 7 mL of sulfuric acid, and heat to 70 °C. Titrate immediately with the permanganate solution to a pink color that persists for 15 s. The temperature at the end of the titration should not be less than 60 °C.

$$N = \frac{g\ Na_2C_2O_4}{mL\ KMnO_4\ solution \times 0.06700}$$

Potassium Thiocyanate, 0.1 N. Weigh exactly 9.7 g of potassium thiocyanate, previously dried for 2 h at 110 °C, transfer to a 1-L volumetric flask, dilute to volume, and mix thoroughly. Standardize as follows: Measure accurately 40 mL of 0.1 N silver nitrate volumetric solution into a 500-mL conical flask and add 100 mL of water, 1 mL of nitric acid, and 2 mL of ferric ammonium sulfate indicator solution. Titrate with the thiocyanate solution, with agitation, to a permanent light pinkish-brown color of the supernatant solution.

$$N = \frac{mL\ AgNO_3 \times N\ AgNO_3}{mL\ KSCN\ solution}$$

Silver Nitrate, 0.1 N. Weigh exactly 16.987 g of silver nitrate, previously dried for 1 h at 100 °C, transfer to a 1-L volumetric flask, dilute to volume, and mix thoroughly. Standardize as follows: Weigh accurately about 0.33 g of potassium chloride (NIST Standard Reference Material No. 999), previously dried for 2 h at 110 °C, place in a 250-mL beaker, and dissolve in 50 mL of water and 1 mL of nitric acid. Pipet 50 mL of the silver nitrate solution into the beaker, slowly, and with agitation. Heat to boiling, cool in the dark, and filter, washing the precipitate thoroughly with 1% nitric acid. To the filtrate, add 2 mL of ferric ammonium sulfate indicator solution, and titrate the excess silver nitrate with 0.1 N potassium thiocyanate volumetric solution, with agitation, to a permanent light pinkish-brown color of the supernatant solution.

$$N = \frac{g \text{ KCl} \times 1000}{74.55(50 - A)}$$

where A = mL of 0.1 N potassium thiocyanate volumetric solution.

Sodium Hydroxide, 1 N. Dissolve 162 g of sodium hydroxide in 150 mL of carbon dioxide-free water, cool the solution to room temperature, and filter through a suitable filter, such as hardened filter paper. (*Note:* A reagent-grade 50% sodium hydroxide solution may be used in place of the solution prepared in this way.) Prepare a 1 N solution by diluting 54.5 mL of the concentrated solution to 1 L with carbon dioxide-free water. Mix thoroughly, and store in a tight polyolefin container. Standardize as follows: Weigh accurately about 8.5 g of potassium hydrogen phthalate, Acidimetric Standard, previously lightly crushed and dried for 2 h at 120 °C. Dissolve in 100 mL of carbon dioxide-free water, add 0.15 mL of phenolphthalein indicator solution, and titrate with the sodium hydroxide solution to a permanent faint pink color.

$$N = \frac{g \text{ KHC}_8\text{H}_4\text{O}_4}{0.20423 \times \text{mL NaOH solution}}$$

Sodium Methoxide in Methanol, 0.01 N. Dilute 20.0 mL of sodium methoxide in methanol, 0.5 M methanolic solution, with methanol to 1 L, and mix thoroughly. Standardize as follows: Measure accurately 40 mL of 0.01 N sulfuric acid volumetric solution into a suitable container, add 0.10 mL of thymol blue indicator solution, and titrate with the sodium methoxide in methanol solution to a blue end point.

$$N = \frac{\text{mL H}_2\text{SO}_4 \times N \text{ H}_2\text{SO}_4}{\text{mL CH}_3\text{ONa solution}}$$

Sodium Nitrite, 0.1 N. Assay a sample of sodium nitrite as described on page 601. Correcting for the assay value obtained, weigh exactly enough of the material to contain 6.900 g of $NaNO_2$. Dissolve in 100 mL of water in a 1-L volumetric flask, dilute to volume, and mix thoroughly.

Sodium Thiosulfate, 0.1 N. Weigh accurately about 210 mg of NIST current standard reference material potassium dichromate, previously crushed and dried at 120 °C for 4 h. Dissolve in 100 mL of water in a glass-stoppered iodine flask. Swirl to dissolve the sample, remove the stopper, and quickly add 3 g of potassium iodide, 2 g of sodium bicarbonate, and 5 mL of hydrochloric acid. Insert the stopper, swirl slightly to release excess carbon dioxide, cover the stopper with water, and allow to stand in dark for 10 min. Rinse the stopper and the inner walls of the flask with water and titrate the liberated iodine with the sodium thiosulfate until solution becomes faintly yellow in color. Add 3 mL of starch indicator solution and continue the titration to the disappearance of the blue color. Restandardize the solution frequently.

$$N = \frac{g \ K_2Cr_2O_7}{0.04904 \times mL \ Na_2S_2O_3}$$

Sulfuric Acid, 1 N. Add slowly (use caution), with stirring, 30 mL of sulfuric acid to about 500 mL of water. Allow the mixture to cool to room temperature, dilute to 1 L, and mix thoroughly. Standardize against sodium carbonate, Alkalimetric Standard, as directed for hydrochloric acid, 1 N.

Zinc Chloride, 0.1 M. Weigh accurately 6.537 g of zinc and dissolve in 80 mL of 10% hydrochloric acid. Warm if necessary to complete dissolution, cool, dilute with water to volume in a 1-L volumetric flask, and mix thoroughly.

Registered Trademarks

The tests in *Reagent Chemicals* employ some products having registered trademarks. These are listed below with the corporate owners.

Amberlite (Rohm and Haas Corp.)
Carbowax (Union Carbide Chemical and Plastics Inc.)
Chromosorb (Johns-Mansville Corp.)
Dowex (Dow Chemical Co.)
Drierite (W. Hammond Drierite Co.)
Eriochrome (Ciba-Geigy Corp.)
Flexol (Union Carbide Chemicals and Plastics Inc.)
GasChrom (Altex Associates, Inc.)
Igepal (GAF Corp.)
Porapak (Waters Associates, Inc.)
Regisil (Regis Chemical Co.)
Supelcoport (Supelco, Inc.)
Teflon (E.I. du Pont de Nemours &a Co., Inc.)
Tenax (Enka Research Institute)
Triton (Union Carbide Chemicals and Plastics Inc.)
Whatman (Whatman)

References

APHA (American Public Health Association). 1998. *Standard Methods for the Examination of Water and Wastewater*, 20th ed.; APHA, American Water Works Association, and Water Pollution Control Federation: Washington, DC.

ASTM (American Society for Testing and Materials). 1995. *Standard Specification for Wire Cloth and Sieves for Testing Purposes*, E11; ASTM: West Conshohocken, PA.

ASTM. 1997a. *Standard Practice for Preparation, Standardization, and Storage of Standard and Reagent Solutions for Chemical Analysis*, E200; ASTM: West Conshohocken, PA.

ASTM. 1997b. *Standard Test Method for Color of Clear Liquids (Platinum–Cobalt Scale)*, D1209; ASTM: West Conshohocken, PA.

ASTM. 1997c. *Standard Test Method for Distillation Range of Volatile Organic Liquids*, D1078; ASTM: West Conshohocken, PA.

ASTM. 1999. *Standard Specification for Reagent Water*, D1193; ASTM: West Conshohocken, PA.

Institute of Medicine. 1996. *Food Chemicals Codex*, 4th ed; National Academy Press: Washington, DC.

Kenkel, J. 1994. *Analytical Chemistry for Technicians;* Lewis Publishers: New York.

Kingston, H.M., and Haswell, S.J., eds. 1997. *Microwave-Enhanced Chemistry: Fundamentals, Sample Preparation, and Applications;* American Chemical Society: Washington, DC.

Lind, J.E., Zwolenik, J.J., and Fuoss, R.M. 1959. Calibration of Conductance Cells at 25 °C with Aqueous Solutions of Potassium Chloride. *J. Am. Chem. Soc.,* 81:1557.

Mitchell, J., and Smith, D.M. 1980. *Aquametry*, 2nd ed.; Wiley-Interscience: New York, p 41.

Moody, J.R. 1982. NBS Clean Laboratories for Trace Element Analysis. *Anal. Chem.* 54(13):1358A–1376A

NIST (National Institute of Standards and Technology). 1974. *The Calibration of Small Volumetric Laboratory Glassware*, NBSIR 74–461; NIST: Gaithersburg, MD.

Schilt, A.A. 1991. *Moisture Measurement by Karl Fischer Titrimetry.* GFS Chemicals: Columbus, OH.

Scholz, E. 1984. *Karl Fischer Titration.* Springer-Verlag: Berlin.

Semiconductor Equipment and Materials International (SEMI). 1999. *Guide for Determination of Method Detection Limits,* Procedure C10-0299; SEMI: Mountain View, California.

Settle, F.A. 1997. *Handbook of Instrumental Techniques for Analytical Chemistry;* Prentice-Hall: Saddle Brook River, NJ.

Skoog, D.A., Holler, F.J., and Nieman, T.A. 1998. *Principles of Instrumental Analysis,* 5th ed.; Saunders College Publishing: Philadelphia.

Smith, H.M., and others. 1950. Measurement of Density of Hydrocarbon Liquids by the Pycnometer. *Anal. Chem.* 22:1452.

United States Pharmacopeia and National Formulary. 1999. USP 24–NF 19; U.S. Pharmacopeial Convention: Rockville, MD.

Willard, H.H., Dean, J.A., Settle, F.A., and Merritt, L.L. 1995. *Instrumental Methods of Analysis,* 7th ed.; Wadsworth: Belmont, CA.

Zief, M., and Mitchell, J.W. 1976. Contamination Control in Trace Element Analysis. *Chemical Analyses;* Elving, J.P., Ed.; Wiley: New York.

Specifications for
Reagent Chemicals

Requirements and Tests

Acetaldehyde
Ethanal

CH₃CHO CH_3CHO **Formula Wt 44.05**

CAS Number 75–07–0

REQUIREMENTS

Assay . . . . ≥99.5% CH_3CHO

MAXIMUM ALLOWABLE

Residue after evaporation 0.005%

Titrable acid . . . . 0.008 meq/g

TESTS

ASSAY. Analyze the sample by gas chromatography using the general parameters cited on page 70. The following specific conditions are also required.

    Column: Type I, methyl silicone

    Detector: Flame ionization

Measure the area under all peaks and calculate the area percent for acetaldehyde.

RESIDUE AFTER EVAPORATION. (Page 16). Evaporate 40 g (51 mL) to dryness in a tared platinum dish on a hot plate (≈100 °C) in a well-ventilated hood, and dry the residue at 105 °C for 30 min.

TITRABLE ACID. Chill 25 g (32 mL) of sample in a graduated cylinder in an ice bath. To a 250-mL Erlenmeyer flask add about 25 mL of deionized water and 75 g of deionized ice. Add 0.5 mL of phenolphthalein indicator solution and titrate with 0.1 N sodium hydroxide to the first perceptible pink color that persists for 15 s. Working in a fume hood, add the chilled sample and titrate immediately to the same faint pink color. Not more than 2.0 mL of 0.1 N sodium hydroxide should be required.

Acetic Acid, Glacial

CH₃COOH   **Formula Wt 60.05**

CAS Number 64–19–7

REQUIREMENTS

Assay . . . . ≥99.7% CH₃COOH

MAXIMUM ALLOWABLE

Color (APHA) 10
Dilution test . . . . Passes test
Residue after evaporation 0.001%
Acetic anhydride [(CH₃CO)₂O] 0.01%
Chloride (Cl) . . . . 1 ppm
Sulfate (SO₄) . . . . 1 ppm
Heavy metals (as Pb) . . . . 0.5 ppm
Iron (Fe) . . . . 0.2 ppm
Substanced reducing dichromate . . . . Passes test
Substances reducing permanganate . . . . Passes test
Titrable base . . . . 0.0004 meq/g

TESTS

ASSAY AND ACETIC ANHYDRIDE. Analyze the sample by gas chromatography using the general parameters cited on page 70. The following specific conditions are also required.

Column: Type I, methyl silicone

Measure the area under all peaks and calculate the area percent for acetic acid and acetic anhydride. Correct for water content.

COLOR (APHA). (Page 19).

DILUTION TEST. Dilute 1 volume of the acid with 3 volumes of water and allow to stand for 1 h. The solution should be as clear as an equal volume of water.

RESIDUE AFTER EVAPORATION. (Page 16). Evaporate 100 g (95 mL) to dryness in a tared dish on a hot plate (≈100 °C) and dry the residue at 105 °C for 30 min.

CHLORIDE. (Page 27). Dilute 10 g (9.5 mL) with 10 mL of water and add 1 mL of silver nitrate reagent solution. Prepare a standard containing 0.01 mg of chloride ion (Cl) in 20 mL of water and add 1 mL of silver nitrate reagent solution. Evaporate the solutions to dryness on a hot plate ($\approx$100 °C). Dissolve the residues with 0.5 mL of ammonium hydroxide, dilute with 20 mL of water, and add 1.5 L of nitric acid. Any turbidity in the solution of the sample should not exceed that of the standard.

SULFATE. To 48 mL (50 g) of sample, add 2 mL of 1% sodium carbonate solution and evaporate to dryness on a hot plate ($\approx$100 °C). Dissolve the residue in 1.0 mL of 10% hydrochloric acid and 15 mL of water. Dilute to 25 mL with water. For the control, take 0.10 mg of sulfate ion (SO_4) in 20 mL of water, add 1.0 mL of 10% hydrochloric acid and dilute to 25 mL. To sample and control solutions, add 1 mL of barium chloride solution. Compare after 10 min. Sample turbidity should not exceed that of the control solution.

HEAVY METALS. (Page 28, Method 1). To 40 g (38 mL) add about 10 mg of sodium carbonate, evaporate to dryness on a hot plate ($\approx$100 °C), dissolve the residue in about 20 mL of water, and dilute with water to 25 mL.

IRON. (Page 30, Method 1). To 50 g (48 mL) add 10 mg of sodium carbonate and evaporate to dryness. Dissolve the residue in 2 mL of hydrochloric acid, dilute with water to 50 mL, and use the solution without further acidification.

SUBSTANCES REDUCING DICHROMATE. To 10 mL, add 1.0 mL of 0.1 N potassium dichromate, and cautiously add 10 mL of sulfuric acid. Cool the solution to room temperature, and allow to stand for 30 min. While the solution is swirled, dilute slowly and cautiously with 50 mL of water, cool, and add 1 mL of freshly prepared 10% potassium iodide reagent solution. Titrate the liberated iodine with 0.1 N thiosulfate, using starch as the indicator. Not more than 0.40 mL of the 0.1 N potassium dichromate should be consumed (not less than 0.60 mL of the 0.1 N thiosulfate should be required). Correct for a complete blank.

SUBSTANCES REDUCING PERMANGANATE. Add 40 g (38 mL) of the sample to 10 mL of water. Cool to 15 °C, add 0.30 mL of 0.1 N potassium permanganate, and allow to stand at 15 °C for 10 min. The pink color should not be entirely discharged.

TITRABLE BASE. To 25 g (24 mL) add 0.10 mL of a solution of 1 g of methyl violet (or crystal violet) in 100 mL of glacial acetic acid. The color should be violet. Titrate the solution with 0.1 N perchloric acid in glacial acetic acid to a green color. Not more than 0.10 mL should be required.

Acetic Acid, Ultratrace
(suitable for use in ultratrace elemental analysis)

CH₃COOH — CH_3COOH **Formula Wt 60.05**

CAS number 64–19–7

NOTE. Reagent must be packaged in a preleached Teflon bottle and used in a clean environment to maintain purity

REQUIREMENTS

Appearance . . . . Clear
Assay . . . . 90% minimum

MAXIMUM ALLOWABLE

Chloride (Cl) . . . . 1 ppm
Sulfate (SO_4) . . . . 1 ppm
Aluminum (Al) . . . . 1 ppb
Barium (Ba) . . . . 1 ppb
Boron (B) . . . . 5 ppb
Cadmium (Cd) . . . . 1 ppb
Calcium (Ca) . . . . 1 ppb
Chromium (Cr) . . . . 1 ppb
Cobalt (Co) . . . . 1 ppb
Copper (Cu) . . . . 1 ppb
Iron (Fe) . . . . 5 ppb
Lead (Pb) . . . . 1 ppb
Lithium (Li) . . . . 1 ppb
Magnesium (Mg) . . . . 1 ppb
Manganese (Mn) . . . . 1 ppb
Molybdenum (Mo) . . . . 1 ppb
Mercury (Hg) . . . . 1 ppb
Potassium (K) . . . . 1 ppb
Silicon (Si) . . . . 5 ppb
Sodium (Na) . . . . 5 ppb
Strontium (Sr) . . . . 1 ppb
Tin (Sn) . . . . 1 ppb
Titanium (Ti) . . . . 1 ppb
Vanadium (V) . . . . 1 ppb
Zinc (Zn) . . . . 1 ppb
Zirconium (Zr) . . . . 1 ppb

TESTS

ASSAY. (By acid–base titrimetry). The assay of acetic acid used for ultratrace metal analysis may vary, depending on the method of purification. Tare a glass-stopped flask containing about 30 mL of water. Quickly add about 3 mL of the sample, stopper, and weigh accurately. Dilute to about 50 mL, add 0.15 mL of methyl orange indicator solution, and titrate with 1 N sodium hydroxide. One mL of 1 N sodium hydroxide corresponds to 0.06005 g of acetic acid.

$$\% \text{ Acetic acid} = (\text{mL} \times \text{N NaOH} \times 6.005) / \text{Sample wt (g)}$$

CHLORIDE. (Page 27). Dissolve 10.0 g in 10 mL of water.

SULFATE. To 48 mL (50 g) of sample, add 2 mL of 1% sodium carbonate solution and evaporate to dryness on hot plate ($\approx$100 °C). Dissolve the residue in 1.0 mL of 10% hydrochloric acid and 15 mL of water. Dilute to 25 mL with water. For the control, take 0.05 mg of sulfate ion (SO_4) in 20 mL of water, add 1.0 mL of 10% hydrochloric acid and dilute to 25 mL. To sample and control solutions, add 1 mL of barium chloride solution. Compare after 10 min. Sample turbidity should not exceed that of the control solution.

TRACE METALS. Determine the aluminum, barium, cadmium, calcium, chromium, cobalt, copper, iron, lead, lithium, magnesium, manganese, molybdenum, nickel, potassium, sodium, strontium, tin, titanium, vanadium, zinc, and zirconium by the ICP–OES method described on page 42.

Determine the mercury by the CVAAS procedure described on page 40.

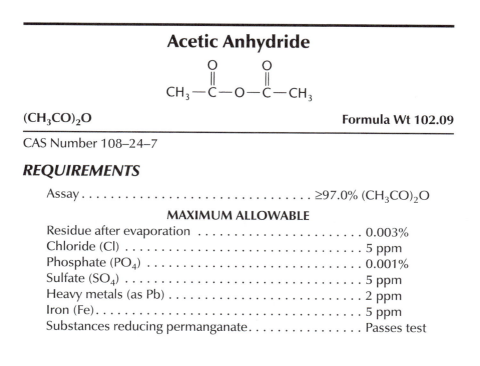

Acetic Anhydride

(CH₃CO)₂O   **Formula Wt 102.09**

CAS Number 108–24–7

REQUIREMENTS

Assay. . . . . $\geq$97.0% $(CH_3CO)_2O$

MAXIMUM ALLOWABLE

Residue after evaporation 0.003%
Chloride (Cl) . . . . 5 ppm
Phosphate (PO_4) 0.001%
Sulfate (SO_4) . . . . 5 ppm
Heavy metals (as Pb) . . . . 2 ppm
Iron (Fe). . . . . 5 ppm
Substances reducing permanganate. . . . . Passes test

TESTS

ASSAY. Analyze the sample by gas chromatography using the general parameters cited on page 70. The following specific conditions are also required.

Column: Type I, methyl silicone

Measure the area under all peaks and calculate the area percent for acetic anhydride.

RESIDUE AFTER EVAPORATION. (Page 16). Evaporate 50 g (46 mL) in a tared dish on a hot plate ($\approx$100 °C), and dry the residue at 105 °C for 30 min.

Sample Solution A for the Determination of Chloride, Phosphate, Sulfate, Heavy Metals, Iron, and Substances Reducing Permanganate. Dilute 40 g (37 mL) of the sample with water to 200 mL (1 mL = 0.2 g).

CHLORIDE. (Page 27). Use 10 mL of sample solution A (2-g sample).

PHOSPHATE. (Page 32, Method 1). Evaporate 10 mL of sample solution A (2-g sample) to dryness on a hot plate ($\approx$100 °C), and dissolve the residue in 25 mL of 0.5 N sulfuric acid.

SULFATE. (Page 33, Method 3). Use 50 mL of sample solution A (10-g sample).

HEAVY METALS. (Page 28, Method 1). To 50 mL of sample solution A (10-g sample) add about 10 mg of sodium carbonate, evaporate to dryness on a hot plate ($\approx$100 °C), dissolve the residue in about 20 mL of water, and dilute with water to 25 mL.

IRON. (Page 30, Method 1). To 10 mL of sample solution A (2-g sample) add 10 mg of sodium carbonate and evaporate to dryness. Dissolve the residue in 2 mL of hydrochloric acid, dilute with water to 50 mL, and use the solution without further acidification.

SUBSTANCES REDUCING PERMANGANATE. To 10 mL of sample solution A (2-g sample) add 0.4 mL of 0.1 N potassium permanganate and allow to stand for 5 min. The pink color should not be entirely discharged.

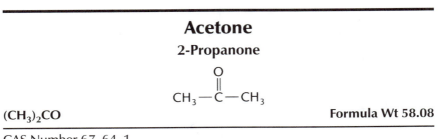

Acetone

2-Propanone

$$CH_3 - \overset{\displaystyle O}{\overset{\displaystyle \|}{C}} - CH_3$$

$(CH_3)_2CO$ **Formula Wt 58.08**

CAS Number 67–64–1

Suitable for use in ultraviolet spectrophotometry or general use. Product labeling shall designate the one or more of those uses for which suitability is represented on the basis of meeting the relevant requirements and tests. The ultraviolet spectrophotometry requirements include all of the requirements for general use.

REQUIREMENTS

General Use

Assay . . . .≥99.5% $(CH_3)_2CO$

MAXIMUM ALLOWABLE

Color (APHA) 10
Residue after evaporation 0.001%
Solubility in water . . . . Passes test
Titrable acid . . . . 0.0003 meq/g
Titrable base . . . . 0.0006 meq/g
Aldehyde (as HCHO) 0.002%
Isopropyl alcohol 0.05%
Methanol 0.05%
Substances reducing permanganate . . . . Passes test
Water 0.5%

Specific Use

Ultraviolet Spectrophotometry

Wavelength (nm)	Absorbance (AU)
400	0.01
350	0.02
340	0.10
330	1.00

TESTS

ASSAY, ISOPROPYL ALCOHOL, AND METHANOL. Analyze the sample by gas chromatography using the general parameters cited on page 70. The following specific conditions are also required.

Column: Type I, methyl silicone

Measure the area under all peaks and calculate the area percent for acetone, isopropyl alcohol, and methanol. Correct for water content.

COLOR (APHA). (Page 19).

RESIDUE AFTER EVAPORATION. (Page 16). Evaporate 100 g (127 mL) to dryness in a tared dish on a hot plate (≈100 °C) and dry the residue at 105 °C for 30 min.

SOLUBILITY IN WATER. Mix 30 g (38 mL) with 38 mL of carbon dioxide-free water. The solution should remain clear for 30 min. Reserve this solution for the test for titrable acid.

TITRABLE ACID. Add 0.10 mL of phenolphthalein indicator solution to the solution prepared in the preceding test. Not more than 1.0 mL of 0.01 N sodium hydroxide should be required to produce a pink color.

TITRABLE BASE. Mix 18 g (23 mL) with 25 mL of water and add 0.05 mL of methyl red indicator solution. Not more than 1.0 mL of 0.01 N hydrochloric acid should be required to produce a red color.

ALDEHYDE. Dilute 2 g (2.5 mL) with water to 10 mL. Prepare a standard containing 0.04 mg of formaldehyde in 10 mL of water and a complete reagent blank in 10 mL of water alone. To each add 0.15 mL of a freshly prepared 5% solution of 5,5-dimethyl-1,3-cyclohexanedione in alcohol. Evaporate each on a hot plate ($\approx$100 °C) until the acetone is volatilized. Dilute each with water to 10 mL and cool quickly in an ice bath while stirring vigorously. Any turbidity in the solution of the sample should not exceed that in the standard. If any turbidity develops in the reagent blank, the 5,5-dimethyl-1,3-cyclohexanedione should be discarded and the entire test rerun with a fresh solution.

SUBSTANCES REDUCING PERMANGANATE. Add 0.05 mL of 0.1 N potassium permanganate to 10 mL of the sample and allow to stand for 15 min at 25 °C. The pink color should not be entirely discharged, when compared with an equal volume of water.

WATER. (Page 55, Method 1). Use 25 mL (20 g) of the sample and a methanol-free system to prevent ketal formation with liberation of water.

ULTRAVIOLET SPECTROPHOTOMETRY. Use the procedure on page 73 to determine the absorbance.

Acetonitrile

CH$_3$CN   **Formula Wt 41.05**

CAS Number 75–05–8

Suitable for use in high-performance liquid chromatography, extraction–concentration analysis, ultraviolet spectrophotometry, or general use. Product labeling shall designate the one or more of these uses for which suitability is represented on the basis of meeting the relevant requirements and tests. The ultraviolet spectrophotometry and liquid chromatography suitability requirements include all of the requirements for general use. The extraction–concentration suitability requirements include only the general use requirement for color.

REQUIREMENTS

General Use

Appearance . . . . Clear
Assay (CH$_3$CN) . . . . ≥99.5%

MAXIMUM ALLOWABLE

Color (APHA). 10
Residue after evaporation 0.005%
Titrable acid. . . . . 8 μeq/g
Titrable base . . . . 0.6 μeq/g
Water. 0.3%

Specific Use

Ultraviolet Spectrophotometry

Wavelength (nm)	Absorbance (AU)
254	0.01
220	0.05
190	1.00

Liquid Chromatography Suitability

Absorbance . . . . Passes test
Gradient elution. . . . . Passes test

Extraction–Concentration Suitability

Absorbance . . . . Passes test
GC–FID . . . . Passes test
GC–ECD . . . . Passes test

TESTS

ASSAY. Analyze the sample by gas chromatography using the general parameters cited on page 70. The following specific conditions are also required.

Column: Type I, methyl silicone

Measure the area under all peaks and calculate the acetonitrile content in area percent. Correct for water content.

COLOR (APHA). (Page 19).

RESIDUE AFTER EVAPORATION. (Page 16). Evaporate 39 g (50 mL) to dryness in a tared platinum dish on a hot plate (≈100 °C) in a well-ventilated hood, and dry the residue at 105 °C for 30 min.

TITRABLE ACID. Mix 10 g (13 mL) with 13 mL of carbon dioxide-free water. Add 0.10 mL of phenolphthalein indicator solution. Not more than 8.0 mL of 0.01 N sodium hydroxide solution should be required in the titration.

TITRABLE BASE. To 78 g (100 mL) in a 250-mL conical flask, add 0.30 mL of 0.05% solution of bromcresol green in alcohol and 0.10 mL of methyl red indicator solution. Titrate with 0.01 N hydrochloric acid to a light orange-pink end point. Not more than 4.5 mL of 0.01 N hydrochloric acid should be required.

WATER. (Page 55, Method 1). Use 10 mL (7.8 g) of the sample.

ULTRAVIOLET SPECTROPHOTOMETRY. Use the procedure on page 73 to determine the absorbance.

LIQUID CHROMATOGRAPHY SUITABILITY. Analyze the sample by using the gradient elution procedure cited on page 71. Use the procedure on page 73 to determine the absorbance.

EXTRACTION–CONCENTRATION SUITABILITY. Analyze the sample by using the general procedure cited on page 72. In the sample preparation step exchange the sample with ACS grade 2,2,4-trimethylpentane suitable for extraction–concentration. Use the procedure on page 73 to determine the absorbance.

Acetyl Chloride
Ethanoic Acid Chloride

CH_3COCl Formula Wt 78.50

CAS Number 75–36–5

REQUIREMENTS

Assay . . . .≥98.5% CH_3COCl

MAXIMUM ALLOWABLE

Color (APHA) 20
Residue after evaporation. 0.005%
Insoluble matter. 0.0025%
Phosphate (PO_4) 0.002%
Heavy metals (as Pb) . . . . .5 ppm
Iron (Fe) . . . . .5 ppm

TESTS

ASSAY. Analyze the sample by gas chromatography using the general parameters cited on page 70. The following specific conditions are also required.

Column: Type I, methyl silicone

Detector: Flame ionization

Column Temperature: Initial 75 °C for 5 min, then increased at a rate of 10 °C per minute to a final temperature of 200 °C

COLOR (APHA). (Page 19).

RESIDUE AFTER EVAPORATION. (Page 16). Evaporate 20 g (18 mL) to dryness in a tared dish on a hot plate (≈100 °C), and dry the residue at 105 °C for 30 min.

INSOLUBLE MATTER. (Page 15). Add 40 g (36 mL) cautiously, with stirring, to 150 mL of water and dilute to 200 mL with water. The solution is clear and colorless. Continue with the procedure as described on page 15. Retain the filtrate separate from the washings for the tests for phosphate, heavy metals, and iron.

PHOSPHATE. (Page 32, Method 1). To 5 mL of the solution from the test for insoluble matter (1.0-g sample) add 5 mg of anhydrous sodium carbonate and evaporate to dryness. Take up the residue in 20 mL of 0.5 N H_2SO_4. For the standard, use 0.02 mg of PO_4 and 5 mg of Na_2CO_3, and treat exactly as the sample.

HEAVY METALS. (Page 28, Method 1). Use 20 mL (4-g sample) of the solution retained from the test for insoluble matter.

IRON. (Page 30, Method 1). Use 10 mL (2.0-g sample) of the solution retained from the test for insoluble matter.

Aluminum

Al   **Atomic Wt 26.98**

CAS Number 7429–90–5

REQUIREMENTS

MAXIMUM ALLOWABLE

Insoluble in dilute hydrochloric acid 0.05%
Copper (Cu). 0.02%
Iron (Fe). 0.1%
Manganese (Mn) 0.002%
Nitrogen compounds (as N). 0.001%
Silicon (Si) 0.1%
Titanium (Ti) 0.03%

TESTS

INSOLUBLE IN DILUTE HYDROCHLORIC ACID. Place a 1 L beaker containing 75 mL of water in a fume hood and add 10.0 g of sample. Cautiously, and with stirring, add 30 mL of hydrochloric acid (exothermic reaction), and then add an additional 70 mL in 10 mL increments. If necessary, one or more additional 5 mL portions of hydrochloric acid may be added to complete the dissolution. Wash the sides of the sample beaker with about 10 mL of water. Boil the solution for 20 minutes, then cool and dilute to 190 mL. Filter through a tared filtering crucible. Retain the filtrate, without washings, for Sample Solution A. Thoroughly wash the contents of the crucible and dry the residue at 105 °C.

Sample Solution A for the Determination of Copper, Iron, Manganese, and Titanium. Dilute the filtrate, without washings, obtained in the test for insoluble in dilute hydrochloric acid with water to 200 mL in a volumetric flask (1 mL = 0.05 g).

COPPER, IRON, AND MANGANESE. (By flame AAS, page 39). Use sample solution A.

Element	Wavelength (nm)	Sample Wt (g)	Standard Added (mg)	Flame Type*	Background Correction
Cu	324.8	0.10	0.02; 0.04	A/A	Yes
Fe	248.3	0.10	0.10; 0.20	A/A	Yes
Mn	279.5	1.00	0.02; 0.04	A/A	Yes

*A/A is air/acetylene.

NITROGEN COMPOUNDS. (Page 31). Use 2.5 g dissolved in 60 mL of freshly boiled 10% sodium hydroxide. For the standard, use 0.5 g of aluminum and 0.02 mg of nitrogen (N).

SILICON.

> *NOTE.* The molybdenum blue test for silicon is extremely sensitive; therefore, all solutions that are more alkaline than pH 4 should be handled in containers other than glass or porcelain. Platinum dishes and plastic graduated cylinders are recommended. The 10% solution of sodium hydroxide is prepared and measured in plastic.

Dissolve 0.25 g in 10 mL of 10% sodium hydroxide solution. Add the aluminum to the sodium hydroxide solution in small portions, particularly if the sample is finely divided. When the vigorous reaction has subsided, digest the covered container on a hot plate ($\approx$100 °C) until dissolution is complete. Uncover and evaporate to a moist residue. Dissolve the residue in 10 mL of water, and add 0.05–0.10 mL of 30% hydrogen peroxide. Boil the solution gently to decompose the excess peroxide, cool, and dilute with water to 100 mL. Adjust the acidity of the solution to pH 2 (external indicator) by adding dilute sulfuric acid (1 + 1). Digest on a hot plate ($\approx$100 °C) until any precipitate is dissolved and the solution is clear. If necessary, add more of the dilute sulfuric acid (1 + 1). When dissolution is complete, cool and dilute with water to 250 mL. Prepare a blank solution by treating 10 mL of the 10% sodium hydroxide solution exactly as described above, but omit the 0.25-g sample. Dilute 5.0 mL of the sample solution with water to 80 mL. For the standard, add 0.01 mg of SiO_2 (0.005 mg of Si) to 5 mL of the blank solution and dilute with water to 80 mL. To each solution add 5.0 mL of a freshly prepared 10% solution of ammonium molybdate. Adjust the pH of each solution to 1.7–1.9 (using a pH meter) with dilute hydrochloric acid (1 + 9) or silica-free ammonium hydroxide. Heat the solutions just to boiling, cool, add 20 mL of hydrochloric acid, and dilute with water to 110 mL. Transfer the solutions to separatory funnels, add 50 mL of butyl alcohol, and shake vigorously. Allow the layers to separate and draw off and discard the aqueous phase. Wash the butyl alcohol three times with 20-mL portions of dilute hydrochloric acid (1 + 99), discarding each aqueous phase. Add 0.5 mL of a freshly prepared 2% solution of stannous chloride in hydrochloric acid and shake. If the butyl alcohol is turbid, wash it with 10 mL of dilute hydrochloric acid (1 + 99). Any blue color in the butyl alcohol from the solution of the sample should not exceed that in the butyl alcohol from the standard.

TITANIUM. To 20 mL of sample solution A (1-g sample) add 3 mL of phosphoric acid and 0.5 mL of 30% hydrogen peroxide. Any color should not exceed that produced by 0.3 mg of titanium in an equal volume of solution containing the quantities of reagents used in the test.

Aluminum Ammonium Sulfate Dodecahydrate
Ammonium Alum

$AlNH_4(SO_4)_2 \cdot 12H_2O$ **Formula Wt 453.33**

CAS Number 7784–26–1

REQUIREMENTS

Assay . . . . .98.0–102.0% $AlNH_4(SO_4)_2 \cdot 12H_2O$

MAXIMUM ALLOWABLE

Insoluble matter.	0.005%
Chloride (Cl)	0.001%
Heavy metals (as Pb)	0.001%
Calcium (Ca)	0.05%
Iron (Fe).	0.001%
Potassium (K).	0.05%
Sodium (Na)	0.01%

TESTS

ASSAY. (By complexometry). Weigh accurately 0.8 g and transfer to a 400-mL beaker. Moisten with 1 mL of glacial acetic acid, add 50 mL of water, 40.0 mL of standard 0.1 M EDTA, and 20 mL of ammonium acetate–acetic acid buffer solution. Warm on a steam bath until solution is complete, and boil gently for 5 min. Cool, add 50 mL of alcohol and 2 mL of dithizone indicator solution, and titrate with standard 0.1 M zinc chloride to a bright rose-pink color. Perform a blank titration of 40.0 mL of standard 0.1 M EDTA, using the same quantities of reagents as for the assay. One milliliter of 0.1 M EDTA consumed by the sample corresponds to 0.04533 g of $AlNH_4(SO_4)_2 \cdot 12H_2O$.

INSOLUBLE MATTER. (Page 15). Use 10 g dissolved in 150 mL of water containing 5 mL of 25% sulfuric acid.

CHLORIDE. (Page 27). Use 1.0 g.

HEAVY METALS. (Page 28, Method 1). Dissolve 2.0 g in 20 mL of water.

CALCIUM, IRON, POTASSIUM, AND SODIUM. (By flame AAS, page 39).

Sample Stock Solution. Dissolve 10.0 g in 100 mL of water (1 mL = 0.1 g).

Element	Wavelength (nm)	Sample Wt (g)	Standard Added (mg)	Flame Type*	Background Correction
Ca	422.7	0.20	0.05; 0.10	N/A	No
Fe	248.3	5.0	0.025; 0.05	A/A	Yes
K	766.5	0.10	0.025; 0.05	A/A	No
Na	589.0	0.20	0.01; 0.02	A/A	No

*A/A is air/acetylene; N/A is nitrous oxide/acetylene.

Aluminum Nitrate Nonahydrate

$Al(NO_3)_3 \cdot 9H_2O$ **Formula Wt 375.13**

CAS Number 7784–27–2

REQUIREMENTS

Assay . . . . .98.0–102.0% $Al(NO_3)_3 \cdot 9H_2O$

MAXIMUM ALLOWABLE

Insoluble matter. 0.005%
Chloride (Cl) 0.001%
Sulfate (SO_4) 0.005%
Substances not precipitated by ammonium
    hydroxide (as SO_4) 0.05%

Heavy metals (as Pb) 0.001%
Iron (Fe). 0.002%

TESTS

ASSAY. (By indirect complexometric titration of Al^{III}). Weigh accurately 5 g and transfer to a 250-mL volumetric flask. Dissolve in about 75 mL of water, add 5 mL of hydrochloric acid, dilute to volume with water, and mix. Place a 25.0-mL aliquot of this solution in a 250-mL beaker, add 50 mL of water, and mix. While stirring, add 40.0 mL of standard 0.1 M EDTA and 20 mL of ammonium acetate–acetic acid buffer solution. Heat the solution to near boiling for 5 min and cool. (The pH should be 4.0–5.0 and should be maintained in that range by adding more buffer during the titration if necessary.) Add 50 mL of alcohol and 2 mL of dithizone indicator solution and titrate the excess of EDTA with 0.1 M zinc sulfate (28.75 g of zinc sulfate heptahydrate diluted to 1 L with water) until the color changes from a green-violet to a rose-pink. Similarly, titrate 40.0 mL of the standard 0.1 M EDTA with 0.1 M zinc sulfate, substituting 25 mL of water for the sample solution. Let V_s and V_b be the volumes of the 0.1 M zinc sulfate required in the two titrations; let V_e be the volume in milliliters and M_e be the molarity of the standard EDTA solution; and let W be the weight of sample, in grams, corresponding to the aliquot taken.

$$\% \ Al(NO_3)_3 \cdot 9H_2O = [1 - (V_s/V_b)](V_e M_e)(37.51/W)$$

INSOLUBLE MATTER. (Page 15). Dissolve 20 g in 150 mL of water and filter immediately without digesting.

CHLORIDE. (Page 27). Use 1.0 g.

SULFATE. Dissolve 4.0 g in 10 mL of hydrochloric acid and evaporate to dryness on a hot plate ($\approx$100 °C). Take up the residue in 10 mL of hydrochloric acid and again evaporate to dryness. Dissolve the residue in 100 mL of water, add a slight excess of ammonium hydroxide, and boil the solution until the odor of ammonia is entirely gone. Dilute with water to 200 mL, filter, and evaporate 50 mL of the filtrate to 10 mL. Retain the remainder of the filtrate for the substances not precipitated by the ammonium hydroxide test. To the evaporated filtrate add 1 mL of dilute hydrochloric acid (1 + 19) and 1 mL of barium chloride reagent solution. Any turbidity should not exceed that produced by 0.05 mg of sulfate ion (SO_4) in an equal volume of solution containing 1 mL of dilute hydrochloric acid (1 + 19) and 1 mL of barium chloride reagent solution. Compare 10 min after adding the barium chloride to the sample and standard solutions.

SUBSTANCES NOT PRECIPITATED BY AMMONIUM HYDROXIDE. To 100
mL of the filtrate retained from the test for sulfate, add 0.5 mL of sulfuric acid and evaporate the solution to dryness. Ignite the residue at 800 ± 25 °C for 15 min.

HEAVY METALS. (Page 28, Method 1). Dissolve 4.0 g in 40 mL of water, and dilute with water to 48 mL. Use 36 mL to prepare the sample solution, and use the remaining 12 mL to prepare the control solution.

IRON. (Page 30, Method 1). Dissolve 1.0 g in 50 mL of water, and use 25 mL of this solution.

Aluminum Potassium Sulfate Dodecahydrate

$AlK(SO_4)_2 \cdot 12H_2O$ **Formula Wt 474.39**

CAS Number 7784–24–9

REQUIREMENTS

Assay . . . . $98.0–102.0\%\ AlK(SO_4)_2 \cdot 12H_2O$

MAXIMUM ALLOWABLE

Insoluble matter. 0.005%
Chloride (Cl) . . . . 5 ppm
Ammonium (NH_4) 0.005%
Heavy metals (as Pb) 0.001%
Iron (Fe) 0.001%
Sodium (Na). 0.02%

TESTS

ASSAY. (By indirect complexometric titration of Al^{III}). Weigh accurately 10 g and transfer to a 250-mL volumetric flask. Dissolve in about 75 mL of water, add 5 mL of hydrochloric acid, dilute to volume with water, and mix. Place a 25.0-mL aliquot of this solution in a 250-mL beaker, add 50 mL of water, and mix. While stirring, add 40.0 mL of standard 0.1 M EDTA and 20 mL of ammonium acetate–acetic acid buffer solution. Heat the solution to near boiling for 5 min and cool. (The pH should be 4.0–5.0 and should be a maintained in that range by adding more buffer during the titration if necessary.) Add 50 mL of alcohol and 2 mL of dithizone indicator solution and titrate the excess of EDTA with 0.1 M zinc sulfate (28.75 g of zinc sulfate heptahydrate diluted to 1 L with water) until the color changes from a green-violet to a rose-pink. Similarly, titrate 40.0 mL of the standard 0.1 M EDTA with 0.1 M zinc sulfate, substituting 25 mL of water for the sample solution. Let V_s and V_b be the volumes of the 0.1 M zinc sulfate required in the two titrations; let V_e be the volume in milliliters and M_e be the molarity of the standard EDTA solution; and let W be the weight of sample, in grams, corresponding to the aliquot taken.

$$\% \ AlK(SO_4)_2 \cdot 12H_2O = [1 - (V_s/V_b)](V_e M_e)(47.44/W)$$

INSOLUBLE MATTER. (Page 15). Use 20 g dissolved in 150 mL of hot water. Filter immediately.

CHLORIDE. (Page 27). Use 2.0 g dissolved in warm water.

AMMONIUM. Dissolve 1.0 g in 100 mL of ammonia-free water. To 20 mL of this solution add 10% sodium hydroxide reagent solution until the precipitate first formed is redissolved. Dilute with water to 50 mL and add 2 mL of Nessler reagent. Any color should not exceed that produced by 0.01 mg of ammonium ion (NH_4) in an equal volume of solution containing the quantities of reagents used in the test.

HEAVY METALS. (Page 28, Method 1). Dissolve 5.0 g in 40 mL of water, and dilute with water to 50 mL. Use 30 mL to prepare the sample solution, and use 10 mL to prepare the control solution.

IRON. (Page 30, Method 1). Use 1.0 g.

SODIUM. (By flame AAS, page 39).

Sample Stock Solution. Dissolve 1.0 g of sample in a 100-mL volumetric flask and dilute to the mark with water (1 mL = 0.01 g).

Element	Wavelength (nm)	Sample Wt (g)	Standard Added (mg)	Flame Type*	Background Correction
Na	589.0	0.05	0.01; 0.02	A/A	No

*A/A is air/acetylene.

Aluminum Sulfate, Hydrated

$Al_2(SO_4)_3 \cdot (14–18)H_2O$

CAS Number 7784–31–8 (for 18-hydrate)

NOTE. This reagent is available in degrees of hydration ranging from 14 to 18 molecules of water.

REQUIREMENTS

Assay (as the labeled hydrate) 98.0–102.0%

MAXIMUM ALLOWABLE

Insoluble matter. 0.01%
Chloride (Cl) 0.005%
Substances not precipitated by ammonium
    hydroxide (as SO_4) 0.2%
Heavy metals (as Pb) 0.001%
Iron (Fe). 0.002%

TESTS

ASSAY. (By indirect complexometric titration of Al^{III}). Weigh accurately 8 g and transfer to a 250-mL volumetric flask. Dissolve in about 75 mL of water, add 5 mL of hydrochloric acid, dilute to volume with water, and mix. Place a 25.0-mL aliquot of this solution in a 250-mL beaker, add 50 mL of water, mix, and then add 40.0 mL of standard 0.1 M EDTA and 20 mL of ammonium acetate–acetic acid buffer solution. Heat the solution to near boiling for 5 min and cool. (The pH should be 4.0–5.0 and should be maintained in that range by adding more buffer during the titration if necessary.) Add 50 mL of alcohol and 2 mL of dithizone indicator solution and titrate the excess of EDTA with 0.1 M zinc sulfate (28.75 g of zinc sulfate heptahydrate diluted to 1 L with water) until the color changes from a green-violet to a rose-pink. Similarly, titrate 40 mL of the standard 0.1 M EDTA with 0.1 M zinc sulfate, substituting 25 mL of water for the sample. Let V_s and V_b be the volumes of the 0.1 M zinc sulfate required in the two titrations; let V_e be the volume in milliliters and M_e be the molarity of the standard EDTA solution; and let W be the weight of sample, in grams, corresponding to the aliquot taken.

$$\% \ Al_2(SO_4)_3 \cdot nH_2O = [1 - (V_s/V_b)](V_e M_e)(K/W)$$

where K has values of 29.72, 31.52, and 33.32 for n values of 14, 16, and 18, respectively.

INSOLUBLE MATTER. (Page 15). Use 20 g dissolved in 150 mL of water containing 5 mL of 25% sulfuric acid.

CHLORIDE. (Page 27). Use 0.20 g.

SUBSTANCES NOT PRECIPITATED BY AMMONIUM HYDROXIDE. Dissolve 2.0 g in 130 mL of boiling water, add 0.15 mL of methyl red indicator solution, then add ammonium hydroxide until the solution is alkaline. Heat to boiling, boil for 5 min, and dilute with water to 150 mL. Filter, transfer 75 mL of the filtrate to a tared evaporating dish, and evaporate to dryness. Ignite the residue at 800 ± 25 °C for 15 min.

HEAVY METALS. (Page 28, Method 1). Dissolve 4.0 g in 40 mL of water, and dilute with water to 48 mL. Use 36 mL to prepare the sample solution, and use the remaining 12 mL to prepare the control solution.

IRON. (Page 30, Method 1). Dissolve 1.0 g in 40 mL of water; use 20 mL of this solution, and only 1 mL of hydrochloric acid.

2-Aminoethanol
Monoethanolamine

$HOCH_2CH_2NH_2$ Formula Wt 61.08

CAS Number 141–43–5

REQUIREMENTS

Assay . . . .≥99.0% $HOCH_2CH_2NH_2$

MAXIMUM ALLOWABLE

Color (APHA) 15
Water 0.30%
Iron (Fe) . . . . 5 ppm
Heavy metals (as Pb) . . . . 5 ppm

TESTS

ASSAY. Analyze the sample by gas chromatography using the general parameters cited on page 70. The following specific conditions are also required.

Column: Type I, methyl silicone

Measure the area under all peaks and calculate the 2-aminoethanol content in area percent. Correct for water content.

COLOR (APHA). (Page 19).

WATER. (Page 56, Method 2). Use 0.10 mL (0.1 g) of the sample.

IRON. (Page 30, Method 1). Use 2.0 g (2.0 mL) in 40 mL of water and 5 mL of hydrochloric acid.

HEAVY METALS. (Page 28, Method 1). Use 4.0 g (4.0 mL) in 30 mL of water and 6 mL of hydrochloric acid.

4-Amino-3-hydroxy-1-naphthalene Sulfonic Acid
1-Amino-2-naphthol-4-sulfonic Acid

$H_2N(HO)C_{10}H_5SO_3H$ Formula Wt 239.25

CAS Number 116–63–2

REQUIREMENTS

Assay . . . .≥98.0% $H_2N(HO)C_{10}H_5SO_3H$

MAXIMUM ALLOWABLE

Solubility in sodium carbonate . . . . Passes test
Residue after ignition 0.1%
Sulfate (SO$_4$) 0.2%
Sensitivity to phosphate . . . . Passes test

TESTS

ASSAY. (By acid–base titrimetry). Weigh accurately about 0.5 g and transfer to a beaker. Add 100 mL of dimethyl sulfoxide and, while stirring, add 25 mL of water in small portions. Insert the electrodes of a suitable pH meter, and titrate with 0.1 N sodium hydroxide, establishing the end point potentiometrically. Correct for a blank determination. One milliliter of 0.1 N sodium hydroxide corresponds to 0.02392 g of $H_2N(HO)C_{10}H_5SO_3H$.

SOLUBILITY IN SODIUM CARBONATE. Dissolve 0.1 g in 3 mL of sodium carbonate reagent solution, and dilute with water to 10 mL. The solution should be clear and dissolution complete, or nearly so.

RESIDUE AFTER IGNITION. (Page 16). Ignite 1.0 g.

SULFATE. Transfer 0.5 g to a beaker, add 25 mL of water and 0.10 mL of hydrochloric acid, and heat on a hot plate ($\approx$100 °C) for 10 min. Cool, dilute with water to 50 mL, and filter. To 10 mL of the clear filtrate add 1 mL of 1 N hydrochloric acid and 2 mL of barium chloride reagent solution. Any turbidity should not exceed that produced by 0.2 mg of sulfate ion (SO$_4$) in an equal volume of solution containing the quantities of reagents used in the test. Compare 10 min after adding the barium chloride to the sample and standard solutions.

SENSITIVITY TO PHOSPHATE. Dissolve 0.10 g in a mixture of 50 mL of water and 10 g of sodium bisulfite, warm if necessary to complete dissolution, and filter. Add 1.0 mL of this solution to a test solution prepared as follows: To 20 mL of water add 0.02 mg of phosphate ion (PO$_4$) and 2 mL of 25% sulfuric acid reagent solution, mix, and then add 1.0 mL of ammonium molybdate reagent solution. A distinct blue color should be produced in the test solution in 5 min.

Ammonium Acetate

CH$_3$COONH$_4$ **Formula Wt 77.08**

CAS Number 631–61–8

REQUIREMENTS

Assay . . . . ≥97%

pH of a 5% solution . . . . 6.7–7.3 at 25 °C

MAXIMUM ALLOWABLE

Insoluble matter. 0.005%
Residue after ignition 0.01%
Chloride (Cl) . . . . 5 ppm
Nitrate (NO_3). 0.001%
Sulfate (SO_4) 0.001%
Heavy metals (as Pb) . . . . 5 ppm
Iron (Fe). . . . . 5 ppm

TESTS

ASSAY. (By alkalimetry for ammonium). Weigh accurately about 2.5 g of sample and dissolve in 50 mL of water in a 500-mL Erlenmeyer flask. Add exactly 50.0 mL of 1 N sodium hydroxide, place a filter funnel loosely in the neck of the flask, and boil until all the ammonia is expelled (about 10–15 min) as determined with litmus paper. Cool, add 0.15 mL of thymol blue indicator solution. Titrate the excess sodium hydroxide with 1 N sulfuric acid. One millimeter of 1 N sodium hydroxide corresponds to 0.07708 g of CH_3COONH_4.

pH OF A 5% SOLUTION. (Page 44). The pH should be 6.7–7.3 at 25 °C.

INSOLUBLE MATTER. (Page 15). Use 20 g dissolved in 200 mL of water.

RESIDUE AFTER IGNITION. (Page 16). Ignite 10 g.

CHLORIDE. (Page 27). Use 2.0 g.

NITRATE. (Page 30). For sample solution A use 1.0 g. For control solution B use 1.0 g and 1 mL of standard nitrate solution.

Sample Solution A and Blank Solution B for the Determination of Sulfate, Heavy Metals, and Iron

Sample Solution A. Dissolve 15 g in 15 mL of water and add about 10 mg of sodium carbonate, 2 mL of hydrogen peroxide, 15 mL of hydrochloric acid, and 25 mL of nitric acid.

Blank Solution B. Prepare a solution containing the quantities of reagents used in sample solution A, omitting only the sample.

Digest each in covered beakers on a hot plate (≈100 °C) until reaction ceases, uncover, and evaporate to dryness. Dissolve each in 3 mL of 1 N acetic acid plus 15 mL of water, filter through a small filter, and dilute with water to 30 mL (1 mL = 0.5 g).

SULFATE. (Page 33, Method 1). Use 10 mL of sample solution A (5-g sample).

HEAVY METALS. (Page 28, Method 1). Dilute 8.0 mL of sample solution A (4-g sample) with water to 25 mL. Use 8.0 mL of blank solution B to prepare the standard solution.

IRON. (Page 30, Method 1). Use 4.0 mL of sample solution A (2-g sample). Prepare the standard from 4.0 mL of blank solution B.

Ammonium Benzoate

Benzoic Acid, Ammonium Salt

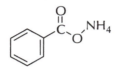

$C_7H_9NO_2$ Formula Wt 139.16

CAS Number 1863–63–4

REQUIREMENTS

Assay . . . . $\geq$98.0% $C_7H_9NO_2$
pH of a 5% solution . . . . 5.0–8.0 at 25 °C

MAXIMUM ALLOWABLE

Insoluble matter. 0.005%
Residue after ignition 0.005%
Substances reducing permanganate . . . . Passes Test
Chlorine compounds (as Cl). 0.01%
Heavy metals (as Pb) 0.001%
Iron (Fe) 0.001%

TESTS

ASSAY. (By alkalimetry for ammonium). Weigh accurately about 5.0 g of sample and dissolve in 50 mL of water in a 500 mL Erlenmeyer flask. Add exactly 50.0 mL of 1 N sodium hydroxide, place a filter funnel loosely in the neck of the flask, and boil until all the ammonia is expelled (about 10–15 min) as determined with litmus paper. Cool, add 0.15 mL of thymol blue indicator solution. Titrate the excess sodium hydroxide with 1 N sulfuric acid. One milliliter of 1 N sodium hydroxide corresponds to 0.13916 g of $C_7H_9NO_2$.

pH OF A 5% SOLUTION. (Page 44). The pH should be 5.0–8.0 at 25 °C.

INSOLUBLE MATTER. (Page 15). Use 20.0 g in 200 mL of water.

RESIDUE AFTER IGNITION. (Page 16). Heat 20.0 g in a platinum dish at a temperature sufficient to volatilize. When volatilized, cool and continue as described.

SUBSTANCES REDUCING PERMANGANATE. Add 1.0 g of the sample to 10 mL of water. Add 0.2 mL of 0.1 N potassium permanganate and let stand for 5 min. The pink color should not be entirely discharged.

CHLORINE COMPOUNDS. (Page 27). Mix 1.0 g with 0.5 g of sodium carbonate and add 10–15 mL of water. Evaporate and ignite until charred. Extract with 20 mL of water and 5.5 mL of nitric acid. Filter through a chloride-free filter and wash with two 10-mL portions of water. Dilute with water to 100 mL. Use 10 mL for the test.

HEAVY METALS. (Page 28, Method 1). Volatilize 2.0 g over a low flame, cool, and add 5 mL of nitric acid and 10 mg of sodium carbonate. Evaporate to dryness on the hot plate, dissolve the residue in water and dilute to 25 mL. For the control, add 0.02 mg of lead to 5 mL of nitric acid and 10 mg of sodium carbonate and evaporate on a hot plate. Dissolve the residue and dilute to 25 mL.

IRON. (Page 30, Method 2). Use 1.0 g.

Ammonium Bromide

NH_4Br Formula Wt 97.94

CAS Number 12124–97–9

REQUIREMENTS

Assay . . . . ≥99.0% NH_4Br
pH of a 5% solution . . . . 4.5–6.0 at 25 °C

MAXIMUM ALLOWABLE

Insoluble matter. 0.005%
Residue after ignition 0.01%
Bromate (BrO_3) 0.002%
Chloride (Cl) 0.2%
Iodide (I) . . . . Passes test
Sulfate (SO_4) 0.005%
Barium (Ba) 0.002%
Heavy metals (as Pb) . . . . 5 ppm
Iron (Fe). . . . . 5 ppm

TESTS

ASSAY. (By indirect argentimetric titration of bromide). Weigh accurately about 0.4 g of sample in 50 mL of water. Add 10 mL of dilute nitric acid and 50.0 mL of 0.1 N silver nitrate, and titrate the excess silver nitrate with 0.1 N ammonium thiocyanate, using 0.5 mL of ferric sulfate as an indicator. One milliliter of 0.1 N silver nitrate consumed corresponds to 0.009794 g of NH_4Br.

pH OF A 5% SOLUTION. (Page 44). The pH should be 4.5–6.0 at 25 °C.

INSOLUBLE MATTER. (Page 15). Use 20 g dissolved in 150 mL of water.

RESIDUE AFTER IGNITION. (Page 16). Ignite 20 g.

BROMATE. (By differential pulse polarography, page 50). Use 5.0 g of sample in 25 mL of solution. For the standard, add 0.10 mg of bromate (BrO_3).

CHLORIDE. (Page 27). Dissolve 0.50 g in 15 mL of dilute nitric acid (1 + 2) in a small flask. Add 3 mL of 30% hydrogen peroxide and digest on a hot plate ($\approx$100 °C) until the solution is colorless. Wash down the sides of the flask with a little water, digest for an additional 15 min, cool, and dilute with water to 200 mL. Dilute 2.0 mL of the solution with water to 20 mL.

IODIDE. Dissolve 5.0 g in 20 mL of water. Add 1 mL of chloroform, 0.15 mL of 10% ferric chloride reagent solution, and 0.25 mL of 10% sulfuric acid, and shake. No violet tint should be produced in the chloroform. (Limit about 0.005%)

SULFATE. (Page 33, Method 1). Allow 30 min for the turbidity to form.

BARIUM. For the sample dissolve 6.0 g in 15 mL of water. For the control, dissolve 1.0 g in 15 mL of water and add 0.1 mg of barium. To each solution add 5 mL of glacial acetic acid, 5 mL of 30% hydrogen peroxide, and 1 mL of hydrochloric acid. Digest in a covered beaker on a hot plate ($\approx$100 °C) until reaction ceases, uncover, and evaporate to dryness. Dissolve each residue in 15 mL of water, filter if necessary, and dilute with water to 23 mL. Add 2 mL of 10% potassium dichromate reagent solution, then add ammonium hydroxide until the orange color is just dissipated and the yellow color persists. Add 25 mL of methanol, stir vigorously, and allow to stand for 10 min. Any turbidity in the solution of the sample should not exceed that in the control.

HEAVY METALS. (Page 28, Method 1). Dissolve 6.0 g in about 20 mL of water, and dilute with water to 30 mL. Use 25 mL to prepare the sample solution, and use the remaining 5.0 mL to prepare the control solution.

IRON. (Page 30, Method 1). Use 2.0 g.

Ammonium Carbonate

CAS Number 8000–73–5

NOTE. This product is a mixture of variable proportions of ammonium bicarbonate and ammonium carbamate.

REQUIREMENTS

Assay. . . . . ≥30.0% NH_3

MAXIMUM ALLOWABLE

Insoluble matter. 0.005%
Nonvolatile matter. 0.01%
Chloride (Cl) . . . . 5 ppm
Sulfur compounds (as SO_4) 0.002%
Heavy metals (as Pb) . . . . 5 ppm
Iron (Fe). . . . . 5 ppm

TESTS

ASSAY. (By acid–base titrimetry). Tare a glass-stoppered flask containing about 25 mL of water. Add 2.0–2.5 g of the sample and weigh accurately. Add slowly 50.0 mL of 1 N hydrochloric acid and titrate the excess with 1 N sodium hydroxide, using methyl orange indicator. One milliliter of 1 N hydrochloric acid corresponds to 0.01703 g of NH_3.

INSOLUBLE MATTER. (Page 15). Use 20 g dissolved in 100 mL of water.

NONVOLATILE MATTER. To 20 g in a tared dish add 10 mL of water, volatilize on a hot plate (≈100 °C), and dry for 1 hour at 105 °C. Retain the residue for the test for heavy metals.

CHLORIDE. (Page 27). Dissolve 2.0 g in 25 mL of hot water, add about 10 mg of sodium carbonate, and evaporate to dryness on a hot plate (≈100 °C). Dissolve the residue in 20 mL of water.

SULFUR COMPOUNDS. Dissolve 2.0 g in 20 mL of water, add about 10 mg of sodium carbonate, and evaporate to dryness. Dissolve the residue in a slight excess of hydrochloric acid, add 2 mL of bromine water, and again evaporate to dryness. Dissolve the residue in 4 mL of water plus 1 mL of dilute hydrochloric acid (1 + 19). Filter through a small filter, wash with two 2-mL portions of water, dilute with water to 10 mL, and add 1 mL of barium chloride reagent solution. Any turbidity should not exceed that produced by 0.04 mg of sulfate ion (SO_4) in an equal volume of solution containing the quantities of reagents used in the test. Compare 10 min after adding the barium chloride to the sample and standard solutions.

HEAVY METALS. (Page 28, Method 1). Dissolve the residue from the test for nonvolatile matter in 5 mL of 1 N acetic acid, and dilute with water to 50 mL. Dilute 10 mL of this solution with water to 25 mL.

IRON. (Page 30, Method 1). To 2.0 g add 5 mL of water and evaporate on a hot plate ($\approx$100 °C). Dissolve the residue in 2 mL of hydrochloric acid, add 20 mL of water, and filter if necessary. Use this solution without further acidification.

Ammonium Chloride

NH_4Cl **Formula Wt 53.49**

CAS Number 12125–02–9

REQUIREMENTS

Assay . . . . $\geq$99.5% NH_4Cl
pH of a 5% solution . . . . 4.5–5.5 at 25 °C

MAXIMUM ALLOWABLE

Insoluble matter. 0.005%
Residue after ignition 0.01%
Calcium (Ca) 0.001%
Magnesium (Mg) . . . . .5 ppm
Heavy metals (as Pb) . . . . .5 ppm
Iron (Fe) . . . . .2 ppm
Phosphate (PO_4) . . . . .2 ppm
Sulfate (SO_4) 0.002%

TESTS

ASSAY. (By argentimetric titration). Dissolve about 200 mg of ammonium chloride sample, accurately weighed, in 20 mL of water. Add about 10 mL of glacial acetic acid, 100 mL of methanol, and 0.25 mL of eosin Y indicator. With stirring, titrate with 0.1 N silver nitrate to a pink end point. One milliliter of 0.1 N silver nitrate corresponds to 0.005349 g of NH_4Cl.

pH OF A 5% SOLUTION. (Page 44). The pH should be 4.5–5.5 at 25 °C.

INSOLUBLE MATTER. (Page 15). Use 20.0 g dissolved in 200 mL of water.

RESIDUE AFTER IGNITION. (Page 16). Ignite 10.0 g.

CALCIUM AND MAGNESIUM. (By flame AAS, page 39).

> *Sample Stock Solution.* Dissolve 10.0 g of sample in 80 mL of water, warm to room temperature. Transfer to a 100 mL volumetric flask and dilute to the mark with water (1 mL = 0.10 g).

Element	Wavelength (nm)	Sample Wt (g)	Standard Added (mg)	Flame Type*	Background Correction
Ca	422.7	2.0	0.02; 0.04	N/A	No
Mg	285.2	2.0	0.01; 0.02	A/A	Yes

*A/A is air/acetylene; N/A is nitrous oxide/acetylene.

Sample Solution A and Blank Solution B for the Determination of Heavy Metals, Iron, Phosphate, and Sulfate

Sample Solution A. Dissolve 25.0 g in 75 mL of water and add about 10 mg of sodium carbonate, 10 mL of hydrogen peroxide, and 50 mL of nitric acid.

Blank Solution B. To 75 mL of water add about 10 mg of sodium carbonate, 10 mL of hydrogen peroxide, and 50 mL of nitric acid.

Digest sample solution A and blank solution B in covered beakers on a hot plate ($\approx$100 °C) until reaction ceases, uncover, and evaporate to dryness. Dissolve the two residues in separate 5-mL portions of 1 N acetic acid and 50 mL of water, filter if necessary, and dilute with water to 100 mL (1 mL of sample solution A = 0.25 g).

HEAVY METALS. (Page 28, Method 1). Dilute 16 mL of sample solution A (4-g sample) with water to 25 mL. Use 16 mL of blank solution B to prepare the standard solution.

IRON. (Page 30, Method 1). Use 20 mL of sample solution A (5-g sample). Prepare the standard with 20 mL of blank solution B.

PHOSPHATE. (Page 32, Method 1). Evaporate 40 mL of sample solution A (10-g sample) to dryness on a hot plate ($\approx$100 °C). Dissolve the residue in 25 mL of approximately 0.5 N sulfuric acid. Use 40 mL of blank solution B to prepare the standard solution.

SULFATE. (Page 33, Method 1). Use 10 mL of sample solution A (2.5-g sample). Use 10 mL of blank solution B to prepare the standard solution.

Ammonium Citrate, Dibasic
Citric Acid, Diammonium Salt
2-Hydroxy-1,2,3-propanetricarboxylic Acid, Diammonium Salt

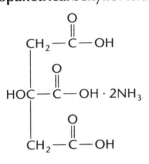

$(NH_4)_2HC_6H_5O_7$ **Formula Wt 226.19**

CAS Number 3012–65–5

REQUIREMENTS

Assay . . . . 98.0–103.0% $(NH_4)_2HC_6H_5O_7$

MAXIMUM ALLOWABLE

Insoluble matter 0.005%
Residue after ignition 0.01%
Chloride (Cl) 0.001%
Heavy metals (as Pb) . . . . 5 ppm
Iron (Fe) 0.001%
Oxalate (C_2O_4) . . . . Passes test
Phosphate (PO_4) . . . . 5 ppm
Sulfur compounds (as SO_4) 0.005%

TESTS

ASSAY. (By alkalimetry for ammonium). Weigh accurately about 3.0 g of sample and dissolve in 50 mL of water in a 500-mL Erlenmeyer flask. Add exactly 50.0 mL of 1 N sodium hydroxide, place a filter funnel loosely in the neck of the flask, and boil until all the ammonia is expelled (about 10–15 min) as determined with litmus paper. Cool, add 0.15 mL of thymol blue indicator solution (0.1% in alcohol). Titrate the excess sodium hydroxide with 1 N sulfuric acid. One milliliter of 1 N sodium hydroxide corresponds to 0.07540 g of $(NH_4)_2 HC_6H_5O_7$.

INSOLUBLE MATTER. (Page 15). Use 20 g dissolved in 200 mL of water.

RESIDUE AFTER IGNITION. (Page 16). Ignite 10 g.

CHLORIDE. (Page 27). Use 1.0 g.

HEAVY METALS. (Page 28, Method 1). Dissolve 6.0 g in about 20 mL of water, add 5 mL of dilute hydrochloric acid (1 + 1), and dilute with water to 30 mL. Use 25 mL to prepare the sample solution, and use the remaining 5.0 mL to prepare the control solution.

IRON. (Page 30, Method 1). Use 1.0 g.

OXALATE. Dissolve 5.0 g in 25 mL of water, and add 3 mL of glacial acetic acid and 2 mL of 10% solution of calcium acetate. No turbidity or precipitate should appear after standing 4 h. (Limit about 0.05%)

PHOSPHATE. (Page 32, Method 1). Mix 4.0 g with 0.5 g of magnesium nitrate in a platinum dish and ignite. Dissolve the residue in 5 mL of water, add 5 mL of nitric acid, and evaporate to dryness. Dissolve the residue in 25 mL of approximately 0.5 N sulfuric acid and continue as described.

SULFUR COMPOUNDS. To 1.0 g add 1 mL of hydrochloric acid and 3 mL of nitric acid. Prepare a standard containing 0.05 mg of sulfate ion (SO_4), 1 mL of hydrochloric acid, and 3 mL of nitric acid. Digest each in a covered beaker on a hot plate ($\approx$100 °C) until the reaction ceases, remove the covers, and evaporate to dryness. Add 0.2–0.5 mg of ammonium vanadate and 10 mL of nitric acid and digest in covered beakers on the hot plate until reactions cease. Remove the covers and evaporate to dryness. Add 10 mL more of nitric acid and repeat the digestions and evaporations. Add 5 mL of dilute hydrochloric acid (1 + 1) and evaporate to dryness. Dissolve the residues in 4 mL of water plus 1 mL of dilute hydrochloric acid (1 + 19), and filter through a small filter. Wash with two 2-mL portions of water, dilute with water to 10 mL, and add 1 mL of barium chloride reagent solution to each. Any turbidity in the solution of the sample should not exceed that in the standard. Compare 10 min after adding the barium chloride to the sample and standard solutions.

Ammonium Dichromate

$(NH_4)_2Cr_2O_7$ **Formula Wt 252.07**

CAS Number 7789–09–5

NOTE. This reagent may be stabilized by the addition of 0.5% to 3.0% water.

REQUIREMENTS

Assay (dried basis). . . . . $\geq$99.5% $(NH_4)_2Cr_2O_7$

MAXIMUM ALLOWABLE

Insoluble matter. 0.005%
Loss on drying at 105 °C. 3.0%
Chloride (Cl) 0.005%
Sulfate (SO$_4$) 0.01%
Calcium (Ca) 0.002%
Iron (Fe). 0.002%
Sodium (Na). 0.005%

TESTS

ASSAY. (By iodometry). Weigh accurately 1.0 g, previously dried to constant weight at 105 °C, transfer to a 100-mL volumetric flask, dissolve in water, dilute to volume, and mix well. Pipet a 10.0-mL aliquot to a glass-stoppered flask, add 50 mL of water, 10 mL of 10% sulfuric acid reagent solution, and 3.0 g of potassium iodide. Mix, stopper, and allow to stand in the dark for 10 min. Titrate the liberated iodine with 0.1 N sodium thiosulfate, adding 3 mL of starch indicator solution near the end point, which is a greenish-blue color. One milliliter of 0.1 N sodium thiosulfate is equivalent to 0.004201 g of $(NH_4)_2Cr_2O_7$.

INSOLUBLE MATTER. Dissolve 20 g in 200 mL of water.

LOSS ON DRYING AT 105 °C. Weigh accurately 2.0 g and dry to constant weight at 105 °C.

CHLORIDE. (Page 27). Dissolve 0.2 g in 10 mL of water, filter if necessary through a small chloride-free filter, and add 1 mL of ammonium hydroxide and 1 mL of silver nitrate reagent solution. Prepare a standard containing 0.01 mg of chloride ion (Cl) in 10 mL of water, and add 1 mL of ammonium hydroxide and 1 mL of silver nitrate reagent solution. Add 2 mL of nitric acid to each. The comparison is best made by the general method for chloride in colored solutions, page 27.

SULFATE. Dissolve 10.0 g in 250 mL of water, filter if necessary, and heat to boiling. Add 25 mL of a solution containing 1.0 g of barium chloride and 2 mL of hydrochloric acid per 100 mL of solution. Digest in a covered beaker on a hot plate ($\approx$100 °C) for 2 hours and allow to stand overnight. If a precipitate is formed, filter, wash thoroughly, and ignite. Fuse the residue with 1.0 g of sodium carbonate. Extract the fused mass with water and filter off the insoluble residue. Add 5 mL of hydrochloric acid to the filtrate, dilute with water to about 200 mL, heat to boiling, and add 10 mL of alcohol. Digest in a covered beaker on the hot plate until reduction of chromate is complete, as indicated by the change to a clear green or colorless solution. Neutralize the solution with ammonium hydroxide and add 2 mL of hydrochloric acid. Heat to boiling and add 10 mL of barium chloride reagent solution. Digest in a covered beaker on a hot plate ($\approx$100 °C) for 2 hours and allow to stand overnight. Filter, wash thoroughly, and ignite. The weight of the precipitate should not exceed the weight obtained in a complete blank test by more than 0.0024 g. If the original precipitate of barium sulfate weighs less than the requirement permits, the fusion with sodium carbonate is not necessary.

CALCIUM, IRON, AND SODIUM. (By flame AAS, page 39).

> **Sample Stock Solution.** Dissolve 20.0 g in water and dilute with water to 100 mL (1 mL = 0.2 g).

Element	Wavelength (nm)	Sample Wt (g)	Standard Added (mg)	Flame Type*	Background Correction
Ca	422.7	2.0	0.02; 0.04	N/A	No
Fe	248.3	4.0	0.04; 0.08	A/A	Yes
Na	589.0	0.20	0.005; 0.01	A/A	No

*A/A is air/acetylene; N/A is nitrous oxide/acetylene.

Ammonium Fluoride

NH$_4$F   **Formula Wt 37.04**

CAS Number 12125–01–8

REQUIREMENTS

Assay . . . . ≥98.0% NH$_4$F

MAXIMUM ALLOWABLE

Insoluble matter. 0.005%
Residue after ignition 0.01%
Chloride (Cl) 0.001%
Sulfate (SO$_4$) 0.005%
Heavy metals (as Pb) . . . . 5 ppm
Iron (Fe). . . . . 5 ppm

TESTS

ASSAY. (By alkalimetry for ammonium). Weigh accurately about 1.0 g of sample and dissolve in 50 mL of water in a 500-mL Erlenmeyer flask. Add exactly 50.0 mL of 1 N sodium hydroxide, place a filter funnel loosely in the neck of the flask, and boil until all the ammonia is expelled (about 10–15 min) as determined with litmus paper. Cool, add 0.15 mL of thymol blue indicator solution (0.1% in alcohol). Titrate the excess sodium hydroxide with 1 N sulfuric acid. One milliliter of 1 N sodium hydroxide corresponds to 0.03704 g of NH$_4$F.

INSOLUBLE MATTER. (Page 15). Use 20 g dissolved in 200 mL of hot water.

RESIDUE AFTER IGNITION. (Page 16). Ignite 10 g in a tared platinum crucible or dish.

CHLORIDE. Dissolve 1.0 g in a mixture of 10 mL of water and 1 mL of nitric acid in a platinum dish. In a test tube or small beaker mix 20 mL of water and 1 mL of silver nitrate reagent solution, add this solution to the sample solution, and mix. Any turbidity should not exceed that produced by 0.01 mg of chloride ion (Cl) in an equal volume of solution containing the quantities of reagents used in the test.

SULFATE. (Page 33, Method 2). Use 10 mL of hydrochloric acid in a platinum dish, and perform four evaporations.

HEAVY METALS. (Page 28, Method 1). Dissolve 5.0 g in about 25 mL of water, add 2 g of sodium acetate, and dilute with water to 30 mL. Adjust the pH with short-range pH paper. Use 1.0 g of sample and 2 g of sodium acetate to prepare the control solution.

IRON. (Page 30, Method 1). Treat 2.0 g in a platinum dish with 10 mL of dilute hydrochloric acid (1 + 1), and evaporate on a hot plate ($\approx$100 °C) to dryness. Repeat the evaporation with a second portion of the dilute acid. Warm the residue with 2 mL of hydrochloric acid, dilute with water to 50 mL, and use this solution without further acidification. Prepare the standard from the residue remaining from the evaporation of 10 mL of hydrochloric acid.

Ammonium Hydroxide
Aqueous Ammonia

NH_4OH   **Formula Wt 35.05**

CAS Number 1336–21–6

REQUIREMENTS

Appearance	Colorless and free from suspended matter or sediment
Assay (as NH_3)	28.0–30.0%

MAXIMUM ALLOWABLE

Residue after ignition	0.002%
Carbon dioxide (CO_2)	0.002%
Chloride (Cl)	0.5 ppm
Phosphate (PO_4)	2 ppm
Nitrate (NO_3)	2 ppm
Sulfate (SO_4)	2 ppm

Heavy metals (as Pb) . . . . 0.5 ppm
Iron (Fe). . . . . 0.2 ppm
Substances reducing permanganate. . . . . Passes test

TESTS

APPEARANCE. Mix the material in the original container, pour 10 mL into a test tube (20 × 150 mm), and compare with distilled water in a similar tube. The liquids should be equally clear and free from suspended matter; looking across the columns by means of transmitted light should reveal no apparent difference in color between the two liquids.

ASSAY. (By acidimetry). Tare a small glass-stoppered Erlenmeyer flask containing 35 mL of water. Using a measuring pipet, without suction, introduce barely under the surface about 2 mL of the ammonium hydroxide, stopper, and weigh accurately. Titrate with 1 N hydrochloric acid, using methyl red as indicator. One milliliter of 1 N acid is equivalent to 0.01703 g of NH_3.

RESIDUE AFTER IGNITION. (Page 16). Evaporate 50 g (56 mL) and ignite.

CARBON DIOXIDE. Dilute 10.0 g (11 mL) of the sample with 10 mL of water free from carbon dioxide, and add 5 mL of clear saturated barium hydroxide solution. Any turbidity should not be greater than is produced when the same quantity of barium hydroxide solution is added to 21 mL of water (free from carbon dioxide) containing 0.3 mg of carbonate standard (0.2 mg of carbon dioxide).

CHLORIDE, PHOSPHATE, NITRATE, AND SULFATE. (By ion chromatography, page 74). Add 2 mg of sodium carbonate to 10 g (11 mL) of sample and evaporate to dryness. Prepare also a blank and a standard with 0.005 mg Cl and 0.02 mg each of NO_3, PO_4, and SO_4. Take up in water and dilute each to 10 mL in a volumetric flask. Analyze 50-μL aliquots by IC and measure the various peaks. The differences between the sample and the blank should not be greater than the corresponding differences between the standard and the sample.

HEAVY METALS. (Page 28, Method 1). To 40.0 g (44 mL) add about 10 mg of sodium carbonate, evaporate to dryness on a hot plate (≈100 °C), dissolve the residue in about 20 mL of water, and dilute with water to 25 mL.

IRON. (Page 30, Method 1). To 50 g (56 mL) add about 10 mg of sodium carbonate, and evaporate to dryness on a hot plate (≈100 °C). Dissolve the residue in 3 mL of hydrochloric acid, dilute with water to 50 mL, and use this solution without further acidification.

SUBSTANCES REDUCING PERMANGANATE. Dilute 3 mL of the sample with 5 mL of water and add 50 mL of 10% sulfuric acid reagent solution. Add 0.05 mL of 0.1 N potassium permanganate, heat to boiling, and keep at this temperature for 5 min. The pink color should not be entirely discharged.

Ammonium Hydroxide, Ultratrace
(suitable for use in ultratrace elemental analysis)

NH_4OH Formula Wt 35.05

CAS number 1336–21–6

> *NOTE.* Reagent must be packaged in a preleached Teflon bottle and used in a clean environment to maintain purity

REQUIREMENTS

Appearance . . . . Clear
Assay 20–30%

MAXIMUM ALLOWABLE

Chloride (Cl) . . . . 0.5 ppm
Phosphate (PO_4) . . . . 2 ppm
Sulphate (SO_4) . . . . 2 ppm
Aluminum (Al) . . . . 1 ppb
Barium (Ba) . . . . 1 ppb
Boron (B) . . . . 5 ppb
Cadmium (Cd) . . . . 1 ppb
Calcium (Ca) . . . . 1 ppb
Chromium (Cr) . . . . 1 ppb
Cobalt (Co) . . . . 1 ppb
Copper (Cu) . . . . 1 ppb
Iron (Fe) . . . . 5 ppb
Lead (Pb) . . . . 1 ppb
Lithium (Li) . . . . 1 ppb
Magnesium (Mg) . . . . 1 ppb
Manganese (Mn) . . . . 1 ppb
Molybdenum (Mo) . . . . 1 ppb
Mercury (Hg) . . . . 1 ppb
Potassium (K) . . . . 1 ppb
Silicon (Si) . . . . 5 ppb
Sodium (Na) . . . . 5 ppb
Strontium (Sr) . . . . 1 ppb
Tin (Sn) . . . . 1 ppb
Titanium (Ti) . . . . 1 ppb
Vanadium (V) . . . . 1 ppb
Zinc (Zn) . . . . 1 ppb
Zirconium (Zr) . . . . 1 ppb

TESTS

ASSAY. (By acidimetry). Tare a small glass-stoppered Erlenmeyer flask containing 35 mL of water. Using a measuring pipet, without suction, introduce barely under the surface about 2 mL of the ammonium hydroxide , stopper, and weigh accurately. Titrate with 1 N hydrochloric acid, using methyl red as indicator. One milliliter of 1 N acid is equivalent to 0.01703 g of NH_3.

$$\% \ NH_4OH = (ml \times N \ HCl \times 1.703) \ / \ Sample \ wt \ (g)$$

CHLORIDE, PHOSPHATE, AND SULFATE. See test for ammonium hydroxide on page 147

TRACE METALS. Determine the aluminum, barium, cadmium, calcium, chromium, cobalt, copper, iron, lead, lithium, magnesium, manganese, molybdenum, nickel, potassium, sodium, strontium, tin, titanium, vanadium, zinc, and zirconium by the ICP–OES method described on page 42.

Determine the mercury by the CVAAS procedure described on page 40.

Ammonium Iodide

NH_4I
                                           Formula Wt 144.94

CAS Number 12027–06–4

NOTE. This compound may turn yellow on storage.

REQUIREMENTS

Assay . . . . ≥99.0% NH_4I
MAXIMUM ALLOWABLE
Insoluble matter. 0.005%
Residue after ignition 0.05%
Chloride and bromide (as Cl) 0.005%
Phosphate (PO_4) 0.001%
Sulfate (SO_4) 0.05%
Barium (Ba) 0.002%
Heavy metals (as Pb) 0.001%
Iron (Fe). . . . . 5 ppm

TESTS

ASSAY. (By oxidation–reduction titration of iodide). Weigh to the nearest 0.1 mg about 0.3 g of sample, and dissolve with about 20 mL of water in a 250-mL glass-stoppered titration flask. Add 30 mL of hydrochloric acid and 5 mL of chloroform. Cool, if necessary, and titrate with 0.05 M potassium iodate solution until the iodine color disappears from the aqueous layer. Stopper, shake vigorously for 30 s, and continue the titration, shaking vigorously after each addition of the iodate until the iodine color in the chloroform is discharged. One milliliter of 0.05 M potassium iodate corresponds to 0.007247 g of NH_4I.

INSOLUBLE MATTER. (Page 15). Use 20 g dissolved in 200 mL of water.

RESIDUE AFTER IGNITION. (Page 16). Ignite 2.0 g.

CHLORIDE AND BROMIDE. Dissolve 1.0 g in 100 mL of water in a distilling flask. Add 1 mL of hydrogen peroxide and 1 mL of phosphoric acid, heat to boiling, and boil gently until all the iodine is expelled and the solution is colorless. Cool, wash down the sides of the flask, and add 0.5 mL of hydrogen peroxide. If an iodine color develops, boil until the solution is colorless and for 10 min longer. If no color develops, boil for 10 min, filter if necessary through a chloride-free filter, and dilute with water to 100 mL. Dilute 20 mL with water to 23 mL, and add 1 mL of nitric acid and 1 mL of silver nitrate reagent solution. Any turbidity should not exceed that produced by 0.01 mg of chloride ion (Cl) in an equal volume of solution containing the quantities of nitric acid and silver nitrate used in the test.

Sample Solution A and Blank Solution B for the Determination of Phosphate, Sulfate, Barium, Heavy Metals, and Iron

Sample Solution A. Dissolve 20.0 g in 40 mL of water in a 600-mL beaker and add about 10 mg of sodium carbonate. Add 16 mL of hydrochloric acid and 32 mL of nitric acid, cover with a watch glass, and warm on a hot plate ($\approx$100 °C). When the rapid evolution of iodine ceases, add an additional 16 mL of nitric acid and 24 mL of hydrochloric acid. Digest in the covered beaker until the bubbling ceases, remove the watch glass, and evaporate to dryness.

Blank Solution B. Evaporate to dryness the quantities of acids and sodium carbonate used to prepare sample solution A.

Dissolve the residues in separate 20-mL portions of water, filter if necessary, and dilute each with water to 100 mL (1 mL of sample solution A = 0.2 g).

PHOSPHATE. (Page 32, Method 1). Evaporate 10 mL of sample solution A (2-g sample) to dryness on a hot plate ($\approx$100 °C). Dissolve the residue in 25 mL of approximately 0.5 N sulfuric acid. Use 10 mL of blank solution B to prepare the standard solution.

SULFATE. (Page 33, Method 1). Use 0.5 mL of sample solution A. Use 0.5 mL of blank solution B to prepare the standard solution.

BARIUM. To 20 mL of sample solution A (4-g sample) add 5 mL of a 1% solution of potassium sulfate. For the standard, add 0.08 mg of barium and 5 mL of a 1% solution of potassium sulfate to 20 mL of blank solution B. Any turbidity in the solution of the sample should not exceed that in the standard. Compare 10 min after adding the potassium sulfate to the sample and standard solutions.

HEAVY METALS. (Page 28, Method 1). Dilute 10 mL of sample solution A (2-g sample) with water to 25 mL. Use 10 mL of blank solution B to prepare the standard solution.

IRON. (Page 30, Method 1). Use 10 mL of sample solution A (2-g sample). Prepare the standard with 10 mL of blank solution B.

Ammonium Metavanadate

NH_4VO_3 Formula Wt 116.98

CAS Number 7803–55–6

REQUIREMENTS

Assay . . . .≥99.0% NH_4VO_3

MAXIMUM ALLOWABLE

Solubility in ammonium hydroxide . . . . Passes test
Carbonate (CO_3) . . . . Passes test
Chloride (Cl) 0.2%
Sulfate (SO_4) 0.05%

TESTS

ASSAY. (By oxidation–reduction titration of vanadium). Weigh accurately about 0.4 g, dissolve in 50 mL of warm water, and add 1 mL of sulfuric acid and 30 mL of sulfurous acid. Boil gently until the sulfur dioxide is expelled, then boil for 5 min longer. Cool, dilute with water to 100 mL, and titrate with 0.1 N potassium permanganate. One milliliter of 0.1 N potassium permanganate corresponds to 0.01170 g of NH_4VO_3.

SOLUBILITY IN AMMONIUM HYDROXIDE. Dissolve 5.0 g in a mixture of ammonium hydroxide and 250 mL of hot water, add ammonium hydroxide until the reagent dissolves. Heat the solution to boiling. The solution should be clear and free from insoluble matter.

CARBONATE. To 0.5 g add 1 mL of water and 2 mL of 10% hydrochloric acid reagent solution. No effervescence should be produced. (Limit about 0.3%)

CHLORIDE. (Page 27). Dissolve 0.5 g in 50 mL of hot water, add 2 mL of nitric acid, and let stand for 1 h. Filter through a chloride-free filter, wash with a few milliliters of water, and dilute with water to 100 mL; use 1.0 mL.

SULFATE. (Page 33, Method 1). Dissolve 0.5 g in 40 mL of hot water, add 2 mL of 10% hydrochloric acid reagent solution and 1.5 g of hydroxylamine hydrochloride, and heat at 60 °C for 5 min. Filter through a suitable filter paper, cool, and dilute with water to 100 mL. To 20 mL add 1 mL of dilute hydrochloric acid (1 + 19), and continue as directed.

Ammonium Molybdate Tetrahydrate
Ammonium Heptamolybdate Tetrahydrate

$(NH_4)_6Mo_7O_{24} \cdot 4H_2O$ Formula Wt 1235.86

CAS Number 12027–67–7

REQUIREMENTS

Assay (as MoO_3) 81.0–83.0%

MAXIMUM ALLOWABLE

Insoluble matter. 0.005%
Chloride (Cl) 0.002%
Nitrate (NO_3) . . . . Passes test
Arsenate, phosphate, and silicate (as SiO_2). 0.001%
Phosphate (PO_4) . . . . 5 ppm
Sulfate (SO_4) 0.02%
Heavy metals (as Pb) 0.001%
Magnesium (Mg) 0.005%
Potassium (K). 0.002%
Sodium (Na). 0.01%

TESTS

ASSAY. (By complexometric titration). Weigh approximately 0.7 g and dissolve in 100 mL of water. Adjust the pH of the solution to 4.0 with dilute nitric acid. Add saturated hexamethylenetetramine solution to a pH of 5–6. Heat the solution to 60 °C and titrate with 0.1 M lead nitrate. Add 0.2 mL of 0.1% PAR (4-[2-pyridyl-azo]resorcinol) indicator solution and titrate from yellow color to the first permanent pink endpoint. One milliliter of 0.1 M lead nitrate corresponds to 0.01439 g of MoO_3.

$$\% \, MoO_3 = (mL \times M \, Pb(NO_3)_2 \times 14.394) \, / \, \text{Sample wt (g)}$$

INSOLUBLE MATTER. (Page 15). Use 20 g dissolved in 200 mL of water.

CHLORIDE. (Page 27). Use 0.5 g.

NITRATE. Dissolve 1.0 g in 10 mL of water containing 5 mg of sodium chloride. Add 0.10 mL of indigo carmine solution and 10 mL of sulfuric acid. The blue color should not be completely discharged in 5 min. (Limit about 0.003%)

ARSENATE, PHOSPHATE, AND SILICATE. Dissolve 2.5 g in 70 mL of water. For the control, dissolve 0.5 g in 70 mL of water and add 0.02 mg of silica (SiO_2). These solutions are prepared in containers other than glass to avoid excessive silica contamination. Adjust the pH to between 3 and 4 (pH paper) with dilute hydrochloric acid (1 + 9), then transfer to glass containers and treat each solution as follows: add 1–2 mL of bromine water and adjust the pH to between 1.7 and 1.9 (using a pH meter) with dilute hydrochloric acid (1 + 9). Heat just to boiling, but do not boil, and cool to room temperature. Dilute with water to 90 mL, add 10 mL of hydrochloric acid, and transfer to a separatory funnel. Add 1 mL of butyl alcohol and 30 mL of 4-methyl-2-pentanone, shake vigorously, and allow the phases to separate. Draw off and discard the aqueous phase and wash the ketone phase three times with 10-mL portions of dilute hydrochloric acid (1 + 99), discarding each aqueous phase. To the washed ketone phase add 10 mL of dilute hydrochloric acid (1 + 99) to which has just been added 0.2 mL of a freshly prepared 2% solution of stannous chloride in hydrochloric acid. Any blue color in the solution of the sample should not exceed that in the control.

PHOSPHATE. (Page 32, Method 3).

SULFATE. (Page 33, Method 2). Use 5 mL of nitric acid.

HEAVY METALS. Dissolve 2.0 g in about 20 mL of water, add 10 mL of 10% sodium hydroxide reagent solution and 2 mL of ammonium hydroxide, and dilute with water to 40 mL. For the control, add 0.01 mg of lead (Pb) to 10 mL of the solution. Add 10 mL of freshly prepared hydrogen sulfide water to each. Any color in the solution of the sample should not exceed that in the control.

MAGNESIUM, POTASSIUM, AND SODIUM. (By flame AAS, page 39).

Sample Stock Solution. Dissolve 5.0 g of sample in 60 mL of water and add 20 mL of 10% sulfuric acid. Transfer to a 100-mL volumetric flask and dilute to the mark with water (1 mL = 0.05 g).

Element	Wavelength (nm)	Sample Wt (g)	Standard Added (mg)	Flame Type*	Background Correction
Mg	285.2	0.20	0.01; 0.02	A/A	Yes
K	766.5	0.50	0.01; 0.02	A/A	No
Na	589.0	0.20	0.01; 0.02	A/A	No

*A/A is air/acetylene.

Ammonium Nitrate

NH$_4$NO$_3$ **Formula Wt 80.04**

CAS Number 6484–52–2

REQUIREMENTS

Assay . . . . ≥98.0% NH$_4$NO$_3$
pH of a 5% solution . . . . 4.5–6.0 at 25 °C

MAXIMUM ALLOWABLE

Insoluble matter. 0.005%
Residue after ignition 0.01%
Chloride (Cl) . . . . 5 ppm
Nitrite (NO$_2$) . . . . Passes test
Phosphate (PO$_4$) . . . . 5 ppm
Sulfate (SO$_4$) 0.002%
Heavy metals (as Pb) . . . . 5 ppm
Iron (Fe) . . . . 2 ppm

TESTS

ASSAY. (By alkalimetry for ammonium). Weigh, to the nearest 0.1 mg, 3.0 g of sample and dissolve in 50 mL of water in a 500-mL Erlenmeyer flask. Add exactly 50.0 mL of 1 N sodium hydroxide, place a filter funnel loosely in the neck of the flask, and boil until all the ammonia is expelled (about 10–15 min) as determined with litmus paper. Cool, add 0.15 mL of thymol blue indicator solution (0.1% in alcohol). Titrate the excess sodium hydroxide with 1 N sulfuric acid. One milliliter of 1 N sodium hydroxide corresponds to 0.08004 g of NH$_4$NO$_3$.

pH OF A 5% SOLUTION. (Page 44). The pH should be 4.5–6.0 at 25 °C.

INSOLUBLE MATTER. (Page 15). Use 20 g dissolved in 200 mL of water.

RESIDUE AFTER IGNITION. (Page 16). Ignite 10 g.

CHLORIDE. (Page 27). Use 2.0 g.

NITRITE. (By differential pulse polarography, page 51). Dissolve 0.150 g of sodium nitrite (NaNO$_2$) in water and dilute with water to 100 mL. Dilute 1.0 mL of this solution with water to 100 mL. This solution should be freshly prepared. Use 1.0 g of sample and 0.005 mg of nitrite. (Limit about 5 ppm)

PHOSPHATE. (Page 32, Method 1). Dissolve 4.0 g in 25 mL of approximately 0.5 N sulfuric acid.

Sample Solution A and Blank Solution B for the Determination of Sulfate, Heavy Metals, and Iron

Sample Solution A. Dissolve 20 g in 20 mL of water, add 10 mg of sodium carbonate, 5 mL of 30% hydrogen peroxide, 20 mL of hydrochloric acid, and 20 mL of nitric acid.

Blank Solution B. Prepare a solution containing the quantities of reagents used in sample solution A.

Digest each in covered beakers on a hot plate ($\approx$100 °C) until reaction ceases, uncover, and evaporate to dryness. Dissolve the residues in 5 mL of 1 N acetic acid and 50 mL of water, filter if necessary, and dilute to 100 mL with water (1 mL = 0.2 g).

SULFATE. (Page 33, Method 1). Use 12.5 mL of sample solution A (2.5-g sample). Use 12.5 mL of blank solution B to prepare the standard solution.

HEAVY METALS. (Page 28, Method 1). Dilute 20 mL of sample solution A (4-g sample) with water to 25 mL. Use 20 mL of blank solution B to prepare the standard solution.

IRON. (Page 30, Method 1). Use 25 mL of sample solution A (5-g sample). Prepare the standard solution with 25 mL of blank solution B.

Ammonium Oxalate Monohydrate
Ethanedioic Acid, Diammonium Salt Monohydrate

$(COONH_4)_2 \cdot H_2O$ **Formula Wt 142.11**

CAS Number 6009–70–7

REQUIREMENTS

Assay . . . . 99.0–101.0% $(COONH_4)_2 \cdot H_2O$

MAXIMUM ALLOWABLE

Insoluble matter. 0.005%
Residue after ignition 0.02%
Chloride (Cl) 0.002%
Sulfate (SO_4) 0.002%
Heavy metals (as Pb) . . . . 5 ppm
Iron (Fe) . . . . 2 ppm

TESTS

ASSAY. (By oxidation–reduction titration of oxalate). Transfer 5.0 g of previously crushed sample, accurately weighed, to a 500-mL volumetric flask. Dissolve it in water, dilute with water to volume, and mix. Pipet 25.0 mL of the solution into 70 mL of water and 5 mL of sulfuric acid in a beaker. Titrate slowly with 0.1 N potassium permanganate until about 25 mL has been added, then heat the mixture to about 70 °C, and complete the titration. One milliliter of 0.1 N potassium permanganate corresponds to 0.007106 g of $(COONH_4)_2 \cdot H_2O$.

INSOLUBLE MATTER. (Page 15). Use 20.0 g dissolved in 400 mL of hot water.

RESIDUE AFTER IGNITION. (Page 16). Ignite 5.0 g.

CHLORIDE. Dissolve 2.0 g in water plus 10 mL of nitric acid, filter if necessary through a chloride-free filter, and dilute with water to 100 mL. To 25 mL of the solution add 1 mL of silver nitrate reagent solution. Any turbidity should not exceed that produced by 0.01 mg of chloride ion (Cl), in an equal volume of solution containing the quantities of reagents used in the test.

Sample Solution A and Blank Solution B for the Determination of Sulfate, Heavy Metals, and Iron

Sample Solution A. Digest 20.0 g of sample plus 10 mg of sodium carbonate with 40 mL of nitric acid plus 35 mL of hydrochloric acid in a covered beaker on a hot plate (≈100 °C) until no more bubbles of gas are evolved. Remove the cover and evaporate until a small amount of crystals form in the beaker. Add 10 mL of 30% hydrogen peroxide, cover the beaker, and digest on the hot plate until reaction ceases. Add an additional 10 mL of 30% hydrogen peroxide, re-cover the beaker, digest on a hot plate (≈100 °C) until reaction ceases, remove the cover, and evaporate to dryness. Add 5 mL of dilute hydrochloric acid (1 + 1), cover, digest on the hot plate for 15 min, remove the cover, and evaporate to dryness. Dissolve the residue in about 50 mL of water, filter if necessary, and dilute with water to 100 mL (1 mL = 0.2 g).

Blank Solution B. Prepare a similar solution containing the quantities of reagents used in preparing sample solution A and treated in the identical manner.

SULFATE. (Page 33, Method 1). Use 12.5 mL of sample solution A (2.5-g sample). Use 12.5 mL of blank solution B to prepare the standard solution.

HEAVY METALS. (Page 28, Method 1). Dilute 20 mL of sample solution A (4-g sample) with water to 25 mL. Use 20 mL of blank solution B to prepare the standard solution.

IRON. (Page 30, Method 1). Use 25 mL of sample solution A (5-g sample). Prepare the standard solution from 25 mL of blank solution B.

Ammonium Peroxydisulfate
Ammonium Persulfate

$(NH_4)_2S_2O_8$ Formula Wt 228.19

CAS Number 7727–54–0

NOTE. Because of inherent instability, this reagent may be expected to decrease in strength and to increase in acidity during storage. After storage for some time the reagent may fail to meet the specified requirements for assay and acidity. Store in a dry, cool place.

REQUIREMENTS

Assay . . . . ≥98.0% $(NH_4)_2S_2O_8$

MAXIMUM ALLOWABLE

Insoluble matter. 0.005%
Residue after ignition 0.05%
Titrable free acid . . . . 0.04 meq/g
Chloride and chlorate (as Cl) 0.001%
Heavy metals (as Pb) 0.005%
Iron (Fe). 0.001%
Manganese (Mn) . . . . 0.5 ppm

TESTS

ASSAY. (By oxidation–reduction titration of persulfate). Weigh accurately 0.5 g and add it to 25.0 mL of standard ferrous sulfate solution (0.25 N) in a glass-stoppered flask. Stopper the flask, allow it to stand for 1 h with frequent shaking, and titrate the excess ferrous sulfate with 0.1 N potassium permanganate. One milliliter of 0.1 N ferrous sulfate corresponds to 0.01141 g of $(NH_4)_2S_2O_8$.

Standard Ferrous Sulfate Solution. Dissolve 7.0 g of clear crystals of ferrous sulfate heptahydrate, $FeSO_4 \cdot 7H_2O$ in 80 mL of freshly boiled and cooled water, and add dilute sulfuric acid (1 + 1) to make 100 mL. Standardize the solution with 0.1 N potassium permanganate. The solution must be prepared and standardized at the time of use.

INSOLUBLE MATTER. (Page 15). Use 20 g dissolved in 200 mL of water.

RESIDUE AFTER IGNITION. (Page 16). Ignite 5.0 g.

TITRABLE FREE ACID. Dissolve 10.0 g in 100 mL of water and add 0.01 mL of methyl red indicator solution. If a red color is produced, not more than 4.0 mL of 0.1 N sodium hydroxide should be required to discharge it.

CHLORIDE AND CHLORATE. (Page 27). Mix 1.0 g with 1 g of sodium carbonate and heat until no more gas is evolved. Dissolve the residue in 20 mL of water, and neutralize with nitric acid.

HEAVY METALS. (Page 28, Method 1). Gently ignite 2.0 g in a porcelain or silica crucible or dish (not in platinum), and to the residue add 1 mL of hydrochloric acid, 1 mL of nitric acid, and about 10 mg of sodium carbonate. Evaporate to dryness on a hot plate ($\approx$100 °C), dissolve the residue in about 20 mL of water, and dilute with water to 50 mL. Dilute 10 mL of the solution with water to 25 mL.

IRON. (Page 30, Method 1). To 1.0 g add 5 mL of water and 10 mL of hydrochloric acid, and evaporate to dryness. Dissolve the residue in 2 mL of hydrochloric acid, dilute with water to 50 mL, and use this solution without further acidification. Prepare the standard solution from the residue remaining from evaporation of 10 mL of hydrochloric acid.

MANGANESE. Gently ignite 20.0 g (not in platinum) until the sample is decomposed, and finally ignite at 600 °C until the sample is volatilized. Dissolve the residue by boiling for 5 min with 35 mL of water plus 10 mL of nitric acid, 5 mL of sulfuric acid, and 5 mL of phosphoric acid, and cool. Prepare a standard containing 0.01 mg of manganese (Mn) in an equal volume of solution containing the quantities of reagents used to dissolve the residue. To each solution add 0.25 g of potassium periodate, boil gently for 5 min, and cool. Any pink color in the solution of the sample should not exceed that in the standard.

Ammonium Phosphate, Dibasic
Diammonium Hydrogen Phosphate

$(NH_4)_2HPO_4$ Formula Wt 132.06

CAS Number 7783–28–0

REQUIREMENTS

Assay . . . . $\geq$98.0% $(NH_4)_2HPO_4$
pH of a 5% solution . . . . 7.7–8.1 at 25 °C

MAXIMUM ALLOWABLE

Insoluble matter. 0.005%
Ammonium hydroxide precipitate 0.005%
Chloride (Cl) 0.001%
Nitrate (NO_3) 0.003%
Sulfate (SO_4) 0.01%
Heavy metals (as Pb) 0.001%

Iron (Fe). 0.001%
Potassium (K). 0.005%
Sodium (Na) 0.005%

TESTS

ASSAY. (By acid–base titration). Weigh accurately about 0.5 g and dissolve in 50 mL of water. Titrate with 0.1 N hydrochloric acid to the potentiometric end point. One milliliter of 0.1 N hydrochloric acid corresponds to 0.01321 g of $(NH_4)_2HPO_4$.

pH OF A 5% SOLUTION. (Page 44). The pH should be 7.7–8.1 at 25 °C.

INSOLUBLE MATTER. (Page 15). Use 20.0 g dissolved in 200 mL of water. Save the filtrate separate from the washings for the test for ammonium hydroxide precipitate.

AMMONIUM HYDROXIDE PRECIPITATE. To the filtrate from the test for insoluble matter add successively enough hydrochloric acid to make the solution acid to methyl red and enough ammonium hydroxide to make the solution again alkaline to methyl red. Heat to boiling, boil gently for 5 min, and cool. Filter, wash, and ignite the precipitate.

CHLORIDE. (Page 27). Use 1.0 g of sample and 3 mL of nitric acid.

NITRATE. (Page 30). For sample solution A use 0.50 g. For control solution B use 0.50 g and 1.5 mL of standard nitrate solution.

SULFATE. Dissolve 1.0 g in 15 mL of water and add 4.0 mL of 10% hydrochloric acid (solution approximately pH 2). Filter through a washed filter paper and add 10 mL of water through the same filter paper. For the control, add 0.10 mg of sulfate ion (SO_4) in 20 mL of water and add 1.0 mL of (1 + 19) hydrochloric acid. Dilute both to 35 mL with water, add 5.0 mL of 30% barium chloride reagent solution, and mix. Observe turbidity after 10 min. Sample solution turbidity should not exceed that of the control solution.

HEAVY METALS. (Page 28, Method 1). Dissolve 4.0 g in 10 mL of water, add 15 mL of 2 N hydrochloric acid, and dilute with water to 32 mL. Use 24 mL to prepare the sample solution, and use the remaining 8.0 mL to prepare the control solution.

IRON. (Page 30, Method 2). Use 1.0 g.

POTASSIUM AND SODIUM. (By flame AAS, page 39).

   Sample Stock Solution. Dissolve 10.0 g of sample with water in a 100-mL volumetric flask and dilute to the mark with water (1 mL = 0.1 g).

Element	Wavelength (nm)	Sample Wt (g)	Standard Added (mg)	Flame Type*	Background Correction
K	766.5	1.0	0.05; 0.10	A/A	No
Na	589.0	0.10	0.005; 0.01	A/A	No

*A/A is air/acetylene.

Ammonium Sulfamate

$NH_4OSO_2NH_2$ Formula Wt 114.13

CAS Number 7773–06–0

REQUIREMENTS

Assay . . . . ≥98.0% $NH_4OSO_2NH_2$
Melting point . . . . Within a range of 2.0 °C, including 133.0 °C

MAXIMUM ALLOWABLE

Insoluble matter. 0.02%
Residue after ignition 0.10%
Heavy metals (as Pb) . . . . 5 ppm

TESTS

ASSAY. (By oxidation–reduction titration). Weigh accurately about 0.35 g and transfer to a 250-mL flask. Add 75 mL of water and 5 mL of sulfuric acid, swirl to dissolve the sample, and titrate slowly with 0.1 M sodium nitrite, shaking the flask vigorously from time to time. Titrate dropwise near the end point, shaking after each addition, until a blue color is produced immediately when a glass rod dipped into the titrated solution is streaked on starch–iodide test paper. One milliliter of 0.1 M sodium nitrite corresponds to 0.01141 g of $NH_4OSO_2NH_2$.

MELTING POINT. (Page 22).

INSOLUBLE MATTER. (Page 15). Use 5.0 g dissolved in 100 mL of water.

RESIDUE AFTER IGNITION. (Page 16). Ignite 1.0 g.

HEAVY METALS. (Page 28, Method 1). Dissolve 6.0 g in 20 mL of water, and dilute with water to 30 mL. Use 25 mL to prepare the sample solution, and use the remaining 5.0 mL to prepare the control solution.

Ammonium Sulfate

(NH$_4$)$_2$SO$_4$ **Formula Wt 132.14**

CAS Number 7783–20–2

REQUIREMENTS

Assay. . . . . ≥99.0% (NH$_4$)$_2$SO$_4$
pH of a 5% solution . . . . 5.0–6.0 at 25 °C

MAXIMUM ALLOWABLE

Insoluble matter. 0.005%
Residue after ignition 0.005%
Chloride (Cl) . . . . 5 ppm
Nitrate (NO$_3$). 0.001%
Phosphate (PO$_4$) . . . . 5 ppm
Heavy metals (as Pb) . . . . 5 ppm
Iron (Fe). . . . . 5 ppm

TESTS

ASSAY. (By alkalimetry for ammonium). Weigh, to the nearest 0.1 mg, 2.5 g of sample and dissolve in 50 mL of water in a 500-mL Erlenmeyer flask. Add exactly 50.0 mL of 1 N sodium hydroxide, place a filter funnel loosely in the neck of the flask, and boil until all the ammonia is expelled (about 10–15 min) as determined with litmus paper. Cool, add 0.15 mL of thymol blue indicator solution (0.1% in alcohol). Titrate the excess sodium hydroxide with 1 N sulfuric acid. One milliliter of 1 N sodium hydroxide corresponds to 0.06607 g of (NH$_4$)$_2$SO$_4$.

pH OF A 5% SOLUTION. (Page 44). The pH should be 5.0–6.0 at 25 °C.

INSOLUBLE MATTER. (Page 15). Use 20 g dissolved in 200 mL of water.

RESIDUE AFTER IGNITION. (Page 16). Ignite 20 g.

CHLORIDE. (Page 27). Use 2.0 g.

NITRATE. (Page 30). For sample solution A use 1.0 g. For control solution B use 1.0 g and 1 mL of standard nitrate solution.

PHOSPHATE. (Page 32, Method 1). Dissolve 4.0 g in 25 mL of 0.5 N sulfuric acid.

HEAVY METALS. (Page 28, Method 1). Dissolve 6.0 g in 20 mL of water, and dilute with water to 30 mL. Use 25 mL to prepare the sample solution, and use the remaining 5.0 mL to prepare the control solution.

IRON. (Page 30, Method 1). Use 2.0 g.

Ammonium Thiocyanate

NH₄SCN   **Formula Wt 76.12**

CAS Number 1762–95–4

REQUIREMENTS

```
Appearance . . . . . . . . . . . . . . . . . . . . . . Colorless or white crystals
Assay . . . . . . . . . . . . . . . . . . . . . . . . . . . . . . . . . . ≥97.5% NH₄SCN
pH of a 5% solution . . . . . . . . . . . . . . . . . . . . . . . 4.5–6.0 at 25 °C
```

MAXIMUM ALLOWABLE

```
Insoluble matter. . . . . . . . . . . . . . . . . . . . . . . . . . . . . . 0.005%
Residue after ignition . . . . . . . . . . . . . . . . . . . . . . . . . . 0.025%
Chloride (Cl) . . . . . . . . . . . . . . . . . . . . . . . . . . . . . . . . 0.005%
Sulfate (SO₄) . . . . . . . . . . . . . . . . . . . . . . . . . . . . . . . . 0.005%
Heavy metals (as Pb) . . . . . . . . . . . . . . . . . . . . . . . . 5 ppm
Iron (Fe) . . . . . . . . . . . . . . . . . . . . . . . . . . . . . . . . . . 3 ppm
Iodine-consuming substances . . . . . . . . . . . . . . . . . . . 0.004 meq/g
```

TESTS

ASSAY. (By argentimetric titration of thiocyanate content). Weigh, to the nearest 0.1 mg, 7 g of sample. Dissolve with 100 mL of water in a 1-L volumetric flask, dilute with water to the mark, and mix well. Transfer a 50.0-mL aliquot to a 250-mL glass-stoppered iodine-type flask, add 5 mL of 5 N nitric acid, then add with agitation exactly 50.0 mL of 0.1 N silver nitrate solution, and shake vigorously. Add 2 mL of ferric ammonium sulfate indicator solution and titrate the excess silver nitrate with 0.1 N ammonium thiocyanate solution. Near the end point, shake after the addition of each drop.

$$\% \, NH_4SCN \; = \; \frac{(50.0 \; mL \times N \, AgNO_3) - (mL \times N \, NH_4SCN)(20)(7.612)}{Sample \, wt \, (g)}$$

pH OF A 5% SOLUTION. (Page 44). The pH should be 4.5–6.0 at 25 °C.

INSOLUBLE MATTER. (Page 15). Use 20 g dissolved in 150 mL of water.

RESIDUE AFTER IGNITION. (Page 16). Ignite 3.3 g. Retain the residue for the test for iron.

CHLORIDE. (Page 27). Dissolve 1.0 g in 20 mL of water in a small flask. Add 10 mL of 25% sulfuric acid reagent solution and 7 mL of 30% hydrogen peroxide. Evaporate to 20 mL by boiling in a well-ventilated hood, add 17 mL of water, and evaporate again. Repeat until all the cyanide has been volatilized. Cool, filter if necessary through a chloride-free filter, and dilute with water to 100 mL. Use 20 mL of this solution.

SULFATE. (Page 33, Method 1). Use 1.0-g sample, and allow 30 min for the turbidity to form.

HEAVY METALS. (Page 28). Dissolve 6.0 g in 20 mL of water, and dilute with water to 30 mL. Use 25 mL to prepare the sample solution, and use the remaining 5.0 mL to prepare the control solution.

IRON. (Page 30, Method 1). To the residue after ignition add 3 mL of dilute hydrochloric acid (1 + 1), cover with a watch glass, and digest on a hot plate ($\approx$100 °C) for 15–20 min. Remove the watch glass and evaporate to dryness. Dissolve the residue in 2 mL of hydrochloric acid, filter if necessary, and dilute with water to 50 mL. Use the solution without further acidification.

IODINE-CONSUMING SUBSTANCES. Dissolve 5.0 g in 50 mL of water plus 1.7 mL of 10% sulfuric acid reagent solution. Add 1 g of potassium iodide and 2 mL of starch indicator solution, and titrate with 0.01 N iodine solution. Not more than 2.0 mL of the iodine solution should be required.

Aniline

Benzenamine

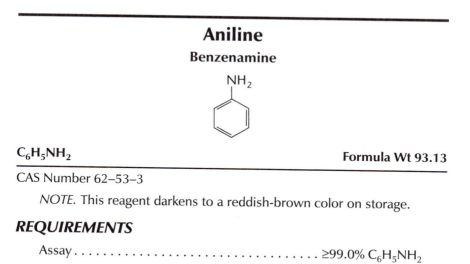

$C_6H_5NH_2$ **Formula Wt 93.13**

CAS Number 62–53–3

NOTE. This reagent darkens to a reddish-brown color on storage.

REQUIREMENTS

Assay. . . . . $\geq$99.0% $C_6H_5NH_2$

MAXIMUM ALLOWABLE

Color (APHA). 250
Residue after ignition. 0.005%
Chlorobenzene (C_6H_5Cl) 0.01%
Hydrocarbons . . . . Passes test
Nitrobenzene ($C_6H_5NO_2$) . . . . Passes test

TESTS

ASSAY AND CHLOROBENZENE. Analyze the sample by gas chromatography using the general parameters cited on page 70. The following specific conditions are also required.

Column: Type I, methyl silicone

Measure the area under all peaks and calculate the area percent for aniline and chlorobenzene.

COLOR (APHA). (Page 19).

RESIDUE AFTER IGNITION. (Page 16). Evaporate 20 g (20 mL) to dryness in the hood and ignite.

HYDROCARBONS. Mix 5 mL with 10 mL of hydrochloric acid. The solution should be clear after dilution with 15 mL of cold water.

NITROBENZENE.

Sample Solution A. Place 10 mL of methanol in a 25-mL glass-stoppered graduated cylinder.

Control Solution B. Place 10 mL of methanol in a 25-mL glass-stoppered graduated cylinder and add 1 mL of a methanol solution containing 0.10 mg of nitrobenzene per mL.

Immerse the cylinders in a beaker of water at or below room temperature, and add 10 mL of the aniline and 2.5 mL of hydrochloric acid to each. Mix well and bring to room temperature.

Transfer a portion of sample solution A to a polarographic cell and deaerate with nitrogen or hydrogen. Record the polarogram from –0.2 to –0.7 V vs. SCE with a current sensitivity of 0.02 µA/mm. Repeat this procedure with control solution B. The diffusion current in sample solution A should not exceed the difference in diffusion current between control solution B and sample solution A. (Limit about 0.001%)

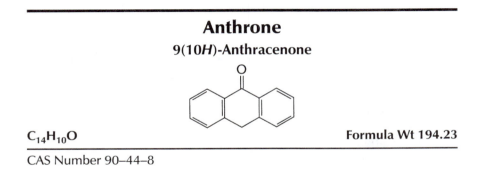

Anthrone
9(10*H*)-Anthracenone

C$_{14}$H$_{10}$O   **Formula Wt 194.23**

CAS Number 90–44–8

REQUIREMENTS

Appearance . . . . .Off-white to light yellow crystals
Melting point . . . . No more than a 5° range, including 156 °C
Sensitivity to carbohydrates . . . . Passes test

MAXIMUM ALLOWABLE

Absorbance of reagent solution . . . . Passes test
Solubility in ethyl acetate . . . . Passes test

TESTS

MELTING POINT. (Page 22).

SENSITIVITY TO CARBOHYDRATES.

Reagent Solution. Slowly add 132 mL of sulfuric acid to 68 mL of ice-cold water in an ice bath. Allow the mixture to cool to room temperature, add 0.100 g of the sample and stir if necessary to dissolve.

Glucose Standard. (1,000 µg in 1 mL). Dissolve 1.00 g of glucose in water, and dilute with water to 100 mL. Pipet 10.0 mL of this solution (1 mL = 10 mg) into a 100-mL volumetric flask, and dilute to volume with water.

Mark four 25-mL volumetric flasks as 0 µg (blank), 250 µg, 500 µg, and 750 µg. Add the glucose standard and water as specified in the following table, and dilute to volume with the reagent solution. Place all flasks in a vigorously boiling water bath for 7.0 min exactly. Chill the flask quickly to room temperature and determine the absorbance of each at 620 nm against the blank in 1.00-cm cells. A plot of the absorbance against concentration should be linear, and the absorbance of the 750-µg flask should not be less than 0.90.

Flask (µg)	Glucose Standard (mL)	Water (mL)
0	0.00	0.75
250	0.25	0.50
500	0.50	0.25
750	0.75	0.00

ABSORBANCE OF REAGENT SOLUTION. Measure the absorbance of the reagent solution (*see* the test for sensitivity to carbohydrates) in a spectrophotometer, using 1.00-cm cells, with sulfuric acid as reference. Record the absorbance at 425 nm and 620 nm, respectively. The absorbance should not exceed 0.75 at 425 nm and 0.045 at 620 nm.

SOLUBILITY IN ETHYL ACETATE. Dissolve 1 g in 50 mL of freshly distilled ethyl acetate in a dry 50-mL glass-stoppered graduate. The mixture should not be heated, dissolution should be complete, and the resulting solution should be clear.

Antimony Trichloride
Antimony(III) Chloride

$SbCl_3$ **Formula Wt 228.12**

CAS Number 10025–91–9

REQUIREMENTS

Assay . . . . ≥99.0% $SbCl_3$

MAXIMUM ALLOWABLE

Insoluble in chloroform 0.05%
Sulfate (SO_4) 0.005%
Arsenic (As) 0.02%
Calcium (Ca) 0.005%
Copper (Cu) 0.001%
Iron (Fe) 0.002%
Lead (Pb) 0.005%
Potassium (K) 0.01%
Sodium (Na) 0.02%

TESTS

ASSAY. (By oxidation–reduction titration). Weigh accurately 0.5 g and, in a glass-stoppered Erlenmeyer flask, dissolve in 5 mL of 10% hydrochloric acid reagent solution. When dissolution is complete, add a solution of 4.0 g of potassium sodium tartrate tetrahydrate in 30 mL of water. Stopper and swirl. Add 50 mL of a cold saturated solution of sodium bicarbonate, and titrate immediately with 0.1 N iodine, adding 3 mL of starch indicator solution near the end of the titration. One milliliter of 0.1 N iodine corresponds to 0.01140 g of $SbCl_3$.

INSOLUBLE IN CHLOROFORM. Dissolve 5.0 g in 25 mL of chloroform, filter through a tared filtering crucible, wash the crucible with several portions of chloroform, and dry at 105 °C.

> *NOTE.* Weigh quickly to avoid moisture absorption and the formation of insoluble oxychlorides. The use of anhydrous, ethanol-free chloroform is recommended.

SULFATE. Dissolve 10.0 g in the minimum volume of hydrochloric acid required to achieve complete dissolution, dilute with water to 75 mL, neutralize with ammonium hydroxide, and filter. Add 2 mL of hydrochloric acid to the filtrate, dilute with water to 100 mL, and heat to boiling. Add 10 mL of barium chloride reagent solution, digest in a covered beaker on the hot plate (≈100 °C) for 2 h, and allow to stand overnight. If a precipitate is formed, filter, wash thoroughly, and ignite. Correct for the weight obtained in a complete blank test.

ARSENIC. (Page 25). Dissolve 1.0 g in 5 mL of hydrochloric acid, pour this solution into a solution of 2 g of stannous chloride dihydrate ($SnCl_2 \cdot 2H_2O$) in 2 mL of hydrochloric acid, and allow to stand overnight. If no precipitate is visible, the arsenic content is less than the limit and the test need not be completed. If a precipitate has formed, filter it on a glass-fiber filter. Wash the beaker and the precipitate with two 5-mL portions of hydrochloric acid and then with water, passing the washings through the filter. Discard the filtrate. Dissolve the precipitate by passing a mixture of 5 mL of hydrochloric acid and 0.15 mL of bromine through the filter, and wash the filter with 5 mL of hydrochloric acid and then with water. Collect the filtrate and washings in the beaker in which the precipitation was made, warm the solution on a hot plate ($\approx$100 °C) to remove the excess bromine, and dilute with water to 1 L. Dilute 15 mL of this solution with water to 35 mL. For the standard, use 0.003 mg of arsenic (As).

CALCIUM, COPPER, IRON, LEAD, POTASSIUM, AND SODIUM. (By flame AAS, page 39).

> *Sample Stock Solution.* Dissolve 20.0 g of sample in 80 mL of hydrochloric acid (1 + 3). Transfer to a 100-mL volumetric flask, and dilute to the mark with hydrochloric acid (1 + 3). (1 mL = 0.20 g)

> *Blank Solution.* Use appropriate amount of hydrochloric acid (1 + 3) as a blank solution to zero the instrument.

Element	Wavelength (nm)	Sample Wt (g)	Standard Added (mg)	Flame Type*	Background Correction
Ca	422.7	0.80	0.02; 0.04	N/A	No
Cu	324.8	4.0	0.02; 0.04	A/A	Yes
Fe	248.3	4.0	0.08; 0.16	A/A	Yes
Pb	217.0	4.0	0.10; 0.20	A/A	Yes
K	766.5	0.10	0.01; 0.02	A/A	No
Na	589.0	0.10	0.01; 0.02	A/A	No

*A/A is air/acetylene; N/A is nitrous oxide/acetylene.

Arsenic Trioxide
Arsenic(III) Oxide, Reductometric Standard

As_2O_3 Formula Wt 197.84

CAS Number 1327–53–3

REQUIREMENTS

Assay . . . . 99.95–100.05% As_2O_3

MAXIMUM ALLOWABLE

Residue after ignition 0.02%

Insoluble in dilute hydrochloric acid. 0.01%

Chloride (Cl) 0.005%

Sulfide (S). . . . . Passes test

Antimony (Sb) 0.05%

Lead (Pb) 0.002%

Iron (Fe) . . . . 5 ppm

TESTS

ASSAY. (By oxidation–reduction titration of arsenic). This assay is based on a direct titration with NIST Standard Reference Material Potassium Dichromate. Accurately weighed portions of the arsenic trioxide are reacted with accurately weighed portions of the NIST potassium dichromate of such size as to provide a small excess of the latter. The excess dichromate is determined by titration with standardized ferrous ammonium sulfate solution, using a potentiometrically determined end point.

> *CAUTION.* Due to the small tolerance in assay limits for this reductometric standard, extreme care must be observed in the weighing, transferring, and titrating operations in the following procedure. Strict adherence to the specified sample weights and final titration volumes is absolutely necessary. It is recommended that the titrations be run at least in duplicate. Duplicate values for the assay of this material should agree within 2 parts in 5000 to be acceptable for averaging.

Procedure. Place 2.00-g portions of the arsenic trioxide in a weighing bottle, dry for 1 h at 105 °C, and cool for 2 h in a desiccator. Weigh accurately to within 0.1 mg and transfer to 400-mL beakers. Weigh each bottle again and determine by difference to the nearest 0.1 mg the weight of each portion of arsenic trioxide.

Weigh accurately to within 0.1 mg an amount of NIST potassium dichromate equivalent to a slight excess of the weight of arsenic trioxide taken, and transfer it to a 50-mL beaker. Dissolve the arsenic trioxide in 20 mL of 20% sodium hydroxide solution, add 100 mL of water and 10 mL of dilute sulfuric acid (1 + 1), and stir. Transfer the potassium dichromate to the solution, using water to loosen any crystals adhering to the beakers. Stir until all of the potassium dichromate is dissolved, and let stand for 10 min.

Add 20 mL of dilute sulfuric acid (1 + 1), dilute with water to 300 mL, and stir. Titrate the excess dichromate with standardized 0.02 N ferrous ammonium sulfate solution, using a platinum indicator electrode and a calomel reference electrode for the measurement of the potential difference in millivolts. The end point of the titration is determined potentiometrically, using the second derivative method (page 52). A correction for "blank" oxidants or reductants must be applied by titrating an accurately weighed 40-mg portion of the NIST potassium dichromate with the standardized ferrous ammonium sulfate in an equal volume of solution containing the quantities of reagents used in this assay. The "blank" is equal to the calculated milliequivalents of $K_2Cr_2O_7$ minus the calculated milliequivalents of $Fe(NH_4)_2(SO_4)_2 \cdot 6H_2O$. A positive "blank" correction results from other oxidants and a negative correction from other reductants.

Calculations:

$$M_1 = \left[\frac{W_1}{49.0307} - VN + B\right] \times 1.00032F$$

where

M_1 = number of milliequivalents of $K_2Cr_2O_7$ equivalent to As_2O_3
W_1 = weight of $K_2Cr_2O_7$ in milligrams
V = volume of $Fe(NH_4)_2(SO_4)_2$ solution in milliliters
N = concentration of $Fe(NH_4)_2(SO_4)_2$ solution in equivalents per liter
B = blank correction in milliequivalents (may be either positive or negative)
F = assay value of the NIST standard in percent divided by 100

(The equivalent formula weight of $K_2Cr_2O_7$ is 49.0307, and the conversion factor for air to vacuum weight of $K_2Cr_2O_7$ is 1.00032.)

Calculate the assay value of the arsenic trioxide from

$$A = \% As_2O_3 = \frac{(49.4603)(M_1)}{(1.00017)W_2}100$$

(W_2 = weight of As_2O_3 in milligrams; 1.00017 is the conversion factor for air to vacuum weight of As_2O_3.)

Preparation and Standardization of Ferrous Ammonium Sulfate Solution, 0.02 N. Weigh 8.0 g of $Fe(NH_4)_2(SO_4)_2 \cdot 6H_2O$ and transfer to a 1-L flask. Add 200 mL of dilute sulfuric acid (1 + 19) and, after the salt is dissolved, dilute to the liter mark with the dilute acid. Mix thoroughly.

Transfer a 40-mg portion of the potassium dichromate, accurately weighed to within 0.1 mg, to a 400-mL beaker. Add 300 mL of dilute sulfuric acid (1 + 19) and stir to dissolve the salt. Titrate the solution with the 0.02 N ferrous ammonium sulfate, using a potentiometrically determined end point. The concentration of the ferrous ammonium sulfate, in equivalents per liter (N), is calculated from the following equation:

$$N = \left[\frac{W}{49.03} \right]\left[\frac{1000}{V} \right]$$

where

$N =$ concentration of ferrous ammonium sulfate in equivalents per liter

$W =$ weight of $K_2Cr_2O_7$ in grams

$V =$ volume of $Fe(NH_4)_2(SO_4)_2$ solution in milliliters

RESIDUE AFTER IGNITION. (Page 16). Ignite 5.0 g in a tared platinum dish. Retain the residue for the test for iron.

INSOLUBLE IN DILUTE HYDROCHLORIC ACID. To 10 g add 90 mL of dilute hydrochloric acid (3 + 7) and 10 mL of hydrogen peroxide. Cover with a watch glass and allow to stand until the sample goes into solution, heating if necessary. Heat to boiling, digest in a covered beaker on a hot plate (≈100 °C) for 1 h, and filter through a tared filtering crucible. Retain the filtrate for the antimony and lead test, wash the insoluble matter thoroughly, and dry at 105 °C.

CHLORIDE. (Page 27). Dissolve 1.0 g in 10 mL of dilute ammonium hydroxide (1 + 2) with the aid of gentle heating, and dilute with water to 100 mL. Neutralize 20 mL with nitric acid.

SULFIDE. Dissolve 1.0 g in 10 mL of 10% sodium hydroxide reagent solution, and add 0.05 mL of lead acetate solution (about 10%). The color should be the same as that of an equal volume of sodium hydroxide solution to which only the lead acetate is added. (Limit about 0.001%)

ANTIMONY AND LEAD. (By flame AAS, page 39).

Sample Stock Solution. Dilute the filtrate retained from the insoluble in hydrochloric acid test with water to the mark in a 200-mL volumetric flask (1 mL = 0.05 g).

Element	Wavelength (nm)	Sample Wt (g)	Standard Added (mg)	Flame Type*	Background Correction
Sb	217.6	1.0	0.5; 1.0	A/A	Yes
Pb	217.0	1.0	0.05; 0.10	A/A	Yes

*A/A is air/acetylene.

IRON. (Page 30, Method 1). To the residue from the test for residue after ignition, add 3 mL of dilute hydrochloric acid (1 + 1), and warm. Add 5 mL of hydrochloric acid, dilute with water to 125 mL, and use 50 mL of this solution without further acidification.

Ascorbic Acid

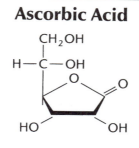

C$_6$H$_8$O$_6$ **Formula Wt 176.13**

CAS Number 50–81–7

REQUIREMENTS

Assay. . . . . ≥99.0% C$_6$H$_8$O$_6$
Specific rotation [α]$_D^{25°}$ . . . . +21.0° ± 0.5°

MAXIMUM ALLOWABLE

Residue after ignition. 0.1%
Heavy metals (as Pb) 0.002%
Iron (Fe). 0.001%

TESTS

ASSAY. (By oxidation–reduction titration). Weigh accurately 0.4 g and in a conical flask dissolve in a mixture of 100 mL of water (made oxygen-free by bubbling
nitrogen gas through it) and 25 mL of 10% sulfuric acid. Swirl and titrate immediately with 0.1 N iodine, adding 3 mL of starch indicator near the end of the
titration. One milliliter of 0.1 N iodine corresponds to 0.008806 g of C$_6$H$_8$O$_6$.

SPECIFIC ROTATION. (Page 23). Weigh accurately about 10 g, dissolve in 90
mL of oxygen-free water in a 100-mL volumetric flask, and dilute with water to
volume. Adjust the temperature of the solution to 25 °C. Observe the optical rotation in a polarimeter at 25 °C using the sodium line, and calculate the specific
rotation.

RESIDUE AFTER IGNITION. (Page 16). Ignite 1.0 g. Moisten the char with 1
mL of sulfuric acid. Retain the residue for the test for iron.

HEAVY METALS. (Page 28, Method 2). Use 1.0 g.

IRON. (Page 30, Method 1). To the residue from the test for Residue after ignition add 3 mL of dilute hydrochloric acid (1 + 1) and 0.10 mL of nitric acid, cover
with a watch glass, and digest on a hot plate (≈100 °C) for 15–20 min. Remove the
watch glass and evaporate to dryness. Dissolve the residue in a mixture of 2 mL of

hydrochloric acid and 10 mL of water, dilute with water to 50 mL, and use without further acidification.

Aurin Tricarboxylic Acid, [tri]Ammonium Salt

5-[(3-Carboxy-4-hydroxyphenyl)(3-carboxy-4-oxo-2,5-cyclohexadien-1-ylidene)methyl]-2-hydroxybenzoic Acid, Triammonium Salt

Aluminon

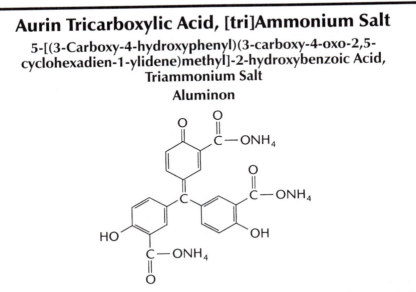

$(HOC_6H_3COONH_4)_2C{:}C_6H_3(COONH_4){:}O$ **Formula Wt 473.44**

CAS Number 569–58–4

REQUIREMENTS

Sensitivity to aluminum . . . . Passes test

MAXIMUM ALLOWABLE

Insoluble matter. 0.1%
Residue after ignition 0.2%

TESTS

SENSITIVITY TO ALUMINUM. Dissolve 25 mg in 20 mL of water and 0.1 mL of dilute ammonium hydroxide (10% NH$_3$). Add 0.1 mL of this solution and 0.1 mL of acetic acid to 10 mL of a solution containing 0.001 mg of aluminum ion (Al). A distinct pink color should appear within 15 min.

INSOLUBLE MATTER. (Page 15). Use 1.0 g dissolved in 100 mL of water plus 0.5 mL of dilute ammonium hydroxide (10% NH$_3$).

RESIDUE AFTER IGNITION. (Page 16). Ignite 0.50 g.

Barium Acetate

(CH₃COO)₂Ba   **Formula Wt 255.42**

$(CH_3COO)_2Ba$

CAS Number 543–80–6

REQUIREMENTS

Assay. . . . . .99.0–102.0% $(CH_3COO)_2Ba$

MAXIMUM ALLOWABLE

Insoluble matter. 0.01%
Chloride (Cl) 0.001%
Oxidizing substances (as NO_3) 0.005%
Calcium (Ca) 0.05%
Potassium (K). 0.003%
Sodium (Na) 0.005%
Strontium (Sr) 0.2%
Heavy metals (as Pb) . . . . 5 ppm
Iron (Fe). 0.001%

TESTS

ASSAY. (By complexometry for barium). Weigh accurately 0.28 g, transfer to a 400-mL beaker, and dissolve in 100 mL of carbon dioxide-free water. Add 100 mL of ethanol, 10 mL of ammonium hydroxide, and 3.0 mL of metalphthalein-screened indicator solution. Titrate immediately with standard 0.1 M EDTA to a color change from magenta to grey-green. One milliliter of 0.1 M EDTA corresponds to 0.02554 g of $(CH_3COO)_2Ba$.

INSOLUBLE MATTER. (Page 15). Use 10 g dissolved in 100 mL of water.

CHLORIDE. (Page 27). Use 1.0 g.

OXIDIZING SUBSTANCES. Place 0.10 g in a dry beaker. Cool the beaker thoroughly in an ice bath and add 22 mL of sulfuric acid that has been cooled to ice-bath temperature. Allow the mixture to warm to room temperature and swirl the beaker at intervals to effect gentle dissolution with slow evolution of the acetic acid vapor. When dissolution is complete, add 3 mL of diphenylamine reagent solution and digest on a hot plate (≈100 °C) for 90 min. Prepare a standard by evaporating to dryness a solution containing 0.005 mg of nitrate (0.5 mL of the nitrate standard solution) and 0.01 g of sodium carbonate. Treat the residue exactly like the sample. Any color produced in the solution of the sample should not exceed that in the standard.

CALCIUM, POTASSIUM, SODIUM, AND STRONTIUM. (By flame AAS, page 39).

Sample Stock Solution A. Dissolve 10.0 g of sample in sufficient water in a beaker, add 5 mL of nitric acid, and transfer to a 100-mL volumetric flask. Dilute to the mark with water (1 mL = 0.10 g).

Sample Stock Solution B. Pipet 10.0 mL of sample stock solution A into a 100-mL volumetric flask, and dilute to the mark with water (1 mL = 0.01 g). Add 2.0 mL of a 2.5% potassium chloride solution to the flask for calcium and strontium.

Element	Wavelength (nm)	Sample Wt (g)	Standard Added (mg)	Flame Type*	Background Correction
Ca	422.7	0.04	0.02; 0.04	N/A	No
K	766.5	1.0	0.02; 0.04	A/A	No
Na	589.0	0.10	0.005; 0.01	A/A	No
Sr	460.7	0.05	0.05; 0.10	N/A	No

*A/A is air/acetylene; N/A is nitrous oxide/acetylene.

HEAVY METALS. (Page 28, Method 1). Dissolve 6.0 g in about 15 mL of water, add 8 mL of dilute hydrochloric acid (1 + 1), and dilute with water to 30 mL. Use 25 mL to prepare the sample solution, and use the remaining 5.0 mL to prepare the control solution.

IRON. Dissolve 1.0 g in 40 mL of water. For the standard, add 0.01 mg of iron (Fe) to 40 mL of water. Add 2 mL of hydrochloric acid to each, dilute with water to 50 mL, and add 0.10 mL of 0.1 N potassium permanganate. Allow to stand for 5 min and add 3 mL of ammonium thiocyanate reagent solution. Any red color in the solution of the sample should not exceed that in the standard.

Barium Carbonate

$BaCO_3$ **Formula Wt 197.34**

CAS Number 513–77–9

REQUIREMENTS

Assay . . . . .99.0–101.0% $BaCO_3$

MAXIMUM ALLOWABLE

Insoluble in dilute hydrochloric acid. 0.015%
Chloride (Cl) 0.002%
Water-soluble titrable base. .0.002 meq/g
Oxidizing substances (as NO_3). 0.005%
Sulfide (S). 0.001%
Calcium (Ca) 0.05%

Potassium (K). 0.005%
Sodium (Na) 0.02%
Strontium (Sr) 0.7%
Heavy metals (as Pb) 0.001%
Iron (Fe). 0.002%

TESTS

ASSAY. (By acid–base titrimetry). Weigh to the nearest 0.1 mg, 4 g of sample, transfer to a 250-mL beaker, add 50 mL of water, and mix. Cautiously add exactly 50.0 mL of 1 N hydrochloric acid, boil, and cool. Add a few drops of bromphenol blue indicator (0.1% in water), and titrate the excess acid with 1 N sodium hydroxide to the blue end point. One milliliter of 1 N hydrochloric acid corresponds to 0.0987 g of $BaCO_3$.

INSOLUBLE IN DILUTE HYDROCHLORIC ACID. Cautiously dissolve 10 g in 100 mL of dilute hydrochloric acid (1 + 9) and dilute with water to 200 mL. Ignore any slight haze that is produced. Filter through a tared filtering crucible, wash thoroughly with dilute hydrochloric acid (1 + 9), and dry at 105 °C.

CHLORIDE. To 1.0 g add 20 mL of water and add dropwise, with stirring, about 2–3 mL of nitric acid. Filter if necessary through a chloride-free filter, wash with a little hot water, and dilute with water to 50 mL. To 25 mL of the solution add 1 mL of silver nitrate reagent solution. Any turbidity should not exceed that produced by 0.01 mg of chloride ion (Cl) in an equal volume of solution containing the quantities of reagents used in the test.

WATER-SOLUBLE TITRABLE BASE. Shake 5.0 g for 5 min with 50 mL of carbon dioxide-free water, cool, and filter. To 25 mL of the filtrate add 0.10 mL of phenolphthalein indicator solution. If a pink color is produced, it should be discharged by 0.50 mL of 0.01 N hydrochloric acid.

OXIDIZING SUBSTANCES. Place 0.10 g in a dry beaker. Cool the beaker thoroughly in an ice bath and add 22 mL of sulfuric acid that has been cooled to ice-bath temperature. Allow the mixture to warm to room temperature and swirl the beaker at intervals to effect gentle dissolution with slow evolution of carbon dioxide. When dissolution is complete, add 3 mL of diphenylamine reagent solution and digest on a hot plate (≈100 °C) for 90 min. Prepare a standard by evaporating to dryness a solution containing 0.005 mg of nitrate (0.5 mL of the nitrate standard solution) and 0.01 g of sodium carbonate. Treat the residue exactly like the sample. Any color produced in the solution of the sample should not exceed that in the standard.

SULFIDE. Dissolve 1.0 g in 10 mL of dilute acetic acid (1 + 4). As soon as dissolution is complete, add 1 mL of silver nitrate reagent solution. Any color after 5 min should not exceed that produced by 0.01 mg of sulfide ion (S) in an equal volume of solution containing the quantities of reagents used in the test.

CALCIUM, POTASSIUM, SODIUM, AND STRONTIUM. (By flame AAS, page 39).

> *Sample Stock Solution A.* Cautiously dissolve 10.0 g of sample in 50 mL of dilute hydrochloric acid (1 + 3) in a beaker. Transfer to a 100-mL volumetric flask and dilute with water to the mark (1 mL = 0.10 g).

> *Sample Stock Solution B.* Pipet 10.0 mL of sample stock solution A into a 100-mL volumetric flask and dilute to the mark with water (1 mL = 0.01 g). Add 2.0 mL of a 2.5% potassium chloride solution to the flask for calcium and strontium.

Element	Wavelength (nm)	Sample Wt (g)	Standard Added (mg)	Flame Type*	Background Correction
Ca	422.7	0.05	0.025; 0.05	N/A	Yes
K	766.5	0.40	0.02; 0.04	A/A	No
Na	589.0	0.05	0.01; 0.02	A/A	No
Sr	460.7	0.01	0.035; 0.07	N/A	Yes

*N/A is nitrous oxide/acetylene; A/A is air/acetylene.

HEAVY METALS. (Page 28, Method 1). Cautiously dissolve 5.0 g in 30 mL of dilute hydrochloric acid (1 + 4), and evaporate to dryness on a hot plate ($\approx$100 °C). Dissolve the residue in about 20 mL of water, and dilute with water to 25 mL. Dilute 15 mL with water to 25 mL, and use to prepare the sample solution. Use 5.0 mL of the remaining solution to prepare the control solution.

IRON. Cautiously dissolve 1.0 g in 15 mL of dilute hydrochloric acid (1 + 2). For the standard, add 0.02 mg of iron (Fe) to 15 mL of dilute hydrochloric acid (1 + 2). Evaporate both solutions to dryness on a hot plate ($\approx$100 °C). Dissolve each residue in about 20 mL of water, add 4 mL of hydrochloric acid, and dilute with water to 100 mL. To 50 mL of each solution add 0.10 mL of 0.1 N potassium permanganate, allow to stand for 5 min, and add 3 mL of ammonium thiocyanate reagent solution. Any red color in the solution of the sample should not exceed that in the standard.

Barium Chloride Dihydrate

$BaCl_2 \cdot 2H_2O$ **Formula Wt 244.26**

CAS Number 10326–27–9

REQUIREMENTS

Assay . . . . $\geq$99.0% $BaCl_2 \cdot 2H_2O$

Loss on drying at 150 °C 14.0–16.0%
pH of a 5% solution . . . . 5.2–8.2 at 25 °C

MAXIMUM ALLOWABLE

Insoluble matter. 0.005%
Oxidizing substances (as NO_3) 0.005%
Calcium (Ca) 0.05%
Potassium (K). 0.0025%
Sodium (Na) 0.005%
Strontium (Sr) 0.1%
Heavy metals (as Pb) . . . . 5 ppm
Iron (Fe). . . . . 2 ppm

TESTS

ASSAY. (By complexometric titration of barium). Weigh accurately 0.8 g of sample and dissolve in 200 mL of water. Add, while stirring, 2.0 mL of hydrochloric acid (1 + 1) and, from 50-mL buret, add 25.0 mL of 0.1 M EDTA. Add 100 mL of methanol, 25 mL of ammonium hydroxide, and about 80–100 mg of metalphthalein indicator mixture. Immediately continue titration until color changes from violet to clear colorless. The color change is gradual.

Calculation:

$$\% \; BaCl_2 \cdot 2H_2O = (mL \times M \; EDTA) \times 24.43) \; / \; Sample \; wt \; (g)$$

LOSS ON DRYING AT 150 °C. Weigh accurately about 1.0 g and dry at 150 °C to constant weight. (The theoretical loss for $BaCl_2 \cdot 2H_2O$ is 14.75%.)

pH OF A 5% SOLUTION. (Page 44). The pH should be 5.2–8.2 at 25 °C.

INSOLUBLE MATTER. (Page 15). Use 20 g dissolved in 200 mL of water. The solution should be clear and colorless.

OXIDIZING SUBSTANCES. Place 0.10 g in a dry beaker. Cool the beaker thoroughly in an ice bath and add 22 mL of sulfuric acid that has been cooled to ice-bath temperature. Allow the mixture to warm to room temperature and swirl the beaker at intervals to effect slow dissolution with gentle evolution of hydrogen chloride. When dissolution is complete, add 3 mL of diphenylamine reagent solution and digest on a hot plate (≈100 °C) for 90 min. Prepare a standard by evaporating to dryness a solution containing 0.005 mg of nitrate (0.5 mL of the standard nitrate solution) and 0.01 g of sodium carbonate. Treat the residue exactly like the sample. Any color produced in the solution of the sample should not exceed that in the standard.

CALCIUM, POTASSIUM, SODIUM, AND STRONTIUM. (By flame AAS, page 39).

Sample Stock Solution A. Dissolve 10.0 g of sample in sufficient water in a beaker, add 5 mL of nitric acid, and transfer to a 100-mL volumetric flask. Dilute to the mark with water (1 mL = 0.10 g).

Sample Stock Solution B. Pipet 10.0 mL of sample stock solution A into a 100-mL volumetric flask and dilute to the mark with water (1 mL = 0.01 g). Add 2.0 mL of a 2.5% potassium chloride solution to the flask for calcium and strontium.

Element	Wavelength (nm)	Sample Wt (g)	Standard Added (mg)	Flame Type*	Background Correction
Ca	422.7	0.04	0.02; 0.04	N/A	No
K	766.5	1.0	0.02; 0.04	A/A	No
Na	589.0	0.10	0.005; 0.01	A/A	No
Sr	460.7	0.05	0.05; 0.10	N/A	No

*N/A is nitrous oxide/acetylene; A/A is air/acetylene.

HEAVY METALS. (Page 28, Method 1). Dissolve 6.0 g in about 20 mL of water, and dilute with water to 30 mL. Use 25 mL to prepare the sample solution, and use the remaining 5.0 mL to prepare the control solution.

IRON. Dissolve 5.0 g in 40 mL of water plus 2 mL of hydrochloric acid. Prepare a standard containing 0.01 mg of iron (Fe) in 40 mL of water plus 2 mL of hydrochloric acid. Add 0.10 mL of 0.1 N potassium permanganate to each, dilute with water to 50 mL, and allow to stand for 5 min. Add 3 mL of ammonium thiocyanate reagent solution to each. Any red color in the solution of the sample should not exceed that in the standard.

Barium Hydroxide Octahydrate

$Ba(OH)_2 \cdot 8H_2O$ **Formula Wt 315.46**

CAS Number 12230–71–6

REQUIREMENTS

Assay . . . . $\geq$98.0% $Ba(OH)_2 \cdot 8H_2O$

MAXIMUM ALLOWABLE

Carbonate (as $BaCO_3$) 2.0%
Insoluble in dilute hydrochloric acid. 0.01%
Chloride (Cl) 0.001%
Sulfide (S). . . . . Passes test
Calcium (Ca) 0.05%
Potassium (K). 0.01%
Sodium (Na). 0.01%

Strontium (Sr) 0.8%
Heavy metals (as Pb) . . . . 5 ppm
Iron (Fe). 0.001%

TESTS

ASSAY. (By acid–base titrimetry). Weigh accurately 4–5 g of clear crystals and dissolve in about 200 mL of carbon dioxide-free water. Add 0.10 mL of phenolphthalein indicator solution and titrate with 1 N hydrochloric acid. Reserve this solution for the determination of carbonate. Each milliliter of 1 N hydrochloric acid corresponds to 0.1577 g of $Ba(OH)_2 \cdot 8H_2O$.

CARBONATE. To the solution reserved from the test for assay add 5.00 mL of 1 N hydrochloric acid, heat to boiling, boil gently to expel all of the carbon dioxide, and cool. Add 0.10 mL of methyl orange indicator and titrate the excess acid with 1 N sodium hydroxide. Each milliliter of 1 N hydrochloric acid consumed corresponds to 0.09867 g of $BaCO_3$.

INSOLUBLE IN DILUTE HYDROCHLORIC ACID. Dissolve 10 g in 100 mL of dilute hydrochloric acid (1 + 9). Heat the solution to boiling, then digest in a covered beaker on a hot plate ($\approx$100 °C) for 1 h. Filter through a tared filtering crucible, wash thoroughly, and dry at 105 °C.

CHLORIDE. Dissolve 2.0 g in 25 mL of water, add 0.05 mL of phenolphthalein indicator solution, and neutralize with nitric acid. If the solution is not perfectly clear, filter it through a chloride-free filter and dilute the filtrate with water to 50 mL. To 25 mL of the solution add 1 mL of nitric acid and 1 mL of silver nitrate reagent solution. Any turbidity should not exceed that produced by 0.01 mg of chloride ion (Cl) in an equal volume of solution containing the quantities of reagents used in the test.

SULFIDE. Dissolve 1.0 g in 8 mL of warm water and add 0.25 mL of alkaline lead solution (prepared by adding 10% sodium hydroxide solution to a 10% lead acetate solution until the precipitate is redissolved) and 2 mL of glacial acetic acid. No darkening should occur. (Limit about 0.001%)

CALCIUM, POTASSIUM, SODIUM, AND STRONTIUM. (By flame AAS, page 39).

Sample Stock Solution A. Dissolve 10.0 g of sample in water in a beaker, add 5 mL of nitric acid, and transfer to a 100-mL volumetric flask. Dilute to the mark with water (1 mL = 0.10 g).

Sample Stock Solution B. Pipet 10.0 mL of Sample Stock Solution A into a 100-ml volumetric flask, and dilute to the mark with water (1 mL = 0.01 g). Add 2.0 mL of a 2.5% potassium chloride solution to the flask for calcium and strontium.

Element	Wavelength (nm)	Sample Wt (g)	Standard Added (mg)	Flame Type*	Background Correction
Ca	422.7	0.04	0.02; 0.04	N/A	No
K	766.5	0.10	0.01; 0.02	A/A	No
Na	589.0	0.05	0.005; 0.01	A/A	No
Sr	460.7	0.01	0.04; 0.08	N/A	No

*N/A is nitrous oxide/acetylene; A/A is air/acetylene.

HEAVY METALS. (Page 28, Method 1). To 5.0 g in a 150-mL beaker, add 25 mL of water, mix, and cautiously add 10 mL of hydrochloric acid. Evaporate to dryness on a hot plate ($\approx$100 °C), dissolve the residue in about 2 mL of water, and dilute with water to 25 mL. For the control, treat 1.0 g of the sample and 0.02 mg of lead exactly as the 5.0 g of sample.

IRON. Dissolve 1.0 g in 40 mL of water. For the standard, add 0.01 mg of iron (Fe) to 40 mL of water. Add 2.5 mL of hydrochloric acid to each, dilute with water to 50 mL, and add 0.10 mL of 0.1 N potassium permanganate solution. Allow to stand for 5 min and add 3 mL of ammonium thiocyanate reagent solution. Any red color in the solution of the sample should not exceed that in the standard.

Barium Nitrate

$Ba(NO_3)_2$　　　　　　　　　　　　　　　　　**Formula Wt 261.35**

CAS Number 10022–31–8

REQUIREMENTS

Assay . . . . $\geq$99.0% $Ba(NO_3)_2$
pH of a 5% solution . . . . 5.0–8.0 at 25 °C

MAXIMUM ALLOWABLE

Insoluble matter. 0.01%
Chloride (Cl) . . . . 5 ppm
Calcium (Ca) 0.05%
Potassium (K). 0.005%
Sodium (Na). 0.005%
Strontium (Sr) 0.1%
Heavy metals (as Pb) . . . . 5 ppm
Iron (Fe) . . . . 2 ppm

TESTS

ASSAY. (By complexometry for barium). Weigh accurately 0.20 g, transfer to a 400-mL beaker, and dissolve in 100 mL of freshly boiled (carbon dioxide-free) water. Add 100 mL of ethanol, 10 mL of ammonium hydroxide, and 0.3 mL of metalphthalein-screened indicator solution. Titrate immediately with standard 0.1 M EDTA to a color change from magenta to grey-green. One milliliter of 0.1 M EDTA corresponds to 0.02614 g of $Ba(NO_3)_2$.

pH OF A 5% SOLUTION. (Page 44). The pH should be 5.0–8.0 at 25 °C.

INSOLUBLE MATTER. (Page 15). Use 10 g dissolved in 150 mL of hot water.

CHLORIDE. Dissolve 2.0 g in 30 mL of warm water, filter if necessary through a chloride-free filter, and add 0.10 mL of nitric acid and 1 mL of silver nitrate reagent solution. Any turbidity should not exceed that produced by 0.01 mg of chloride ion (Cl) in an equal volume of solution containing the quantities of reagents used in the test.

CALCIUM, POTASSIUM, SODIUM, AND STRONTIUM. (By flame AAS, page 39).

> *Sample Stock Solution.* Dissolve 10.0 g of sample in sufficient water in a beaker, add 5 mL of nitric acid, and transfer to a 100-mL volumetric flask. Dilute to the mark with water (1 mL = 0.10 g). Add 2.0 mL of a 2.5% potassium chloride solution to the flask for calcium and strontium.

Element	Wavelength (nm)	Sample Wt (g)	Standard Added (mg)	Flame Type*	Background Correction
Ca	422.7	0.10	0.05; 0.10	N/A	No
K	766.5	0.40	0.02; 0.04	A/A	No
Na	589.0	0.10	0.005; 0.01	A/A	No
Sr	460.7	0.10	0.10; 0.20	N/A	No

*N/A is nitrous oxide/acetylene; A/A is air/acetylene.

HEAVY METALS. Dissolve 4.0 g in about 50 mL of water and dilute with water to 60 mL. For the control, add 0.01 mg of lead to 15 mL of the solution and dilute with water to 45 mL. For the sample use the remaining 45-mL portion. Adjust the pH of the control and sample solutions to between 3 and 4 (using a pH meter) with 1 N acetic acid or ammonium hydroxide (10% NH_3), dilute with water to 48 mL, and mix. Add 10 mL of freshly prepared hydrogen sulfide water to each and mix. Any color in the solution of the sample should not exceed that in the control.

IRON. Dissolve 5.0 g in 45 mL of hot water plus 2 mL of hydrochloric acid and allow to cool. Prepare a standard containing 0.01 mg of iron (Fe) in 45 mL of water plus 2 mL of hydrochloric acid. Add 0.10 mL of 0.1 N potassium permanganate to each, allow to stand for 5 min, and add 3 mL of ammonium thiocyanate reagent solution. Any red color in the solution of the sample should not exceed that in the standard.

Benzene

C₆H₆   **Formula Wt 78.11**

CAS Number 71–43–2

Suitable for use in ultraviolet spectrophotometry or general use. Product labeling shall designate one or both of these uses for which suitability is represented on the basis of meeting the relevant requirements and tests. The ultraviolet spectrophotometry requirements include all the requirements for general use.

REQUIREMENTS

General Use

Assay . . . . ≥99.0% C₆H₆

MAXIMUM ALLOWABLE

Color (APHA) 10
Residue after evaporation. 0.001%
Substances darkened by sulfuric acid . . . . Passes test
Thiophene (limit about 1 ppm). . . . . Passes test
Sulfur compounds (as S) 0.005%
Water (H₂O). 0.05%

Specific Use

Ultraviolet Spectrophotometry

Wavelength (nm)	Absorbance (AU)
380 to 400	0.01
350	0.02
330	0.04
300	0.10
290	0.30
280	1.00

TESTS

ASSAY. Analyze the sample by gas chromatography using the general parameters cited on page 70. The following specific conditions are also required.

Column: Type I, methyl silicone

Measure the area under all peaks and calculate the benzene content in area percent. Correct for water content.

COLOR (APHA). (Page 19).

RESIDUE AFTER EVAPORATION. (Page 16). Evaporate 100 g (115 mL) to dryness in a tared dish on a hot plate ($\approx$100 °C) and dry the residue at 105 °C for 30 min.

SUBSTANCES DARKENED BY SULFURIC ACID. Shake 25 mL with 15 mL of sulfuric acid for 15–20 s and allow to separate. Neither the benzene nor the acid should be darkened.

THIOPHENE. Add 5.0 mL of freshly prepared isatin reagent solution to a dry, clean 50-mL porcelain crucible. Carefully overlay with 5.0 mL of the sample, allow to stand undisturbed for 1 h, and compare with a blank containing 5.0 mL of isatin reagent solution in a 50-mL porcelain crucible. No bluish-green color should be present.

Solution A. Dissolve 0.25 g of isatin in 25 mL of sulfuric acid.

Solution B. Dissolve 0.25 g of ferric chloride in 1 mL of water. Dilute to 50 mL with sulfuric acid and allow the evolution of gas to cease.

To 2.5 mL of solution A add 5.0 mL of solution B and dilute to 100 mL with sulfuric acid.

SULFUR COMPOUNDS. Place 30 mL of 0.5 N potassium hydroxide in methanol in a conical flask, add 5.2 g (6.0 mL) of the sample, and boil the mixture gently for 30 min under a reflux condenser, avoiding the use of a rubber stopper or connection. Detach the condenser, dilute with 50 mL of water, and heat on a hot plate ($\approx$100 °C) until the benzene and methanol are evaporated. Add 50 mL of bromine water and heat for 15 min longer. Transfer the solution to a beaker, neutralize with dilute hydrochloric acid (1 + 3), add an excess of 1 mL of the acid, and concentrate to about 50 mL. Filter if necessary, heat the filtrate to boiling, add 5 mL of barium chloride reagent solution, digest in a covered beaker on a hot plate ($\approx$100 °C) for 2 hours, and allow to stand overnight. If a precipitate is formed, filter, wash thoroughly, ignite, and weigh. Correct for the weight obtained in a complete blank test.

WATER. (Page 56, Method 2). Use 100 μL (88 mg) of the sample.

ULTRAVIOLET SPECTROPHOTOMETRY. Use the procedure on page 73 to determine the absorbance.

1,3-Benzenediol

Resorcinol

1,3-$C_6H_4(OH)_2$	**Formula Wt 110.11**

CAS Number 108–46–3

REQUIREMENTS

Assay . . . . 99.0–100.5% $C_6H_4(OH)_2$
Melting point . . . . 110.0–112.0 °C

MAXIMUM ALLOWABLE

Insoluble matter. 0.005%
Residue after ignition 0.01%
Titrable acid . . . . .0.004 meq/g

TESTS

ASSAY. (By bromination). Weigh accurately 1.5 g, transfer to a 500-mL volumetric flask, dissolve in water, and dilute with water to volume. Stopper and mix thoroughly. Place a 25-mL aliquot in a 500-mL iodine flask. Add 50.0 mL of 0.1 N bromine, 50 mL of water, and 5 mL of hydrochloric acid. Stopper, shake, and allow to stand in the dark for 15 min, shaking occasionally. Add 5 mL of potassium iodide reagent solution (16.5 g/100 mL), taking care to prevent loss of bromine. Stopper and shake. Rinse the stopper, neck, and inner walls of the flask with about 25 mL of water. Titrate the liberated iodine with 0.1 N sodium thiosulfate, adding 3 mL of starch indicator solution near the end of the titration. Correct for a complete blank. One milliliter of 0.1 N bromine corresponds to 0.001835 g of $C_6H_4(OH)_2$

MELTING POINT. (Page 22).

INSOLUBLE MATTER. (Page 15). Use 20 g dissolved in 150 mL of water.

RESIDUE AFTER IGNITION. (Page 16). Ignite 10 g in a tared crucible or dish.

TITRABLE ACID. Dissolve 1 g in 50 mL of water, add 0.15 mL of methyl orange indicator solution, and neutralize with 0.02 N sodium hydroxide. Add 5.0 g of sample to the neutralized solution, and titrate with 0.02 N sodium hydroxide. Not more than 1.0 mL of 0.02 N sodium hydroxide should be consumed in the titration.

Benzoic Acid

C₆H₅COOH $\quad$ **Formula Wt 122.12**

CAS Number 65–85–0

REQUIREMENTS

Assay. . . . .≥99.5% C₆H₅COOH
Freezing point . . . . 122–123 °C

MAXIMUM ALLOWABLE

Residue after ignition. 0.005%
Insoluble in methanol 0.005%
Chlorine compounds (as Cl) 0.005%
Sulfur compounds (as S) 0.002%
Heavy metals (as Pb) . . . . 5 ppm
Substances reducing permanganate. . . . . Passes test

TESTS

ASSAY. (By acid–base titrimetry). Dissolve about 0.5 g of the sample, accurately weighed, in 25 mL of 50% alcohol previously neutralized with 0.1 N sodium hydroxide, add 0.15 mL of phenolphthalein indicator solution, and titrate with 0.1 N sodium hydroxide. One milliliter of 0.1 N sodium hydroxide corresponds to 0.01221 g of C₆H₅COOH.

FREEZING POINT. Place 12–15 g in test tube (20 × 150 mm), and heat gently to melt the benzoic acid. Insert an accurate thermometer in the test tube so that the bulb of the thermometer is centrally located, and heat gently to about 130 °C. Place the tube containing the melted sample in a larger test tube (38 × 200 mm) and keep it centered by means of corks. Allow the sample to cool slowly. If crystallization does not start when the temperature is about 120 °C, stir gently with the thermometer to start freezing, and read the thermometer every half minute. The temperature that remains constant for 2–3 min is the freezing point.

RESIDUE AFTER IGNITION. (Page 16). Heat 20 g in a platinum dish at a temperature sufficient to volatilize the acid slowly, but do not char or ignite it. When nearly all the acid has been volatilized, cool, and continue as described.

INSOLUBLE IN METHANOL. Dissolve 20 g in 200 mL of methanol and digest under complete reflux for 30 min. Filter through a tared filtering crucible, wash thoroughly with methanol, and dry at 105 °C.

CHLORINE COMPOUNDS. (Page 27). Mix 1.0 g with 0.5 g of sodium carbonate and add 10–15 mL of water. Evaporate on a hot plate (≈100 °C) and ignite until the mass is thoroughly charred, avoiding an unduly high temperature. Extract the fusion with 20 mL of water and 5.5 mL of nitric acid. Filter through a chloride-free filter, wash with two 10-mL portions of water, and dilute with water to 50 mL. Dilute 10 mL with water to 20 mL.

SULFUR COMPOUNDS. Mix 2.0 g with 1.0 g of sodium carbonate and add in small portions 15 mL of water. Evaporate and thoroughly ignite, protected from sulfur in the flame. Treat the residue with 20 mL of water and 2 mL of 30% hydrogen peroxide and heat on a hot plate (≈100 °C) for 15 min. Add 5 mL of hydrochloric acid and evaporate to dryness on the hot plate. Dissolve the residue in 10 mL of water, filter, wash with two 5-mL portions of water, and dilute with water to 25 mL. Add to this solution 0.5 mL of 1 N hydrochloric acid and 2 mL of barium chloride reagent solution. Any turbidity should not exceed that in a standard prepared as follows: treat 1 g of sodium carbonate with 2 mL of 30% hydrogen peroxide and 5 mL of hydrochloric acid and evaporate to dryness on a hot plate (≈100 °C). Dissolve the residue and 0.12 mg of sulfate ion (SO_4) in sufficient water to make 25 mL and add 0.5 mL of 1 N hydrochloric acid and 2 mL of barium chloride reagent solution. Compare 10 min after adding the barium chloride to the sample and standard solutions.

HEAVY METALS. (Page 28, Method 1). Volatilize 4.0 g over a low flame, cool, and add 5 mL of nitric acid and about 10 mg of sodium carbonate. Evaporate to dryness on a hot plate (≈100 °C), dissolve the residue in 20 mL of water, and dilute with water to 25 mL. For the control, add 0.02 mg of lead to 5 mL of nitric acid and about 10 mg of sodium carbonate, evaporate to dryness on a hot plate (≈100 °C), dissolve the residue in 20 mL of water, and dilute with water to 25 mL.

SUBSTANCES REDUCING PERMANGANATE. Dissolve 1.0 g in 100 mL of water containing 1 mL of sulfuric acid. Heat to 85 °C (steam bath temperature) and add 0.50 mL of 0.1 N potassium permanganate solution. The pink color should not be entirely discharged in 5 min.

Benzoyl Chloride

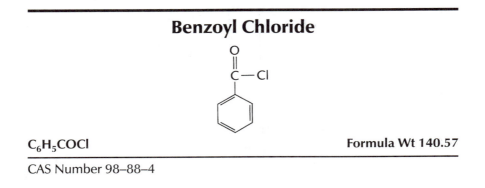

C_6H_5COCl Formula Wt 140.57

CAS Number 98–88–4

REQUIREMENTS

Assay . . . . .98.0–100.5% C_6H_5COCl
Freezing point . . . . –2.0 to 0.0 °C

MAXIMUM ALLOWABLE

Residue after ignition 0.005%
Phosphorus compounds (as P) 0.002%
Heavy metals (as Pb) 0.001%
Iron (Fe) 0.001%

TESTS

ASSAY. (By acid–base titrimetry). Weigh accurately about 2 mL in a glass-stoppered flask, add 50.0 mL of 1 N sodium hydroxide, and stopper the flask. Allow to stand, with frequent agitation, until the sample has dissolved. Add 0.15 mL of phenolphthalein indicator solution and titrate the excess sodium hydroxide with 1 N hydrochloric acid. One milliliter of 1 N sodium hydroxide corresponds to 0.07028 g of C_6H_5COCl.

FREEZING POINT. Place 15 mL in a test tube (20 × 150 mm) in which is centered an accurate thermometer. The sample tube is centered by corks in an outer tube about 38 × 200 mm. Cool the whole apparatus, without stirring, in a bath of shaved or crushed ice mixed with sufficient salt and water to provide a temperature of –5 °C. When the temperature is about –2 °C, stir to start the freezing and read the thermometer every 30 s. The temperature that remains constant for 1–2 min is the freezing point.

RESIDUE AFTER IGNITION. (Page 16). Evaporate 40 g (33 mL) to dryness in a tared dish, adding sufficient sulfuric acid to wet the sample. Add to the carbonized mass 2 mL of nitric acid and 0.25 mL of sulfuric acid, and heat cautiously until white fumes of sulfur trioxide are no longer evolved. Ignite at 500–600 °C until the carbon is burned off. Retain the residue to prepare sample solution A.

PHOSPHORUS COMPOUNDS. To 1 g (0.85 mL) add 7 mL of water and 3 mL of nitric acid. Heat the mixture to boiling, boil for 2 min, cool, dilute with water to 20 mL, and filter. To 10 mL of the filtrate add 5 mL of 10% sulfuric acid reagent solution, and evaporate to fumes of sulfur trioxide. Cool, dilute with water to 25 mL, add 1 mL of ammonium molybdate reagent solution and 1 mL of 4-(methylamino)phenol sulfate reagent solution, and allow to stand for 2 h at room temperature. Any blue color should not exceed that produced by 0.03 mg of phosphate ion (PO_4) in an equal volume of solution containing the quantities of reagents used in the test.

Sample Solution A for the Determination of Heavy Metals and Iron.

Add 10 mL of 10% hydrochloric acid reagent solution to the residue obtained in the test for residue after ignition and evaporate to dryness on a hot plate (≈100 °C). Dissolve the residue in 0.5 mL of hydrochloric acid and 25 mL of hot water, filter if necessary, and dilute with water to 200 mL (1 mL = 0.2 g).

HEAVY METALS. (Page 28, Method 1). Use 10 mL of sample solution A (2-g sample).

IRON. (Page 30, Method 1). Use 5.0 mL of sample solution A (1-g sample).

Benzyl Alcohol
Benzenemethanol

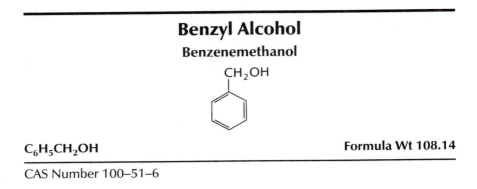

$C_6H_5CH_2OH$ **Formula Wt 108.14**

CAS Number 100–51–6

REQUIREMENTS

Assay . . . . ≥99.0% $C_6H_5CH_2OH$

MAXIMUM ALLOWABLE

Color (APHA) 20
Residue after ignition 0.005%
Acetophenone ($C_6H_5COCH_3$) 0.02%
Benzaldehyde (C_6H_5CHO) 0.01%

TESTS

ASSAY, ACETOPHENONE, AND BENZALDEHYDE. Analyze the sample by gas chromatography using the parameters cited on page 70. The following specific conditions are also required.

Column: Type III (polyethylene glycol)

Measure the area under all peaks and calculate the area percent for benzyl alcohol, acetophenone, and benzaldehyde.

COLOR (APHA). (Page 19).

RESIDUE AFTER IGNITION. (Page 16). Use 25 g (24 mL).

Bismuth Nitrate Pentahydrate
Bismuth(III) Nitrate Pentahydrate

$Bi(NO_3)_3 \cdot 5H_2O$ Formula Wt 485.07

CAS Number 10035–06–0

REQUIREMENTS

Assay. . . . . $\geq$98.0% $Bi(NO_3)_3 \cdot 5H_2O$

MAXIMUM ALLOWABLE

Insoluble matter. 0.005%
Chloride (Cl) 0.001%
Sulfate (SO_4) 0.005%
Arsenic (As) 0.001%
Calcium (Ca) 0.005%
Copper (Cu). 0.002%
Iron (Fe). 0.001%
Lead (Pb) 0.002%
Potassium (K). 0.01%
Sodium (Na) 0.02%
Silver (Ag) 0.001%

TESTS

ASSAY. (By complexometric titration for bismuth). Weigh accurately 1.8 g and transfer to a 400-mL beaker. Dissolve in 5 mL of nitric acid and add about 250 mL of water. Add about 50 mg of ascorbic acid and 50 mg of xylenol orange indicator mixture. Titrate with standard 0.1 M EDTA to canary yellow. (The solution color changes from purple to salmon pink just before the canary yellow end point). One milliliter of 0.1 M EDTA corresponds to 0.04851 g of $Bi(NO_3)_3 \cdot 5H_2O$.

INSOLUBLE MATTER. (Page 15). Use 20 g dissolved in 100 mL of dilute nitric acid (1 + 4).

CHLORIDE. (Page 27). Use 1.0 g.

SULFATE. Dissolve 3.0 g in 5 mL of warm hydrochloric acid, dilute with water to 100 mL, and nearly neutralize with 10% ammonium hydroxide. Filter, and wash the residue and the filter with hot water to 150 mL. Transfer 100.0 mL to an evaporating dish, evaporate to about 15 mL, add 0.5 g of sodium carbonate, evaporate to dryness, and ignite gently. Cool, add 15 mL of hot water, neutralize with 10% hydrochloric acid, filter, and wash with water to 20 mL. Add to the combined filtrate and washings 1 mL of 1 N hydrochloric acid and 2 mL of barium chloride reagent solution, and allow to stand for 10 min. Any turbidity should not exceed that of a standard prepared as follows: To 0.5 g of sodium carbonate add 0.1 mg of sulfate ion (SO_4) and 2 mL of hydrochloric acid, and evaporate to dryness on a hot plate ($\approx$100 °C). Dissolve the residue in 20 mL of water, and add 1 mL of 1 N hydrochloric acid and 2 mL of barium chloride reagent solution.

ARSENIC. (Page 25). Mix 0.5 g with 5 mL of water, and cautiously add 2 mL of sulfuric acid. Heat the mixture until white fumes of sulfur trioxide are evolved copiously. Cool, cautiously add 10 mL of water, and evaporate to strong fumes of sulfur trioxide. Cool, cautiously add a second 10 mL of water, and again evaporate to strong fumes of sulfur trioxide. Cool, and cautiously wash the solution into a generator flask with sufficient water to make the volume 35 mL. The solution should show not more than 0.005 mg of arsenic (As).

CALCIUM, COPPER, IRON, LEAD, POTASSIUM, AND SODIUM. (By flame AAS, page 39).

> **Sample Stock Solution.** Dissolve 10.0 g of sample in 20 mL of nitric acid and 20 mL of water. Transfer to a 100-mL volumetric flask and dilute to the mark with water (1 mL = 0.1 g).

Element	Wavelength (nm)	Sample Wt (g)	Standard Added (mg)	Flame Type*	Background Correction
Ca	422.7	2.0	0.05; 0.10	N/A	No
Cu	324.7	2.0	0.02; 0.04	A/A	Yes
Fe	248.3	2.0	0.02; 0.04	A/A	Yes
Pb	217.0	2.0	0.04; 0.08	A/A	Yes
K	766.5	0.10	0.01; 0.02	A/A	No
Na	589.0	0.10	0.01; 0.02	A/A	No

*N/A is nitrous oxide/acetylene; A/A is air/acetylene.

SILVER. Dissolve 1.0 g in a mixture of 5 mL of nitric acid and 15 mL of water, and add 0.05 mL of hydrochloric acid. Any turbidity should not exceed that of a standard containing 0.01 mg of silver.

Boric Acid

H_3BO_3   **Formula Wt 61.83**

CAS Number 10043–35–3

REQUIREMENTS

Assay . . . . ≥99.5% H_3BO_3

MAXIMUM ALLOWABLE

Insoluble in methanol. 0.005%
Nonvolatile with methanol 0.05%

Chloride (Cl) 0.001%
Phosphate (PO$_4$) 0.001%
Sulfate (SO$_4$) 0.01%
Calcium (Ca) 0.005%
Heavy metals (as Pb) 0.001%
Iron (Fe). 0.001%

TESTS

ASSAY. (By acid–base titrimetry). Weigh, to the nearest 0.1 mg, 2.5 g of sample and transfer to a 250-mL glass-stoppered flask. Dissolve in a mixture of 50 mL of warm water and 75 mL of glycerol, previously neutralized to a pH of 9.0 with 0.1 N sodium hydroxide. Titrate with 1 N sodium hydroxide to a pH of 9.0. One milliliter of 1 N sodium hydroxide corresponds to 0.06183 g of H$_3$BO$_3$.

INSOLUBLE IN METHANOL. Heat 20 g with 200 mL of methanol under complete reflux until the acid is dissolved, then reflux for 30 min. Filter through a tared filtering crucible, wash thoroughly with hot methanol, and dry at 105 °C.

NONVOLATILE WITH METHANOL. To 2.0 g of the sample in a platinum dish, add 25 mL of methanol and 0.5 mL of hydrochloric acid and evaporate to dryness. Repeat the evaporation twice, each with 15 mL of methanol and 0.3 mL of hydrochloric acid. Add to the residue 0.10 mL of sulfuric acid and ignite at 800 ± 25 8C for 15 min.

CHLORIDE. (Page 27). Use 1.0 g dissolved in 20 mL of warm water.

Sample Solution A for the Determination of Phosphate, Sulfate, Heavy Metals, and Iron. To 10 g add 10 mg of sodium carbonate, 100 mL of methanol, and 5 mL of hydrochloric acid. Digest in a covered beaker on a hot plate (≈100 °C) to effect dissolution, then uncover and evaporate to dryness. Add 75 mL of methanol and 5 mL of hydrochloric acid and repeat the digestion and evaporation. Dissolve the residue in 5 mL of 1 N acetic acid and digest in a covered beaker on a hot plate (≈100 °C) for 15 min. Filter and dilute with water to 100 mL (1 mL = 0.1 g).

PHOSPHATE. (Page 32, Method 1). Add 0.5 mL of nitric acid to 20 mL of sample solution A (2-g sample), and evaporate to dryness. Dissolve the residue in 25 mL of approximately 0.5 N sulfuric acid and continue as described.

SULFATE. (Page 33, Method 1). Use 5.0 mL of sample solution A (0.5-g sample).

CALCIUM. (By flame AAS, page 39).

Sample Stock Solution. To 20 g add 20 mg of sodium carbonate, 200 mL of methanol, and 10 mL of hydrochloric acid. Digest in a covered beaker on a hot plate (≈100 °C) to effect solution, then uncover and evaporate to dryness. Add 75 mL of methanol and 5 mL of hydrochloric acid, and repeat the digestion and evaporation. Dissolve the residue in 5 mL of 1 N acetic acid, and digest in a covered beaker on a hot plate (≈100 °C) for 15 min. Filter and dilute with water to volume in a 200-mL volumetric flask (1 mL = 0.1 g).

Element	Wavelength (nm)	Sample Wt (g)	Standard Added (mg)	Flame Type*	Background Correction
Ca	422.7	1.0	0.05; 0.10	N/A	No

*N/A is nitrous oxide/acetylene.

HEAVY METALS. (Page 28, Method 1). Dilute 20 mL of sample solution A (2-g sample) with water to 25 mL.

IRON. (Page 30, Method 1). Use 10 mL of sample solution A (1-g sample).

Bromcresol Green

3′,3″,5′,5″-Tetrabromo-*m*-cresolsulfonphthalein
4,4′-(3*H*-2,1-Benzoxathiol-3-ylidene)bis-[(2,6-dibromo-3-methyl)phenol] *S,S*-Dioxide

$C_{21}H_{14}Br_4O_5S$ **Formula Wt 698.02**

CAS Numbers 76–60–8; 62625–32–5 (sodium salt)

NOTE. This specification applies both to the sultone form and to the salt form of this indicator.

REQUIREMENTS

Clarity of solution. . . . . Passes test
Visual transition interval. . . . . .pH 3.8 (yellow) to pH 5.4 (blue)

TESTS

CLARITY OF SOLUTION. If the indicator is the acid form, dissolve 0.1 g in 100 mL of alcohol. If the indicator is a salt form, dissolve 0.1 g in 100 mL of water. Not more than a faint trace of turbidity or insoluble matter should remain. Reserve the solution for the test for visual transition interval.

VISUAL TRANSITION INTERVAL. Dissolve 1 g of potassium chloride in 100 mL of water. Adjust the pH of the solution to 3.8 (using a pH meter as described on page 43) with 0.01 N hydrochloric acid. Add 0.10–0.30 mL of the 0.1% solution reserved from the test for clarity of solution. The color of the solution should be yellow. Titrate the solution with 0.01 N sodium hydroxide to a pH of 4.5 (using the pH meter). The color of the solution should be green. Continue the titration to a pH of 5.4. The color of the solution should be blue. Not more than 2.0 mL of 0.01 N sodium hydroxide should be consumed in the entire titration.

Bromine

Br$_2$ **Formula Wt 159.808**

CAS Number 7726–95–6

REQUIREMENTS

Assay . . . . ≥99.5% Br$_2$

MAXIMUM ALLOWABLE

Residue after evaporation 0.005%
Chlorine (Cl) 0.05%
Iodine (I) 0.001%
Organic bromine compounds . . . . Passes test
Sulfur compounds (as S) 0.001%
Heavy metals (as Pb) . . . . 2 ppm
Nickel (Ni) . . . . 5 ppm

TESTS

ASSAY. (By iodometric titration). Weigh a 250-mL volumetric flask containing 15 g of potassium iodide in 45 mL of water. Add 1 mL of sample. Stopper promptly, mix by shaking, and reweigh. Dilute to volume with water and mix thoroughly. Place a 25.0-mL aliquot of this solution in a conical flask and titrate the liberated iodine with 0.1 N sodium thiosulfate, adding 3 mL of starch indicator solution near the end of the titration. One milliliter of 0.1 N sodium thiosulfate corresponds to 0.007990 g of Br$_2$.

RESIDUE AFTER EVAPORATION. (Page 16). In a fume hood, evaporate 31 g (10 mL) to dryness from a tared container on a hot plate (≈100 °C), and dry the residue at 105 °C for 30 min. Reserve the residue for the test for heavy metals.

CHLORINE. To each of two 250-mL wide-mouth conical flasks add 0.5 mL of dilute sulfuric acid (1 + 4), 5 mL of potassium bromide solution (1.50 g of KBr per liter), and 35 mL of water. For the sample add 3 g (1.0 mL) of the bromine to one of the flasks. For the standard, add 1.5 mg of chloride ion (3.2 mg of KCl) to the other. Digest each on a hot plate ($\approx$100 °C) until the sample solution is colorless, to each add 4 mL of potassium peroxydisulfate solution (1.0 g per 100 mL), and wash down the sides of the flasks with a little water. Digest again on a hot plate ($\approx$100 °C) for 30–40 min. Cool and dilute each with water to 100 mL. Transfer 1.0 mL of each solution into separate 50-mL beakers and add 22 mL of water, 1 mL of nitric acid, and 1 mL of silver nitrate reagent solution. Any turbidity in the solution of the sample should not exceed that in the standard.

IODINE. To each of two 250-mL wide-mouth conical flasks, add 1.0 mL of 10% potassium chloride solution, 50 mL of water, and a few silicon carbide boiling chips. For the sample add 10.5 g (3.5 mL) of the bromine to one of the flasks. For the control, add 0.3 mL of the bromine and 0.1 mg of iodide ion (0.13 mg of KI) to the other flask. Boil cautiously in a hood to remove the excess bromine, adding water as required to maintain a volume of not less than 50 mL. When no trace of yellow color remains in either flask, cool, and add 5 mL of 10% potassium iodide solution and 5 mL of dilute sulfuric acid (1 + 4) to each. Any color in the solution of the sample should not exceed that in the control.

ORGANIC BROMINE COMPOUNDS. Use matching infrared cells consisting of rock salt windows separated by 1/2-inch Teflon spacers. Fill one of the cells with bromine from a pipet inserted through an opening drilled in the plastic. Scan the sample from 2.5 to 15 μm, using the empty matching cell for reference. The absorbance of the sample should not exceed 0.10 in the region from 2.5 to 12 μm (except for possible water bands at 2.6 to 2.9 μm and 6.3 μm) or 0.20 in the region from 12 to 15 μm.

SULFUR COMPOUNDS. To 8 g (2.6 mL) add 5 mL of water and evaporate to dryness on a hot plate ($\approx$100 °C). To the residue add 5 mL of dilute hydrochloric acid (1 + 19), filter if necessary, dilute to 50 mL with water, and add 1 mL of barium chloride reagent solution to 10 mL of the solution. Any turbidity should not exceed that produced by 0.05 mg of sulfate ion (SO_4) in an equal volume of solution containing the quantities of reagents used in the test. Compare 10 min after adding the barium chloride to the sample and standard solutions. Reserve the remaining solution for the test for nickel.

HEAVY METALS. For the sample use the residue reserved from the test for residue after evaporation. For the standard, use a solution containing 0.06 mg of lead (Pb). To each add 3 mL of nitric acid, 1 mL of water, and about 10 mg of sodium carbonate. Evaporate to dryness on a hot plate ($\approx$100 °C), dissolve the residues in about 20 mL of water, and dilute with water to 25 mL. Adjust the pH of the standard and sample solutions to between 3 and 4 (using a pH meter) with 1 N acetic acid or ammonium hydroxide (10% NH₃), dilute with water to 40 mL, and mix. Add 10 mL of freshly prepared hydrogen sulfide water to each and mix. Any color in the solution of the sample should not exceed that in the standard.

NICKEL. To 25 mL of the solution reserved from the test for sulfur compounds, add 5 mL of bromine water. Stir and add ammonium hydroxide (1 + 1) until the bromine color is discharged. Add 5 mL of 1% dimethylglyoxime solution in ethyl alcohol and 5 mL of 10% sodium hydroxide reagent solution. Any red color should not exceed that produced by 0.02 mg of nickel (Ni) in an equal volume of solution containing the quantities of reagents used in the test. Compare 10 min after adding the dimethylglyoxime to the sample and standard solutions.

Bromphenol Blue

3′,3″,5′,5″-Tetrabromophenolsulfonphthalein
4,4′-(3H-2,1-Benzoxathiol-3-ylidene)bis-[2,6-dibromophenol] S,S-Dioxide

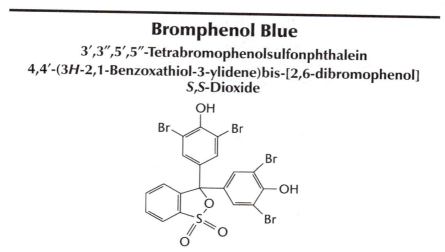

$C_{19}H_{10}Br_4O_5S$ **Formula Wt 669.97**

CAS Numbers 115–39–9; 62625–28–9 (sodium salt)

NOTE. This specification applies both to the sultone form and to the salt form of this indicator.

REQUIREMENTS

Clarity of solution . . . . Passes test
Visual transition interval. . . . . pH 3.0 (yellow) to pH 4.6 (blue)

TESTS

CLARITY OF SOLUTION. If the indicator is the acid form, dissolve 0.1 g in 100 mL of alcohol. If the indicator is the salt form, dissolve 0.1 g in 100 mL of water.

Not more than a faint trace of turbidity or insoluble matter should remain. Reserve the solution for the test for visual transition interval.

VISUAL TRANSITION INTERVAL. Dissolve 1 g of potassium chloride in 100 mL of water. Adjust the pH of the solution to 3.0 (using a pH meter as described on page 43) with 0.01 N hydrochloric acid. Add 0.10–0.30 mL of the 0.1% solution of the indicator reserved from the test for clarity of solution. The color of the solution should be yellow with a slight greenish hue. Titrate the solution with 0.01 N sodium hydroxide to a pH of 3.4 (using the pH meter). The color of the solution should be green. Continue the titration to pH 4.6. The color of the solution should be blue. Not more than 13.0 mL of 0.01 N sodium hydroxide should be consumed in the entire titration.

Bromthymol Blue

3′,3″-Dibromothymolsulfonphthalein
4,4′-(3H-2,1-Benzoxathiol-3-ylidene)bis-[2-bromo-3-methyl-6-(1-methylethyl)phenol] S,S-Dioxide

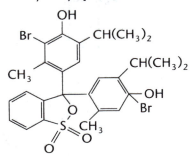

$C_{27}H_{28}Br_2O_5S$ **Formula Wt 624.39**

CAS Numbers 76–59–5; 34722–90–2 (sodium salt)

NOTE. This specification applies both to the sultone form and to the salt form of this indicator.

REQUIREMENTS

Clarity of solution. . . . . Passes test
Visual transition interval. . . . . .pH 6.0 (yellow) to pH 7.6 (blue)

TESTS

CLARITY OF SOLUTION. If the indicator is the acid form, dissolve 0.1 g in 100 mL of alcohol. If the indicator is a salt form, dissolve 0.1 g in 100 mL of water. Not more than a faint trace of turbidity or insoluble matter should remain. Reserve the solution for the test for visual transition interval.

VISUAL TRANSITION INTERVAL. Dissolve 1 g of potassium chloride in 100 mL of water. Adjust the pH of the solution to 6.0 (using a pH meter as described on page 43) with 0.01 N hydrochloric acid or sodium hydroxide. Add 0.10–0.30 mL of the 0.1% solution reserved from the test for clarity of solution. The color of the solution should be yellow, with not more than a faint trace of green color. Titrate the solution with 0.01 N sodium hydroxide to a pH of 6.7 (using a pH meter). The color of the solution should be green. Continue the titration to pH 7.6. The color of the solution should be blue. Not more than 0.30 mL of 0.01 N sodium hydroxide should be consumed in the entire titration.

Brucine Sulfate Heptahydrate
2,3-Dimethoxystrychnidin-10-one Sulfate Heptahydrate

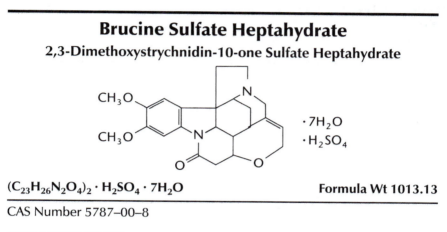

$(C_{23}H_{26}N_2O_4)_2 \cdot H_2SO_4 \cdot 7H_2O$

Formula Wt 1013.13

CAS Number 5787–00–8

REQUIREMENTS

Sensitivity to nitrate . . . . Passes test

MAXIMUM ALLOWABLE

Clarity of solution . . . . Passes test
Loss on drying at 105 °C 13.0%
Residue after ignition 0.1%

TESTS

SENSITIVITY TO NITRATE.

Test Solutions 1, 2, 3, and 4. Prepare a sample solution by dissolving 0.6 g in dilute sulfuric acid (2 + 1), as described on page 30. Place 50 mL of this sample solution in each of four test tubes, and add nitrate ion (NO_3) to each as follows: No. 1, none; No. 2, 0.01 mg; No. 3, 0.02 mg; and No. 4, 0.03 mg.

Heat the four test tubes in a preheated (boiling) water bath for 10 min. Cool rapidly in an ice bath to room temperature. Set a spectrophotometer at 410 nm and, using 1-cm cells, adjust the instrument to read 0 absorbance with solution 1 in the light path. Determine the absorbance for solutions 2, 3, and 4 at this adjustment, using similar cells. The absorbances for solutions 2, 3, and 4 should not be less than 0.025, 0.050, and 0.075, respectively, and the plot of absorbances versus nitrate concentrations should be linear.

CLARITY OF SOLUTION. Dissolve 1 g in 100 mL of water, and digest in a covered beaker on a hot plate ($\approx$100 °C) for 1 h. The solution should be as colorless and clear as an equal volume of water.

LOSS ON DRYING AT 105 °C. Weigh accurately about 1 g and dry on a tared dish for 6 h at 105 °C.

RESIDUE AFTER IGNITION. (Page 16). Ignite 1.0 g. Omit the addition of sulfuric acid.

2-Butanone

Methyl Ethyl Ketone

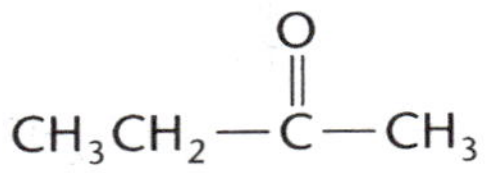

$$CH_3CH_2-\overset{\overset{\displaystyle O}{\|}}{C}-CH_3$$

CH₃COCH₂CH₃   **Formula Wt 72.11**

CAS Number 78–93–3

REQUIREMENTS

Assay . . . .$\geq$99.0% $CH_3COCH_2CH_3$

MAXIMUM ALLOWABLE

Color (APHA) 15
Residue after evaporation. 0.0025%
Titrable acid . . . . 0.0005 meq/g
Water (H₂O). 0.20%

TESTS

ASSAY. Analyze the sample by gas chromatography using the general parameters cited on page 70. The following specific conditions are also required.

Column: Type I, methyl silicone

Measure the area under all peaks and calculate the 2-butanone content in area percent. Correct for water content.

COLOR (APHA). (Page 19).

RESIDUE AFTER EVAPORATION. (Page 16). Use 80 g (100 mL) in a tared beaker on a hot plate (≈100 °C). When the liquid has evaporated, complete the drying at 105 °C to constant weight after cooling in a desiccator.

TITRABLE ACID. Mix 40 g (50 mL) with 50 mL of carbon dioxide-free water, and add 0.15 mL of phenolphthalein indicator solution. Not more than 2.0 mL of 0.01 N sodium hydroxide should be required to produce a pink color.

WATER. (Page 55, Method 1). Use 25 mL (20 g) of the sample and a methanol-free system to prevent ketal formation with liberation of water.

---

n-Butyl Acetate

$CH_3COO(CH_2)_3CH_3$ **Formula Wt 116.16**

CAS Number 123–86–4

REQUIREMENTS

Assay . . . . ≥99.5% $CH_3COO(CH_2)_3CH_3$

MAXIMUM ALLOWABLE

Color (APHA) 10
Residue after evaporation 0.001%
Titrable acid . . . . 0.0016 meq/g
Substances darkened by sulfuric acid . . . . Passes test
Water (H_2O) 0.1%
n-Butyl alcohol (C_4H_9OH) 0.2%

TESTS

Assay and *n*-Butyl Alcohol. Analyze the sample by gas chromatography using the general parameters cited on page 70. The following specific conditions are also required.

Column: Type I, methyl silicone

Measure the area under all peaks and calculate the area percent for *n*-butyl acetate and *n*-butyl alcohol. Correct for water content.

COLOR (APHA). (Page 19).

RESIDUE AFTER EVAPORATION. (Page 16). Evaporate 100 g (114 mL) to dryness in a tared dish on a hot plate (≈100 °C) and dry the residue at 105 °C for 30 min.

TITRABLE ACID. Neutralize 25 mL of reagent alcohol, to which 0.05 mL of bromthymol blue indicator solution has been added, with 0.01 N sodium hydroxide to a blue-green color. Add 25 g (28 mL) of the sample, mix, and titrate with 0.01 N sodium hydroxide to the same color. Not more than 4.0 mL of 0.01 N sodium hydroxide should be required.

SUBSTANCES DARKENED BY SULFURIC ACID. Cool 5 mL of the sample to below 10 °C. Cautiously add, dropwise, 5 mL of 95.0 ± 0.5% sulfuric acid. The color of the solution should not exceed that of a color standard composed of 9 volumes of water plus one volume of a color standard containing 3.25 g of $CoCl_2 \cdot 6H_2O$, 2.25 g of $FeCl_3 \cdot 6H_2O$, and 3 mL of hydrochloric acid per 100 mL of solution.

WATER. (Page 55, Method 1). Use 25 g (28 mL) of the sample.

Butyl Alcohol
1-Butanol

$CH_3(CH_2)_2CH_2OH$ **Formula Wt 74.12**

CAS Number 71–36–3

REQUIREMENTS

Assay . . . .≥99.4% $CH_3(CH_2)_2CH_2OH$

MAXIMUM ALLOWABLE

Color (APHA) 10
Residue after evaporation. 0.005%
Titrable acid . . . . .0.0008 meq/g
Carbonyl compounds (as butyraldehyde). 0.01%
Butyl ether 0.2%
Water (H_2O). 0.1%

TESTS

ASSAY AND BUTYL ETHER. Analyze the sample by gas chromatography using the general parameters cited on page 70. The following specific conditions are also required.

Column: Type I, methyl silicone

Measure the areas under all peaks and calculate the butyl alcohol and butyl ether content in area percent. Correct for water content.

COLOR (APHA). (Page 19).

RESIDUE AFTER EVAPORATION. (Page 16). Evaporate 20 g (25 mL) to dryness in a tared platinum dish on a hot plate ($\approx$100 °C) and dry the residue at 105 °C for 30 min.

TITRABLE ACID. To 60 g (74 mL) in a glass-stoppered flask add 0.10 mL of phenolphthalein indicator solution, and titrate with 0.01 N alcoholic potassium hydroxide to a pink end point that persists for at least 15 s. Not more than 5.0 mL of 0.01 N potassium hydroxide should be consumed.

CARBONYL COMPOUNDS. (By differential pulse polarography, page 50). Use 1.0 mL (0.80 g) sample and 0.08 mg of butaraldehyde.

WATER. (Page 55, Method 1). Use 30 mL (24 g) of the sample.

---

tert-Butyl Alcohol
2-Methyl-2-propanol

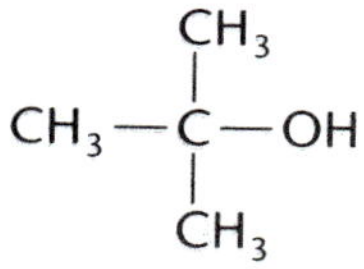

$$CH_3-\underset{\underset{CH_3}{|}}{\overset{\overset{CH_3}{|}}{C}}-OH$$

(CH$_3$)$_3$COH   **Formula Wt 74.12**

CAS Number 75–65–0

NOTE. Since the melting point of this reagent is close to 25 °C, the sample may be warmed to about 30 °C for testing.

REQUIREMENTS

Assay. . . . . $\geq$99.0% (CH$_3$)$_3$COH

MAXIMUM ALLOWABLE

Color (APHA). 20
Residue after evaporation 0.003%
Titrable acid. . . . . 0.001 meq/g
Carbonyl compounds (as formaldehyde) 0.01%
Water (H$_2$O) 0.1%

TESTS

ASSAY. Analyze the sample by gas chromatography using the parameters cited on page 70. The following specific conditions are also required.

Column: Type I, methyl silicone

Measure the areas under all peaks and calculate the *tert*-butyl alcohol content in area percent. Correct for water content.

COLOR (APHA). (Page 19).

RESIDUE AFTER EVAPORATION. (Page 16). Evaporate 60 g (77 mL) of sample to dryness in a tared evaporating dish on a hot plate (≈100 °C) and dry the residue at 105 °C for 30 min.

TITRABLE ACID. To 35 g (45 mL) in a glass-stoppered Erlenmeyer flask, add 0.15 mL of phenolphthalein indicator solution and shake for 1 min. No pink color should appear. Titrate the solution with 0.01 N alcoholic potassium hydroxide to a faint pink color that persists for at least 15 s. Not more than 3.5 mL should be required.

CARBONYL COMPOUNDS. (By differential pulse polarography, page 50). Use 1.3 mL (1.0 g) sample. For the standard, use 0.10 mg of formaldehyde.

WATER. (Page 55, Method 1). Use 45 mL (35 g) of the sample.

Cadmium Chloride, Anhydrous

$CdCl_2$ **Formula Wt 183.32**

CAS Number 10108–64–2

REQUIREMENTS

Assay . . . . ≥99.0% $CdCl_2$

MAXIMUM ALLOWABLE

Insoluble matter. 0.01%
Nitrate and nitrite (as NO_3) 0.003%
Sulfate (SO_4) 0.01%
Ammonium (NH_4) 0.01%
Calcium (Ca) 0.01%
Copper (Cu). 0.001%
Lead (Pb) 0.005%
Potassium (K). 0.02%
Sodium (Na). 0.05%
Zinc (Zn) 0.05%
Iron (Fe). 0.001%

TESTS

ASSAY. (By complexometric titration of cadmium). Weigh accurately about 0.4 g of sample and dissolve in about 200 mL of water. Add 15 mL of a saturated aqueous hexamethylenetetramine solution and 50 mg of xylenol orange indicator mixture. Titrate with standard 0.1 M EDTA to a color change of red-purple to lemon-yellow. One milliliter of 0.1 M EDTA corresponds to 0.01833 g of $CdCl_2$.

INSOLUBLE MATTER. (Page 15). Use 10 g dissolved in 150 mL of water. Retain the filtrate and washings for the test for sulfate.

NITRATE AND NITRITE.

Sample Solution A. Dissolve 0.30 g in a 50-mL centrifuge tube in 3 mL of water by heating on a hot plate ($\approx$100 °C). Add 20 mL of nitrate-free sulfuric acid (2 + 1) and cool in an ice bath. Centrifuge for 10 min, decant the solution into a 200-mm test tube, and dilute to 50 mL with brucine sulfate reagent solution.

Control Solution B. Dissolve 0.30 g in a 50-mL centrifuge tube in 2 mL of water and 1 mL of the standard nitrate solution containing 0.01 mg of nitrate ion (NO_3) per mL by heating on a hot plate ($\approx$100 °C). Add 20 mL of nitrate-free sulfuric acid (2 + 1) and cool in an ice bath. Centrifuge for 10 min, decant the solution into a 200-mm test tube, and dilute to 50 mL with brucine sulfate reagent solution.

Continue with the procedure described on page 30, starting with the preparation of blank solution C.

SULFATE. (Page 33, Method 1). Use 1/20 of the filtrate from the test for insoluble matter. Allow 30 min for the turbidity to form.

AMMONIUM. Dissolve 1.0 g in 80 mL of water and add with stirring 20 mL of freshly boiled 10% sodium hydroxide reagent solution. After allowing the precipitate to settle, dilute 10 mL of the clear supernatant liquid with water to 50 mL, and add 2 mL of Nessler reagent. Any color should not exceed that produced by 0.01 mg of ammonium ion (NH_4) in an equal volume of solution containing 4 mL of the sodium hydroxide solution and 2 mL of the Nessler reagent.

CALCIUM, COPPER, LEAD, POTASSIUM, SODIUM, AND ZINC. (By flame AAS, page 39).

Sample Stock Solution. Dissolve 10.0 g of sample in a 100-mL volumetric flask, add 5 mL of nitric acid, and dilute to the mark with water (1 mL = 0.1 g).

Element	Wavelength (nm)	Sample Wt (g)	Standard Added (mg)	Flame Type*	Background Correction
Ca	422.7	0.20	0.02; 0.04	N/A	No
Cu	324.7	2.0	0.02; 0.04	A/A	Yes
Pb	217.0	2.0	0.10; 0.20	A/A	Yes
K	766.5	0.20	0.02; 0.04	A/A	No
Na	589.0	0.04	0.01; 0.02	A/A	No
Zn	213.9	0.04	0.01; 0.02	A/A	Yes

*A/A is air/acetylene; N/A is nitrous oxide/acetylene.

IRON. (Page 30, Method 1). Use 1.0 g.

Cadmium Chloride, Crystals

$CdCl_2 \cdot 21{,}2\,H_2O$ **Formula Wt 228.35**

CAS Number 7790–78–5

REQUIREMENTS

Assay . . . . 79.5–81.0% $CdCl_2$

MAXIMUM ALLOWABLE

Insoluble matter. 0.005%
Nitrate and nitrite (as NO_3) 0.003%
Sulfate (SO_4) 0.005%
Ammonium (NH_4) 0.005%
Calcium (Ca) 0.005%
Copper (Cu). . . . . 5 ppm
Lead (Pb) 0.005%
Potassium (K). 0.02%
Sodium (Na). 0.05%
Zinc (Zn) 0.05%
Iron (Fe). . . . . 5 ppm

TESTS

ASSAY. (By complexometric titration of cadmium). Weigh accurately about 0.8 g of sample and dissolve in about 200 mL of water. Add 15 mL of a saturated aqueous hexamethylenetetramine solution and 50 mg of xylenol orange indicator mixture. Titrate with standard 0.1 M EDTA to a color change of red-purple to lemon-yellow. One milliliter of 0.1 M EDTA corresponds to 0.01832 g of $CdCl_2$.

INSOLUBLE MATTER. (Page 15). Use 20 g dissolved in 150 mL of water. Reserve the filtrate for the test for sulfate.

NITRATE AND NITRITE.

Sample Solution A. Dissolve 0.30 g in a 50-mL centrifuge tube in 3 mL of water by heating on a hot plate (≈100 °C). Add 20 mL of nitrate-free sulfuric acid (2 + 1) and cool in an ice bath. Centrifuge for 10 min, decant the solution into a 200-mm test tube, and dilute to 50 mL with brucine sulfate reagent solution.

Control Solution B. Dissolve 0.30 g in a 50-mL centrifuge tube in 2 mL of water and 1 mL of the standard nitrate solution containing 0.01 mg of nitrate ion (NO_3) per mL by heating on a hot plate (≈100 °C). Add 20 mL of nitrate-free sulfuric acid $(2 + 1)$ and cool in an ice bath. Centrifuge for 10 min, decant the solution into a 200-mm test tube, and dilute to 50 mL with brucine sulfate reagent solution.

Continue with the procedure described on page 30, starting with the preparation of blank solution C.

SULFATE. (Page 33, Method 1). Use 2.0 g of sample. Allow 30 min for the turbidity to form.

AMMONIUM. Dissolve 1.0 g in 80 mL of water and add, with stirring, 20 mL of freshly boiled 10% sodium hydroxide reagent solution. After allowing the precipitate to settle, dilute 20 mL of the clear supernatant liquid with water to 50 mL, and add 2 mL of Nessler reagent. Any color should not exceed that produced by 0.01 mg of ammonium ion (NH_4) in an equal volume of solution containing 4 mL of the sodium hydroxide solution and 2 mL of the Nessler reagent.

CALCIUM, COPPER, LEAD, POTASSIUM, SODIUM, AND ZINC. (By flame AAS, page 39).

Sample Stock Solution. Dissolve 10.0 g of sample in a 100-mL volumetric flask, add 5 mL of nitric acid, and dilute to the mark with water (1 mL = 0.10 g).

Element	Wavelength (nm)	Sample Wt (g)	Standard Added (mg)	Flame Type*	Background Correction
Ca	422.7	2.0	0.05; 0.10	N/A	No
Cu	324.7	2.0	0.01; 0.02	A/A	Yes
Pb	217.0	2.0	0.10; 0.20	A/A	Yes
K	766.5	0.20	0.02; 0.04	A/A	No
Na	589.0	0.04	0.01; 0.02	A/A	No
Zn	213.9	0.04	0.01; 0.02	A/A	Yes

*A/A is air/acetylene; N/A is nitrous oxide/acetylene.

IRON. (Page 30, Method 1). Use 2.0 g.

Cadmium Sulfate

Cadmium Sulfate, Anhydrous

CdSO$_4$ **Formula Wt 208.47**

CAS Number 10124–36–4

REQUIREMENTS

Assay . . . . ≥99.0% CdSO$_4$

MAXIMUM ALLOWABLE

Insoluble Matter.	0.005%
Loss on drying at 150 °C.	1.0%
Chloride (Cl)	0.001%
Nitrate and nitrite (as NO_3)	0.003%
Calcium (Ca)	0.01%
Copper (Cu).	0.002%
Lead (Pb)	0.003%
Potassium (K).	0.02%
Sodium (Na).	0.05%
Zinc (Zn)	0.05%
Iron (Fe).	0.001%

TESTS

ASSAY. (By complexometric titration of cadmium). Weigh accurately 0.9 g, transfer to a 400-mL beaker, and dissolve in 200 mL of water. Add 15 mL of a saturated aqueous hexamethylenetetramine solution and 50 mg of xylenol orange indicator mixture. Titrate with standard 0.1 M EDTA to a color change of red-purple to lemon-yellow. One milliliter of 0.1 M EDTA corresponds to 0.02085 g of $CdSO_4$.

INSOLUBLE MATTER. (Page 15). Use 20 g dissolved in 150 mL of water.

LOSS ON DRYING AT 150 °C. Weigh accurately about 1 g and dry to constant weight at 150 °C.

CHLORIDE. (Page 27). Use 1.0 g.

NITRATE AND NITRITE. (Page 30). For sample solution A use 0.50 g. For control solution B use 0.50 g and 1.5 mL of standard nitrate solution.

CALCIUM, COPPER, LEAD, POTASSIUM, SODIUM, AND ZINC. (By flame AAS, page 39).

Sample Stock Solution A. Dissolve 10.0 g of sample with water in a 100-mL volumetric flask, and dilute to the mark with water (1 mL = 0.10 g).

Sample Stock Solution B. Transfer 10.0 mL of sample stock solution A to a 100-mL volumetric flask, and dilute to the mark with water (1 mL = 0.01 g).

Element	Wavelength (nm)	Sample Wt (g)	Standard Added (mg)	Flame Type*	Background Correction
Ca	422.7	0.20	0.02; 0.04	N/A	No
Cu	324.7	2.0	0.04; 0.08	A/A	Yes
Pb	217.0	2.0	0.04; 0.08	A/A	Yes
K	766.5	0.20	0.02; 0.04	A/A	No
Na	589.0	0.04	0.01; 0.02	A/A	No
Zn	213.9	0.04	0.01; 0.02	A/A	Yes

*A/A is air/acetylene; N/A is nitrous oxide/acetylene.

IRON. (Page 30, Method 1). Dissolve 1.0 g in 30 mL of water, add 5 mL of hydrochloric acid, and dilute with water to 50 mL. Use this solution without further acidification.

Cadmium Sulfate $\frac{8}{3}$-Hydrate
Cadmium Sulfate, Crystals

$CdSO_4 \cdot \frac{8}{3}H_2O$ **Formula Wt 256.52**

CAS Number 7790–84–3

REQUIREMENTS

Assay . . . . .98.0–102.0% $CdSO_4 \cdot \frac{8}{3}H_2O$

MAXIMUM ALLOWABLE

Insoluble matter. 0.005%
Chloride (Cl) 0.001%
Nitrate and nitrite (as NO_3) 0.003%
Calcium (Ca) 0.005%
Copper (Cu). 0.002%
Lead (Pb) 0.003%
Potassium (K). 0.01%
Sodium (Na) 0.02%
Zinc (Zn) 0.05%
Iron (Fe). 0.001%

TESTS

ASSAY. (By complexometric titration of cadmium). Weigh accurately 1.0 g, transfer to a 400-mL beaker, and dissolve in 175 mL of water. Add 15 mL of a saturated aqueous hexamethylenetetramine solution and 50 mg of xylenol orange indicator mixture. Titrate with standard 0.1 M EDTA to a color change of red-purple to lemon-yellow. One milliliter of 0.1 M EDTA corresponds to 0.02565 g of $CdSO_4 \cdot \frac{8}{3}H_2O$.

INSOLUBLE MATTER. (Page 15). Use 20 g dissolved in 150 mL of water.

CHLORIDE. (Page 27). Use 1.0 g.

NITRATE AND NITRITE. (Page 30). For sample solution A use 0.50 g. For control solution B use 0.50 g and 1.5 mL of standard nitrate solution.

CALCIUM, COPPER, LEAD, POTASSIUM, SODIUM, AND ZINC. (By flame AAS, page 39).

Sample Stock Solution A. Dissolve 20.0 g of sample with water in a 100-mL volumetric flask, and dilute to the mark with water (1 mL = 0.20 g).

Sample Stock Solution B. Transfer 5.0 mL of sample stock solution A to a 100-mL volumetric flask, and dilute to the mark with water (1 mL = 0.01 g).

Element	Wavelength (nm)	Sample Wt (g)	Standard Added (mg)	Flame Type*	Background Correction
Ca	422.7	0.40	0.02; 0.04	N/A	No
Cu	324.7	4.0	0.04; 0.08	A/A	Yes
Pb	217.0	4.0	0.12; 0.24	A/A	Yes
K	766.5	0.40	0.02; 0.04	A/A	No
Na	589.0	0.05	0.005; 0.01	A/A	No
Zn	213.9	0.04	0.01; 0.02	A/A	Yes

*A/A is air/acetylene; N/A is nitrous oxide/acetylene.

IRON. (Page 30, Method 1). Dissolve 1.0 g in 30 mL of water, add 5 mL of hydrochloric acid, and dilute with water to 50 mL. Use this solution without further acidification.

Calcium Acetate Monohydrate

Ca(CH₃COO)₂ · H₂O   **Formula Wt 176.18**

$Ca(CH_3COO)_2 \cdot H_2O$

CAS Number 5743–26–0

REQUIREMENTS

Assay . . . . ≥99.0% Ca(CH₃COO)₂ · H₂O

MAXIMUM ALLOWABLE

Insoluble matter. 0.005%
Alkalinity . . . . Passes test
Titrable acid . . . . 0.035 meq/g
Chloride (Cl) 0.001%
Sulfate (SO₄) 0.01%
Barium (Ba) 0.01%
Heavy metals (as Pb) 0.005%
Iron (Fe) 0.001%
Magnesium (Mg) 0.05%
Potassium (K). 0.01%
Sodium (Na). 0.02%
Strontium (Sr) 0.05%

TESTS

ASSAY. (By complexometric titration for calcium). Weigh accurately 0.70 g, transfer to a 250-mL beaker, and dissolve in 150 mL of water. While stirring, add from a 50-mL buret about 30 mL of standard 0.1 M EDTA. Add 10% sodium hydroxide solution until the pH is above 12 (using a pH meter). Add about 300 mg of hydroxy naphthol blue indicator mixture and continue the titration with the EDTA solution to a blue color. One milliliter of 0.1 M EDTA corresponds to 0.01762 g of $Ca(CH_3COO)_2 \cdot H_2O$.

INSOLUBLE MATTER. (Page 15). Use 20 g dissolved in 200 mL of water.

ALKALINITY AND TITRABLE ACID. Dissolve 2.0 g in 25 mL of carbon dioxide-free water, and add 0.10 mL of phenolphthalein indicator solution. The sample solution should show no pink color due to excess alkalinity. Titrate the solution with 0.10 N sodium hydroxide until a pink color is produced after shaking. Not more than 0.70 mL of 0.10 N sodium hydroxide should be required in the titration.

CHLORIDE. (Page 27). Dissolve 1.0 g in 20 mL of water.

SULFATE. (Page 33, Method 1). Use 0.5 g.

BARIUM. Dissolve 2.0 g in 15 mL of water, add 2 drops of glacial acetic acid, filter, and add to the filtrate 0.3 mL of potassium dichromate solution (1 in 10). Compare the turbidity to a standard containing 0.2 mg of barium treated exactly as the sample.

HEAVY METALS. (Page 28, Method 1). Dissolve 0.4 g in 20 mL of water.

IRON. (Page 30, Method 1). Dissolve 1.0 g in 20 mL of water.

MAGNESIUM, POTASSIUM, SODIUM, AND STRONTIUM. (By flame AAS, page 39).

Sample Stock Solution. Dissolve 2.0 g in 80 mL of water containing 2 mL of nitric acid. Transfer to a 100-mL volumetric flask and dilute to the mark with water (1 mL = 0.02 g).

Element	Wavelength (nm)	Sample Wt (g)	Standard Added (mg)	Flame Type*	Background Correction
Mg	285.2	0.02	0.01; 0.02	A/A	Yes
Na	589.0	0.05	0.01; 0.02	A/A	No
K	766.5	0.20	0.02; 0.04	A/A	No
Sr	460.7	0.20	0.10; 0.20	N/A	No

*A/A is air/acetylene; N/A is nitrous oxide/acetylene.

Calcium Carbonate

CaCO₃

Formula Wt 100.09

CAS Number 471–34–1

REQUIREMENTS

Assay (dried basis) . . . .≥99.0% CaCO₃

MAXIMUM ALLOWABLE

Insoluble in dilute hydrochloric acid. 0.01%
Ammonium hydroxide precipitate 0.01%
Chloride (Cl) 0.001%
Fluoride (F). 0.0015%
Sulfate (SO₄) 0.01%
Ammonium (NH₄) 0.003%
Barium (Ba) 0.01%
Heavy metals (as Pb) 0.001%
Iron (Fe) 0.003%
Magnesium (Mg) 0.02%
Potassium (K). 0.01%
Sodium (Na). 0.1%
Strontium (Sr) 0.1%

TESTS

ASSAY. (By complexometric titration of calcium). Weigh accurately 0.4 g, previously dried at 300 °C for 4 h, and transfer to a 400-mL beaker. Cover the beaker with a watch glass. Add 2 mL of dilute hydrochloric acid from a pipet placed between the lip of the beaker and the watch glass. Swirl the beaker to aid dissolution. With water, wash down the inner wall of the beaker, the outer surface of the pipet, and the watch glass. Dilute to about 100 mL with water. While stirring, add from a 50-mL buret about 30 mL of standard 0.1 M EDTA. Adjust the solution to above pH 12 (pH meter with suitable high-pH glass electrode) with 10% sodium hydroxide solution. Add 300 mg of hydroxy naphthol blue indicator mixture and continue the titration immediately with the EDTA solution to a blue color. One milliliter of 0.1 M EDTA corresponds to 0.01001 g of CaCO₃.

INSOLUBLE IN DILUTE HYDROCHLORIC ACID. Add 10 g to about 100 mL of water, swirl, slowly and carefully add 20 mL of hydrochloric acid, and dilute with water to 150 mL. Heat the solution to boiling, boil gently to expel the carbon dioxide, and digest in a covered beaker on a hot plate (≈100 °C) for 1 h. Filter through a tared filtering crucible (reserve the filtrate for the determination of ammonium hydroxide precipitate), wash thoroughly, and dry at 105 °C.

AMMONIUM HYDROXIDE PRECIPITATE. Add 0.10 mL of methyl red indicator solution to the filtrate reserved from the test for insoluble in dilute hydrochloric acid and make slightly alkaline with ammonium hydroxide. Heat the solution to boiling and boil gently for 5 min to coagulate the precipitate. If necessary, add more ammonium hydroxide to maintain the alkalinity throughout this operation. Filter through a small filter paper and wash with a small amount of hot water. Redissolve the precipitate from the filter paper with hot dilute hydrochloric acid (1 + 3) and wash the filter paper free of acid. Heat the filtrate plus washings, which should amount to about 25 mL, to boiling and boil gently for 1 to 2 min. Add 0.05 mL of methyl red indicator solution, make slightly alkaline with ammonium hydroxide, and again boil gently to coagulate the precipitate. Filter through the same small filter paper, wash thoroughly, and ignite in a tared crucible.

CHLORIDE. (Page 27). Dissolve 1.0 g in 10 mL of water plus 2 mL of nitric acid, filter if necessary through a small chloride-free filter, wash with hot water, and dilute with water to 20 mL.

FLUORIDE. Dissolve 1.0 g in 50 mL of water and 20 mL of 1 N hydrochloric acid in a 250-mL beaker. For the standards add 1.0, 2.0, 3.0, 5.0, 10.0, and 15.0 mL of standard fluoride solution to 250-mL glass beakers containing 50 mL of water and 5 mL of 1 N hydrochloric acid. Boil the sample and six standard solutions for a few seconds, cool rapidly, and transfer to plastic beakers. To each solution add 10 mL of 1 M sodium citrate reagent solution and 10 mL of 0.2 M disodium (ethylenedinitrilo)tetraacetate reagent solution and mix. If necessary, adjust the pH to 5.5 ± 0.1 with dilute hydrochloric acid or dilute sodium hydroxide. Transfer to 100-mL volumetric flasks, dilute to the mark with water, mix, and pour into plastic beakers. For each solution, measure immediately the potential of a fluoride electrode versus a reference electrode using a pH meter with an expanded scale or an appropriate potentiometer. Plot a 2-cycle semilogarithmic calibration curve with micrograms of fluoride per 100 mL of solution on the logarithmic scale. Determine the concentration of fluoride from the calibration curve.

SULFATE. Cautiously dissolve 1.0 g of sample with 10 mL of (1 + 1) hydrochloric acid in a 100-mL beaker. Add 0.05 mL of bromine water, and evaporate to dryness. Dissolve the residue in 5 mL of water, and evaporate to dryness. Dissolve the residue in 15 mL of water, add 1 mL of (1 + 19) hydrochloric acid, and filter through a small, washed filter paper. Add two 3-mL portions of water to the same filter. For the control, take 0.10 mg of sulfate ion (SO_4) in 15 mL of water, and add 1 mL of (1 + 19) hydrochloric acid. Dilute both solutions to 25 mL, add 1 mL of barium chloride reagent solution, and mix. Compare after 10 min. Sample turbidity should not exceed that of the control solution.

Sample Solution A for the Determination of Ammonium, Heavy Metals, and Iron. Cautiously dissolve 20 g in 100 mL of dilute hydrochloric acid (1 + 1) and evaporate on a hot plate ($\approx$100 °C) to dryness or a moist residue.

Dissolve the residue in about 100 mL of water, filter, and dilute with water to 200 mL (1 mL = 0.1 g).

AMMONIUM. Dilute 10 mL (1-g sample) of sample solution A with water to 80 mL, add 20 mL of 10% sodium hydroxide reagent solution, stopper, mix well, and allow to stand for 1 h. Decant 50 mL through a filtering crucible that has been washed with 10% sodium hydroxide solution, and add to the filtrate 2 mL of Nessler reagent. Any color should not exceed that produced by 0.015 mg of ammonium ion (NH_4) in an equal volume of solution containing the quantities of reagents used in the test.

BARIUM. For the sample dissolve 2.0 g in 15 mL of water and sufficient nitric acid (about 2.5 mL) to effect dissolution. For the control, dissolve 1.0 g in 15 mL of water and sufficient nitric acid (about 1.5 mL) to effect dissolution and add 0.1 mg of barium ion (Ba). Heat each solution to boiling, boil gently for 5 min, cool, filter if necessary, and dilute with water to 23 mL. Add 2 mL of potassium dichromate reagent solution and neutralize with ammonium hydroxide until the orange color is just dissipated and the yellow color persists. To each solution slowly add, with constant stirring, 25 mL of methanol. Any turbidity in the solution of the sample should not exceed that in the control.

HEAVY METALS. (Page 28, Method 1). Use 30 mL (3-g sample) of sample solution A for the sample solution, and use 10 mL (1-g sample) of sample solution A to prepare the control solution.

IRON. (Page 30, Method 1). Use 3.3 mL of sample solution A (0.33-g sample).

MAGNESIUM, POTASSIUM, SODIUM, AND STRONTIUM. (By flame AAS, page 39).

Sample Stock Solution A. Cautiously dissolve 5.0 g of sample in 10 mL of nitric acid (1 + 1). Transfer to a 100-mL volumetric flask and dilute to the mark with water (1 mL = 0.05 g).

Sample Stock Solution B. Transfer 10.0 mL (0.5 g) to a 100-mL volumetric flask, and dilute to the mark with water (1 mL = 0.005 g).

Element	Wavelength (nm)	Sample Wt (g)	Standard Added (mg)	Flame Type*	Background Correction
Mg	285.2	0.10	0.01; 0.02	A/A	Yes
K	766.5	0.50	0.05; 0.10	A/A	No
Na	589.0	0.01	0.01; 0.02	A/A	No
Sr	460.7	0.10	0.10; 0.20	N/A	No

*A/A is air/acetylene; N/A is nitrous oxide/acetylene.

Calcium Carbonate, Low in Alkalies

$CaCO_3$ **Formula Wt 100.09**

CAS Number 471–34–1

REQUIREMENTS

Assay (dried basis) . . . . ≥99.0% $CaCO_3$

MAXIMUM ALLOWABLE

Insoluble in dilute hydrochloric acid 0.01%
Ammonium hydroxide precipitate 0.01%
Chloride (Cl) 0.001%
Fluoride (F) 0.0015%
Sulfate (SO_4) 0.005%
Ammonium (NH_4) 0.003%
Barium (Ba) 0.01%
Heavy metals (as Pb) 0.001%
Iron (Fe). 0.002%
Magnesium (Mg) 0.01%
Potassium (K). 0.01%
Sodium (Na) 0.01%
Strontium (Sr) 0.1%

TESTS

ASSAY. (By complexometric titration of calcium). Weigh accurately 0.4 g, previously dried at 300 °C for 4 h, and transfer to a 400-mL beaker. Cover the beaker with a watch glass. Add 2 mL of dilute hydrochloric acid from a pipet placed between the lip of the beaker and the watch glass. Swirl the beaker to aid dissolution. With water, wash down the inner wall of the beaker, the outer surface of the pipet, and the watch glass. Dilute to about 100 mL with water. While stirring, add, from a 50-mL buret, about 30 mL of standard 0.1 M EDTA. Adjust the solution to above pH 12 (pH meter with suitable high-pH glass electrode) with 10% sodium hydroxide solution. Add 300 mg of hydroxy naphthol blue indicator mixture and continue the titration immediately with the EDTA solution to a blue color. One milliliter of 0.1 M EDTA corresponds to 0.01001 g of $CaCO_3$.

INSOLUBLE IN DILUTE HYDROCHLORIC ACID. Add 20 g to about 200 mL of water, swirl, slowly and carefully add 40 mL of hydrochloric acid, and dilute with water to 300 mL. Heat the solution to boiling, boil gently to expel the carbon dioxide, and digest in a covered beaker on a hot plate (≈100 °C) for 1 h. Filter through a tared filtering crucible (reserve the filtrate for the determination of ammonium hydroxide precipitate), wash thoroughly, and dry at 105 °C.

AMMONIUM HYDROXIDE PRECIPITATE. Add 0.20 mL of methyl red indicator solution to the filtrate reserved from the test for insoluble in dilute hydrochloric acid and make slightly alkaline with ammonium hydroxide. Heat the solution to boiling and boil gently for 5 min to coagulate the precipitate. If necessary, add more ammonium hydroxide to maintain the alkalinity throughout the operation. Filter through a small filter paper and wash with a small amount of hot water. (Reserve this filtrate for the determination of sulfate.) Redissolve the precipitate from the filter paper with hot dilute hydrochloric acid (1 + 3) and wash the filter paper free of acid. Heat the filtrate plus washings, which should amount to about 25 mL, to boiling and boil gently for 1 to 2 min. Add 0.05 mL of methyl red indicator solution, make slightly alkaline with ammonium hydroxide, and again boil gently to coagulate the precipitate. Filter through the same small filter paper, wash thoroughly, and ignite in a tared crucible.

CHLORIDE. (Page 27). Dissolve 1.0 g in 10 mL of water plus 2 mL of nitric acid, filter if necessary through a small chloride-free filter, wash with hot water, and dilute with water to 20 mL.

FLUORIDE. Dissolve 1.0 g in 50 mL of water and 25 mL of 1 N hydrochloric acid in a 250-mL beaker. For the standards add 1.0, 2.0, 3.0, 5.0, 10.0, and 15.0 mL of standard fluoride solution to 250-mL glass beakers containing 50 mL of water and 5 mL of 1 N hydrochloric acid. Boil the sample and six standard solutions for a few seconds, cool rapidly, and transfer to plastic beakers. To each solution add 20 mL of 1 M sodium citrate reagent solution and 10 mL of 0.2 M disodium (ethylenedinitrilo)tetraacetate reagent solution and mix. If necessary, adjust the pH to 5.5 ± 0.1 with dilute hydrochloric acid or dilute sodium hydroxide. Transfer to 100-mL volumetric flasks, dilute to the mark with water, mix, and pour into plastic beakers. For each solution measure immediately the potential of a fluoride electrode versus a reference electrode using a pH meter with an expanded scale or an appropriate potentiometer. Plot a 2-cycle semilogarithmic calibration curve with micrograms of fluoride per 100 mL of solution on the logarithmic scale. Determine the concentration of fluoride from the calibration curve.

SULFATE. Cautiously dissolve 2.0 g of sample with 15 mL of (1 + 1) hydrochloric acid in a 100-mL beaker. Add 0.05 mL of bromine water, and evaporate to dryness. Dissolve the residue in 5 mL of water, and evaporate to dryness. Dissolve the residue in 15 mL of water, add 1 mL of (1 + 19) hydrochloric acid, and filter through a small, washed filter paper. Add two 3-mL portions of water to the same filter. For the control, take 0.10 mg of sulfate ion (SO_4) in 15 mL of water, and add 1 mL of (1 + 19) hydrochloric acid. Dilute both solutions to 25 mL, add 1 mL of barium chloride reagent solution, and mix. Compare after 10 min. Sample turbidity should not exceed that of the control solution.

Sample Solution A for the Determination of Ammonium, Heavy Metals, and Iron. Cautiously dissolve 20 g in 100 mL of dilute hydrochloric acid (1 + 1) and evaporate on a hot plate (≈100 °C) to dryness or a moist residue. Dissolve the residue in about 100 mL of water, filter, and dilute with water to 200 mL in a volumetric flask (1 mL = 0.10 g).

AMMONIUM. Dilute 10 mL (1-g sample) of sample solution A with water to 80 mL, add 20 mL of 10% sodium hydroxide reagent solution, stopper, mix well, and allow to stand for 1 h. Decant 50 mL through a filtering crucible that has been washed with 10% sodium hydroxide solution, and add to the filtrate 2 mL of Nessler reagent. Any color should not exceed that produced by 0.015 mg of ammonium ion (NH_4) in an equal volume of solution containing the quantities of reagents used in the test.

BARIUM. For the sample dissolve 2.0 g in 15 mL of water and sufficient nitric acid (about 2.5 mL) to effect dissolution. For the control, dissolve 1.0 g in 15 mL of water and sufficient nitric acid (about 1.5 mL) to effect dissolution and add 0.1 mg of barium. Heat each to boiling, boil gently for 5 min, cool, filter if necessary, and dilute with water to 23 mL. Add 2 mL of potassium dichromate reagent solution and neutralize with ammonium hydroxide until the orange color is just dissipated and the yellow color persists. To each solution slowly add, with constant stirring, 25 mL of methanol. Any turbidity in the solution of the sample should not exceed that in the control.

HEAVY METALS. (Page 28, Method 1). Use 30 mL (3-g sample) of sample solution A to prepare the sample solution, and use 10 mL of sample solution A (1-g sample) to prepare the control solution.

IRON. (Page 30, Method 1). Use 5.0 mL of sample solution A (0.5-g sample).

MAGNESIUM, POTASSIUM, SODIUM, AND STRONTIUM. (By flame AAS, page 39).

Sample Stock Solution. Cautiously dissolve 5.0 g of sample in 10 mL of nitric acid (1 + 1). Transfer to a 100-mL volumetric flask, and dilute to the mark with water (1 mL = 0.05 g).

Element	Wavelength (nm)	Sample Wt (g)	Standard Added (mg)	Flame Type*	Background Correction
Mg	285.2	0.10	0.01; 0.02	A/A	Yes
K	766.5	0.50	0.05; 0.10	A/A	No
Na	589.0	0.10	0.01; 0.02	A/A	No
Sr	460.7	0.10	0.10; 0.20	N/A	No

*A/A is air/acetylene; N/A is nitrous oxide/acetylene.

Calcium Carbonate, Chelometric Standard

$CaCO_3$ **Formula Wt 100.09**

CAS Number 471–34–1

REQUIREMENTS

Assay (dried basis) . . . . 99.95–100.05% $CaCO_3$

MAXIMUM ALLOWABLE

Insoluble in dilute hydrochloric acid. 0.01%
Ammonium hydroxide precipitate 0.01%
Chloride (Cl) 0.001%
Fluoride (F). 0.0015%
Sulfate (SO_4) 0.005%
Ammonium (NH_4) 0.003%
Barium (Ba) 0.01%
Heavy metals (as Pb) 0.001%
Iron (Fe). 0.002%
Magnesium (Mg) 0.01%
Potassium (K). 0.01%
Sodium (Na). 0.01%
Strontium (Sr) 0.1%

TESTS

ASSAY. (By complexometric titration for calcium). This method is a comparative procedure in which the calcium carbonate to be assayed is compared with NIST Standard Reference Material Calcium Carbonate (SRM 915). A minimum of two samples and two standards should be run according to this procedure.

Procedure. Dry the sample and NIST standard at 110 °C for 2 h. Weigh (to the nearest 0.1 mg) 0.4000 ± 0.0040 g of calcium carbonate and place in a 400-mL beaker with a magnetic stirring bar and 50 mL of water. Cover with a watch glass and add 12 mL of 10% hydrochloric acid with the pipet tip under the watch glass. Wash down the watch glass and the sides of the beaker. All of the sample must be dissolved.

Weigh (to the nearest 0.1 mg) 1.4400 ± 0.0100 g of (ethylenedinitrilo)tetraacetic acid disodium salt (dihydrate) (NOTE 1). Add the EDTA directly to the calcium solution, dilute to about 125 mL, and add 15 mL of 5 M potassium hydroxide. This addition should dissolve all the EDTA and leave a clear solution. Add 0.05–0.10 mL of 0.02% calcein indicator solution and titrate the excess calcium with 0.0200 M EDTA under UV illumination in a dark room or box (NOTE 2). The end point is extremely sharp from a bright yellow-green fluorescence to black.

$$\% \, CaCO_3 = \frac{\left[\dfrac{(Wt \, EDTA) + (0.37226 \times mL \times M)}{Wt \, sample \, CaCO_3}\right]}{\left[\dfrac{(Wt \, EDTA) + (0.37226 \times mL \times M)}{Wt \, NIST \, CaCO_3 \times NIST \, assay \, factor}\right]}$$

NOTE 1. This method is independent of the quality of the disodium EDTA used, but it is recommended that reagent-quality material be selected, then crushed, and mixed thoroughly prior to use. Do not dry the EDTA because water loss will change the proportions needed for the test. A disodium dihydrogen EDTA dihydrate having an assay of 99.9–100.1% is commercially available. ACS Reagent is 99–101% and suitable for this test.

NOTE 2. The UV lamp should have dark blue glass (little visible light) and should be placed at a right angle to the viewer, shielded from the viewer's eyes.

NOTE 3. Buoyancy corrections cancel.

Potassium Hydroxide, 5 M.   Dissolve 56 g of KOH in water and dilute with water to 200 mL.

Calcein Indicator Solution, 0.02%.   Dissolve 20 mg of calcein in 10 mL of water by addition of a few drops of 5 M potassium hydroxide and dilute with water to 100 mL. This solution is not stable at room temperature for more than 1 day. Frozen solutions are stable, so the indicator solution may be split into portions in plastic bottles and frozen, and a fresh bottle may be thawed as needed.

EDTA, 0.02 M.   Dissolve 7.446 g of reagent-quality (ethylenedinitrilo)tetra-acetic acid disodium salt dihydrate in water and dilute with water to 1 L. Standardize against a calcium solution prepared by dissolving 2.002 g (weighed to the nearest 0.1 mg) of NIST calcium carbonate in water by addition of 6 mL of concentrated hydrochloric acid, followed by dilution to exactly 1 L. The molarity of this solution should be calculated from the weight taken. The standardization of a 25-mL aliquot of the calcium solution will require 10 mL of 5 M potassium hydroxide and 0.05–0.10 mL of calcein indicator solution, as in the procedure cited.

Insoluble in dilute hydrochloric acid and other tests except assay are the same as for calcium carbonate, low in alkalies, page 215.

Calcium Chloride Desiccant

CaCl₂   **Formula Wt 110.98**

CAS Number 10043–52–4

REQUIREMENTS

Assay . . . . ≥96.0% CaCl₂

MAXIMUM ALLOWABLE

Titrable base. . . . . 0.006 meq/g

TESTS

ASSAY. (By complexometric titration of calcium). Weigh accurately about 2.5 g in a 250-mL beaker. Dissolve in 100 mL of water, add 5 mL of 10% hydrochloric acid. Transfer to a 250-mL volumetric flask, dilute to the mark with water, and mix. Pipet 50.0 mL of the sample solution to a 400-mL beaker, add 75 m L of water, and while stirring add 20 mL of 0.1 M EDTA from 50-mL buret. Adjust the pH using a pH meter to pH 12–13 with 10% sodium hydroxide solution. Add 300 mg of hydroxy naphthol blue indicator mixture and continue the titration immediately with EDTA solution to a blue color. One milliliter of 0.1 M EDTA corresponds to 0.0111 g of CaCl₂.

$$\% \, CaCl_2 = \frac{mL \times M \, EDTA \times 11.1}{Sample \, wt \, (g)/5}$$

TITRABLE BASE. Dissolve 5.0 g in 50 mL of water and add 0.10 mL of phenolphthalein indicator solution. If any pink color is produced, it should be discharged by not more than 3.0 mL of 0.01 N hydrochloric acid.

Calcium Chloride Dihydrate

CaCl₂ · 2H₂O   **Formula Wt 147.01**

CAS Number 10035–04–8

REQUIREMENTS

Assay . . . . 99.0–105.0% CaCl₂ · 2H₂O
pH of a 5% solution . . . . 4.5–8.5 at 25 °C

MAXIMUM ALLOWABLE
Insoluble matter. 0.01%
Oxidizing substances (as NO_3) 0.003%
Sulfate (SO_4) 0.01%
Ammonium (NH_4) 0.005%
Barium (Ba) 0.005%
Heavy metals (as Pb) . . . . 5 ppm
Iron (Fe). 0.001%
Magnesium (Mg) 0.005%
Potassium (K). 0.01%
Sodium (Na) 0.02%
Strontium (Sr) 0.1%

TESTS

ASSAY. (By complexometric titration of calcium). Weigh accurately about 2.5 g in a 250-mL beaker. Dissolve in 100 mL of water, add 5 mL of 10% hydrochloric acid. Transfer to a 250-mL volumetric flask, dilute to the mark with water, and mix. Pipet 50.0 mL of the sample solution to a 400 mL beaker, add 75 mL of water, and while stirring add 20 mL of 0.1 M EDTA from a 50-mL buret. Adjust the pH using a pH meter to pH 12–13 with 10% sodium hydroxide solution. Add 300 mg of hydroxy naphthol blue indicator mixture and continue the titration immediately with EDTA solution to a blue color. One milliliter of 0.1 M EDTA corresponds to 0.014701 g of $CaCl_2 \cdot 2H_2O$.

$$\% \, CaCl_2 \cdot 2H_2O = \frac{mL \times M \, EDTA \times 14.701}{Sample \, wt \, (g) \, / \, 5}$$

pH OF A 5% SOLUTION. (Page 44). The pH should be 4.5–8.5 at 25 °C.

INSOLUBLE MATTER. (Page 15). Dissolve 20 g in 200 mL of water.

OXIDIZING SUBSTANCES. Place 0.20 g in a dry beaker. Cool the beaker thoroughly in an ice bath and add 22 mL of sulfuric acid that has been cooled to ice-bath temperature. Allow the mixture to warm to room temperature and swirl the beaker at intervals to effect gentle dissolution with slow evolution of hydrogen chloride. When dissolution is complete, add 3 mL of diphenylamine reagent solution and digest on a hot plate ($\approx$100 °C) for 90 min. Prepare a standard by evaporating to dryness a solution containing 0.006 mg of nitrate (0.6 mL of the nitrate standard solution) and 0.01 g of sodium carbonate. Treat the residue exactly like the sample. Any color produced in the solution of the sample should not exceed that in the standard.

SULFATE. Dissolve 1.0 g of sample in 20 mL of water, and add 1.0 mL of 10% hydrochloric acid. Filter through a small, washed filter paper. Add two 3-mL portions of water through filter paper. For the control, take 0.10 mg of sulfate (SO_4)

in 25 mL of water, and add 1.0 mL of 10% hydrochloric acid. Dilute both solutions to 35 mL with water, and add 1 mL of barium chloride solution. Compare turbidity after 10 min. Sample turbidity should not exceed that of the control solution.

Sample Solution A for Determining Ammonium, Barium, Heavy Metals, and Iron.
Dissolve 50 g in about 200 mL of water, filter if necessary, and dilute with water to 250 mL in a volumetric flask (1 mL = 0.2 g).

AMMONIUM. Dilute 10 mL (1-g sample) of sample solution A with water to 80 mL, add 20 mL of 10% sodium hydroxide reagent solution, stopper, mix well, and allow to stand for 1 h. Decant 50 mL through a filtering crucible that has been washed with 10% sodium hydroxide solution and add to the filtrate 2 mL of Nessler reagent. Any color should not exceed that produced by 0.015 mg of ammonium ion (NH_4) in an equal volume of solution containing the quantities of reagents used in the test.

BARIUM. For the sample add 2 g of sodium acetate and 0.05 mL of glacial acetic acid to 15 mL (3-g sample) of sample solution A. For the control, add 2 g of sodium acetate, 0.05 mL of glacial acetic acid, and 0.1 mg of barium ion to 5 mL (1-g sample) of sample solution A and dilute with water to 15 mL. To each solution add 2 mL of potassium dichromate reagent solution and allow to stand for 15 min. Any turbidity in the solution of the sample should not exceed that in the control.

HEAVY METALS. (Page 28, Method 1). Use 30 mL (6-g sample) of sample solution A to prepare the sample solution, and use 10 mL (2-g sample) of sample solution A to prepare the control solution.

IRON. (Page 30, Method 1). Use 5.0 mL (1-g sample) of sample solution A.

MAGNESIUM, POTASSIUM, SODIUM, AND STRONTIUM. (By flame AAS, page 39).

Sample Stock Solution.
Dissolve 2.0 g of sample in 50 mL of water and 5 mL of nitric acid in a 200-mL volumetric flask, and dilute to the mark with water (1 mL = 0.01 g).

Element	Wavelength (nm)	Sample Wt (g)	Standard Added (mg)	Flame Type*	Background Correction
Mg	285.2	0.20	0.01; 0.02	A/A	Yes
K	766.5	0.20	0.02; 0.04	A/A	No
Na	589.0	0.05	0.01; 0.02	A/A	No
Sr	460.7	0.05	0.05; 0.10	N/A	No

*A/A is air/acetylene; N/A is nitrous oxide/acetylene.

Calcium Hydroxide

Ca(OH)$_2$ **Formula Wt 74.09**

CAS Number 1305–62–0

REQUIREMENTS

Assay . . . .≥95.0% Ca(OH)$_2$ and ≤3.0% CaCO$_3$

MAXIMUM ALLOWABLE

Insoluble in hydrochloric acid 0.03%
Chloride (Cl) 0.03%
Sulfur compounds (as SO$_4$) 0.1%
Heavy metals (as Pb) 0.003%
Iron (Fe). 0.05%
Magnesium (Mg) 0.5%
Potassium (K). 0.05%
Sodium (Na) 0.05%
Strontium (Sr) 0.05%

TESTS

ASSAY. (By acid–base titrimetry). Weigh accurately 1.4 g of sample, mix with 100 mL of carbon dioxide-free water, and add 0.15 mL of phenolphthalein indicator solution. Titrate with 1.0 N hydrochloric acid to the disappearance of the pink color (end point = A mL). (The sample is initially suspended in the solution, but dissolves as the acid is added.) Add 0.20 mL of methyl orange indicator solution, and continue the titration to a pinkish-orange color (end point = B mL).

$$\% \, Ca(OH)_2 = \frac{A \times N \times 3.705}{\text{Sample wt (g)}}$$

$$\% \, CaCO_3 = \frac{(B - A) \times N \times 5.005}{\text{Sample wt (g)}}$$

INSOLUBLE IN HYDROCHLORIC ACID. Add 5.0 g to 100 mL of water, add 25 mL of hydrochloric acid, heat the solution to boiling, and digest in a covered beaker on a hot plate (≈100 °C) for 1 h. Filter through a tared filtering crucible, wash thoroughly, and dry at 105 °C.

CHLORIDE. (Page 27). Dissolve 0.17 g in 10 mL of water plus 1 mL of nitric acid, dilute with water to 50 mL, and dilute 10 mL of this solution with water to 20 mL.

SULFUR COMPOUNDS. Mix 4.0 g with 100 mL of water, add 5 mL of bromine water, and digest on a hot plate ($\approx$100 °C) for 10 min. Add 6 mL of hydrochloric acid, boil to expel the excess bromine, filter, and wash the filter with hot water. Heat the solution to boiling, add 5 mL of barium chloride reagent solution, digest in a covered beaker on a hot plate ($\approx$100 °C) for 2 h, and allow to stand overnight. If a precipitate is formed, filter, wash thoroughly, and ignite. Correct for the weight obtained in a complete blank test.

Sample Solution A for the Determination of Heavy Metals and Iron.
To 3.0 g, add 40 mL of water, mix, cautiously add 10 mL of hydrochloric acid and 3 mL of nitric acid, and evaporate on a hot plate ($\approx$100 °C) to dryness. Take up the residue with 1 mL of 10% hydrochloric acid and 30 mL of hot water, filter, and wash with a few mL of water. Add 0.10 mL of phenolphthalein indicator solution and sufficient 1 N sodium hydroxide to produce a pink color, then add sufficient 1 N hydrochloric acid to discharge the pink color, and dilute with water to 60 mL (10 mL = 0.5 g).

HEAVY METALS. (Page 28, Method 1). Use 20 mL (1-g sample) of sample solution A to prepare the sample solution, and use 6.7 mL (0.33-g sample) of sample solution A to prepare the control solution.

IRON. (Page 30, Method 1). Dilute 4.0 mL of sample solution A (0.2-g sample) with water to 100 mL, and use 10 mL of this dilution.

MAGNESIUM, POTASSIUM, SODIUM, AND STRONTIUM. (By flame AAS, page 39).

Stock Solution A.
Dissolve carefully with stirring 2.0 g of sample in 30 mL of water containing 5 mL of nitric acid. Cool to room temperature. Transfer to a 100 mL volumetric flask and dilute to the mark with water (1 mL = 0.02 g).

Sample Stock Solutin B.
Transfer 5.0 mL of stock solution A to a 100-mL volumetric flask and dilute to the mark with water (1 mL = 0.001 g).

Element	Wavelength (nm)	Sample Wt (g)	Standard Added (mg)	Flame Type*	Background Correction
Mg	285.2	0.002	0.01; 0.02	A/A	Yes
K	766.5	0.02	0.01; 0.02	A/A	No
Na	589.0	0.02	0.01; 0.02	A/A	No
Sr	460.7	0.20	0.10; 0.20	N/A	No

*A/A is air/acetylene; N/A is nitrous oxide/acetylene.

Calcium Nitrate Tetrahydrate

Ca(NO₃)₂ · 4H₂O

$Ca(NO_3)_2 \cdot 4H_2O$

Formula Wt 236.15

CAS Number 13477–34–4

REQUIREMENTS

Assay . . . . 99.0–103.0% $Ca(NO_3)_2 \cdot 4H_2O$
pH of a 5% solution . . . . 5.0–7.0 at 25 °C

MAXIMUM ALLOWABLE

Insoluble matter. 0.005%
Chloride (Cl) 0.005%
Nitrite (NO_2) 0.001%
Sulfate (SO_4) 0.002%
Barium (Ba) 0.005%
Heavy metals (as Pb) . . . . 5 ppm
Iron (Fe). . . . . 5 ppm
Magnesium (Mg) 0.05%
Potassium (K). 0.005%
Sodium (Na) 0.01%
Strontium (Sr) 0.05%

TESTS

ASSAY. (By complexometric titration of calcium). Weigh accurately about 0.90 g, transfer to a 250-mL beaker, and dissolve in 150 mL of water. While stirring, add from a 50-mL buret about 30 mL of standard 0.1 M EDTA. Then adjust the solution to above pH 12 with 10% sodium hydroxide solution. Add about 300 mg of hydroxy naphthol blue indicator mixture and continue the titration immediately with the EDTA solution to a blue color. One milliliter of 0.1 M EDTA corresponds to 0.02362 g of $Ca(NO_3)_2 \cdot 4H_2O$.

pH OF A 5% SOLUTION. (Page 44). The pH should be 5.0–7.0 at 25 °C.

INSOLUBLE MATTER. (Page 15). Dissolve 20 g in 200 mL of water.

CHLORIDE. (Page 27). Dissolve 1.0 g in 50 mL of water, and dilute 10 mL of the solution with water to 20 mL.

NITRITE. (page 51). Use 1.0 g of sample and 0.01 mg of nitrite.

SULFATE. (Page 33, Method 2). Use 6 mL of hydrochloric acid.

BARIUM. For the sample dissolve 2.0 g in 15 mL of water, add 2 g of sodium acetate and 0.05 mL of glacial acetic acid, and filter if necessary. For the standard, add

0.1 mg of barium to 15 mL of water, then add 2 g of sodium acetate and 0.05 mL of glacial acetic acid. Add to each solution 2 mL of 10% potassium dichromate reagent solution, mix, and let stand for 15 min. Any turbidity in the solution of the sample should not exceed that in the standard.

HEAVY METALS. (Page 28, Method 1). Dissolve 6.0 g in about 20 mL of water, and dilute with water to 30 mL. Use 25 mL to prepare the sample solution, and use the remaining 5.0 mL to prepare the control solution.

IRON. (Page 30, Method 1). Use 2.0 g.

MAGNESIUM, POTASSIUM, SODIUM, AND STRONTIUM. (By flame AAS, page 39).

> *Sample Stock Solution.* Dissolve 4.0 g of sample in 80 mL of water. Transfer to a 100 mL volumetric flask and dilute to the mark with water (1 mL = 0.04 g).

Element	Wavelength (nm)	Sample Wt (g)	Standard Added (mg)	Flame Type*	Background Correction
Mg	285.2	0.02	0.01; 0.02	A/A	Yes
K	766.5	0.20	0.01; 0.02	A/A	No
Na	589.0	0.10	0.01; 0.02	A/A	No
Sr	460.7	0.20	0.10; 0.20	N/A	No

*A/A is air/acetylene; N/A is nitrous oxide/acetylene.

Calcium Sulfate Dihydrate

$CaSO_4 \cdot 2H_2O$ Formula Wt 172.17

CAS Number 10101–41–4

REQUIREMENTS

Assay . . . . .98.0–102.0% $CaSO_4 \cdot 2H_2O$

MAXIMUM ALLOWABLE

Insoluble in dilute hydrochloric acid. 0.02%
Chloride (Cl) 0.005%
Nitrate (NO_3) . . . . Passes test
Carbonate (CO_3) . . . . Passes test
Heavy metals (as Pb) 0.002%
Iron (Fe). 0.001%
Magnesium (Mg) 0.02%
Potassium (K). 0.005%

Sodium (Na) 0.02%
Strontium (Sr) 0.05%

TESTS

ASSAY. (By complexometric titration for calcium). Dissolve 0.30 g, accurately weighed, in 100 mL of water and 5 mL of hydrochloric acid; boil if necessary to effect dissolution. Cool and stir, preferably with a magnetic stirrer, while adding the following reagents in the cited order: 0.5 mL of 2,2′,2″-nitrilotriethanol (i.e., triethanolamine), 300 mg of hydroxy naphthol blue indicator mixture, and (from a 50-mL buret) about 30 mL of standard 0.05 M EDTA. Now add 10% sodium hydroxide solution until the initial red color changes to clear blue; then continue the addition dropwise until the pH is above 12. Continue the titration immediately with the EDTA solution to a blue color. One milliliter of 0.05 M EDTA corresponds to 0.008608 g of $CaSO_4 \cdot 2H_2O$.

INSOLUBLE IN DILUTE HYDROCHLORIC ACID. Dissolve 4.0 g in a mixture of 100 mL of water and 20 mL of hydrochloric acid, heating on a hot plate ($\approx$100 °C) if necessary until dissolution is complete. Filter through a tared filtering crucible, wash thoroughly with hot 10% hydrochloric acid, followed by water, and dry at 105 °C.

CHLORIDE. (Page 27). Dissolve 0.20 g in a mixture of 10 mL of water and 7 mL of nitric acid, warming if necessary to complete dissolution. Filter, if necessary, through a chloride-free filter, and dilute with water to 20 mL.

NITRATE. Mix 1.0 g with 9 mL of water and 1 mL of 0.5% sodium chloride solution. Add 0.10 mL of indigo carmine reagent solution, mix, and add 10 mL of sulfuric acid. The blue color should not be completely discharged in 10 min. (Limit about 0.005%)

CARBONATE. Mix 1 g with 5 mL of water and slowly add 2 mL of hydrochloric acid. The evolution of a gas should not be observed.

HEAVY METALS. (Page 28, Method 1). Add 2.0 g to a mixture of 25 mL of water and 5 mL of hydrochloric acid, heat the solution to boiling, and nearly neutralize with ammonium hydroxide. Filter, wash with water to about 40 mL, complete the neutralization with ammonium hydroxide (10% NH_3), and dilute with water to 50 mL. Use 25 mL for the sample solution. To prepare the standard use 25 mL of the following solution: add 0.04 mg of lead to 5 mL of hydrochloric acid, neutralize with ammonium hydroxide, and dilute with water to 50 mL.

IRON. (Page 30, Method 1). Boil 1.0 g with 6 mL of hydrochloric acid and 40 mL of water, cool, filter, and wash with water to 50 mL. Test the solution without further acidification.

MAGNESIUM, POTASSIUM, SODIUM, AND STRONTIUM. (By flame AAS, page 39).

Sample Stock Solution. Dissolve 2.0 g in a mixture of 50 mL of water and 10 mL of hydrochloric acid by heating on a hot plate until dissolution is complete. Cool to room temperature. Transfer to a 100-mL volumetric flask and dilute to the mark with water (1 mL = 0.02 g).

Element	Wavelength (nm)	Sample Wt (g)	Standard Added (mg)	Flame Type*	Background Correction
Mg	285.2	0.05	0.01; 0.02	A/A	Yes
K	766.5	0.20	0.01; 0.02	A/A	No
Na	589.0	0.05	0.01; 0.02	A/A	No
Sr	460.7	0.20	0.10; 0.20	N/A	No

*A/A is air/acetylene; N/A is nitrous oxide/acetylene.

Carbon Disulfide

CS$_2$ Formula Wt 76.13

CAS Number 75–15–0

NOTE. Carbon disulfide should be supplied and stored in amber glass containers and protected from direct sunlight.

REQUIREMENTS

Assay . . . . ≥99.9% CS$_2$

MAXIMUM ALLOWABLE

Color (APHA) 10
Residue after evaporation. 0.002%
Hydrogen sulfide (H$_2$S) . . . . Passes test
Sulfur dioxide (SO$_2$) . . . . Passes test
Water (H$_2$O). 0.05%

TESTS

ASSAY. Analyze the sample by gas chromatography using the general parameters cited on page 70. The following specific conditions are also required.

Column: Type I, methyl silicone

Measure the area under all peaks and calculate the carbon disulfide content in area percent. Correct for water content.

COLOR (APHA). (Page 19).

RESIDUE AFTER EVAPORATION. (Page 16). Evaporate 50 g (40 mL) to dryness in a tared dish at 50 to 60 °C, and heat at 60 °C for 1 h.

HYDROGEN SULFIDE AND SULFUR DIOXIDE. Shake vigorously 25 g (20 mL) with 0.2 mL of 0.01 N iodine for 15 s in a glass-stoppered cylinder. The pink color should not disappear. (Limit for H_2S is about 1.5 ppm; limit for SO_2 is about 2.5 ppm.)

WATER. (Page 56, Method 2). Use 20 μL (25 mg) of the sample.

Ceric Ammonium Nitrate
Ammonium Hexanitratocerate(IV)

$(NH_4)_2Ce(NO_3)_6$ Formula Wt 548.22

CAS Number 16774–21–3

REQUIREMENTS

Assay . . . .≥98.5% $(NH_4)_2Ce(NO_3)_6$

MAXIMUM ALLOWABLE

Insoluble in dilute sulfuric acid 0.05%
Chloride (Cl) 0.01%
Phosphate (PO$_4$) 0.02%
Iron (Fe). 0.005%

TESTS

ASSAY. (By titration of oxidative capacity of cerium-IV). Weigh accurately 2.4 to 2.5 g and dissolve in 50 mL of water. From a pipet add 50 mL of 0.1 N ferrous sulfate solution, and swirl until the precipitate that forms is redissolved. Add 10 mL of phosphoric acid and 0.10 mL of diphenylaminesulfonic acid, sodium salt indicator solution. Titrate at once with 0.1 N standard potassium dichromate solution to a change from faint green to violet. Record this titration volume as A mL. Pipet 25 mL of 0.1 N ferrous sulfate solution into 50 mL of water, and treat it in the same way (with phosphoric acid and indicator but without sample). Titrate to a change from green to gray-blue and record this titration volume as B mL.

$$\% (NH_4)_2Ce(NO_3)_6 = \frac{(2B - A) \times 0.05482 \times 100}{\text{Sample wt (g)}}$$

Ferrous Sulfate Solution, 0.1 N. Dissolve 27.8 g of clear ferrous sulfate heptahydrate crystals in about 500 mL of water containing 50 mL of sulfuric acid, and dilute with water to 1 L.

Diphenylaminesulfonic Acid Sodium Salt Indicator Solution. Dissolve 0.10 g of the salt in 100 mL of water.

INSOLUBLE IN DILUTE SULFURIC ACID. To 5.0 g add 10 mL of sulfuric acid, stir, then cautiously add 90 mL of water to dissolve. Heat to boiling and digest in a covered beaker on a hot plate ($\approx$100 °C) for 1 hour. Filter through a tared filtering crucible, wash thoroughly, and dry at 105 °C.

CHLORIDE. (Page 27). Use 0.10 g dissolved in 10 mL of water. The comparison is best made by the general method for chloride in colored solutions.

PHOSPHATE. (Page 32, Method 2). Dissolve 0.25 g in 30 mL of dilute sulfuric acid (1 + 9), add hydrogen peroxide until the solution just turns colorless, then boil to destroy excess peroxide. Cool and dilute with water to 50 mL. Dilute 10 mL with water to 60 mL, and adjust the pH between 2 and 3 (pH paper) with ammonium hydroxide.

> *NOTE.* This neutralization must be made with care to avoid formation of a permanent precipitate that would void the test. If this happens, discard the solution and start with another 10 mL.

Add 0.5 g of ammonium molybdate, and adjust the pH to 1.8 (using a pH meter) with dilute hydrochloric acid (1 + 9). Heat the solution to boiling and cool to room temperature. Dilute with water to 90 mL, add 10 mL of hydrochloric acid, and continue as described. Carry along a standard containing 0.01 mg of phosphate ion (PO_4) in 10 mL of dilute sulfuric acid (3 + 47) treated exactly as the 10 mL of sample.

IRON. (Page 30, Method 2). Dissolve 1.0 g in 30 mL of dilute sulfuric acid (1 + 9) and add, dropwise, 3% hydrogen peroxide solution until the yellow color disappears. For the standard, add 0.05 mg of iron (Fe) to 30 mL of dilute sulfuric acid (1 + 9) and the same volume of 3% hydrogen peroxide solution used for the sample. To each solution, add ammonium hydroxide until the pH is between 1 and 3, cool to room temperature, and adjust the pH to 3.5 (pH meter). Dilute each solution with water to 50 mL, and use 10 mL of each.

Ceric Ammonium Sulfate Dihydrate
Ammonium Tetrasulfatocerate Dihydrate

$(NH_4)_4Ce(SO_4)_4 \cdot 2H_2O$ **Formula Wt 632.58**

CAS Number 10378–47–9

REQUIREMENTS

Assay . . . . $\geq$94% $(NH_4)_4Ce(SO_4)_4 \cdot 2H_2O$

MAXIMUM ALLOWABLE

Insoluble in dilute sulfuric acid. 0.05%
Phosphate (PO_4) 0.03%
Iron (Fe). 0.01%

TESTS

ASSAY. (By titration of the oxidative capacity of cerium IV). Weigh accurately 2.4–2.5 g and place in a beaker. Cover with 10 mL of water and add cautiously 5 mL of sulfuric acid. Stir thoroughly and add water to complete dissolution. Dilute to 75–100 mL with water, add 0.05 mL of ferroin indicator, and titrate with 0.1 N ferrous ammonium sulfate to the red end point. One milliliter of 0.1 N ferrous ammonium sulfate corresponds to 0.06326 g of $(NH_4)_4Ce(SO_4)_4 \cdot 2H_2O$

INSOLUBLE IN DILUTE SULFURIC ACID. To 2.5 g add 5 mL of sulfuric acid, stir, and then cautiously add 90 mL of water to dissolve. Heat to boiling and digest on a hot plate ($\approx$100 °C) for 1 h. Filter through a tared filtering crucible and wash thoroughly, first with 2% sulfuric acid, then with water. Dry at 105 °C.

PHOSPHATE. (Page 32, Method 2). Dissolve 0.75 g by adding 1 mL of sulfuric acid and stirring in 30 mL of water. Add hydrogen peroxide dropwise until just colorless, boil to destroy the excess peroxide, cool, and dilute to 50 mL. Dilute a 2-mL aliquot to 60 mL and adjust the pH to between 2 and 3 with ammonium hydroxide. (NOTE: This neutralization must avoid the formation of a precipitate, which would void the test.) Add 0.5 g of ammonium molybdate and adjust the pH to 1.8 (using a pH meter) with dilute hydrochloric acid. Heat to boiling and cool to room temperature. Dilute with water to 90 mL, add 10 mL of hydrochloric acid, and continue as described. Carry along a standard containing 0.01 mg of phosphate ion (PO_4) in 2.0 mL of sulfuric acid (1 + 29) treated as the 2.0-mL aliquot of the sample.

IRON. (Page 30, Method 2). Dissolve 1 g by adding 2 mL of sulfuric acid and stirring in 30 mL of water. Add dropwise 3% hydrogen peroxide until the color disappears. For the standard, add 0.1 mg of iron to 30 mL of water containing 2 mL of sulfuric acid and the same volume of hydrogen peroxide used in the test. Dilute the standard and sample each to 100 mL, and use 10-mL aliquots.

Chloramine-T Trihydrate
N-Chloro-4-toluenesulfonamide, Sodium Salt, Trihydrate

$H_3CC_6H_4SO_2NClNa \cdot 3H_2O$ **Formula Wt 281.69**

CAS Number 127–65–1

REQUIREMENTS

Assay . . . . .98.0–103.0% $C_7H_7ClNNaO_2S \cdot 3H_2O$
pH of a 5% solution . . . . 8.0–10.0 at 25 °C
Suitability for determination of bromide . . . . Passes test
Clarity of aqueous solution . . . . Passes test

MAXIMUM ALLOWABLE

Insoluble in alcohol 1.5%

TESTS

ASSAY. (By iodometric titration). Weigh accurately 0.5 g, transfer to a glass-stoppered Erlenmeyer flask, and dissolve in 100 mL of water. Add 5 mL of acetic acid and 2 g of potassium iodide, and allow to stand in the dark for 10 min. Titrate the liberated iodine with standardized 0.1 N sodium thiosulfate, using starch indicator solution. One milliliter of 0.1 N sodium thiosulfate corresponds to 0.01408 g of $C_7H_7ClNNaO_2S \cdot 3H_2O$.

pH OF A 5% SOLUTION. (Page 44). The pH should be 8.0–10.0 at 25 °C. Reserve the solution for the clarity test of an aqueous solution.

SUITABILITY FOR DETERMINATION OF BROMIDE. Transfer 0.01 mg and 0.02 mg of bromide ion (Br) (2.0 mL and 4.0 mL of diluted bromide standard solution) to two color-comparison tubes, and dilute each to 50 mL with water. For the blank, use 50 mL of water. To each add 2.0 mL of acetate buffer solution, 2.0 mL of phenol red solution, and 0.5 mL of chloramine-T solution, mixing thoroughly immediately after each addition. Twenty minutes after adding the chloramine-T solution, add, with mixing, 0.5 mL of sodium thiosulfate solution. The color of the solution containing 0.01 mg of bromide should be distinctly more reddish purple than the blank, and the purple color should increase with the higher bromide concentration.

> *Chloramine-T Solution.* Dissolve 0.50 g of the sample in water, and dilute to 100 mL with water.

> *Diluted Bromide Standard Solution, 0.005 mg of Br per mL.* Dilute 5 mL of bromide standard (0.1 mg of Br per mL) to 100 mL with water.

> *Acetate Buffer Solution.* Dissolve 9.0 g of sodium chloride and 6.8 g of sodium acetate trihydrate in about 50 mL of water. Add 3.0 mL of glacial acetic acid, and dilute to 100 mL. The pH should be 4.6–4.7.

> *Phenol Red Solution.* Dissolve 21 mg of the sodium salt of phenol red in water, and dilute to 100 mL.

> *Sodium Thiosulfate Solution, 2 M.* Dissolve 49.6 g of sodium thiosulfate pentahydrate in water, and dilute to 100 mL.

CLARITY OF AQUEOUS SOLUTION. The solution prepared for the test for pH of a 5% solution should be free from turbidity and insolubles, and not more than slightly yellow in color.

INSOLUBLE IN ALCOHOL. Dissolve 5.0 g in 100 mL of reagent alcohol, and stir for 30 min. Filter the solution through a tared medium-porosity sintered-glass filter, and wash with 5 mL of reagent alcohol. Dry the filter at 105 °C to constant weight.

Chloroacetic Acid

Monochloroacetic Acid
Chloroethanoic Acid

$ClCH_2COOH$ **Formula Wt 94.50**

CAS Number 79–11–8

REQUIREMENTS

Assay . . . . ≥99.0% $ClCH_2COOH$

MAXIMUM ALLOWABLE

Insoluble matter. 0.01%
Residue after ignition 0.02%
Carbonyl compounds (as acetone). 0.02%
Other carbonyl compounds. 0.01%
Chloride (Cl) 0.01%
Sulfate (SO_4) 0.02%
Heavy metals (as Pb) 0.001%
Iron (Fe). 0.002%
Substances darkened by sulfuric acid. . . . . Passes test

TESTS

ASSAY. (By acid–base titrimetry). Weigh accurately 3 g, transfer to a conical flask, and dissolve in about 50 mL of water. Add 0.15 mL of phenolphthalein indicator solution, and titrate with 1 N sodium hydroxide. One milliliter of 1 N sodium hydroxide corresponds to 0.0945 g of $ClCH_2COOH$.

INSOLUBLE MATTER. (Page 15). Use 10 g in 100 mL of water.

RESIDUE AFTER IGNITION. (Page 16). Use 5.0 g. Retain the residue for the test for iron.

CARBONYL COMPOUNDS. (Page 50). Use 1.0 g of sample diluted with 4 mL of water and neutralize to pH between 6.5 and 6.8 with 10% sodium hydroxide solution. For the standard, use 0.20 mg of acetone and 0.10 mg of acetylaldehyde.

CHLORIDE. (Page 27). Use 0.1 g.

SULFATE. Dissolve 5.0 g in 30 mL of water, add 25 mg of anhydrous sodium carbonate, evaporate to dryness, and heat over a low flame until the chloroacetic acid is volatilized. Dissolve the residue in 50 mL of water and neutralize with hydrochloric acid. Filter if necessary, wash, and dilute with water to 100 mL. To 5 mL (0.25-g sample), add 1 mL of dilute hydrochloric acid (1 + 19). Dilute to 10 mL and add 1 mL of barium chloride reagent solution. Any turbidity produced in the sample solution after 10 min should not exceed that produced by 0.05 mg of sulfate ion (SO_4) in an equal volume of solution containing the quantities of reagents used in the test.

HEAVY METALS. (Page 28, Method 1). Dissolve 5.0 g in 30 mL of water, neutralize with (1 + 1) ammonium hydroxide, and dilute with water to 50 mL. For the test solution, use 30 mL. For the standard–control solution, add 0.02 mg of lead to 10 mL.

IRON. (Page 30, Method 1). To the residue after ignition add 3 mL of hydrochloric acid, cover with a watch glass, and digest on a hot plate (≈100 °C) for 15 min. Remove the cover and evaporate to dryness. Dissolve the residue with 1 mL of hydrochloric acid and dilute with water to 100 mL. Use 10 mL (0.5-g sample) as the sample solution.

SUBSTANCES DARKENED BY SULFURIC ACID. Heat 1.0 g with 10 mL of 95.0 ± 0.5% sulfuric acid at 50 °C for 5 min. The solution should not have more than a faintly brown color.

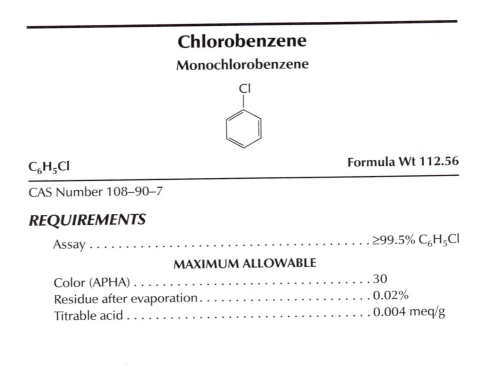

Chlorobenzene

Monochlorobenzene

C_6H_5Cl **Formula Wt 112.56**

CAS Number 108–90–7

REQUIREMENTS

Assay . . . .≥99.5% C_6H_5Cl

MAXIMUM ALLOWABLE

Color (APHA) 30
Residue after evaporation. 0.02%
Titrable acid . . . . .0.004 meq/g

TESTS

ASSAY. Analyze the sample by gas chromatography using the general parameters cited on page 70. The following specific conditions are also required.

Column: Type I, methyl silicone

Measure the area under all peaks and calculate the chlorobenzene content in area percent.

COLOR (APHA). (Page 19).

RESIDUE AFTER EVAPORATION. (Page 16). Evaporate 20 g (18 mL) in a tared dish on a hot plate (≈100 °C), and dry the residue at 105 °C for 30 min.

TITRABLE ACID. Neutralize 200 mL of methanol to a methyl red end point with 0.1 N sodium hydroxide. Add 25 g (23 mL) of sample, and titrate with 0.1 N sodium hydroxide to the same methyl red end point. Not more than 1.0 mL should be required.

Chloroform
Trichloromethane

$CHCl_3$ **Formula Wt 119.38**

CAS Number 67–66–3

Suitable for use in high-performance liquid chromatography, extraction–concentration analysis, ultraviolet spectrophotometry, or general use. Product labeling shall designate the one or more of these uses for which suitability is represented on the basis of meeting the relevant requirements and tests. The ultraviolet spectrophotometry and liquid chromatography suitability requirements include all of the requirements for general use. The extraction–concentration suitability requirements include only the general use requirement for color.

NOTE. Chloroform should be supplied and stored in amber glass containers and protected from direct sunlight. This solvent usually contains additives to retard decomposition. Typical additives include alcohol or mixed amylenes. The substances darkened by sulfuric acid test is not required for chloroform containing mixed amylenes.

REQUIREMENTS

General Use

Assay . . . . ≥99.8% $CHCl_3$

MAXIMUM ALLOWABLE

Color (APHA) 10
Residue after evaporation. 0.001%
Acetone and aldehyde . . . . Passes test
Acid and chloride . . . . Passes test
Free chlorine (Cl) . . . . Passes test
Lead (Pb) . . . . 0.05 ppm
Substances darkened by sulfuric acid . . . . Passes test
Suitability for use in dithizone tests . . . . Passes test

Specific Use

Ultraviolet Spectrophotometry

Wavelength (nm)	Absorbance (AU)
290 to 400	0.01
270	0.05
260	0.15
255	0.25
245	1.00

Liquid Chromatography Suitability

Absorbance . . . . Passes test

Extraction–Concentration Suitability

Absorbance . . . . Passes test
GC–FID . . . . Passes test
GC–ECD . . . . Passes test

TESTS

ASSAY. Analyze the sample by gas chromatography using the general parameters cited on page 70. The following specific conditions are also required.

Column: Type I, methyl silicone

Measure the area under all peaks and calculate the chloroform and active additive content in area percent. The assay is the sum of the chloroform content plus all known active additives.

COLOR (APHA). (Page 19).

RESIDUE AFTER EVAPORATION. (Page 16). Evaporate 100 g (67 mL) to dryness in a tared dish on a hot plate (≈100 °C) and dry the residue at 105 °C for 30 min.

ACETONE AND ALDEHYDE. Shake 15 mL with 20 mL of ammonia-free water for 5 min in a separatory funnel. Allow the layers to separate and transfer 10 mL of the aqueous layer to a 125-mL glass-stoppered flask containing 40 mL of ammonia-free water. Adjust and maintain the temperature of this solution at 25 ± 1 °C. Add 5 mL of Nessler reagent solution and allow to stand for 5 min. No turbidity or precipitate should develop. [Limit about 0.005% as $(CH_3)_2CO$]

ACID AND CHLORIDE. Shake 25 g (17 mL) with 25 mL of water for 5 min, allow the liquids to separate, and draw off the aqueous phase. Add a small piece of blue litmus paper to 10 mL of the aqueous phase. The blue litmus paper should not change color. Add 0.25 mL of silver nitrate reagent solution to another 10 mL of the aqueous phase. No turbidity should be produced in this solution.

FREE CHLORINE. Shake 10 mL for 2 min with 10 mL of water to which 0.10 mL of 10% potassium iodide reagent solution has been added, and allow to separate. The lower layer should not show a violet tint.

LEAD. For the solution, of the sample transfer 40 g (27 mL) to a separatory funnel, add 20 mL of dilute nitric acid (1 + 99), and shake vigorously for 1 min. Allow the phases to separate, draw off and discard the chloroform phase, and use the aqueous phase for the solution of the sample. For the standard, prepare a solution containing 0.002 mg of lead ion (Pb) in 20 mL of dilute nitric acid (1 + 99) in another separatory funnel. To each solution add 4 mL of ammonium cyanide, lead-free, reagent solution (CAUTION. Do this in a well-ventilated hood.) and 5 mL of dithizone standard solution in chloroform, and shake vigorously for one-half min. Allow the phases to separate, draw off the dithizone–chloroform phase into a clean, dry comparison tube, and compare the color to a tube containing only the chloroform using a white background. The purplish hue in the solution of the sample due to any red lead dithizonate present should not exceed that in the standard.

> *NOTE.* When doing this test, all glassware must be carefully cleaned and thoroughly rinsed with warm dilute nitric acid (1 + 1) to remove any adsorbed lead, and finally rinsed with distilled water. All glassware used in the preparation and storage of reagents and in performing the test must be made of lead-free glass.

SUBSTANCES DARKENED BY SULFURIC ACID. To 40 mL in a separatory funnel add 5 mL of sulfuric acid, shake the mixture vigorously for 5 min, and allow the liquids to separate completely. The chloroform layer should be colorless. The acid layer should have no more color than 5 mL of a color standard of the following composition: 0.4 mL of cobaltous chloride solution (5.95 g of $CoCl_2 \cdot 6H_2O$ and 2.5 mL of hydrochloric acid in 100 mL), 1.6 mL of ferric chloride solution (4.50 g of $FeCl_3 \cdot 6H_2O$ and 2.5 mL of hydrochloric acid in 100 mL), 0.4 mL of cupric sulfate solution (6.24 g of $CuSO_4 \cdot 5H_2O$ and 2.5 mL of hydrochloric acid in 100 mL), and 17.6 mL of water.

SUITABILITY FOR USE IN DITHIZONE TESTS.

> *Solution A.* Dissolve 4 mg of dithizone in 200 mL of chloroform.

Alkaline Extractions. To 50 mL of a 5% solution of sodium hydroxide in a 200-mL glass-stoppered flask add 0.025 mg of cadmium ion (Cd) and 25 mL of solution A, shake, and allow to stand for 10 min. The color after standing 10 min should be identical in hue and intensity with that in a similar, freshly prepared solution.

Acid Extractions. Dilute 25 mL of solution A with the chloroform to 100 mL, and transfer 25 mL of this dilution to each of two 200-mL glass-stoppered flasks. Add 25 mL of 0.1 N hydrochloric acid to one flask and 25 mL of water to the other flask, and shake each intermittently for 10 min. The green color in the chloroform layers should be identical. Add 0.5 mg of mercury to the flask to which 0.1 N hydrochloric acid was added, shake, and allow to stand for 10 min. The orange color should be identical in hue and intensity with that in a similar, freshly prepared solution that is allowed to stand for only 1 min instead of 10 min. Since mercury dithizonate is slightly light-sensitive, both flasks should be shaken for about 15 s before final comparison is made.

> *NOTE.* Dithizone is extremely sensitive to various metal ions. Therefore, the distilled water, 0.1 N hydrochloric acid, and the 5% solution of sodium hydroxide must be free of metal ions that react with dithizone. In case of doubt, the solutions must be freed of such metals before use by shaking with portions of chloroform–dithizone solution until the color of the dithizone remains unchanged.

ULTRAVIOLET SPECTROPHOTOMETRY. Use the procedure on page 73 to determine the absorbance.

LIQUID CHROMATOGRAPHY SUITABILITY. Use the procedure on page 73 to determine the absorbance.

EXTRACTION–CONCENTRATION SUITABILITY. Analyze the sample, using the general procedure cited on page 72. In the sample preparation step, exchange the sample with ACS-grade isooctane suitable for extraction–concentration. Use the procedure on page 73 to determine the absorbance.

Chloroplatinic Acid Hexahydrate
Hexachloroplatinic Acid Hexahydrate
Platinic Chloride Hexahydrate

$H_2PtCl_6 \cdot 6H_2O$ **Formula Wt 517.90**

CAS Number 16941–12–1

REQUIREMENTS

Assay . . . . ≥37.50% Pt

MAXIMUM ALLOWABLE

Solubility in alcohol . . . . Passes test
Alkali and other salts (as sulfates) 0.05%
Suitability for potassium determinations . . . . Passes test

TESTS

ASSAY. (By electrolytic determination of platinum content). Using precautions to prevent absorption of moisture, weigh accurately 1 g, and transfer to a tared platinum dish of about 100-mL capacity. Dissolve in 80 mL of water and add 2 mL of sulfuric acid. Cover the dish with a split watch glass and electrolyze the solution for 4 hours at a current of 0.5 A and a temperature 55 to 60 °C and 1½ hours at 1 A, using a rotating platinum anode and the platinum dish as the cathode. Wash the cover several times during the electrolysis. When the electrolysis is complete, transfer the solution to a second tared platinum dish, and reserve for the alkali and other salts test. Rinse the platinum dish and deposited platinum with water, dry, and ignite at 600 °C for 5 min. The weight of the deposit should not be less than 37.50% of the sample weight. All weighings must be determined to ±0.00002 g.

SOLUBILITY IN ALCOHOL. Dissolve 1 g in 10 mL of alcohol and allow to stand, with occasional stirring, for 15 min. The solution should contain no more than traces of insoluble matter.

ALKALI AND OTHER SALTS. Evaporate the solution retained from the assay in a tared dish and ignite at 800 ± 25 °C for 30 min.

SUITABILITY FOR POTASSIUM DETERMINATIONS. Dissolve 7.456 g of potassium chloride, previously dried at 105 °C, in water and dilute with water to 1 L in a volumetric flask. To a 10-mL aliquot of this solution add 0.100 to 0.110 g of sodium chloride and dilute with water to 100 mL. Heat on a hot plate (≈100 °C) and add between 0.90 and 1.00 g of the chloroplatinic acid crystals dissolved in about 1 mL of water. Evaporate the solution on a hot plate (≈100 °C) to a moist residue. (Do not evaporate to dryness, as this will render the sodium salt insoluble.) Cool and add 10 mL of absolute ethyl alcohol, crush the crystals with a glass stirring rod flattened on the end, and allow the mixture to stand for 30 min. Filter by decantation through a previously dried and weighed fritted-glass crucible. Add 10 mL of absolute ethyl alcohol to the residue, grind the residue in the beaker, and filter again by decantation. When the filtrate runs clear, transfer the precipitate to the crucible, and wash with three 10-mL portions of alcohol. Discard the filtrate. Dry the filtering crucible in an oven at 105 °C for 1 h, cool, and weigh. The weight of the residue must not be less than 0.2410 g nor more than 0.2450 g.

Chromium Potassium Sulfate Dodecahydrate
Chromium(III) Potassium Sulfate Dodecahydrate

$CrK(SO_4)_2 \cdot 12H_2O$ Formula Wt 499.40

CAS Number 7788–99–0

REQUIREMENTS

Assay .98.0–102.0% as $CrK(SO_4)_2 \cdot 12H_2O$

MAXIMUM ALLOWABLE

Insoluble matter. 0.01%
Chloride (Cl) 0.002%
Aluminum (Al) 0.02%
Iron (Fe). 0.01%
Ammonium (NH_4) 0.01%
Heavy metals (as Pb) 0.01%

TESTS

ASSAY. (By iodometric titration for oxidative capacity of Cr-VI). Weigh accurately 0.65 g and transfer to a 500-mL iodine flask with 50 mL of water. Add 5 mL of 30% hydrogen peroxide and 15 mL of 10% sodium hydroxide solution. Dilute with 100 mL of water and boil until the solution color is yellow. Add dropwise 5 mL of 5% nickel sulfate hexahydrate solution. When the vigorous evolution of oxygen has ceased, boil for 3 min and cool. Add 1.0 g of potassium iodide and 5 mL each of phosphoric and hydrochloric acids. Mix well and titrate the liberated iodine with 0.1 N sodium thiosulfate, adding 3 mL of starch indicator solution near the end of the titration. One milliliter of 0.1 N sodium thiosulfate corresponds to 0.01665 g of $CrK(SO_4)_2 \cdot 12H_2O$.

INSOLUBLE MATTER. (Page 15). Use 10 g dissolved in 100 mL of water.

CHLORIDE. Dissolve 1.0 g in 10 mL of water, heat the solution to boiling, and add 2 mL of ammonium hydroxide to the hot, constantly stirred solution. Boil gently to expel excess ammonia, filter through a small chloride-free filter, wash with hot water until the volume of filtrate and washings is about 45 mL, dilute with water to 50 mL, and mix. To 25 mL of this solution add 1.5 mL of nitric acid and 1 mL of silver nitrate reagent solution. Any turbidity should not exceed that produced by 0.01 mg of chloride ion (Cl) in an equal volume of solution containing the quantities of reagents used in the test.

ALUMINUM AND IRON. (By flame AAS, page 39).

Sample Stock Solution. Dissolve 20.0 g of sample in 60 mL of water, transfer to a 100-mL volumetric flask, and dilute to the mark with water (1 mL = 0.2 g).

Element	Wavelength (nm)	Sample Wt (g)	Standard Added (mg)	Flame Type*	Background Correction
Al	309.3	4.0	1.0; 2.0	N/A	Yes
Fe	248.3	4.0	0.20; 0.40	A/A	Yes

*N/A is nitrous oxide/acetylene, A/A is air/acetylene.

AMMONIUM. (Page 25). Dilute the distillate from a 1.0 g sample with water to 50 mL, and use 5 mL of the dilution. For the standard, use 0.01 mg of ammonium ion (NH_4).

HEAVY METALS. (Page 28, Method 1). Dissolve 1.0 g and 30 mg of mercuric chloride in 40 mL of water. Add 15 mL of hydrogen sulfide water, stir, and allow to stand for 30 min. Filter and wash thoroughly with hydrogen sulfide water containing 1% potassium sulfate. Ignite the precipitate at low temperature in a small porcelain dish in a well-ventilated hood to char the paper, then at 525 ± 25 °C for 30 min. To the residue add 1 mL of hydrochloric acid, 1 mL of nitric acid, and about 30 mg of sodium carbonate, and evaporate to dryness on a hot plate (≈ 100 °C). Dissolve the residue in about 20 mL of water, and dilute with water to 100 mL. Use 20 mL to prepare the sample solution.

Chromium Trioxide
Chromium(VI) Oxide

CrO_3 **Formula Wt 99.99**

CAS Number 1333–82–0

REQUIREMENTS

Assay . . . . ≥98.0% CrO_3

MAXIMUM ALLOWABLE

Insoluble matter. 0.01%
Chloride (Cl) 0.005%
Nitrate (NO_3). 0.05%
Sulfate (SO_4) 0.005%
Sodium (Na) 0.2%
Aluminum, barium, and iron 0.03%

TESTS

ASSAY. (By iodometric titration for oxidative capacity). Weigh accurately 5 g, transfer to a 1000-mL volumetric flask, dissolve in water, dilute with water to volume, and mix thoroughly. Place a 25-mL aliquot of this solution in a glass-stoppered conical flask and dilute with 100 mL of water. Add 5 mL of dilute sulfuric acid (1 + 1) and 3 g of potassium iodide and allow to stand in the dark for 15 min. Dilute with 100 mL of water and titrate the liberated iodine with 0.1 N sodium thiosulfate, adding 3 mL of starch indicator solution near the end of the titration. Correct for a complete blank. One milliliter of 0.1 N sodium thiosulfate corresponds to 0.003333 g of CrO_3.

INSOLUBLE MATTER. (Page 15). Use 10.0 g dissolved in 100 mL of water.

CHLORIDE. Dissolve 1.0 g in water, filter if necessary through a chloride-free filter, and dilute with water to 50 mL. To 10 mL of this solution, add 1.5 mL of ammonium hydroxide and 1 mL of silver nitrate reagent solution, mix, and add 2 mL of nitric acid. Any turbidity should not exceed that produced in a standard containing 1 mL of ammonium hydroxide, 1 mL of silver nitrate reagent solution, 2 mL of nitric acid, and 0.01 mg of added chloride ion (Cl). The comparison is best made by the general method for chloride in colored solutions, page 27.

NITRATE.

Sample Stock Solution. Dissolve 1.0 g in about 20 mL of water. Add 10% sodium hydroxide reagent solution until the color of the solution just turns light yellow, then add 0.2 mL of glacial acetic acid to change the color to orange (pH about 6.0–6.5). Slowly add a solution of 5 g of lead acetate in about 10 mL of water, dilute to 50 mL with water, mix, and allow to stand for 15 min. Transfer to a 50-mL centrifuge tube, centrifuge, and decant through a filter. *This solution must be clear and colorless.*

NOTE. Two filter papers may be required to obtain a clear, colorless filtrate. Use a Nessler tube to check.

Sample Solution A. Dissolve 0.1 g of mercuric acetate in 2.0 mL of sample stock solution plus 1.0 mL of water in a dry test tube. Add 0.05 g of urea and shake to dissolve. Put the tube in an ice bath and immediately, but slowly, add 7 mL of cold chromotropic acid reagent while swirling. Keep the tube in the ice bath an additional 2–3 min, remove, and let stand for 30 min, swirling occasionally.

Control Solution B. Dissolve 0.1 g of mercuric acetate in 2.0 mL of sample stock solution in a dry test tube. Add 1.0 mL of a solution containing 0.02 mg of nitrate ion (NO_3) per mL, then add 0.05 g of urea and shake to dissolve. Put the tube in an ice bath and immediately, but slowly, add 7 mL of cold chromotropic acid reagent while swirling. Keep the tube in the ice bath an additional 2–3 min, remove, and let stand for 30 min, swirling occasionally.

Chromotropic Acid Reagent. Dissolve 0.02 g of recrystallized chromotropic acid, disodium salt, in concentrated sulfuric acid and dilute to 200 mL with the acid. Store in an amber bottle.

Recrystallization of Reagent. Add solid sodium sulfate to a saturated aqueous solution of the foregoing salt. Filter the resultant crystals, using suction with a Buchner funnel, wash with ethyl alcohol, and air dry. Transfer the crystals to a beaker, add sufficient water to redissolve, and then add just enough sodium sulfate to reprecipitate the salt. Filter as before, wash with ethyl alcohol, and dry at a temperature no higher than 80 °C. Store in an amber bottle.

NOTE. If an orange or red precipitate forms in the solution of the sample before the addition of the reagent, discard the solution and start again. Immediate addition of the reagent should prevent formation of such a precipitate.

Blank Solution C. Dissolve 0.1 g of mercuric acetate in 3.0 mL of water in a dry test tube. Add 0.05 g of urea, shake to dissolve, and put the tube in an ice bath. Add 7 mL of cold chromotropic acid reagent while swirling, remove from the bath, and let stand for 30 min.

Transfer sample solution A, control solution B, and blank solution C to dry 15-mL centrifuge tubes and centrifuge until the supernatant liquid is clear. Set a spectrophotometer at 405 nm and, using 1-cm cells, adjust the instrument to read zero (absorbance units) with blank solution C in the light path, then determine the absorbance of sample solution A. Adjust the instrument to read zero (absorbance units) with sample solution A in the light path and determine the absorbance of control solution B. The absorbance of sample solution A should not exceed that of control solution B.

SULFATE. Dissolve 10.0 g in 350 mL of water and add 5.0 g of sodium carbonate. Heat to boiling and add 35 mL of a solution containing 1 g of barium chloride and 2 mL of hydrochloric acid in 100 mL of solution. Digest in a covered beaker on a hot plate ($\approx$100 °C) for 2 h and allow to stand overnight. If a precipitate is formed, filter, wash thoroughly, and ignite. Fuse the ignited precipitate with 1 g of sodium carbonate. Extract the fused mass with water and filter out the insoluble residue. Add 5 mL of hydrochloric acid to the filtrate, dilute with water to about 200 mL, heat to boiling, and add 10 mL of alcohol. Digest on a hot plate ($\approx$100 °C) until the reduction of chromate is complete, as indicated by the change to a clear green or colorless solution. Neutralize the solution with ammonium hydroxide and add 2 mL of hydrochloric acid. Heat to boiling and add 10 mL of barium chloride reagent solution. Digest in a covered beaker on a hot plate ($\approx$100 °C) for 2 h and allow to stand overnight. Filter, wash thoroughly, and ignite. Correct for the weight obtained in a complete blank test.

SODIUM. (By flame AAS, page 39).

Sample Stock Solution. Dissolve 1.0 g of sample in a 100-mL volumetric flask, and dilute to the mark with water (1 mL = 0.01 g).

Element	Wavelength (nm)	Sample Wt (g)	Standard Added (mg)	Flame Type*	Background Correction
Na	589.0	0.01	0.01; 0.02	A/A	No

*A/A is air/acetylene.

ALUMINUM, BARIUM, AND IRON. Dissolve 5.0 g in 100 mL of water, and add 10 mL of ammonium hydroxide and 1 mL of 30% hydrogen peroxide. Heat the solution to boiling and boil gently for 5 min. Cool, filter, and wash the precipitate with water. Dissolve the precipitate in warm dilute hydrochloric acid (1 + 1), collecting the filtrate in the original beaker. Dilute the filtrate with water to about 30 mL, adjust to pH 7 with ammonium hydroxide, and add 0.5 mL of 30% hydrogen peroxide. Heat the solution to boiling and boil gently for 5 min, cool, filter, and ignite.

Chromotropic Acid Disodium Salt
4,5-Dihydroxy-2,7-naphthalenedisulfonic Acid, Disodium Salt

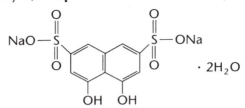

$(HO)_2C_{10}H_4(SO_3Na)_2 \cdot 2H_2O$ **Formula Wt 400.29**

CAS Number 5808–22–0

> *NOTE.* Two forms of this reagent are available: the disodium salt dihydrate and the free acid, $4,5\text{-}(OH)_2C_{10}H_4\text{-}2,7\text{-}(SO_3H)_2$ (formula wt 320.29). This specification applies to both forms, but the sample weight of the acid should be changed to 80% of that for the salt.

REQUIREMENTS

Appearance . . . . Off-white powder
Clarity of solution. . . . . Passes test
Sensitivity to nitrate . . . . Passes test
Sensitivity to formaldehyde . . . . Passes test

TESTS

CLARITY OF SOLUTION. Use 5.0 g of the salt or 4.0 g of the acid dissolved in 100 mL of hot water. Add a few drops of sodium hydroxide reagent solution if necessary. The solution should be free of turbidity.

> *Sample Solution A.* Dissolve 0.20 g of the salt (0.16 g of the acid) in 100 mL of sulfuric acid.

SENSITIVITY TO NITRATE. Into three 50-mL volumetric flasks pipet, respectively, 1.0, 2.0, and 3.0 mL of a standard solution containing 0.01 mg/mL of nitrate ion (NO_3). For the blank add 1.0 mL of water to another 50-mL volumetric flask. To each flask, add approximately 5 mL of sulfuric acid, swirl, and cool in an ice bath. Add 2.0 mL of sample solution A to each flask, swirl again, and cool. Dilute to 50 mL with sulfuric acid and shake well. Set a spectrophotometer at 410 nm and, using 5-cm cells, determine the absorbance of the standard solutions against the blank. A plot of absorbance versus concentration of nitrate should be linear, with a minimum absorbance of 0.20 for the solution containing 0.03 mg of nitrate.

SENSITIVITY TO FORMALDEHYDE. Into three 50-mL volumetric flasks pipet, respectively, 1.0, 2.0, and 3.0 mL of standard formaldehyde solution. For the blank, add 1.0 mL of water to another 50-mL volumetric flask. To each flask, add 5.0 mL of sample solution A, stopper, and shake. Dilute to the mark with sulfuric acid and let stand in a hot-water bath (70–80 °C) for 15 min. Cool in an ice bath to room temperature, fill to volume with sulfuric acid, stopper, and shake. Set a spectrophotometer at 580 nm and, using 1-cm cells, determine the absorbance of the standard solutions against the blank. A plot of absorbance versus concentration of formaldehyde should be linear, with a minimum absorbance of 0.30 for the solution containing 0.03 mg of formaldehyde.

Citric Acid, Anhydrous, and Citric Acid, Monohydrate
2-Hydroxy-1,2,3-propanetricarboxylic Acid

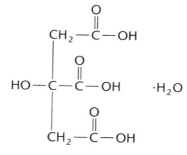

HOCOCH$_2$C(OH)(COOH)CH$_2$COOH   **Formula Wt. 192.13**

HOCOCH$_2$C(OH)(COOH)CH$_2$COOH · H$_2$O   **Formula Wt. 210.14**

CAS Number 77–92–9 (Anhydrous); 5949–29–1 (Monohydrate)

NOTE: This reagent is available as either the anhydrous or monohydrate form. The label should identify the specific form.

REQUIREMENTS

Assay . . . . ≥99.5% $C_6H_8O_7$
99.0–102.0% $C_6H_8O_7 \cdot H_2O$

MAXIMUM ALLOWABLE

Insoluble matter. 0.005%
Residue after ignition 0.02%
Chloride (Cl) 0.001%
Oxalate (C_2O_4). . . . . Passes test
Phosphate (PO_4) 0.001%
Sulfate (SO_4) 0.002%
Iron (Fe). . . . . 3 ppm
Lead (Pb) . . . . 2 ppm
Substances carbonizable by hot sulfuric acid . . . . Passes test

TESTS

ASSAY. (By acid–base titrimetry). Weigh accurately 2.6–2.9 g, dissolve in about 40 mL of water, and add 0.15 mL of phenolphthalein indicator solution. Titrate with 1 N sodium hydroxide to a pink color that persists for 5 min. One milliliter of 1 N sodium hydroxide corresponds to 0.06404 g of $C_6H_8O_7$ or 0.07005 g of $C_6H_8O_7 \cdot H_2O$.

INSOLUBLE MATTER. (Page 15). Use 20 g dissolved in 150 mL of water.

RESIDUE AFTER IGNITION. (Page 16). Ignite 5.0 g. Use 2 mL of sulfuric acid to moisten the residue.

CHLORIDE. (Page 27). Use 1.0 g.

OXALATE. Dissolve 5 g in 25 mL of water and add 2 mL of 10% calcium acetate solution. No turbidity or precipitate should appear after standing 4 h. (Limit about 0.05%)

PHOSPHATE. (Page 32, Method 1). Mix 2.0 g with 0.5 g of magnesium nitrate in a platinum dish and ignite. Dissolve the residue in 5 mL of water, add 5 mL of nitric acid, and evaporate to dryness. Dissolve the residue in 25 mL of approximately 0.5 N sulfuric acid and continue as described.

SULFATE. To 3.0 g add 0.2–0.5 mg of ammonium metavanadate and 10 mL of nitric acid. Digest in a covered beaker on a hot plate (≈100 °C) until the reaction ceases, remove the cover, and evaporate to dryness. Add 10 mL of nitric acid and repeat the digestion and evaporation. Add 5 mL of dilute hydrochloric acid (1 + 1) and evaporate to dryness. Dissolve the residue in 4 mL of water plus 1 mL of dilute hydrochloric acid (1 + 19), filter if necessary through a small filter, wash with two 2-mL portions of water, and dilute with water to 10 mL. Add 1 mL of barium chloride reagent solution. Any turbidity should not exceed that produced by 0.06 mg of sulfate ion (SO_4) treated in the same manner as the sample. Compare 10 min after adding the barium chloride to the sample and standard solutions.

IRON. (Page 30, Method 1). Use 3.3 g.

LEAD. Dissolve 1.5 g in 10 mL of water, adjust the pH to between 9 and 10 (using a pH meter) with ammonium hydroxide, and dilute with water to 15 mL. Prepare a control containing 0.003 mg of lead and the amount of ammonium hydroxide required to neutralize the sample, and evaporate it to dryness on a hot plate ($\approx$100 °C). Moisten the residue from the control with 0.1 mL of nitric acid, dissolve in 5 mL of water, and add 5 mL of ammonium citrate (lead-free) reagent solution. Adjust the pH to between 9 and 10 (using a pH meter) with ammonium hydroxide and dilute with water to 15 mL. Transfer the sample solution and the control solution to separatory funnels, add to each 5 mL of dithizone extraction reagent solution in chloroform, and shake vigorously for 2 min. Let the layers separate, draw off the dithizone–chloroform layers into clean, dry comparison tubes, and dilute to 10 mL with chloroform. Any red color producing a purplish hue in the chloroform from the sample solution should not exceed that in the chloroform from the control.

SUBSTANCES CARBONIZABLE BY HOT SULFURIC ACID (TARTRATES, ETC.). Carefully powder about 1.0 g of the sample and take precautions to prevent contamination during the powdering. Mix 0.30 g with 10 mL of sulfuric acid in a test tube previously rinsed with the acid. Heat the mixture at 110 °C (in a brine bath) for 30 min and keep the test tube covered during the heating. Any color should not exceed that produced in a color standard of the following composition: 1 part of cobaltous chloride solution (5.95 g of $CoCl_2 \cdot 6H_2O$ and 2.5 mL of hydrochloric acid in 100 mL), 2.4 parts of ferric chloride solution (4.5 g of $FeCl_3 \cdot 6H_2O$ and 2.5 mL of hydrochloric acid in 100 mL), 0.4 part of cupric sulfate solution (6.24 g of $CuSO_4 \cdot 5H_2O$ and 2.5 mL of hydrochloric acid in 100 mL), and 6.2 parts of water.

Cobalt Acetate Tetrahydrate
Cobalt(II) Acetate Tetrahydrate
Cobaltous Acetate Tetrahydrate

$Co(CH_3COO)_2 \cdot 4H_2O$

Formula Wt 249.08

CAS Number 6147–53–1

REQUIREMENTS

Assay . . . . 98.0–102.0% $Co(CH_3COO)_2 \cdot 4H_2O$

MAXIMUM ALLOWABLE

Insoluble matter. 0.01%
Chloride (Cl) 0.002%
Nitrate (NO_3) 0.01%
Sulfate (SO_4) 0.005%
Calcium (Ca) 0.005%
Copper (Cu). 0.002%
Iron (Fe) 0.001%
Lead (Pb) 0.001%
Magnesium (Mg) 0.005%
Nickel (Ni) 0.1%
Potassium (K). 0.01%
Sodium (Na). 0.05%
Zinc (Zn) 0.01%

TESTS

ASSAY. (By complexometric titration of cobalt). Weigh accurately about 1.0 g and transfer to a 400-mL beaker with the aid of 200 mL of water. Add 0.2 g of ascorbic acid and about 100 mg of murexide indicator mixture. Titrate with 0.1 M EDTA to a color change from yellow to violet that persists for at least 30 seconds. One milliliter of 0.1 M EDTA corresponds to 0.02491 g of $Co(CH_3COO)_2 \cdot 4H_2O$.

INSOLUBLE MATTER. (Page 15). Use 10.0 g dissolved in 100 mL of water containing 1 mL of glacial acetic acid. Reserve the filtrate without washings for the test for sulfate.

CHLORIDE. (Page 27). Use 0.5 g. The comparison is best made by the general method for chloride in colored solutions on page 27.

NITRATE. (Page 30). Dissolve 1.0 g in 10 mL of water. Add this solution in small portions and with constant stirring to 10 mL of 10% sodium hydroxide reagent solution. Digest in a covered beaker on a hot plate ($\approx$100 °C) for 15 min. Cool, dilute with water to 20 mL, and filter.

> **Sample Solution A.** To 4 mL of the preceding filtrate, add 6 mL of water. Dilute to 50 mL with brucine sulfate reagent solution.

> **Control Solution B.** To 4 mL of the preceding filtrate, add 4 mL of water and 2 mL of the standard nitrate solution containing 0.01 mg of nitrate per mL. Dilute to 50 mL with brucine sulfate reagent solution.

Continue with the procedure described on page 30, starting with the preparation of blank solution C.

SULFATE. (Page 33, Method 1). Use 1/10 of the filtrate from the insoluble matter test. Allow 30 min for the turbidity to develop.

CALCIUM, COPPER, IRON, LEAD, MAGNESIUM, NICKEL, POTASSIUM, SODIUM, AND ZINC. (By flame AAS, page 39).

Sample Stock Solution. Dissolve 25.0 g in water in a 250-mL volumetric flask, add 2.5 mL of nitric acid, and dilute with water to volume (1 mL = 0.10 g).

Element	Wavelength (nm)	Sample Wt (g)	Standard Added (mg)	Flame Type*	Background Correction
Ca	422.7	0.80	0.02; 0.04	N/A	No
Cu	324.8	2.0	0.02; 0.04	A/A	Yes
Fe	248.3	2.50	0.025; 0.05	A/A	Yes
Pb	217.0	2.0	0.02; 0.04	A/A	Yes
Mg	285.2	0.20	0.005; 0.01	A/A	Yes
Ni	232.0	0.10	0.05; 0.10	A/A	Yes
K	766.5	0.20	0.01; 0.02	A/A	No
Na	589.0	0.04	0.01; 0.02	A/A	No
Zn	213.9	0.20	0.01; 0.02	A/A	Yes

*A/A is air/acetylene; N/A is nitrous oxide/acetylene.

Cobalt Chloride Hexahydrate
Cobalt(II) Chloride Hexahydrate
Cobaltous Chloride Hexahydrate

$CoCl_2 \cdot 6H_2O$ **Formula Wt 237.93**

CAS Number 7791–13–1

REQUIREMENTS

Assay . . . . 98.0–102.0% $CoCl_2 \cdot 6H_2O$

MAXIMUM ALLOWABLE

Insoluble matter. 0.01%
Nitrate (NO_3) 0.01%
Sulfate (SO_4) 0.01%
Ammonium (NH_4) 0.005%
Calcium (Ca) 0.005%
Copper (Cu). 0.002%
Iron (Fe). 0.005%
Magnesium (Mg) 0.005%
Nickel (Ni) 0.1%
Potassium (K). 0.01%
Sodium (Na) 0.05%
Zinc (Zn) 0.03%

TESTS

ASSAY. (By complexometric titration of cobalt). Weigh, to the nearest 0.1 mg, 1.0 g of sample and dissolve in 200 mL of water. Add about 0.2 g of ascorbic acid and 5 g of ammonium acetate followed by 10 mL of ammonium hydroxide. Add about 0.1 g of murexide indicator mixture, and titrate with 0.1 M EDTA until the end point is signaled by a color change to violet. The color change must persist for 30 s. One milliliter of 0.1 M EDTA corresponds to 0.02379 g of $CoCl_2 \cdot 6H_2O$.

INSOLUBLE MATTER. (Page 15). Use 10.0 g dissolved in 100 mL of water.

NITRATE. Dissolve 1.0 g in 10 mL of water. Add this solution in small portions and with constant stirring to 10 mL of 10% sodium hydroxide reagent solution. Digest in a covered beaker on a hot plate ($\approx$100 °C) for 15 min. Cool, dilute with water to 20 mL, and filter.

> **Sample Solution A.** To 4 mL of the preceding filtrate add 6 mL of water. Dilute to 50 mL with brucine sulfate reagent solution.

> **Control Solution B.** To 4 mL of the preceding filtrate add 4 mL of water and 2 mL of the standard nitrate solution containing 0.01 mg of nitrate ion per mL. Dilute to 50 mL with brucine sulfate reagent solution.

Continue with the procedure described on page 30, starting with preparation of blank solution C.

SULFATE. (Page 33, Method 1). Use 1/20 of the filtrate from the test for insoluble matter. Allow 30 min for the turbidity to develop.

AMMONIUM. (Page 25). Dilute the distillate from a 1.0-g sample with water to 50 mL, and use 10 mL of the dilution. For the standard, use 0.01 mg of ammonium ion (NH_4).

CALCIUM, COPPER, IRON, MAGNESIUM, NICKEL, POTASSIUM, SODIUM, AND ZINC. (By flame AAS, page 39).

> **Sample Stock Solution.** Dissolve 20.0 g of sample in water, add 2 mL of nitric acid, and dilute to the mark with water in a 200-mL volumetric flask (1 mL = 0.10 g).

Element	Wavelength (nm)	Sample Wt (g)	Standard Added (mg)	Flame Type*	Background Correction
Ca	422.7	0.80	0.02; 0.04	N/A	No
Cu	324.8	2.0	0.04; 0.08	A/A	Yes
Fe	248.3	2.0	0.05; 0.10	A/A	Yes
Mg	285.2	0.20	0.005; 0.01	A/A	Yes
K	766.5	0.20	0.01; 0.02	A/A	No
Na	589.0	0.04	0.01; 0.02	A/A	No
Ni	232.0	0.10	0.05; 0.10	A/A	Yes
Zn	213.9	0.10	0.01; 0.02	A/A	Yes

*A/A is air/acetylene; N/A is nitrous oxide/acetylene.

Cobalt Nitrate Hexahydrate
Cobalt(II) Nitrate Hexahydrate
Cobaltous Nitrate Hexahydrate

$Co(NO_3)_2 \cdot 6H_2O$

Formula Wt 291.03

CAS Number 10026–22–9

REQUIREMENTS

Assay . . . . 98.0–102.0% $Co(NO_3)_2 \cdot 6H_2O$

MAXIMUM ALLOWABLE

Insoluble matter. 0.01%
Chloride (Cl) 0.002%
Sulfate (SO_4) 0.005%
Ammonium (NH_4) 0.2%
Calcium (Ca) 0.005%
Copper (Cu). 0.002%
Iron (Fe). 0.001%
Lead (Pb) 0.002%
Magnesium (Mg) 0.005%
Nickel (Ni) 0.15%
Potassium (K). 0.01%
Sodium (Na) 0.05%
Zinc (Zn) 0.01%

TESTS

ASSAY. (By complexometric titration of cobalt). Weigh, to the nearest 0.1 mg, 1.0 g of sample and dissolve in 200 mL of water. Add about 0.2 g of ascorbic acid and 5 g of ammonium acetate followed by 10 mL of ammonium hydroxide. Add about 0.1 g of murexide indicator mixture, and titrate with 0.1 M EDTA until the end point is signaled by a color change to violet. The color change must persist for 30 s. One milliliter of 0.1 M EDTA corresponds to 0.02910 g of $Co(NO_3)_2 \cdot 6H_2O$.

INSOLUBLE MATTER. (Page 15). Use 10 g dissolved in 100 mL of water.

CHLORIDE. Dissolve 0.50 g in 15 mL of water plus 1 mL of nitric acid, filter if necessary through a small chloride-free filter, and add 1 mL of silver nitrate reagent solution. Any turbidity should not exceed that produced by 0.01 mg of chloride in an equal volume of solution containing the quantities of reagents used in the test. The comparison is best made by the general method for chloride in colored solutions, page 27.

SULFATE. (Page 33, Method 2). Use 4 mL of dilute hydrochloric acid $(1 + 1)$. Evaporate twice.

AMMONIUM. (Page 25). Dilute the distillate from a 1.0-g sample with water to 100 mL, and use 5 mL of the dilution. For the standard, use 0.1 mg of ammonium ion (NH_4).

CALCIUM, COPPER, IRON, LEAD, MAGNESIUM, NICKEL, POTASSIUM, SODIUM, AND ZINC. (By flame AAS, page 39).

Sample Stock Solution. Dissolve 20.0 g of sample in water, add 2 mL of nitric acid, and dilute to the mark with water in a 200-mL volumetric flask (1 mL = 0.1 g).

Element	Wavelength (nm)	Sample Wt (g)	Standard Added (mg)	Flame Type*	Background Correction
Ca	422.7	0.80	0.02; 0.04	N/A	No
Cu	324.8	2.0	0.04; 0.08	A/A	Yes
Fe	248.3	2.0	0.02; 0.04	A/A	Yes
Pb	217.0	2.0	0.04; 0.08	A/A	Yes
Mg	285.2	0.20	0.005; 0.01	A/A	Yes
Ni	232.0	0.10	0.075; 0.15	A/A	Yes
K	766.5	0.20	0.01; 0.02	A/A	No
Na	589.0	0.04	0.01; 0.02	A/A	No
Zn	213.9	0.10	0.01; 0.02	A/A	Yes

*A/A is air/acetylene; N/A is nitrous oxide/acetylene.

Copper

Cu   **Atomic Wt 63.55**

CAS Number 7440–50–8

REQUIREMENTS

Assay . . . . ≥99.90% Cu

MAXIMUM ALLOWABLE

Insoluble in dilute nitric acid 0.02%
Antimony and tin (as Sn) 0.01%
Arsenic (As) . . . . 5 ppm
Iron (Fe) 0.005%
Lead (Pb) 0.005%
Manganese (Mn) 0.001%

Silver (Ag) 0.002%

Phosphorus (P) 0.001%

TESTS

ASSAY. (By electrolytic determination of copper content). Accurately weigh 5.0050–5.0070 g of the sample and transfer this weighed sample to a tall-form lip-less beaker provided with a close-fitting cover glass. Add 42 mL of a solution that contains 10 mL of sulfuric acid plus 7 mL of nitric acid and 25 mL of water, cover, and allow to stand until the active reactions have subsided. Heat at 80–90 °C until the copper is completely dissolved and brown nitrous fumes are completely expelled. Wash down the cover glass and the sides of the beaker and dilute the solution sufficiently with water so that it will cover a cathode cylinder. Insert a tared cathode and an anode and electrolyze at a current density of about 0.6 A per dm^2 for about 16 h. If electrolysis of the copper is not completed, as indicated by copper plating on a new surface of the cathode stem when the level of the solution is raised, continue the electrolysis until all of the copper is deposited.

When all of the copper is deposited, slowly lower the beaker, without inter-rupting the current, and wash the electrodes continuously with a stream of water from a wash bottle while removing the beaker. Reserve this solution for preparing sample solution A. Rinse the electrodes by raising and lowering a beaker of water around them. Remove the beaker of rinse water, discontinue the current, and remove the electrodes. Wash the cathode in acetone and dry it in an oven at 110 °C for 3–5 min. Cool, weigh, and calculate the percentage of copper from the weight of the deposit. This value may have to be corrected for a small amount of undeposited copper as determined in preparing sample solution A.

Sample Solution A. Place the anode in a 150-mL beaker and add 10 mL of dilute nitric acid (1 + 1) containing 0.20 mL of alcohol. Heat for 5 min on a hot plate (≈100 °C), rotating the anode so that any deposit is dissolved in the dilute nitric acid. Remove the anode and wash it with a stream of water. Add this solution and washings to the electrolyte from which the copper was deposited. Evaporate the electrolyte to fumes of sulfur trioxide and continue until the volume is reduced to approximately 5 mL. Cool, add 50 mL of water, cool, and transfer the solution to a 100-mL volumetric flask. Dilute with water to 100 mL and mix (1 mL = 0.05 g). To 20 mL of sample solution A (1-g sample) add an excess of ammonium hydroxide. If the presence of copper is indicated by a blue color, determine the amount colorimetrically, calculate the amount in sample solution A, and add to the amount deter-mined electrolytically. Retain the remainder of sample solution A for suc-ceeding tests.

INSOLUBLE IN DILUTE NITRIC ACID. Dissolve 30.0 g in 200 mL of dilute nitric acid (1 + 1) and dilute with water to 225 mL. Filter through a tared filtering crucible and reserve the filtrate separate from the washings (sample solution B). Wash thoroughly with hot water and dry at 105 °C.

ANTIMONY AND TIN. Dissolve 4.0 g in 35 mL of dilute nitric acid (1 + 1) in a 150-mL beaker and evaporate to between 6 and 10 mL. Dilute to 30 mL with water and digest on a hot plate ($\approx$100 °C) for 15 min. Any insoluble residue should not be greater than that obtained when a control solution containing 0.4 mg of tin dissolved in 35 mL of dilute nitric acid (1 + 1) is treated exactly like the 4.0 g of the sample.

ARSENIC. (Page 25). Dissolve 0.60 g in 20 mL of dilute nitric acid (1 + 1) in a 250-mL beaker. Add 6 mL of sulfuric acid, evaporate in a hood to dense fumes of sulfur trioxide, and continue the fuming for 15 min. Cool, cautiously add 15 mL of water to dissolve the salts, and fume again in the same way. Repeat the addition of water and the fuming. Cautiously transfer the solution to a generator flask, using a total of 30 mL of water. Add 20 mL of hydrochloric acid containing 3 g of $SnCl_2 \cdot 2H_2O$. Cool in an ice bath, add 15 g of zinc shot, connect the scrubber–absorber assembly, and allow the hydrogen evolution and the color development to proceed for 45 min, keeping the flask in the ice bath. Any red color in the silver diethyldithiocarbamate solution of the sample should not exceed that in a standard containing 0.003 mg of arsenic (As) similarly treated.

IRON, LEAD, MANGANESE, AND SILVER. (By flame AAS, page 39).

> **Sample Stock Solution.** Dissolve 10.0 g of sample in 80 mL of (1 + 1) nitric acid. Boil a few minutes to remove the oxides of nitrogen, cool, and dilute with water to the mark in a 100-mL volumetric flask (1 mL = 0.1 g).

Element	Wavelength (nm)	Sample Wt (g)	Standard Added (mg)	Flame Type*	Background Correction
Fe	248.3	1.0	0.05; 0.10	A/A	Yes
Pb	217.0	1.0	0.05; 0.10	A/A	Yes
Mn	279.5	1.0	0.01; 0.02	A/A	Yes
Ag	328.1	1.0	0.02; 0.04	A/A	Yes

*A/A is air/acetylene.

PHOSPHORUS. Dilute 10 mL of sample solution A (0.5-g sample) to 25 mL, add 1 mL of ammonium molybdate reagent solution and 1 mL of 4-(methylamino)phenol sulfate reagent solution, and allow to stand at room temperature for 2 h. Any blue color should not exceed that produced by 1.5 mL of phosphate standard solution (0.005 mg of P) in 25 mL of approximately 0.5 N sulfuric acid when treated with the reagents used with the 25-mL dilution of the 10 mL of sample solution A.

Crystal Violet
Hexamethylpararosaniline Chloride

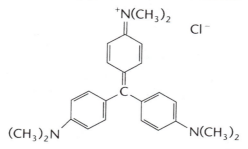

C₂₅H₃₀ClN₃   **Formula Wt 407.99**

$C_{25}H_{30}ClN_3$

CAS Number 548–62–9

REQUIREMENTS

Assay . . . . ≥90.0%
Sensitivity as indicator . . . . Passes test
Loss on drying . . . . ≤2.5%
Absorbance characteristics . . . . Passes test

TESTS

ASSAY. (By spectrophotometry). Weigh accurately 50.0 mg of sample previously dried for 4 h at 110 °C, dissolve in about 100 mL of water in a 250-mL volumetric flask, and dilute with water to volume. Dilute 10 mL of this solution with water to 1 L, and determine the absorbance in a 1-cm cell at the wavelength of maximum absorption at about 590 nm, against water as the blank. Reserve the remainder of the solution for the determination of absorbance characteristics.

$$\% \text{ Crystal Violet} = \text{Absorbance at 590 nm} \times 209$$

SENSITIVITY.

Solution A. Dissolve 100 mg of sample in 10 mL of glacial acetic acid.

Solution B. Dilute 0.30 mL of 70% perchloric acid with dioxane to 25 mL.

To 50 mL of glacial acetic acid add 0.10 mL of solution A. The purplish blue color of the solution is changed to bluish-green by the addition of not more than 0.05 mL of solution B.

LOSS ON DRYING. Weigh accurately 1.0 g and dry for 4 h at 110 °C.

ABSORBANCE CHARACTERISTICS. Using the solution retained from the assay, determine the absorbance in a 1-cm cell at $A - 15$ nm and at $A + 15$ nm, against water as the blank. The ratio of the absorbance at $A - 15$ nm to the absorbance at $A + 15$ nm should be between 0.98 and 1.9 (A is the wavelength of maximum absorption obtained in the assay).

Cupferron

N-Hydroxy-*N*-dinitrosobenzenamine, Ammonium Salt
Ammonium Salt of Nitrosophenylhydroxylamine

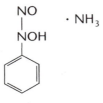

$C_6H_5N(NCO)ONH_4$ **Formula Wt 155.16**

CAS Number 135–20–6

NOTE. Packages of this reagent usually contain a lump or a small cloth-wrapped package of ammonium carbonate as a preservative.

REQUIREMENTS

Appearance . . . . Passes test
Suitability for precipitation of iron, etc.. . . . . Passes test

MAXIMUM ALLOWABLE

Solubility in water . . . . Passes test
Residue after ignition 0.05%

TESTS

APPEARANCE. The material should consist of crystalline flakes, white to light buff in color. It should be free flowing and contain no tarry matter.

SUITABILITY FOR PRECIPITATION OF IRON, ETC. Accurately weigh about 3.0 g, dissolve in water, and dilute with water to 100 mL. Prepare a ferric chloride solution containing approximately 0.0008 g of iron per mL. [Dissolve about 4 g of ferric chloride hexahydrate in 1 L of dilute hydrochloric acid (1 + 99).] To 40 mL of the iron solution add 20 mL of hydrochloric acid and 75 mL of water. Cool in an ice bath for about 15 min and add, from a buret, with constant stirring, 14.0 mL of the cupferron solution. Allow to stand for 15 min at room temperature and filter. (The addition of a small amount of filter paper pulp to the solution prior to filtration aids in obtaining a clear filtrate.) To the filtrate add 2 mL of the cupferron solution. No further precipitate of the brown iron compound should form. (A colorless precipitate of the acid form of the reagent may appear.)

SOLUBILITY IN WATER. Dissolve 5.0 g in 100 mL of water. The solution should be practically clear and have no more than a pale yellow color.

RESIDUE AFTER IGNITION. Decompose 5.0 g in a tared platinum dish by means of an infrared lamp placed above the sample. Cool, moisten the residue with 1 mL of sulfuric acid, and heat gently to remove the sulfuric acid. Finally, ignite at 800 ± 25 °C for 15 min.

Cupric Acetate Monohydrate
Copper(II) Acetate Monohydrate

$(CH_3COO)_2Cu \cdot H_2O$ **Formula Wt 199.65**

CAS Number 6046–93–1

REQUIREMENTS

Assay . . . . 98.0–102.0% $(CH_3COO)_2Cu \cdot H_2O$

MAXIMUM ALLOWABLE

Insoluble matter. 0.01%
Chloride (Cl) 0.003%
Sulfate (SO_4) 0.01%
Calcium (Ca) 0.005%
Iron (Fe). 0.002%
Nickel (Ni) 0.01%
Potassium (K). 0.01%
Sodium (Na) 0.05%

TESTS

ASSAY. (By iodometric titration of oxidative capacity of Cu^{II}). Weigh accurately 0.80 g and dissolve in 50 mL of water. Add 4 mL of glacial acetic acid, 1 mL of 10% sulfuric acid, and 3 g of potassium iodide. Swirl. Titrate the liberated iodine with 0.1 N sodium thiosulfate, adding about 2 g of potassium thiocyanate and 3 mL of starch indicator solution near the end of the titration. Correct for a complete blank. One milliliter of 0.1 N sodium thiosulfate corresponds to 0.01996 g of $(CH_3COO)_2Cu \cdot H_2O$.

INSOLUBLE MATTER. (Page 15). Use 10 g dissolved in 150 mL of water containing 1 mL of glacial acetic acid. Omit heating. Reserve the filtrate without washings for the sulfate test.

CHLORIDE. Dissolve 0.50 g in 15 mL of water, filter if necessary through a small chloride-free filter, and add 1 mL of nitric acid and 1 mL of silver nitrate reagent solution. Any turbidity should not exceed that produced by 0.015 mg of chloride ion (Cl) in an equal volume of solution containing the quantities of reagents used in the test. The comparison is best made by the general method for chloride in colored solutions, page 27.

SULFATE. (Page 33, Method 1). Use 1/20 of the filtrate from the test for insoluble matter. Allow 30 min for the turbidity to form.

CALCIUM, IRON, NICKEL, POTASSIUM, AND SODIUM. (By flame AAS, page 39).

> **Sample Stock Solution A.** Dissolve by warming 25.0 g of sample in a mixture of 60 mL of water and 15 mL of nitric acid. Cool to room temperature, transfer to a 100-mL volumetric flask, and dilute to the mark with water (1 mL = 0.25 g).

> **Sample Stock Solution B.** Dilute 4.0 mL of sample stock solution A with water to 100 mL in a volumetric flask (1 mL = 0.01 g).

Element	Wavelength (nm)	Sample Wt (g)	Standard Added (mg)	Flame Type*	Background Correction
Ca	422.7	2.0	0.05; 0.10	N/A	No
Fe	248.3	5.0	0.05; 0.10	A/A	Yes
Ni	232.0	2.0	0.10; 0.20	A/A	Yes
K	766.5	0.20	0.02; 0.04	A/A	No
Na	589.0	0.04	0.01; 0.02	A/A	No

*A/A is air/acetylene; N/A is nitrous oxide/acetylene.

Cupric Chloride Dihydrate
Copper(II) Chloride Dihydrate

$CuCl_2 \cdot 2H_2O$ **Formula Wt 170.48**

CAS Number 10125–13–0

REQUIREMENTS

Assay . . . .≥99.0% $CuCl_2 \cdot 2H_2O$

MAXIMUM ALLOWABLE
Insoluble matter. 0.01%
Nitrate (NO_3) 0.015%

Sulfate (SO₄) 0.005%
Calcium (Ca) 0.005%
Iron (Fe). 0.005%
Nickel (Ni) 0.01%
Potassium (K). 0.01%
Sodium (Na) 0.02%

TESTS

ASSAY. (By iodometric titration of oxidative capacity of Cu^{II}). Weigh accurately 0.7 g, transfer to a glass-stoppered conical flask, and dissolve in 100 mL of water. Add 2 mL of glacial acetic acid and 3 g of potassium iodide. Stopper and swirl. Titrate the liberated iodine with 0.1 N sodium thiosulfate, adding about 2.0 g of potassium thiocyanate and 3 mL of starch indicator solution near the end of the titration. One milliliter of 0.1 N sodium thiosulfate corresponds to 0.01705 g of $CuCl_2 \cdot 2H_2O$.

INSOLUBLE MATTER. (Page 15). Use 10 g in 100 mL of dilute hydrochloric acid (1 + 50). Omit heating. Retain the filtrate for the test for sulfate.

NITRATE.

Sample Stock Solution. Dissolve 1.0 g in 10 mL of water, add the solution with stirring to 50 mL of 10% sodium hydroxide reagent solution, and digest on a hot plate (≈100 °C) for 15 min. Cool, and filter through a retentive filter. Neutralize the filtrate with sulfuric acid, and dilute with water to 50 mL.

For sample solution A, dilute 10 mL of sample stock solution with brucine sulfate reagent solution to 50 mL. For control solution B, add 0.03 mg of nitrate ion (NO_3), and dilute with brucine sulfate reagent solution to 50 mL. Continue with the procedure described on page 30, starting with the preparation of blank solution C.

SULFATE. Heat the filtrate from the test for insoluble matter to boiling, add 10 mL of barium chloride reagent solution, digest in a covered beaker on a hot plate (≈100 °C) for 2 hours, and allow to stand overnight. If a precipitate is formed, filter, wash thoroughly, and ignite it. The weight of the precipitate should not be 0.0012 g more than the weight obtained in a complete blank test.

CALCIUM, IRON, NICKEL, POTASSIUM, AND SODIUM. (By flame AAS, page 39).

Sample Stock Solution. Dissolve 10.0 g of sample in water and add 5 mL of nitric acid. Transfer to a 100-mL volumetric flask and dilute to the mark with water (1 mL = 0.10 g).

Element	Wavelength (nm)	Sample Wt (g)	Standard Added (mg)	Flame Type*	Background Correction
Ca	422.7	2.0	0.05; 0.10	N/A	No
Fe	248.3	2.0	0.05; 0.10	A/A	Yes
Ni	232.0	2.0	0.10; 0.20	A/A	No
K	766.5	0.10	0.01; 0.02	A/A	No
Na	589.0	0.10	0.01; 0.02	A/A	No

*A/A is air/acetylene; N/A is nitrous oxide/acetylene.

Cupric Nitrate Hydrate

Copper(II) Nitrate Hydrate

$Cu(NO_3)_2 \cdot 2.5H_2O$ **Formula Wt 232.59**

$Cu(NO_3)_2 \cdot 3H_2O$ **Formula Wt 241.60**

CAS Number 10031–43–3

NOTE. This reagent is available containing either 2.5 or 3 molecules of water.

REQUIREMENTS

Assay . . . . 98.0–102.0% $Cu(NO_3)_2 \cdot 3H_2O$

MAXIMUM ALLOWABLE

Insoluble matter. 0.01%
Chloride (Cl) 0.002%
Sulfate (SO$_4$) 0.01%
Calcium (Ca) 0.005%
Iron (Fe) 0.005%
Lead (Pb) 0.001%
Nickel (Ni) 0.01%
Potassium (K). 0.005%
Sodium (Na). 0.01%

TESTS

ASSAY. (By iodometric titration of oxidative capacity of Cu^{II}). Weigh accurately 0.9–1.0 g, dissolve in 5 mL of hydrochloric acid in a 100-mL beaker, and evaporate to dryness on a steam bath. Add 2 mL of sulfuric acid and heat on a hot plate to dense fumes of sulfur trioxide. Cool, add 5 mL of water and 5 mL of hydrochloric acid, and again evaporate to dryness on a hot plate ($\approx$100 °C). Dissolve the residue in water, transfer to a conical flask, and dilute with water to about 75 mL. Add 2 mL of glacial acetic acid and 3 g of potassium iodide. Titrate the liberated iodine with 0.1 N sodium thiosulfate, adding about 2 g of potassium thiocyanate and 3 mL of starch indicator solution near the end of the titration. Correct for a complete blank. One milliliter of 0.1 N sodium thiosulfate corresponds to 0.02326 g of $Cu(NO_3)_2 \cdot 2.5H_2O$ or 0.02416 g of $Cu(NO_3)_2 \cdot 3H_2O$.

INSOLUBLE MATTER. (Page 15). Use 10.0 g dissolved in 100 mL of dilute nitric acid (1 + 200). Omit heating.

CHLORIDE. Dissolve 0.50 g in 15 mL of dilute nitric acid (1 + 15), filter if necessary through a small chloride-free filter, and add 1 mL of silver nitrate reagent solution. Any turbidity should not exceed that produced by 0.01 mg of chloride ion (Cl) in an equal volume of solution containing the quantities of reagents used in the test. The comparison is best made by the general method for chloride in colored solution, page 27.

SULFATE. (Page 33, Method 2). Use 1.5 mL of hot dilute hydrochloric acid (1 + 1). Do two evaporations. Allow 30 min for the turbidity to form.

CALCIUM, IRON, LEAD, NICKEL, POTASSIUM, AND SODIUM. (By flame AAS, page 39).

Sample Stock Solution. Dissolve 25.0 g of sample in 50 mL of water and 10 mL of nitric acid. Transfer to a 100-mL volumetric flask and dilute to the mark with water (1 mL = 0.25 g).

Element	Wavelength (nm)	Sample Wt (g)	Standard Added (mg)	Flame Type*	Background Correction
Ca	422.7	2.0	0.05; 0.10	N/A	No
Fe	248.3	2.0	0.10; 0.20	A/A	Yes
Pb	217.0	5.0	0.05; 0.10	A/A	Yes
Ni	232.0	2.0	0.10; 0.20	A/A	Yes
K	766.5	0.20	0.01; 0.02	A/A	No
Na	589.0	0.20	0.01; 0.02	A/A	No

*A/A is air/acetylene; N/A is nitrous oxide/acetylene.

Cupric Oxide, Powdered
Copper(II) Oxide

CuO

Formula Wt 79.55

CAS Number 1317–38–0

REQUIREMENTS

Assay. . . . . ≥99.0% CuO

MAXIMUM ALLOWABLE

Insoluble in dilute hydrochloric acid.	0.02%
Carbon compounds (as C)	0.01%
Chloride (Cl)	0.005%
Nitrogen compounds (as N).	0.002%
Sulfate (SO_4)	0.02%
Free alkali.	Passes test
Calcium (Ca)	0.01%
Iron (Fe)	0.05%
Potassium (K).	0.02%
Sodium (Na).	0.05%

TESTS

ASSAY. (By complexometric titration of copper). Weigh, to the nearest 0.1 mg, 0.2 g of sample. Dissolve in a mixture of 20 mL of water and 5 mL of nitric acid, and boil for a few min to dissolve the sample completely. Cool, dilute with 50 mL of water in a 400-mL beaker, and add 50 mL of alcohol and 50.0 mL of 0.1 M EDTA. Then adjust the pH to about 5 (using a pH meter) with pH 5 buffer, add 0.2 mL of PAN indicator, and titrate with 0.1 M copper sulfate while maintaining the pH at 5 by the addition of more buffer during the titration. The color change is from green to deep blue.

$$\% \text{ CuO} = \frac{[(\text{mL} \times \text{M EDTA}) - (\text{mL} \times \text{M CuSO}_4)] \times 7.954}{\text{Sample wt (g)}}$$

INSOLUBLE IN DILUTE HYDROCHLORIC ACID. Dissolve 5.0 g by warming on a hot plate ($\approx$100 °C) with 30 mL of dilute hydrochloric acid (2 + 1). Dilute with about 100 mL of water, heat to boiling, and digest in a covered beaker on a hot plate ($\approx$100 °C) for 1 h. Filter through a tared filtering crucible, wash thoroughly, and dry at 105 °C.

CARBON COMPOUNDS. Ignite 1.2 g in a stream of carbon dioxide-free air or oxygen, and pass the effluent gases into 20 mL of dilute ammonium hydroxide (2.5% NH_3) plus 2 mL of water. Prepare a standard of 20 mL of the dilute ammonium hydroxide and 1.06 mg of sodium carbonate (Na_2CO_3) (0.12 mg of C). Add 2 mL of barium chloride reagent solution to each and compare promptly. Any turbidity in the solution of the sample should not exceed that in the standard.

CHLORIDE. Shake 1.0 g with 25 mL of dilute nitric acid (1 + 3) for 10 min, filter through a chloride-free filter and dilute with water to 50 mL (sample solution A). To 10 mL of sample solution A add 5 mL of water and 1 mL of silver nitrate reagent solution. Any turbidity should not exceed that produced by 0.01 mg of chloride ion (Cl) in an equal volume of solution containing 1 mL of nitric acid and 1 mL of silver nitrate reagent solution. The comparison is best made by the general method for chloride in colored solutions, page 27.

NITROGEN COMPOUNDS. (Page 31). Use 3.0 g dissolved in 30 mL of freshly boiled 10% sodium hydroxide. For the standard, use 0.06 mg of nitrogen (N).

SULFATE. (Page 33, Method 2). Use 1.5 mL of dilute aqua regia (1 volume nitric acid + 4 volumes hydrochloric acid + 6 volumes water). Allow 30 min for the turbidity to form.

FREE ALKALI. Boil gently 3 g with 30 mL of water in a flask for 10 min. Allow to cool, add water to restore the original volume, mix well, and allow to settle. Decant 20 mL of the liquid and add 0.10 mL of phenolphthalein indicator solution. No red color should be produced.

CALCIUM, IRON, POTASSIUM, AND SODIUM. (By flame AAS, page 39).

Sample Stock Solution A. Dissolve 4.0 g of sample by heating with 30 mL of dilute hydrochloric acid (2 + 1). Cool to room temperature, transfer to a 100-mL volumetric flask, and dilute to the mark with water (1 mL = 0.04 g).

Sample Stock Solution B. Dilute 10.0 mL of sample stock solution A with water to 50 mL in a volumetric flask (1 mL = 0.01 g).

Element	Wavelength (nm)	Sample Wt (g)	Standard Added (mg)	Flame Type*	Background Correction
Ca	422.7	0.40	0.02; 0.04	N/A	No
Fe	248.3	0.40	0.10; 0.20	A/A	Yes
K	766.5	0.20	0.02; 0.04	A/A	No
Na	589.0	0.04	0.01; 0.02	A/A	No

*A/A is air/acetylene; N/A is nitrous oxide/acetylene.

Cupric Oxide, Wire
Copper(II) Oxide, Wire

CuO   **Formula Wt 79.55**

CAS Number 1317–38–0

REQUIREMENTS

MAXIMUM ALLOWABLE

Carbon compounds (as C) 0.002%
Nitrogen compounds (as N). 0.002%
Sulfate (SO_4) 0.012%

TESTS

CARBON COMPOUNDS. Ignite 6.0 g in a stream of carbon dioxide-free air or oxygen and pass the effluent gases into 20 mL of dilute ammonium hydroxide

(2.5% NH_3) plus 2 mL of water. Prepare a standard of 20 mL of the dilute ammonium hydroxide and 1.06 mg of sodium carbonate (Na_2CO_3) (0.12 mg of C). Add 2 mL of barium chloride reagent solution to each and compare promptly. Any turbidity in the solution representing the sample should not exceed that in the standard.

NITROGEN COMPOUNDS. (Page 31). Use 3.0 g dissolved in 30 mL of freshly boiled 10% sodium hydroxide. For the standard, use 0.06 mg of nitrogen (N).

SULFATE. (Page 33, Method 2). Use 2 mL of dilute aqua regia (1 volume nitric acid + 4 volumes hydrochloric acid + 6 volumes water). Allow 30 min for the turbidity to form.

Cupric Sulfate Pentahydrate
Copper(II) Sulfate Pentahydrate

$CuSO_4 \cdot 5H_2O$ Formula Wt 249.68

CAS Number 7758–99–8

REQUIREMENTS

Assay . . . . 98.0–102.0% $CuSO_4 \cdot 5H_2O$

MAXIMUM ALLOWABLE

Insoluble matter. 0.005%
Chloride (Cl) 0.001%
Nitrogen compounds (as N). 0.002%
Calcium (Ca) 0.005%
Iron (Fe). 0.003%
Nickel (Ni) 0.005%
Potassium (K). 0.01%
Sodium (Na). 0.02%

TESTS

ASSAY. (By iodometric titration of oxidative capacity of Cu^{II}). Weigh accurately 0.9–1.0 g, dissolve in water, transfer to a conical flask, and dilute with water to about 50 mL. Add 4 mL of glacial acetic acid, 1 mL of 10% sulfuric acid, and 3 g of potassium iodide. Titrate the liberated iodine with 0.1 N sodium thiosulfate, adding about 2 g of potassium thiocyanate and 3 mL of starch indicator solution near the end of the titration. Correct for a blank. One milliliter of 0.1 N sodium thiosulfate corresponds to 0.02497 g of $CuSO_4 \cdot 5H_2O$.

INSOLUBLE MATTER. (Page 15). Use 20 g dissolved in 200 mL of dilute sulfuric acid (1 + 40). Omit heating.

CHLORIDE. Dissolve 1.0 g in 15 mL of water, filter if necessary through a small chloride-free filter, and add 1 mL of nitric acid and 1 mL of silver nitrate reagent solution. Any turbidity should not exceed that produced by 0.01 mg of chloride ion (Cl) in an equal volume of solution containing the quantities of reagents used in the test. The comparison is best made by the general method for chloride in colored solutions, page 27.

NITROGEN COMPOUNDS. (Page 31). Use 1.0 g. For the standard, use 0.02 mg of nitrogen (N).

CALCIUM, IRON, NICKEL, POTASSIUM, AND SODIUM. (By flame AAS, page 39).

Sample Stock Solution. Dissolve 20.0 g of sample in water and add 5 mL of nitric acid. Transfer to a 100-mL volumetric flask, and dilute to the mark with water (1 mL = 0.20 g).

Element	Wavelength (nm)	Sample Wt (g)	Standard Added (mg)	Flame Type*	Background Correction
Ca	422.7	0.80	0.02; 0.04	N/A	No
Fe	248.3	4.0	0.12; 0.24	A/A	Yes
Ni	232.0	4.0	0.10; 0.20	A/A	Yes
K	766.5	0.40	0.02; 0.04	A/A	No
Na	589.0	0.05	0.005; 0.01	A/A	No

*A/A is air/acetylene; N/A is nitrous oxide/acetylene.

Cuprous Chloride
Copper(I) Chloride

CuCl   **Formula Wt 99.00**

CAS Number 7758–89–6

REQUIREMENTS

Assay. . . . . ≥90.0% CuCl

MAXIMUM ALLOWABLE

```
Insoluble in acid . . . . . . . . . . . . . . . . . . . . . . . . . . . . . . 0.02%
Sulfate (SO₄) . . . . . . . . . . . . . . . . . . . . . . . . . . . . . . . 0.1%
Calcium (Ca) . . . . . . . . . . . . . . . . . . . . . . . . . . . . . . . 0.01%
Iron (Fe) . . . . . . . . . . . . . . . . . . . . . . . . . . . . . . . . . 0.005%
Potassium (K) . . . . . . . . . . . . . . . . . . . . . . . . . . . . . . . 0.02%
Sodium (Na) . . . . . . . . . . . . . . . . . . . . . . . . . . . . . . . . 0.05%
```

TESTS

ASSAY. (By iodometric titration of reductive capacity of Cu^I). Weigh accurately 0.4 g and dissolve in the cold in 30 mL of ferric ammonium sulfate solution (10.0 g of ferric ammonium sulfate dissolved in 100 mL of 20% hydrochloric acid reagent solution). Add 5 mL of phosphoric acid, dilute with 200 mL of water, and titrate with 0.1 N potassium permanganate. The solution should be prepared and titrated in a flask in which the air is replaced with carbon dioxide. A stream of carbon dioxide should be passed through the flask during the titration. One milliliter of 0.1 N permanganate corresponds to 0.009900 g of CuCl.

INSOLUBLE IN ACID. Heat 5.0 g with 50 mL of dilute hydrochloric acid (1 + 1) and add nitric acid in small portions until the sample is dissolved. Dilute with 50 mL of water. If insoluble matter is present, filter through a tared filtering crucible, wash thoroughly, and dry at 105 °C.

> *Sample Solution A for the Determination of Sulfate.* Dilute the combined filtrate and washings from the test for insoluble in acid with water to 200 mL (1 mL = 0.025 g).

SULFATE. (Page 33, Procedure A, Method 1). Use 2 mL of sample solution A. Allow 30 min for the turbidity to form.

CALCIUM, IRON, POTASSIUM, AND SODIUM. (By flame AAS, page 39).

> *Sample Stock Solution.* To 10.0 g of sample, add 45 mL of dilute hydrochloric acid (1 + 1), and heat on a hot plate. Cautiously add 5 mL of nitric acid and keep warming and stirring until dissolved. Cool to room temperature, transfer to a 100-mL volumetric flask, and dilute to the mark with water (1 mL = 0.10 g).

Element	Wavelength (nm)	Sample Wt (g)	Standard Added (mg)	Flame Type*	Background Correction
Ca	422.7	0.20	0.02; 0.04	N/A	No
Fe	248.3	1.0	0.05; 0.10	A/A	Yes
K	766.5	0.20	0.02; 0.04	A/A	No
Na	589.0	0.02	0.005; 0.01	A/A	No

*A/A is air/acetylene; N/A is nitrous oxide/acetylene.

Cyclohexane

C₆H₁₂

C_6H_{12}

Formula Wt 84.16

CAS Number 110–82–7

REQUIREMENTS

General Use

Appearance . . . . Clear
Assay . . . . ≥99.0% C_6H_{12}

MAXIMUM ALLOWABLE

Color (APHA) 10
Residue after evaporation 0.002%
Substances darkened by sulfuric acid . . . . Passes test
Water (H_2O) 0.02%

Specific Use

ULTRAVIOLET SPECTROPHOTOMETRY.

Wavelength (nm)	Absorbance (AU)
300–400	0.01
260	0.02
250	0.03
240	0.08
230	0.20
220	0.50
210	1.00

TESTS

ASSAY. Analyze the sample by gas chromatography using the general parameters cited on page 70. The following specific conditions are also required.

Column: Type I, methyl silicone

Measure the area under all peaks and calculate the cyclohexane content in area percent. Correct for water content.

COLOR (APHA). (Page 19).

RESIDUE AFTER EVAPORATION. (Page 16). Evaporate 100.0 g (129 mL) to dryness in a tared dish on a hot plate (≈100 °C) and dry the residue at 105 °C for 30 min.

SUBSTANCES DARKENED BY SULFURIC ACID. Shake 25 mL with 15 mL of sulfuric acid for 15–20 s and allow to separate. Neither the cyclohexane nor the acid should be darkened.

WATER. (Page 56, Method 2). Use 200 µL (160 mg) of the sample.

ULTRAVIOLET SPECTROPHOTOMETRY. Use the procedure on page 73 to determine the absorbance.

Cyclohexanone

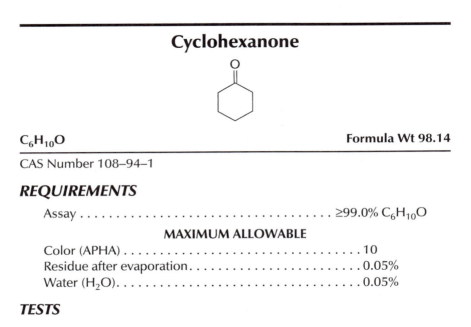

$C_6H_{10}O$ **Formula Wt 98.14**

CAS Number 108–94–1

REQUIREMENTS

Assay . . . . ≥99.0% $C_6H_{10}O$

MAXIMUM ALLOWABLE

Color (APHA) 10
Residue after evaporation. 0.05%
Water (H_2O). 0.05%

TESTS

ASSAY. Analyze the sample by gas chromatography using the parameters cited on page 70. The following specific conditions are also required.

Column: Type I, methyl silicone

Measure the area under all peaks and calculate the cyclohexanone content in area percent. Correct for water content.

COLOR (APHA). (Page 19).

RESIDUE AFTER EVAPORATION. (Page 16). Evaporate 100 g (105 mL) to dryness in a tared dish on a hot plate (≈100 °C) and dry the residue at 105 °C for 30 min.

WATER. (Page 55, Method 1). Use 30 mL (24 g) of the sample and a methanol-free system to prevent ketal formation with liberation of water.

(1,2-Cyclohexylenedinitrilo)tetraacetic Acid

$$N(CH_2CO_2H)_2$$

$$\cdot\ H_2O$$

$$N(CH_2CO_2H)_2$$

$C_{14}H_{22}N_2O_8 \cdot H_2O$ **Formula Wt 364.35**

CAS Number 13291–61–7

REQUIREMENTS

Assay . . . . 97.5–100.5% $C_{14}H_{22}N_2O_8 \cdot H_2O$

MAXIMUM ALLOWABLE

Residue after ignition 0.2%
Heavy metals (as Pb) 0.001%
Iron (Fe). 0.005%

TESTS

ASSAY. (By complexometry). Accurately weigh 0.7–0.8 g, and dissolve it in 50 mL of 0.1 N sodium hydroxide in a 250-mL beaker. Add 25 mL of water, 10 mL of ammonium acetate buffer solution, and 0.15 mL of phenolphthalein indicator solution. Neutralize the solution to the pink color of phenolphthalein by adding dilute ammonium hydroxide (10% NH_3) dropwise. Add 1 mL of glacial acetic acid and 0.10 mL of xylenol orange indicator solution. Titrate with 0.1 M zinc chloride standard solution to the change from yellow to off-yellow orange that remains for at least 1 min.

$$\% \, C_{14}H_{22}N_2O_8 \cdot H_2O = \frac{3.644 \times (\text{ml of 0.1 M ZnCl}_2)}{\text{Sample wt (g)}}$$

Ammonium Acetate Buffer Solution. Dissolve 200 g of ammonium acetate in water and dilute with water to 1 L.

Xylenol Orange Indicator Solution. Dissolve 0.50 g of the tetrasodium salt of xylenol orange in 100 mL of water.

RESIDUE AFTER IGNITION. (Page 16). Use 3.0 g.

Sample Solution A for the Determination of Heavy Metals and Iron. Ignite 3.0 g thoroughly and heat in an oven at 500 °C until most of the carbon is volatilized. Cool, add 0.15 mL of nitric acid, and heat at 500 °C until all the carbon is volatilized. Dissolve the residue in 2 mL of dilute hydrochloric acid (1 + 1), digest in a covered dish on a hot plate (≈100 °C) for 10 min, remove the cover, and evaporate to dryness. Dissolve in 1 mL of 1 N acetic acid and 20 mL of hot water, digest for 5 min, cool, and dilute with water to 30 mL (1 mL = 0.1 g).

HEAVY METALS. (Page 28, Method 1). Dilute 20 mL of sample solution A (2-g sample) with water to 25 mL.

IRON. (Page 30, Method 1). Use 2.0 mL of sample solution A (0.2-g sample).

1,2-Dichloroethane
Ethylene Dichloride

CH_2ClCH_2Cl **Formula Wt 98.96**

CAS Number 107–06–2

Suitable for use in ultraviolet spectrophotometry or general use. Product labeling shall designate one or both of these uses for which suitability is represented on the basis of meeting the relevant requirements and tests. The ultraviolet spectrophotometry requirements include all of the requirements for general use.

REQUIREMENTS

General Use

Appearance . . . . Clear
Assay . . . . ≥99.0% CH_2ClCH_2Cl

MAXIMUM ALLOWABLE

Color (APHA) 10
Residue after evaporation. 0.002%
Titrable acid . . . . 0.0003 meq/g
Water (H_2O). 0.03%

Specific Use

Ultraviolet Spectrophotometry

Wavelength (nm)	*Absorbance (AU)*
255–400.	0.01
250	0.02
245	0.05
240	0.10
235	0.20
230	0.50
228	1.00

TESTS

ASSAY. Analyze the sample by gas chromatography using the general parameters cited on page 70. The following specific conditions are also required.

Column: Type I, methyl silicone

Measure the area under all peaks and calculate the 1,2-dichloroethane content in area percent. Correct for water content.

COLOR (APHA). (Page 19).

RESIDUE AFTER EVAPORATION. (Page 16). Evaporate 100 g (82 mL) to dryness in a tared dish on a hot plate (≈100 °C) and dry the residue at 105 °C for 30 min.

TITRABLE ACID. To 25 mL of alcohol in a 100-mL glass-stoppered flask, add 0.10 mL of phenolphthalein indicator solution and 0.01 N sodium hydroxide until a faint pink color persists after shaking for 30 s. Add 31 g (25 mL) of sample, mix well, and titrate with 0.01 N sodium hydroxide until the pink color is restored. Not more than 0.93 mL of 0.01 N sodium hydroxide should be required.

> *NOTE.* Special care should be taken during the addition of the sample and titration to avoid contamination from carbon dioxide.

WATER. (Page 56, Method 2). Use 100 µL (125 mg) of the sample.

ULTRAVIOLET SPECTROPHOTOMETRY. Use the procedure on page 73 to determine the absorbance.

2',7'-Dichlorofluorescein
2',7'-Dichloro-3',6'-dihydroxyspiro[isobenzofuran-1(3*H*),9'-[9*H*]xanthen]-3-one
2',7'-Dichloro-3,6-fluorandiol

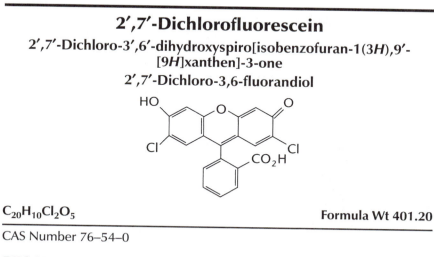

$C_{20}H_{10}Cl_2O_5$ **Formula Wt 401.20**

CAS Number 76–54–0

REQUIREMENTS

Appearance . . . . Orange to red-brown powder
Clarity of alcohol solution . . . . Passes test
Suitability as adsorption indicator . . . . Passes test

TESTS

CLARITY OF ALCOHOL SOLUTION. Dissolve 100 mg in 100 mL of 70% alcohol. The solution should be clear and greenish yellow. Reserve the solution for the suitability test.

SUITABILITY AS ABSORPTION INDICATOR. Prepare a 0.10 M solution of sodium chloride by dissolving 0.58 g of NaCl in water, and diluting with water to 100.0 mL. Dilute 10.0 mL of the 0.10 M sodium chloride with 30 mL of water, add 0.5 mL of the solution reserved from the preceding test, and titrate with 0.10 N silver nitrate, protecting the titration vessel from direct sunlight. The color change from greenish yellow to rose should require between 9.9 and 10.1 mL of 0.10 N silver nitrate.

2,6-Dichloroindophenol Sodium Salt
2,6-Dichloro-N-[(4-hydroxyphenyl)imino]-2,5-cyclohexadien-1-one, Sodium Salt

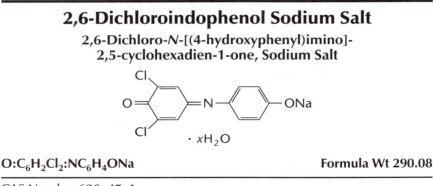

$\cdot\, xH_2O$

$O{:}C_6H_2Cl_2{:}NC_6H_4ONa$ **Formula Wt 290.08**

CAS Number 620–45–1

REQUIREMENTS

MAXIMUM ALLOWABLE
Loss on drying at 120 °C 12.0%
Interfering dyes . . . . Passes test

TESTS

LOSS ON DRYING AT 120 °C. Weigh accurately about 1.0 g, and dry at 120 °C to constant weight.

INTERFERING DYES.

Solution A. Dissolve 50 mg of the sample and 42 mg of sodium bicarbonate in water, and dilute with water to 200 mL. Filter through a dry filter, rejecting the first 20 mL of the filtrate.

Solution B. Dissolve 50 mg of ascorbic acid in 50 mL of a solution composed of 1.5 g of metaphosphoric acid plus 4 mL of glacial acetic acid, and dilute to 50 mL with water.

To 15 mL of solution A, add 2.5 mL of solution B. The mixture should become nearly colorless, with no trace of red or blue.

***

Dichloromethane
Methylene Chloride

CH_2Cl_2 Formula Wt 84.93

CAS Number 75–09–2

Suitable for use in high-performance liquid chromatography, extraction–concentration analysis, ultraviolet spectrophotometry, or general use. Product labeling shall designate the one or more of these uses for which suitability is represented on the basis of meeting the relevant requirements and tests. The ultraviolet spectrophotometry and liquid chromatography suitability requirements include all of the requirements for general use. The extraction–concentration suitability requirements include only the general use requirement for color.

REQUIREMENTS

General Use

Appearance . . . . Clear
Assay . . . . ≥99.5% CH_2Cl_2

MAXIMUM ALLOWABLE

Color (APHA) 10
Residue after evaporation 0.002%
Titrable acid . . . . 0.0003 meq/g
Free halogens . . . . Passes test
Water (H_2O) 0.02%

Specific Use

Ultraviolet Spectrophotometry

Wavelength (nm)	Absorbance (AU)
340–400	0.01
260	0.04
250	0.10
240	0.35
235	1.0

Liquid Chromatography Suitability

Absorbance . . . . Passes test

Extraction–Concentration Suitability

Absorbance . . . . Passes test

GC–FID . . . . Passes test

GC–ECD . . . . Passes test

TESTS

ASSAY. Analyze the sample by gas chromatography using the parameters cited on page 70. The following specific conditions are also required.

Column: Type I, methyl silicone

Measure the area under all peaks and calculate the dichloromethane content in area percent. Correct for water content.

COLOR (APHA). (Page 19).

RESIDUE AFTER EVAPORATION. (Page 16). Evaporate 100 g (76 mL) in a tared dish on a hot plate (≈100 °C) and dry the residue at 105 °C for 30 min.

TITRABLE ACID.

NOTE. Special care should be taken in the test during the addition of the sample and the titration to avoid contamination from carbon dioxide.

To 25 mL of alcohol in a 100-mL glass-stoppered flask, add 0.10 mL of phenolphthalein indicator solution and 0.01 N sodium hydroxide solution until a pink color persists for at least 30 s after vigorous shaking. Add 25 mL (33 g) of sample from a pipet and mix thoroughly with the neutralized alcohol. If no pink color remains, titrate with 0.01 N sodium hydroxide to the end point where the pink color persists for at least 30 s. Not more than 0.90 mL of 0.01 N sodium hydroxide should be required.

FREE HALOGENS. Shake 10 mL for 2 min with 10 mL of water to which 0.10 mL of 10% potassium iodide reagent solution has been added and allow to separate. The lower layer should not show a violet tint.

WATER. (Page 56, Method 2). Use 250 μL (330 mg) of the sample.

ULTRAVIOLET SPECTROPHOTOMETRY. Use the procedure on page 73 to determine the absorbance.

LIQUID CHROMATOGRAPHY SUITABILITY. Use the procedure on page 73 to determine the absorbance.

EXTRACTION–CONCENTRATION SUITABILITY. Analyze the sample, using the general procedure cited on page 72. In the sample preparation step, exchange the sample with ACS-grade hexanes suitable for extraction–concentration.

Diethanolamine
2,2'-Iminodiethanol

$(HOCH_2CH_2)_2NH$ **Formula Wt 105.14**

CAS Number 111–42–2

REQUIREMENTS

Assay. . . . . ≥98.5% $(HOCH_2CH_2)_2NH$
Apparent equivalent weight 104.0–106.0

MAXIMUM ALLOWABLE

Color (APHA). 15
Residue after ignition. 0.005%
Monoethanolamine 1.0%
Triethanolamine. 1.0%
Water (H_2O) 0.15%

TESTS

ASSAY, MONOETHANOLAMINE, AND TRIETHANOLAMINE. Analyze   the sample by gas chromatography using the general parameters cited on page 70. The following specific conditions are also required.

Column: Type I, methyl silicone

Measure the area under all peaks and calculate the diethanolamine, monoethanolamine, and triethanolamine content in weight percent. Correct for water content and subtract the ethanol blank.

APPARENT EQUIVALENT WEIGHT. To 50 mL of water in a 250-mL conical flask, add 0.3 mL of mixed indicator and neutralize by adding 0.1 N hydrochloric acid just to disappearance of the green color. From a suitable weighing pipet introduce 1.8–2.0 g of sample, weighing to within 0.5 mg. Mix and titrate with standard 0.5 N hydrochloric acid to disappearance of the green color.

$$\text{Apparent equivalent weight} = \frac{\text{grams of sample} \times 1000}{\text{mL of HCl} \times N}$$

Mixed Indicator Solution. Prepare separate solutions of bromcresol green and methyl red in methanol (0.1 g/100 mL). When needed, mix freshly each day 5 parts of the bromcresol green solution with 1 part of the methyl red solution.

COLOR (APHA). (Page 19).

RESIDUE AFTER IGNITION. (Page 16). Evaporate 50.0 g (45 mL) to dryness in a tared dish. Add 0.10 mL of sulfuric acid, again take to dryness, and finally ignite at 800 ± 25 °C.

WATER. (Page 56, Method 2). Use 0.1 g of the sample.

Diethylamine
N-Ethylethanamine

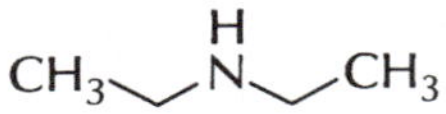

$$CH_3 \diagdown \overset{\overset{\displaystyle H}{|}}{N} \diagup CH_3$$

C₄H₁₁N   **Formula Wt 73.14**

CAS number 109-89-7

REQUIREMENTS

Assay . . . . $\geq$99.0% $C_4H_{11}N$

MAXIMUM ALLOWABLE

Color (APHA) 20
Monoethylamine 0.5%
Triethylamine 0.5%
Water 0.1%

TESTS

ASSAY, MONOETHYLAMINE, AND TRIETHYLAMINE. Analyze the sample by gas chromatography using the general parameters cited on page 70. The following specific conditions are also required.

Column: Type I, methyl silicone

Measure the area under each peak and calculate the diethylamine, monoethylamine, and triethylamine content in area percent. Correct for water content.

COLOR (APHA). (Page 19).

WATER. (Page 56, Method 2). Use 50 µL (35mg) of the sample.

4-(Dimethylamino)benzaldehyde

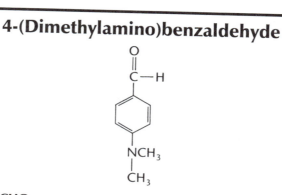

$(CH_3)_2NC_6H_4CHO$

Formula Wt 149.19

CAS Number 100–10–7

REQUIREMENTS

Melting point. . . . . 73–75 °C

MAXIMUM ALLOWABLE

Solubility in alcohol . . . . Passes test
Color (APHA) of alcohol solution 60
Solubility in hydrochloric acid . . . . Passes test
Color of hydrochloric acid solution . . . . Passes test
Residue after ignition 0.1%

TESTS

MELTING POINT. (Page 22).

SOLUBILITY IN ALCOHOL. Dissolve 1.0 g in 25 mL of alcohol. The solution should be clear.

COLOR (APHA) OF ALCOHOL SOLUTION. (Page 19). Dissolve 4.0 g in alcohol and dilute with alcohol to 100 mL.

SOLUBILITY IN HYDROCHLORIC ACID. Dissolve 1.0 g in 20 mL of dilute hydrochloric acid (1 + 10). Dissolution should be complete; the solution, clear. Reserve the solution for the test for color of hydrochloric acid solution.

COLOR OF HYDROCHLORIC ACID SOLUTION. Transfer a portion of the solution prepared in the test for solubility in hydrochloric acid into a 1 cm curette. Measure the absorbance at 450 nm vs. water. The absorbance should not exceed 0.1 AU.

RESIDUE AFTER IGNITION. (Page 16). Ignite 1.0 g.

N,N-Dimethylformamide

HCON(CH₃)₂   **Formula Wt 73.09**

CAS Number 68–12–2

Suitable for use in ultraviolet spectrophotometry or general use. Product labeling shall designate the one or both of these uses for which suitability is represented on the basis of meeting the relevant requirements and tests. The ultraviolet spectrophotometry requirements include all of the requirements for general use.

REQUIREMENTS

General Use

Appearance . . . . Clear
Assay . . . . ≥99.8% HCON(CH₃)₂

MAXIMUM ALLOWABLE

Color (APHA) 15
Residue after evaporation. 0.005%
Titrable base. . . . . 0.003 meq/g
Titrable acid . . . . 0.0005 meq/g
Water (H₂O). 0.15%

Specific Use

Ultraviolet Spectrophotometry

Wavelength (nm)	Absorbance (AU)
340–400.	0.01
310	0.05
295	0.10
275	0.30
270	1.00

TESTS

ASSAY. Analyze the sample by gas chromatography using the general parameters cited on page 70. The following specific conditions are also required.

Column: Type I, methyl silicone

Measure the area under all peaks and calculate the *N,N*-dimethylformamide content in area percent. Correct for water content.

COLOR (APHA). (Page 19).

RESIDUE AFTER EVAPORATION. (Page 16). Evaporate 20.0 g (21 mL) to dryness in a tared dish on a hot plate (≈100 °C) and dry the residue at 105 °C for 30 min.

NOTE. In the tests for titrable base–acid, take care to minimize exposure of the sample to the atmosphere because of rapid absorption of carbon dioxide. Flush the flask with nitrogen during the titrations.

TITRABLE BASE. Mix 10.0 g (10.6 mL) with 25 mL of carbon dioxide-free water in a glass-stoppered flask, and add 0.05 mL of methyl red indicator solution. If the solution becomes yellow, titrate with 0.01 N hydrochloric acid until a red color appears. Not more than 3.0 mL of the hydrochloric acid solution should be required.

TITRABLE ACID. Quickly add 19.0 g (20 mL) of the sample from a graduated cylinder to a 125-mL conical flask. Add 0.10 mL of thymol blue indicator solution. The color of the solution should be yellow. Titrate with 0.01 N sodium methoxide solution to a blue end point. Not more than 1.0 mL should be consumed in the titration.

> *Thymol Blue Indicator Solution.* Dissolve 0.30 g of thymol blue in 100 mL of dry methanol.

> *Sodium Methoxide in Methanol, 0.01 N.* Dissolve 0.540 g of sodium methylate in dry methanol and dilute with the methanol to 1 L in a volumetric flask. Standardize by titrating 40 mL of 0.01 N sulfuric acid, using thymol blue indicator.

WATER. (Page 55, Method 1). Use 25.0 mL (24.0 g) of the sample.

ULTRAVIOLET SPECTROPHOTOMETRY. Use the procedure on page 73 to determine the absorbance.

Dimethylglyoxime
2,3-Butanedione Dioxime

$CH_3C:NOHC:NOHCH_3$ **Formula Wt 116.12**

CAS Number 95–45–4

REQUIREMENTS

Melting point. . . . . About 240 °C
Suitability for nickel determination. . . . . Passes test

MAXIMUM ALLOWABLE
Insoluble in alcohol 0.05%
Residue after ignition. 0.05%

TESTS

MELTING POINT. (Page 22).

SUITABILITY FOR NICKEL DETERMINATION. Dissolve 0.665 g of nickel sulfate hexahydrate ($NiSO_4 \cdot 6H_2O$) in water and dilute with water to 50 mL. Dilute 20 mL of this solution with water to 100 mL, heat to boiling, and add 0.25 g of the dimethylglyoxime in 25 mL of alcohol. Add dilute ammonium hydroxide (1 + 4) dropwise until the solution is alkaline to litmus, cool, and filter. To the filtrate add 1 mL of the nickel sulfate solution and heat to boiling. A substantial precipitate of red nickel dimethylglyoxime should appear.

INSOLUBLE IN ALCOHOL. Gently boil 2.0 g with 100 mL of alcohol under a reflux condenser until no more dissolves. Filter through a tared filtering crucible, wash with 50 mL of alcohol in small portions, and dry at 105 °C.

RESIDUE AFTER IGNITION. (Page 16). Ignite 2.0 g.

Dimethyl Sulfoxide
Methyl Sulfoxide
Sulfinylbismethane

$(CH_3)_2SO$ **Formula Wt 78.13**

CAS Number 67–68–5

Suitable for use in ultraviolet spectrophotometry or general use. Product labeling shall designate the one or both of these uses for which suitability is represented on the basis of meeting the relevant requirements and tests. The ultraviolet spectrophotometry requirements include all of the requirements for general use.

REQUIREMENTS
General Use
Appearance . . . . Clear, colorless liquid
Assay . . . . ≥99.9% $(CH_3)_2SO$

MAXIMUM ALLOWABLE
Residue after evaporation. 0.01%
Titrable acid . . . . 0.001 meq/g
Water (H_2O). 0.1%

Specific Use
Ultraviolet Spectrophotometry

Wavelength (nm)	Absorbance (AU)
350–400.	0.01
330	0.02
310	0.06
290	0.18
270	0.40

TESTS

ASSAY. Analyze the sample by gas chromatography using the general parameters cited on page 70. The following specific conditions are also required.

Column: Type III, polyethylene glycol

Measure the area under all peaks and calculate the dimethylsulfoxide content in area percent. Correct for water content.

RESIDUE AFTER EVAPORATION. (Page 16). Evaporate 100.0 g (91 mL) to dryness in a tared dish on a hot plate.

TITRABLE ACID. Dissolve 50.0 g in 100 mL of water and add 0.1 mL of phenolphthalein indicator solution. If the solution remains colorless, titrate with 0.01 N sodium hydroxide until a pink color appears. Not more than 5.0 mL of 0.01 N sodium hydroxide should be consumed.

WATER. (Page 56, Method 2). Use 50 µL (55 mg) of the sample.

ULTRAVIOLET SPECTROPHOTOMETRY. Use the procedure on page 73 to determine the absorbance.

Dioxane

1,4-Dioxane

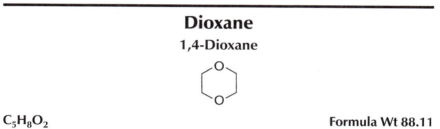

C₅H₈O₂ $C_5H_8O_2$

Formula Wt 88.11

CAS Number 123–91–1

CAUTION. Dioxane tends to form explosive peroxides, especially when anhydrous. It should not be allowed to evaporate to dryness unless the absence of peroxides has been shown.

NOTE. Dioxane usually contains a stabilizer. If a stabilizer is present, its identity and quantity must be stated on the label.

REQUIREMENTS

Assay . . . . ≥99.0% $C_5H_8O_2$
Freezing point . . . . Not below 11.0 °C

MAXIMUM ALLOWABLE

Color (APHA) 20
Peroxide (as H_2O_2) 0.005%

Residue after evaporation. 0.005%
Titrable acid . . . . .0.0016 meq/g
Carbonyl (as HCHO) 0.01%
Water (H_2O). 0.05%

TESTS

ASSAY. Analyze the sample by gas chromatography using the general parameters cited on page 70. The following specific conditions are also required.

Column: Type I, methyl silicone

Measure the area under all peaks and calculate the dioxane content in area percent. Correct for water content.

FREEZING POINT.

Apparatus: The sample container is a test tube (25 × 100 mm) supported by a cork in a water-tight glass cylinder (50 × 110 mm). The cylinder is mounted in a water bath that provides at least a 37-mm layer of water surrounding the sides and bottom of the cylinder. An accurate thermometer with 0.1 °C subdivisions is centered in the test tube and a thermometer with 1 °C subdivisions is mounted in the water bath. The stirrer, about 30 cm long, is a wire with a loop at the bottom.

Pour sufficient sample into the test tube to make a 50-mm column. Assemble the apparatus with the bulb of the 0.1 ° thermometer immersed halfway between the top and bottom of the sample in the test tube. Fill the water bath with a mixture of ice and water to within 12 mm of the top, and adjust the temperature to 0 °C by the addition of more ice, if necessary. Stir the sample continuously during the test by moving the wire loop up and down throughout the entire depth of the sample at a regular rate of 20 cycles per min. Record the reading of the 0.1 ° thermometer every 30 s. Anticipate that the temperature will fall gradually at first, then become constant for 1–2 min at the freezing point, and finally fall gradually again. The freezing point is the average of 4 consecutive readings that lie within a range of 0.2 °C.

COLOR (APHA). (Page 19).

PEROXIDE. Dilute 10 mL of the dioxane with water to 50 mL and use 5.0 mL of the solution (1.0-g sample). For the standard, dilute a solution containing 0.05 mg of hydrogen peroxide standard with water to 5.0 mL. Add 5.0 mL of titanium tetrachloride reagent solution to each and mix. After standing for 5 ± 1 min, any yellow color in the solution of the sample should not exceed that in the standard. The color intensities may be determined with a spectrophotometer in 1-cm cells at a wavelength of 410 nm.

CAUTION. If peroxide is present, do not perform the test for residue after evaporation.

RESIDUE AFTER EVAPORATION. (Page 16). Evaporate 20.0 g (19.0 mL) to dryness in a tared platinum dish on a hot plate ($\approx$100 °C), and dry the residue at 105 °C for 30 min.

TITRABLE ACID. Mix 31.0 g (30 mL) with 30 mL of carbon dioxide-free water, and titrate with 0.01 N sodium hydroxide, using phenolphthalein as the indicator. Not more than 5.0 mL of the titrant is required.

CARBONYL. To 1.0 mL of sample in a polarograph cell add 2.0 mL of water, 5.0 mL of pH 6.5 buffer solution, 1.0 mL of 0.20% Triton X-100 (polyethylene glycol ether of isooctylphenol) solution, and 1.0 mL of 2.0% hydrazine sulfate solution. Prepare a similar mixture in which 1.0 mL of the water is replaced by 1.0 mL of a water solution containing 0.10 mg of formaldehyde. Deaerate each solution with nitrogen for 10 min and record the polarograms from −0.6 to −1.6 V versus SCE at a sensitivity of 20 μA full scale. The total diffusion current for the sample should not be greater than one-half the diffusion current for sample plus formaldehyde.

The approximate half-wave potential for the formaldehyde hydrazone is −1.06 V.

> **Buffer Solution.** Dissolve 10.1 g of anhydrous dibasic sodium phosphate (Na_2HPO_4) and 3.05 g of citric acid monohydrate in water and dilute to 500 mL.

WATER. (Page 56, Method 2). Use 100 μL (103 mg) of the sample.

Diphenylamine
N-Phenylbenzeneamine

$(C_6H_5)_2NH$ **Formula Wt 169.23**

CAS Number 122–39–4

NOTE. This reagent discolors on exposure to light.

REQUIREMENTS

Melting point. . . . . 52.5–54.0 °C
Sensitivity to nitrate . . . . Passes test

MAXIMUM ALLOWABLE

Solubility in alcohol . . . . Passes test
Residue after ignition 0.03%
Nitrate (NO_3). . . . . Passes test

TESTS

MELTING POINT. (Page 22).

SENSITIVITY TO NITRATE. To 8 mL of the solution retained from the test for nitrate add 0.01 mg of nitrate ion (NO_3) and allow to stand at 60 °C for 5 min. A blue color should be produced.

SOLUBILITY IN ALCOHOL. Dissolve 1.0 g in 50 mL of alcohol. The solution should be clear and colorless.

RESIDUE AFTER IGNITION. (Page 16). Ignite 4.0 g. Moisten the char with 2 mL of sulfuric acid.

NITRATE. To 20 mL of water add 60 mL of nitrate-free sulfuric acid, adjust the temperature to about 60 °C, and add 0.5 mL of hydrochloric acid and 0.0100 g of the sample. No blue color should be produced in 5 min. Retain the solution for the test for sensitivity to nitrate.

Diphenylaminesulfonic Acid Sodium Salt
Sodium Diphenylaminesulfonate

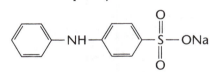

$C_6H_5NHC_6H_4SCO_3Na$ **Formula Wt 271.27**

CAS Number 6152–67–6

REQUIREMENTS

Sensitivity as indicator . . . . Passes test

TESTS

SENSITIVITY AS INDICATOR. Dissolve 0.15 g in 100 mL of water. To 25 mL of water in a 50-mL test tube, add 10 mL of 4 N sulfuric acid, 5 mL of phosphoric acid, 0.05 mL of 0.01 N ferrous ammonium sulfate, and 0.05 mL of the sample solution. The addition of 0.10 mL of 0.01 N potassium dichromate should produce a violet color. The color should be completely discharged by the addition of 0.10 mL of 0.01 N ferrous ammonium sulfate.

Diphenylcarbazone Compound with *s*-Diphenylcarbazide

(Phenylazo)formic Acid 2-Phenylhydrazide Compound with
1,5-Diphenylcarbohydrazide
"*s*-Diphenylcarbazone"

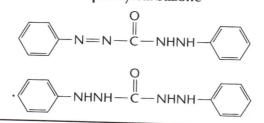

CAS Number 538–62–5 with 140–22–7

REQUIREMENTS

Sensitivity . . . . Passes test

MAXIMUM ALLOWABLE

Residue after ignition 0.1%
Solubility in acetone. . . . . Passes test

TESTS

SENSITIVITY. Dissolve 0.135 g of mercuric chloride in 500 mL of water. To 20 mL of this solution add 0.25 mL of the solution reserved from the test for solubility. A violet color should develop.

RESIDUE AFTER IGNITION. (Page 16). Ignite 2.0 g. Moisten the char with 2 mL of sulfuric acid.

SOLUBILITY IN ACETONE. Dissolve 0.2 g in 25 mL of acetone. The sample should dissolve completely, and the solution should be clear and red. Reserve this solution for the sensitivity test.

1,5-Diphenylcarbohydrazide

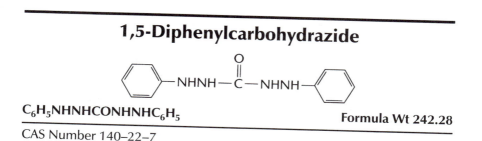

$C_6H_5NHNHCONHNHC_6H_5$ **Formula Wt 242.28**

CAS Number 140–22–7

REQUIREMENTS

Melting point . . . . 173–176 °C
Sensitivity to chromate. . . . . Passes test

MAXIMUM ALLOWABLE

Solubility in aqueous acetone. Passes test
Residue after ignition 0.05%

TESTS

MELTING POINT. (Page 22).

SENSITIVITY TO CHROMATE. Dissolve 0.25 g in a mixture of 50 mL of acetone and 50 mL of water. To each of two 50-mL volumetric flasks transfer 20 mL of water and 10 mL of 1 N sulfuric acid. To one flask add 5.8 mL of standard chromate solution (1 mL = 0.01 mg of CrO_4) and to both add 2.0 mL of the sample solution, mix, and allow to stand for 1 min. To both flasks add 5 mL of 4 M monobasic sodium phosphate, dilute with water to the mark, and allow to stand for 5 min. Determine the absorbance of the solution containing chromate in a 1-cm cell at 540 nm, against the other solution in a similar matched cell set at zero absorbance as the blank. The absorbance should be not less than 0.42.

SOLUBILITY IN AQUEOUS ACETONE. Mix 250 mL of acetone and 50 mL of water, and adjust the pH to 6.0 ± 0.2 (using a pH meter) with 0.02 M acetic acid or 0.02 M sodium hydroxide. Dissolve 10.0 g in the mixture without heating. Dissolution should be complete and the solution clear, free of any red color, and not more than faintly yellow.

RESIDUE AFTER IGNITION. (Page 16). Ignite 2.0 g. Moisten the char with 1 mL of sulfuric acid.

Dithizone
(Phenylazo)thioformic Acid 2-Phenylhydrazide
Diphenylthiocarbazone

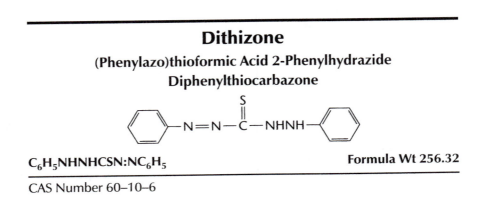

$C_6H_5NHNHCSN:NC_6H_5$ **Formula Wt 256.32**

CAS Number 60–10–6

REQUIREMENTS

Assay . . . .$\geq$85.0% $C_6H_5NHNHCSN:NC_6H_5$
Ratio of absorbances . . . . $\geq$1.55

MAXIMUM ALLOWABLE

Residue after ignition 0.3%
Heavy metals (as Pb) 0.002%

TESTS

ASSAY. (By spectrophotometry). Weigh accurately 10.0 mg, transfer to a 100-mL volumetric flask, dissolve in about 75 mL of chloroform, and dilute to volume with chloroform. Dilute 5.0 mL of this solution to 100.0 mL with chloroform, and determine the absorbance of this solution in 1.00-cm cells at 620 nm, using chloroform as the blank. Reserve the remainder of the solution for determining the ratio of absorbances. Determine the molar absorptivity by dividing the absorbance by 1.95×10^{-5} (the molar concentration of the dithizone) and calculate the assay value as follows:

$$\% \text{ Dithizone} = \frac{\text{Molar absorptivity at 620 nm}}{34,600} \times 100$$

RATIO OF ABSORBANCES. Determine the molar absorptivity at 450 nm of the solution prepared in the test for assay, using the method described in the test. The ratio of the value at 620 nm to that at 450 nm should not be less than 1.55.

RESIDUE AFTER IGNITION. (Page 16). Ignite 1.0 g. Moisten the char with 1 mL of nitric acid and 1 mL of sulfuric acid.

HEAVY METALS. (Page 28, Method 2). Use 1.0 g.

Dodecyl Alcohol
Lauryl Alcohol

$CH_3(CH_2)_{11}OH$ **Formula Wt 186.34**

CAS Number 112–53–8

REQUIREMENTS

Assay (GC). . . . .$\geq$98.0% $C_{12}H_{26}O$
MAXIMUM ALLOWABLE
Color (APHA). 10
Water. 0.05%

TESTS

ASSAY. Analyze the sample by gas chromatography by using the general parameters cited on page 70. The following specific conditions are also required.

Sample Solution. Accurately weigh and transfer 0.10 g of the dodecyl alcohol into a 50-mL volumetric flask. Dissolve and dilute to volume with hexanes. Mix well and stopper.

Column: Type 1, methyl silicone

Column Temperature: 80 °C–230 °C programmed at 10 °C/min, 5 min hold at 230 °C

Injector Temperature: 220 °C

Detector Temperature: 250 °C

Sample Size: 0.2 μL

Carrier Gas: Helium at 6 mL/min

Detector: Flame ionization

Measure the area under all the peaks after the solvent peaks and calculate the dodecyl alcohol content in area percent. Relative retention for dodecyl alcohol to hexanes is approximately 8.8. Correct for water content.

COLOR (APHA). (Page 19).

WATER. (Page 56, Method 2). Use 1.0 mL (0.832 g) of the sample.

Eosin Y

2′,4′,5′,7′-Tetrabromofluorescein, Disodium Salt

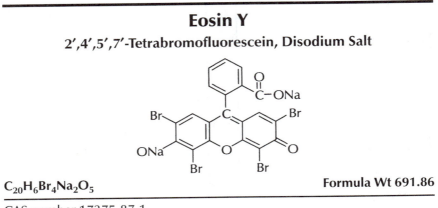

$C_{20}H_6Br_4Na_2O_5$ **Formula Wt 691.86**

CAS number 17375-87-1

REQUIREMENTS

Clarity of aqueous solution. . . . . Passes test
Suitability as adsorption indicator. . . . . Passes test

TESTS

CLARITY OF AQUEOUS SOLUTION. Dissolve 0.05 g in 100 mL of water. The solution should be clear and complete. Reserve the solution for the suitability test.

SUITABILITY AS ADSORPTION INDICATOR. Prepare a 0.10 N solution of sodium chloride by dissolving 0.58 g of sodium chloride in water and dilute to 100.0 mL. Dilute 10.0 mL of the 0.10 N sodium chloride with 30.0 mL of methanol and 10.0 mL of acetic acid. Add 0.15 mL of the solution reserved from the preceding test and, with vigorous stirring, titrate with 0.10 N silver nitrate, protecting the titration vessel from direct sunlight. The color change from orange to bright pink should require between 9.9 and 10.1 mL of the silver nitrate solution.

Eriochrome Black T
3-Hydroxy-4-[(1-hydroxy-2-naphthalenyl)azo]-7-nitro-1-naphthalenesulfonic Acid, Monosodium Salt
C.I. Mordant Black 11

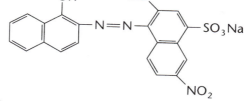

$C_{20}H_{12}N_3NaO_7S$

Formula Wt 461.38

CAS Number 1787–61–7

REQUIREMENTS

Clarity of solution . . . . Passes test
Suitability as complexometric indicator . . . . Passes test

TESTS

CLARITY OF SOLUTION. Dissolve 0.1 g in 100 mL of water, warming if necessary. Not more than a faint trace of turbidity or insoluble matter should remain. Reserve the solution for the test for suitability.

SUITABILITY AS COMPLEXOMETRIC INDICATOR. Add 10 mL of ammoniacal buffer (pH 10) to 100 mL of water. Add 0.5 mL of the solution reserved from the test for clarity of solution and 10.0 mL of magnesium standard solution (0.01 mg of Mg per mL). The color of the solution should be red-violet. Add 0.10 mL of 0.1 M EDTA solution. The color of the solution should change to blue.

Ethyl Acetate

$$CH_3-\overset{\displaystyle O}{\overset{\displaystyle \|}{C}}-OCH_2CH_3$$

$CH_3COOCH_2CH_3$ **Formula Wt 88.11**

CAS Number 141–78–6

Suitable for use in ultraviolet spectrophotometry or general use. Product labeling shall designate the one or both of these uses for which suitability is represented on the basis of meeting the relevant requirements and tests. The ultraviolet spectrophotometry requirements include all of the requirements for general use.

REQUIREMENTS

General Use

Assay . . . . ≥99.5% $CH_3COOCH_2CH_3$

MAXIMUM ALLOWABLE

Color (APHA) 10
Residue after evaporation. 0.003%
Water (H_2O). 0.2%
Titrable acid . . . . .0.0009 meq/g
Substances darkened by sulfuric acid . . . . .Passes test

Specific Use

Ultraviolet Spectrophotometry

Wavelength (nm)	Absorbance (AU)
330–400.	0.01
275	0.05
263	0.10
257	0.50
255	1.00

TESTS

ASSAY. Analyze the sample by gas chromatography using the general parameters cited on page 70. The following specific conditions are also required.

Column: Type I, methyl silicone

Measure the area under all peaks and calculate the ethyl acetate content in area percent. Correct for water content.

COLOR (APHA). (Page 19).

RESIDUE AFTER EVAPORATION. (Page 16). Evaporate 40.0 g (45.0 mL) to dryness in a tared dish on a hot plate ($\approx$100 °C) and dry the residue at 105 °C for 30 min.

WATER. (Page 55, Method 1). Use 20 mL (18 g) of the sample.

TITRABLE ACID. To 10 mL of alcohol add 0.10 mL of phenolphthalein indicator solution, and neutralize with 0.01 N sodium hydroxide. Add 9.0 g (10.0 mL) of the sample, mix gently, and titrate with 0.01 N sodium hydroxide until the pink color persists for 15 s. Not more than 0.80 mL should be required.

SUBSTANCES DARKENED BY SULFURIC ACID. Superimpose 5 mL of ethyl acetate upon 5 mL of sulfuric acid. No dark coloration should be produced at zone of impact.

ULTRAVIOLET SPECTROPHOTOMETRY. Use the procedure on page 73 to determine the absorbance.

Ethyl Alcohol
Ethanol

CH_3CH_2OH **Formula Wt 46.07**

CAS Number 64–17–5

Suitable for use in ultraviolet spectrophotometry or general use. Product labeling shall designate the one or both of these uses for which suitability is represented on the basis of meeting the relevant requirements and tests. The ultraviolet spectrophotometry requirements include all of the requirements for general use.

REQUIREMENTS

General Use

Assay . . . . ≥95.0% by volume

MAXIMUM ALLOWABLE

Color (APHA) 10
Solubility in water . . . . Passes test
Residue after evaporation 0.001%
Acetone, isopropyl alcohol. . . . . Passes test
Titrable acid. . . . . 0.0005 meq/g
Titrable base . . . . 0.0002 meq/g
Methanol (CH_3OH) 0.1%
Substances darkened by sulfuric acid. . . . . Passes test
Substances reducing permanganate. . . . . Passes test

Specific Use

Ultraviolet Spectrophotometry

Wavelength (nm)	Absorbance (AU)
270–400	0.01
240	0.05
230	0.15
220	0.25
210	0.40

TESTS

ASSAY. Analyze the sample by gas chromatography using the general parameters cited on page 70. The following specific conditions are also required.

Column: Type I, methyl silicone

Measure the area under all peaks and calculate the ethyl alcohol content in area percent. Correct for water content.

COLOR (APHA). (Page 19).

SOLUBILITY IN WATER. Mix 15 mL with 45 mL of water and allow to stand for 1 h. The mixture should be as clear as an equal volume of water.

RESIDUE AFTER EVAPORATION. (Page 16). Evaporate 100.0 g (124 mL) to dryness in a tared dish on a hot plate ($\approx$100 °C) and dry the residue at 105 °C for 30 min.

ACETONE, ISOPROPYL ALCOHOL. Dilute 1.0 mL with 1 mL of water. Add 1 mL of a saturated solution of disodium hydrogen phosphate and 3 mL of a saturated solution of potassium permanganate. Warm the mixture to between 45 and 50 °C and allow to stand until the permanganate color is discharged. Add 3 mL of 10% sodium hydroxide reagent solution and filter, without washing, through glass. Prepare a control containing 1 mL of the saturated solution of disodium hydrogen phosphate, 3 mL of 10% sodium hydroxide reagent solution, and 0.08 mg of acetone in 9 mL. To each solution add 1 mL of 1% furfural reagent solution, and allow to stand for 10 min. To 1.0 mL of each solution add 3 mL of hydrochloric acid. Any pink color produced in the solution of the sample should not exceed that in the control. (Limit about 0.001% acetone, 0.003% isopropyl alcohol)

TITRABLE ACID. Add 10 mL to 25 mL of water in a glass-stoppered flask and add 0.15 mL of phenolphthalein indicator solution. Add 0.01 N sodium hydroxide until a slight pink color persists after shaking for 30 s. Add 25 mL of the sample, mix well, and titrate with 0.01 N sodium hydroxide until the pink color is restored. Not more than 1.0 mL of the hydroxide solution should be required.

TITRABLE BASE. Dilute 25 mL with 25 mL of water and add 0.15 mL of methyl red indicator solution. Not more than 0.40 mL of 0.01 N sulfuric acid should be required to produce a pink color.

METHANOL. Dilute 5 mL with water to 100 mL. To 1 mL of the solution add 0.2 mL of dilute phosphoric acid (1 + 9) and 0.25 mL of a 5% potassium permanganate solution. Allow to stand for 15 min, add 0.3 mL of 10% sodium bisulfite solution, and shake until colorless. Add slowly 5 mL of ice-cold 80% sulfuric acid (3 volumes of acid plus 1 volume of water), keeping the mixture cool during the addition. Add 0.1 mL of a 1% aqueous solution of chromotropic acid, mix, and digest on a hot plate ($\approx$100 °C) for 20 min. Any violet color should not exceed that produced by 0.04 mg of methanol in 1 mL of water treated in the same way as the 1 mL of the diluted sample.

SUBSTANCES DARKENED BY SULFURIC ACID. Cool 10 mL of sulfuric acid, contained in a small conical flask, to 10 °C. Add dropwise, with constant agitation, 10 mL of the sample, while keeping the temperature of the mixture below 20 °C. The resulting solution should have no more color than either of the two liquids before mixing.

SUBSTANCES REDUCING PERMANGANATE. Cool 20 mL to 15 °C, add 0.10 mL of 0.1 N potassium permanganate, and allow to stand at 15 °C for 5 min. The pink color should not be entirely discharged.

ULTRAVIOLET SPECTROPHOTOMETRY. Use the procedure on page 73 to determine the absorbance.

Ethyl Alcohol, Absolute
Ethanol, Absolute

CH_3CH_2OH   **Formula Wt 46.07**

CAS Number 64–17–5

REQUIREMENTS

Assay . . . . $\geq$99.5% CH_3CH_2OH by volume
(about 99.2% by weight)

MAXIMUM ALLOWABLE

Water (H_2O) 0.2%
Color (APHA) 10
Solubility in water . . . . Passes test
Residue after evaporation 0.001%
Acetone, isopropyl alcohol. . . . . Passes test

Titrable acid . . . . 0.0005 meq/g
Titrable base. . . . . 0.0002 meq/g
Methanol (CH₃OH) 0.1%
Substances darkened by sulfuric acid . . . . Passes test
Substances reducing permanganate . . . . Passes test

TESTS

WATER. (Page 56, Method 2). Use 20 mL (15.8 g).

ASSAY AND OTHER TESTS . These are the same as for ethyl alcohol, page 291.

Ethyl Ether
Diethyl Ether

(CH₃CH₂)₂O   **Formula Wt 74.12**

CAS Number 60–29–7

CAUTION. Ethyl ether tends to form explosive peroxides, especially when anhydrous. It should not be allowed to evaporate to dryness unless the absence of peroxides has been shown. The presence of water or appropriate reducing agents lessens peroxide formation.

NOTE. Ethyl ether normally contains about 2% of alcohol and 0.5% of water as stabilizers.

REQUIREMENTS

Assay . . . . ≥98.0% (CH₃CH₂)₂O

MAXIMUM ALLOWABLE

Color (APHA) 10
Peroxide (as H_2O_2) . . . . 1 ppm
Residue after evaporation 0.001%
Titrable acid . . . . 0.0002 meq/g
Carbonyl (as HCHO) 0.001%

TESTS

ASSAY. Assay the sample by gas chromatography using the general parameters cited on page 70. The following specific conditions are also required.

Column: Type I, methyl silicone

Measure the area under all peaks and calculate the ethyl ether content in area percent. The assay value is the combination of ethyl ether and added stabilizers.

COLOR (APHA). (Page 19).

PEROXIDE. To 35.0 g (50 mL) of the ethyl ether in a separatory funnel, add 5.0 mL of titanium tetrachloride reagent solution. Shake vigorously, allow the layers to separate, and drain the lower layer into a 25-mL glass-stoppered graduated cylinder. For the standard, transfer 5.0 mL of titanium tetrachloride reagent solution to a similar graduated cylinder and add a volume of solution containing 0.035 mg of hydrogen peroxide standard. Dilute both solutions with water to 10.0 mL and mix. Allow the mixture to stand at least 30 min. Any yellow color in the solution of the sample should not exceed that in the standard. The color intensities may be determined with a spectrophotometer in 1-cm cells at a wavelength of 410 nm.

> *NOTE.* Fresh ethyl ether should meet this test, but after storage for several months peroxide may be formed.

> *CAUTION.* If peroxide is present, do not make the test for residue after evaporation.

RESIDUE AFTER EVAPORATION. (Page 16). Evaporate 100.0 g (141 mL) to dryness in a tared dish on a water bath, and dry the residue at 105 °C for 30 min.

TITRABLE ACID.

> *NOTE.* Great care should be taken in the test during the addition of the sample and the titration to avoid contamination from carbon dioxide.

To 10 mL of water in a glass-stoppered flask, add 0.10 mL of bromthymol blue indicator solution and 0.01 N sodium hydroxide until a blue color persists after vigorous shaking. Add 25 mL of sample from a pipet and shake briskly to mix the two layers. If no blue color remains, titrate with 0.01 N sodium hydroxide until the blue color is restored and persists for several minutes. Not more than 0.30 mL of 0.01 N sodium hydroxide should be required.

CARBONYL. To 5.0 g (7.0 mL) of sample in a 30-mL separatory funnel add 5.0 mL of pH 6.5 buffer solution, 1.0 mL of 2.0% hydrazine sulfate solution, 1.0 mL of 0.20% Triton X-100 (polyethylene glycol ether of isooctylphenol) solution, and 3.0 mL of water. Prepare a similar mixture in which part of the water is replaced by an equal volume of aqueous solution containing 0.05 mg of formaldehyde. Shake each funnel well to mix the contents, allow the phases to separate, and draw off each lower layer into a polarographic cell. Deaerate with nitrogen for 10 min. Record the polarogram from -0.6 to -1.6 V versus SCE at a sensitivity of 5 μA full scale. The total diffusion current for the sample should not be greater than one-half the diffusion current for sample plus aldehyde.

Approximate half-wave potentials for the hydrazones of known carbonyl compounds are: formaldehyde, -1.06 V; acetaldehyde, -1.21 V; and acetone, -1.4 V.

Buffer Solution. Dissolve 10.1 g of anhydrous dibasic sodium phosphate (Na_2HPO_4) and 3.05 g of citric acid monohydrate in water and dilute with water to 500 mL.

Ethyl Ether, Anhydrous
Diethyl Ether, Anhydrous

$(CH_3CH_2)_2O$ **Formula Wt 74.12**

CAS Number 60–29–7

CAUTION. Ethyl ether tends to form explosive peroxides, especially when anhydrous. It should not be allowed to evaporate to dryness or near dryness unless the absence of peroxides has been shown. The formation of peroxides is more rapid in ethyl ether kept in containers that have been opened and partly emptied. Some ethyl ether may contain a stabilizer. If it does, the amount and type should be marked on the label.

REQUIREMENTS

Assay . . . . ≥99.0% $(CH_3CH_2)_2O$

MAXIMUM ALLOWABLE

Color (APHA) 10
Peroxide (as H_2O_2) . . . . 1 ppm
Residue after evaporation 0.001%
Titrable acid . . . . 0.0002 meq/g
Carbonyl (as HCHO) 0.001%
Alcohol (CH_3CH_2OH) . . . . Passes test
Water (H_2O) 0.03%

TESTS

ASSAY. Analyze the sample by gas chromatography using the general parameters cited on page 70. The following specific conditions are also required.

    Column: Type I, methyl silicone

Measure the area under all peaks and calculate the ethyl ether content in area percent. Correct for water content.

COLOR (APHA). (Page 19).

PEROXIDE. To 35.0 g (50 mL) of the ethyl ether in a separatory funnel, add 5.0 mL of titanium tetrachloride reagent solution. Shake vigorously, allow the layers to separate, and drain the lower layer into a 25-mL glass-stoppered graduated cylinder. For the standard, transfer 5.0 mL of titanium tetrachloride reagent solution to a similar graduated cylinder, and add a volume of solution containing 0.035 mg of hydrogen peroxide standard. Dilute both solutions with water to 10.0 mL and mix. Allow the mixture to stand at least 30 min. Any yellow color in the solution of the sample should not exceed that in the standard. The color intensities may be determined with a spectrophotometer in 1-cm cells at a wavelength of 410 nm.

NOTE. Fresh ethyl ether should meet this test, but after storage for several months peroxide may be formed.

CAUTION. If peroxide is present, do not make the test for residue after evaporation.

RESIDUE AFTER EVAPORATION. (Page 16). Evaporate 100.0 g (141 mL) to dryness in a tared dish on a water bath and dry the residue at 105 °C for 30 min.

TITRABLE ACID.

NOTE. Great care should be taken in the test during the addition of the sample and the titration to avoid contamination from carbon dioxide.

To 10 mL of water in a glass-stoppered flask, add 0.10 mL of bromthymol blue indicator solution and 0.01 N sodium hydroxide until a blue color persists after vigorous shaking. Add 25 mL of sample from a pipet and shake briskly to mix the two layers. If no blue color remains, titrate with 0.01 N sodium hydroxide until the blue color is restored and persists for several minutes. Not more than 0.30 mL of 0.01 N sodium hydroxide should be required.

CARBONYL. To 5.0 g (7.0 mL) of sample in a 30-mL separatory funnel add 5.0 mL of pH 6.5 buffer solution, 1.0 mL of 2.0% hydrazine sulfate solution, 1.0 mL of 0.20% Triton X-100 (polyethylene glycol ether of isooctylphenol) solution, and 3.0 mL of water. Prepare a similar mixture in which part of the water is replaced by an equal volume of aqueous solution containing 0.05 mg of formaldehyde. Shake each funnel well to mix the contents, allow the phases to separate, and draw off each lower layer into a polarographic cell. Deaerate with nitrogen for 10 min. Record the polarogram from –0.6 to –1.6 V versus SCE at a sensitivity of 5 μA full scale. The total diffusion current for the sample should not be greater than one-half the diffusion current for sample plus aldehyde.

Approximate half-wave potentials for the hydrazones of known carbonyl compounds are: formaldehyde, –1.06 V; acetaldehyde, –1.21 V; and acetone, –1.4 V.

Buffer Solution. Dissolve 10.1 g of anhydrous dibasic sodium phosphate (Na_2HPO_4) and 3.05 g of citric acid monohydrate in water and dilute with water to 500 mL.

ALCOHOL. Transfer 100 mL to a separatory funnel and shake with five successive portions, 20 mL, 10 mL, 10 mL, 5 mL, and 5 mL, respectively, of distilled water at about 25 °C. Shake each portion for 2 min and separate the water layer carefully. Finally, pour the combined water extract from one flask to another six times to assure minimum contamination with ether. Transfer 1 mL of the water extract with a pipet to a comparison tube and add 4 mL of water. For a standard take 5 mL of a solution of 0.2 mL of absolute alcohol in a liter of water. Add 10 mL of nitrochromic acid reagent to each solution, mix, and allow to stand for 1 h. At the end of this time the color of the sample solution should show no more change from yellow to green or blue than is shown by the standard. The test and the standard must be kept at the same temperature. (Limit about 0.05%)

> **Nitrochromic Acid Reagent.** Mix 1 volume of 5% potassium chromate solution with 133 volumes of water and 66 volumes of colorless nitric acid. This reagent should not be used if more than 1 month old.

WATER. (Page 56, Method 2). Use 0.2 mL (0.14 g) of the sample.

(Ethylenedinitrilo)tetraacetic Acid
N,N,N´,N´-Ethylenediaminetetraacetic Acid
N,N´-1,2-Ethanediylbis[*N*-carboxymethyl]glycine
EDTA, Free Acid

$C_{10}H_{16}N_2O_8$ **Formula Wt 292.25**

CAS Number 60–00–4

REQUIREMENTS

Assay . . . . 99.4–100.6% $C_{10}H_{16}N_2O_8$

MAXIMUM ALLOWABLE

Insoluble in dilute ammonium hydroxide 0.005%
Residue after ignition 0.2%
Nitrilotriacetic acid [$(HOCOCH_4)_3N$] 0.1%
Calcium (Ca) 0.001%
Magnesium (Mg) . . . . 5 ppm
Heavy metals (as Pb) 0.001%
Iron (Fe) 0.005%

TESTS

ASSAY. (By complexometric titration). Weigh accurately 4.0 g and transfer to a 250-mL volumetric flask. Dissolve in 25 mL of 1 M sodium hydroxide, dilute to volume with water, and mix thoroughly. Fill a 50-mL buret with this sample solution. Weigh accurately 0.2 g of calcium carbonate, chelometric standard, dried in an oven at 300 °C for 3 h and cooled in a charged desiccator for 2 h, and transfer to a 400-mL beaker. Cover the beaker with a watch glass. Add 2 mL of dilute hydrochloric acid from a pipet placed between the lip of the beaker and the watch glass. Swirl the beaker to aid dissolution. With water, wash down the inner wall of the beaker, the outer surface of the pipet, and the watch glass. Dilute to about 100 mL with water. While stirring, add from the buret about 30 mL of the sample solution. Add 15 mL of 1 M sodium hydroxide and 300 mg of hydroxy naphthol blue indicator mixture. Continue the titration immediately with the sample solution to a blue color. Let W_c be the weight, in grams, of calcium carbonate; let W_s be the weight, in grams, of the sample corresponding to the aliquot of the sample solution taken; and let V be the volume, in milliliters, of the sample solution required in the titration.

$$C_{10}H_{16}N_2O_8 = (W_c \times 73,000) / (V \times W_s)$$

INSOLUBLE IN DILUTE AMMONIUM HYDROXIDE. To 20.0 g sample add 190 mL of water and 0.10 mL of methyl red indicator solution. Slowly add ammonium hydroxide, keeping the solution acidic, until all the sample is dissolved. Warm if necessary. Heat to boiling and digest in a covered beaker on a hot plate ($\approx$100 °C) for 1 h. Filter through a tared filtering crucible, wash thoroughly, and dry at 105 °C.

RESIDUE AFTER IGNITION. (Page 16). Ignite 3.0 g. Moisten the char with 1 mL of sulfuric acid.

NITRILOTRIACETIC ACID. For a stock solution transfer 10.0 g of the sample to a 100-mL volumetric flask. Dissolve in 87 mL of potassium hydroxide solution (100 g of KOH per liter), dilute with water to volume, and mix.

> **Sample Solution A.** Dilute 10 mL of the stock solution with water to 100 mL in a volumetric flask.

> **Control Solution B.** Add 1.0 mL (10 mg) of the nitrilotriacetic acid standard to 10 mL of the stock solution and dilute with water to 100 mL in a volumetric flask.

Transfer 20 mL of sample solution A to a 150-mL beaker and 20 mL of control solution B to a second 150-mL beaker. Add to each beaker 1 mL of potassium hydroxide solution (100 g of KOH per L) and 2 mL of ammonium nitrate solution (100 g of NH_4NO_3 per L). Add approximately 0.05 g of Eriochrome Black T indicator and titrate with an aqueous cadmium nitrate solution [30 g of $Cd(NO_3)_2 \cdot 4H_2O$ per L] to a red end point.

Sample Solution C. Transfer 20 mL of sample solution A to a 100-mL volumetric flask.

Control Solution D. Transfer 20 mL of control solution B to a 100-mL volumetric flask.

Add to each flask a volume of the cadmium nitrate solution equal to the volume determined by the titration plus 0.05 mL in excess. Add 1.5 mL of potassium hydroxide solution (100 g of KOH per L), 10 mL of ammonium nitrate solution (100 g of NH_4NO_3 per L), and 0.5 mL of a 0.1% solution of methyl red in alcohol. Dilute each solution with water to 100 mL and mix.

Transfer a portion of control solution D to an H-type polarographic cell equipped with a saturated calomel electrode and deaerate with nitrogen for 10 min. Record the polarogram from −0.6 to −1.2 V versus SCE at a sensitivity of 0.006, μA/mm. Repeat with sample solution C. The diffusion current in sample solution C should not exceed 0.1 times the difference in diffusion current between control solution D and sample solution C.

CALCIUM AND MAGNESIUM. (By flame AAS, page 39).

Sample Stock Solution. Dissolve 5.0 g of sample in 50 mL of water and 5 mL of ammonium hydroxide in a 150-mL beaker, and heat if necessary. Transfer to a 100-mL volumetric flask and dilute to the mark with water (1 mL = 0.05 g).

Element	Wavelength (nm)	Sample Wt (g)	Standard Added (mg)	Flame Type*	Background Correction
Ca	422.7	1.0	0.01; 0.02	N/A	No
Mg	285.2	1.0	0.005; 0.01	A/A	Yes

*N/A is nitrous oxide/acetylene; A/A is air/acetylene.

Sample Solution E for the Determination of Heavy Metals and Iron. Char 3.0 g thoroughly and heat in an oven at 500 °C until most of the carbon is volatilized. Cool, add 0.15 mL of nitric acid, and heat at 500 °C until all of the carbon is volatilized. Dissolve the residue in 2 mL of dilute hydrochloric acid (1 + 1), digest in a covered dish on a hot plate (≈100 °C) for 10 min, remove the cover, and evaporate to dryness. Dissolve in 1 mL of 1 N acetic acid and 20 mL of hot water, digest for 5 min, cool, and dilute with water to 30 mL (1 mL = 0.1 g).

HEAVY METALS. (Page 28, Method 1). Dilute 20 mL of sample solution E (2.0-g sample) with water to 25 mL.

IRON. (Page 30, Method 1). Use 2.0 mL of sample solution E (0.2-g sample).

(Ethylenedinitrilo)tetraacetic Acid, Disodium Salt Dihydrate

N,N,N′,N′-Ethylenediaminetetraacetic Acid, Disodium Salt, Dihydrate
N,N′-1,2-Ethanediylbis[*N*-carboxymethyl]glycine, Disodium Salt, Dihydrate
EDTA (Dihydrate)

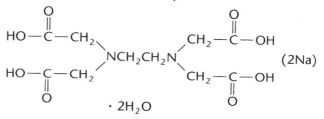

$C_{10}H_{14}N_2O_8Na_2 \cdot 2H_2O$

Formula Wt 372.24

CAS Number 6381–92–6

REQUIREMENTS

Assay. . . . . 99.0–101.0% $C_{10}H_{14}N_2O_8Na_2 \cdot 2H_2O$
pH of a 5% solution . . . . 4.0–6.0 at 25 °C

MAXIMUM ALLOWABLE

Insoluble matter. 0.005%
Nitrilotriacetic acid [(HOCOCH$_2$)$_2$N]. 0.1%
Heavy metals (as Pb) 0.005%
Iron (Fe). 0.01%

TESTS

ASSAY. (By complexometric titration). Weigh accurately 5.0 g of sample and transfer to a 250-mL volumetric flask. Dissolve in water, dilute to volume with water, and mix thoroughly. Fill a 50-mL buret with this sample solution. Weigh accurately 0.2 g of calcium carbonate, chelometric standard, dried in an oven at 300 °C for 3 hours and cooled in a charged desiccator for 2 h, and transfer to a 400-mL beaker. Cover the beaker with a watch glass. Add 2 mL of dilute hydrochloric acid from a pipet placed between the lip of the beaker and the watch glass. Swirl the beaker to aid dissolution. With water, wash down the inner wall of the beaker, the outer surface of the pipet, and the watch glass. Dilute to about 100 mL with water. While stirring, add from the buret about 30 mL of the sample solution. Add 15 mL of 1 M sodium hydroxide and 300 mg of hydroxy naphthol blue indicator mixture. Continue the titration immediately with the sample solution to a blue color. Let W_c be the weight, in grams, of calcium carbonate; let W_s be the weight, in grams, of the sample corresponding to the aliquot of the sample solution taken; and let V be the volume, in milliliters, of the sample solution required in the titration.

$$\% \ C_{10}H_{14}N_2O_8Na_2 \cdot 2H_2O = (W_c \times 92{,}980) / (V \times W_s)$$

pH OF A 5% SOLUTION.   (Page 44). The pH should be 4.0–6.0 at 25 °C.

INSOLUBLE MATTER.   (Page 15). Use 20.0 g dissolved in 200 mL of hot water.

NITRILOTRIACETIC ACID.   For a stock solution, transfer 10.0 g of the sample to a 100-mL volumetric flask. Dissolve in 40 mL of potassium hydroxide solution (100 g of KOH per L), dilute with water to volume, and mix.

> **Sample Solution A.**   Dilute 10 mL of the stock solution with water to 100 mL in a volumetric flask.

> **Control Solution B.**   Add 1.0 mL (10 mg) of nitrilotriacetic acid standard to 10 mL of the stock solution and dilute with water to 100 mL in a volumetric flask.

Transfer 20 mL of sample solution A to a 150-mL beaker and 20 mL of control solution D to a second 150-mL beaker. Add to each beaker 1 mL of potassium hydroxide solution (100.0 g of KOH per L) and 2 mL of ammonium nitrate solution (100.0 g of NH_4NO_3 per L). Add approximately 0.05 g of Eriochrome Black T indicator and titrate with an aqueous cadmium nitrate solution [30.0 g of $Cd(NO_3)_2 \cdot 4H_2O$ per liter] to a red end point.

> **Sample Solution C.**   Transfer 20 mL of sample solution A to a 100-mL volumetric flask.

> **Control Solution D.**   Transfer 20 mL of control solution B to a 100-mL volumetric flask.

Add to each flask a volume of the cadmium nitrate solution equal to the volume determined by the titration plus 0.05 mL in excess. Add 1.5 mL of potassium hydroxide solution (100.0 g of KOH per L), 10 mL of ammonium nitrate solution (100.0 g of NH_4NO_3 per L), and 0.5 mL of a 0.1% solution of methyl red in alcohol. Dilute each solution with water to 100 mL and mix.

Transfer a portion of control solution D to an H-type polarographic cell equipped with a saturated calomel electrode and deaerate with nitrogen for 10 min. Record the polarogram from –0.6 to –1.2 V versus SCE at a sensitivity of 0.006 µA/mm. Repeat with sample solution C. The diffusion current in sample solution C should not exceed 0.1 times the difference in diffusion current between control solution D and sample solution C.

> **Sample Solution E for the Determination of Heavy Metals and Iron.**   To 1.0 g add 1 mL of sulfuric acid, heat cautiously until the sample is charred, and ignite in an oven at 500 °C until most of the carbon is volatilized. Cool,

add 1 mL of nitric acid, heat until the acid is evaporated, and ignite again at 500 °C until all of the carbon is volatilized. Cool, add 4 mL of dilute hydrochloric acid (1 + 1), digest on a hot plate (≈100 °C) for 10 min, and evaporate to dryness. Add 10 mL of 10% ammonium acetate solution and digest on a hot plate (≈100 °C) for 30 min. Filter, wash thoroughly, and dilute with water to 50 mL (1 mL = 0.02 g).

HEAVY METALS. (Page 28, Method 1). Dilute 20 mL of sample solution E (0.4-g sample) with water to 25 mL.

IRON. (Page 30, Method 1). Use 5.0 mL of sample solution E (0.1-g sample).

Ferric Ammonium Sulfate Dodecahydrate
Ammonium Iron(III) Sulfate Dodecahydrate
Ammonium Ferric Sulfate Dodecahydrate

$NH_4Fe(SO_4)_2 \cdot 12H_2O$ **Formula Wt 482.20**

CAS Number 7783–83–7

REQUIREMENTS

Appearance . . . . .Pale violet crystals
Assay . . . . .98.5–102.0% $NH_4Fe(SO_4)_2 \cdot 12(H_2O)$

MAXIMUM ALLOWABLE

Insoluble matter. 0.01%
Chloride (Cl) 0.001%
Nitrate (NO_3). 0.01%
Copper (Cu). 0.003%
Zinc (Zn) 0.003%
Ferrous iron (Fe^{++}) . . . . Passes test
Substances not precipitated by ammonium hydroxide. . 0.05%

TESTS

ASSAY. (By iodometric titration of oxidative capacity of Fe^{III}). Weigh accurately 1.8 g and transfer to an iodine flask with 50 mL of water. Add 3 mL of hydrochloric acid and 3.0 g of potassium iodide. Stopper, swirl, and allow to stand in the dark for 30 min. Dilute with 100 mL of water and titrate the liberated iodine with 0.1 N sodium thiosulfate, adding 3 mL of starch indicator solution near the end of the titration. Correct for a blank. One milliliter of 0.1 N sodium thiosulfate corresponds to 0.04822 g of $NH_4Fe(SO_4)_2 \cdot 12H_2O$.

INSOLUBLE MATTER. (Page 15). Use 10.0 g of sample dissolved in 100 mL of dilute hydrochloric acid (1 + 99).

CHLORIDE. Dissolve 4.0 g in 40 mL of dilute nitric acid (1 + 9), filter if necessary through a chloride-free filter, and dilute with water to 60 mL. For the sample use 30 mL of the solution. For the control, add 0.01 mg of chloride ion (Cl) and 1 mL of nitric acid to 15 mL of the solution and dilute with water to 30 mL. To each solution add 1 mL of silver nitrate reagent solution. Any turbidity in the solution of the sample should not exceed that in the control.

NITRATE. Dissolve 2.0 g in 1 mL of water by swirling in a test tube. Add 10 mL of dilute ammonium hydroxide (1 + 1), mix well, and filter through a dry filter paper. Do not wash the precipitate.

> **Sample Solution A.** Add 1.0 mL of the filtrate to 2 mL of water, dilute to 50 mL with brucine sulfate reagent solution, and mix.

> **Control Solution B.** Add 1.0 mL of the filtrate to 2 mL of the standard nitrate solution containing 0.01 mg of nitrate ion (NO_3) per mL, dilute to 50 mL with brucine sulfate reagent solution, and mix.

Continue with the procedure described on page 30, starting with the preparation of blank solution C.

COPPER AND ZINC. (By flame AAS, page 39).

> **Sample Stock Solution.** Dissolve 10.0 g of sample in 90 mL of nitric acid (1 + 19) in a 100-mL volumetric flask, and dilute to the mark with water (1 mL = 0.10 g).

Element	Wavelength (nm)	Sample Wt (g)	Standard Added (mg)	Flame Type*	Background Correction
Cu	324.7	1.0	0.03; 0.06	A/A	Yes
Zn	213.9	0.20	0.005; 0.01	A/A	Yes

*A/A is air/acetylene.

FERROUS IRON. Dissolve 1.0 g in a mixture of 20 mL of water and 1 mL of hydrochloric acid and add 0.05 mL of a freshly prepared 5% solution of potassium ferricyanide in a 25-mL glass-stoppered flask. Stopper immediately. No blue or green color should be produced in 1 min. (Limit about 0.001%)

SUBSTANCES NOT PRECIPITATED BY AMMONIUM HYDROXIDE. Dissolve 2.0 g in 50 mL of water. Heat to boiling and pour slowly with constant stirring into 20 mL of dilute ammonium hydroxide (1 + 4). Filter while hot and wash with hot water until the filtrate measures 100 mL. Evaporate the filtrate to dryness in a tared dish and ignite at 800 ± 25 °C for 15 min.

Ferric Chloride Hexahydrate
Iron(III) Chloride Hexahydrate

$FeCl_3 \cdot 6H_2O$ **Formula Wt 270.30**

CAS Number 10025–77–1

REQUIREMENTS

Assay . . . . 97.0–102.0% $FeCl_3 \cdot 6H_2O$

MAXIMUM ALLOWABLE

Insoluble matter. 0.01%
Nitrate (NO_3) 0.01%
Phosphorus compounds (as PO_4) 0.01%
Sulfate (SO_4) 0.01%
Copper (Cu). 0.003%
Zinc (Zn) 0.003%
Ferrous iron (Fe^{++}) . . . . Passes test
Substances not precipitated by ammonium
        hydroxide (as sulfates) 0.1%

TESTS

ASSAY. (By iodometric titration of oxidative capacity of Fe^{III}). Weigh accurately 1.0 g and transfer to an iodine flask with 50 mL of water. Add 3 mL of hydrochloric acid and 3.0 g of potassium iodide. Stopper, swirl, and allow to stand in the dark for 30 min. Dilute with 100 mL of water and titrate the liberated iodine with 0.1 N sodium thiosulfate, adding 3 mL of starch indicator solution near the end of the titration. Correct for a blank. One milliliter of 0.1 N sodium thiosulfate corresponds to 0.02703 g of $FeCl_3 \cdot 6H_2O$.

INSOLUBLE MATTER. (Page 15). Use 10.0 g dissolved in 50 mL of dilute hydrochloric acid (1 + 49).

> ***Sample Solution A for the Determination of Nitrate, Sulfate, and Substances Not Precipitated by Ammonium Hydroxide.*** Dissolve 10.0 g in 75 mL of water, heat to boiling, and pour slowly with constant stirring into 200 mL of dilute ammonium hydroxide (1 + 3). Filter through a folded filter while still hot, wash with hot water until the volume of the filtrate plus washings measures 300 mL, and thoroughly mix the solution (1 mL = 0.033 g).

NITRATE.

> ***Sample Solution A-1.*** Add 42 mL of brucine sulfate reagent solution and 2 mL of water to 6 mL of sample solution A (0.2-g sample).

Control Solution B. Add 42 mL of brucine sulfate reagent solution and 2 mL of the standard nitrate solution containing 0.01 mg of nitrate ion (NO_3) per mL to 6 mL of sample solution A.

Blank Solution C. Add 42 mL of brucine sulfate reagent solution to 8 mL of water.

Continue with the procedure described on page 30.

PHOSPHORUS COMPOUNDS. Dissolve 5.0 g in 40 mL of dilute nitric acid (1 + 1) and evaporate on a hot plate ($\approx$100 °C) to a syrupy residue. Dissolve the residue in about 80 mL of water, dilute with water to 100 mL, and dilute 20 mL of this solution with water to 100 mL. For the test dilute 10 mL (0.1-g sample) with water to 70 mL. Prepare a standard containing 0.01 mg of phosphate ion (PO_4) in 70 mL of water. To each solution add 5 mL of ammonium molybdate solution (5 g in 50 mL) and adjust the pH to 1.8 (using a pH meter) by adding dilute hydrochloric acid (1 + 1) or dilute ammonium hydroxide (1 + 1). Cautiously heat the solutions to boiling but do not boil, and cool to room temperature. If a precipitate forms, it will dissolve when the solution is acidified in the next step. To each solution add 10 mL of hydrochloric acid and dilute each with water to 100 mL. Transfer the solutions to separatory funnels, add 35 mL of ether to each, shake vigorously, and allow to separate. Draw off and discard the aqueous phases. Wash the ether phases twice with 10-mL portions of dilute hydrochloric acid (1 + 9), discarding the washings each time. To the washed ether phases add 10 mL of dilute hydrochloric acid (1 + 9) to which has just been added 0.2 mL of a freshly prepared 2% solution of stannous chloride in hydrochloric acid. Shake the solutions and allow the phases to separate. Any blue color in the ether phase from the solution of the sample should not exceed that in the ether phase from the standard.

SULFATE. (Page 33, Method 1). Evaporate 15 mL of sample solution A to 10 mL. Allow 30 min for the turbidity to form.

COPPER AND ZINC. (By flame AAS, page 39).

Sample Stock Solution. Dissolve 10.0 g of sample with water in a 100-mL volumetric flask, and dilute to the mark with water (1 mL = 0.10 g).

Element	Wavelength (nm)	Sample Wt (g)	Standard Added (mg)	Flame Type*	Background Correction
Cu	324.7	1.0	0.05; 0.10	A/A	Yes
Zn	213.9	0.20	0.005; 0.01	A/A	Yes

*A/A is air/acetylene.

FERROUS IRON. Dissolve 0.5 g in 20 mL of dilute hydrochloric acid (1 + 19) in a 25-mL glass-stoppered flask and add 0.05 mL of a freshly prepared 5% solution of potassium ferricyanide. Stopper immediately. No blue or green color should be produced in 1 min. (Limit about 0.002%)

SUBSTANCES NOT PRECIPITATED BY AMMONIUM HYDROXIDE. To 30 mL of sample solution A (1-g sample) add 0.50 mL of sulfuric acid, evaporate in a tared dish or crucible, ignite cautiously until ammonium salts are volatilized, and finally ignite at 800 ± 25 °C for 15 min.

Ferric Nitrate Nonahydrate
Iron(III) Nitrate Nonahydrate

$Fe(NO_3)_3 \cdot 9H_2O$

Formula Wt 404.00

CAS Number 7782–61–8

REQUIREMENTS

Assay. . . . . .98.0–101.0% $Fe(NO_3)_3 \cdot 9H_2O$

MAXIMUM ALLOWABLE

Insoluble matter. 0.005%
Chloride (Cl) . . . . 5 ppm
Sulfate (SO_4) 0.01%
Substances not precipitated by ammonium
    hydroxide (as sulfates) 0.1%

TESTS

ASSAY. (By complexometric titration of iron). Weigh accurately 1.6 g and transfer to a 400-mL beaker. Dissolve with 200 mL of water and add 2 mL of 20% hydrochloric acid. From a 50 mL buret, add 30.0 mL of 0.1 M EDTA. Heat the solution to 40–50 °C and adjust the pH to 2.5 ± 0.2 with 10% sodium acetate solution. Add 0.25 mL of Variamine Blue B indicator solution and continue the addition of 0.1 M EDTA until burgundy color changes to canary yellow. One milliliter of 0.1 M EDTA corresponds to 0.04040 g of $Fe(NO_3)_3 \cdot 9H_2O$

INSOLUBLE MATTER. (Page 15). Use 20.0 g dissolved in 200 mL of dilute nitric acid (1 + 99).

CHLORIDE. Dissolve 2.0 g in 20 mL of water, filter if necessary through a chloride-free filter, and add 1 mL of nitric acid plus 1 mL of phosphoric acid and 1 mL of silver nitrate reagent solution. Any turbidity should not exceed that produced by 0.01 mg of chloride ion (Cl) in an equal volume of solution containing the quantities of reagents used in the test.

Sample Solution A for the Determination of Sulfate and Substances Not Precipitated by Ammonium Hydroxide. Dissolve 5.0 g in 70 mL of water, heat to boiling, and pour slowly with constant stirring into 100 mL of dilute ammonium hydroxide (1 + 6). Filter through a folded filter while still hot and wash with hot water until the filtrate (sample solution A) measures 250 mL. Mix thoroughly (1 mL = 0.02 g).

SULFATE. To 25 mL (0.5-g sample) of sample solution A add 5 mL of hydrochloric acid and 10 mL of nitric acid. For the standard, add 0.05 mg of sulfate ion (SO_4) and 1.5 mL of ammonium hydroxide to 25 mL of water, and add 5 mL of hydrochloric acid and 10 mL of nitric acid. Digest each solution in a covered beaker on a hot plate ($\approx$100 °C) until reaction ceases, uncover, and evaporate to dryness. Dissolve the residues in 4 mL of water plus 1 mL of dilute hydrochloric acid (1 + 19), filter if necessary through a small filter, wash with two 2-mL portions of water, and dilute with water to 10 mL. To each solution add 1 mL of barium chloride reagent solution. Any turbidity in the solution of the sample should not exceed that in the standard. Compare 10 min after adding the barium chloride to the sample and standard solutions.

SUBSTANCES NOT PRECIPITATED BY AMMONIUM HYDROXIDE. Evaporate 50 mL of sample solution A to dryness in a tared dish and add 0.50 mL of sulfuric acid. Heat carefully to volatilize the ammonium salts and excess sulfuric acid, and finally ignite at 800 ± 25 °C for 15 min.

Ferrous Ammonium Sulfate Hexahydrate
Ammonium Ferrous Sulfate Hexahydrate
Ammonium Iron(II) Sulfate Hexahydrate

$Fe(NH_4)_2(SO_4)_2 \cdot 6H_2O$ Formula Wt 392.14

CAS Number 7783–85–9

NOTE. Due to inherent oxidation, ferric ion (Fe^{3+}) content may increase on storage.

REQUIREMENTS

Assay . . . . .98.5–101.5% $Fe(NH_4)_2(SO_4)_2 \cdot 6H_2O$

MAXIMUM ALLOWABLE

Insoluble matter. 0.01%
Phosphate (PO_4) 0.003%
Copper (Cu). 0.003%
Manganese (Mn) 0.01%
Zinc (Zn) 0.003%
Ferric ion (Fe^{3+}) 0.01%
Substances not precipitated
    by ammonium hydroxide. 0.05%

TESTS

ASSAY. (By titration of reductive capacity of Fe^{II}). Weigh accurately 1.6 g and dissolve in a mixture of 100 mL of water and 3 mL of sulfuric acid that has previously been freed from oxygen by bubbling with an inert gas. Titrate with 0.1 N potassium permanganate to a permanent faint pink end point. One milliliter of 0.1 N permanganate corresponds to 0.03921 g of $Fe(NH_4)_2(SO_4)_2 \cdot 6H_2O$.

INSOLUBLE MATTER. (Page 15). Use 10.0 g dissolved in 100 mL of freshly boiled dilute sulfuric acid (1 + 99).

PHOSPHATE. Dissolve 1.0 g in 10 mL of water and add 2 mL of nitric acid. Boil to oxidize the iron and expel the excess gases. Dilute with water to 90 mL. Dilute 30 mL with water to 80 mL, add 5 mL of ammonium molybdate solution (5.0 g in 50 mL), and adjust the pH to 1.8 (using a pH meter) with dilute ammonium hydroxide (1 + 1). Prepare a standard containing 0.01 mg of phosphate ion (PO_4), the residue from evaporation of the amount of ammonium hydroxide used in adjusting the pH of the sample, and 5 mL of the ammonium molybdate solution in 85 mL. Adjust the pH to 1.8 (using a pH meter) with dilute hydrochloric acid (1 + 9). Cautiously heat the solutions to boiling, but do not boil, and cool to room temperature. If a precipitate forms, it will dissolve when acidified in the next operation. Add 10 mL of hydrochloric acid and dilute with water to 100 mL. Transfer the solutions to separatory funnels, add 35 mL of ether to each, shake vigorously, and allow to separate. Draw off and discard the aqueous phase. Wash the ether phase twice with 10-mL portions of dilute hydrochloric acid (1 + 9), discarding the washings each time. Add 10 mL of dilute hydrochloric acid (1 + 9) to which has just been added 0.2 mL of a freshly prepared 2% solution of stannous chloride in hydrochloric acid, shake, and allow to separate. Any blue color in the ether extract from the sample should not exceed that in the ether extract from the standard.

COPPER, MANGANESE, AND ZINC. (By flame AAS, page 39).

Sample Stock Solution. Dissolve 10.0 g of sample with water in a 100-mL volumetric flask and dilute to the mark with water (1 mL = 0.10 g).

Element	Wavelength (nm)	Sample Wt (g)	Standard Added (mg)	Flame Type*	Background Correction
Cu	324.7	1.0	0.03; 0.06	A/A	Yes
Mn	279.5	0.20	0.02; 0.04	A/A	Yes
Zn	213.9	0.20	0.005; 0.01	A/A	Yes

*A/A is air/acetylene.

FERRIC ION. Place 0.20 g of the sample and 0.5 g of sodium bicarbonate in a dry 125-mL glass-stoppered conical flask. For the control, place 0.10 g of the sample and 0.5 g of sodium bicarbonate in a similar flask. To each flask add 94 mL of a freshly boiled and cooled solution of dilute sulfuric acid (1 + 25) and loosely stopper the flasks until effervescence stops. To the control add 0.01 mg of ferric ion (Fe^{3+}). Add 6 mL of ammonium thiocyanate reagent solution to each flask. Any red color in the solution of the sample should not exceed that in the control.

SUBSTANCES NOT PRECIPITATED BY AMMONIUM HYDROXIDE.
Dissolve 2.0 g in 25 mL of dilute nitric acid (1 + 4) and boil gently to oxidize the iron. Pour the solution slowly with constant stirring into 50 mL of dilute ammonium hydroxide (1 + 4). Filter, wash moderately with hot water, and evaporate the solution to dryness in a tared dish. Heat gently to volatilize the excess ammonium salts and finally ignite at 800 ± 25 °C for 15 min.

Ferrous Sulfate Heptahydrate
Iron(II) Sulfate Heptahydrate

$FeSO_4 \cdot 7H_2O$ **Formula Wt 278.01**

CAS NUMBER 7782–63–0

NOTE. Due to inherent oxidation, ferric ion (Fe^{3+}) content may increase on storage.

REQUIREMENTS

Assay . . . . ≥99.0% $FeSO_4 \cdot 7H_2O$

MAXIMUM ALLOWABLE

Insoluble matter. 0.01%
Chloride (Cl) 0.001%
Phosphate (PO_4) 0.001%
Copper (Cu). 0.005%
Manganese (Mn) 0.05%
Zinc (Zn) 0.005%
Ferric ion (Fe^{3+}) 0.1%
Substances not precipitated
    by ammonium hydroxide. 0.05%

TESTS

ASSAY. (By titration of reductive capacity of Fe^{II}). Weigh accurately 1.0 g, and dissolve in a mixture of 100 mL of water and 3 mL of sulfuric acid that has previously been freed from oxygen by bubbling with an inert gas. Titrate with 0.1 N potassium permanganate to a permanent faint pink end point. One milliliter of 0.1 N potassium permanganate corresponds to 0.02780 g of $FeSO_4 \cdot 7H_2O$.

INSOLUBLE MATTER. (Page 15). Use 10.0 g dissolved in 100 mL of freshly boiled dilute sulfuric acid (1 + 99).

CHLORIDE. Dissolve 1.0 g in 10 mL of water and add slowly 2 mL of nitric acid. After the evolution of nitrogen oxides has ceased, dilute with water to 15 mL, filter if necessary through a small chloride-free filter, and add 1 mL of silver nitrate reagent solution. Any turbidity should not exceed that produced by 0.01 mg of chloride ion (Cl) in an equal volume of solution containing the quantities of reagents used in the test. The comparison is best made by the general method for chloride in colored solutions, page 27.

PHOSPHATE. (Page 32, Method 2). Dissolve 1.0 g in 10 mL of dilute nitric acid (1 + 5) and boil to expel the excess gases. Dilute the solution with water to 80 mL, add 5 mL of ammonium molybdate solution (5 g in 50 mL), and adjust the pH to 1.8 (using a pH meter) with dilute ammonium hydroxide (1 + 1). Cautiously heat the solution to boiling, but do not boil, and cool to room temperature. If a precipitate forms, it will dissolve in the next operation. Add 10 mL of hydrochloric acid, dilute with water to 100 mL, and continue as described.

Prepare a standard containing 0.01 mg of phosphate ion (PO_4), the residue from evaporation of the amount of ammonium hydroxide used in adjusting the pH of the sample, and 5 mL of the ammonium molybdate solution in 85 mL. Adjust the pH to 1.8 (using a pH meter) with dilute hydrochloric acid (1 + 9). Cautiously heat the solutions to boiling, but do not boil; cool to room temperature. Add 10 mL of hydrochloric acid, and dilute with water to 100 mL.

Transfer the solutions to separatory funnels, add 35 mL of ether to each, shake vigorously, and allow to separate. Draw off and discard the aqueous phase. Wash the ether phase twice with 100-mL portions of dilute hydrochloric acid (1 + 9), discarding the washings each time. Add 10 mL of dilute hydrochloric acid (1 + 9) to which has just been added 0.2 mL of a freshly prepared 2% solution of stannous chloride in hydrochloric acid, shake, and allow to separate. Any blue color in the ether extract from the sample should not exceed that in the ether extract from the standard.

COPPER, MANGANESE, AND ZINC. (By flame AAS, page 39).

Sample Stock Solution. Dissolve 10.0 g of sample with water in a 100-mL volumetric flask and dilute to the mark with water (1 mL = 0.10 g).

Element	Wavelength (nm)	Sample Wt (g)	Standard Added (mg)	Flame Type*	Background Correction
Cu	324.7	1.0	0.05; 0.10	A/A	Yes
Mn	279.5	0.10	0.025; 0.05	A/A	Yes
Zn	213.9	0.10	0.005; 0.01	A/A	Yes

*A/A is air/acetylene.

FERRIC ION. Place 0.20 g of the sample and 0.5 g of sodium bicarbonate in a dry 125-mL glass-stoppered conical flask. For the control, place 0.10 g of the sample and 0.5 g of sodium bicarbonate in a similar flask. To each flask add 94 mL of a freshly boiled and cooled solution of dilute sulfuric acid (1 + 25) and loosely stopper the flasks until effervescence stops. To the control add 0.1 mg of ferric ion (Fe^{3+}). Add 6 mL of ammonium thiocyanate reagent solution to each flask. Any red color in the solution of the sample should not exceed that in the control.

SUBSTANCES NOT PRECIPITATED BY AMMONIUM HYDROXIDE.
Dissolve 2.0 g in 25 mL of dilute nitric acid (1 + 4) and boil gently to oxidize the iron. Pour the solution slowly with constant stirring into 50 mL of dilute ammonium hydroxide (1 + 4). Filter, wash moderately with hot water, and evaporate the solution to dryness in a tared dish. Heat gently to volatilize the excess ammonium salts and finally ignite at 800 ± 25 °C for 15 min.

Formaldehyde Solution
(with stabilizer)

HCHO   **Formula Wt 30.03**

CAS Number 50–00–0

NOTE. This reagent contains 10–15% of methanol as a stabilizer.

REQUIREMENTS

Assay . . . . 36.5–38.0% HCHO

MAXIMUM ALLOWABLE

Color (APHA) 10
Residue after ignition 0.005%
Titrable acid . . . . 0.006 meq/g
Chloride (Cl) . . . . 5 ppm
Sulfate (SO_4) 0.002%
Heavy metals (as Pb) . . . . 5 ppm
Iron (Fe) . . . . 5 ppm

TESTS

NOTE. The assay procedure specified herein is intended to determine both formaldehyde and its paraformaldehyde polymerization product that may form upon standing. A different procedure (*see*, for example, *Reagent Chemicals*, 5th edition) should be used if only the formaldehyde content is desired.

ASSAY. (By bisulfite addition and acid–base titrimetry). Transfer a glass-stoppered weighing vial containing 2.0 g of sample, accurately weighed, to 100 mL of sodium sulfite solution in a 500-mL conical flask. Avoid getting any of the sample on the wall of the flask. Add 50 mL of 1 N sodium hydroxide and 0.15 mL of thymolphthalein indicator, and then titrate to a colorless end point with 1 N sulfuric acid. Correct for a reagent blank and any significant quantity of acid (as HCOOH) in the sample. One milliliter of 1 N sodium hydroxide corresponds to 0.03003 g of HCHO.

COLOR (APHA). (Page 19).

RESIDUE AFTER IGNITION. (Page 16). Evaporate 20.0 g (20 mL) to dryness in a tared dish on a hot plate ($\approx$100 °C), add 0.05 mL of sulfuric acid to the residue, and ignite gently to volatilize the excess sulfuric acid. Finally, ignite at 800 ± 25 °C for 15 min.

TITRABLE ACID. To 10 mL in a flask containing 20 mL of water, add 0.15 mL of bromthymol blue indicator solution, and titrate with 0.1 N sodium hydroxide. Not more than 0.60 mL of 0.1 N sodium hydroxide should be required.

CHLORIDE. (Page 27). Dilute 2.0 mL with 20 mL of water.

SULFATE. (Page 33, Method 1). Use 2.5 g (2.5 mL).

HEAVY METALS. (Page 28, Method 1). To 4.0 mL add about 10 mg of sodium carbonate, evaporate to dryness, and heat gently to volatilize any organic matter. Dissolve the residue in about 20 mL of water and dilute with water to 25 mL.

IRON. (Page 30, Method 2). Dilute 2.0 mL with water to 15 mL. Adjust the pH to 5 with ammonium hydroxide after addition of the 1,10-phenanthroline reagent solution.

Formamide

$HCONH_2$ **Formula Wt 45.04**

CAS Number 75–12–7

REQUIREMENTS

Assay. . . . . $\geq$99.5% $HCONH_2$
Freezing point . . . . .2.0–3.0 °C

MAXIMUM ALLOWABLE

Color (APHA). 10

TESTS

ASSAY. Analyze the sample by gas chromatography using the general parameters cited on page 70. The following specific conditions are also required.

Column: Type I, methyl silicone

Measure the area under all peaks and calculate the formamide content in area percent.

FREEZING POINT. Place 15 mL in a test tube (20 × 150 mm) in which is centered an accurate thermometer. The sample tube is centered by corks in an outer tube about 38 × 200 mm. Cool the whole apparatus, without stirring, in a bath of shaved or crushed ice with enough water to wet the outer tube. When the temperature is about 0 °C, stir to start the freezing and read the thermometer every 30 s. The temperature that remains constant for 1–2 min is the freezing point.

COLOR (APHA). (Page 19).

Formic Acid, 96%

HCOOH   **Formula Wt 46.03**

CAS Number 64–18–6

> *CAUTION.* Slow decomposition of this reagent may produce pressure in the bottle. Loosen cap occasionally to vent the gas.

REQUIREMENTS

Assay . . . . ≥96.0% HCOOH

MAXIMUM ALLOWABLE

Color (APHA) 15
Dilution test . . . . Passes test
Residue after evaporation 0.003%
Acetic acid (CH_3COOH) 0.4%
Ammonium (NH_4) 0.005%
Chloride (Cl) 0.001%
Sulfate (SO_4) 0.003%
Sulfite (SO_3) . . . . Passes test
Heavy metals (as Pb) 0.001%
Iron (Fe) 0.001%

TESTS

ASSAY. (By acid–base titrimetry). Tare a small glass-stoppered conical flask containing 15 mL of water. Quickly introduce 1.0–1.5 mL of the sample and weigh. Dilute with water to about 50 mL and titrate with 1 N sodium hydroxide, using 0.15 mL of phenolphthalein indicator. One milliliter of 1 N sodium hydroxide corresponds to 0.04603 g of HCOOH.

COLOR (APHA). (Page 19).

DILUTION TEST. Dilute 1 volume of the acid with 3 volumes of water. No turbidity should be observed within 1 h.

RESIDUE AFTER EVAPORATION. (Page 16). Evaporate 50.0 g (42 mL) to dryness in a tared dish on a hot plate (≈100 °C) and dry the residue at 105 °C for 30 min.

ACETIC ACID. Analyze the sample by gas chromatography as described on page 70. The following specific conditions are satisfactory.

> **Column:** 18 m × 64 mm, stainless steel, packed with 80/100-mesh Porapak Q
>
> **Column Temperature:** 150 °C
>
> **Injection Port Temperature:** 150 °C
>
> **Detector Temperature:** 175 °C
>
> **Detector Current:** 200 mA
>
> **Carrier Gas:** Helium at 40 mL/min
>
> **Sample Size:** 5 μL
>
> **Detector:** Thermal conductivity
>
> **Approximate Retention Times (min):** Water, 12; formic acid, 35; acetic acid, 81

Measure the area under all peaks and calculate the acetic acid content in area percent.

AMMONIUM. (By differential pulse polarography, page 49). Dilute 10.0 g (8.2 mL) of sample to 50 mL with water; use 5.0 mL in each test, neutralizing with 3.2 mL of ammonia-free 6 N sodium hydroxide. For the standard, use 0.05 mg of ammonium (NH_4).

CHLORIDE. (Page 27). Use 1.0 g (0.83 mL).

SULFATE. (Page 33, Method 3). Use 1.7 g (1.4 mL).

SULFITE. Dilute 25 mL with 25 mL of water; add 0.10 mL of 0.1 N iodine solution. The mixture should retain a distinct yellow color.

HEAVY METALS. (Page 28, Method 1). To 2 g (1.6 mL) in a beaker add about 10 mg of sodium carbonate and evaporate to dryness. Dissolve the residue in about 20 mL of water and dilute with water to 25 mL.

IRON. (Page 30, Method 1). To 6.0 g (5.0 mL) in a beaker, add about 10 mg of sodium carbonate, and evaporate to dryness. Dissolve the residue in 6 mL of hydrochloric acid, dilute with water to 60 mL, and use 10 mL of this solution without further acidification.

Formic Acid, 88%

HCOOH   **Formula Wt 46.03**

CAS Number 64–18–6

REQUIREMENTS

Assay . . . . ≥88.0% HCOOH

MAXIMUM ALLOWABLE

Color (APHA) 15
Dilution test . . . . Passes test
Residue after evaporation 0.002%
Acetic acid (CH_3COOH) 0.4%
Ammonium (NH_4) 0.005%
Chloride (Cl) 0.001%
Sulfate (SO_4) 0.002%
Sulfite (SO_3) . . . . Passes test
Heavy metals (as Pb) . . . . 5 ppm
Iron (Fe) . . . . 5 ppm

TESTS

Assay and other tests are the same as for Formic Acid, 96%, page 314.

2-Furancarboxyaldehyde

2-Furaldehyde
Furfural

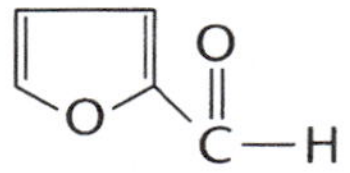

$C_5H_4O_2$ **Formula Wt 96.08**

CAS Number 98–01–1

REQUIREMENTS

Assay . . . . ≥98.0% C_4H_3OCHO

MAXIMUM ALLOWABLE

Titrable acid . . . . 0.02 meq/g
Residue after evaporation 0.5%

TESTS

ASSAY. Analyze the sample by gas chromatography using the general parameters cited on page 70. The following specific conditions are also required.

   Column: Type I, methyl silicone

Measure the area under all peaks and calculate the 2-furancarboxyaldehyde content in area percent.

TITRABLE ACID. To 300 mL of water in an Erlenmeyer flask, add 0.15 mL of phenolphthalein indicator solution, and titrate with 0.1 N sodium hydroxide to a faint pink color that is stable for about 1 min. Add 10.0 g (9.0 mL) of sample to the flask, mix well, and titrate with 0.1 N sodium hydroxide to the original faint pink color. Not more than 2.0 mL of 0.1 N sodium hydroxide should be consumed.

RESIDUE AFTER EVAPORATION. (Page 16). Evaporate 25.0 g (22 mL) to dryness in a tared dish on a hot plate ($\approx$100 °C), and dry the residue at 105 °C for 30 min.

Gallic Acid

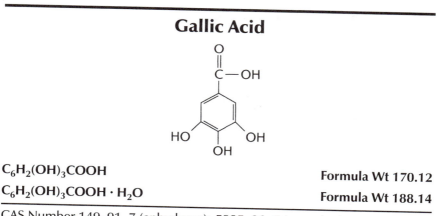

| $C_6H_2(OH)_3COOH$ | **Formula Wt 170.12** |
| $C_6H_2(OH)_3COOH \cdot H_2O$ | **Formula Wt 188.14** |

CAS Number 149–91–7 (anhydrous); 5995–86–8 (monohydrate)

   NOTE. This reagent is available in both the monohydrate and the anhydrous form. The label should identify the degree of hydration.

REQUIREMENTS

Assay . . . . $\geq$98.0% $C_6H_2(OH)_3COOH$
or $C_6H_2(OH)_3COOH \cdot H_2O$

MAXIMUM ALLOWABLE

Insoluble matter. 0.01%
Residue after ignition 0.05%
Sulfate (SO_4) 0.02%

TESTS

ASSAY. (By acid–base titrimetry). Weigh accurately about 0.7 g, transfer to a beaker, add 100 mL of water, and stir to dissolve. When dissolution is complete, titrate potentiometrically with 0.1 N sodium hydroxide. Correct for a blank determination. One milliliter of 0.1 N sodium hydroxide corresponds to 0.017012 g of $C_6H_2(OH)_3COOH$ or 0.018814 g of $C_6H_2(OH)_3COOH \cdot H_2O$.

INSOLUBLE MATTER. (Page 15). Use 10.0 g dissolved in 300 mL of water.

RESIDUE AFTER IGNITION. (Page 16). Ignite 4.0 g.

SULFATE. (Page 33, Method 1). Dissolve 2.5 g in 50 mL of hot water, cool in ice water while stirring, and filter. Dilute the filtrate with water to 50 mL and use 5 mL.

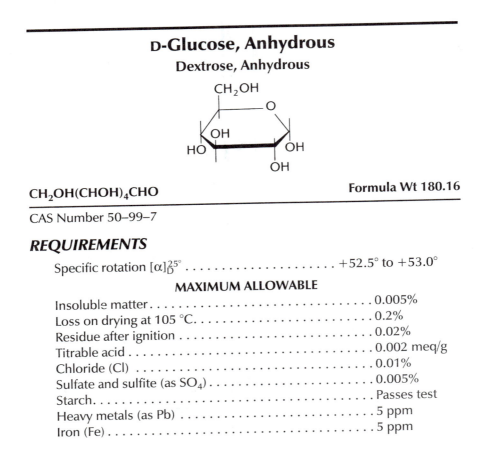

D-Glucose, Anhydrous
Dextrose, Anhydrous

$CH_2OH(CHOH)_4CHO$ | **Formula Wt 180.16**

CAS Number 50–99–7

REQUIREMENTS

Specific rotation $[\alpha]_D^{25°}$ . . . . +52.5° to +53.0°

MAXIMUM ALLOWABLE

Insoluble matter.	0.005%
Loss on drying at 105 °C.	0.2%
Residue after ignition	0.02%
Titrable acid	0.002 meq/g
Chloride (Cl)	0.01%
Sulfate and sulfite (as SO_4)	0.005%
Starch.	Passes test
Heavy metals (as Pb)	5 ppm
Iron (Fe)	5 ppm

TESTS

SPECIFIC ROTATION. (Page 23). Weigh accurately about 10 g and dissolve in 90 mL of water in a 100-mL volumetric flask. Add 0.2 mL of ammonium hydroxide and dilute with water to volume at 25 °C. Observe the optical rotation in a polarimeter at 25 °C using sodium light, and calculate the specific rotation. It should not be less than +52.5° nor more than +53.0°.

INSOLUBLE MATTER. (Page 15). Use 20 g dissolved in 150 mL of water.

LOSS ON DRYING. Weigh accurately about 1 g and dry at 105 °C for 6 h.

RESIDUE AFTER IGNITION. (Page 16). Ignite 5.0 g until charred. Moisten the char with 2 mL of sulfuric acid.

TITRABLE ACID. To 100 mL of carbon dioxide-free water add 0.15 mL of phenolphthalein indicator solution and 0.02 N sodium hydroxide until a pink color is produced. Dissolve 10.0 g of the sample in this solution and titrate with 0.01 N sodium hydroxide to the same end point. Not more than 2.5 mL should be required.

CHLORIDE. (Page 27). Dissolve 0.50 g in 50 mL of water, and use 10 mL of this solution.

SULFATE AND SULFITE. (Page 33, Method 1). After sample dissolution, add 1 mL of bromine water and boil.

STARCH. Dissolve 1.0 g in 10 mL of water and add 0.05 mL of 0.1 N iodine solution. No blue color should appear.

HEAVY METALS. (Page 28, Method 1). Dissolve 6.0 g in about 20 mL of water, and dilute with water to 30 mL. Use 25 mL to prepare the sample solution, and use the remaining 5.0 mL to prepare the control solution.

IRON. (Page 30, Method 1). Use 2.0 g.

Glycerol
1,2,3-Propanetriol

$$\overset{\displaystyle OH}{\underset{\displaystyle |}{HOCH_2CHCH_2OH}}$$

CH₂OHCHOHCH₂OH

$CH_2OHCHOHCH_2OH$

Formula Wt 92.09

CAS Number 56–81–5

REQUIREMENTS

Assay (by volume) . . . .≥99.5% $C_3H_5(OH)_3$

MAXIMUM ALLOWABLE

Color (APHA) 10
Residue after ignition 0.005%
Neutrality . . . . Passes test
Chlorinated compounds (as Cl) 0.003%
Sulfate (SO_4) 0.001%
Acrolein and glucose . . . . Passes test
Fatty acid esters (as butyric acid) 0.05%
Substances darkened by sulfuric acid . . . . Passes test
Heavy metals (as Pb) . . . . .2 ppm
Water (H_2O). 0.5%

TESTS

ASSAY. Analyze the sample by gas chromatography using the general parameters cited on page 70. The following specific conditions are also required.

Column: Type II, mixed cyano, phenyl methyl silicone

Measure the area under all peaks and calculate the glycerol content in area percent. Correct for water content.

COLOR (APHA). (Page 19).

RESIDUE AFTER IGNITION. (Page 16). Heat 20 g in a tared open dish, ignite the vapors, and when the glycerol has been entirely consumed, ignite at 800 ± 25 °C for 15 min.

NEUTRALITY. A 10% aqueous solution of glycerol should not affect the color of either red or blue litmus paper in 1 min.

CHLORINATED COMPOUNDS. Transfer 5.0 g of the glycerol to a dry 100-mL round-bottom flask fitted with a ground joint. Add 15 mL of morpholine, connect a matching standard taper reflux condenser, and reflux the mixture gently for 3 h. Rinse the condenser with 10 mL of water, collecting the washing in the flask. Cool. Cautiously acidify the solution with nitric acid, add 1 mL of silver nitrate reagent solution, and dilute with water to 50 mL. Any turbidity should not exceed that produced by 0.15 mg of chloride ion (Cl) in an equal volume of solution containing the quantities of reagents used in the test, omitting the refluxing.

SULFATE. (Page 33, Method 1). Use 5.0 g.

ACROLEIN AND GLUCOSE. Heat a mixture of 5 mL of the glycerol and 5 mL of 10% potassium hydroxide solution at 60 °C for 5 min. No yellow color should develop.

FATTY ACID ESTERS. To 40 g in a 250-mL conical flask add 50 mL of hot, freshly boiled water. Add 10.0 mL of 0.1 N sodium hydroxide, cover with a loosely fitting pear-shaped bulb, and digest on a hot plate ($\approx$100 °C) for 45 min. Cool and titrate the excess alkali with 0.1 N hydrochloric acid, using 0.15 mL of bromthymol blue indicator solution and titrating to a bluish green end point. Run a blank with 50 mL of the same water and 10.0 mL of 0.1 N sodium hydroxide solution, heating for the same length of time and titrating to the same end point. The difference between the volumes of acid used in the titration of the blank and of the sample should be less than 2.3 mL.

SUBSTANCES DARKENED BY SULFURIC ACID. Vigorously shake 5 mL of glycerol with 5 mL of 94.5–95.5% sulfuric acid in a glass-stoppered 25-mL cylinder for 1 min and allow the liquid to stand for 1 h. The liquid should not be darker than a standard made up of 0.4 mL of cobaltous chloride color solution, 3.0 mL of ferric chloride color solution, and 6.6 mL of water.

> *Ferric Chloride Color Solution.* Dissolve 1.80 g of ferric chloride in 40 mL of (1 + 39) hydrochloric acid. (1 mL = 45.0 mg $FeCl_3 \cdot 6H_2O$).

> *Cobaltous Chloride Color Solution.* Dissolve 2.38 g of cobaltous chloride in 40 mL of (1 + 39) hydrochloric acid (1 mL = 59.5 mg $CoCl_2 \cdot 6H_2O$).

HEAVY METALS. (Page 28, Method 1). Dilute 13 g (10 mL) with water to 36 mL. Use 32 mL to prepare the sample solution, and use the remaining 4.0 mL to prepare the control solution.

WATER. (Page 55, Method 1). Use 4.0 mL (5.0 g) of sample.

Glycine
Aminoacetic Acid

H_2NCH_2COOH Formula Wt 75.07

CAS Number 56–40–6

REQUIREMENTS

Assay . . . . $\geq$98.5% H_2NCH_2COOH

MAXIMUM ALLOWABLE

Residue after ignition 0.1%
Heavy metals (as Pb) 0.002%
Chloride (Cl) 0.005%
Sulfate (SO$_4$) 0.005%
Ammonium (NH$_4$) 0.005%
Substances darkened by sulfuric acid . . . . Passes test
Hydrolyzable substances . . . . Passes test

TESTS

ASSAY. (Total alkalinity by nonaqueous titration). Weigh accurately 0.25 g and dissolve in 100 mL of glacial acetic acid. Add 0.5 mL of crystal violet indicator solution and titrate with 0.1 N perchloric acid in glacial acetic acid to a green end point. Correct for a blank. One milliliter of 0.1 N perchloric acid corresponds to 0.007507 g of H_2NCH_2COOH.

RESIDUE AFTER IGNITION. (Page 16). Use 2.0 g.

HEAVY METALS. (Page 28, Method 1). Use 1.0 g.

CHLORIDE. (Page 27). Use 0.2 g.

SULFATE. (Page 33, Method 1). Use 1.0 g.

AMMONIUM. (Page 25). Use 0.2 g. For the standard, use 0.01 mg of ammonium ion (NH_4) treated exactly as the sample.

SUBSTANCES DARKENED BY SULFURIC ACID. Dissolve 0.5 g in 5 mL of sulfuric acid. Allow the solution to stand for 15 min. The color of the sample solution should not exceed that of an equal volume of the acid.

HYDROLYZABLE SUBSTANCES. Boil 10 mL of a 1 in 10 solution of glycine for 1 min and set aside for 2 h. The solution, when compared to an equal volume of an untreated sample solution, appears as clear and as mobile.

Gold Chloride
Tetrachloroauric(III) Acid Trihydrate

$HAuCl_4 \cdot 3H_2O$ Formula Wt 393.83

CAS Number 16961–25–4

REQUIREMENTS

Assay . . . . 49.0% Au

MAXIMUM ALLOWABLE

Insoluble in ether 0.1%
Alkalies and other metals (as sulfates) 0.2%

TESTS

ASSAY. (Gold content by gravimetry). Evaporate the solution from the test for insoluble in ether to dryness or a sirupy residue. Dissolve the residue in 100 mL of dilute hydrochloric acid (1 + 19), disregarding any undissolved material. Add 10 mL of sulfurous acid and digest on a hot plate (≈100 °C) until the precipitate is well coagulated. Filter, wash with dilute hydrochloric acid (1 + 99), and ignite in a tared crucible. Retain the filtrate and washings. The weight of the residue should be not less than 49.0% of the weight of the sample.

NOTE. Sulfurous acid assay must be >6% as SO_2.

INSOLUBLE IN ETHER. Weigh accurately about 1 g. Add 10 mL of ether, allow to stand for 10 min with occasional stirring, filter through a tared filtering crucible of fine porosity, wash with small portions of ether, and dry at 105 °C. Retain the filtrate and washings.

ALKALIES AND OTHER METALS (AS SULFATES). Evaporate the filtrate and washings reserved from the assay, in a tared dish, to dryness. Ignite the residue cautiously, and finally ignite at 800 ± 25 °C for 15 min.

Hexamethylenetetramine

$C_6H_{12}N_4$ **Formula Wt 140.19**

CAS Number 100-97-0

REQUIREMENTS

Assay (dried basis). . . . .≥99.0% $C_6H_{12}N_4$
MAXIMUM ALLOWABLE
Loss on drying 2.0%
Residue after ignition 0.1%
Heavy metals (as Pb) 0.001%

TESTS

ASSAY. Transfer 1 g of previously dried sample, accurately weighed, to a 100-mL beaker. Add 40 mL of 1 N sulfuric acid and boil gently for at least 1 h to remove formaldehyde. Cool, add 20 mL of water, and titrate the excess acid with 1 N sodium hydroxide, using 0.15 mL of methyl red indicator solution. One milliliter of 1 N sulfuric acid corresponds to 0.03505 g of $C_6H_{12}N_4$.

LOSS ON DRYING. Accurately weigh about 1 g of sample and dry over phosphorus pentoxide for 4 h. Save the dried sample for the assay determination.

RESIDUE AFTER IGNITION. (Page 16). Ignite 1–2 g, accurately weighed.

HEAVY METALS. (Page 28, Method 1). Dissolve 2 g in 10 mL of water, add 2 mL of 3 N hydrochloric acid, and dilute with water to 25 mL. Proceed as directed, except use glacial acetic acid to adjust the pH.

Hexanes

Suitable for use in high-performance liquid chromatography, extraction–concentration analysis, ultraviolet spectrophotometry, or general use. Product labeling shall designate the one or more of these uses for which suitability is represented on the basis of meeting the relevant requirements and tests. The ultraviolet spectrophotometry and liquid chromatography suitability requirements include all of the requirements for general use. The extraction–concentration suitability requirements include only the general use requirement for color.

NOTE. This reagent is generally a mixture of several isomers of hexane (C_6H_{14}), predominantly n-hexane, 2-methylpentane, and 3-methylpentane, plus methylcyclopentane (C_6H_{12}).

REQUIREMENTS

General Use

Assay . . . . ≥98.5% hexanes (sum of 5 isomers, total hexanes, plus methylcyclopentane)

MAXIMUM ALLOWABLE

Color (APHA) 10
Residue after evaporation. 0.001%
Water-soluble titrable acid . . . . 0.0003 meq/g
Sulfur compounds (as S) 0.005%
Thiophene . . . . Passes test

Specific Use

Ultraviolet Spectrophotometry

Wavelength (nm)	Absorbance (AU)
280–400.	0.01
250	0.02
240	0.04
230	0.10
220	0.20
210	1.0

Liquid Chromatography Suitability

Absorbance . . . . Passes test

Extraction–Concentration Suitability

Absorbance . . . . Passes test
GC–FID . . . . Passes test
GC–ECD . . . . Passes test

TESTS

ASSAY. Analyze the sample by gas chromatography using the general parameters cited on page 70. The following specific conditions are also required.

Column: Type I, methyl silicone

Measure the area under all peaks and calculate the hexanes and methylcyclopentane in area percent. The retention times are (min): 2-methylpentane, 7.5; 3-methylpentane, 8.5; *n*-hexane, 9.2; methylcyclopentane, 10.2.

COLOR (APHA). (Page 19).

RESIDUE AFTER EVAPORATION. (Page 16). Evaporate 100 g (145 mL) to dryness in a tared dish on a hot plate ($\approx$100 °C) and dry the residue at 105 °C for 30 min.

WATER-SOLUBLE TITRABLE ACID. To 30 g (44 mL) in a separatory funnel, add 50 mL of water and shake vigorously for 2 min. Allow the layers to separate, draw off the aqueous layer, and add 0.15 mL of phenolphthalein indicator solution to the aqueous layer. Not more than 1.0 mL of 0.01 N sodium hydroxide should be required to produce a pink color.

SULFUR COMPOUNDS. To 30 mL of 0.5 N potassium hydroxide in methanol in a conical flask, add 5.5 g (8 mL) of the sample, and boil the mixture gently for 30 min under a reflux condenser, avoiding the use of a rubber stopper or connection. Detach the condenser, dilute with 50 mL of water, and heat on a hot plate ($\approx$100 °C) until the hexanes and methanol are evaporated. Add 50 mL of bromine water, heat for an additional 15 min, and transfer to a beaker. Neutralize with dilute hydrochloric acid (1 + 3), add an excess of 1 mL of the acid, evaporate to about 50 mL, and filter if necessary. Heat the filtrate to boiling, add 5 mL of barium chloride reagent solution, digest in a covered beaker on the hot plate ($\approx$100 °C) for 2 h, and allow to stand overnight. If a precipitate is formed, filter, wash thoroughly, and ignite. Correct for the weight obtained in a complete blank test.

THIOPHENE. To 25 mL in a glass-stoppered flask add 15 mL of sulfuric acid to which has been added a few milligrams of isatin, shake the mixture for 15–20 s, and allow to stand for 1 h. The acid layer should not be colored blue or green.

ULTRAVIOLET SPECTROPHOTOMETRY. Use the procedure on page 73 to determine the absorbance.

LIQUID CHROMATOGRAPHY SUITABILITY. Use the procedure on page 73 to determine the absorbance.

EXTRACTION–CONCENTRATION SUITABILITY. Analyze the sample by using the general procedure cited on page 72. Use the procedure on page 73 to determine the absorbance.

Hydrazine Sulfate

$(NH_2)_2 \cdot H_2SO_4$ **Formula Wt 130.12**

CAS Number 10034–93–2

REQUIREMENTS

Assay . . . . ≥99.0% $(NH_2)_2 \cdot H_2SO_4$

MAXIMUM ALLOWABLE

Insoluble matter. 0.005%
Residue after ignition 0.05%
Chloride (Cl) 0.005%
Heavy metals (as Pb) 0.002%
Iron (Fe) 0.001%

TESTS

ASSAY. (By titration of reducing power). Weigh accurately about 1 g, dissolve in water, and dilute with water to 500 mL in a volumetric flask. To 50.0 mL add 1 g of sodium bicarbonate, dissolve the reagent, and add 50.0 mL of 0.1 N iodine solution. Titrate the excess iodine with 0.1 N sodium thiosulfate, adding 3 mL of starch indicator near the end point. One milliliter of 0.1 N iodine corresponds to 0.003253 g of $(NH_2)_2 \cdot H_2SO_4$.

INSOLUBLE MATTER. (Page 15). Use 20 g dissolved in 300 mL of water.

RESIDUE AFTER IGNITION. (Page 16). Ignite 2.0 g. Reserve the residue for the test for iron.

CHLORIDE. (Page 27). Dissolve 1.0 g in water, dilute with water to 100 mL, and use 20 mL of this solution.

HEAVY METALS. (Page 28, Method 1). Dissolve 2.0 g in 40 mL of warm water. Use 30 mL to prepare the sample solution, and use the remaining 10 mL to prepare the control solution.

IRON. (Page 30, Method 1). To the residue after ignition add 3 mL of dilute hydrochloric acid (1 + 1) and 0.10 mL of nitric acid, cover with a watch glass, and digest on a hot plate ($\approx$100 °C) for 15–20 min. Remove the watch glass and evaporate to dryness. Dissolve the residue in a mixture of 4 mL of hydrochloric acid and 10 mL of water and dilute with water to 100 mL. Use 50 mL of this solution without further acidification.

Hydriodic Acid, 47%
(with stabilizer)

HI   **Formula Wt 127.91**

CAS Number 10034–85–2

NOTE. To avoid danger of explosions, this acid should be distilled only in an inert atmosphere. The reagent may have a slight yellow color. It contains hypophosphorus acid as a stabilizer.

REQUIREMENTS

Assay . . . . $\geq$47.0% HI

MAXIMUM ALLOWABLE

Chloride and bromide (as Cl) 0.05%
Sulfate (SO_4) 0.005%
Stabilizer (H_3PO_2) 1.5%
Heavy metals (as Pb) 0.001%
Iron (Fe) 0.001%

TESTS

ASSAY. (By acid–base titrimetry) Tare a glass-stoppered flask containing about 25 mL of water. Under the water surface, quickly add about 3 mL of the sample, then stopper, and weigh accurately to the nearest 0.1 mg. Dilute to about 75 mL, add 0.15 mL of phenolphthalein indicator solution, and titrate with 1 N sodium hydroxide.

$$\% \ HI = (mL \times N \ NaOH \times 12.79) / Sample \ wt \ (g)$$

CHLORIDE AND BROMIDE. Dilute 1 g (0.67 mL) with water to 100 mL in a 250-mL conical flask. Add 1 mL of hydrogen peroxide and 1 mL of phosphoric acid. Heat to boiling and boil gently until all the iodine is expelled and the solution is colorless. Cool, wash down the sides of the flask, and add 0.5 mL of hydrogen peroxide. If an iodine color develops, boil until the solution is colorless and for 10 min longer. If no color develops, boil for 10 min. Dilute with water to 100 mL, take a 2-mL portion of the solution, and dilute with water to 23 mL. Prepare a standard containing 0.01 mg of chloride ion (Cl) in 23 mL of water. Add 1 mL of nitric acid and 1 mL of silver nitrate reagent solution to each. Any turbidity produced in the solution of the sample should not exceed that in the standard.

SULFATE. For sample, dilute 2.0 mL (3.0 g) to 25 mL with water. For control, take 0.7 mL (1.0 g) of sample and dilute to 25 mL with water. Neutralize both with (1 + 1) ammonium hydroxide and add 1 mL of (1 + 19) hydrochloric acid to each. Heat to boil and cool both. To the control solution, add 0.10 mg of sulfate ion (SO_4) and dilute both solutions to 25 mL, add 1 mL of barium chloride. Any turbidity in sample after 1 h should not exceed that of the control solution.

> *Sample Solution A for the Determination of Stabilizer, Heavy Metals, and Iron.* To 5 g (3.4 mL) add 10 mL of water, 7 mL of nitric acid, and 9 mL of hydrochloric acid. Evaporate to dryness on a hot plate ($\approx$100 °C), dissolve in water, and dilute with water to 50 mL in a volumetric flask (1 mL = 0.1 g).

STABILIZER. Dilute 1.0 mL of sample solution A (0.1-g sample) with water to 1 L. Dilute 4.8 mL of the solution with water to 85 mL. Add 5 mL of a 10% solution of ammonium molybdate and adjust the pH to 1.8 (using a pH meter) with dilute hydrochloric acid (1 + 9). Heat to boiling and cool to room temperature. Add 10 mL of hydrochloric acid and transfer to a separatory funnel. Add 35 mL of ether and shake vigorously. Allow the layers to separate and draw off and discard the aqueous layer. Wash the ether layer twice with 10-mL portions of dilute hydrochloric acid (1 + 9), drawing off and discarding the aqueous portion each time. Add 10 mL of dilute hydrochloric acid (1 + 9) to which has just been added 0.2 mL of a 2% stannous chloride solution in hydrochloric acid and shake. Any blue color in the ether should not exceed that produced by 0.01 mg of phosphate ion (PO_4) treated exactly like the 4.8-mL portion of the sample.

HEAVY METALS. (Page 28, Method 1). Dilute 20 mL of sample solution A (2-g sample) with water to 25 mL.

IRON. (Page 30, Method 1). Use 10 mL of sample solution A (1-g sample).

Hydriodic Acid, 55%
(suitable for use in methoxyl determinations)

HI	**Formula Wt 127.91**

CAS Number 10034–85–2

NOTE. This reagent is uninhibited and may turn yellow on standing.

REQUIREMENTS

Assay . . . . 55.0–58.0% HI

MAXIMUM ALLOWABLE

Free iodine (I$_2$) 0.75%
Residue after ignition 0.01%
Chloride and bromide (as Cl) 0.05%
Phosphate (PO$_4$) 0.001%
Sulfate (SO$_4$) 0.005%
Heavy metals (as Pb) 0.001%
Iron (Fe) 0.001%

TESTS

ASSAY. (By acid–base titrimetry). Tare a glass-stoppered flask containing about 25 mL of water. Under the water surface, quickly add about 3 mL of the sample, then stopper, and weigh accurately to the nearest 0.1 mg. Dilute to about 75 mL, add 0.15 mL of phenolphthalein indicator solution, and titrate with 1 N sodium hydroxide.

$$\% \text{ HI} = (\text{mL} \times \text{N NaOH} \times 12.79) / \text{Sample wt (g)}$$

FREE IODINE. Weigh a glass-stoppered flask containing 15 mL of water. Add about 3 mL of sample, stopper, and reweigh. Dilute with 40 mL of water and titrate with 0.1 N sodium thiosulfate to the disappearance of the yellow color. Not more than 0.60 mL of the thiosulfate should be required per gram of the acid.

RESIDUE AFTER IGNITION. (Page 16). Evaporate 10 g (6.0 mL) to dryness in a tared platinum crucible and ignite.

CHLORIDE AND BROMIDE. Dilute 1 g (0.6 mL) with water to 100 mL in a 250-mL conical flask. Add 1 mL of hydrogen peroxide and 1 mL of phosphoric acid. Heat to boiling and boil gently until all the iodine is expelled and the solution is colorless. Cool, wash down the sides of the flask, and add 0.5 mL of hydrogen peroxide. If an iodine color develops, boil until the solution is colorless and for 10 min longer. If no color develops, boil for 10 min. Dilute with water to 100 mL, take a 2-mL portion, and dilute with water to 23 mL. Prepare a standard containing 0.01 mg of chloride ion (Cl) in 23 mL of water. Add 1 mL of nitric acid and 1 mL of silver nitrate reagent solution to each. Any turbidity produced in the solution of the sample should not exceed that in the standard.

Sample Solution A for the Determination of Phosphate, Heavy Metals, and Iron. Evaporate 5 g (3.0 mL) of sample with 2 mL of nitric acid to dryness on a hot plate ($\approx$100 °C). Re-evaporate with several portions of water until all the iodine has been volatilized. Warm the residue with 2 mL of 1 N hydrochloric acid and take up with 40 mL of hot water. Cool, dilute with water to 50 mL, and mix well (1 mL = 0.1 g).

PHOSPHATE. To 20 mL of sample solution A (2-g sample) add 2 mL of 25% sulfuric acid, 10 mL of water, 1 mL of ammonium molybdate–sulfuric acid reagent solution, and 1 mL of 4-(triethylamino)phenol sulfate reagent solution. Heat to 60 °C for 10 min. Any blue color should not exceed that produced by 0.02 mg of phosphate ion (PO_4) in an equal volume of solution containing the quantities of reagents used in the test.

SULFATE. For sample, dilute 2.0 mL (3.0 g) to 25 mL with water. For control, take 0.7 mL (1.0 g) of sample and dilute to 25 mL with water. Neutralize both with (1 + 1) ammonium hydroxide and add 1 mL of (1 + 19) hydrochloric acid to each. Heat to boil and cool both. To the control solution, add 0.10 mg of sulfate ion (SO_4), dilute both solutions to 35 mL, and add 1 mL of barium chloride. Any turbidity in sample after 1 h should not exceed that of the control solution.

HEAVY METALS. (Page 28, Method 1). Dilute 20 mL of sample solution A (2-g sample) with water to 25 mL.

IRON. (Page 30, Method 1). Use 10 mL of sample solution A (1-g sample).

Hydrobromic Acid, 48%

HBr   **Formula Wt 80.91**

CAS Number 10035–10–6

REQUIREMENTS

Assay . . . . 47.0–49.0% HBr

MAXIMUM ALLOWABLE

Residue after ignition 0.002%
Chloride (Cl) 0.05%
Iodide (I) 0.003%
Phosphate (PO_4) 0.001%
Sulfate and sulfite (as SO_4) 0.003%
Heavy metals (as Pb) . . . . 5 ppm
Iron (Fe) . . . . 1 ppm
Selenium (Se) . . . . 0.01 ppm

TESTS

ASSAY. (By acid–base titrimetry). Tare a glass-stoppered conical flask containing about 15 mL of water. Quickly add about 4 mL of the acid and weigh. Dilute with water to 50 mL and titrate with 1 N sodium hydroxide, using 0.15 mL of phenolphthalein indicator. One milliliter of 1 N sodium hydroxide corresponds to 0.08091 g of HBr.

RESIDUE AFTER IGNITION. (Page 16). To 50 g (34 mL) in a tared dish (not platinum) add 0.05 mL of sulfuric acid, evaporate as far as possible on a hot plate ($\approx$100 °C), then heat gently to volatilize the solution. Finally, ignite at 800 $\pm$ 25 °C for 15 min.

CHLORIDE. To 1 g (0.67 mL) in a 150-mL conical flask add 50 mL of dilute nitric acid (1 + 3) and digest on a hot plate ($\approx$100 °C) until the solution is color-less. Wash down the sides of the flask with a little water and digest for an addi-tional 15 min. Cool, filter if necessary through a chloride-free filter, and dilute with water to 100 mL. Dilute 2.0 mL of the solution with water to 23 mL. Prepare a standard containing 0.01 mg of chloride ion (Cl) in 23 mL of water. Add 1 mL of nitric acid and 1 mL of silver nitrate reagent solution to each. Any turbidity produced in the solution of the sample should not exceed that in the standard.

IODIDE. Dilute 6 g (4.0 mL) with 20 mL of water. Add 5 mL of chloroform, 0.2 mL of 10% ferric chloride solution, and 0.1 mL of dilute sulfuric acid (1 + 15) and mix gently in a separatory funnel. Draw off the chloroform layer into a com-parison tube of about 1.5-cm diameter. No violet color should be observed on looking down through the chloroform.

PHOSPHATE. (Page 32, Method 1). Add 5 mL of nitric acid to 6 g (4.0 mL) of the sample, and evaporate to dryness on a hot plate ($\approx$100 °C). Dissolve the resi-due in 75 mL of approximately 0.5 N sulfuric acid, and use 25 mL of the solution.

SULFATE AND SULFITE. Dilute 2.7 mL (4.0 g) of the sample with water to 20 mL. Add 0.5 mL of bromine water and gently heat to boil for a few minutes to expel excess bromine. Add 1.0 mL of 1% sodium carbonate solution and evapo-rate to dryness on a hot plate ($\approx$100 °C). Dissolve the residue in 1 mL of (1 + 19) hydrochloric acid and 20 mL of water. Filter through small, washed filter paper, add two 3-mL portions of water through, and dilute to 35 mL with water. For the control, take 0.12 mg of sulfate ion (SO_4) in 30 mL of water, add 1 mL of (1 + 19) hydrochloric acid, dilute to 35 mL with water, add 1 mL of barium chloride solu-tion, and compare turbidity after 10 min. Sample turbidity should not exceed that of control solution.

HEAVY METALS. (Page 28, Method 1). To 4 g (2.7 mL) in a beaker, add about 10 mg of sodium carbonate, and evaporate to dryness on a hot plate ($\approx$100 °C). Dis-solve the residue in about 20 mL of water, and dilute with water to 25 mL.

IRON. (Page 30, Method 1). To 10 g (6.6 mL) add about 10 mg of sodium car-bonate, and evaporate to dryness on a hot plate ($\approx$100 °C). Dissolve the residue in 2 mL of hydrochloric acid, dilute with water to 50 mL, and use the solution with-out further acidification.

SELENIUM. Place 150 g (100 mL) in an all-glass distilling apparatus, add 2 mL of bromine, and distill off approximately 25 mL. Dilute the distillate with an equal volume of water and decolorize with sulfurous acid. Add 0.1 g of hydroxylamine hydrochloride and warm gently on a hot plate ($\approx$100 °C) for 30 min. Cool to room temperature, filter through a white glass-wool mat in a small Gooch crucible, without washing, and examine immediately. No pink or red color should be observed on the mat.

Hydrochloric Acid

HCl   **Formula Wt 36.46**

CAS Number 7647–01–0

REQUIREMENTS

Appearance . . . . Free from suspended matter or sediment
Assay . . . . 36.5–38.0% HCl

MAXIMUM ALLOWABLE

Color (APHA) 10
Residue after ignition . . . . 5 ppm
Bromide (Br) 0.005%
Sulfate (SO$_4$) . . . . 1 ppm
Sulfite (SO$_3$) . . . . 1 ppm
Extractable organic substances. . . . . 5 ppm
Free chlorine (Cl) . . . . 1 ppm
Ammonium (NH$_4$) . . . . 3 ppm
Arsenic (As) . . . . 0.01 ppm
Heavy metals (as Pb) . . . . 1 ppm
Iron (Fe) . . . . 0.2 ppm

TESTS

APPEARANCE. Mix the material in the original container, pour 10 mL into a test tube (20 × 150 mm), and compare with distilled water in a similar tube. The liquids should be equally clear and free from suspended matter.

ASSAY. (By acid–base titrimetry). Tare a glass-stoppered flask containing about 30 mL of water. Quickly add about 3 mL of the sample, stopper, and weigh accurately. Dilute to about 50 mL, add 0.15 mL of methyl orange indicator solution, and titrate with 1 N sodium hydroxide. One milliliter of 1 N sodium hydroxide corresponds to 0.03646 g of HCl.

COLOR (APHA). (Page 19).

RESIDUE AFTER IGNITION. (Page 16). To 200 g (170 mL) in a tared platinum dish, add 0.05 mL of sulfuric acid, and evaporate as far as possible on a hot plate ($\approx$100 °C). Heat gently to volatilize the excess sulfuric acid, and ignite.

BROMIDE. Pipet 1 mL into 25 mL of water in a 100-mL beaker, add 2 mL of phenol red indicator solution (0.020 g of the sodium salt in 100 mL of water), and titrate with 1 N sodium hydroxide. Add just sufficient titrant to change the yellow color to red. Transfer the solution to a 100-mL volumetric flask with about 25 mL of water. For the standard, add identical amounts of indicator and water to 0.06 mg of bromide ion (Br) in a 100-mL volumetric flask. To each flask add 5 mL of acetate buffer solution (68 g of sodium acetate, $CH_3COONa \cdot 3H_2O$, and 30 mL of glacial acetic acid per liter) and 2.0 mL of freshly prepared chloramine-T solution (0.125 g per 100 mL). Mix, allow to stand for 20.0 min, and add 20 mL of 0.1 N sodium thiosulfate solution. Swirl, dilute with water to volume, and mix. Compare visually or spectrophotometrically at 590 nm in 1.0-cm cells. Any blue-violet color in the sample should not exceed that in the standard.

SULFATE. To 42 mL (50 g) of the sample, add 2 mL of 1% sodium carbonate solution and evaporate to dryness. Take up the residue with 10 mL of water, filter through washed filter paper, and wash with 5 mL of water. Add 1 mL of (1 + 9) hydrochloric acid and dilute to 20 mL with water. For the control, use 0.05 mg of sulfate ion (SO_4) in 15 mL of water, add 1 mL (1 + 9) hydrochloric acid, and dilute to 20 mL with water. To both, add 1 mL of barium chloride solution. Compare after 10 min. Sample turbidity should not exceed that of the control solution.

SULFITE. Transfer 400 mL of freshly boiled and cooled water to each of two 500-mL conical flasks. Adjust the temperature of one flask (for the reagent blank) to 25 °C and the other flask (for the sample solution) to 15 °C. To each flask add 3 mL of 10% potassium iodide reagent solution, 5 mL of hydrochloric acid, and 2 mL of starch indicator solution. To the sample flask add 85 mL of hydrochloric acid sample, and adjust the temperature of the contents of the two flasks to within 1 °C of each other. Using a magnetic stirrer and a Teflon-coated stirring bar, titrate each with 0.01 N iodine until a faint permanent blue color is produced. Subtract the reagent blank titrant volume from the sample titrant volume to obtain the net sample titration. The net titration should not exceed 0.25 mL.

EXTRACTABLE ORGANIC SUBSTANCES. Place 120 g (100 mL) of sample in each of two 125-mL separatory funnels. To one funnel, add 7.0 mL of 2,2,4-trimethylpentane as a sample blank. To the second funnel, add 5.0 mL of 2,2,4-trimethylpentane and 2.0 mL of the internal standard solution. Stopper the funnels, shake for 2 min, allow the layers to separate, and drain the hydrochloric acid layer. Prepare a standard solution by adding 2.0 mL of the internal standard solution to 5.0 mL of the chloroform standard solution.

   Internal Standard Solution. Add by syringe 30 mg (22.7 µL) of 1,1,1-trichloroethane to 100 mL of 2,2,4-trimethylpentane.

Chloroform Standard Solution. Add by syringe 12 mg (8.1 µL) of chloroform to 100 mL of 2,2,4-trimethylpentane. The 2,2,4-trimethylpentane used for both the standard and the analysis should be free from impurities that interfere with the chromatographic analysis.

Analyze the sample blank and the sample 2,2,4-trimethylpentane layers and the standard solution by gas chromatography as described on page 70. With a flame ionization detector, most organic compounds can be detected at levels of 1 ppm or less. The use of an electron-capture or other selective detector would make it possible to detect chlorinated aliphatics or other selected compounds at much lower levels. The following parameters have given satisfactory results.

Column: Type I, methyl silicone, 30 m × 0.53 mm i.d., 5.0-µm film thickness

Column Temperature: 35 °C initial hold, 5 min, then programmed at 10 °C per min to 230 °C, final hold 21 min. Total run time, 45.5 min

Injector Temperature: 230 °C

Detector Temperature: 250 °C

Carrier Gas: Helium, with flow rate 3–4 mL per min

Detector: Flame ionization, range 10^{-11} A

Sample Size: 0.5 µL. Adjust the attenuation to produce a peak height of at least 10% of full scale for the internal standard peak

Approximate Retention Times (min.): Dichloromethane, 3.8; chloroform, 6.8; 1,2-dichloroethane, 7.7; 1,1,1-trichloroethane, 8.0; benzene, 8.6; carbon tetrachloride, 8.8; 2,2,4-trimethylpentane, 10–13; chlorobenzene, 13.8; 1,2,4-trichlorobenzene, 20.5; lindane, 37

Use the sample blank extraction to correct for impurities in the solvent or sample that interfere with the internal standard peak for 1,1,1-trichloroethane. The ratio of the total area of any peaks in the extraction (other than the internal standard and solvent) to the area of the internal standard peak in the sample should not exceed the ratio of the area of the chloroform peak to the area of the internal standard peak in the standard solution.

FREE CHLORINE. To 50 mL of the sample add 50 mL of freshly boiled water and cool. Add 0.1 mL of 2% potassium iodide solution and 1 mL of carbon disulfide, and mix. The carbon disulfide should not acquire a pink color within 30 s. The potassium iodide should be free from iodate.

AMMONIUM. (By differential pulse polarography, page 49). Use 3.3 g (2.8 mL) of sample and 5.5 mL of ammonia-free 6 N sodium hydroxide. For the standard, use 0.010 mg of ammonium (NH_4).

ARSENIC. (Page 25). Mix 300 g (255 mL) with 10 mL of dilute sulfuric acid (1 + 1) in a generator flask. Add potassium chlorate crystals in small increments until a yellow color is produced, and evaporate on a hot plate just to fumes of sulfur trioxide, adding a few more crystals of potassium chlorate whenever the yellow color starts to fade. Cool, cautiously wash down the flask with 5–10 mL of water, and again heat just to fumes of sulfur trioxide. Cool, cautiously wash down the flask with 5–10 mL of water, and again heat just to fumes of sulfur trioxide. Cool, cautiously wash down the flask with 5–10 mL of water, and repeat the fuming. (The volume of sulfuric acid at this point should be about 4 mL.) Cool, cautiously dilute with water to 55 mL, and use this solution without further acidification. For the standard, use 0.003 mg of arsenic (As).

HEAVY METALS. (Page 28, Method 1). To 20 g (17 mL) in a 150-mL beaker, add about 10 mg of sodium carbonate, and evaporate to dryness on a hot plate ($\approx$100 °C). Dissolve the residue in about 20 mL of water, and dilute with water to 25 mL.

IRON. (Page 30, Method 1). To 50 g (42 mL) in a porcelain or glass vessel add about 10 mg of sodium carbonate and evaporate to dryness on a hot plate ($\approx$100 °C). Dissolve in 2 mL of hydrochloric acid, dilute with water to 50 mL, and use the solution without further acidification.

Hydrochloric Acid, Ultratrace
(suitable for use in ultratrace elemental analysis)

HCl   **Formula Wt 36.46**

CAS number 7647–01–0

NOTE. Reagent must be packaged in a preleached teflon bottle and used in a clean environment to maintain purity

REQUIREMENTS

Appearance . . . . Clear
Assay 20–38%

MAXIMUM ALLOWABLE

Sulfate . . . . 1 ppm
Aluminum (Al) . . . . 1 ppb
Barium (Ba) . . . . 1 ppb
Boron (B) . . . . 5 ppb
Cadmium (Cd) . . . . 1 ppb
Calcium (Ca) . . . . 1 ppb
Chromium (Cr) . . . . 1 ppb
Cobalt (Co) . . . . 1 ppb
Copper (Cu) . . . . 1 ppb
Iron (Fe) . . . . 5 ppb

Lead (Pb) . . . . .1 ppb
Lithium (Li). . . . . .1 ppb
Magnesium (Mg) . . . . .1 ppb
Manganese (Mn) . . . . .1 ppb
Molybdenum (Mo). . . . . .1 ppb
Mercury (Hg) . . . . .1 ppb
Potassium (K). . . . . .1 ppb
Selenium (Se). . . . . .10 ppb
Silicon (Si) . . . . .5 ppb
Sodium (Na). . . . . .5 ppb
Strontium (Sr) . . . . .1 ppb
Tin (Sn) . . . . .1 ppb
Titanium (Ti). . . . . .1 ppb
Vanadium (V) . . . . .1 ppb
Zinc (Zn) . . . . .1 ppb
Zirconium (Zr) . . . . .1 ppb

TESTS

ASSAY. (By acid–base titrimetry).   The assay of hydrochloric acid used for ultratrace metal analysis may vary from 20% to 38%, depending on the method of purification.   Tare a glass-stopped flask containing about 30 mL of water. Quickly add about 3 mL of the sample, stopper, and weigh accurately.   Dilute to about 50 mL, add 0.15 mL of methyl orange indicator solution, and titrate with 1 N sodium hydroxide.   One milliliter of 1 N sodium hydroxide corresponds to 0.03646 g of HCl.

$$\% \text{ HCl} = (\text{ml} \times \text{N NaOH} \times 3.646) / \text{Sample wt (g)}$$

SULFATE. See test for hydrochloric acid, page 333.

TRACE METALS. Determine the aluminum, barium, cadmium, calcium, chromium, cobalt, copper, iron, lead, lithium,   magnesium, manganese, molybdenum, nickel, potassium, sodium,   strontium, tin, titanium, vanadium, zinc, and zirconium by the ICP–OES method described on page 42.

Determine the mercury by the CVAAS procedure described on page 40.

Determine the selenium by the HGAAS method described on page 41. To a set of 3 100-mL volumetric flasks, transfer 50 g (42 mL) of sample. To two flasks, add the specified amount of selenium standard from the following table. Dilute each to 100 mL with (1 + 1) hydrochloric acid. Mix. Prepare fresh working standard before use.

Element	Wavelength (nm)	Sample Wt (g)	Standard Addition (μg)
Se	196.0	50.0	0.50; 0.10 (= 10 and 20 ppb)

Hydrofluoric Acid

HF   **Formula Wt 20.01**

CAS Number 7664–39–3

REQUIREMENTS

Assay . . . . 48.0–51.0% HF

MAXIMUM ALLOWABLE

Fluosilicic acid (H_2SiF_4) 0.01%
Residue after ignition . . . . 5 ppm
Chloride (Cl) . . . . 5 ppm
Phosphate (PO_4) . . . . 1 ppm
Sulfate and sulfite (as SO_4). . . . . 5 ppm
Arsenic (As) . . . . 0.05 ppm
Copper (Cu). . . . . 0.1 ppm
Iron (Fe). . . . . 1 ppm
Heavy metals (as Pb) . . . . 0.5 ppm

TESTS

ASSAY. (By acid–base titrimetry). Tare a stoppered plastic flask, containing about 25 mL of water, and deliver about 1.5 mL of sample under the water surface, using a plastic pipet. Weigh accurately, wash down the sides, and dilute the solution to 50–60 mL with water. Add 0.15 mL of phenolphthalein indicator solution, and titrate with 1 N sodium hydroxide.

$$\% \text{ HF} = (\text{mL} \times \text{N NaOH} \times 2.001) / \text{Sample wt (g)}$$

FLUOSILICIC ACID. Weigh about 37.5 g (33 mL) into a large platinum dish. Add 2 g of potassium chloride and 3 mL of hydrochloric acid and evaporate to dryness on a hot plate ($\approx$100 °C) in the hood. Wash down the sides of the dish with a small amount of water, add 3 mL of hydrochloric acid, and repeat the evaporation. Dissolve the residue in about 100 mL of water, cool to 0 °C, and add 0.15 mL of phenolphthalein indicator solution. Neutralize any free acid with 0.1 N sodium hydroxide solution to a colorless end point, keeping the temperature of the solution near 0 °C. Heat the solution to boiling and titrate with 0.1 N sodium hydroxide solution. One milliliter of 0.1 N sodium hydroxide corresponds to 0.0036 g of H_2SiF_6.

RESIDUE AFTER IGNITION. (Page 16). To 200 g (180 mL) in a tared platinum dish, add 0.05 mL of sulfuric acid, evaporate as far as possible on a hot plate ($\approx$100 °C) in the hood, heat gently to volatilize the excess sulfuric acid, and ignite.

CHLORIDE. Add 1.8 mL to 45 mL of water, filter if necessary through a chloride-free filter, and add 1 mL of nitric acid and 1 mL of silver nitrate reagent solution.

Any turbidity should not exceed that produced by 0.01 mg of chloride ion (Cl) in an equal volume of solution containing the quantities of reagents used in the test.

> *Sample Solution A for the Determination of Phosphate, Sulfate and Sulfite, Arsenic, Copper, Iron, and Heavy Metals.* In a fume hood, to 500.0 g (425 mL) sample in a platinum or teflon dish, add 10 mg of sodium carbonate and evaporate to dryness on a hot plate ($\approx$125 °C). Cool, add 5 mL of hydrochloric acid, evaporate to dryness. Cool, add 5 mL of water, evaporate to dryness. To the residue, add three 10 mL portions of (1 + 9) hydrochloric acid and transfer to a 50-mL volumetric flask and dilute to the mark with water (1 mL = 10.0 g).

PHOSPHATE. To 2.0 mL (20.0 g) of sample solution A, add 25 mL of approximately 0.5 N sulfuric acid, 1 mL of ammonium molybdate–sulfuric acid reagent solution, and 1 mL of 4-(methylamino)phenol sulfate reagent solution. Allow to stand for 2 h at room temperature. Any blue color should not exceed that produced by 0.02 mg of phosphate ion (PO_4) in an equal volume of solution containing the quantities of reagents used in the test.

SULFATE AND SULFITE. To 1.0 mL (10.0 g) of sample solution A in a teflon beaker, add 1 mL of 30% hydrogen peroxide. Evaporate to dryness on a hot plate ($\approx$100 °C) in the hood, wash down the sides of the dish with a small volume of water, and add 3 mL of perchloric acid. Evaporate to about 1 mL, dilute with about 15 mL of water, and add 0.15 mL of phenolphthalein indicator solution. Neutralize with ammonium hydroxide, dilute with water to 20 mL, and add 2 mL of dilute hydrochloric acid (1 + 19) and 2 mL of barium chloride reagent solution. Any turbidity should not exceed that produced by 0.1 mg of sulfate ion (SO_4) in an equal volume of solution containing the quantities of reagents used in the test. Compare 10 min after adding the barium chloride to the sample and standard solutions.

ARSENIC. To 4.0 mL (40.0 g sample) of sample solution A in a platinum dish, add 5 mL of sulfuric acid and 5 mL of hydrochloric acid, swirling the beaker gently after each addition. Allow to stand at room temperature for 5 min. Evaporate in a hood in a sand bath to dense fumes of sulfur trioxide. Cool, wash down the sides of the beaker with 15–20 mL of water, and evaporate again in a sand bath to dense fumes of sulfur trioxide. Cool and transfer to a 125-mL generator flask, rinsing the beaker thoroughly with water. Place the flask on a hot plate, evaporate to dense fumes of sulfur trioxide, add sulfuric acid if necessary to make a volume of about 4 mL, and cool. Add 50 mL of water, 2 mL of 16.5% potassium iodide solution, and 0.5 mL of stannous chloride solution (40% $SnCl_2 \cdot 2H_2O$ in hydrochloric acid), and mix. Proceed as described in the general method for arsenic on page 25 under Procedure, starting with the second sentence which begins "Allow the mixture to stand…" Any red color in the silver diethyldithiocarbamate solution of the sample should not exceed that in a standard containing 0.002 mg of arsenic (As).

COPPER AND IRON. (By flame AAS, page 39). To 3 10.0 mL of sample solution A portions (100.0 g sample) in a 25-mL volumetric flask, add 5 mL of (1 +1) hydrochloric acid to each and add standards as per following table to second and third flasks respectively.

Element	Wavelength (nm)	Sample Wt (g)	Standard Added (mg)	Flame Type*	Background Correction
Cu	324.8	100.0	0.01; 0.02	A/A	Yes
Fe	248.3	100.0	0.10; 0.20	A/A	Yes

*A/A is air/acetylene.

HEAVY METALS. (Page 28, Method 1). Take 4.0 mL (40.0 g) of sample solution A and dilute to 25 mL with water. Solution color should not exceed the control solution of 0.02 mg of lead.

Hydrofluoric Acid, Ultratrace
(suitable for use in ultratrace elemental analysis)

HF   **Formula Wt 20.01**

CAS number 7664–39–3

NOTE. Reagent must be packaged in a preleached Teflon bottle and used in a clean environment to maintain purity.

REQUIREMENTS

Appearance . . . . Clear
Assay 46–51.0%

MAXIMUM ALLOWABLE

Chloride (Cl) . . . . 0.5 ppm
Phosphate (PO_4) . . . . 1 ppm
Sulfate and Sulfite (SO_4). . . . . 1 ppm
Aluminum (Al) . . . . 1 ppb
Barium (Ba) . . . . 1 ppb
Boron (B). . . . . 5 ppb
Cadmium (Cd). . . . . 1 ppb
Calcium (Ca) . . . . 1 ppb
Chromium (Cr) . . . . 1 ppb
Cobalt (Co) . . . . 1 ppb
Copper (Cu). . . . . 1 ppb
Iron (Fe). . . . . 5 ppb
Lead (Pb) . . . . 1 ppb
Lithium (Li) . . . . 1 ppb
Magnesium (Mg) . . . . 1 ppb

Manganese (Mn) . . . . 1 ppb
Molybdenum (Mo). . . . . 1 ppb
Mercury (Hg) . . . . 1 ppb
Potassium (K). . . . . 1 ppb
Silicon (Si) . . . . 5 ppb
Sodium (Na). . . . . 5 ppb
Strontium (Sr) . . . . 1 ppb
Tin (Sn) . . . . 1 ppb
Titanium (Ti). . . . . 1 ppb
Vanadium (V) . . . . 1 ppb
Zinc (Zn) . . . . 1 ppb
Zirconium (Zr) . . . . 1 ppb

TESTS

ASSAY. (By acid–base titrimetry).   Tare a stoppered plastic flask containing about 25 mL of water, and deliver, using a plastic pipet, about 1.5 mL of sample under the water surface. Weigh accurately, wash down the sides, and dilute the solution to 50–60 mL with water. Add 0.15 mL of phenolphthalein indicator solution, and titrate with 1 N sodium hydroxide.

$$\% \text{ HF} = (\text{mL} \times \text{N NaOH} \times 2.001) / \text{Sample wt (g)}$$

CHLORIDE. See test for hydrofluoric acid, page 337. Use 18 mL of sample and add it to 45 mL of water.

PHOSPHATE. See test for hydrofluoric acid, page 338.

SULFATE AND SULFITE. See test for hydrofluoric acid, page 338. Use 5.0 mL (50.0 g) of sample solution A.

TRACE METALS. Determine the aluminum, barium, cadmium, calcium, chromium, cobalt, copper, iron, lead, lithium, magnesium, manganese, molybdenum, nickel, potassium, sodium, strontium, tin, titanium, vanadium, zinc, and zirconium by the ICP–OES method described on page 42.

Determine the mercury by the CVAAS procedure described on page 40.

Hydrogen Peroxide

H₂O₂   **Formula Wt 34.01**

CAS Number 7722–84–1

NOTE. This reagent should be stored in a cool place in containers with a vent in the stopper. The assay requirement applies to material stored properly for a reasonable time.

REQUIREMENTS

Assay . . . . .29.0–32.0% H_2O_2

MAXIMUM ALLOWABLE

Color (APHA) 10
Residue after evaporation 0.002%
Titrable acid . . . . 0.0006 meq/g
Chloride (Cl) . . . . 3 ppm
Nitrate (NO_3) . . . . 2 ppm
Phosphate (PO_4) . . . . 2 ppm
Sulfate (SO_4) . . . . 5 ppm
Ammonium (NH_4) . . . . 5 ppm
Heavy metals (as Pb) . . . . 1 ppm
Iron (Fe) . . . . 0.5 ppm

TESTS

ASSAY. (By titration of the reductive capacity). Weigh accurately about 1 mL in a tared 100-mL volumetric flask, dilute to volume with water, and mix thoroughly. To 20.0 mL of this solution add 20 mL of dilute sulfuric acid (1 + 15) and titrate with 0.1 N potassium permanganate. One milliliter of 0.1 N potassium permanganate corresponds to 0.001700 g of H_2O_2.

COLOR (APHA). (Page 19).

RESIDUE AFTER EVAPORATION. (Page 16). Evaporate 50 g (45 mL) to dryness in a tared porcelain or silica dish on a hot plate ($\approx$100 °C) and dry the residue at 105 °C for 30 min.

TITRABLE ACID. Dilute 10 g (9 mL) with 90 mL of carbon dioxide-free water. Add 0.15 mL of methyl red indicator solution and titrate with 0.01 N sodium hydroxide. The volume of sodium hydroxide solution consumed should not be more than 0.6 mL greater than the volume required for a blank test on 90 mL of the water used for dilution.

CHLORIDE, NITRATE, PHOSPHATE, AND SULFATE. (By ion chromatography, page 74). Obtain three 100-mL beakers (blank, test, standard) and add 10 mL water and 2.0 mg of sodium carbonate to each. Add 10.5 g (9.45 mL) of the sample to the standard and test. To the standard, add 0.02 mg each of chloride, nitrate, and phosphate ions and 0.05 mg of sulfate ion. To the blank, add 0.5 g (0.45 mL) of the sample. Evaporate each to dryness on a steam bath or low-temperature hot plate. Dissolve the residues in a little water and dilute each to 25 mL. Analyze 50-mL or 100-mL aliquots by ion chromatography and measure the various peaks. The differences between the sample and the blank should not be greater than the corresponding differences between the standard and the sample.

AMMONIUM. Add 0.1 mL of sulfuric acid to 2 g (1.8 mL) of the sample and evaporate on a hot plate ($\approx$100 °C). Dissolve the residue in 45 mL of water and add 3 mL of 10% sodium hydroxide reagent solution and 2 mL of Nessler reagent. Any color should not exceed that produced by 0.01 mg of ammonium ion (NH_4) in an equal volume of solution containing the quantities of reagents used in the test.

HEAVY METALS. (Page 28, Method 1). To 20 g (18 mL) in a 100-mL beaker add about 10 mg of sodium chloride and evaporate to dryness on a hot plate ($\approx$100 °C). Dissolve the residue in about 20 mL of water and dilute with water to 25 mL.

IRON. (Page 30, Method 1). To 20 g (18 mL) in a 100-mL beaker, add about 10 mg of sodium chloride, and evaporate to dryness on a hot plate ($\approx$100 °C). Dissolve in 2 mL of hydrochloric acid, dilute with water to 50 mL, and use the solution without further acidification.

Hydrogen Peroxide, Ultratrace
(suitable for use in ultratrace elemental analysis)

H_2O_2 Formula Wt 34.01

CAS number 7722–84–1

> *NOTE.* Package in a pre-cleaned polyethylene bottle and use in a clean environment to maintain purity.

REQUIREMENTS

Appearance . . . . Clear
Assay . . . . 25–35% H_2O_2

MAXIMUM ALLOWABLE

Chloride (Cl) . . . . 3 ppm
Nitrate (NO_3) . . . . 2 ppm
Phosphate (PO_4) . . . . 2 ppm
Sulfate (SO_4) . . . . 5 ppm
Aluminum (Al) . . . . 1 ppb
Barium (Ba) . . . . 1 ppb
Boron (B) . . . . 5 ppb
Cadmium (Cd) . . . . 1 ppb
Calcium (Ca) . . . . 1 ppb
Chromium (Cr) . . . . 1 ppb
Cobalt (Co) . . . . 1 ppb
Copper (Cu). . . . . 1 ppb
Iron (Fe). . . . . 5 ppb

Lead (Pb) . . . . 1 ppb
Lithium (Li) . . . . 1 ppb
Magnesium (Mg) . . . . 1 ppb
Manganese (Mn) . . . . 1 ppb
Molybdenum (Mo) . . . . 1 ppb
Mercury (Hg) . . . . 1 ppb
Potassium (K) . . . . 1 ppb
Silicon (Si) . . . . 5 ppb
Sodium (Na) . . . . 5 ppb
Strontium (Sr) . . . . 1 ppb
Tin (Sn) . . . . 1 ppb
Titanium (Ti) . . . . 1 ppb
Vanadium (V) . . . . 1 ppb
Zinc (Zn) . . . . 1 ppb
Zirconium (Zr) . . . . 1 ppb

TESTS

ASSAY. (By titration of the reductive capacity). Weigh accurately about 1 mL of sample in a tared 100-mL volumetric flask containing about 20 mL of water. Dilute to volume with water to the mark, and mix thoroughly. To 20.0 mL of this solution, add 20 mL of dilute sulfuric acid (1 + 15) and titrate with 0.1 N potassium permanganate.

$$\% \ H_2O_2 = (mL \times N \ KMnO_4 \times 17.00 \times 5) \ / \ Sample \ wt \ (g)$$

CHLORIDE, NITRATE, PHOSPHATE, AND SULFATE. See test for hydrogen peroxide, page 341.

TRACE METALS. Determine the aluminum, barium, cadmium, calcium, chromium, cobalt, copper, iron, lead, lithium, magnesium, manganese, molybdenum, nickel, potassium, sodium, strontium, tin, titanium, vanadium, zinc, and zirconium by the ICP–OES method described on page 42.

Determine the mercury by the CVAAS procedure described on page 40.

Hydroxylamine Hydrochloride

NH$_2$OH · HCl   **Formula Wt 69.49**

CAS Number 5470–11–1

REQUIREMENTS

Assay . . . . ≥96.0% NH$_2$OH · HCl

MAXIMUM ALLOWABLE

Clarity of alcohol solution . . . . Passes test
Residue after ignition 0.05%
Titrable free acid. . . . . 0.25 meq/g
Ammonium (NH$_4$) . . . . Passes test
Sulfur compounds (as SO$_4$) 0.005%
Heavy metals (as Pb) . . . . 5 ppm
Iron (Fe) . . . . 5 ppm

TESTS

ASSAY. (By titration of the reductive capacity). Weigh accurately 1.5 g, previously dried for 24 h over magnesium perchlorate or phosphorus pentoxide, and transfer to a 250-mL volumetric flask. Dissolve in oxygen-free water, dilute to volume with oxygen-free water, and take 20.0 mL of the sample solution. Dissolve 5 g of ferric ammonium sulfate in 30 mL of oxygen-free dilute sulfuric acid (1 + 50) and add this solution to the sample solution. Protect the solution from oxygen in the air, boil gently for 5 min, and cool. Dilute with 150 mL of oxygen-free water and titrate with 0.1 N potassium permanganate. One milliliter of 0.1 N potassium permanganate corresponds to 0.003474 g of NH$_2$OH · HCl.

CLARITY OF ALCOHOL SOLUTION. Dissolve 1.0 g in 25 mL of alcohol and compare with an equal volume of alcohol. The sample solution should have no more color or turbidity than the alcohol. Retain the solution for the test for ammonium.

RESIDUE AFTER IGNITION. (Page 16). Ignite 2.0 g.

TITRABLE FREE ACID. Dissolve 10 g in 50 mL of water, add 0.15 mL of bromphenol blue, and titrate with 1 N sodium hydroxide to the neutral (green) end point. Not more than 2.5 mL of 1 N sodium hydroxide should be required.

AMMONIUM. To the solution obtained in the test for insoluble in alcohol, add 1 mL of a solution of chloroplatinic acid (2.6 g in 20 mL). The solution should remain clear for 10 min.

SULFUR COMPOUNDS. Dissolve 2.0 g in 10 mL of water containing 10 mg of sodium carbonate and add 4 mL of nitric acid. Slowly add 12 mL of 10% hydrogen peroxide solution (mix 4 mL of 30% hydrogen peroxide with 8 mL of water). Digest in a covered beaker until reaction ceases, uncover, and evaporate to dryness on a hot plate (≈100 °C). Dissolve the residue in 10 mL of water, add 2 mL of dilute hydrochloric acid (1 + 19), filter if necessary, and dilute with water to 20 mL. To 10 mL add 1 mL of barium chloride reagent solution. Any turbidity should not exceed that produced by 0.05 mg of sulfate ion (SO$_4$) in an equal volume of solution containing the quantities of reagents used in the test. Compare sample 10 min after adding the barium chloride to the standard.

HEAVY METALS. (Page 28, Method 1). Dissolve 6.0 g in about 20 mL of water, and dilute with water to 30 mL. Use 25 mL to prepare the sample solution, and use the remaining 5.0 mL to prepare the control solution.

IRON. (Page 30, Method 1). To the residue after ignition add 3 mL of dilute hydrochloric acid (1 + 1), cover the dish, and digest on a hot plate ($\approx$100 °C) for 15–20 min. Remove the cover, evaporate to dryness, dissolve the residue in 2 mL of hydrochloric acid, and dilute with water to 50 mL. Use the solution without further acidification.

Hydroxy Naphthol Blue
1-(2-Naphtholazo-3,6-disulfonic acid)-2-naphthol-4-sulfonic Acid, Disodium Salt

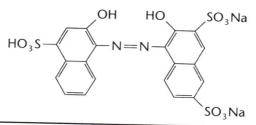

CAS Number 165660–27–5

NOTE. This reagent is deposited on crystals of sodium chloride. The small, blue crystals are freely water-soluble. In the pH range between 12 and 13, the solution of the indicator is reddish pink in the presence of calcium ion and deep blue in the presence of excess (ethylenedinitrilo)tetraacetate.

REQUIREMENTS

Suitability for calcium determination . . . . Passes test

TEST

SUITABILITY FOR CALCIUM DETERMINATION. Dissolve 0.3 g in 100 mL of water, add 10 mL of 1 N sodium hydroxide and 1.0 mL of calcium chloride solution (1 g in 200 mL), and dilute with water to 165 mL. The solution is reddish pink in color. Add 1.0 mL of 0.05 M (ethylenedinitrilo)tetraacetic acid disodium salt. The solution becomes deep blue.

Imidazole

1,3-Diazole

NHCH:NCH:CH   **Formula Wt 68.08**

CAS Number 288–32–4

REQUIREMENTS

Assay . . . . ≥99% $C_3H_4N_2$
pH of a 5% solution . . . . 9.5–11.0 at 25 °C

MAXIMUM ALLOWABLE

Residue after ignition 0.1%
Iron (Fe) 0.001%
Water (H_2O) 0.2%

TESTS

ASSAY. (By acid–base titrimetry). Weigh accurately 0.15 g, and dissolve in 25 mL of glacial acetic acid. Titrate with 0.1 N perchloric acid in acetic acid by using 0.1 mL of crystal violet indicator solution. One milliliter of 0.1 N perchloric acid corresponds to 0.006808 g of $C_3H_4N_2$. Correct for a blank.

pH OF A 5% SOLUTION. (Page 44). The pH should be 9.5–11.0 at 25 °C.

RESIDUE AFTER IGNITION. (Page 16). Use 1.0 g.

IRON. (By flame AAS, page 39).

Sample Stock Solution. Dissolve 25.0 g of sample with water in a 100-mL volumetric flask, and dilute to the mark with water (1 mL = 0.25 g).

Element	Wavelength (nm)	Sample Wt (g)	Standard Added (mg)	Flame Type*	Background Correction
Fe	248.3	5.0	0.05; 0.10	A / A	No

*A/A is air/acetylene.

WATER. (Page 55, Method 1). Use 2.0 g of sample.

Iodic Acid

Iodic(V) Acid

HIO$_3$ Formula Wt 175.91

CAS Number 7782–68–5

REQUIREMENTS

Assay . . . . ≥99.5% HIO$_3$

MAXIMUM ALLOWABLE

Insoluble matter. 0.01%
Residue after ignition 0.02%
Chloride and bromide (as Cl) 0.02%
Iodide (I) 0.01%
Nitrogen compounds (as N). 0.1%
Sulfate (SO$_4$) 0.015%
Heavy metals (as Pb) 0.001%
Iron (Fe). 0.002%

TESTS

ASSAY. (By indirect iodometric titration of iodate). Weigh accurately 0.5 g, transfer to a 250-mL volumetric flask, dissolve in water, dilute with water to volume, and mix thoroughly. Place a 50.0-mL aliquot of this solution in a glass-stoppered conical flask and add 2 g of potassium iodide and 5 mL of 15% hydrochloric acid. Stopper, swirl, allow to stand in the dark for 10 min, and add 100 mL of cold water. Titrate the liberated iodine with 0.1 N sodium thiosulfate, adding 3 mL of starch indicator solution near the end of the titration. Correct for a blank. One milliliter of 0.1 N sodium thiosulfate corresponds to 0.02932 g of HIO$_3$.

INSOLUBLE MATTER. (Page 15). Use 10 g dissolved in 100 mL of water.

RESIDUE AFTER IGNITION. (Page 16). Ignite 5.0 g.

CHLORIDE AND BROMIDE. Dissolve 1.0 g in 100 mL of water in a distilling flask. Add 1 mL of hydrogen peroxide and 1 mL of phosphoric acid, heat to boiling, and boil gently until all the iodine is expelled and the solution is colorless. Cool, wash down the sides of the flask, and add 0.5 mL of hydrogen peroxide. If an iodine color develops, boil until the solution is colorless and for 10 min longer. If no color develops, boil for 10 min, filter if necessary through a chloride-free filter, and dilute with water to 100 mL. Dilute 5.0 mL of this solution with 18 mL of water, and add 1 mL of nitric acid and 1 mL of silver nitrate reagent solution. Any turbidity should not exceed that produced by 0.01 mg of chloride ion (Cl) in an equal volume of solution containing the quantities of nitric acid and silver nitrate used in the test.

IODIDE. Dissolve 1.0 g in 20 mL of water. Add 1 mL of chloroform and 0.5 mL of 1 N sulfuric acid. No violet color should appear in the chloroform within 1 min.

NITROGEN COMPOUNDS. Dissolve 0.50 g in water and dilute with water to 200 mL. Transfer 40 mL to a flask connected through a spray trap to a condenser, the end of which dips beneath the surface of 10 mL of 0.1 N hydrochloric acid. Add to the flask 10 mL of 10% sodium hydroxide reagent solution and 0.5 g of aluminum wire in small pieces, allow to stand for 1 h, and slowly distill about 35 mL. To the distillate add 1 mL of 10% sodium hydroxide reagent solution, dilute with water to 50 mL, and add 2 mL of Nessler reagent. Any color should not exceed that produced when a quantity of an ammonium salt containing 0.1 mg of nitrogen (N) is treated exactly like the sample.

Sample Solution A and Blank Solution B for the Determination of Sulfate, Heavy Metals, and Iron

Sample Solution A. Dissolve 10 g in 20 mL of water and add about 10 mg of sodium carbonate. Add 20 mL of hydrochloric acid and evaporate to dryness on a hot plate ($\approx$100 °C). Repeat the evaporation twice using 10 mL of hydrochloric acid.

Blank Solution B. Evaporate to dryness the quantities of acid and sodium carbonate used to prepare sample solution A.

Dissolve the residues in separate 20-mL portions of water, filter if necessary, and dilute each with water to 100 mL (1 mL of sample solution A = 0.1 g).

SULFATE. (Page 33, Method 1). Use 3.3 mL of sample solution A (0.33-g sample).

HEAVY METALS. (Page 28, Method 1). Dilute 20 mL of sample solution A (2-g sample) to 25 mL. Use 20 mL of blank solution B to prepare the control solution.

IRON. (Page 30, Method 1). Use 5.0 mL of sample solution A (0.5-g sample).

Iodine

I_2 **Formula Wt 253.81**

CAS Number 7553–56–2

REQUIREMENTS

Assay . . . . $\geq$99.8% I_2

MAXIMUM ALLOWABLE

Nonvolatile matter 0.01%
Chlorine and bromine (as Cl) 0.005%

TESTS

ASSAY. (By titration of oxidizing power). Weigh accurately 0.5 g and transfer to a glass-stoppered conical flask. Add 50 mL of water containing 3 g of potassium iodide and 1 mL of sulfuric acid. Stopper, swirl, and allow to stand for 2–3 min. Titrate the iodine with 0.1 N sodium thiosulfate, adding 3 mL of starch indicator solution near the end of the titration. One milliliter of 0.1 N sodium thiosulfate corresponds to 0.01269 g of I_2.

NONVOLATILE MATTER. Transfer 10 g to a tared porcelain crucible or dish. Heat on a hot plate ($\approx$100 °C) until the sample is volatilized. Heat to 105 °C for 1 h, cool in a desiccator, and weigh.

CHLORINE AND BROMINE. Add 1.0 g to 100 mL of hot water containing 0.6 g of hydrazine sulfate in a 250-mL Erlenmeyer flask. Heat on a steam bath until dissolution is effected. Cool and neutralize with 1 N sodium hydroxide solution, using an external indicator. Add 2 mL of hydrogen peroxide and 1 mL of phosphoric acid. Heat to boiling and boil gently until the solution is colorless. Cool, wash down the sides of the flask, and add 0.5 mL of hydrogen peroxide. If an iodine color develops, boil until the solution is colorless and for 10 min longer. If no color develops, boil for 10 min. Filter if necessary through a chloride-free filter, dilute with water to 100 mL, and take a 20-mL portion. For the standard, dilute a solution containing 0.01 mg of chloride ion (Cl) with water to 20 mL. Add 1 mL of nitric acid and 1 mL of silver nitrate reagent solution to each. Any turbidity in the solution of the sample should not exceed that in the standard.

Iodine Monochloride
(suitable for use in Wijs solution)

ICl

Formula Wt 162.36

CAS Number 7790–99–0

REQUIREMENTS

Assay . . . . The I/Cl ratio of the Wijs solution prepared
from the reagent must be 1.10 ± 0.1.
Insoluble matter. . . . . ≤0.005%

TESTS

INSOLUBLE MATTER.

Stock Solution. Dissolve 79.3 ± 0.1 g in 250 mL of glacial acetic acid, and filter through a tared filtering crucible. Store the solution in a dry, low-actinic glass bottle.

Wash the stock solution filter thoroughly with acetic acid and dry at 105 °C.

ASSAY. (By selective titration of reducing and oxidizing power). Prepare Wijs solution (1.4–1.5% ICl in acetic acid) by diluting 11.7 ± 0.1 mL of the stock solution with 217 mL of glacial acetic acid. Determine the I/Cl ratio of this solution as follows:

Iodine. To 150 mL of saturated chlorine water in a 500-mL conical flask, add some glass beads, and pipet 5 mL of the Wijs solution into the flask. Shake, heat to boiling, and boil briskly for 10 min. Cool, add 30 mL of 2% sulfuric acid and 15 mL of 15% potassium iodide solution, and mix well. Titrate immediately with 0.1 N sodium thiosulfate, adding 3 mL of starch indicator solution near the end point.

Total Halogen. Transfer 150 mL of freshly boiled water to a dry 500-mL conical flask. Add 15 mL of 15% potassium iodide solution, pipet 20 mL of Wijs solution into the flask, and mix well. Titrate immediately with 0.1 N sodium thiosulfate, adding 3 mL of starch indicator solution near the end point. Calculate the halogen ratio as follows:

$$\frac{I}{Cl} = \frac{2A}{3B - 2A}$$

where

A = volume, in mL, of thiosulfate used to titrate iodine
B = volume, in mL, of thiosulfate used to titrate total halogen (as iodine)

Iron, Low in Magnesium and Manganese

Fe   Atomic Wt 55.847

CAS Number 7439–89–6

REQUIREMENTS

MAXIMUM ALLOWABLE

Magnesium (Mg) . . . . .5 ppm
Manganese (Mn) 0.002%

TESTS

MAGNESIUM AND MANGANESE. (By flame AAS, page 39).

Sample Stock Solution. In a fume hood, cautiously dissolve 10.0 g in 60 mL of dilute hydrochloric acid (1 + 1) and 20 mL of nitric acid in a covered 400-mL beaker. Substitute a ribbed cover glass, evaporate to dryness, and bake at moderate heat for 5 min. Add 30 mL of hydrochloric acid, heat gently until the salts are dissolved, and dilute with water to 100 mL in a volumetric flask (1 mL = 0.10 g).

Element	Wavelength (nm)	Sample Wt (g)	Standard Added (mg)	Flame Type*	Background Correction
Mg	285.2	1.0	0.005; 0.01	A/A	Yes
Mn	279.5	1.0	0.02; 0.04	A/A	Yes

*A/A is air/acetylene.

Isobutyl Alcohol

2-Methyl-1-propanol

CH$_3$
|
CH$_3$CHCH$_2$OH

(CH$_3$)$_2$CHCH$_2$OH   **Formula Wt 74.12**

CAS Number 78–83–1

Suitable for use in ultraviolet spectrophotometry or general use. Product labeling shall designate the one or more of these uses for which suitability is represented on the basis of meeting the relevant requirements and tests. The ultraviolet spectrophotometry requirements include all of the requirements for general use.

REQUIREMENTS

General Use

Assay . . . . ≥99.0% (CH$_3$)$_2$CHCH$_2$OH

MAXIMUM ALLOWABLE

Color (APHA) 10
Residue after evaporation 0.001%
Solubility in water . . . . Passes test
Titrable acid . . . . 0.0005 meq/g
Water (H$_2$O) 0.1%
Carbonyl compounds . . . . 0.01% butyraldehyde
and 0.02% 2-butanone

Specific Use

Ultraviolet Spectrophotometry

Wavelength (nm)	Absorbance (AU)
400–330	0.01
260	0.05
250	0.07
240	0.25
230	1.00

ASSAY. Analyze the sample by gas chromatography using the parameters cited on page 70. The following specific conditions are also required.

Column: Type I, methyl silicone

Measure the area under all peaks and calculate the isobutyl alcohol content in area percent. Correct for water content.

COLOR (APHA). (Page 19).

RESIDUE AFTER EVAPORATION. (Page 16). Evaporate 100 g (125 mL) to dryness in a tared dish on a hot plate ($\approx$100 °C) and dry the residue at 105 °C for 30 min.

SOLUBILITY IN WATER. Dilute 1 mL with 16 mL of water. The isobutyl alcohol should dissolve completely and the solution should be clear.

TITRABLE ACID. To 60 g (75 mL) in a glass-stoppered flask add 0.15 mL of phenolphthalein indicator solution and shake for 1 min. No pink color should form. Then titrate the solution with 0.01 N alcoholic potassium hydroxide to a faint pink color that persists for at least 15 s. Not more than 3.0 mL of 0.01 N potassium hydroxide should be consumed.

WATER. (Page 55, Method 1). Use 50 mL (40 g) of the sample.

CARBONYL COMPOUNDS. (By differential pulse polarography, page 50). Use 1.25 mL (1.0 g) of sample. For the standards, use 0.10 mg of butyraldehyde and 0.20 mg of 2-butanone.

ULTRAVIOLET SPECTROPHOTOMETRY. Use the procedure on page 73 to determine the absorbance.

Isopentyl Alcohol
Isoamyl Alcohol
3-Methyl-1-butanol

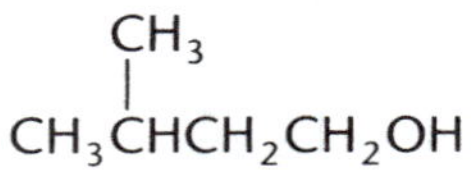

$$CH_3CHCH_2CH_2OH$$
$$(CH_3)$$

(CH$_3$)$_2$CHCH$_2$CH$_2$OH

Formula Wt 88.15

CAS Number 123–51–3

REQUIREMENTS

Assay . . . . $\geq$98.5% $C_5H_{11}OH$

MAXIMUM ALLOWABLE

Water (H_2O) 0.5%
Titrable acid . . . . 0.002 meq/g
Residue after evaporation 0.003%
Acids and esters (as amyl acetate) 0.2%
Carbonyl compounds (as HCHO) 0.1%

TESTS

ASSAY. Analyze the sample by gas chromatography using the general parameters cited on page 70. The following specific conditions are also required.

Column: Type I, methyl silicone

Measure the area under all peaks and calculate the isopentyl alcohol content in area percent. Correct for water content.

WATER. (Page 55, Method 1). Use 25 mL (20 g) of the sample.

TITRABLE ACID. To 20 g (25 mL) of sample in a glass-stoppered flask, add 0.15 mL of phenolphthalein indicator solution and shake for 1 min. No pink color should form. Titrate the solution with 0.01 N alcoholic potassium hydroxide to a faint pink color that persists for at least 15 s. Not more than 4.0 mL of the alcoholic potassium hydroxide should be consumed.

RESIDUE AFTER EVAPORATION. (Page 16). Evaporate 40 g (50 mL) to dryness in a tared dish on a steam bath and dry the residue at 105 °C for 30 min.

ACIDS AND ESTERS. Dilute 100 mL of sample with 100 mL of anhydrous isopropyl alcohol, add 25.0 mL of 0.1 N methanolic potassium hydroxide, and heat gently under a reflux condenser for 10–15 min. Cool and add 20 mL of carbon dioxide-free water through the reflux condenser. Detach the flask, add 0.15 mL of phenolphthalein indicator solution, and titrate the excess potassium hydroxide with 0.1 N hydrochloric acid to a colorless end point. The volume of 0.1 N potassium hydroxide consumed in the test should not be more than 12.3 mL greater than the volume consumed in a complete blank test.

CARBONYL COMPOUNDS. (By differential pulse polarography, page 50). Use 0.25 g (0.30 mL) of sample. For the standard, use 0.25 mg of formaldehyde.

Isopropyl Alcohol

2-Propanol

$$CH_3\overset{\overset{\displaystyle OH}{|}}{C}HCH_3$$

$CH_3CHOHCH_3$ **Formula Wt 60.10**

CAS Number 67–63–0

Suitable for use in ultraviolet spectrophotometry or general use. Product labeling shall designate the one or both of these uses for which suitability is represented on the basis of meeting the relevant requirements and tests. The ultraviolet spectrophotometry requirements include all of the requirements for general use.

REQUIREMENTS

General Use

Assay . . . . ≥99.5% $CH_3CHOHCH_3$

MAXIMUM ALLOWABLE

Carbonyl Compounds . . . . . . 0.002% propionaldehyde or acetone
Color (APHA) 10
Residue after evaporation. 0.001%
Solubility in water . . . . Passes test
Water (H_2O). 0.2%
Titrable acid or base . . . . 0.0001 meq/g

Specific Use

Ultraviolet Spectrophotometry

Wavelength (nm)	Absorbance (AU)
400–330.	0.01
300	0.02
275	0.03
260	0.04
245	0.08
230	0.20
220	0.40
210	1.00

TESTS

ASSAY. Analyze the sample by gas chromatography using the general parameters cited on page 70. The following specific conditions are also required.

Column: Type I, methyl silicone

Measure the area under all peaks and calculate the isopropyl alcohol content in area percent. Correct for water content.

CARBONYL COMPOUNDS. (By differential pulse polarography, page 50). Use 2.6 mL (2.0 g) of sample. Prepare separate standards with 0.04 mg of propionaldehyde and 0.04 mg of acetone.

COLOR (APHA). (Page 19).

RESIDUE AFTER EVAPORATION. (Page 16). Evaporate 100 g (128 mL) to dryness in a tared dish on a hot plate ($\approx$100 °C) and dry the residue at 105 °C for 30 min.

SOLUBILITY IN WATER. Mix 10 mL with 40 mL of water and allow to stand 1 h. The solution should be as clear as an equal volume of water.

WATER. (Page 55, Method 1). Use 10 mL (7.8 g) of the sample.

TITRABLE ACID OR BASE. Mix 10 mL with 10 mL of carbon dioxide-free water and add 0.15 mL of bromthymol blue indicator solution. The solution should be neutral or require not more than 0.10 mL of 0.01 N hydrochloric acid or 0.10 mL of 0.01 N sodium hydroxide to render it so.

ULTRAVIOLET SPECTROPHOTOMETRY. Use the procedure on page 73 to determine the absorbance.

Isopropyl Ether
Diisopropyl Ether

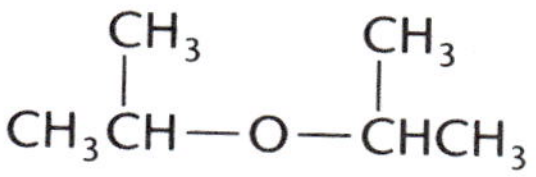

$$CH_3CH-O-CHCH_3$$

with CH_3 groups on each CH

$(C_3H_7)_2O$

Formula Wt 102.18

CAS Number 108–20–3

WARNING. Isopropyl ether is a volatile and flammable liquid with a tendency to form an explosive peroxide. No evaporation or distillation should be performed unless peroxide is shown to be absent. Generally, a stabilizer is present to retard peroxide formation.

REQUIREMENTS

Assay . . . . $\geq$99.0% $(C_3H_7)_2O$

MAXIMUM ALLOWABLE

Color (APHA) 25
Peroxide (as $C_6H_{14}O_2$) 0.05%
Residue after evaporation. 0.01%
Titrable acid . . . . 0.0007 meq/g

TESTS

ASSAY. Analyze the sample by gas chromatography using the general parameters cited on page 70. The following specific conditions are also required.

Column: Type I, methyl silicone

Measure the area under all peaks and calculate the isopropyl ether content in area percent. Correct for water content.

COLOR (APHA). (Page 19).

PEROXIDE. Place 150 mL of 10% sulfuric acid in each of two 500-mL glass-stoppered Erlenmeyer flasks. Reserve one for a blank determination. Into the other, pipet 7.2 g (10 mL) of sample and mix well. To each flask, add 0.15–0.25 mL of 1% ammonium molybdate solution and 15 mL of 10% potassium iodide solution. Mix and allow to stand in the dark for 15 min at room temperature. Titrate each with 0.05 N sodium thiosulfate until the yellow color begins to fade. Add 3 mL of starch indicator solution, and continue the titration just to disappearance of the blue color. Correct for a blank determination.

$$\% \ C_6H_{14}O_2 = (ml \times N \ NaS_2O_3 \times 5.91) \ / \ Sample \ wt \ (g)$$

RESIDUE AFTER EVAPORATION. (Page 16). See the warning above. Evaporate 70 mL (50 g) of sample.

TITRABLE ACID. Titrate 36 g (50 mL) with 0.02 N alcoholic potassium hydroxide, with 0.15 mL of phenolphthalein solution as the indicator, to a pink end point that remains for at least 15 s. Not more than 1.25 mL should be required.

Lactic Acid, 85%
2-Hydroxypropanoic Acid

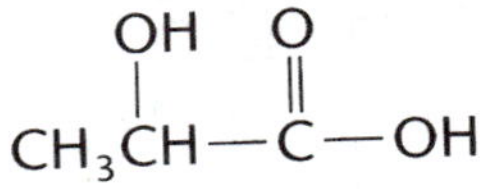

CAS Number 50–21–5

NOTE. The reagent generally available is a mixture of lactic acid, $CH_3CHOHCOOH$, and lactic acid lactate, $C_6H_{10}O_5$.

REQUIREMENTS

Assay . . . . 85.0–90.0% $C_3H_6O_3$
Substances darkened by sulfuric acid . . . . Passes test

MAXIMUM ALLOWABLE

Residue after ignition 0.02%
Chloride (Cl) 0.001%
Sulfate (SO_4) 0.002%
Heavy metals (as Pb) . . . . 5 ppm
Iron (Fe) . . . . 5 ppm

TESTS

ASSAY. (By indirect acid–base titrimetry). Weigh accurately 3 mL in a tared, glass-stoppered flask. Add 50.0 mL of 1 N sodium hydroxide, boil gently for 20 min, and titrate the excess alkali with 1 N sulfuric acid, using 0.15 mL of phenolphthalein indicator solution. Correct for a complete blank test. One milliliter of 1 N sodium hydroxide corresponds to 0.09008 g of $C_3H_6O_3$.

SUBSTANCES DARKENED BY SULFURIC ACID. Cool 5 mL of sulfuric acid to 15 °C and transfer to a clean, dry test tube. Cool 5 mL of the sample to 15 °C and add to the test tube in such a way as to form a distinct layer over the sulfuric acid. A dark color should not develop at the interface of the two layers within 10 min.

RESIDUE AFTER IGNITION. (Page 16). Ignite 10 g (8 mL) and moisten the char with 1 mL of sulfuric acid. Retain the final residue for the preparation of sample solution A.

CHLORIDE. (Page 27). Dilute 12 g (10 mL) with water to 60 mL, mix, and use 5 mL of this solution. Retain the remainder of the solution for the test for sulfate.

SULFATE. (Page 33, Method 1). Use 12.5 mL of the solution prepared in the test for chloride.

> **Sample Solution A.** To the residue from the test for residue after ignition, add 2 mL of water and 2 mL of hydrochloric acid, and evaporate to dryness on a hot plate (≈100 °C). Take up the residue in 1 mL of hydrochloric acid and 20 mL of water, filter if necessary, and dilute with water to 50 mL in a volumetric flask (1 mL = 0.2 g).

HEAVY METALS. (Page 28, Method 2). Test a 4.0-g sample.

IRON. (Page 30, Method 1). Use 10 mL of sample solution A.

Lactose Monohydrate

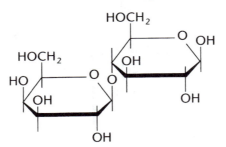

C$_{12}$H$_{22}$O$_{11}$ · H$_2$O   **Formula Wt 360.32**

CAS Number 64044–51–5

REQUIREMENTS

Water (H$_2$O). 4.0–6.0%

MAXIMUM ALLOWABLE

Insoluble matter. 0.005%
Residue after ignition 0.03%
Dextrose . . . . Passes test
Sucrose . . . . Passes test
Heavy metals (as Pb) . . . . 5 ppm
Iron (Fe). . . . . 5 ppm

TESTS

WATER. (Page 55, Method 1). Use 2.0 g of the sample.

INSOLUBLE MATTER. (Page 15). Use 20 g dissolved in 150 mL of water.

RESIDUE AFTER IGNITION. (Page 16). Ignite 5.0 g. Moisten the char with 2 mL of sulfuric acid.

DEXTROSE. To 10 g, finely powdered, add 50 mL of 70% ethyl alcohol. Shake frequently for 30 min and filter through a small, dry filter paper. Evaporate 10 mL of the filtrate to dryness on a hot plate (≈100 °C), and reserve the remainder of the filtrate for the test for sucrose. Dissolve the residue in 5 mL of water, filter if necessary, and transfer the filtrate to a test tube. Add 5 mL of cupric acetate solution, mix, immerse the test tube in a boiling-water bath for 3 min, and allow the solution to stand at room temperature for 25 min. No red precipitate should appear in the test tube.

Cupric Acetate Solution. Dissolve 13.3 g of cupric acetate monohydrate, $(CH_3COO)_2Cu \cdot H_2O$, in a mixture of 195 mL of water and 5 mL of glacial acetic acid.

SUCROSE. Evaporate 25 mL of the filtrate reserved from the test for dextrose to dryness on a hot plate ($\approx$100 °C), and dissolve the residue in 9 mL of water. Add 1 mL of dilute hydrochloric acid (20 + 8) and 100 mg of resorcinol, transfer the mixture to a small test tube, and immerse the tube in a boiling-water bath for 8 min. The solution should remain colorless or acquire not more than a slight yellow coloration.

HEAVY METALS. (Page 28, Method 1). Dissolve 6.0 g in about 20 mL of water, and dilute with water to 30 mL. Use 25 mL to prepare the sample solution, and use the remaining 5.0 mL to prepare the control solution.

IRON. (Page 30, Method 1). Use 2.0 g.

Lanthanum Chloride, Hydrated

$LaCl_3 \cdot (6–7)H_2O$

CAS Number 10099–58–8

NOTE. This reagent is available in degrees of hydration ranging from 6 to 7 molecules of water.

REQUIREMENTS

Assay. . . . . 64.5–70.0% $LaCl_3$

MAXIMUM ALLOWABLE

Insoluble matter. 0.01%
Calcium (Ca) . . . . 10 ppm
Magnesium (Mg) . . . . 10 ppm

TESTS

ASSAY. (By gravimetric determination of La). Weigh 0.7–0.8 g accurately, dissolve in 25 mL of water, filter any undissolved material, and wash with water. Add 100 mL of 5% oxalic acid solution, digest on a hot plate ($\approx$100 °C) until the supernatant solution is clear. Filter through a fine ashless filter paper, wash the precipitate with 10-mL portions of 1% oxalic acid solution, and dry. Ignite in a muffle furnace at 900 °C for 30 min. Cool and weigh as La_2O_3. One gram of lanthanum oxide corresponds to 1.506 g of $LaCl_3$.

INSOLUBLE MATTER. (Page 15). Use 20.0 g dissolved in 100 mL of water.

CALCIUM AND MAGNESIUM. (By flame AAS, page 39).

> *Sample Stock Solution.* Dissolve 5.0 g of sample in 40 mL of water, transfer to a 50-mL volumetric flask, and dilute to the mark with water (1 mL = 0.10 g).

Element	Wavelength (nm)	Sample Wt (g)	Standard Added (mg)	Flame Type*	Background Correction
Ca	422.7	1.0	0.01; 0.02	N/A	No
Mg	285.2	1.0	0.005; 0.01	A/A	Yes

*A/A is air/acetylene; N/A is nitrous oxide/acetylene.

Lead Acetate Trihydrate
Lead(II) Acetate Trihydrate

$(CH_3COO)_2Pb \cdot 3H_2O$ **Formula Wt 379.3**

CAS Number 6080–56–4

REQUIREMENTS

Assay . . . . 99.0–103.0% $(CH_3COO)_2Pb \cdot 3H_2O$

MAXIMUM ALLOWABLE

Insoluble matter. 0.01%
Chloride (Cl) . . . . .5 ppm
Nitrate and nitrite (as NO_3) 0.005%
Calcium (Ca) 0.005%
Copper (Cu). 0.002%
Iron (Fe) 0.001%
Potassium (K). 0.005%
Sodium (Na). 0.01%

TESTS

ASSAY. (By complexometric titration of Pb). Weigh accurately 1.5 g, transfer to a 400-mL beaker, and dissolve in 250 mL of water containing 0.15 mL of acetic acid. Add 15 mL of a saturated aqueous solution of hexamethylenetetramine. Add a few milligrams of xylenol orange indicator mixture and titrate with standard 0.1 M EDTA to a yellow color. One milliliter of 0.1 M EDTA corresponds to 0.03793 g of $(CH_3COO)_2Pb \cdot 3H_2O$.

INSOLUBLE MATTER. (Page 15). Use 10 g dissolved in 100 mL of dilute glacial acetic acid (1 + 99) free from carbon dioxide.

CHLORIDE. (Page 27). Use 2.0 g.

NITRATE AND NITRITE.

Sample Solution A. Dissolve 0.20 g in 1 mL of water. Add 20 mL of nitrate-free sulfuric acid (2 + 1), centrifuge, and decant the liquid portion into a 200-mm test tube. Dilute to 50 mL with brucine sulfate reagent solution.

Control Solution B. Dissolve 0.20 g in 1 mL of the standard nitrate solution containing 0.01 mg of nitrate ion (NO_3) per mL. Add 20 mL of nitrate-free sulfuric acid (2 + 1), centrifuge, and decant the liquid portion into a 200-mm test tube. Dilute to 50 mL with brucine sulfate reagent solution.

Continue with the procedure described on page 30, starting with the preparation of blank solution C.

CALCIUM, COPPER, IRON, POTASSIUM, AND SODIUM. (By flame AAS, page 39).

Sample Stock Solution A. Dissolve 25.0 g of sample in 75 mL of water and 5 mL of nitric acid. Transfer to a 100-mL volumetric flask and dilute to the mark with water (1 mL = 0.25 g).

Sample Stock Solution B. Transfer 4.0 mL of sample stock solution A to a 100-mL volumetric flask and dilute to the mark with water (1 mL = 0.01 g).

Element	Wavelength (nm)	Sample Wt (g)	Standard Added (mg)	Flame Type*	Background Correction
Ca	422.7	1.0	0.025; 0.05	N/A	No
Cu	324.7	1.0	0.02; 0.04	A/A	Yes
Fe	248.3	5.0	0.05; 0.10	A/A	Yes
K	766.5	1.0	0.025; 0.05	A/A	No
Na	589.0	0.10	0.005; 0.01	A/A	No

*A/A is air/acetylene; N/A is nitrous oxide/acetylene.

Lead Carbonate

CAS Number 598–63–0 (normal); 1344–36–1 (basic)

NOTE. This specification applies to both the normal lead carbonate and the basic lead carbonate.

REQUIREMENTS

MAXIMUM ALLOWABLE

Insoluble in dilute acetic acid. 0.02%
Chloride (Cl) 0.002%
Nitrate and nitrite (as NO_3) . . . . Passes test
Cadmium (Cd). 0.002%

Calcium (Ca) 0.01%
Iron (Fe) 0.005%
Potassium (K). 0.02%
Sodium (Na). 0.05%
Zinc (Zn) 0.003%

TESTS

INSOLUBLE IN DILUTE ACETIC ACID. Digest 5.0 g in 50 mL of water plus 7 mL of glacial acetic acid in a covered beaker on a steam bath for 1 h. Filter through a tared filtering crucible, wash thoroughly with dilute acetic acid (2 + 98), and dry at 105 °C.

CHLORIDE. (Page 27). Dissolve 1.0 g in 20 mL of 10% nitric acid reagent solution, filter if necessary through a chloride-free filter, and dilute with water to 50 mL. Use 25 mL of this solution.

NITRATE AND NITRITE.

Sample Solution A. Dissolve 0.20 g in 1 mL of water and 8 mL of dilute acetic acid (1 + 7) by heating in a boiling-water bath. Add 20 mL of nitrate-free sulfuric acid (2 + 1), cool, centrifuge, and decant the liquid portion into a 200-mm test tube. Dilute to 50 mL with brucine sulfate reagent solution.

Control Solution B. Dissolve 0.20 g in 8 mL of dilute acetic acid (1 + 7) and 1 mL of the standard nitrate solution containing 0.01 mg of nitrate ion (NO_3) per mL by heating in a boiling-water bath. Add 20 mL of nitrate-free sulfuric acid (2 + 1), cool, centrifuge, and decant the liquid portion into a 200-mm test tube. Dilute to 50 mL with brucine sulfate reagent solution.

Continue with the procedure described on page 30, starting with the preparation of blank solution C. (Limit about 0.005%)

CADMIUM, CALCIUM, IRON, POTASSIUM, SODIUM, AND ZINC. (By flame AAS, page 39).

Sample Stock Solution. Dissolve 10.0 g of sample in 50 mL of water and sufficient nitric acid to achieve complete dissolution. Transfer the solution to a 100-mL volumetric flask and dilute to the mark with water (1 mL = 0.1 g).

Element	Wavelength (nm)	Sample Wt (g)	Standard Added (mg)	Flame Type*	Background Correction
Cd	228.8	1.0	0.02; 0.04	A/A	Yes
Ca	422.7	0.40	0.02; 0.04	N/A	No
Fe	248.3	1.0	0.05; 0.10	A/A	Yes
K	766.5	0.20	0.02; 0.04	A/A	No
Na	589.0	0.04	0.01; 0.02	A/A	No
Zn	213.9	0.40	0.01; 0.02	A/A	Yes

*A/A is air/acetylene; N/A is nitrous oxide/acetylene.

Lead Chromate

PbCrO$_4$

Formula Wt 323.18

CAS Number 7758–97–6

REQUIREMENTS

Assay . . . . $\geq$98.0% PbCrO$_4$

MAXIMUM ALLOWABLE

Soluble matter 0.15%
Carbon compounds (as C) 0.01%

TESTS

ASSAY. (By titration of oxidizing power of chromate). Weigh accurately 0.5 g of powdered sample and dissolve by warming with 40 mL of 10% sodium hydroxide reagent solution in a glass-stoppered conical flask. Add 2 g of potassium iodide; when it is dissolved, dilute with 80 mL of water and 15 mL of hydrochloric acid. Stopper, swirl, and allow to stand in the dark for 5 min. Titrate the liberated iodine with 0.1 N sodium thiosulfate, adding 3 mL of starch indicator solution near the end of the titration. Correct for a blank. One milliliter of 0.1 N sodium thiosulfate corresponds to 0.01077 g of PbCrO$_4$.

$$\% \ PbCrO_4 = (mL \times N \ Na_2S_2O_3 \times 10.77) \ / \ Sample \ wt \ (g)$$

SOLUBLE MATTER. Boil 0.50 g of the powdered sample for 5 min with 100 mL of dilute acetic acid (1 + 20), stirring well during the heating. Cool and filter. Evaporate 25 mL of the filtrate to dryness in a tared dish on a hot plate ($\approx$100 °C) and dry at 105 °C. Heat 50 mL of the filtrate with 2 g of the powdered sample for 5 min at 80–90 °C, cool, dilute with water to 50 mL, and filter. Evaporate 25 mL of this filtrate in a tared dish on a hot plate ($\approx$100 °C) and dry at 105 °C. The difference between the weights of the two residues should not be more than 0.0015 g.

CARBON COMPOUNDS. This procedure uses induction heating of the sample and a conductometric measurement of evolved carbon dioxide.

Transfer 1.0 g of sample and one tin capsule to a combustion crucible. Add 1.5 g each of iron and copper accelerator. Ignite in an induction furnace in a stream of carbon dioxide-free oxygen until combustion is complete. Pass the effluent gas through an approximately 0.1% solution of reagent-grade barium hydroxide. Measure the change in conductance, in mhos, using a suitable cell and detection system. Subtract the reading from a blank and calculate the carbon content of the sample from a calibration curve.

Prepare the calibration curve as follows: transfer 0, 10, 25, 75, and 100 μL of carbon standard solution (1 μL = 4 μg of carbon) to tin capsules. Evaporate the solutions to dryness in the capsules at 80 °C. Transfer the capsules to combustion crucibles, add the accelerators, ignite, and measure the conductance under the same conditions as described for running the sample. Subtract the reading of the blank from that of each of the standards and plot the calibration curve as mhos versus micrograms of carbon.

Lead Dioxide
Lead(IV) Oxide

PbO_2 **Formula Wt 239.19**

CAS Number 1309–60–0

REQUIREMENTS

Assay . . . . ≥97.0% PbO_2

MAXIMUM ALLOWABLE

Acid-insoluble matter 0.2%
Carbon compounds (as C) 0.04%
Chloride (Cl) 0.002%
Nitrate (NO_3) 0.02%
Sulfate (SO_4) 0.05%
Manganese (Mn) . . . . 5 ppm
Calcium (Ca) 0.02%
Copper (Cu) 0.05%
Potassium (K) 0.05%
Sodium (Na) 0.1%

TESTS

ASSAY. (By titration of oxidizing power). Dissolve 25 g of sodium chloride in 100 mL of water in a stoppered flask. Add 2 g of potassium iodide and 20 mL of hydrochloric acid. Add, accurately weighed, 0.25 g of finely powdered sample and stir until the sample is dissolved. Titrate the liberated iodine with 0.1 N sodium thiosulfate, using 3 mL of starch indicator added near the end of the titration. Run a complete blank determination using the same quantities of reagents, diluting the sample, and stirring for the same time. One milliliter of 0.1 N sodium thiosulfate corresponds to 0.01196 g of PbO_2.

ACID-INSOLUBLE MATTER. To 1.0 g add 25 mL of water and 3 mL of nitric acid. Add carefully with stirring 5 mL of 30% hydrogen peroxide or more, if needed, to dissolve all the lead dioxide. Filter through a tared filtering crucible, wash thoroughly, and dry at 105 °C. Reserve the filtrate for the test for substances not precipitated by hydrogen sulfide.

$$\% \ PbO_2 = (mL \times N \ Na_2S_2O_3 \times 11.96) / Sample \ wt \ (g)$$

CARBON COMPOUNDS. This procedure uses induction heating of the sample and a conductometric measurement of the evolved carbon dioxide.

Transfer 1.0 g of sample and one tin capsule to a combustion crucible. Add 1.5 g each of iron and copper accelerator. Ignite in an induction furnace in a stream of carbon dioxide-free oxygen until combustion is complete. Pass the effluent gas through an approximately 0.1% solution of barium hydroxide. Measure the change in conductance, in mhos, by using a suitable cell and detection system. Subtract the reading from a blank and calculate the carbon content of the sample from a calibration curve.

Prepare the calibration curve as follows: transfer 0, 10, 25, 75, and 100 μL of carbon standard solution (1 μL = 4 μg of carbon) to tin capsules. Evaporate the solutions to dryness in the capsules at 80 °C. Transfer the capsules to combustion crucibles, add the accelerators, ignite, and measure the conductance under the same conditions as described for running the sample. Subtract the reading of the blank from that of each of the standards and plot the calibration curve as mhos versus micrograms of carbon.

CHLORIDE. (Page 27). Dissolve 2.0 g in 20 mL of water plus 2 mL of glacial acetic acid and 2 mL of 30% hydrogen peroxide. Filter through a chloride-free filter, dilute with water to 100 mL, and use 25 mL of this solution.

NITRATE. Dissolve 0.25 g in a test tube containing 10 mL of water, 1 mL of acetic acid, and 0.5 mL of 30% hydrogen peroxide. Pass sulfur dioxide through the solution to precipitate the lead, filter, wash with two 2-mL portions of water, and dilute the filtrate with water to 20 mL.

Sample Solution A. To 4.0 mL of the filtrate in a test tube add 1 mL of water, dilute to 50 mL with brucine sulfate reagent solution, and mix.

Control Solution B. To 4.0 mL of the filtrate in a test tube add 1 mL of the standard nitrate solution containing 0.01 mg of nitrate ion (NO_3) per mL, dilute to 50 mL with brucine sulfate reagent solution, and mix.

Continue with the procedure described on page 30, starting with the preparation of blank solution C.

SULFATE. Decompose 2.0 g with 10 mL of hydrochloric acid and 3 mL of nitric acid and evaporate to dryness on a hot plate ($\approx$100 °C). Add 25 mL of water and 2 g of sodium carbonate and digest on a hot plate ($\approx$100 °C) for several hours, with occasional stirring. Dilute with water to 100 mL and filter. Neutralize the filtrate with hydrochloric acid and add an excess of 1 mL of the acid. Heat to boiling, add 5 mL of barium chloride reagent solution, digest in a covered beaker on the hot plate for 2 h, and allow to stand overnight. Filter, using a fine ashless paper, wash thoroughly, and ignite. Correct for the weight obtained in a complete blank test.

MANGANESE. Add 10 g to 15 mL of dilute nitric acid (1 + 1). Add carefully, with stirring, about 10 mL of 30% hydrogen peroxide—or more, if necessary—to dissolve all of the lead dioxide. Evaporate nearly to dryness, cool, add 20 mL of dilute sulfuric acid (1 + 1), and evaporate to dense fumes of sulfur trioxide. Cool, cautiously dilute to 100 mL with water, allow to settle, and filter. Dilute 20 mL of the filtrate with water to 50 mL, add 10 mL of nitric acid and 5 mL of phosphoric acid, and boil gently for 5 min. Cool slightly, add 0.25 g of potassium periodate, and again boil gently for 5 min. Any pink color should not exceed that produced by 0.01 mg of manganese in 50 mL of dilute sulfuric acid (1 + 9) treated exactly like the 50 mL of filtrate.

CALCIUM, COPPER, POTASSIUM, AND SODIUM. (By flame AAS, page 39).

Sample Stock Solution. Dissolve by warming 10.0 g of sample with a mixture of 20 mL of water and 15 mL of nitric acid. Cool to room temperature, transfer to a 100-mL volumetric flask, and dilute to the mark with water (1 mL = 0.1 g).

Element	Wavelength (nm)	Sample Wt (g)	Standard Added (mg)	Flame Type*	Background Correction
Ca	422.7	0.40	0.04; 0.08	N/A	No
Cu	324.8	0.10	0.025; 0.05	A/A	Yes
K	766.5	0.10	0.025; 0.05	A/A	No
Na	589.0	0.02	0.01; 0.02	A/A	No

*A/A is air/acetylene; N/A is nitrous oxide/acetylene.

Lead Monoxide
Lead(II) Oxide
Litharge

PbO   **Formula Wt 223.19**

CAS Number 1317–36–8

REQUIREMENTS

Assay . . . . ≥99.0% PbO

MAXIMUM ALLOWABLE

Insoluble in dilute acetic acid 0.02%
Chloride (Cl) 0.002%
Nitrate (NO_3) 0.01%

Calcium (Ca) 0.005%
Copper (Cu). 0.005%
Iron (Fe). 0.002%
Potassium (K). 0.005%
Silver (Ag) . . . . 5 ppm
Sodium (Na) 0.02%

TESTS

ASSAY. (By complexometric titration of Pb). Weigh accurately 0.8 g, transfer to a 400-mL beaker, and dissolve by warming with a mixture of 1 mL of acetic acid and 10 mL of water. After solution is complete, dilute to 250 mL with water and add 15 mL of saturated aqueous hexamethylenetetramine solution and 50 mg of xylenol orange indicator mixture. Titrate with 0.1 M EDTA to a color change of purple-red to lemon yellow. One milliliter of 0.1 M EDTA corresponds to 0.02232 g of PbO.

$$\% \text{ PbO} = (\text{mL} \times \text{M EDTA} \times 22.32) / \text{Sample wt (g)}$$

INSOLUBLE IN DILUTE ACETIC ACID. Heat 5.0 g with a mixture of 50 mL of water and 15 mL of acetic acid until solution is complete. Filter through a tared sintered-glass crucible, wash thoroughly with hot water, dry at 105 °C for 1 h, cool, and weigh.

CHLORIDE. (Page 27). Use 0.5 g.

NITRATE. (Page 30). Dissolve 0.5 g in a mixture of 10 mL of water and 1 mL of acetic acid. Add 5 mL of 25% sulfuric acid reagent solution, mix well, and allow to stand for 15 min, mixing occasionally. Filter, wash with 5 mL of water, and dilute the filtrate to 25 mL.

> **Sample Solution A.** Dilute 5.0 mL of the nitrate-test filtrate to 50 mL with brucine sulfate reagent solution.

> **Control Solution B.** To 5.0 mL of the nitrate-test filtrate add 1.0 mL of the standard nitrate solution containing 0.01 mg of nitrate ion per mL. Dilute to 50 mL with brucine sulfate reagent solution.

Continue with the procedure described on page 30, starting with the preparation of blank solution C.

CALCIUM, COPPER, IRON, POTASSIUM, SILVER, AND SODIUM. (By flame AAS, page 39).

> **Sample Stock Solution.** Dissolve by warming 20.0 g of sample with a mixture of 50 mL of water and 30 mL of nitric acid. Cool to room temperature and transfer to a 100-mL volumetric flask, and dilute to the mark with water (1 mL = 0.2 g).

Element	Wavelength (nm)	Sample Wt (g)	Standard Added (mg)	Flame Type*	Background Correction
Ca	422.7	0.80	0.02; 0.04	N/A	No
Cu	324.8	0.80	0.02; 0.04	A/A	Yes
Fe	248.3	4.0	0.08; 0.16	A/A	Yes
K	766.5	0.80	0.02; 0.04	A/A	No
Ag	328.1	4.0	0.02; 0.04	A/A	Yes
Na	589.0	0.10	0.01; 0.02	A/A	No

*A/A is air/acetylene; N/A is nitrous oxide/acetylene.

Lead Nitrate

Lead(II) Nitrate

$Pb(NO_3)_2$ **Formula Wt 331.20**

CAS Number 10099–74–8

REQUIREMENTS

Assay . . . . ≥99.0% $Pb(NO_3)_2$

MAXIMUM ALLOWABLE

Insoluble matter. 0.005%
Chloride (Cl) 0.001%
Calcium (Ca) 0.005%
Copper (Cu). 0.002%
Iron (Fe) 0.001%
Potassium (K). 0.005%
Sodium (Na). 0.02%

TESTS

ASSAY. (By complexometric titration of Pb). Weigh accurately 1.3 g, transfer to a 400-mL beaker, and dissolve in 250 mL of water containing 0.15 mL of acetic acid. Add 15 mL of a saturated aqueous solution of hexamethylenetetramine. Add a few milligrams of xylenol orange indicator mixture and titrate with standard 0.1 M EDTA to a yellow color. One milliliter of 0.1 M EDTA corresponds to 0.03312 g of $Pb(NO_3)_2$.

$$\% \ Pb(NO_3)_2 = (mL \times M \ EDTA \times 33.12) / Sample \ wt \ (g)$$

INSOLUBLE MATTER. (Page 15). Use 20 g dissolved in 200 mL of dilute nitric acid (1 + 99).

CHLORIDE. (Page 27). Use 1.0 g.

CALCIUM, COPPER, IRON, POTASSIUM, AND SODIUM. (By flame AAS, page 39).

> *Sample Stock Solution.* Dissolve 20.0 g of sample with 60 mL of water and add 10 mL of nitric acid. Transfer to a 100-mL volumetric flask, and dilute to the mark with water (1 mL = 0.2 g).

Element	Wavelength (nm)	Sample Wt (g)	Standard Added (mg)	Flame Type*	Background Correction
Ca	422.7	1.0	0.05; 0.10	N/A	No
Cu	324.7	1.0	0.02; 0.04	A/A	Yes
Fe	248.3	4.0	0.04; 0.08	A/A	Yes
K	766.5	1.0	0.025; 0.05	A/A	No
Na	589.0	0.10	0.01; 0.02	A/A	No

*A/A is air/acetylene; N/A is nitrous oxide/acetylene.

Lead Perchlorate Trihydrate

$Pb(ClO_4)_2 \cdot 3H_2O$ **Formula Wt 460.15**

CAS Number 13637–76–8

REQUIREMENTS

Assay. . . . . .97.0%–102.0% $Pb(ClO_4)_2 \cdot 3H_2O$
pH of a 5% solution . . . . 3.0–5.0 at 25 °C

MAXIMUM ALLOWABLE

Insoluble matter. 0.005%
Chloride (Cl) 0.01%
Iron (Fe). 0.001%

TESTS

ASSAY. (By complexometric titration of lead). While protecting from moisture, weigh accurately 1.8 g, transfer to a 400-mL beaker, and dissolve in 250 mL of water. Add 15 mL of a saturated solution of hexamethylenetetramine and a few milligrams of xylenol orange indicator mixture, and titrate with standard 0.1 M EDTA solution to a yellow color. A precipitate that dissolves readily during the titration will form. One milliliter of 0.1 M EDTA corresponds to 0.04602 g of $Pb(ClO_4)_2 \cdot 3H_2O$.

$$\% \, PbClO_4 \cdot 3H_2O = (mL \times M \, EDTA \times 46.02) \, / \, \text{Sample wt (g)}$$

pH OF A 5% SOLUTION. (Page 44). The pH should be 3.0–5.0 at 25 °C.

INSOLUBLE MATTER. (Page 15). Use 20 g dissolved in water.

CHLORIDE. (Page 27). Use 1 g in 100 mL and take a 10-mL aliquot.

IRON. (Page 30, Method 2). Use 1 g, and filter after addition of hydroxylamine hydrochloride. Wash in the funnel and use the filtrate.

Lead Subacetate
(for sugar analysis)

CAS Number 1335–32–6

REQUIREMENTS

Basic lead (PbO) . . . .≥33.0%
MAXIMUM ALLOWABLE
Loss on drying at 105 °C 1.5%
Insoluble in dilute acetic acid 0.02%
Insoluble in water 1.0%
Chloride (Cl) 0.003%
Nitrate and nitrite (as NO_3) . . . . .Passes test
Calcium (Ca) 0.01%
Copper (Cu) 0.002%
Iron (Fe) 0.002%
Potassium (K) 0.02%
Sodium (Na) 0.05%

TESTS

BASIC LEAD. (By indirect acidimetry). Weigh accurately 5 g of sample and dissolve in 100 mL of carbon dioxide-free water in a 500-mL volumetric flask. Add 50.0 mL of 1 N acetic acid and 100 mL of carbon dioxide-free 3% solution of sodium oxalate. Mix thoroughly, dilute to volume with carbon dioxide-free water, and allow the precipitate to settle. Titrate 100.0 mL of the clear supernatant liquid with 1 N sodium hydroxide, using 0.15 mL of phenolphthalein indicator. Each milliliter of 1 N acetic acid consumed is equivalent to 0.1116 g of PbO.

LOSS ON DRYING. Weigh accurately about 0.5 g in a tared dish or crucible and dry for 2 h at 105 °C. Cool and reweigh.

INSOLUBLE IN DILUTE ACETIC ACID. Dissolve 5.0 g in 100 mL of dilute acetic acid (1 + 19), warming if necessary to effect complete dissolution. Filter through a tared filtering crucible, wash with dilute acetic acid (1 + 19) until the washings are no longer darkened by hydrogen sulfide, and dry at 105 °C.

INSOLUBLE IN WATER. Agitate 1.0 g in a small stoppered flask with 50 mL of carbon dioxide-free water, just until dissolved, and filter at once through a tared filtering crucible. Wash with carbon dioxide-free water and dry at 105 °C.

CHLORIDE. (Page 27). Dissolve 1.0 g in 30 mL of dilute nitric acid (1 + 9). Use 10 mL of this solution.

NITRATE AND NITRITE.

Sample Solution A. Suspend 0.50 g in 3 mL of water. Dilute to 50 mL with brucine sulfate reagent solution.

Control Solution B. Suspend 0.50 g in 1.5 mL of water and 1.5 mL of the standard nitrate solution containing 0.01 mg of nitrate ion (NO_3) per mL. Dilute to 50 mL with brucine sulfate reagent solution.

Blank Solution C. Use 50 mL of brucine sulfate reagent solution.

Heat sample solution A, control solution B, and blank solution C in a preheated (boiling) water bath until the lead subacetate in A and B is dissolved, then heat for 10 min more. Cool rapidly in an ice bath to room temperature, centrifuge sample solution A and control solution B for 10 min, and decant the liquids into 200-mm test tubes. Set a spectrophotometer at 410 nm and, using 1-cm cells, adjust the instrument to read zero absorbance with blank solution C in the light path, then determine the absorbance of sample solution A. Adjust the instrument to read 0 absorbance with sample solution A in the light path and determine the absorbance of control solution B. The absorbance of sample solution A should not exceed that of control solution B. (Limit about 0.003%)

CALCIUM, COPPER, IRON, POTASSIUM, AND SODIUM. (By flame AAS, page 39).

Sample Stock Solution. Dissolve 20.0 g of sample in a mixture of 50 mL of water and 10 mL of nitric acid, warming to achieve complete dissolution. Cool to room temperature and transfer to a 100-mL volumetric flask, and dilute to the mark with water (1 mL = 0.2 g).

Element	Wavelength (nm)	Sample Wt (g)	Standard Added (mg)	Flame Type*	Background Correction
Ca	422.7	1.0	0.05; 0.10	N/A	No
Cu	324.8	1.0	0.02; 0.04	A/A	Yes
Fe	248.3	4.0	0.08; 0.16	A/A	Yes
K	766.5	0.20	0.02; 0.04	A/A	No
Na	589.0	0.04	0.01; 0.02	A/A	No

*A/A is air/acetylene; N/A is nitrous oxide/acetylene.

Lithium Carbonate

Li_2CO_3   **Formula Wt 73.89**

CAS Number 554–13–2

> *NOTE.* The formula weight of this reagent is likely to deviate from the value cited, since the natural distribution of ^6Li and ^7Li isotopes is often altered in current sources of lithium compounds.

REQUIREMENTS

Assay . . . . ≥99.0% Li_2CO_3

MAXIMUM ALLOWABLE

Insoluble in dilute hydrochloric acid. 0.01%
Chloride (Cl) 0.005%
Nitrate (NO_3) . . . . 5 ppm
Sulfur compounds (as SO_4) 0.2%
Ammonium (NH_4) . . . . 5 ppm
Heavy metals (as Pb) 0.002%
Iron (Fe) 0.002%
Calcium (Ca) 0.01%
Potassium (K). 0.01%
Sodium (Na). 0.1%

TESTS

ASSAY. (By indirect acid–base titrimetry for carbonate). Weigh accurately 1.3 g of sample, add 50 mL of water, and then add cautiously 50.0 mL of 1 N hydrochloric acid. Boil gently to expel carbon dioxide, cool, add 0.15 mL of methyl red indicator solution, and titrate the excess hydrochloric acid with 1 N sodium hydroxide solution. One milliliter of 1 N hydrochloric acid corresponds to 0.03694 g of Li_2CO_3.

$$\% \ Li_2CO_3 = (mL \times N \ HCl \times 3.694) \ / \ Sample \ wt \ (g)$$

INSOLUBLE IN DILUTE HYDROCHLORIC ACID. Suspend 10 g in 75 mL of water, then add cautiously and slowly 25 mL of hydrochloric acid. Heat to boiling and digest in a covered beaker on a hot plate ($\approx$100 °C) for 1 h. Filter through a tared filtering crucible, wash thoroughly with hot water, and dry at 105 °C.

CHLORIDE. (Page 27). Dissolve 2.0 g in 25 mL of dilute nitric acid (1 + 5), dilute with water to 50 mL, and use 5.0 mL of this solution.

NITRATE.

> *Sample Solution A.* Add 2.0 g to 4 mL of water in a test tube. Add brucine sulfate reagent solution, drop by drop, until the sample is completely dissolved. Then dilute to 50 mL with the same reagent solution and mix.

Control Solution B. Add 2.0 g to 3 mL of water and 1 mL of the standard nitrate solution containing 0.01 mg of nitrate ion (NO_3) per mL. Add brucine sulfate reagent solution, drop by drop, until the sample is completely dissolved. Then dilute to 50 mL with the same reagent solution and mix.

Continue with the procedure described on page 30, starting with the preparation of blank solution C.

SULFUR COMPOUNDS. Suspend 0.10 g in 10 mL of water and cautiously neutralize with dilute hydrochloric acid (1 + 9). Add 0.10 mL of bromine water and boil gently until the bromine is completely expelled. Add 2 mL of dilute hydrochloric acid (1 + 19), filter, and dilute with water to 40 mL. To 10 mL add 1 mL of barium chloride reagent solution. Any turbidity should not exceed that produced by 0.05 mg of sulfate ion (SO_4) in an equal volume of solution containing 1 mL of dilute hydrochloric acid (1 + 19) and 1 mL of barium chloride reagent solution. Compare 10 min after adding the barium chloride to the sample and standard solutions.

AMMONIUM. (Page 25). Suspend 2.0 g in 50 mL of water and cautiously add 5 mL of hydrochloric acid. The distillate shows not more than 0.01 mg of ammonium ion (NH_4) plus the residue from 5 mL of hydrochloric acid.

HEAVY METALS. (Page 28, Method 1). Suspend 1.5 g in 10 mL of water and cautiously add 10 mL of hydrochloric acid. Evaporate the solution to a moist residue on a hot plate (≈ 100 °C), dissolve the residue in 20 mL of water, and dilute with water to 25 mL. For the control, suspend 0.5 g in 10 mL of water, add 0.02 mg of lead, and treat exactly as the 1.5 g of sample.

IRON. (Page 30, Method 1). Suspend 0.5 g in 30 mL of water, cautiously add 3 mL of hydrochloric acid, and dilute with water to 50 mL. Use this solution without further acidification. Use only 2 mL of hydrochloric acid for the control solution.

CALCIUM, POTASSIUM, AND SODIUM. (By flame AAS, page 39).

Sample Stock Solution. Suspend 5 g of sample in 50 mL of water and cautiously add 25 mL of hydrochloric acid. Tranfer the solution to a 100-mL volumetric flask and dilute to the mark with water (1 mL = 0.05 g).

Element	Wavelength (nm)	Sample Wt (g)	Standard Added (mg)	Flame Type*	Background Correction
Ca	422.7	0.10	0.02; 0.04	N/A	No
K	766.5	1.0	0.05; 0.10	A/A	No
Na	589.0	0.01	0.01; 0.02	A/A	No

*A/A is air/acetylene; N/A is nitrous oxide/acetylene.

Lithium Chloride

LiCl   **Formula Wt 42.39**

CAS Number 7447–41–8

NOTE. The formula weight of this reagent is likely to deviate from the value cited because the natural distribution of ^6Li and ^7Li isotopes is often altered in current sources of lithium compounds.

REQUIREMENTS

Assay . . . . ≥99% LiCl

MAXIMUM ALLOWABLE

Insoluble matter. 0.01%
Titrable base. . . . . .0.008 meq/g
Loss on drying at 105 °C. 1.0%
Nitrate (NO_3) 0.001%
Sulfate (SO_4) 0.01%
Barium (Ba) 0.003%
Heavy metals (as Pb) 0.002%
Iron (Fe). 0.001%
Calcium (Ca) 0.01%
Potassium (K). 0.01%
Sodium (Na). 0.20%

TESTS

ASSAY. (By indirect argentimetric titration of chloride). Weigh accurately 0.8 g, dissolve in water in a 100-mL volumetric flask, and dilute with water to volume. Transfer 20.0 mL to a 250-mL glass-stoppered flask, add 50 mL of water, and mix. Add, slowly and with constant stirring, 50.0 mL of 0.1 N silver nitrate. Mix, add 3 mL of nitric acid and 10 mL of toluene, and mix thoroughly. Add 2 mL of ferric ammonium sulfate indicator solution, and titrate the excess silver nitrate with 0.1 N thiocyanate solution. One milliliter of 0.1 N silver nitrate corresponds to 0.004239 g of LiCl.

$$\% \text{ LiCl} = \frac{\text{mL} \times \text{N AgNO}_3 \times 4.239}{[\text{Sample wt (g)}] \ / \ 5}$$

INSOLUBLE MATTER. (Page 15). Use 10.0 g dissolved in 150 mL of water.

TITRABLE BASE. Transfer 10.0 g to a 150-mL beaker. Dissolve in 100 mL of carbon dioxide-free water, add 0.15 mL of methyl red indicator solution, and titrate with 0.01 N hydrochloric acid. Correct for a blank determination. Not more than 8.0 mL of 0.01 N hydrochloric acid should be required to produce a pink color.

LOSS ON DRYING AT 105 °C. Weigh accurately 1.0 g in a tared dish or crucible and dry to constant weight at 105 °C.

NITRATE. Mix 1.0 g with 5 mL of phenoldisulfonic acid reagent solution, and cautiously dilute with water to 25 mL. Cautiously add ammonium hydroxide until the maximum yellow color develops. Any yellow color should not exceed that produced in a standard, treated similarly, containing 0.01 mg of added nitrate (NO_3).

SULFATE. (Page 33, Method 1). Use 0.5 g.

BARIUM. Dissolve 3.3 g in 25 mL of water, filter if necessary, and add 2 mL of potassium dichromate reagent solution. Add ammonium hydroxide until the orange color dissipates and the yellow color persists. Any turbidity produced within 15 min should not exceed that in a standard containing 0.10 mg of added barium.

HEAVY METALS. (Page 28, Method 1). Dissolve 2.0 g in 40 mL of water. Use 30 mL to prepare the sample solution, and use the remaining 10 mL to prepare the control solution.

IRON. (Page 30, Method 1). Use 1.0 g.

CALCIUM, POTASSIUM, AND SODIUM. (By flame AAS, page 39).

Sample Stock Solution A. Dissolve 10.0 g of sample with water in a 100-mL volumetric flask, and dilute to the mark with water (1 mL = 0.1 g).

Sample Stock Solution B. Pipet 10.0 mL of sample stock solution A into a 100-mL volumetric flask, and dilute to the mark with water (1 mL = 0.01 g).

Element	Wavelength (nm)	Sample Wt (g)	Standard Added (mg)	Flame Type*	Background Correction
Ca	422.7	0.50	0.05; 0.10	N/A	No
K	766.5	0.50	0.05; 0.10	A/A	No
Na	589.0	0.005	0.01; 0.02	A/A	No

*A/A is air/acetylene; N/A is nitrous oxide/acetylene.

Lithium Hydroxide Monohydrate

LiOH · H₂O

Formula Wt 41.96

CAS Number 1310–66–3

NOTE. The formula weight of this reagent is likely to deviate from the value cited, since the natural distribution of ^6Li and ^7Li is often altered in current sources of lithium compounds.

REQUIREMENTS

Assay . . . . ≥98.0% LiOH · H$_2$O

MAXIMUM ALLOWABLE

Lithium carbonate 2.0%
Insoluble matter. 0.01%
Chloride (Cl) 0.01%
Sulfate (SO$_4$) 0.05%
Heavy metals (as Pb) 0.002%
Iron (Fe). 0.002%

TESTS

ASSAY AND LITHIUM CARBONATE. (By acidimetry). Weigh accurately 1.6 g and dissolve in 50 mL of carbon dioxide-free water. Add 0.15 mL of phenolphthalein indicator solution and titrate with 1 N hydrochloric acid just to a colorless end point (A mL). Add 0.15 mL of bromphenol blue indicator solution and titrate with 0.1 N hydrochloric acid to a yellow end point (B mL). Calculate the assay using (A − 0.1 B) mL. One milliliter of 1 N hydrochloric acid corresponds to 0.04196 g of LiOH · H$_2$O. Calculate the carbonate content using B mL. One milliliter of 0.1 N hydrochloric acid corresponds to 0.007389 g of Li$_2$CO$_3$.

$$\% \text{ Li}_2\text{CO}_3 = [(A - B)\text{mL} \times \text{N HCl} \times 4.196] / \text{Sample wt (g)}$$

INSOLUBLE MATTER. (Page 15). Use 10.0 g in 100 mL of water. Do not boil the sample.

CHLORIDE. (Page 27). Use 0.1 g.

SULFATE. (Page 33, Method 1). Use 0.10 g sample and add hydrochloric acid (1 + 1) until just acid. Then proceed as directed in the general procedure. For the standard, add the same amount of hydrochloric acid (1 + 1) as used for the sample.

HEAVY METALS. (Page 28, Method 1). Use 1.0 g dissolved in 10 mL of water. Neutralize with hydrochloric acid (1 + 1) and dilute to 25 mL.

IRON. (Page 30, Method 1). Use 0.5 g in 10 mL of water, add 3 mL of hydrochloric acid and dilute to 40 mL.

Lithium Metaborate

LiBO$_2$ **Formula Wt 49.75**

CAS Number 13453–69–5

NOTE. The formula weight of this reagent is likely to deviate from the value cited, since the natural distribution of ^{6}Li and ^{7}Li isotopes is often altered in current sources of lithium compounds.

REQUIREMENTS

Assay. . . . . .98.0–102.0% $LiBO_2$
Bulk density. . . . . ≥0.25 g/mL

MAXIMUM ALLOWABLE

Insoluble matter. 0.01%
Loss on fusion at 950 °C. 2.0%
Phosphorus compounds (as PO_4) 0.004%
Silicon (Si) 0.01%
Aluminum (Al) 0.001%
Calcium (Ca) 0.01%
Iron (Fe). 0.002%
Magnesium (Mg) . . . . 5 ppm
Potassium (K). 0.005%
Sodium (Na) 0.005%
Heavy metals (as Pb) 0.001%

TESTS

ASSAY. (By acidimetry). Weigh accurately 1 g, dissolve in 80 mL of water in a 125-mL conical flask, and heat to boiling. Cool to room temperature, add 0.15 mL of 0.1% bromphenol blue indicator solution, and titrate with 1 N hydrochloric acid until the first appearance of yellow.

$$\%LiBO_2 = \frac{V \times N \times 497.5}{W \times (100 - L)}$$

where

$V =$ volume, in mL, of HCl
$N =$ normality of HCl
$W =$ weight, in grams, of $LiBO_2$
$L =$ % loss on fusion at 950 °C

BULK DENSITY. Weigh a dry 50-mL glass-stoppered cylinder, calibrated to contain, then add approximately 50 mL of sample. Stopper and reweigh to the nearest 0.1 g. Grasp the cylinder above its base and from a height of 1 inch lower the base sharply against a No. 12 rubber stopper. Level the surface of the powder with a gentle back and forth motion of the cylinder in mid-air. Record the volume. Repeat two more times, starting with the rotation of the cylinder. Use the mean of the three volume readings to calculate the apparent density.

INSOLUBLE MATTER. (Page 15). Use 10 g dissolved in 400 mL of water.

LOSS ON FUSION AT 950 °C. Heat 1 g, accurately weighed, in a platinum crucible or dish at 950 °C for 15 min. Cool in a desiccator.

PHOSPHORUS COMPOUNDS. Dissolve 5.0 g in 40 mL of dilute nitric acid (1 + 1) and evaporate on a steam bath to a syrupy residue. Dissolve the residue in about 80 mL of water, dilute with water to 100 mL, and dilute 20 mL of this solution with water to 100 mL. For the test dilute 10 mL (0.1-g sample) with water to 70 mL. Prepare a standard containing 0.01 mg of phosphate ion (PO_4) in 70 mL of water. To each solution add 5 mL of ammonium molybdate solution (5 g in 50 mL) and adjust the pH to 1.8, using a pH meter, by adding dilute hydrochloric acid (1 + 1) or dilute ammonium hydroxide (1 + 1). Cautiously heat the solutions to near boiling, but do not boil, and cool to room temperature. If a precipitate forms, it will dissolve when the solution is acidified in the next steps. To each solution add 10 mL of hydrochloric acid and dilute each with water to 100 mL. Transfer the solutions to separatory funnels, add 35 mL of ether to each, shake vigorously, and allow to separate. Draw off and discard the aqueous phases. Wash the ether phases twice with 10-mL portions of dilute hydrochloric acid (1 + 9) and discard the washings each time. To the washed ether phases add 10 mL of dilute hydrochloric acid (1 + 9) to which has been added 0.2 mL of a freshly prepared 2% solution of stannous chloride dihydrate in hydrochloric acid. Shake the solutions and allow the phases to separate. Any blue color in the ether phase from the solution of the sample should not exceed that in the ether phase from the standard.

SILICON. Fuse 1.00 g at 950 °C for 15 min in a 15-mL platinum crucible. Remove the crucible from the furnace, swirl to spread the melt around the sides of the crucible, and cool. Add a small magnetic stirring bar to the crucible, stir while adding 6.0 mL of 3 N hydrochloric acid, and stir until the melt disintegrates. Transfer the slurry with 50 mL of water to a 150-mL beaker. Adjust the pH to 2.0 ± 0.2 with 0.06 N hydrochloric acid (or dilute ammonium hydroxide), using a pH meter. The solution should be clear at the final pH adjustment. Add 2.0 mL of ammonium molybdate solution, mix and let stand for 10 min. Add 4.0 mL of tartaric acid solution, then 1.0 mL of reducing solution. Mix the solution after each addition. Finally, dilute the solution with water to 100 mL. After 30 min the blue color should not exceed that produced by 0.21 mg of silica in an equal volume of solution containing 10.0 mL of 0.06 N hydrochloric acid, 50 mL of water, and the quantities of reagents used in the test. If desired, the absorbances can be measured with a spectrophotometer at 650 nm, using water as a reference.

> ***Ammonium Molybdate Solution.*** Dissolve 7.5 g of ammonium molybdate tetrahydrate, $(NH_4)_6Mo_7O_{24} \cdot 4H_2O$, in 75 mL of water. Add 10.0 mL of 18 N sulfuric acid, dilute with water to 100 mL, filter, and store in a plastic bottle.

> ***Tartaric Acid Solution.*** Dissolve 50 g of tartaric acid, dilute with water to 500 mL, and filter into a plastic bottle.

Reducing Solution. Dissolve 0.7 g of sodium sulfite in 20 mL of water, then dissolve in this solution 0.15 g of 4-amino-3-hydroxy-1-naphthalenesulfonic acid. Add with constant stirring a solution of 9 g of sodium bisulfite in 80 mL of water. Filter the resulting solution into a plastic bottle. Discard after 3 days.

ALUMINUM. Add 25 mL of methanol and 1.8 mL of hydrochloric acid to 1.0 g in a 100-mL plastic beaker. Evaporate to dryness on a hot plate ($\approx$100 °C), add 15 mL of methanol and 0.3 mL of hydrochloric acid, and evaporate again to dryness. Add another 15 mL of methanol and 0.3 mL of hydrochloric acid, repeat the evaporation, and dissolve the residue in 15.0 mL of 0.06 N hydrochloric acid. Transfer the solution of the sample to a 100-mL volumetric flask with 45 mL of water. For the blank and the standard, respectively, add 15.0 mL of 0.06 N hydrochloric acid to two 100-mL volumetric flasks containing 0 and 10 µg of aluminum, then add 45 mL of water. To each of the three flasks add 5.0 mL of lithium chloride solution, 1.0 mL of hydroxylamine hydrochloride reagent solution, and 4.0 mL of ascorbic acid solution. Mix and let stand for 5 min. Add 10.0 mL of buffer solution, mix, and let stand for 10 min. Add 5.0 mL of alizarin red S solution and dilute with water to 100 mL. The final pH should be 4.6 ± 0.1. After 1 h measure the absorbance of the sample and the standard at 500 nm in 5-cm cells with the blank as the reference. The absorbance of the sample should not exceed that of the standard.

Lithium Chloride Solution. Dissolve 61 g of lithium chloride in water and dilute with water to 200 mL. Prepare fresh solution daily.

Ascorbic Acid Solution. Dissolve 5.0 g of ascorbic acid in water and dilute with water to 100 mL. Prepare fresh solution daily.

Sodium Acetate–Acetic Acid Buffer. Dissolve 100 g of sodium acetate trihydrate, $CH_3COONa \cdot 3H_2O$, in water. Add 30 mL of glacial acetic acid and dilute with water to 500 mL. Filter if necessary.

Alizarin Red S Solution. Dissolve 0.250 g of alizarin red S in water, dilute with water to 250 mL, filter, and store in glass.

CALCIUM, IRON, MAGNESIUM, POTASSIUM, AND SODIUM. (By flame AAS, page 39).

Sample Stock Solution and Blank Solution. Weigh 10.0 g of sample into a 2-L beaker, add 400 mL of methanol and 5 mL of hydrochloric acid, and evaporate to dryness. Repeat the evaporation, using 200 mL of methanol and 2 mL of hydrochloric acid, then add 100 mL of methanol and 1 mL of hydrochloric acid, and perform a third evaporation. Prepare a blank in the same manner, using the quantities of reagents used for preparation of the sample. Dissolve each residue in 50 mL of water and 1 mL of hydrochloric acid, transfer to a 100-mL volumetric flask, and dilute to the mark with water (1 mL = 0.1 g).

Element	Wavelength (nm)	Sample Wt (g)	Standard Added (mg)	Flame Type*	Background Correction
Ca	422.7	0.20	0.02; 0.04	N/A	No
Fe	248.3	1.0	0.02; 0.04	A/A	Yes
Mg	285.2	1.0	0.005; 0.01	A/A	Yes
K	766.5	1.0	0.05; 0.10	A/A	No
Na	589.0	0.10	0.005; 0.01	A/A	No

*A/A is air/acetylene; N/A is nitrous oxide/acetylene.

HEAVY METALS. Dilute 20 mL of glacial acetic acid with water to 80 mL, add 5.0 g of sample while stirring, and heat to about 80 °C to dissolve. Prepare a control with 1.0 g of sample, 5.0 mL of glacial acetic acid, and 0.04 mg of lead in a volume of 80 mL. The pH of the solutions should be between 3 and 4. Keep the solutions at 80 °C, add 10 mL of freshly prepared hydrogen sulfide water to each, and mix. Transfer to 100-mL Nessler tubes. Any color in the solution of the sample should not exceed that in the control. (If the solutions are permitted to cool, a white crystalline precipitate forms.)

Lithium Perchlorate

LiClO$_4$ **Formula Wt 106.39**

CAS Number 7791–03–9

> *NOTE.* The formula weight of this reagent is likely to deviate from the value cited, since the natural distribution of ^{6}Li and ^{7}Li isotopes is often altered in current sources of lithium compounds.

REQUIREMENTS

Assay . . . . ≥95.0% LiClO$_4$
pH of a 5% solution . . . . 6.0–7.5 at 25 °C

MAXIMUM ALLOWABLE

Insoluble matter. 0.005%
Chloride (Cl) 0.003%
Sulfate (SO$_4$) 0.001%
Heavy metals (as Pb) . . . . 5 ppm
Iron (Fe) . . . . 5 ppm

TESTS

ASSAY. (By argentimetric titration of chloride after reduction). Weigh accurately about 0.25–0.30 g of sample into a platinum dish or crucible. Add 3 g of sodium carbonate, Na$_2$CO$_3$, and heat gently at first, then to fusion until a clear melt is obtained. Leach with minimal dilute nitric acid and boil gently to expel carbon dioxide. Cool to room temperature, add 10 mL of 30% ammonium acetate solution, and titrate with 0.1 N silver nitrate to the dichlorofluorescein end point. One milliliter of 0.1 N silver nitrate corresponds to 0.01064 g of $LiClO_4$.

$$\% \ LiClO_4 = (mL \times N \ AgNO_3 \times 10.64) \ / \ Sample \ wt \ (g)$$

pH OF A 5% SOLUTION. (Page 44). The pH should be 6.0–7.5 at 25 °C.

INSOLUBLE MATTER. (Page 15). Use 20 g dissolved in 200 mL of water.

CHLORIDE. (Page 27). Use 1.0 g.

SULFATE. Dissolve 40 g in 300 mL of water, add 2 mL of hydrochloric acid, filter, and heat to boiling. Add 5 mL of barium chloride reagent solution, digest in a covered beaker on a hot plate ($\approx$100 °C) for 2 h, and allow to stand overnight. If a precipitate is formed, filter through a fine ashless paper, wash thoroughly, and ignite. Correct for the weight obtained in a complete blank test.

HEAVY METALS. (Page 28, Method 1). Dissolve 6.0 g in about 20 mL of water and dilute with water to 30 mL. Use 25 mL to prepare the sample solution, and use the remaining 5.0 mL to prepare the control solution.

IRON. Dissolve 1.0 g in 20 mL of water. Add 1 mL of hydroxylamine hydrochloride reagent solution, 4 mL of 1,10-phenanthroline reagent solution, and 1 mL of 10% sodium acetate solution, and mix. Any red color should not exceed that produced by 0.005 mg of iron in an equal volume of solution containing the quantities of reagents used in the test. Compare 1 h after adding the reagents to the sample and standard solutions.

Lithium Sulfate Monohydrate

$Li_2SO_4 \cdot H_2O$

Formula Wt 127.96

CAS Number 10102–25–7

NOTE. The formula weight of this reagent is likely to deviate from the value cited, since the natural distribution of ^6Li and ^7Li isotopes is often altered in current sources of lithium compounds.

REQUIREMENTS

Assay (dried basis). . . . . $\geq$99.0% Li_2SO_4
Loss on drying at 150 °C 13.0–15.0%

MAXIMUM ALLOWABLE
Insoluble matter. 0.01%
Chloride (Cl) 0.002%

Nitrate (NO_3) 0.001%
Heavy metals (as Pb) 0.001%
Iron (Fe) 0.001%
Potassium (K) 0.05%
Sodium (Na) 0.05%

TESTS

ASSAY. (By indirect acid–base titrimetry). Weigh, to the nearest 0.1 mg, 0.3 g of previously dried sample (*see* loss on drying test) and dissolve in 50 mL of water. Pass the solution through a cation-exchange column (see page 56) at the rate of about 5 mL/min, and collect the eluate in a 500-mL titration flask. Then wash the resin in the column with water at a rate of about 10 mL/min, and collect in the same titration flask. Add 0.15 mL of phenolphthalein indicator solution to the flask and titrate with 0.1 N sodium hydroxide. Continue the titration, as the washing proceeds, until 50 mL of eluate requires no further titration. One milliliter of 0.1 N sodium hydroxide corresponds to 0.06398 g of $Li_2SO_4 \cdot H_2O$.

$$\% \ Li_2SO_4 \cdot H_2O = (mL \times N \ NaOH \times 63.98) / \text{Sample wt (g)}$$

LOSS ON DRYING AT 150 °C. Weigh accurately about 1 g, and dry at 150 °C to constant weight. (The theoretical loss for $Li_2SO_4 \cdot H_2O$ is 14.08%). Save the dried sample for the assay.

INSOLUBLE MATTER. (Page 15). Use 10 g dissolved in 100 mL of water.

CHLORIDE. (Page 27). Use 0.50 g.

NITRATE.

Sample Solution A. Dissolve 1.0 g in 5 mL of water by heating in a boiling-water bath. Dilute to 50 mL with brucine sulfate reagent solution.

Control Solution B. Dissolve 1.0 g in 4 mL of water and 1 mL of the standard nitrate solution containing 0.01 mg of nitrate ion (NO_3) per mL by heating in a boiling-water bath. Dilute to 50 mL with brucine sulfate reagent solution.

Continue with the procedure described on page 30, starting with the preparation of blank solution C.

HEAVY METALS. (Page 28, Method 1). Dissolve 4.0 g in 40 mL of water. Use 30 mL to prepare the sample solution, and use the remaining 10 mL to prepare the control solution.

IRON. (Page 30, Method 1). Use 1.0 g.

POTASSIUM AND SODIUM. (By flame AAS, page 39).

Sample Stock Solution. Dissolve 1.0 g of sample with water in a 100-mL volumetric flask and dilute to the mark with water (1 mL = 0.01 g).

Element	Wavelength (nm)	Sample Wt (g)	Standard Added (mg)	Flame Type*	Background Correction
K	766.5	0.10	0.05; 0.10	A/A	No
Na	589.0	0.02	0.01; 0.02	A/A	No

*A/A is air/acetylene.

Lithium Tetraborate
(suitable for use in X-ray spectroscopy)

$Li_2B_4O_7$ **Formula Wt 169.12**

CAS Number 12007–60–2

NOTE. The formula weight of this reagent is likely to deviate from the value cited, since the natural distribution of ^6Li and ^7Li isotopes is often altered in current sources of lithium compounds.

REQUIREMENTS

Assay. . . . . 98.0–102.0% $Li_2B_4O_7$
Bulk density. . . . . ≥0.25 g/mL

MAXIMUM ALLOWABLE

Insoluble matter. 0.01%
Loss on fusion at 950 °C. 2.0%
Phosphorus compounds (as PO_4) 0.004%
Silicon (Si) 0.01%
Aluminum (Al) 0.001%
Calcium (Ca) 0.01%
Iron (Fe). 0.001%
Magnesium (Mg) . . . . 5 ppm
Potassium (K). 0.005%
Sodium (Na) 0.005%
Heavy metals (as Pb) 0.001%

TESTS

ASSAY. (By acidimetry). Weigh accurately 1 g, dissolve in 80 mL of water in a 125-mL conical flask, and heat to boiling. Cool to room temperature, add 0.15 mL of 0.1% bromphenol blue indicator solution, and titrate with 1 N hydrochloric acid until the first appearance of yellow.

$$\%Li_2B_4O_7 = \frac{V \times N \times 845.6}{W \times (100 - L)}$$

where

$V =$ volume, in mL, of HCl
$N =$ normality of HCl
$W =$ weight, in grams, of $Li_2B_4O_7$
$L =$ % loss on fusion at 950 °C

BULK DENSITY. Weigh a dry 50-mL glass-stoppered cylinder, calibrated to contain, then add approximately 50 mL of sample. Stopper and reweigh to the nearest 0.1 g. Grasp the cylinder above its base and from a height of 1 inch lower the base sharply against a No. 12 rubber stopper. Level the surface of the powder with a gentle back and forth motion of the cylinder in mid-air. Record the volume. Repeat two more times, starting with the rotation of the cylinder. Use the mean of the three volume readings to calculate the apparent density.

INSOLUBLE MATTER. (Page 15). Use 10 g dissolved in 400 mL of water.

LOSS ON FUSION AT 950 °C. Heat 1 g, accurately weighed, in a platinum crucible or dish at 950 °C for 15 min.

PHOSPHORUS COMPOUNDS. Dissolve 5.0 g in 40 mL of dilute nitric acid (1 + 1) and evaporate on a steam bath to a syrupy residue. Dissolve the residue in about 80 mL of water, dilute with water to 100 mL, and dilute 20 mL of this solution with water to 100 mL. For the test dilute 10 mL (0.1-g sample) with water to 70 mL. Prepare a standard containing 0.01 mg of phosphate ion (PO_4) in 70 mL of water. To each solution add 5 mL of ammonium molybdate solution (5 g in 50 mL) and adjust the pH to 1.8, using a pH meter, by adding dilute hydrochloric acid (1 + 1) or dilute ammonium hydroxide (1 + 1). Cautiously heat the solutions to near boiling, but do not boil, then cool to room temperature. If a precipitate forms, it will dissolve when the solution is acidified in the next steps. To each solution add 10 mL of hydrochloric acid and dilute each with water to 100 mL. Transfer the solutions to separatory funnels, add 35 mL of ether to each, shake vigorously, and allow to separate. Draw off and discard the aqueous phases. Wash the ether phases twice with 10-mL portions of dilute hydrochloric acid (1 + 9) and discard the washings each time. To the washed ether phases add 10 mL of dilute hydrochloric acid (1 + 9) to which has been added 0.2 mL of a freshly prepared 2% solution of stannous chloride dihydrate in hydrochloric acid. Shake the solutions and allow the phases to separate. Any blue color in the ether phase from the solution of the sample should not exceed that in the ether phase from the standard.

SILICON. Fuse 1.00 g at 950 °C for 15 min in a 15-mL platinum crucible. Remove the crucible from the furnace, swirl to spread the melt around the sides of the crucible, and cool. Add a small magnetic stirring bar to the crucible, stir while adding 6.0 mL of 3 N hydrochloric acid, and stir until the melt disintegrates. Transfer the slurry with 50 mL of water to a 150-mL beaker. Adjust the pH to 2.0 ± 0.2 with 0.06 N hydrochloric acid (or dilute ammonium hydroxide) using a pH meter. The solution should be clear at the final pH adjustment. Add 2.0 mL of ammonium molybdate solution, mix and let stand for 10 min. Add 4.0 mL of tartaric acid solution, then 1.0 mL of reducing solution. Mix the solution after each addition. Finally, dilute the solution with water to 100 mL. After 30 min the blue color should not exceed that produced by 0.21 mg of silica in an equal volume of solution containing 10.0 mL of 0.06 N hydrochloric acid, 50 mL of water, and the quantities of reagents used in the test. If desired, the absorbances can be measured with a spectrophotometer at 650 nm, using water as a reference.

Ammonium Molybdate Solution. Dissolve 7.5 g of ammonium molybdate tetrahydrate, $(NH_4)_6Mo_7O_{24} \cdot 4H_2O$, in 75 mL of water. Add 10.0 mL of 18 N sulfuric acid, dilute with water to 100 mL, filter, and store in a plastic bottle.

Tartaric Acid Solution. Dissolve 50 g of tartaric acid in about 400 mL of water, dilute with water to 500 mL, and filter into a plastic bottle.

Reducing Solution. Dissolve 0.7 g of sodium sulfite in 20 mL of water, then dissolve into this solution 0.15 g of 4-amino-3-hydroxy-1-naphthalenesulfonic acid. Add with constant stirring a solution of 9 g of sodium bisulfite in 80 mL of water. Filter the resulting solution into a plastic bottle. Discard after 3 days.

ALUMINUM. Add 25 mL of methanol and 1.8 mL of hydrochloric acid to 1.0 g in a 100-mL plastic beaker. Evaporate to dryness on a hot plate ($\approx$100 °C), add 15 mL of methanol and 0.3 mL of hydrochloric acid, and evaporate again to dryness. Add another 15 mL of methanol and 0.3 mL of hydrochloric acid, repeat the evaporation, and dissolve the residue in 15.0 mL of 0.06 N hydrochloric acid. Transfer the solution of the sample to a 100-mL volumetric flask with 45 mL of water. For the blank and the standard, respectively, add 15.0 mL of 0.06 N hydrochloric acid to two 100-mL volumetric flasks containing 0 and 10 µg of aluminum, then add 45 mL of water. To each of the three flasks add 5.0 g of lithium chloride solution, 1.0 mL of hydroxylamine hydrochloride reagent solution, and 4.0 mL of ascorbic acid solution. Mix and let stand for 5 min. Add 10.0 mL of buffer solution, mix, and let stand for 10 min. Add 5.0 mL of alizarin red S solution and dilute with water to 100 mL. The final pH should be 4.6 ± 0.1. After 1 h measure the absorbance of the sample and the standard at 500 nm in 5-cm cells with the blank as the reference. The absorbance of the sample should not exceed that of the standard.

Lithium Chloride Solution. Dissolve 61 g of lithium chloride in water and dilute with water to 200 mL. Prepare fresh solution daily.

Ascorbic Acid Solution. Dissolve 5.0 g of ascorbic acid in water and dilute with water to 100 mL. Prepare fresh solution daily.

Sodium Acetate–Acetic Acid Buffer. Dissolve 100 g of sodium acetate trihydrate, $CH_3COONa \cdot 3H_2O$, in water. Add 30 mL of glacial acetic acid and dilute with water to 500 mL. Filter if necessary.

Alizarin Red S Solution. Dissolve 0.250 g of alizarin red S in water, dilute with water to 250 mL, filter, and store in glass.

CALCIUM, IRON, MAGNESIUM, POTASSIUM, AND SODIUM. (By   flame AAS, page 39).

Sample Stock Solution and Blank Solution. Weigh 10 g of sample into a 2-L beaker, add 400 mL of methanol and 5 mL of hydrochloric acid, and evaporate to dryness. Repeat the evaporation using 200 mL of methanol and 2 mL of hydrochloric acid, then add 100 mL of methanol and 1 mL of hydrochloric acid, and perform a third evaporation. Prepare a blank in the same manner, using the quantities of reagent used for preparation of the sample. Dissolve each residue in 50 mL of water and 1 mL of hydrochloric acid, transfer to a 100-mL volumetric flask, and dilute to the mark with water (1 mL = 0.1 g).

Element	Wavelength (nm)	Sample Wt (g)	Standard Added (mg)	Flame Type*	Background Correction
Ca	422.7	0.20	0.02; 0.04	N/A	No
Fe	248.3	1.0	0.01; 0.02	A/A	Yes
Mg	285.2	1.0	0.005; 0.01	A/A	Yes
K	766.5	1.0	0.05; 0.10	A/A	No
Na	589.0	0.10	0.005; 0.01	A/A	No

*A/A is air/acetylene; N/A is nitrous oxide/acetylene.

HEAVY METALS. Dilute 20 mL of glacial acetic acid with water to 80 mL, add 5.0 g of sample while stirring, and heat to about 80 °C to dissolve. Prepare a control with 1.0 g of sample, 5.0 mL of glacial acetic acid, and 0.04 mg of lead in a volume of 80 mL. The pH of the solutions should be between 3 and 4. Keep the solutions at 80 °C, add 10 mL of freshly prepared hydrogen sulfide water to each, and mix. Transfer to 100-mL Nessler tubes. Any color in the solution of the sample should not exceed that in the control. (If the solutions are permitted to cool, a white crystalline precipitate forms.)

Litmus Paper

NOTE. When litmus paper is used in a solution, the length taken of a 0.6-cm-wide strip should not exceed 0.5 cm.

REQUIREMENTS

MAXIMUM ALLOWABLE

Ash. . . . . .0.4 mg per strip (about 3 sq cm)
Phosphate (PO_4) . . . . Passes test
Rosin acids (for blue paper only) . . . . Passes test
Sensitivity. . . . . Passes test

TESTS

ASH. Carefully ignite 10 strips in a tared crucible, and finally ignite at 800 ± 25 °C for 15 min.

PHOSPHATE. (Page 32, Method 1). Cut five strips into small pieces, mix with 0.5 g of magnesium nitrate in a porcelain crucible, and ignite. To the ignited residue add 5 mL of nitric acid and evaporate to dryness. Dissolve in 25 mL of approximately 0.5 N sulfuric acid, and continue as described.

ROSIN ACIDS. Immerse a strip of blue paper in a solution of 0.10 g of silver nitrate in 50 mL of water. The color of the paper should not change in 30 s.

SENSITIVITY. Drop the pieces from six strips of paper into 100 mL of test solution in a beaker and stir continuously. Blue paper should change color in 45 s in 0.0005 N acid. Red paper should change color in 30 s in 0.0005 N alkali.

Magnesium Acetate Tetrahydrate

$(CH_3COO)_2Mg \cdot 4H_2O$ **Formula Wt 214.46**

CAS Number 16674–78–5

REQUIREMENTS

Assay . . . . 98.0–102.0% $(CH_3CO_2)_2Mg \cdot 4H_2O$

MAXIMUM ALLOWABLE

Insoluble matter.	0.005%
Chloride (Cl)	0.001%
Nitrogen compounds (as N).	0.001%
Sulfate (SO_4)	0.005%
Barium (Ba)	0.001%
Calcium (Ca)	0.01%
Manganese (Mn)	0.001%
Potassium (K).	0.005%
Sodium (Na)	0.005%
Strontium (Sr)	0.005%
Heavy metals (as Pb)	5 ppm
Iron (Fe).	5 ppm

TESTS

ASSAY. (By complexometric titration of Mg). Weigh accurately 0.85 g and transfer to a 250-mL beaker with 50 mL of water. Add 5 mL of pH 10 ammoniacal buffer solution and about 50 mg of Eriochrome Black T indicator mixture. Titrate with standard 0.1 M EDTA to a clear blue color (and disappearance of the last trace of red). One milliliter of 0.1 M EDTA corresponds to 0.02145 g of $(CH_3COO)_2Mg \cdot 4H_2O$.

$$\% \ (CH_3COO)_2Mg \cdot 4H_2O = (mL \times M \ EDTA \times 21.45) \ / \ Sample \ wt \ (g)$$

INSOLUBLE MATTER. (Page 15). Use 20 g dissolved in 200 mL of dilute acetic acid (1 + 99). Retain the filtrate, without washings, for the test for sulfate.

> **Sample Solution A for the Determination of Chloride, Nitrogen Compounds, and Iron.** Dissolve 10 g in water, filter if necessary through a chloride-free filter, and dilute with water to 100 mL (1 mL = 0.10 g).

CHLORIDE. (Page 27). Use 10 mL of sample solution A (1-g sample).

NITROGEN COMPOUNDS. (Page 31). Use 10 mL of sample solution A (1-g sample). For the standard, use 0.01 mg of nitrogen (N).

SULFATE. (Page 33, Method 1). Use 1/20 of the filtrate from the test for insoluble matter. Allow 30 min for the turbidity to form.

BARIUM. Dissolve 6.0 g in 25 mL of dilute nitric acid (1 + 1). For the control, dissolve 1.0 g in 25 mL of dilute nitric acid (1 + 1) and add 0.05 mg of barium. Evaporate each solution on a hot plate ($\approx$100 °C) to a syrupy consistency and add 20 mL of water. If the solutions are not clear, add 10% nitric acid until clear. Add 1 mL of 1 N acetic acid and 2 mL of potassium dichromate reagent solution to each, and adjust the pH to 7 (using a pH meter) with ammonium hydroxide (10% NH_3). Add 25 mL of methanol to each and stir vigorously. Any turbidity in the solution of the sample should not exceed that in the control.

CALCIUM, MANGANESE, POTASSIUM, SODIUM, AND STRONTIUM. (By flame AAS, page 39).

> **Sample Stock Solution.** Dissolve 10.0 g of sample in 75 mL of water, transfer to a 100-mL volumetric flask, and dilute to the mark with water (1 mL = 0.10 g).

Element	Wavelength (nm)	Sample Wt (g)	Standard Added (mg)	Flame Type*	Background Correction
Ca	422.7	0.20	0.02; 0.04	N/A	No
Mn	279.5	2.0	0.02; 0.04	A/A	Yes
K	766.5	1.0	0.05; 0.10	A/A	No
Na	589.0	0.20	0.01; 0.02	A/A	No
Sr	460.7	1.0	0.05; 0.10	N/A	No

*A/A is air/acetylene; N/A is nitrous oxide/acetylene.

HEAVY METALS. (Page 28, Method 1). Dissolve 6.0 g in 20 mL of water, and dilute with water to 30 mL. Use 25 mL to prepare the sample solution, and use the remaining 5.0 mL to prepare the control solution.

IRON. (Page 30, Method 1). Use 20 mL of sample solution A (2.0 g sample).

Magnesium Chloride Hexahydrate

$MgCl_2 \cdot 6H_2O$

Formula Wt 203.30

CAS Number 7791–18–6

REQUIREMENTS

Assay . . . . 99.0–102.0% $MgCl_2 \cdot 6H_2O$

MAXIMUM ALLOWABLE

Insoluble matter.	0.005%
Nitrate (NO_3).	0.001%
Phosphate (PO_4)	5 ppm
Sulfate (SO_4)	0.002%
Ammonium (NH_4)	0.002%
Barium (Ba)	0.005%
Calcium (Ca)	0.01%
Manganese (Mn)	5 ppm
Potassium (K).	0.005%
Sodium (Na)	0.005%
Strontium (Sr)	0.005%
Heavy metals (as Pb)	5 ppm
Iron (Fe).	5 ppm

TESTS

ASSAY. (By complexometric titration of Mg). Weigh accurately 0.8 g and transfer to a 250-mL beaker with 50 mL of water. Add 5 mL of pH 10 ammoniacal buffer solution and about 50 mg of Eriochrome Black T indicator mixture. Titrate with standard 0.1 M EDTA to a clear blue color (and disappearance of the last trace of red). One milliliter of 0.1 M EDTA corresponds to 0.02033 g of $MgCl_2 \cdot 6H_2O$.

$$\% \ MgCl_2 \cdot 6H_2O = (mL \times M \ EDTA \times 20.33) \ / \ Sample \ wt \ (g)$$

INSOLUBLE MATTER. (Page 15). Use 20 g dissolved in 200 mL of water.

NITRATE.

Sample Solution A. Dissolve 1.0 g in 3 mL of water by heating in a boiling-water bath. Dilute to 50 mL with brucine sulfate reagent solution.

Control Solution B. Dissolve 1.0 g in 2 mL of water and 1 mL of the standard nitrate solution containing 0.01 mg of nitrate ion (NO_3) per mL by heating in a boiling-water bath. Dilute to 50 mL with brucine sulfate reagent solution.

Continue with the procedure described on page 30, starting with the preparation of blank solution C.

PHOSPHATE. (Page 32, Method 1). Dissolve 4.0 g in 25 mL of approximately 0.5 N sulfuric acid, and continue as described.

SULFATE. (Page 33, Method 1). Use 2.5 g of sample.

AMMONIUM. Dissolve 1.0 g in 90 mL of water and add 10 mL of freshly boiled 10% sodium hydroxide reagent solution. Allow to settle, decant 50 mL and add 2 mL of Nessler reagent. Any color should not exceed that produced by 0.01 mg of ammonium ion (NH_4) in an equal volume of solution containing 5 mL of 10% sodium hydroxide solution and 2 mL of Nessler reagent.

BARIUM. Dissolve 1.5 g in 20 mL of water. For the control, dissolve 0.5 g in 15 mL of water, add 0.05 mg of barium ion (Ba), and dilute with water to 20 mL. To each solution add 1 mL of 1 N acetic acid and 2 mL of potassium dichromate reagent solution. Adjust the pH of the control and sample solutions to 7 (using a pH meter) with ammonium hydroxide (10% NH_3), add 25 mL of methanol, and stir vigorously. Any turbidity in the solution of the sample should not exceed that in the control.

CALCIUM, MANGANESE, POTASSIUM, SODIUM, AND STRONTIUM. (By flame AAS, page 39).

> **Sample Stock Solution.** Dissolve 10.0 g of sample with water in a 100-mL volumetric flask and dilute to the mark with water (1 mL = 0.10 g).

Element	Wavelength (nm)	Sample Wt (g)	Standard Added (mg)	Flame Type*	Background Correction
Ca	422.7	0.20	0.02; 0.04	N/A	No
Mn	279.5	2.0	0.01; 0.02	A/A	Yes
K	766.5	1.0	0.05; 0.10	A/A	No
Na	589.0	0.20	0.01; 0.20	A/A	No
Sr	460.7	1.0	0.05; 0.10	N/A	No

*A/A is air/acetylene; N/A is nitrous oxide/acetylene.

HEAVY METALS. (Page 28, Method 1). Dissolve 6.0 g in 20 mL of water, and dilute with water to 30 mL. Use 25 mL to prepare the sample solution, and use the remaining 5.0 mL to prepare the control solution.

IRON. (Page 30, Method 1). Use 2.0 g.

Magnesium Nitrate Hexahydrate

$Mg(NO_3)_2 \cdot 6H_2O$ **Formula Wt 256.41**

CAS Number 13446–18–9

REQUIREMENTS

Assay . . . . 98.0–102.0% $Mg(NO_3)_2 \cdot 6H_2O$
pH of a 5% solution . . . . 5.0–8.2 at 25 °C

MAXIMUM ALLOWABLE

Insoluble matter. 0.005%
Chloride (Cl) 0.001%
Phosphate (PO_4) . . . . 5 ppm
Sulfate (SO_4) 0.005%
Ammonium (NH_4) 0.003%
Barium (Ba) 0.005%
Calcium (Ca) 0.01%
Manganese (Mn) . . . . 5 ppm
Potassium (K). 0.005%
Sodium (Na) 0.005%
Strontium (Sr) 0.005%
Heavy metals (as Pb) . . . . 5 ppm
Iron (Fe). . . . . 5 ppm

TESTS

ASSAY. (By complexometric titration of Mg). Weigh accurately 1 g and transfer to a 250-mL beaker with 50 mL of water. Add 5 mL of pH 10 ammoniacal buffer solution and about 50 mg of Eriochrome Black T indicator mixture. Titrate with standard 0.1 M EDTA to a clear blue color (and disappearance of the last trace of red). One milliliter of 0.1 M EDTA corresponds to 0.02564 g of $Mg(NO_3)_2 \cdot 6H_2O$.

$$\% \ Mg(NO_3)_2 \cdot 6H_2O = (mL \times M \ EDTA \times 25.64) \ / \ Sample \ wt \ (g)$$

pH OF A 5% SOLUTION. (Page 44). The pH should be 5.0–8.2 at 25 °C.

INSOLUBLE MATTER. (Page 15). Use 20 g dissolved in 200 mL of water.

CHLORIDE. (Page 27). Use 1.0 g.

PHOSPHATE. (Page 32, Method 1). Dissolve 4.0 g in 25 mL of approximately 0.5 N sulfuric acid, and continue as described.

SULFATE. (Page 33, Method 2). Use 10 mL of dilute hydrochloric acid (1 + 1), do two evaporations, and allow 30 min for the turbidity to form.

AMMONIUM. Dissolve 1.0 g in 100 mL of water, add 10 mL of freshly boiled 10% sodium hydroxide reagent solution, and dilute with water to 150 mL. Allow the precipitate to settle. Filter through sintered-glass or porous porcelain filter. To 50 mL of the filtrate, add 2 mL of Nessler reagent solution. Any color should not exceed that produced by 0.01 mg of ammonium ion (NH_4) in an equal volume of solution containing the quantities of reagents used in the test.

BARIUM. Dissolve 1.5 g in 20 mL of water. For the control, dissolve 0.5 g in 15 mL of water, add 0.05 mg of barium ion (Ba), and dilute with water to 20 mL. To each solution add 1 mL of 1 N acetic acid and 2 mL of potassium dichromate reagent solution. Adjust the pH of the control and sample solutions to 7 (using a pH meter) with ammonium hydroxide (10% NH_3), add 25 mL of methanol, and stir vigorously. Any turbidity in the solution of the sample should not exceed that in the control.

CALCIUM, MANGANESE, POTASSIUM, SODIUM, AND STRONTIUM. (By flame AAS, page 39).

Sample Stock Solution. Dissolve 10.0 g of sample with water in a 100-mL volumetric flask and dilute to the mark with water (1 mL = 0.10 g).

Element	Wavelength (nm)	Sample Wt (g)	Standard Added (mg)	Flame Type*	Background Correction
Ca	422.7	0.20	0.02; 0.04	N/A	No
Mn	279.5	2.0	0.01; 0.02	A/A	Yes
K	766.5	1.0	0.05; 0.10	A/A	No
Na	589.0	0.20	0.01; 0.02	A/A	No
Sr	460.7	1.0	0.05; 0.10	A/A	No

*A/A is air/acetylene; N/A is nitrous oxide/acetylene.

HEAVY METALS. (Page 28, Method 1). Dissolve 6.0 g in 20 mL of water, and dilute with water to 30 mL. Use 25 mL to prepare the sample solution, and use the remaining 5.0 mL to prepare the control solution.

IRON. (Page 30, Method 1). Use 2.0 g.

Magnesium Oxide

MgO   **Formula Wt 40.30**

CAS Number 1039–48–4

REQUIREMENTS

Assay (dried basis) . . . . ≥95.0% MgO

MAXIMUM ALLOWABLE

Insoluble in dilute hydrochloric acid. 0.02%
Water-soluble substances. 0.4%
Loss on ignition 2.0%
Ammonium hydroxide precipitate 0.02%
Chloride (Cl) 0.01%
Nitrate (NO_3) 0.005%
Sulfate and sulfite (as SO_4) 0.02%
Barium (Ba) 0.005%

Calcium (Ca) 0.05%
Manganese (Mn) . . . . 5 ppm
Potassium (K). 0.005%
Sodium (Na) 0.5%
Strontium (Sr) 0.005%
Heavy metals (as Pb) 0.003%
Iron (Fe). 0.01%

TESTS

ASSAY. (By complexometric titration of Mg). Weigh accurately about 0.5 g of the sample, previously ignited at 800 ± 25 °C. Dissolve cautiously in water and a slight excess of hydrochloric acid, transfer to a 250-mL volumetric flask, dilute to the mark with water, and mix thoroughly. Pipet 50.0 mL of this solution into a 250-mL beaker, add 50 mL of water and 0.10 mL of methyl red indicator solution, and neutralize with 10% sodium hydroxide. Add 15 mL of pH 10 buffer, about 0.1 g of ascorbic acid, and a few milligrams of Eriochrome Black T indicator. Titrate with 0.1 M EDTA to a blue end point. One milliliter of 0.1 M EDTA corresponds to 0.004030 g of MgO.

$$\% \ MgO = \frac{mL \times M \ EDTA \times 4.03}{[Sample \ wt \ (g)]/5}$$

INSOLUBLE IN DILUTE HYDROCHLORIC ACID. Dissolve 5.0 g in 125 mL of dilute hydrochloric acid (1 + 4), heat to boiling, and boil gently for 5 min. Digest the solution in a covered beaker on a hot plate ($\approx$100 °C) for 1 h. Filter through a tared filtering crucible, wash thoroughly, and dry at 105 °C. Retain the filtrate for the determination of ammonium hydroxide precipitate.

WATER-SOLUBLE SUBSTANCES. Suspend 2.0 g in 50 mL of water, heat to boiling, and filter while hot. Evaporate 25 mL of the clear filtrate to dryness in a tared dish. Moisten the residue with 0.10 mL of sulfuric acid, ignite gently to remove the excess acid, and finally ignite at 800 ± 25 °C for 15 min.

LOSS ON IGNITION. Weigh accurately about 0.25 g in a tared covered platinum crucible. Ignite at 800 ± 25 °C for 15 min. Cool and reweigh.

AMMONIUM HYDROXIDE PRECIPITATE. To the filtrate from the determination of insoluble in dilute hydrochloric acid add ammonium hydroxide until neutral to methyl red and add an excess of 1 mL. Boil the solution for 5 min, allow to cool, and filter. Dissolve the precipitate on the filter with 5 mL of hot dilute hydrochloric acid (1 + 1) and wash to a volume of 25 mL. Add ammonium hydroxide until neutral to methyl red, add an excess of 0.5 mL, and boil for 5 min. Filter through the same filter, wash with hot water, and ignite.

CHLORIDE. (Page 27). Use 0.10 g dissolved in 10 mL of dilute nitric acid (1 + 9).

NITRATE.

Sample Solution A. Suspend 0.20 g in 3 mL of water. Dilute to 50 mL with brucine sulfate reagent solution.

Control Solution B. Suspend 0.20 g in 2 mL of water and 1 mL of the standard nitrate solution containing 0.01 mg of nitrate ion (NO_3) per milliliter. Dilute to 50 mL with brucine sulfate reagent solution.

Blank Solution C. Use 50 mL of brucine sulfate reagent solution.

Heat sample solution A, control solution B, and blank solution C in a preheated (boiling) water bath until the magnesium oxide is dissolved in A and B, then heat for 10 min more. Continue with the procedure described on page 30.

SULFATE AND SULFITE. Dissolve 0.50 g of sample in $(1 + 1)$ hydrochloric acid and evaporate to dryness on a hot plate ($\approx 100\ °C$). Cool, dissolve the residue in 5 mL of water, and evaporate to dryness. Cool and dissolve the residue in 20 mL of water. Filter through a small, washed filter paper. Add 2 3-mL portions of water through the filter paper, and dilute with water to 30 mL. For control, take 0.10 mg of sulfate ion (SO_4) in 30 mL of water. To both, add 1 mL of $(1 + 19)$ hydrochloric acid and 1 mL of barium chloride solution. After 30 min, sample turbidity should not exceed that of the control solution.

BARIUM. Dissolve 1.5 g in 15 mL of nitric acid. For the control, dissolve 0.5 g in 15 mL of nitric acid and add 0.05 mg of barium. Evaporate each solution on a hot plate ($\approx 100\ °C$) to a syrupy consistency and add 20 mL of water. If the solutions are not clear, add 10% nitric acid dropwise until clear. Add 1 mL of 1 N acetic acid and 2 mL of potassium dichromate reagent solution to each, and adjust the pH to 7 (using a pH meter) with ammonium hydroxide (10% NH_3). Add 25 mL of methanol to each and stir vigorously. Any turbidity in the solution of the sample should not exceed that in the control.

CALCIUM, MANGANESE, POTASSIUM, SODIUM, AND STRONTIUM. (By flame AAS, page 39).

Sample Stock Solution. Dissolve 10.0 g of sample with dilute nitric acid $(1 + 4)$, and digest in a covered beaker on a steam bath for 20 min. Cool, transfer to a 100-mL volumetric flask, and dilute to the mark with water (1 mL = 0.10 g).

Element	Wavelength (nm)	Sample Wt (g)	Standard Added (mg)	Flame Type*	Background Correction
Ca	422.7	0.20	0.05; 0.10	N/A	No
Mn	279.5	2.0	0.01; 0.02	A/A	Yes
K	766.5	1.0	0.05; 0.10	A/A	No
Na	589.0	0.002	0.01; 0.02	A/A	No
Sr	460.7	1.0	0.05; 0.10	N/A	No

*A/A is air/acetylene; N/A is nitrous oxide/acetylene.

HEAVY METALS. (Page 28, Method 1). Dissolve 5.0 g in 50 mL of dilute hydrochloric acid (1 + 1), heat if necessary to obtain complete dissolution, and dilute with water to 150 mL. Evaporate 30 mL to dryness on a hot plate (≈100 °C), dissolve the residue in 20 mL of water, and dilute with water to 25 mL. For the control, add 0.02 mg of lead to 10 mL of the sample solution, evaporate to dryness on a hot plate (≈100 °C), dissolve the residue in 20 mL of water, and dilute with water to 25 mL.

IRON. (Page 30, Method 1). Add 1.0 g to 50 mL of dilute hydrochloric acid (1 + 1), and boil gently for 5 min. Cool, dilute with water to 50 mL, and use 5 mL of this solution without further acidification.

Magnesium Perchlorate, Desiccant

$Mg(ClO_4)_2$ **Formula Wt 223.21**

CAS Number 10034–81–8

REQUIREMENTS

Suitability for moisture absorption . . . . Passes test

MAXIMUM ALLOWABLE

Titrable free acid . . . . 0.005 meq/g
Titrable base . . . . 0.025 meq/g
Loss on drying at 190 °C 8%

TESTS

SUITABILITY FOR MOISTURE ABSORPTION. Weigh about 2 g in a tared, 50-mL weighing bottle, having a diameter of 50 mm and a height of 30 mm. Place the bottle, with cover removed, for 16 h or overnight in a closed container in which the atmosphere possesses a relative humidity of 85%, maintained by equilibrium with sulfuric acid having a specific gravity of 1.16. The increase in weight should not be less than 25%.

TITRABLE FREE ACID AND TITRABLE BASE. Dissolve 2.0 g in 25 mL of water and add 0.15 mL of methyl orange indicator solution. Any red color produced should be changed to yellow by the addition of not more than 1.0 mL of 0.01 N sodium hydroxide (titrable free acid). If a yellow color is produced, not more than 5.0 mL of 0.01 N hydrochloric acid should be required to produce a red color (titrable base).

LOSS ON DRYING AT 190 °C.

CAUTION. Do not dry the sample in the presence of any organic material.

Weigh accurately about 1 g and dry at 190 ± 10 °C to constant weight. Calculate the loss in weight as percent water.

Magnesium Sulfate Heptahydrate

$MgSO_4 \cdot 7H_2O$ Formula Wt 246.48

CAS Number 10034–99–8

REQUIREMENTS

Assay . . . . 98.0–102.0% $MgSO_4 \cdot 7H_2O$
pH of a 5% solution . . . . 5.0–8.2 at 25 °C

MAXIMUM ALLOWABLE

Insoluble matter. 0.005%
Chloride (Cl) . . . . 5 ppm
Nitrate (NO_3) 0.002%
Ammonium (NH_4) 0.002%
Calcium (Ca) 0.02%
Manganese (Mn) . . . . 5 ppm
Potassium (K). 0.005%
Sodium (Na). 0.005%
Strontium (Sr) 0.005%
Heavy metals (as Pb) . . . . 5 ppm
Iron (Fe). . . . . 5 ppm

TESTS

ASSAY. (By complexometric titration of Mg). Weigh accurately 1 g and transfer to a 250-mL beaker with 50 mL of water. Add 5 mL of pH 10 ammoniacal buffer solution and about 50 mg of Eriochrome Black T indicator mixture. Titrate with standard 0.1 M EDTA to a clear blue color (and disappearance of the last trace of red). One milliliter of 0.1 M EDTA corresponds to 0.02465 g of $MgSO_4 \cdot 7H_2O$.

$$\% \ MgSO_4 \cdot 7H_2O = (mL \times M \ EDTA \times 24.65) \ / \ Sample \ wt \ (g)$$

pH OF A 5% SOLUTION. (Page 44). The pH should be 5.0–8.2 at 25 °C.

INSOLUBLE MATTER. (Page 15). Use 20 g dissolved in 200 mL of water.

CHLORIDE. (Page 27). Use 2.0 g.

NITRATE.

Sample Solution A. Dissolve 1.0 g in 5 mL of water by heating in a water bath. Dilute to 50 mL with brucine sulfate reagent solution.

Control Solution B. Dissolve 1.0 g in 3 mL of water and 2 mL of the standard nitrate solution containing 0.01 mg of nitrate ion (NO_3) per mL by heating in a boiling-water bath. Dilute to 50 mL with brucine sulfate reagent solution.

Continue with the procedure described on page 30, starting with the preparation of blank solution C.

AMMONIUM. Dissolve 2.0 g in 90 mL of water, add 10 mL of 10% sodium hydroxide reagent solution and allow to settle. Draw off 25 mL of the clear solution, dilute with water to 50 mL, and add 2 mL of Nessler reagent. Any color should not exceed that produced by 0.01 mg of ammonium ion (NH_4) in an equal volume of solution containing 2.5 mL of 10% sodium hydroxide solution and 2 mL of Nessler reagent.

CALCIUM, MANGANESE, POTASSIUM, SODIUM, AND STRONTIUM. (By flame AAS, page 39).

Sample Stock Solution. Dissolve 10.0 g of sample with water in a 100-mL volumetric flask and dilute to the mark with water (1 mL = 0.10 g).

Element	Wavelength (nm)	Sample Wt (g)	Standard Added (mg)	Flame Type*	Background Correction
Ca	422.7	0.20	0.04; 0.08	N/A	No
Mn	279.5	2.0	0.01; 0.02	A/A	Yes
K	766.5	0.50	0.025; 0.05	A/A	No
Na	589.0	0.20	0.01; 0.02	A/A	No
Sr	460.7	2.0	0.10; 0.20	N/A	No

*A/A is air/acetylene; N/A is nitrous oxide/acetylene.

HEAVY METALS. (Page 28, Method 1). Dissolve 6.0 g in about 20 mL of water, and dilute with water to 30 mL. Use 25 mL to prepare the sample solution, and use the remaining 5.0 mL to prepare the control solution.

IRON. (Page 30, Method 1). Use 2.0 g.

Manganese Chloride Tetrahydrate
Manganese(II) Chloride Tetrahydrate
Manganous Chloride Tetrahydrate

$MnCl_2 \cdot 4H_2O$ **Formula Wt 197.90**

CAS Number 13446–34–9

REQUIREMENTS

Assay . . . . 98.0–101.0% $MnCl_2 \cdot 4H_2O$
pH of a 5% solution . . . . 3.5–6.0 at 25 °C

MAXIMUM ALLOWABLE

Insoluble matter. 0.005%
Sulfate (SO_4) 0.005%
Heavy metals (as Pb) . . . . 5 ppm
Iron (Fe). . . . . 5 ppm
Calcium (Ca) 0.005%
Magnesium (Mg) 0.005%
Potassium (K). 0.01%
Sodium (Na). 0.05%
Zinc (Zn) 0.005%

TESTS

ASSAY. (By complexometric titration of Mn). Weigh accurately 0.8 g, transfer to a 500-mL beaker, and dissolve in 200 mL of water. Add a few milligrams of ascorbic acid to prevent oxidation. From a buret, add about 30 mL of standard 0.1 M EDTA. Now add 10 mL of pH 10 ammoniacal buffer solution and about 50 mg of Eriochrome Black T indicator mixture. Continue the titration with EDTA to a clear blue color (and disappearance of the last trace of red). One milliliter of 0.1 M EDTA corresponds to 0.01979 g of $MnCl_2 \cdot 4H_2O$.

$$\% \ MnCl_2 \cdot 4H_2O = (mL \times M \ EDTA \times 19.79) \ / \ Sample \ wt \ (g)$$

pH OF A 5% SOLUTION. (Page 44). The pH should be 3.5–6.0 at 25 °C.

INSOLUBLE MATTER. (Page 15). Use 20 g dissolved in 150 mL of water.

SULFATE. Dissolve 10 g in 100 mL of water, add 1 mL of hydrochloric acid, filter, and heat the filtrate to boiling. Add 10 mL of barium chloride reagent solution, digest in a covered beaker on a hot plate ($\approx$100 °C) for 2 h, and allow to stand overnight. If a precipitate is formed, filter through a fine ashless paper, wash thoroughly, and ignite. Correct for the weight obtained in a complete blank test.

HEAVY METALS. (Page 28, Method 1). Dissolve 6.0 g in about 20 mL of water, and dilute with water to 30 mL. Use 25 mL to prepare the sample solution, and use the remaining 5.0 mL to prepare the control solution. Adjust the pH with ammonium acetate solution.

IRON. (Page 30, Method 1). Use 2.0 g.

CALCIUM, MAGNESIUM, POTASSIUM, SODIUM, AND ZINC. (By   flame AAS, page 39).

> *Sample Stock Solution.* Dissolve 5.0 g in water in a 100-mL volumetric flask, add 0.5 mL of hydrochloric acid, and dilute with water to the mark (1 mL = 0.05 g).

Element	Wavelength (nm)	Sample Wt (g)	Standard Added (mg)	Flame Type*	Background Correction
Ca	422.7	0.80	0.02; 0.04	N/A	No
Mg	285.2	0.20	0.005; 0.01	A/A	Yes
K	766.5	0.20	0.01; 0.02	A/A	No
Na	589.0	0.04	0.01; 0.02	A/A	No
Zn	213.9	0.20	0.01; 0.02	A/A	Yes

*A/A is air acetylene; N/A is nitrous oxide/acetylene.

Manganese Sulfate Monohydrate
Manganese(II) Sulfate Monohydrate
Manganous Sulfate Monohydrate

$MnSO_4 \cdot H_2O$ Formula Wt 169.02

CAS Number 10034–96–5

REQUIREMENTS

Assay . . . . 98.0–101.0% $MnSO_4 \cdot H_2O$
Loss on ignition 10.0–12.0%
Substances reducing permanganate. . . . . Passes test

MAXIMUM ALLOWABLE

Insoluble matter. 0.01%
Chloride (Cl) 0.005%
Calcium (Ca) 0.005%
Magnesium (Mg) 0.005%
Nickel (Ni) 0.02%
Potassium (K). 0.01%
Sodium (Na) 0.05%
Zinc (Zn) 0.005%
Heavy metals (as Pb) 0.002%
Iron (Fe). 0.002%

TESTS

ASSAY. (By complexometric titration of Mn^{II}). Weigh accurately 0.7 g, transfer to a 500-mL beaker, and dissolve in 200 mL of water. Add a few milligrams of ascorbic acid to prevent oxidation. From a buret, add about 30 mL of standard 0.1

M EDTA, followed by 10 mL of pH 10 ammoniacal buffer solution and about 50 mg of Eriochrome Black T indicator mixture. Continue the titration with 0.1 M EDTA to a clear blue color. One milliliter of 0.1 M EDTA corresponds to 0.01690 g of $MnSO_4 \cdot H_2O$.

LOSS ON IGNITION. Weigh accurately about 2 g. Ignite to constant weight at 400–500 °C.

SUBSTANCES REDUCING PERMANGANATE. Dissolve 7.5 g in 200 mL of water containing 3 mL of sulfuric acid and 3 mL of phosphoric acid. To this solution add 0.10 mL of 0.1 N potassium permanganate in excess of the amount required to produce a pink color in 200 mL of water containing 3 mL of sulfuric acid and 3 mL of phosphoric acid, and allow to stand for 1 min. The pink color should not be entirely discharged.

INSOLUBLE MATTER. (Page 15). Use 10 g dissolved in 130 mL of water.

CHLORIDE. (Page 27). Dissolve 1.0 g in 100 mL of water, and use 20 mL of this solution.

CALCIUM, MAGNESIUM, NICKEL, POTASSIUM, SODIUM, AND ZINC. (By flame AAS, page 39).

Sample Stock Solution. Dissolve 10.0 g in water in a 100-mL volumetric flask, add 0.5 mL of hydrochloric acid, and dilute with water to the mark (1 mL = 0.10 g).

Element	Wavelength (nm)	Sample Wt (g)	Standard Added (mg)	Flame Type*	Background Correction
Ca	422.7	0.80	0.02; 0.04	N/A	No
Mg	285.2	0.20	0.005; 0.01	A/A	Yes
Ni	232.0	0.50	0.05; 0.10	A/A	Yes
K	766.5	0.20	0.01; 0.02	A/A	No
Na	589.0	0.02	0.005; 0.01	A/A	No
Zn	213.9	0.20	0.005; 0.01	A/A	Yes

*A/A is air/acetylene; N/A is nitrous oxide/acetylene.

HEAVY METALS. (Page 28, Method 1). Dissolve 1.0 g in about 20 mL of water, and dilute with water to 25 mL.

IRON. (Page 30, Method 1). Dissolve 1.0 g in water, dilute with water to 50 mL, and use 25 mL of this solution without further acidification.

Mannitol

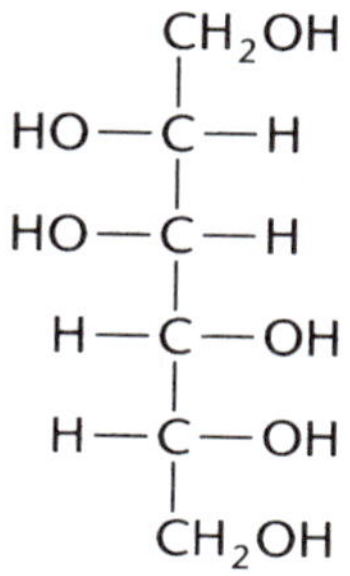

$$CH_2OH$$
$$HO-C-H$$
$$HO-C-H$$
$$H-C-OH$$
$$H-C-OH$$
$$CH_2OH$$

HOCH$_2$(CHOH)$_4$CH$_2$OH   **Formula Wt 182.17**

CAS Number 69–65–8

REQUIREMENTS

Specific rotation $[\alpha]_D^{25°}$ . . . . +23.3° to +24.3°
Reducing sugars. . . . . Passes test

MAXIMUM ALLOWABLE

Insoluble matter. 0.01%
Loss on drying at 105 °C 0.05%
Residue after ignition. 0.01%
Titrable acid. . . . . 0.0008 meq/g
Heavy metals (as Pb) . . . . 5 ppm

TESTS

SPECIFIC ROTATION. (Page 23). Weigh accurately about 10 g, transfer to a 100-mL volumetric flask, and add 12.8 g of sodium borate. Dissolve the mixture in sufficient water to make about 90 mL of solution, allow to stand with occasional shaking for 1 h, dilute with water to volume at 25 °C, and mix. Observe the optical rotation in a polarimeter at 25 °C using sodium line, and calculate the specific rotation.

REDUCING SUGARS. Add 1 mL of a saturated solution of the mannitol (about 0.2 g per mL) to 5 mL of Benedict's solution on a hot plate ($\approx$100 °C) and heat for 5 min. No more than a slight precipitate should form.

INSOLUBLE MATTER. (Page 15). Use 20 g dissolved in 100 mL of water.

LOSS ON DRYING AT 105 °C. Weigh accurately 5.0 g in a tared dish or crucible and dry at 105 °C for 2 h.

RESIDUE AFTER IGNITION. (Page 16). Ignite 10 g. Moisten the char with 1 mL of sulfuric acid.

TITRABLE ACID. To 100 mL of carbon dioxide-free water add 0.15 mL of phenolphthalein indicator solution and 0.01 N sodium hydroxide until a pink color is produced. Dissolve 10 g of the sample in this solution and titrate with 0.01 N sodium hydroxide to the same end point. Not more than 0.8 mL should be required.

HEAVY METALS. (Page 28, Method 1). Dissolve 4.0 g in about 25 mL of water, and dilute with water to 30 mL.

Mercuric Acetate
Mercury(II) Acetate

$(CH_3COO)_2Hg$ **Formula Wt 318.68**

CAS Number 1600–27–7

REQUIREMENTS

Assay . . . .≥98.0% $(CH_3COO)_2Hg$

MAXIMUM ALLOWABLE

Insoluble matter. 0.01%
Residue after reduction 0.02%
Chloride (Cl) 0.005%
Nitrate (NO_3) 0.005%
Sulfate (SO_4) 0.005%
Other heavy metals (as Pb) 0.002%
Iron (Fe) 0.001%
Mercurous mercury (as Hg) 0.4%

TESTS

ASSAY. (By thiocyanate precipitation titration of mercuric ion). Weigh accurately 0.7 g of sample and dissolve in 100 mL of dilute nitric acid (1 + 19). Add 2 mL of ferric ammonium sulfate indicator solution and titrate with 0.1 N ammonium thiocyanate. One milliliter of 0.1 N ammonium thiocyanate corresponds to 0.01593 g of $(CH_3COO)_2Hg$.

INSOLUBLE MATTER. (Page 15). Use 10.0 g dissolved in 100 mL of dilute acetic acid (5 + 95). Also use dilute acetic acid (1 + 99) as the wash.

Sample Solution A for the Determination of Residue after Reduction, Chloride, Sulfate, Other Heavy Metals, and Iron. Dissolve 10.0 g in 15 mL of water, 2 mL of nitric acid, and 10 mL of formic acid (96%). Digest under total reflux until all the mercury is reduced to metal and the solution is clear. Cool, filter through a thoroughly washed filter paper, and wash with a small quantity of water. Dilute the filtrate and washings with water to 100 mL in a volumetric flask (1 mL = 0.1 g).

RESIDUE AFTER REDUCTION. Evaporate 50 mL (5-g sample) of sample solution A plus 0.10 mL of sulfuric acid to dryness in a tared dish in a well-ventilated hood. Continue heating until the excess sulfuric acid has been volatilized. Finally, ignite at 800 ± 25 °C for 15 min. Correct for the weight obtained in a complete blank test.

CHLORIDE. (Page 27). Use 2.0 mL of sample solution A (0.2-g sample) for the determination of chloride.

NITRATE.

Sample Solution A-1. Dissolve 0.50 g completely in 3.0 mL of water in a dry test tube. Put the tube in an ice bath and immediately, but slowly, add 7 mL of cold chromotropic acid reagent while swirling. Keep the tube in the ice bath an additional 2–3 min, remove, and let stand for 30 min, swirling occasionally.

Chromotropic Acid Reagent. Dissolve 0.02 g of recrystallized 4,5-dihydroxy-2,7-naphthalenedisulfonic acid, disodium salt, in concentrated, nitrate-free sulfuric acid and dilute to 200 mL with the acid. Store in an amber bottle.

Recrystallization of Reagent. Add solid sodium sulfate to a saturated aqueous solution of the reagent. Filter the resultant crystals using suction with a Buchner funnel, wash with ethyl alcohol, and air dry. Transfer the crystals to a beaker, add sufficient water to redissolve, then add just enough sodium sulfate to reprecipitate the salt. Filter as before, wash with ethyl alcohol, and dry at a temperature no higher than 80 °C. Store in an amber bottle.

NOTE. If an orange or red precipitate forms in the solution of the sample before the addition of the reagent, discard the solution and start again. Immediate addition of the reagent should prevent formation of such a precipitate.

Control Solution B. Dissolve 0.50 g of sample in 0.5 mL of water plus 2.5 mL of a solution containing 0.01 mg of nitrate ion (NO_3) per milliliter in a dry test tube. Put the tube in an ice bath and immediately, but slowly, add 7 mL of cold chromotropic acid reagent while stirring. Keep the tube in the ice bath an additional 2 to 3 min, remove, and let stand for 30 min, swirling occasionally.

Blank Solution C. Add 7 mL of cold chromotropic acid reagent to 3.0 mL of water in a dry test tube immersed in an ice bath. Remove and let stand for 30 min.

Transfer sample solution A-1 and control solution B to dry 15-mL centrifuge tubes and centrifuge until the supernatant liquid is clear. Set a spectrophotometer at 405 nm and, using 1-cm cells, adjust the instrument to read zero absorbance with blank solution C in the light path, then determine the absorbance of sample solution A-1. Adjust the instrument to read zero absorbance with sample solution A-1 in the light path and determine the absorbance of control solution B. The absorbance of sample solution A-1 versus blank solution C should not exceed that of control solution B versus sample solution A-1.

SULFATE. (Page 33, Method 3). Use 10 mL of sample solution A (1-g sample) for the determination of sulfate.

OTHER HEAVY METALS. (Page 28, Method 1). Add 10 mg of sodium carbonate to 10 mL of sample solution A (1-g sample) for the determination of other heavy metals and evaporate to dryness. Dissolve the residue in about 20 mL of water, filter, and dilute with water to 25 mL.

IRON. (Page 30, Method 1). To 10 mL of sample solution A (1-g sample) for the determination of iron, add about 10 mg of sodium carbonate, and evaporate to dryness. Dissolve the residue in 2 mL of hydrochloric acid, dilute with water to 50 mL, and use the solution without further acidification.

MERCUROUS MERCURY. Dissolve 5.0 g in 100 mL of a 12.5% solution of potassium iodide. Add 5.0 mL of 0.1 N iodine and 2 mL of 1 N hydrochloric acid. Allow to stand, protected from the light, for 1 h with frequent agitation. Titrate the iodine with 0.1 N sodium thiosulfate, adding 3 mL of starch indicator near the end point, and correct for a complete blank. Not more than 1.0 mL of 0.1 N iodine should have been consumed.

Mercuric Bromide
Mercury(II) Bromide

$HgBr_2$ **Formula Wt 360.40**

CAS Number 7789–47–1

REQUIREMENTS

Appearance . . . . . White or having at most a very faint yellow tinge.
If a mercurous salt is present, the
product darkens on exposure to light.

MAXIMUM ALLOWABLE

Residue after reduction 0.02%
Insoluble in methanol 0.05%
Chloride (Cl) 0.25%

TESTS

RESIDUE AFTER REDUCTION. Dissolve 5.0 g in 10 mL of water plus 10 mL of ammonium hydroxide. Add 40 mL of formic acid (96%) and reflux until all the mercury is reduced to metal. Cool, filter through a well-washed filter paper, and wash with a small quantity of water. Add 0.10 mL of sulfuric acid to the combined filtrate and washings. Evaporate in a tared dish in a well-ventilated hood. Continue heating until the excess sulfuric acid has been volatilized. Finally, ignite at 800 ± 25 °C for 15 min. Correct for the weight obtained in a complete blank test.

INSOLUBLE IN METHANOL. Dissolve 2.0 g in 30 mL of methanol. Filter through a tared filtering crucible, wash with 100 mL of methanol, and dry at 105 °C.

CHLORIDE. To 0.70 g add 20 mL of water, 5 mL of 10% sodium hydroxide reagent solution, and 4 mL of 30% hydrogen peroxide, and digest on a hot plate ($\approx$100 °C) until all reaction ceases. Cool, filter through a chloride-free filter, wash with water, and dilute the filtrate and washings with water to 100 mL. To 5.0 mL in a small conical flask, add 20 mL of water and 1.5 mL of ammonium carbonate solution [20 g of ammonium carbonate and 20 mL of dilute ammonium hydroxide (10% NH_3) in sufficient water to make 100 mL]. Add, with agitation, 5 mL of silver nitrate reagent solution and allow to stand, with frequent agitation, for 10 min. Filter through a chloride-free filter, wash with water, and dilute the filtrate and washings with water to 100 mL. To 25 mL add 0.5 mL of nitric acid. Any turbidity should not exceed that in a control obtained by treating 0.20 g of sample and 1.25 mg of added chloride ion (Cl) in 20 mL of water exactly like the solution of the sample.

Mercuric Chloride
Mercury(II) Chloride

$HgCl_2$ Formula Wt 271.50

CAS Number 7487–94–7

REQUIREMENTS

Assay . . . . $\geq$99.5% $HgCl_2$
Solution in ethyl ether . . . . Passes test

MAXIMUM ALLOWABLE

Residue after reduction 0.02%
Iron (Fe) 0.002%

TESTS

ASSAY. (By complexometric titration of Hg^{II}). Weigh accurately about 0.7 g, and dissolve completely in 250 mL of water. From a buret add 20 mL of 0.1 M EDTA, then add 10 mL of a solution containing 2 g of silver nitrate. Add 2 g of ammonium nitrate, 20 mL of a saturated aqueous solution of hexamethylenetetramine, and a few milligrams of xylenol orange indicator. Continue the titration with 0.1 M EDTA to a yellow end point. One milliliter of 0.1 M EDTA corresponds to 0.02715 g of $HgCl_2$.

SOLUTION IN ETHYL ETHER. Dissolve 2.0 g in 60 mL of ethyl ether in a stoppered flask. Shake the flask to agitate the sample and ether, but do not expose the contents to atmospheric moisture. Not more than a faint trace of insoluble residue should remain.

RESIDUE AFTER REDUCTION. Suspend 5.0 g in 10 mL of water plus 10 mL of ammonium hydroxide. Add 40 mL of formic acid (96%) and reflux until all the mercury is reduced to metal. Cool, filter through a well-washed filter paper, and wash with a small quantity of water. Add 0.10 mL of sulfuric acid to the combined filtrate and washings. Evaporate in a tared dish in a well-ventilated hood. Continue heating until the excess sulfuric acid has been volatilized. Finally, ignite at 800 ± 25 °C for 15 min. Correct for the weight obtained in a complete blank test.

IRON. (Page 30, Method 1). To the residue after reduction add 3 mL of dilute hydrochloric acid $(1 + 1)$, cover the dish, and digest on a hot plate (≈ 100 °C) for 15–20 min. Remove the cover and evaporate to dryness. Dissolve the residue in 10 mL of dilute hydrochloric acid $(1 + 9)$, dilute with water to 50 mL, and use 5.0 mL of this solution.

Mercuric Iodide, Red

Mercury(II) Iodide, Red

HgI_2 Formula Wt 454.40

CAS Number 7774–29–0

REQUIREMENTS

Assay (dried basis) . . . . $\geq$99.0% HgI_2
Solubility in potassium iodide solution . . . . Passes test

MAXIMUM ALLOWABLE

Mercurous mercury (as Hg) 0.1%
Soluble mercury salts (as Hg) 0.05%

TESTS

ASSAY. (By iodometry). Dry about 1 g of sample over magnesium perchlorate overnight. Weigh accurately 0.5 g of the dried sample, place in a glass-stoppered conical flask, and add 30 mL of hydrochloric acid and 20 mL of water. Rotate the flask until the mecuric iodide is dissolved, add 5 mL of chloroform, and titrate the solution with 0.05 M potassium iodate until the iodine color is discharged from the aqueous layer. Stopper the flask, shake well for 30 s, then continue the titration, shaking vigorously after each addition of the potassium iodate until the chloroform is free of iodine color. One milliliter of 0.05 M potassium iodate corresponds to 0.02272 g of HgI_2.

SOLUBILITY IN POTASSIUM IODIDE SOLUTION. Dissolve 10.0 g of the sample in 100 mL of 10% potassium iodide reagent solution in a glass-stoppered flask. A complete, or practically complete, solution results. Retain the solution for the test for mercurous mercury.

MERCUROUS MERCURY. To the solution reserved from the test for solubility in potassium iodide solution add 5.0 mL of 0.1 N iodine and 3 mL of 1 N hydrochloric acid. Allow to stand in a dark place for 1 h with frequent agitation. Titrate the excess iodine with 0.1 N sodium thiosulfate, adding 3 mL of starch indicator near the end point, and correct for a complete blank. Not more than 0.50 mL of the 0.1 N iodine should be consumed.

SOLUBLE MERCURY SALTS. Shake 1.0 g of the sample with 20 mL of water for 2 min and filter. Dilute 10 mL of the filtrate with water to 40 mL and add 10 mL of hydrogen sulfide water. Any color should not exceed that produced by 0.25 mg of mercury ion (Hg) in an equal volume of solution containing the quantities of reagents used in the test.

Mercuric Nitrate Monohydrate or Dihydrate
Mercury(II) Nitrate Monohydrate or Dihydrate

$Hg(NO_3)_2 \cdot H_2O$	**Formula Wt 342.62**
$Hg(NO_3)_2 \cdot 2H_2O$	**Formula Wt 360.63**

CAS Number 22852–67–1 (dihydrate); 7783–34–8 (monohydrate)

NOTE. This reagent is available as either the mono- or dihydrate. The label should identify the degree of hydration.

REQUIREMENTS

Assay. . . . .≥98.0% $Hg(NO_3)_2 \cdot H_2O$
or $Hg(NO_3)_2 \cdot 2H_2O$

MAXIMUM ALLOWABLE

Residue after reduction 0.01%
Chloride (Cl) 0.002%
Sulfate (SO$_4$) 0.002%
Iron (Fe) 0.001%
Mercurous mercury (as Hg) 0.2%

TESTS

ASSAY. (By thiocyanate precipitation titration of Hg^{II}). Weigh accurately about 0.6 g of sample, and transfer with 100 mL of water to a 250-mL flask. Add 5 mL of nitric acid and 2 mL of ferric ammonium sulfate indicator solution. Cool to 15 °C, and titrate with 0.1 N ammonium thiocyanate to a permanent reddish brown color. One milliliter of 0.1 N ammonium thiocyanate corresponds to 0.01713 g of Hg(NO$_3$)$_2$ · H$_2$O or 0.01803 g of Hg(NO$_3$)$_2$ · 2H$_2$O.

RESIDUE AFTER REDUCTION. Dissolve 10 g in 15 mL of water, 2 mL of nitric acid, and 10 mL of formic acid (96%). Digest under total reflux until all the mercury is reduced to metal and the solution is clear. Cool, filter through a well-washed filter paper, and wash with a small quantity of water. Add 0.10 mL of sulfuric acid to the combined filtrate and washings. Evaporate in a tared dish in a well-ventilated hood. Continue heating until the excess sulfuric acid has been volatilized. Finally, ignite at 800 ± 25 °C for 15 min. Correct for the weight obtained in a complete blank test. Retain the residue for the determination of iron.

CHLORIDE. Dissolve 2.0 g in 50 mL of water and 2 mL of formic acid (96%). Add, dropwise, 10% sodium hydroxide solution until a small amount of permanent precipitate is formed. Digest under a reflux condenser until the mercury is all reduced to the metal and the supernatant liquid is clear. Cool, filter through a chloride-free filter, wash thoroughly, and dilute the filtrate with water to 100 mL. To 50 mL of the dilution add 1 mL of nitric acid and 1 mL of silver nitrate reagent solution. Any turbidity should not exceed that produced by 0.02 mg of chloride ion (Cl) in an equal volume of solution containing the quantities of reagents used in the test.

SULFATE. Dissolve 5.0 g in 50 mL of water and 5 mL of formic acid. Digest under a reflux condenser until the mercury is all reduced to the metal and the supernatant liquid is clear. Cool, filter, and wash thoroughly. To the combined filtrate and washings add 20 mg of sodium carbonate, and evaporate to dryness on a hot plate (≈100 °C). Dissolve the residue in 10 mL of water and 1 mL of dilute hydrochloric acid (1 + 19), and filter if necessary. To the filtrate add 1 mL of barium chloride reagent solution. Any turbidity should not exceed that produced by 0.1 mg of sulfate ion (SO$_4$) in an equal volume of solution containing the quantities of reagents used in the test. Compare 10 min after adding the barium chloride to the sample and standard solutions.

IRON. (Page 30, Method 1). To the residue obtained in the test for residue after reduction, add 3 mL of dilute hydrochloric acid (1 + 1), cover with a watch glass, and digest on a hot plate (≈100 °C) for 20 min. Remove the cover and evaporate to dryness. Dissolve the residue in 2 mL of dilute hydrochloric acid (1 + 1), add about 40 mL of water, filter if necessary, and dilute with water to 100 mL. Use 10 mL of this solution.

MERCUROUS MERCURY. Dissolve 5.0 g in 100 mL of 10% potassium iodide solution, and add 5 mL of 0.1 N iodine and 3 mL of dilute hydrochloric acid (1 + 19). Allow to stand in the dark for 1 h with frequent agitation. Titrate the excess iodine with 0.1 N sodium thiosulfate, adding 3 mL of starch indicator solution near the end point. Perform a complete blank test. The volume of 0.1 N iodine consumed should not exceed 0.50 mL.

Mercuric Oxide, Red
Mercury(II) Oxide, Red

HgO   Formula Wt 216.59

CAS Number 21908–53–2

REQUIREMENTS

Assay . . . . ≥99.0% HgO

MAXIMUM ALLOWABLE
Insoluble in dilute hydrochloric acid 0.03%
Residue after reduction 0.025%
Chloride (Cl) 0.025%
Sulfate (SO$_4$) 0.015%
Nitrogen compounds (as N). 0.005%
Iron (Fe). 0.005%

TESTS

ASSAY. (By complexometric titration of Hg^{II}). Weigh accurately 0.8 g of sample. Dissolve in 3.5 mL of nitric acid, dilute to about 100 mL with water, and add 50 mg of xylenol orange indicator. Add a saturated solution of hexamethylenetetramine until the color changes to purple, followed by 3–4 mL in excess, then titrate with 0.1 M EDTA until the color changes to yellow. One milliliter of 0.1 M EDTA corresponds to 0.02166 g of HgO.

INSOLUBLE IN DILUTE HYDROCHLORIC ACID. Dissolve 4.0 g in 40 mL of dilute hydrochloric acid (1 + 3), heat to boiling, and digest in a covered beaker on a hot plate (≈100 °C) for 1 h. Filter through a tared filtering crucible, wash well with water, and dry at 105 °C.

Sample Solution A for the Determination of Residue After Reduction and Sulfate. Dissolve 10.0 g in 15 mL of water, 2 mL of nitric acid, and 10 mL of formic acid (96%). Digest under total reflux until all the mercury is reduced to metal and the solution is clear. Cool, filter through a well-washed filter paper, and wash with a small quantity of water. Dilute the filtrate and washings with water to 200 mL in a volumetric flask (1 mL = 0.05 g).

RESIDUE AFTER REDUCTION. Evaporate 80 mL (4-g sample) of sample solution A plus 0.10 mL of sulfuric acid to dryness in a tared dish in a well-ventilated hood. Continue heating until the excess sulfuric acid has been volatilized. Finally, ignite at 800 ± 25 °C for 15 min. Correct for the weight obtained in a complete blank test. Retain the residue for the test for iron.

CHLORIDE. Dissolve 1.0 g in 50 mL of water plus 1 mL of formic acid (96%). Add, dropwise, 10% sodium hyroxide reagent solution until a small amount of permanent precipitate is formed. Digest under a reflux condenser until all the mercury is reduced to metal and the solution is clear. Cool, filter through a chloride-free filter, and dilute with water to 100 mL. Dilute 4.0 mL of this solution with water to 20 mL and add 1 mL of nitric acid and 1 mL of silver nitrate reagent solution. Any turbidity should not exceed that produced by 0.01 mg of chloride ion (Cl) in an equal volume of solution containing the quantities of reagents used in the test.

SULFATE. (Page 33, Method 3). Use 6.7 mL of sample solution A (0.33-g sample).

NITROGEN COMPOUNDS. (Page 31). Dissolve 0.2 g in 5 mL of dilute hydrochloric acid (1 + 1). For the standard, use 0.01 mg of nitrogen (N) and 5 mL of the dilute acid.

IRON. (Page 30, Method 1). To the residue obtained in the test for residue after reduction, add 3 mL of dilute hydrochloric acid (1 + 1), cover with a watch glass, and digest on a hot plate (≈100 °C) for 20 min. Remove the cover and evaporate to dryness. Dissolve in 2 mL of dilute hydrochloric acid (1 + 1), add about 40 mL of water, filter if necessary, and dilute with water to 100 mL. Use 5 mL of this solution.

Mercuric Oxide, Yellow
Mercury(II) Oxide, Yellow

HgO   **Formula Wt 216.59**

CAS Number 21908–53–2

REQUIREMENTS

Assay . . . . ≥99.0% HgO

MAXIMUM ALLOWABLE

Insoluble in dilute hydrochloric acid 0.03%
Residue after reduction 0.05%
Chloride (Cl) 0.025%
Sulfate (SO$_4$) 0.01%
Nitrogen compounds (as N). 0.005%
Iron (Fe). 0.003%

TESTS

ASSAY. (By complexometric titration of Hg^{II}). Weigh accurately 0.8 g of sample. Dissolve in 3.5 mL of nitric acid, dilute to about 100 mL with water, and add 50 mg of xylenol orange indicator. Add a saturated solution of hexamethylenetetramine until the color changes to purple, followed by 3–4 mL in excess, then titrate with 0.1 M EDTA until the color changes to yellow. One milliliter of 0.1 M EDTA corresponds to 0.02166 g of HgO.

INSOLUBLE IN DILUTE HYDROCHLORIC ACID. Dissolve 3.0 g in 30 mL of dilute hydrochloric acid (1 + 3), heat to boiling, and digest in a covered beaker on a hot plate (≈100 °C) for 1 h. Filter through a tared filtering crucible, wash thoroughly, and dry at 105 °C.

Sample Solution A for the Determination of Residue After Reduction and Sulfate. Dissolve 10.0 g in 15 mL of water, 2 mL of nitric acid, and 10 mL of formic acid (96%). Digest under total reflux until all the mercury is reduced to metal and the solution is clear. Cool, filter through a well-washed filter paper, and wash with a small quantity of water. Dilute the filtrate and washings with water to 200 mL in a volumetric flask (1 mL = 0.05 g).

RESIDUE AFTER REDUCTION. Evaporate 60 mL (3 g) of sample solution A plus 0.10 mL of sulfuric acid to dryness in a tared dish in a well-ventilated hood. Continue heating until the excess sulfuric acid has been volatilized. Finally, ignite at 800 ± 25 °C for 15 min. Correct for the weight obtained in a complete blank test. Retain the residue for the test for iron.

CHLORIDE. Dissolve 1.0 g in 50 mL of water and 1 mL of formic acid (96%). Add, dropwise, 10% sodium hydroxide reagent solution until a small amount of permanent precipitate is formed. Digest under a reflux condenser until all the mercury is reduced to metal and the solution is clear. Cool, filter through a chloride-free filter, and dilute with water to 100 mL. Dilute 4.0 mL of this solution with water to 20 mL and add 1 mL of nitric acid and 1 mL of silver nitrate reagent solution. Any turbidity should not exceed that produced by 0.01 mg of chloride ion (Cl) in an equal volume of solution containing the quantities of reagents used in the test.

SULFATE. (Page 33, Method 3). Use 10 mL of sample solution A (0.5-g sample).

MAXIMUM ALLOWABLE

Residue after reduction 0.02%
Mercuric chloride ($HgCl_2$) 0.01%
Sulfate (SO_4) 0.01%

TESTS

ASSAY. (By iodometric titration of mercury). Weigh accurately 0.9 g and transfer to a 250-mL glass-stoppered conical flask. Add 50.0 mL of 0.1 N iodine and 2 g of potassium iodide. Stopper and swirl until the precipitate redissolves. Titrate the excess of iodine with 0.1 N sodium thiosulfate, adding 3 mL of starch indicator solution near the end point. One milliliter of 0.1 N iodine corresponds to 0.02360 g of Hg_2Cl_2.

RESIDUE AFTER REDUCTION. Dissolve 5.0 g in 10 mL of water plus 10 mL of ammonium hydroxide. Add 40 mL of formic acid (96%) and reflux until all the mercury is reduced to metal. Cool, filter through a thoroughly washed filter paper, and wash with a small quantity of water. Add 0.10 mL of sulfuric acid to the combined filtrate and washings. Evaporate in a tared dish in a well-ventilated hood. Continue heating until the excess sulfuric acid has been volatilized. Finally, ignite at 800 ± 25 °C for 15 min. Correct for the weight obtained in a complete blank test.

MERCURIC CHLORIDE. Shake 1.0 g with 10 mL of alcohol for 5 min, and filter. To 5 mL of the filtrate add 0.10 mL of hydrochloric acid and 5 mL of hydrogen sulfide water. Any darkening should not be more than that produced by 0.05 mg of mercuric chloride ($HgCl_2$) in a mixture of 5 mL of alcohol, 0.10 mL of hydrochloric acid, and 5 mL of hydrogen sulfide water.

SULFATE. (Page 33, Method 3). Digest for 10 min with 10 mL of dilute hydrochloric acid (1 + 1), and filter.

Mercurous Nitrate Dihydrate
Mercury(I) Nitrate Dihydrate

$Hg_2(NO_3)_2 \cdot 2H_2O$ **Formula Wt 561.22**

CAS Number 14836–60–3

NOTE. This reagent is commercially available in a degree of hydration ranging from 0 to 2 molecules of water per molecule of $Hg_2(NO_3)_2$. The specifications refer to both anhydrous and hydrated reagents, and the assay is expressed as the dihydrate.

REQUIREMENTS

Assay . . . . ≥97.0% $Hg_2(NO_3)_2 \cdot 2H_2O$

MAXIMUM ALLOWABLE

Insoluble matter. 0.005%
Residue after reduction 0.01%
Chloride (Cl) 0.005%
Sulfate (SO$_4$) 0.005%
Iron (Fe). 0.001%
Mercuric mercury (as Hg) 0.5%

TESTS

ASSAY. (By iodometric titration of mercury). Weigh accurately 1 g of sample and transfer to a 250-mL iodine flask. Dissolve by stirring with 20 mL of acetic acid (1 + 2) and 10 mL of 20% sodium acetate trihydrate solution. Dilute with 50 mL of water. Add 50.0 mL of 0.1 N iodine and 2 g of potassium iodide. Stopper and swirl until the precipitate redissolves. Titrate the excess iodine with 0.1 N sodium thiosulfate, adding 3 mL of starch indicator solution near the end point. One milliliter of 0.1 N iodine corresponds to 0.02806 g of $Hg_2(NO_3)_2 \cdot 2H_2O$.

$$\% \ Hg_2(NO_3)_2 \cdot 2H_2O = \frac{[(V \times N \ I_2) - (V \times N \ Na_2S_2O_3)]28.061}{Sample \ wt \ (g)}$$

INSOLUBLE MATTER. (Page 15). Use 20 g dissolved in 200 mL of dilute nitric acid (5 + 95).

> **Sample Solution A for the Determination of Residue After Reduction, Chloride, and Sulfate.** Dissolve 25 g in 30 mL of water, 4 mL of nitric acid, and 20 mL of formic acid (96%). Digest under total reflux until all the mercury(I) is reduced to mercury and the solution is clear. Cool, filter through a well-washed filter paper, and wash with a small quantity of water. Dilute the filtrate and washings with water to 500 mL in a volumetric flask (1 mL = 0.05 g).

RESIDUE AFTER REDUCTION. Evaporate 400 mL (20-g sample) of sample solution A plus 0.10 mL of sulfuric acid to dryness in a tared dish in a well-ventilated hood. Continue heating until the excess sulfuric acid has been volatilized. Finally, ignite at 800 ± 25 °C for 15 min. Correct for the weight obtained in a complete blank test. Retain the residue for the test for iron.

CHLORIDE. (Page 27). Use 4.0 mL of sample solution A (0.2-g sample).

SULFATE. (Page 33, Method 1). Use 20 mL of sample solution A (1-g sample).

IRON. (Page 30, Method 1). To the residue obtained in the test for residue after reduction, add 3 mL of dilute hydrochloric acid (1 + 1), cover with a watch glass, and digest on a hot plate (≈100 °C) for 20 min. Remove the cover, and evaporate to dryness. Dissolve the residue in 2 mL of dilute hydrochloric acid (1 + 1) and 30–40 mL of water, filter if necessary, and dilute with water to 100 mL. Use 5.0 mL of this solution.

MERCURIC MERCURY. Dissolve 1.0 g in a mixture of 5 mL of water and 0.25 mL of nitric acid in a glass-stoppered, 50-mL cylinder. When dissolution is complete, add 10 mL of water and 5 mL of 1 N hydrochloric acid, and mix well. Dilute with water to 50 mL, shake well, and allow to stand for 15 min. Filter the solution, and dilute 1 mL of the filtrate with water to 25 mL. For the standard, mix 0.25 mL of nitric acid with about 10 mL of water, add 5 mL of 1 N hydrochloric acid, dilute with water to 50 mL, and mix. To 1 mL of the standard solution, add 0.1 mg of mercury ion (Hg) prepared from mercury chloride, and dilute with water to 25 mL. Add 10 mL of hydrogen sulfide water to both the sample and the standard solution. Any color in the sample solution should not exceed that in the standard.

Mercury

Hg   **Atomic Wt 200.59**

CAS Number 7439–97–6

REQUIREMENTS

Appearance . . . . Passes test

MAXIMUM ALLOWABLE

Nonvolatile matter . . . . 5 ppm

TESTS

APPEARANCE. The mercury should have a bright mirrorlike surface, free from film or scum. It should pour freely from a thoroughly clean, dry glass container without leaving any mercury adhering to the glass.

NONVOLATILE MATTER. Perform all tests in a hood. Transfer 200 g to a tared boat of 25–30-mL capacity. Place the boat in the approximate center of a glass combustion tube, about 350 mm long and 35 mm in diameter, and mount in a horizontal position. The tube is fitted at one end with a removable closure and the exit end is bent downward at 90 ° to the horizontal and is drawn out so as to fit into a one-hole rubber stopper held in the mouth of a suction flask. The suction flask is cooled in an ice bath. After the boat and sample are placed in the tube, close the entrance end and reduce the pressure to less than 15 mm of mercury. Heat gently until all the mercury has distilled from a quiet surface (without ebullition). Remove the boat, heat it in a muffle furnace at 600 °C for 15 min, and cool.

Metalphthalein
Phthalein Purple
o-Cresolphthalein complexone

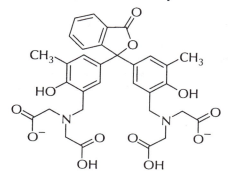

$C_{32}H_{32}N_2O_{12}$

Formula Wt 636.62

CAS Number 2411–89–4

REQUIREMENTS

Clarity of solution . . . . Passes test
Suitability as a mixed indicator for complexometry . . . . Passes test

TESTS

CLARITY OF SOLUTION. Dissolve 0.18 g in 100 mL of water containing 0.5 mL of ammonium hydroxide. Not more than a trace of turbidity or insoluble matter should remain.

SUITABILITY AS A MIXED INDICATOR FOR COMPLEXOMETRY. Using the test sample, prepare metalphthalein-screened indicator as defined on page 87. Mix 50 mL of water and 50 mL of alcohol. Add 10.0 mL of 0.1 M EDTA, 10 mL of ammonium hydroxide, and 0.3 mL of the metalphthlein-screened indicator solution. Titrate with 0.1 M barium nitrate. The color change is from gray-green to magenta. Add back 0.05 mL of EDTA. The color should return to gray-green.

> **Barium Nitrate, 0.1 M.** Dissolve 0.654 g of barium nitrate in water and dilute to 25.0 mL in a volumetric flask.

Methanol
Methyl Alcohol

CH₃OH   **Formula Wt 32.04**

CAS Number 67–56–1

Suitable for use in high-performance liquid chromatography, extraction–concentration analysis, ultraviolet spectrophotometry, or general use. Product labeling shall designate the one or more of these uses for which suitability is represented based on meeting the relevant requirements and tests. The ultraviolet spectrophotometry and liquid chromatography suitability requirements include all of the requirements for general use. The extraction–concentration suitability requirements include only the general use requirement for color.

REQUIREMENTS

General Use

Appearance . . . . Clear
Assay . . . . ≥99.8% CH₃OH
Substances darkened by sulfuric acid . . . . Passes test
Substances reducing permanganate . . . . Passes test

MAXIMUM ALLOWABLE

Color (APHA) 10
Water (H₂O) 0.1%
Residue after evaporation 0.001%
Solubility in water . . . . Passes test
Carbonyl compounds . . . . 0.001% each of acetone, formaldehyde, and acetaldehyde
Titrable acid . . . . 0.0003 meq/g
Titrable base . . . . 0.0002 meq/g

Specific Use

Ultraviolet Spectrophotometry

Wavelength (nm)	Absorbance (AU)
280–400	0.01
260	0.04
240	0.10
230	0.20
220	0.40
210	0.80
205	1.00

Liquid Chromatography Suitability

Absorbance . . . . Passes test
Gradient elution. . . . . Passes test

Extraction–Concentration Suitability

Absorbance . . . . Passes test
GC–FID . . . . Passes test
GC–ECD . . . . Passes test

TESTS

ASSAY. Analyze the sample by gas chromatography using the general parameters cited on page 70. The following specific conditions are also required.

Column: Type I, methyl silicone

Measure the area under all peaks and calculate the methanol content in area percent. Correct for water content.

SUBSTANCES DARKENED BY SULFURIC ACID. Cool 10 mL of sulfuric acid, contained in a small conical flask, to 10 °C and add dropwise with constant agitation 10 mL of the sample, keeping the temperature of the mixture below 20 °C. The mixture should be colorless and have no more color than the acid or methanol before mixing.

SUBSTANCES REDUCING PERMANGANATE. Cool 20 mL to 15 °C, add 0.10 mL of 0.1 N potassium permanganate, and allow to stand at 15 °C for 15 min. The pink color should not be entirely discharged.

COLOR (APHA). (Page 19).

WATER. (Page 55, Method 1). Use 25 mL (19.8 g) of the sample.

RESIDUE AFTER EVAPORATION. (Page 16). Evaporate 100 g (125 mL) in a tared dish on a hot plate (≈100 °C) and dry the residue at 105 °C for 30 min.

SOLUBILITY IN WATER. Dilute 15 mL with 45 mL of water, mix, and allow to stand for 1 h. The solution should be as clear as an equal volume of water.

CARBONYL COMPOUNDS. (By differential pulse polarography, page 50). Use 5.0 g (6.4 mL) of sample. For the standard, use 0.05 mg each of acetone, formaldehyde, and acetaldehyde.

TITRABLE ACID. To 25 mL of water and 10 mL of ethyl alcohol in a glass-stoppered flask, add 0.50 mL of phenolphthalein indicator solution and 0.01 N sodium hydroxide until a slight pink color persists after shaking for 30 s. Add 15 g (19 mL) of sample, mix well, and titrate with 0.01 N sodium hydroxide until the pink color is restored. Not more than 0.50 mL should be required.

NOTE. Special care should be taken during the addition of the sample and titration to avoid contamination from carbon dioxide.

TITRABLE BASE. Dilute 22.6 g (28.6 mL) with 25 mL of water and add 0.15 mL of methyl red indicator solution. Not more than 0.40 mL of 0.01 N hydrochloric acid should be required to produce a pink color.

ULTRAVIOLET SPECTROPHOTOMETRY. Use the procedure on page 73 to determine the absorbance.

LIQUID CHROMATOGRAPHY SUITABILITY. Analyze the sample, using the gradient elution procedure cited on page 71. Use the procedure on page 73 to determine the absorbance.

EXTRACTION–CONCENTRATION SUITABILITY. Analyze the sample using the general procedure cited on page 72. Use the procedure on page 73 to determine the absorbance.

2-Methoxyethanol
Ethylene Glycol Monomethyl Ether

$CH_3OCH_2CH_2OH$ **Formula Wt 76.10**

CAS Number 109–86–4

REQUIREMENTS

Assay . . . . ≥99.3% $CH_3OCH_2CH_2OH$
MAXIMUM ALLOWABLE
Color (APHA) 10
Titrable acid . . . . 0.002 meq/g
Water 0.1%

TESTS

ASSAY. Analyze the sample by gas chromatography using the general parameters cited on page 70. The following specific conditions are also required.

Column: Type III, polyethylene glycol

Measure the area under all peaks and calculate the 2-methoxyethanol content in area percent. Correct for water content.

COLOR (APHA). (Page 19).

TITRABLE ACID. To 50 mL of water in a conical flask, add 0.15 mL of phenol red indicator, and adjust to a pink color with 0.01 N sodium hydroxide. Add 25 g (26 mL) of sample, and titrate with 0.01 N sodium hydroxide to the same color. Not more than 5.0 mL should be required.

WATER. (Page 55, Method 1). Use 25 mL (24 g) of sample.

4-(Methylamino)phenol Sulfate

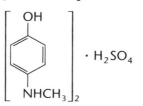

$(CH_3NHC_6H_4OH)_2 \cdot H_2SO_4$ **Formula Wt 344.39**

CAS Number 1936–57–8

REQUIREMENTS

Assay 99.0–101.5%
Residue after ignition . . . . ≤0.1%
Suitability for determination of phosphate . . . . Passes test

TESTS

ASSAY. (By titration of reductive capacity). Weigh about 250 mg to the nearest 0.1 mg. Transfer to a 500-mL conical flask containing 100 mL of water and 10 mL of 0.1 N sulfuric acid. Dissolve, add 0.15 mL of ferroin indicator solution, and titrate with 0.1 N ceric ammonium nitrate to a light green color that persists for 15 s.

% $(CH_3NHC_6H_4OH)_2 \cdot H_2SO_4 =$

$$[mL \times N\ (NH_4)_2Ce(NO_3)_6 \times 8.61]\ /\ \text{Sample wt (g)}$$

Standard Ceric Ammonium Nitrate Solution, 0.1 N. Mix 59 g of ceric ammonium nitrate, $(NH_4)_2Ce(NO_3)_6$, with 31 mL of sulfuric acid in a 500-mL beaker with stirring. Cautiously add water in 20-mL portions, with stirring, allowing 2–3 min between additions. Continue the addition of water until the ceric ammonium nitrate is completely dissolved. Filter, if necessary, through a sintered glass crucible or funnel, and dilute to 1 L with water in a volumetric flask.

Standardization: Weigh about 200 mg of dry primary standard arsenic trioxide to the nearest 0.1 mg. Transfer to a 500-mL conical flask, add 15 mL of 10% sodium hydroxide reagent solution, and warm the mixture gently to hasten dissolution. When dissolution is complete, cool to room temperature and add 25 mL of dilute sulfuric acid (1 + 5). Dilute to 100 mL with water and add, as catalyst, 0.15 mL of 0.01 M osmium tetroxide (0.25 g of OsO_4 in 100 mL of 0.1 N sulfuric acid). Add 0.15 mL of ferroin indicator solution, and titrate with the ceric ammonium nitrate solution until the reddish orange color changes to colorless or very pale blue. (*NOTE:* As the osmium tetroxide catalyst ages, 0.15–0.65 mL may be needed to achieve the desired action. A sluggish end point indicates insufficient osmium tetroxide.)

$$\text{Normality of Ceric Ammonium Nitrate} = \frac{W \times 1000}{V \times 49.95}$$

where

$W =$ weight, in mg, of As_2O_3
$V =$ volume, in mL, of $(NH_4)_2Ce(NO_3)_6$ solution

RESIDUE AFTER IGNITION. (Page 16). Ignite 5.0 g.

SUITABILITY FOR DETERMINATION OF PHOSPHATE. Dissolve 2 g in 100 mL of water. To 10 mL of this solution add 90 mL of water and 20 g of sodium bisulfite, dissolve, and mix. Transfer 1 mL of this solution to each of two solutions containing 25 mL of 0.5 N sulfuric acid and 1 mL of ammonium molybdate–sulfuric acid reagent solution. For the control, add 0.005 mg of phosphate ion (PO_4) to one of the solutions. For the sample use the remaining solution. Allow the solutions to stand at room temperature for 2 h. The control should turn perceptibly darker blue than the sample.

Methyl Orange
4-(Dimethylamino)azobenzenesulfonic Acid, Sodium Salt
4-[[4-(Dimethylamino)phenyl]azo]benzenesulfonic Acid, Sodium Salt
C.I. Acid Orange 52

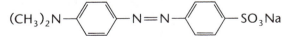

$C_{14}H_{14}N_3NaO_3S$ **Formula Wt 327.34**

CAS Number 547–58–0

REQUIREMENTS

Clarity of solution. . . . . Passes test
Visual transition interval. . . . From pH 3.2 (pink) to pH 4.4 (yellow)

TESTS

CLARITY OF SOLUTION. Dissolve 0.1 g in 100 mL of water. Not more than a faint trace of turbidity or insoluble matter should remain. Reserve the solution for the test for visual transition interval.

VISUAL TRANSITION INTERVAL. Dissolve 1 g of potassium chloride in 100 mL of water. Adjust the pH of the solution to 3.0 (using a pH meter as described on page 43) with 0.01 N hydrochloric acid. Add 0.15 mL of the 0.1% solution reserved from the test for clarity of solution. The solution should have a definite pink color. Titrate the solution with 0.01 N sodium hydroxide to a pH of 3.2 (using a pH meter). A small amount of yellow color should appear, producing a pinkish orange color. Continue the titration to pH 4.4; the solution should have a definite yellow color. No more than 12.5 mL of 0.01 N sodium hydroxide should be consumed in the entire titration.

4-Methyl-2-pentanone
Methyl Isobutyl Ketone

$$CH_3-\overset{\overset{\displaystyle O}{\|}}{C}CH_2\overset{\overset{\displaystyle CH_3}{|}}{C}HCH_3$$

$(CH_3)_2CHCH_2COCH_3$ **Formula Wt 100.16**

CAS Number 108–10–1

REQUIREMENTS

Appearance . . . . Clear
Assay . . . . ≥98.5% $(CH_3)_2CHCH_2COCH_3$

MAXIMUM ALLOWABLE

Color (APHA) 15
Residue after evaporation 0.005%
Titrable acid . . . . 0.002 meq/g
Water 0.1%

TESTS

ASSAY. Analyze the sample by gas chromatography using the general parameters cited on page 70. The following specific conditions are also required.

Column: Type I, methyl silicone

Measure the area under all peaks and calculate the 4-methyl-2-pentanone content in area percent. Correct for water content.

COLOR (APHA). (Page 19).

RESIDUE AFTER EVAPORATION. (Page 16). Evaporate 100 g (125 mL) to dryness in a tared dish on a hot plate (≈100 °C) and dry the residue at 125 °C for 30 min.

TITRABLE ACID. To 25 mL of alcohol in a glass-stoppered flask add 0.50 mL of phenolphthalein indicator solution. Add 0.01 N sodium hydroxide solution until a slight pink color persists after shaking for 30 s. Add 25 mL of the sample, mix, and titrate with 0.01 N sodium hydroxide until the pink color is restored. Not more than 4.0 mL of the sodium hydroxide solution should be required.

> *NOTE.* Great care should be taken in the test during the addition of the sample and the titration to avoid contamination from carbon dioxide.

WATER. (Page 55, Method 1). Use 25 mL (20 g) of the sample and a methanol-free system to prevent ketal formation with liberation of water.

1-Methyl-2-pyrrolidone
N-Methyl Pyrrolidone

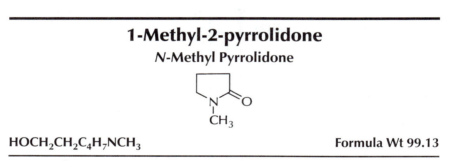

HOCH$_2$CH$_2$C$_4$H$_7$NCH$_3$ **Formula Wt 99.13**

CAS Number 872–50–4

Suitable for use in ultraviolet spectrophotometry or general use. Product labeling shall designate the one or both of these uses for which suitability is represented on the basis of meeting the relevant requirements and tests. The ultraviolet spectrophotometry requirements include all of the requirements for general use.

REQUIREMENTS

General Use

Assay . . . . ≥99.0% C$_5$H$_9$ON

MAXIMUM ALLOWABLE

Color (APHA) 50
Water 0.05%
Free amines (as CH$_3$NH$_2$) 0.01%
Chloride (Cl) . . . . 1 ppm

Specific Use

Ultraviolet Spectrophotometry

Wavelength (nm)	Absorbance (AU)
400	0.01
350	0.01
300	0.05
285	0.15
275	1.00

TESTS

ASSAY. Analyze the sample by gas chromatography by using the general parameters cited on page 70. The following specific conditions are also required.

Column: Type 1, methyl silicone

Measure the area under all the peaks, and calculate the 1-methyl-2-pyrrolidone content in area percent. Correct for water content.

COLOR (APHA). (Page 19).

WATER. (Page 55, Method 1). Use 24.3 mL (25 g) of the sample and a methanol-free system to prevent ketal formation with liberation of water.

FREE AMINES. Accurately weigh a 10.0 g sample in a tared, glass-stoppered flask. Add 100 mL of water and 0.1 mL of methyl red solution. Titrate with standardized 0.01 N hydrochloric acid to the first permanent pink color. No more than 3.2 mL should be required.

$$\% \ CH_3NH_2 = (mL \ HCl \times N \ HCl \times 3.106) \ / \ Sample \ wt \ (g)$$

CHLORIDE. Dilute 24.3 mL (25 g) of sample to the mark in a 100-mL volumetric flask with deionized water. Transfer 10.0 mL of sample solution to each of two 50-mL Nessler tubes and 30.0 mL to each of two additional Nessler tubes. To one tube containing 10 mL, add 0.5 mL of 0.01 mg/mL chloride standard (0.5-ppm standard), and add 1.0 mL of chloride standard to the other tube with 10 mL of solution (1.0-ppm standard). Also add 0.5 mL of standard to one tube containing 30 mL (spiked sample). Dilute all tubes to about 45 mL with deionized water and mix. Add 1 mL of nitric acid and 1 mL of 0.1 N silver nitrate to each tube. Dilute to 50 mL and mix thoroughly. After 10 min, observe the opalescence. Any opalescence in the sample must be no greater than that in the 0.5-ppm standard, and any opalescence in the spiked sample must be greater than that in the 1-ppm standard.

ULTRAVIOLET SPECTROPHOTOMETRY. Use the procedure on page 73 to determine the absorbance.

Methyl Red
2-[4-(Dimethylamino)phenylazo]benzoic Acid
C.I. Acid Red 2

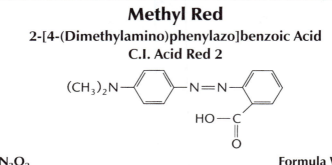

$C_{15}H_{15}N_3O_2$ **Formula Wt 269.30**

CAS Number 845–10–3 (sodium salt); 493–52–7 (free acid)

NOTE. Three forms of this indicator are available: the compound named above (I), the sodium salt of the compound (II), and the hydrochloride of the compound (III). I must meet the requirements for melting point (range), clarity of alcohol solution, and visual transition interval. This form is recommended for nonaqueous titrations, particularly when an aprotic solvent is used. II must meet the requirements for clarity of alcohol solution, clarity of aqueous solution, and visual transition interval. This form is the choice for titrations in aqueous media and is also suitable for nonaqueous titrations where the medium is an amphiprotic solvent. III must meet the requirements for clarity of alcohol solution and visual transition interval. This form can be used as an indicator for titrations in aqueous media and amphiprotic solvents.

REQUIREMENTS

Melting point . . . . .179–182 °C(I)
Clarity of alcohol solution . . . . .Passes test (I, II, III)
Clarity of aqueous solution. .Passes test (II)
Visual transition interval. . . . . .From pH 4.2 (pink) to
pH 6.2 (yellow) (I,II,III)

TESTS

MELTING POINT. (Page 22).

CLARITY OF ALCOHOL AND AQUEOUS SOLUTIONS. Dissolve 0.1 g of II in 100 mL of water. For all three forms (I, II, and III), dissolve 0.1 g in 100 mL of alcohol. Not more than a faint trace of turbidity or insoluble matter should remain. Reserve the alcohol solutions for the test for visual transition interval.

VISUAL TRANSITION INTERVAL. Dissolve 1 g of potassium chloride in 100 mL of water. Adjust the pH of the solution to 4.2 (using a pH meter as described on page 43) with 0.01 N hydrochloric acid. Add 0.15 mL of the 0.1% solution reserved from the test for clarity of solution. The color of the solution should be pink. Titrate the solution with 0.01 N sodium hydroxide to a pH of 5.5 (using the pH meter). The solution should be orange in color. Continue the titration to a pH of 6.2. The solution should be yellow in color. Not more than 1.0 mL of the 0.01 N sodium hydroxide should be consumed in the entire titration.

Methyl *tert*-Butyl Ether

$CH_3OC(CH_3)_3$ **Formula Wt 88.14**

CAS Number 1634–04–4

Suitable for use in ultraviolet spectrophotometry, extraction–concentration, or general use. Product labeling shall designate the one or more of these uses for which suitability is represented on the basis of meeting the relevant requirements and tests. The ultraviolet spectrophotometry requirements include all of the requirements for general use. The extraction–concentration suitability requirements include only the general use requirement for color.

REQUIREMENTS

General Use

Assay (GC). . . . . ≥99.0% $C_5H_{12}O$

MAXIMUM ALLOWABLE

Color (APHA). 10
Peroxide (as H_2O_2) . . . . 1 ppm
Residue after evaporation 0.001%
Water. 0.05%

Specific Use

Ultraviolet Spectrophotometry

Wavelength (nm)	Absorbance (AU)
350	0.01
300	0.01
250	0.10
225	0.50
210	1.0

Extraction–Concentration Suitability

Absorbance . . . . Passes test
GC–FID . . . . Passes test
GC–ECD . . . . Passes test

TESTS

ASSAY. Analyze the sample by gas chromatography by using the general parameters cited on page 70. The following specific conditions are also required.

   Column: Type 1, methyl silicone

Measure the area under all the peaks, and calculate the methyl *t*-butyl ether content in area percent. Correct for water content.

COLOR (APHA). (Page 19).

PEROXIDE. To 47 mL (35 g) of sample in a separatory funnel, add 5.0 mL of titanium tetrachloride solution. Shake vigorously, let separate, and drain the lower layer into a 25-mL glass-stoppered, graduated cylinder. For the standard, transfer 5.0 mL of titanium tetrachloride solution to a similar graduated cylinder, and add 0.035 mg of freshly prepared hydrogen peroxide standard. Dilute both solutions with water to 10.0 mL and mix. Let mixtures stand for 30 min. Any yellow color in the sample solution should not exceed that in the standard. The color intensities may be determined with a spectrophotometer in 1-cm cells at a wavelength of 410 nm.

   CAUTION: If the sample fails the peroxide test, do not perform the test for residue after evaporation.

RESIDUE AFTER EVAPORATION. Evaporate 136 mL (100 g) to dryness in a tared platinum dish on a hot plate ($\approx$100 °C), and dry the residue at 105 °C for 30 min. The weight of the residue should not exceed 0.001 g.

WATER. (Page 55, Method 1). Use 25 mL (18.4 g) of the sample.

ULTRAVIOLET SPECTROPHOTOMETRY. Use the procedure on page 73 to determine the absorbance.

EXTRACTION–CONCENTRATION SUITABILITY. Analyze the sample by using the general procedure cited on page 72.

Methylthymol Blue, Sodium Salt
3,3′-Bis[*N*,*N*-di(carboxymethyl)-aminomethyl]thymolsulfonphthalein, Sodium Salt

$C_{37}H_{40}N_2O_{13}Na_4S$ **Formula Wt 844.76**

CAS Number 1945–77–3

REQUIREMENTS

   Clarity of solution. . . . . Passes test

Suitability as a metal indicator . . . . Passes test
Visual transition interval. . . . . From pH 6.5 (yellow)
to pH 8.5 (blue)

TESTS

CLARITY OF SOLUTION. Dissolve 0.1 g in 100 mL of water. Not more than a trace of turbidity or insoluble matter should remain. Reserve this solution for the test for visual transition interval.

SUITABILITY AS METAL INDICATOR. Mix 50 mL of water with 5 mL of diethylamine. Add 25–50 mg of methylthymol blue indicator (page 91) and 10.0 mL of 0.1 M EDTA. The solution is colorless or gray. Titrate with 0.1 M strontium chloride. The color change is from colorless-gray to blue. Add back 0.05 mL of EDTA. The color should return to the colorless-gray.

> **Strontium Chloride, 0.1 M.** Dissolve 0.666 g of strontium chloride hexahydrate in water and dilute to 25.0 mL in a volumetric flask.

VISUAL TRANSITION INTERVAL. Dissolve 1.0 g of potassium chloride in 100 mL of water. Adjust the pH of the solution to 6.5 (using a pH meter as described on page 43) with 0.01 N hydrochloric acid or 0.01 N sodium hydroxide. Add 0.15 mL of the 0.1% solution reserved from the test for clarity of solution. The color of solution should be yellow. Titrate the solution with 0.01 N sodium hydroxide to a pH of 8.5. The color of the solution should be blue. Not more than 0.3 mL of 0.01 N sodium hydroxide should be consumed in the entire titration.

Molybdenum Trioxide
Molybdenum(VI) Oxide
Molybdic Acid Anhydride

MoO_3 Formula Wt 143.94

CAS Number 1313–27–5

REQUIREMENTS

Assay . . . . $\geq$99.5% MoO_3

MAXIMUM ALLOWABLE

Insoluble in dilute ammonium hydroxide. 0.01%
Chloride (Cl) 0.002%
Nitrate (NO_3). . . . . Passes test
Arsenate, phosphate, and silicate (as SiO_2) 0.001%
Phosphate (PO_4) . . . . 5 ppm
Sulfate (SO_4) 0.02%

Ammonium (NH_4) 0.002%

Heavy metals (as Pb) 0.005%

TESTS

ASSAY. (By complexometric titration of molybdenum). Weigh approximately 0.6 g and dissolve in 100 mL of dilute NH_4OH. Adjust the pH of the solution to 4.0 with dilute nitric acid. Add saturated hexamethylenetetramine solution to a pH of 5–6. Heat the solution to 60 °C and titrate with 0.1 M lead nitrate. Using 0.1% PAR (4-[2-pyridylazo]resorcinol) indicator solution, titrate from yellow color to the first permanent pink endpoint. One milliliter of 0.1 N lead nitrate corresponds to 0.01439 g of MoO_3.

$$\% \ MoO_3 = [mL \times M \ Pb(NO_3)_2 \times 14.394] \ / \ Sample \ wt \ (g)$$

INSOLUBLE IN DILUTE AMMONIUM HYDROXIDE. Dissolve 10 g in 120 mL of dilute ammonium hydroxide (1 + 6), heat to boiling, and digest in a covered beaker on a hot plate (≈100 °C) for 2 h. Filter through a tared filtering crucible, wash thoroughly, and dry at 105 °C.

CHLORIDE. Dissolve 1.0 g in 4 mL of ammonium hydroxide, heat on a hot plate (≈100 °C) until completely dissolved, and evaporate to dryness. Dissolve the residue in 20 mL of water, filter if necessary through a chloride-free filter, add 4 mL of nitric acid, and dilute with water to 30 mL. To 15 mL add 1 mL of silver nitrate reagent solution. Any turbidity should not exceed that produced by 0.01 mg of chloride ion (Cl) in an equal volume of solution containing the quantities of reagents used in the test. The comparison is best made by the general method for chloride in colored solutions, page 27.

NITRATE. Grind 1.0 g with 9 mL of water and 1 mL of sodium chloride solution containing 5 mg of sodium chloride. Add 0.20 mL of indigo carmine reagent solution and 10 mL of sulfuric acid. The blue color should not be completely discharged in 5 min. (Limit about 0.003%)

ARSENATE, PHOSPHATE, AND SILICATE. Dissolve 2.5 g in 60 mL of water and enough silica-free ammonium hydroxide to effect dissolution (about 4 mL) in a platinum dish. For the control, dissolve 0.5 g in 60 mL of water and enough silica-free ammonium hydroxide to effect dissolution and add 0.02 mg of silica (SiO_2). Heat on a hot plate (≈100 °C) until the solutions are neutral as determined with an external indicator (at least 30 min). Cool and adjust the pH to between 3 and 4 with an external indicator. Transfer to beakers and dilute with water to 80 mL. Add enough bromine water to impart a distinct yellow color to the solutions. Adjust the pH of each solution to 1.8 with dilute hydrochloric acid (1 + 9) (using a pH meter). Heat just to boiling and allow to cool to room temperature. (If a precipitate forms, it will dissolve when the solution is acidified in the next operation.) Add 10 mL of hydrochloric acid and dilute with water to 100 mL. Transfer the solutions to separatory funnels and add 30 mL of 4-methyl-2-pentanone and 1 mL of butyl alcohol. Shake vigorously and allow to separate. Draw off and discard the aqueous phase. Wash the ketone phase three times with 10-mL portions of dilute hydrochloric acid $(1 + 99)$, discarding the aqueous phase each time. To the washed ketone phase add 10 mL of dilute hydrochloric acid $(1 + 99)$ to which has just been added 0.2 mL of a freshly prepared 2% solution of stannous chloride dihydrate in hydrochloric acid. Any blue color in the solution of the sample should not exceed that in the control.

PHOSPHATE. (Page 32, Method 3).

SULFATE. Boil 1.0 g with 15 mL of dilute nitric acid $(1 + 2)$ for 5 min. Cool thoroughly, dilute with water to 50 mL, mix well, and filter. Evaporate 10 mL of the filtrate to dryness on a hot plate ($\approx$100 °C). Dissolve the residue in 4 mL of water plus 1 mL of dilute hydrochloric acid $(1 + 19)$, filter if necessary through a small filter, wash with two 2-mL portions of water, and dilute with water to 10 mL. Add 1 mL of barium chloride reagent solution. Any turbidity should not exceed that produced by 0.04 mg of sulfate ion (SO_4) in an equal volume of solution containing the quantities of reagents used in the test. Compare 10 min after adding the barium chloride to the sample and standard solutions.

AMMONIUM. Dissolve 1.0 g in 15 mL of freshly boiled 10% sodium hydroxide reagent solution and dilute with water to 100 mL. To 50 mL add 2 mL of Nessler reagent. Any color should not exceed that produced by 0.01 mg of ammonium ion (NH_4) in an equal volume of solution containing 1.5 mL of 10% sodium hydroxide reagent solution and 2 mL of Nessler reagent.

HEAVY METALS. Dissolve 1.0 g in 15 mL of 10% sodium hydroxide solution, add 2 mL of ammonium hydroxide, and dilute with water to 40 mL. For the control, add 0.025 mg of lead ion (Pb) to 10 mL of the solution and dilute with water to 40 mL. For the sample dilute the remaining solution with water to 40 mL. Add 10 mL of freshly prepared hydrogen sulfide water to each and mix. Any color in the solution of the sample should not exceed that in the control.

Molybdic Acid, 85%

CAS Number 7782–91–4

NOTE. This reagent consists largely of an ammonium molybdate.

REQUIREMENTS

Assay . . . . $\geq$85.0% MoO_3

MAXIMUM ALLOWABLE
Insoluble in dilute ammonium hydroxide 0.01%

Chloride (Cl) 0.002%
Arsenate, phosphate, and silicate (as SiO_2). 0.001%
Phosphate (PO_4) . . . . 5 ppm
Sulfate (SO_4) 0.2%
Heavy Metals (as Pb) 0.003%

TESTS

ASSAY. (By complexometric titration of molybdenum). Weigh approximately 0.6 g and dissolve in 100 mL of dilute NH_4OH. Adjust the pH of the solution to 4.0 with dilute nitric acid. Add saturated hexamethylenetetramine solution to a pH of 5–6. Heat the solution to 60 °C and titrate with 0.1 M lead nitrate. Using 0.1% PAR (4-[2-pyridylazo]resorcinol) indicator solution, titrate from yellow color to the first permanent pink endpoint. One milliliter of 0.1 N lead nitrate corresponds to 0.01439 g of MoO_3.

$$\% \text{ MoO}_3 = [\text{mL} \times M \text{ Pb(NO}_3)_2 \times 14.394] / \text{Sample wt (g)}$$

INSOLUBLE IN DILUTE AMMONIUM HYDROXIDE. Dissolve 10 g in 100 mL of dilute ammonium hydroxide (1 + 9), heat to boiling, and digest in a covered beaker on a hot plate (≈100 °C) for 2 h. Filter through a tared filtering crucible, wash thoroughly, and dry at 105 °C.

CHLORIDE. Dissolve 1.0 g in 4 mL of ammonium hydroxide, heat on a hot plate (≈100 °C) until completely dissolved, and evaporate to dryness. Dissolve the residue in 20 mL of water, add 4 mL of nitric acid, and dilute with water to 30 mL. To 15 mL add 1 mL of silver nitrate reagent solution. Any turbidity should not exceed that produced by 0.01 mg of chloride ion (Cl) in an equal volume of solution containing the quantities of reagents used in the test. The comparison is best made by the general method for chloride in colored solutions, page 27.

ARSENATE, PHOSPHATE, AND SILICATE. Dissolve 2.5 g in 60 mL of water and enough silica-free ammonium hydroxide to effect dissolution (about 4 mL) in a platinum dish. For the control, dissolve 0.5 g in 60 mL of water and enough silica-free ammonium hydroxide to effect dissolution and add 0.02 mg of silica (SiO_2). Heat on a hot plate (≈100 °C) until the solutions are neutral as determined with an external indicator (at least 30 min). Cool and adjust the pH to between 3 and 4 with an external indicator. Transfer to beakers and dilute with water to 80 mL. Add enough bromine water to impart a distinct yellow color to the solutions. Adjust the pH of each solution to 1.8 with dilute hydrochloric acid (1 + 9) (using a pH meter). Heat just to boiling and allow to cool to room temperature. (If a precipitate forms, it will dissolve when the solution is acidified in the next operation.) Add 10 mL of hydrochloric acid and dilute with water to 100 mL. Transfer the solutions to separatory funnels and add 30 mL of 4-methyl-2-pentanone and 1 mL of butyl alcohol. Shake vigorously and allow to separate.

Draw off and discard the aqueous phase. Wash the ketone phase 3 times with 10-mL portions of dilute hydrochloric acid (1 + 99), discarding the aqueous phase each time. To the washed ketone phase add 10 mL of dilute hydrochloric acid (1 + 99), to which has just been added 0.2 mL of a freshly prepared 2% solution of stannous chloride in hydrochloric acid. Any blue color in the solution of the sample should not exceed that in the control.

PHOSPHATE. (Page 32, Method 3).

SULFATE. Boil 1.0 g with 15 mL of dilute nitric acid (1 + 2) for 5 min. Cool thoroughly, dilute with water to 100 mL, and filter. Evaporate 2 mL of the filtrate to dryness on a hot plate (≈100 °C), dissolve the residue in 4 mL of water plus 1 mL of dilute hydrochloric acid (1 + 19), and filter if necessary through a small filter. Wash with two 2-mL portions of water, dilute with water to 10 mL, and add 1 mL of barium chloride reagent solution. Any turbidity should not exceed that produced by 0.04 mg of sulfate ion (SO_4) in an equal volume of solution containing the quantities of reagents used in the test. Compare 10 min after adding the barium chloride to the sample and standard solutions.

HEAVY METALS. Dissolve 1.3 g in 15 mL of 10% sodium hydroxide, add 2 mL of ammonium hydroxide, and dilute with water to 40 mL. For the control, add 0.02 mg of lead ion (Pb) to 10 mL of the solution and dilute with water to 40 mL. For the sample dilute the remaining portion with water to 40 mL. Add 10 mL of freshly prepared hydrogen sulfide water to each and mix. Any color in the solution of the sample should not exceed that in the control.

Morpholine

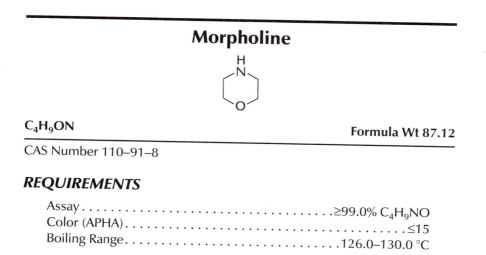

C₄H₉ON

Formula Wt 87.12

CAS Number 110–91–8

REQUIREMENTS

Assay. . . . . ≥99.0% C₄H₉NO
Color (APHA). . . . .≤15
Boiling Range. . . . . .126.0–130.0 °C

TESTS

ASSAY. (By acid–base titrimetry). Add 0.30–0.40 mL of mixed indicator solution to 50 mL of water in a 250-mL flask, and neutralize by the dropwise addition of 0.1 N hydrochloric acid just to the disappearance of the green color. Weigh accurately 1.4–1.6 g of the sample into the flask and swirl to effect complete dissolution. Titrate with 0.5 N hydrochloric acid to the disappearance of the green color. One milliliter of 0.5 N hydrochloric acid corresponds to 0.04356 g of C_4H_9NO.

Mixed Indicator Solution. Prepare separate 0.1% solutions of bromcresol green in methanol and the sodium salt of methyl red in water. Mix 5 parts by volume of the bromcresol green solution with 1 part of the methyl red solution.

COLOR (APHA). (Page 19).

BOILING RANGE. Distill 100 mL by the method described on page 18. The corrected initial boiling point should not lie below 126 °C, and the corrected dry point should not be above 130 °C. The barometric pressure correction for morpholine is + 0.045 °C per mm under 760 mm and –0.045 °C per mm above 760 mm. (The boiling point for pure morpholine at 760-mm mercury is 128.2 °C.)

Murexide

5,5'-Nitrilobarbituric Acid Monoammonium Salt

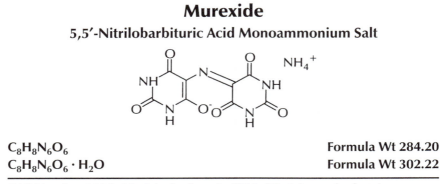

$C_8H_8N_6O_6$ **Formula Wt 284.20**
$C_8H_8N_6O_6 \cdot H_2O$ **Formula Wt 302.22**

CAS Number 3051–09–0 (anhydrous); 6032–80–0 (monohydrate)

NOTE. This material is available in anhydrous and monohydrate forms. The tests are identical for both forms.

REQUIREMENTS

Suitability as complexometric indicator . . . . Passes test

TESTS

SUITABILITY AS COMPLEXOMETRIC INDICATOR. Dissolve 0.1 g in 100 mL of water. Add 1 mL to 100 mL of water, then add 10 mL of pH 10 ammoniacal buffer, and divide into 2 equal portions, A and B. Titrate A with nickel standard solution (0.01 mg Ni/mL) until the blue component of the color is discharged. The color of the solution changes from purple to red. Use portion B as a control. The purple color is restored by adding 0.1 mL of 0.1 M EDTA. Portion A matches portion B.

Naphthol Green B
Acid Green 1

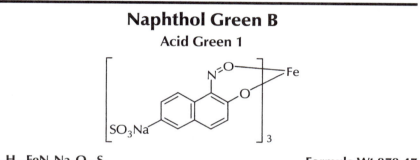

$$C_{30}H_{15}FeN_3Na_3O_{15}S_3 \qquad\qquad \textbf{Formula Wt 878.47}$$

CAS Number 19381–50–1

REQUIREMENTS

Clarity of solution . . . . Passes test
Suitability as a mixed indicator for complexometry . . . . Passes test

TESTS

CLARITY OF SOLUTION. Dissolve 0.02 g in 100 mL water which contains 0.5 mL of ammonium hydroxide. Not more than a trace of turbidity or insoluble matter should remain.

SUITABILITY AS A MIXED INDICATOR FOR COMPLEXOMETRY. Using the test sample, prepare the metalphthalein-screened indicator solution as defined on page 87. Mix 50 mL of water and 50 mL of alcohol. Add 10.0 mL of 0.1 M EDTA, 10 mL of ammonium hydroxide, and 0.3 mL of the metalphthalein-screened indicator solution. Titrate with 0.1 M barium nitrate. The color change is from gray-green to magenta. Add back 0.05 mL of 0.1 M EDTA. The color should return to a gray-green.

> **Barium nitrate, 0.1 M.** Dissolve 0.654 g of barium nitrate in water and dilute to 25.0 mL in a volumetric flask.

N-(1-Naphthyl)ethylenediamine Dihydrochloride

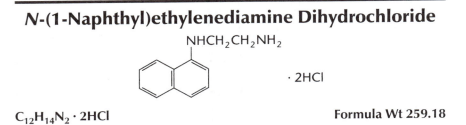

$$NHCH_2CH_2NH_2$$

· 2HCl

$C_{12}H_{14}N_2 · 2HCl$ **Formula Wt 259.18**

CAS Number 1465–25–4

REQUIREMENTS

Sensitivity to sulfanilamide . . . . Passes test
Solubility . . . . Passes test

MAXIMUM ALLOWABLE

Water (H_2O) 5%

TESTS

SENSITIVITY TO SULFANILAMIDE. In two different Nessler tubes dilute 1.0 and 2.5 mL of sulfanilamide solution with water to 100 mL. For the blank, omit the sulfanilamide from a third tube and fill with water to 100 mL. To each tube add 8 mL of trichloroacetic acid solution and 1 mL of sodium nitrite solution. After 3 min add 1 mL of ammonium sulfamate solution, then after 2 additional min add 1 mL of sample solution. The tubes containing the sulfanilamide should turn pink to red, the higher concentration producing the deeper color.

> *Sulfanilamide Solution.* Dissolve 0.02 g of sulfanilamide in 100 mL of water.

> *Trichloroacetic Acid Solution.* Dissolve 15 g of trichloroacetic acid in 100 mL of water.

> *Sodium Nitrite Solution.* Dissolve 0.1 g of sodium nitrite in 100 mL of water.

> *Ammonium Sulfamate Solution.* Dissolve 0.5 g of ammonium sulfamate in 100 mL of water.

> *Sample Solution.* Dissolve 0.1 g of *N*-(1-naphthyl)ethylenediamine dihydrochloride in 100 mL of water.

SOLUBILITY. A solution of 1 g in 50 mL of water is clear and dissolution is complete.

WATER. (Page 55, Method 1). Use 3.0 g of the sample.

Nickel Sulfate Hexahydrate
Nickel(II) Sulfate Hexahydrate
Nickelous Sulfate Hexahydrate

$NiSO_4 \cdot 6H_2O$

Formula Wt 262.85

CAS Number 10101–97–0

REQUIREMENTS

Assay . . . . 98.0–102.0% $NiSO_4 \cdot 6H_2O$

MAXIMUM ALLOWABLE

Insoluble matter. 0.005%
Chloride (Cl) 0.001%
Nitrogen compounds (as N). 0.002%
Calcium (Ca) 0.005%
Cobalt (Co) 0.002%
Copper (Cu). 0.005%
Iron (Fe). 0.001%
Magnesium (Mg) 0.005%
Manganese (Mn) 0.002%
Potassium (K). 0.01%
Sodium (Na) 0.05%

TESTS

ASSAY. (By complexometric titration of nickel). Weigh accurately 1 g, transfer to a 400-mL beaker, and dissolve in 250 mL of water. Add 10 mL of pH 10 ammonical buffer solution and about 50 mg of murexide indicator mixture. Titrate immediately with standard 0.1 M EDTA to a purple color. One milliliter of 0.1 M EDTA corresponds to 0.02628 g of $NiSO_4 \cdot 6H_2O$.

INSOLUBLE MATTER. (Page 15). Use 20 g dissolved in 150 mL of water.

CHLORIDE. Dissolve 1.0 g in 15 mL of water, filter if necessary through a small chloride-free filter, and add 1 mL of nitric acid and 1 mL of silver nitrate reagent solution. Any turbidity should not exceed that produced by 0.01 mg of chloride ion (Cl) in an equal volume of solution containing the quantities of reagents used in the test. The comparison is best made by the general method for chloride in colored solutions, page 27.

NITROGEN COMPOUNDS. (Page 31). Use 1.0 g. For the standard, use 0.02 mg of nitrogen (N).

CALCIUM, COBALT, COPPER, IRON, MAGNESIUM, MANGANESE, POTASSIUM, AND SODIUM. (By flame AAS, page 39).

Sample Stock Solution. Dissolve 20.0 g of sample in 75 mL of water and 2 mL of nitric acid. Transfer to a 100-mL volumetric flask and dilute to the mark with water (1 mL = 0.20 g).

Element	Wavelength (nm)	Sample Wt (g)	Standard Added (mg)	Flame Type*	Background Correction
Ca	422.7	2.0	0.05; 0.10	N/A	No
Co	240.7	2.0	0.02; 0.04	A/A	Yes
Cu	324.8	2.0	0.05; 0.10	A/A	Yes
Fe	248.3	4.0	0.04; 0.08	A/A	Yes
Mg	285.2	0.20	0.005; 0.01	A/A	Yes
Mn	279.5	2.0	0.02; 0.04	A/A	Yes
K	766.5	0.20	0.01; 0.02	A/A	No
Na	589.0	0.04	0.01; 0.02	A/A	No

*A/A is air/acetylene; N/A is nitrous oxide/acetylene.

Ninhydrin
1*H*-Indene-1,2,3-trione Monohydrate
1,2,3-Triketohydrindene Monohydrate

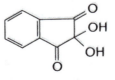

C$_9$H$_6$O$_4$ **Formula Wt 178.14**

CAS Number 485–47–2

REQUIREMENTS

Appearance . . . . White to brownish-white crystals
Identification and melting point . . . . Passes test
Solubility . . . . Passes test
Sensitivity to amino acids. . . . . Passes test

TESTS

IDENTIFICATION AND MELTING POINT. A sample heated above 150 °C becomes purple and melts with decomposition at approximately 250 °C.

SOLUBILITY. A solution of 0.10 g of sample in 10 mL of water is clear and pale yellow in color. Dissolution is complete. Retain the solution for the sensitivity test.

SENSITIVITY TO AMINO ACIDS. Dissolve 0.10 g of beta-alanine in 10 mL of warm water and heat to boiling for 5 min. Add 0.50 mL of the ninhydrin solution reserved from the solubility test and continue heating. A rose color starts to develop within a minute and intensifies to a deep violet color on continued heating.

Nitric Acid

HNO$_3$ **Formula Wt 63.01**

CAS Number 7697–37–2

NOTE. This material may darken during storage due to photochemical reactions.

REQUIREMENTS

Appearance . . . . .Colorless and free from suspended matter or sediment

Assay. . . . . 68.0–70.0% HNO$_3$

MAXIMUM ALLOWABLE

Color (APHA). 10
Residue after ignition . . . . 5 ppm
Chloride (Cl) . . . . 0.5 ppm
Sulfate (SO$_4$) . . . . 1 ppm
Arsenic (As) . . . . 0.01 ppm
Heavy metals (as Pb) . . . . 0.2 ppm
Iron (Fe). . . . . 0.2 ppm

TESTS

APPEARANCE. Mix the acid in the original container, transfer 10 mL to a test tube (20 mm × 150 mm) and compare with distilled water in a similar tube. The liquids should be equally clear and free from suspended matter.

ASSAY. (By acid–base titrimetry). Tare a small glass-stoppered flask containing about 15 mL of water. Quickly pipet about 2 mL of the sample under the water surface, stopper, cool, and weigh accurately. Dilute with about 40 mL of water, add 0.15 mL of methyl orange indicator solution, and titrate with 1 N sodium hydroxide. One milliliter of 1 N sodium hydroxide corresponds to 0.06301 g of HNO$_3$.

COLOR (APHA). (Page 19).

RESIDUE AFTER IGNITION. To 200 g (140 mL) in a tared dish add 0.05 mL of sulfuric acid, evaporate as far as possible on low heat, and heat gently to volatilize the excess sulfuric acid. Finally, ignite at 800 ± 25 °C for 15 min.

CHLORIDE. Dilute 20 g (14 mL) with 10 mL of water and add 1 mL of silver nitrate reagent solution. Prepare a standard containing 0.01 mg of chloride ion (Cl) in 20 mL of water and add 1 mL of silver nitrate reagent solution. Evaporate the solutions to dryness on a hot plate ($\approx$100 °C). Dissolve the residues in 0.5 mL of ammonium hydroxide, dilute with water to 20 mL, and add 1.5 mL of nitric acid. Any turbidity in the solution of the sample should not exceed that in the standard.

SULFATE. (Page 33, Method 3).

ARSENIC. (Page 25). To 300 g (210 mL) in a 250-mL beaker add 5 mL of sulfuric acid, and evaporate on a hot plate ($\approx$100 °C) to a volume of about 5 mL. Transfer the beaker to a hot plate in a hood and heat to dense fumes of sulfur trioxide. Cool, cautiously wash down the beaker with about 10 mL of water, evaporate again to dense fumes of sulfur trioxide, and continue the fuming for 15 min. Cool, wash down the beaker with about 10 mL of water, and repeat the fuming. Cool, cautiously wash the solution into a generator flask with sufficient water to make 35 mL, add 0.5 g of hydrazine sulfate, and proceed as described. For the standard, use 0.003 mg of arsenic (As).

HEAVY METALS. (Page 28, Method 1). To 100 g (70 mL) in a beaker add about 10 mg of sodium carbonate, evaporate to dryness on a hot plate ($\approx$100 °C), dissolve the residue in about 20 mL of water, and dilute with water to 25 mL.

IRON. (Page 30, Method 1). Evaporate 50 g (35 mL) to dryness, dissolve the residue in 2 mL of hydrochloric acid, dilute with water to 50 mL, and use the solution without further acidification.

Nitric Acid, 90%

HNO$_3$ **Formula Wt 63.01**

CAS Number 7697–37–2

REQUIREMENTS

Assay . . . . $\geq$90.0% HNO$_3$
Dilution test . . . . Passes test

MAXIMUM ALLOWABLE

Residue after ignition 0.002%
Dissolved oxides . . . . 0.1% as N$_2$O$_3$
Chloride (Cl) . . . . 0.7 ppm
Sulfate (SO$_4$) . . . . 5 ppm
Arsenic (As) . . . . 0.3 ppm
Heavy metals (as Pb) . . . . 5 ppm
Iron (Fe) . . . . 2 ppm

TESTS

ASSAY. (By acid–base titrimetry). Tare a small glass-stoppered flask containing about 15 mL of water. Quickly pipet about 2 mL of the sample under the water surface, stopper, cool, and weigh accurately. Dilute with about 40 mL of water, add 0.15 mL of methyl orange indicator solution, and titrate with 1 N sodium hydroxide. One milliliter of 1 N sodium hydroxide corresponds to 0.06301 g of HNO_3.

DILUTION TEST. Dilute 1 volume of acid with 3 volumes of water; mix and allow to stand for 1 h. No turbidity or precipitate should be observed.

RESIDUE AFTER IGNITION. To 50 g (33 mL) in a tared dish add 0.10 mL of sulfuric acid, evaporate as far as possible on low heat, and heat gently to volatilize the excess sulfuric acid. Finally, ignite at 800 ± 25 °C for 15 min.

DISSOLVED OXIDES. Dilute 10 g (6.6 mL) to 150 mL with cold water in a casserole. Add 20 mL of dilute sulfuric acid (1 + 4) and titrate with 0.1 N potassium permanganate, rapidly at first, then slowly till the pink color lasts 3 min. Not more than 5.0 mL of the permanganate solution should be required. One milliliter of 0.1 N permanganate solution corresponds to 0.0019 g of N_2O_3.

CHLORIDE. Dilute 15 g (10 mL) with 10 mL of water and add 1 mL of silver nitrate reagent solution. Prepare a standard of 0.01 mg of chloride ion (Cl) in 20 mL of water and add 1 mL of silver nitrate reagent solution. Evaporate the solutions to dryness on a hot plate ($\approx$100 °C). Dissolve the residues in 0.5 mL of ammonium hydroxide, dilute with water to 20 mL, and add 1.5 mL of nitric acid. Any turbidity in the solution of the sample should not exceed that in the standard.

SULFATE. (Page 33, Method 3).

ARSENIC. (Page 25). To 10 g (6.6 mL) in a 150-mL beaker add 5 mL of sulfuric acid and evaporate on a hot plate ($\approx$100 °C) to a volume of about 5 mL. Transfer the beaker to a hot plate in a hood and heat to dense fumes of sulfur trioxide. Cool, cautiously wash down the sides of the beaker with about 10 mL of water, evaporate again to dense fumes of sulfur trioxide, and continue the fuming for 15 min. Cool, cautiously wash down the sides of the beaker with about 10 mL of water, and repeat the fuming. Cool, cautiously wash the solution into an arsine generator flask with 35 mL of water, add 0.5 g of hydrazine sulfate, and proceed as described. For the standard, use 0.003 mg of arsenic (As).

HEAVY METALS. (Page 28, Method 1). To 4 g (2.7 mL) add about 10 mg of sodium carbonate, evaporate to dryness on a hot plate ($\approx$100 °C), dissolve the residue in about 20 mL of water, and dilute with water to 25 mL.

IRON. (Page 30, Method 1). Evaporate 5 g (3.3 mL) to dryness, dissolve the residue in 2 mL of hydrochloric acid, dilute with water to 50 mL, and use the solution without further acidification.

Nitric Acid, Ultratrace
(suitable for use in ultratrace elemental analysis)

HNO_3 **Formula Wt 63.01**

CAS number 7697–37–2

NOTE. This material may darken during storage due to photochemical reactions. Reagent must be packaged in a preleached Teflon bottle and used in a clean environment to maintain purity.

REQUIREMENTS

Appearance . . . . Clear
Assay . . . . 65–70% HNO_3

MAXIMUM ALLOWABLE

Chloride (Cl) . . . . 0.5 ppm
Sulfate (SO_4) . . . . 1 ppm
Aluminum (Al) . . . . 1 ppb
Barium (Ba) . . . . 1 ppb
Boron (B) . . . . 5 ppb
Cadmium (Cd) . . . . 1 ppb
Calcium (Ca) . . . . 1 ppb
Chromium (Cr) . . . . 1 ppb
Cobalt (Co) . . . . 1 ppb
Copper (Cu) . . . . 1 ppb
Iron (Fe) . . . . 5 ppb
Lead (Pb) . . . . 1 ppb
Lithium (Li) . . . . 1 ppb
Magnesium (Mg) . . . . 1 ppb
Manganese (Mn) . . . . 1 ppb
Molybdenum (Mo) . . . . 1 ppb
Mercury (Hg) . . . . 1 ppb
Potassium (K) . . . . 1 ppb
Silicon (Si) . . . . 5 ppb
Sodium (Na) . . . . 5 ppb
Strontium (Sr) . . . . 1 ppb
Tin (Sn) . . . . 1 ppb
Titanium (Ti) . . . . 1 ppb
Vanadium (V) . . . . 1 ppb
Zinc (Zn) . . . . 1 ppb
Zirconium (Zr) . . . . 1 ppb

TESTS

ASSAY. (By acid–base titrimetry). Tare a small, glass-stoppered flask containing about 15 mL of water. Quickly pipet about 2 mL of the sample under the water surface, stopper, and weigh accurately. Dilute with about 40 mL of water, add 0.15 mL of methyl orange indicator solution, and titrate with 1 N sodium hydroxide. One milliliter of 1 N sodium hydroxide corresponds to 0.06301 g of HNO_3.

$$\% \ HNO_3 = (mL \times N \ NaOH \times 6.30) \ / \ Sample \ wt \ (g)$$

CHLORIDE. See test for nitric acid, page 440

SULFATE. (Page 33, Method 3).

TRACE METALS. Determine the aluminum, barium, cadmium, calcium, chromium, cobalt, copper, iron, lead, lithium, magnesium, manganese, molybdenum, nickel, potassium, sodium, strontium, tin, titanium, vanadium, zinc, and zirconium by the ICP–OES method described on page 42.

Determine the mercury by the CVAAS procedure described on page 40.

Nitrilotriacetic Acid
N,N-Bis(carboxymethyl)glycine

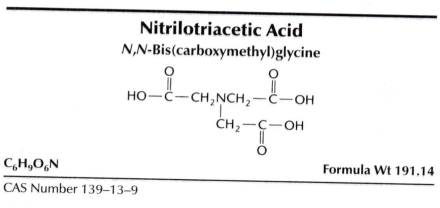

$C_6H_9O_6N$ **Formula Wt 191.14**

CAS Number 139–13–9

REQUIREMENTS

Assay. . . . .≥98.0% $C_6H_9O_6N$
Clarity of solution . . . . Passes test

TESTS

ASSAY (BY ALKALIMETRY) AND CLARITY OF SOLUTION. To check first for clarity of solution, weigh accurately 0.35 g, transfer to a beaker, and dissolve in 100 mL of water. The sample should dissolve completely, with heating and stirring if necessary, to produce a clear solution. Add 0.15 mL of phenolphthalein indicator solution, and titrate with 0.1 N sodium hydroxide. One milliliter of 0.1 N sodium hydroxide corresponds to 0.006372 g of $C_6H_9O_6N$.

Nitrobenzene

NO₂

C₆H₅NO₂   **Formula Wt 123.11**

CAS Number 98–95–3

REQUIREMENTS

Assay . . . . ≥99.0% C₆H₅NO₂

MAXIMUM ALLOWABLE

Residue after evaporation. 0.005%
Water-soluble titrable acid . . . . 0.0005 meq/g
Chloride (Cl) . . . . 5 ppm

TESTS

ASSAY. Analyze the sample by gas chromatography using the general parameters cited on page 70. The following specific conditions are also required.

Column: Type I, methyl silicone

Measure the area under all peaks and calculate the nitrobenzene content in area percent. Correct for water content.

RESIDUE AFTER EVAPORATION. (Page 16). Evaporate 20 g (16.7 mL) to dryness in a tared dish on a hot plate (≈100 °C) and dry the residue at 105 °C for 30 min.

WATER-SOLUBLE TITRABLE ACID. Shake 25 g (21 mL) with 60 mL of water for 1 min and allow the layers to separate. Draw off and discard the nitrobenzene. Titrate 48 mL of the water extract with 0.01 N sodium hydroxide, using 0.15 mL of bromphenol blue indicator solution. Not more than 1.0 mL of 0.01 N sodium hydroxide should be required.

CHLORIDE. Dilute the remaining 12 mL of water from the test for acidity with water to 25 mL, filter through a chloride-free filter, and dilute 10 mL of the filtrate with water to 20 mL. Add 1 mL of nitric acid and 1 mL of silver nitrate reagent solution. Any turbidity should not exceed that produced by 0.01 mg of chloride ion (Cl) in an equal volume of solution containing the quantities of reagents used in the test.

Nitromethane

CH₃NO₂

CH_3NO_2

Formula Wt 61.04

CAS Number 75–52–5

REQUIREMENTS

Appearance . . . . Clear
Assay . . . . ≥95% CH_3NO_2

MAXIMUM ALLOWABLE

Color (APHA) 10
Water (H_2O) 0.05%

TESTS

ASSAY. Analyze the sample by gas chromatography using the general parameters cited on page 70. The following specific conditions are also required.

Column: Type III, polyethylene glycol

Measure the area under all peaks and calculate the nitromethane content in area percent. Correct for water content.

COLOR (APHA). (Page 19).

WATER. (Page 56, Method 2). Use 100 µL (100 ng) of the sample.

1-Octanol

n-Octyl Alcohol
(suitable for use in determining TSCA partition coefficients)

$C_8H_{18}O$

Formula Wt 130.23

CAS Number 111–87–5

REQUIREMENTS

General Use

Appearance . . . . Clear
Assay . . . . ≥99% $C_8H_{18}O$

MAXIMUM ALLOWABLE

Color (APHA) 10
Residue after evaporation 0.004%
Titrable acid . . . . 0.0002 meq/g

Specific Use

Ultraviolet Spectrophotometry

Wavelength (nm)	Absorbance (AU)
280–400	0.01
260	0.05
240	0.10
230	0.5
220	1.0

TESTS

ASSAY. Analyze the sample by gas chromatography using the general parameters cited on page 70. The following specific conditions are also required.

Column: Type I, methyl silicone

Measure the area under all peaks and calculate the 1-octanol content in area percent. Correct for water content.

COLOR (APHA). (Page 19).

RESIDUE AFTER EVAPORATION. (Page 16). Evaporate 100 g (120.5 mL) to dryness in a tared platinum dish on a hot plate (≈ 100 °C) in a well-ventilated hood, and dry the residue at 105 °C for 30 min.

TITRABLE ACID. To 25 mL of ethyl alcohol add 0.15 mL of phenolphthalein indicator, neutralize the solution with either 0.01 N sodium hydroxide or 0.01 N hydrochloric acid. Add 21.0 g (25 mL) of sample and titrate with 0.01 N sodium hydroxide to a pink end point. Not more than 0.4 mL should be required.

ULTRAVIOLET SPECTROPHOTOMETRY. Use the procedure on page 73 to determine the absorbance.

2-Octanol
Methyl Hexyl Carbinol
2-Capryl Alcohol

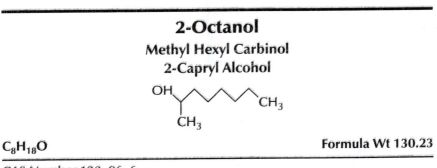

$C_8H_{18}O$ **Formula Wt 130.23**

CAS Number 123–96–6

REQUIREMENTS

Appearance . . . . Clear
Assay . . . . ≥99% $C_8H_{18}O$

MAXIMUM ALLOWABLE

Color (APHA) 10
Residue after evaporation 0.004%
Titrable acid . . . . 0.0002 meq/g

TESTS

ASSAY. Analyze the sample by gas chromatography using the general parameters cited on page 70. The following specific conditions are also required.

Column: Type I, methyl silicone

Measure the area under each peak and calculate the 2-octanol content in area percent.

COLOR (APHA). (Page 19).

RESIDUE AFTER EVAPORATION. (Page 16). Evaporate 100 g (122 mL) to dryness in a tared platinum dish on a hot plate (≈100 °C) in a fume hood and dry the residue at 105 °C for 30 min.

TITRABLE ACID. To 25 mL of ethyl alcohol add 0.15 mL of phenolphthalein indicator and neutralize the solution with either 0.01 N sodium hydroxide or 0.01 N hydrochloric acid. Add 20.0 g (24.4 mL) of sample and titrate with 0.01 N sodium hydroxide to a pink endpoint. Not more than 0.4 mL is required.

Osmium Tetroxide
Osmic Acid

OsO_4   **Formula Wt 254.20**

CAS Number 20816–12–0

CAUTION. Osmium tetroxide vapors are poisonous and highly irritating to the eyes and respiratory membranes. Use only in a well-ventilated fume hood.

REQUIREMENTS

Assay . . . . ≥98.0% OsO_4
Clarity of solution . . . . Passes test
Identification . . . . Passes test

TESTS

CLARITY OF SOLUTION. Accurately weigh 0.20 g and add to 5 mL of reagent alcohol. The resulting solution is clear and not more than slightly yellow, and no appreciable amount of insoluble residue remains. Reserve the solution for the assay test.

ASSAY AND IDENTIFICATION. (Assay by iodometry). To the solution from the clarity of solution test, add 5–7 mL of 1 N sodium hydroxide in a glass-stoppered flask. A dark red precipitate is produced (identification). Add 3 g of potassium iodide, mix to dissolve, slowly add 40 mL of 25% sulfuric acid, stopper, and let stand in a dark place for 20 min. Titrate the liberated iodine with 0.1 N sodium thiosulfate, and use starch solution as an external indicator (the reduced osmium solution is green, and the endpoint may be difficult to determine with indicator internally). Correct for a blank. One milliliter of 0.1 N sodium thiosulfate corresponds to 0.006355 g of OsO_4.

Oxalic Acid Dihydrate
Ethanedioic Acid Dihydrate

$$HO-\overset{\overset{\displaystyle O}{\|}}{C}-\overset{\overset{\displaystyle O}{\|}}{C}-OH \cdot 2H_2O$$

$C_2H_2O_4 \cdot 2H_2O$ **Formula Wt 126.07**

CAS Number 6153–56–6

REQUIREMENTS

Assay . . . . .99.5–102.5% $H_2C_2O_4 \cdot 2H_2O$
Substances darkened by hot sulfuric acid . . . . Passes test

MAXIMUM ALLOWABLE

Insoluble matter. 0.005%
Residue after ignition 0.01%
Chloride (Cl) 0.002%
Sulfate (SO_4) 0.005%
Calcium (Ca) 0.001%
Nitrogen compounds (as N). .0.001%
Heavy metals (as Pb) . . . . .5 ppm
Iron (Fe) . . . . .2 ppm

TESTS

ASSAY. (By titration of reductive capacity). Weigh accurately 0.25 g and add 50 mL of water and 5 mL of sulfuric acid. Titrate slowly with 0.1 N potassium permanganate until about 25 mL has been added. Heat the solution to about 70 °C, and complete the titration to a permanent faint pink color. One milliliter of 0.1 N potassium permanganate consumed corresponds to 0.006303 g of $H_2C_2O_4 \cdot 2H_2O$.

SUBSTANCES DARKENED BY HOT SULFURIC ACID.
Dissolve 2.0 g in 20 mL of sulfuric acid and heat at 150 °C for 30 min. Any color should not exceed that produced in a color standard of the following composition: 0.2 mL of cobalt chloride solution (5.95 g of $CoCl_2 \cdot 6H_2O$ and 2.5 mL of hydrochloric acid in 100 mL), 1.0 mL of ferric chloride solution (4.50 g of $FeCl_3 \cdot 6H_2O$ and 2.5 mL of hydrochloric acid in 100 mL), and 0.2 mL of cupric sulfate solution (6.24 g of $CuSO_4 \cdot 5H_2O$ and 2.5 mL of hydrochloric acid in 100 mL), in 20 mL of solution.

INSOLUBLE MATTER. (Page 15). Use 20.0 g dissolved in 250 mL of water.

RESIDUE AFTER IGNITION. (Page 16). Dry 10.0 g in a tared crucible or dish for 1 h at 105 °C, slowly ignite, and continue as described. Retain the residue for the test for iron.

CHLORIDE. (Page 27). Dissolve 0.5 g in 20 mL of nitric acid.

SULFATE. Digest 1.0 g of sample plus about 10 mg of sodium carbonate with 3 mL of 30% hydrogen peroxide and 0.5 mL of nitric acid in a covered beaker on a hot plate ($\approx$100 °C) until reaction ceases. Remove the cover and evaporate to dryness. Dissolve the residue in 6 mL of dilute hydrochloric acid (1 + 1) and again evaporate to dryness. Dissolve the residue in 4 mL of water plus 1 mL of dilute hydrochloric acid (1 + 19), filter if necessary through a small filter, wash the precipitate with two 2-mL portions of water, and dilute with water to 10 mL. Add 1 mL of barium chloride reagent solution. Any turbidity should not exceed that produced when a solution containing 0.05 mg of sulfate ion (SO_4) is treated exactly like the 1.0 g of sample. Compare 10 min after adding the barium chloride to the sample and standard solutions.

CALCIUM. (By flame AAS, page 39).

Sample Stock Solution. Suspend 10.0 g of sample in 50 mL of nitric acid (1 + 24) and add 25 mL of 30% hydrogen peroxide. Cover and digest on a hot plate ($\approx$100 °C) until reaction ceases. Add 5 mL more of the hydrogen peroxide and when reaction ceases, uncover and evaporate to dryness. Dissolve the residue in 80 mL of nitric acid (1 + 99), transfer to a 100-mL volumetric flask, and dilute to the mark with nitric acid (1 + 99) (1 mL = 0.10 g).

Element	Wavelength (nm)	Sample Wt (g)	Standard Added (mg)	Flame Type*	Background Correction
Ca	422.7	2.0	0.02; 0.04	N/A	No

*N/A is nitrous oxide/acetylene.

NITROGEN COMPOUNDS. (Page 31). Use 1.0 g. For the standard, use 0.01 mg of nitrogen (N).

HEAVY METALS. (Page 28, Method 1). Place 4.0 g in a covered beaker, and add about 10 mg of sodium carbonate, 5 mL of 30% hydrogen peroxide, and 0.5 mL of nitric acid. Digest on a hot plate (≈100 °C) until reaction ceases, remove the cover, and evaporate to dryness. Digest the residue with 0.5 mL of hydrochloric acid and 0.1 mL of nitric acid in a covered beaker for 15 min, remove the cover, and again evaporate to dryness. Dissolve the residue in about 20 mL of water, and dilute with water to 25 mL. For the standard, solution treat 0.02 mg of lead exactly as the 4.0 g of sample.

IRON. (Page 30, Method 1). To the residue after ignition add 3 mL of dilute hydrochloric acid (1 + 1), cover with a watch glass, and digest on a hot plate (≈100 °C) for 15 min. Remove the cover, and evaporate to dryness. Dissolve the residue in 2 mL of dilute hydrochloric acid (1 + 1), add about 40 mL of water, filter if necessary, and dilute with water to 50 mL. Use 25.0 mL (5.0 g) of the solution.

PAN
1-(2-Pyridylazo)-2-naphthol

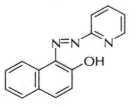

$C_{15}H_{11}N_3O$ **Formula Wt 249.27**

CAS Number 85–85–8

REQUIREMENTS

Suitability as complexometric ion indicator (Cu). . . . . . . Passes test

TESTS

SUITABILITY AS METAL ION INDICATOR (Cu). Weigh 0.01 g of sample, and dilute to 10 mL with reagent alcohol. Mix 50 mL of water, 50 mL of alcohol, 10.0

mL of 0.1 M EDTA, and 5 mL of ammonium acetate–acetic acid buffer. Add 0.2 mL of the PAN sample indicator solution, and titrate with 0.1 M cupric sulfate. The color change is from blue–green to deep blue. Add back 0.05 mL of EDTA. The color should return to blue–green.

Cupric Sulfate, 0.1 M. Dissolve 0.50 g of cupric sulfate pentahydrate in 20.0 mL of water.

PAR
4-(2′-Pyridylazo)resorcinol

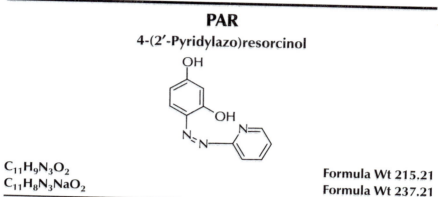

$C_{11}H_9N_3O_2$ Formula Wt 215.21
$C_{11}H_8N_3NaO_2$ Formula Wt 237.21

CAS Number 1141-59-9 free acid; 16593-81-0 monosodium salt

NOTE. This specification includes both the free acid and the salt form of the indicator.

REQUIREMENTS

Clarity of solution . . . . Passes test
Suitability as a complexometric ion indicator . . . . Passes test

TESTS

CLARITY OF SOLUTION. If the indicator is the salt form, weigh 0.10 g of sample and dissolve in 100 mL of water. If the indicator is the acid form, dissolve 0.1 g in 1 N NaOH and dilute to 100 ml maintaining the pH at 8.9. Not more than a trace of turbidity or insoluble matter should remain. Reserve the solution for the test for suitability as a complexometric ion indictor.

SUITABILITY AS A COMPLEXOMETRIC ION INDICATOR. Add 10.0 mL of 0.1 M EDTA to 100 mL of water. Adjust the pH of the solution to 4.0 with 1% nitric acid. Add saturated hexamethylenetetramine solution to a pH of 5–6. Add 0.2 mL of the 0.1% solution reserved from the clarity of solution test. Heat the solution to 60 °C and titrate with 0.1 M lead nitrate. The color of the solution should change from yellow to pink. Add back 0.05 mL of EDTA. The solution should return to the original yellow color.

Pararosaniline Hydrochloride

4-[(4-Aminophenyl)(-4-imino-2,5-cyclohexadien-1-ylidene)methyl]benzenamine Monohydrochloride

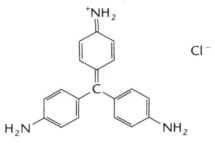

$C_{19}H_{17}N_3 \cdot HCl$ **Formula Wt 323.82**

CAS Number 569–61–9

REQUIREMENTS

Assay . . . . $\geq$99.0% $C_{19}H_{18}N_3Cl$
Suitability for SO_2 determination:
    Absorbance of reagent blank . . . . $\leq$0.170
    Calibration curve slope (µg/mL) 0.706–0.786

TESTS

ASSAY. (By spectrophotometry at 540 nm).

> **Sample Stock Solution.** Weigh accurately 0.2 g of sample. Dissolve completely by shaking with 1 N hydrochloric acid in a 100-mL volumetric flask and dilute to volume.

Dilute 1.0 mL of sample stock solution with water to the mark in a 100-mL volumetric flask. Pipet 5.0 mL into a 50-mL volumetric flask, add 5 mL of 1 M sodium acetate–acetic acid buffer, and dilute with water to volume. Allow to stand for 1 h at room temperature, and determine the absorbance at 540 nm with a spectrophotometer, using 1-cm cells at a slit width of 0.04 mm.

$$\% \ C_{19}H_{18}N_3Cl = (A \times 21.30) \ / \ W$$

where

    A = absorbance
    W = weight, in grams, of sample in 1.0 mL of sample stock solution.

> **Buffer Solution.** Dissolve 13.61 g of sodium acetate trihydrate, $CH_3COONa \cdot 3H_2O$, in water in a 100-mL volumetric flask. Add 5.7 mL of glacial acetic acid, and dilute to volume with water.

SUITABILITY FOR SO₂ DETERMINATION.

NOTE. All absorbance measurements using this procedure are temperature-dependent (0.015 absorbance units/°C), and should be performed at 22 °C.

Absorbance of Reagent Blank. Pipet 10 mL of 0.04 M potassium tetrachloromercurate and 1 mL of 0.6% sulfamic acid into a 25-mL volumetric flask. Allow to stand for 10 min, then add from a pipet 2.0 mL of 0.2% formaldehyde solution and 5.0 mL of pararosaniline reagent. Dilute to volume with water, allow to stand for 30 min, and determine the absorbance at 548 nm in 1-cm cells, using water as the reference.

Calibration Curve Slope. Pipet 1.0 mL of standard sulfite solution into a 25-mL volumetric flask. Add 9 mL of 0.04 M potassium tetrachloromercurate and 1 mL of 0.6% sulfamic acid, allow to stand for 10 min, then add from a pipet 2.0 mL of 0.2% formaldehyde solution and 5.0 mL of pararosaniline reagent. Dilute to volume with water, allow to stand for 30 min and determine the absorbance at 548 nm in 1-cm cells, using water as the reference.

$$\text{Absorbance (g/mL)} = (A_s - A_b) \,/\, W$$

where

$A_s =$ absorbance of sample
$A_b =$ absorbance of reagent blank
$W =$ weight, in micrograms, of SO_2 in 1 mL

Reagents. All six reagents cited should be freshly prepared.

Potassium Tetrachloromercurate, K_2HgCl_4, 0.04 M. Dissolve 10.86 g of mercuric chloride ($HgCl_2$) in water in a 1-L volumetric flask. Add 5.96 g of potassium chloride and 0.066 g of (ethylenedinitrilo)tetraacetic acid $[(HOCOCH_2)_2NCH_2CH_2N(CH_2COOH)]$, dissolve, and dilute with water to volume. The pH of this reagent should not be less than 5.2.

Sulfite Stock Solution. Dissolve 0.4 g of sodium sulfite (Na_2SO_3) or 0.3 g of sodium metabisulfite ($Na_2S_2O_5$) in 500 mL of deaerated water (purged with high-purity nitrogen gas or by boiling). This solution contains 320–400 µg/mL as SO_2. The actual concentration is determined by adding excess iodine solution and back-titrating with sodium thiosulfate solution that has been standardized against potassium iodate or potassium dichromate. Sulfite solutions are unstable.

For blank solution A in the back-titration, transfer 25 mL of water to a 500-mL iodine flask, and add from a pipet 50.0 mL of 0.01 N iodine. For sample solution B in a second 500-mL iodine flask, add successively from a pipet 25.0 mL of stock sulfite solution and 50.0 mL of 0.01 N iodine. Stopper the flasks, allow to react for 5 min, and titrate each with 0.01 N sodium thiosulfate to a pale yellow color. Add 5 mL of starch indicator solution, and continue the titration to the disappearance of the blue color. Calculate the concentration of SO_2 as follows:

$$SO_2 \text{ concentration, } \mu g/mL = [(A - B) \times N \times 32,030] / V$$

where

$A =$ volume, in milliliters, of thiosulfate solution consumed by blank
$B =$ volume, in milliliters, of thiosulfate solution consumed by sample
$N =$ normality of thiosulfate solution
$V =$ volume, in milliliters, of sample

Standard Sulfite Solution. Immediately after its standardization, pipet a 2-mL aliquot of stock sulfite solution into a 100-mL volumetric flask, and dilute to volume with 0.04 M potassium tetrachloromercurate. This solution is stable for 30 days if stored at 5 °C.

Sulfamic Acid, 0.6%. Dissolve 0.6 g of sulfamic acid in 100 mL of water. This solution can be kept for a few days if protected from air.

Formaldehyde, 0.2%. Dilute 5 mL of 36–38% formaldehyde with water to 1 L. Prepare this solution daily.

Pararosaniline Reagent. Pipet 20.0 mL of sample stock solution of pararosaniline (refer to assay) into a 250-mL volumetric flask. Add an additional 0.2 mL of stock for each percent the pararosaniline assay is below 100%. Add 25 mL of 3 M phosphoric acid, and dilute to volume with water.

1-Pentanol
n-Amyl Alcohol

$CH_3(CH_2)_3CH_2OH$ **Formula Wt 88.15**

CAS Number 71–41–0

REQUIREMENTS

Assay . . . .≥98.0% $CH_3(CH_2)_3CH_2OH$

MAXIMUM ALLOWABLE

Water (H_2O). 0.5%
Color (APHA) 30
Residue after evaporation. 0.003%
Acids and esters. . . . . .0.075 meq/g
Carbonyl compounds (as HCHO) 0.1%

TESTS

ASSAY. Analyze the sample by gas chromatography using the parameters cited on page 70. The following specific conditions are also required.

   Column: Type I, methyl silicone

Measure the area under all peaks and calculate the 1-pentanol content in area percent. Correct for water content.

WATER. (Page 55, Method 1). Use 25 mL (20 g) of the sample.

COLOR (APHA). (Page 19).

RESIDUE AFTER EVAPORATION. (Page 16). Use 40.0 g (50 mL) of sample, evaporate to dryness on a hot plate ($\approx$100 °C), and complete drying at 105 °C.

ACIDS AND ESTERS. Dilute 8.1 g (10 mL) with 20 mL of ethyl alcohol, add 10 mL of 0.1 N sodium hydroxide, and reflux gently for 10 min. Cool, add 0.15 mL of phenolphthalein indicator solution, and titrate the excess sodium hydroxide with 0.1 N hydrochloric acid. Run a reagent blank. The net volume of 0.1 N sodium hydroxide consumed by the sample should not exceed 6.1 mL.

CARBONYL COMPOUNDS. (By differential pulse polarography, page 50). Use 0.25 g (0.30 mL) of sample. For the standard, use 0.25 mg of formaldehyde.

Perchloric Acid

$HClO_4$ **Formula Wt 100.46**

CAS Number 7601–90–3

   NOTE. This specification applies to perchloric acid with nominal concentrations of 50%, 60%, or 70%.

REQUIREMENTS

Assay . . . . 48–50% $HClO_4$
                                           60–62% $HClO_4$
                                       69.0–72.0% $HClO_4$

MAXIMUM ALLOWABLE

Color (APHA) 10
Residue after ignition 0.003%
Silicate and phosphate (as SiO_2) . . . . 5 ppm
Chloride (Cl) 0.001%
Nitrogen compounds (as N). 0.001%
Sulfate (SO_4) 0.001%

Heavy metals (as Pb) . . . . .1 ppm
Iron (Fe) . . . . .1 ppm

TESTS

NOTE. All tests are written for 70% concentration. The densities of the 60% acid (1.54 g/mL) and the 50% acid (1.40 g/mL) are less than the density of the 70% reagent (1.67 g/mL), and proper allowance should be made for this when measured volumes are taken for relevant tests.

ASSAY. (By alkalimetric titration). Weigh accurately 3 mL, dilute with water to 60 mL, add 0.15 mL of phenolphthalein indicator solution, and titrate with 1 N sodium hydroxide. One milliliter of 1 N sodium hydroxide corresponds to 0.1005 g of $HClO_4$.

COLOR (APHA). (Page 19).

RESIDUE AFTER IGNITION. (Page 16). To 67 g (42 mL) in a tared dish add 1 mL of nitric acid, evaporate to dryness, and ignite at 800 ± 25 °C for 15 min.

SILICATE AND PHOSPHATE. Dilute 2.0 g (1.2 mL) with 50 mL of water in a platinum dish, add 1.0 g of sodium carbonate, and digest on a hot plate ($\approx$100 °C) for 15–30 min. Cool, add 5 mL of a 10% solution of ammonium molybdate, acidify with hydrochloric acid to about pH 4 (indicator paper), and transfer to a beaker. Adjust the pH to 1.8 (using a pH meter) with dilute hydrochloric acid (1 + 9). Dilute with water to 90 mL, heat just to boiling, and cool to room temperature. Add 10 mL of hydrochloric acid and dilute with water to 100 mL. Transfer to a separatory funnel, add 50 mL of 4-methyl-2-pentanone, shake vigorously, and allow to separate. Draw off and discard the aqueous phase. Wash the ketone layer three times with 20-mL portions of dilute hydrochloric acid (1 + 99), drawing off and discarding the aqueous layer each time. Add 10 mL of dilute hydrochloric acid (1 + 99), to which has just been added 0.2 mL of a freshly prepared 2% solution of stannous chloride dihydrate in hydrochloric acid. Shake vigorously and allow to separate. Any blue color should not exceed that produced by 0.01 mg of silica, SiO_2, and 1 g of sodium carbonate, Na_2CO_3, treated exactly like the sample.

CHLORIDE. (Page 27). Dilute 1 g (0.6 mL) with 25 mL of water.

NITROGEN COMPOUNDS. (Page 31). Use 5 g (3.0 mL). For the standard, use 0.05 mg of nitrogen (N).

SULFATE. Dilute 32.0 g (20 mL) with water to 160 mL and neutralize with ammonium hydroxide, using litmus paper as an indicator. Add 1 mL of hydrochloric acid, heat to boiling, add 5 mL of barium chloride reagent solution, digest in a covered beaker on the hot plate ($\approx$100 °C) for 2 h, and allow to stand overnight. If a precipitate is formed, filter, wash thoroughly, and ignite. The weight of the precipitate should not be more than 0.0008 g greater than the weight obtained from a complete blank test, including the evaporated residue from a volume of ammonium hydroxide equal to that used to neutralize the sample.

Sample Solution A for the Determination of Heavy Metals and Iron. To 40.0 g (25 mL) add 1 mL of nitric acid and about 20 mg of sodium carbonate. Evaporate to dryness on a hot plate in a well-ventilated hood, dissolve the residue in 2 mL of 1 N acetic acid, and dilute with water to 40 mL (1 mL = 1.0 g).

HEAVY METALS. (Page 28, Method 1). Dilute 20 mL of sample solution A (20 g sample) with water to 25 mL.

IRON. (Page 30, Method 1). Use 10 mL of sample solution A (10 g sample).

Perchloric Acid, Ultratrace
(suitable for use in ultratrace elemental analysis)

$HCIO_4$ **Formula Wt 100.46**

CAS number 7601–90–3

NOTE. Reagent must be packaged in a preleached Teflon bottle and used in a clean environment to maintain purity.

REQUIREMENTS

Appearance . . . . Clear
Assay . . . . .65–72% $HCIO_4$

MAXIMUM ALLOWABLE

Chloride (Cl) . . . . 3 ppm
Silicate and phosphate (as SiO_2) . . . . 5 ppm
Sulfate (SO_4) . . . . 5 ppm
Aluminum (Al) . . . . 1 ppb
Barium (Ba) . . . . 1 ppb
Boron (B) . . . . 5 ppb
Cadmium (Cd) . . . . 1 ppb
Calcium (Ca) . . . . 1 ppb
Chromium (Cr) . . . . 1 ppb
Cobalt (Co) . . . . 1 ppb
Copper (Cu) . . . . 1 ppb
Iron (Fe) . . . . 5 ppb
Lead (Pb) . . . . 1 ppb
Lithium (Li) . . . . 1 ppb
Magnesium (Mg) . . . . 1 ppb
Manganese (Mn) . . . . 1 ppb

Molybdenum (Mo) . . . . 1 ppb
Mercury (Hg) . . . . 1 ppb
Potassium (K) . . . . 1 ppb
Silicon (Si) . . . . 5 ppb
Sodium (Na) . . . . 5 ppb
Strontium (Sr) . . . . 1 ppb
Tin (Sn) . . . . 1 ppb
Titanium (Ti) . . . . 1 ppb
Vanadium (V) . . . . 1 ppb
Zinc (Zn) . . . . 1 ppb
Zirconium (Zr) . . . . 1 ppb

TESTS

ASSAY. (By alkalimetric titration). Weigh accurately 3 mL, dilute with water to 60 mL, add 0.15 mL of phenolphthalein indicator solution, and titrate with 1 N sodium hydroxide. One milliliter of 1 N sodium hydroxide corresponds to 0.1005 g of $HClO_4$.

$$\% \ HClO_4 = (mL \times N \ NaOH \times 10.05) \ / \ Sample \ wt \ (g)$$

CHLORIDE. (Page 27). Dilute 3.3 g with 25 mL of water.

SILICATE AND PHOSPHATE. Use test for perchloric acid, page 456.

SULFATE. Use test for perchloric acid, page 456. Precipitate weight <0.0004 g.

TRACE METALS. Determine the aluminum, barium, cadmium, calcium, chromium, cobalt, copper, iron, lead, lithium, magnesium, manganese, molybdenum, nickel, potassium, sodium, strontium, tin, titanium, vanadium, zinc, and zirconium by the ICP–OES method described on page 42.

Determine the mercury by the CVAAS procedure described on page 40.

Perchloroethylene
Tetrachloroethylene

$Cl_2C:CCl_2$ **Formula Wt 165.85**

CAS Number 127–18–4

NOTE. Perchloroethylene usually contains a stabilizer. If a stabilizer is present, its identity and quantity must be stated on the label.

REQUIREMENTS

General Use
Assay . . . . ≥99.0% C_2Cl_4

MAXIMUM ALLOWABLE

Color (APHA). 10
Residue after evaporation . . . . 5 ppm
Water. 0.05%

Specific Use

Ultraviolet Spectrophotometry

Wavelength (nm)	Absorbance (AU)
400	0.03
350	0.05
305	0.2
300	1.0

TESTS

ASSAY. Analyze the sample by gas chromatography using the general parameters cited on page 70. The following specific conditions are also required.

Column: Type 1, methyl silicone

Measure the area under all the peaks and calculate the perchloroethylene content in area percent. Correct for water content. The assay value is the combination of perchloroethylene and added stabilizers.

COLOR (APHA). (Page 19).

RESIDUE AFTER EVAPORATION. (Page 16). Evaporate 124 mL (200 g) to dryness in a tared dish on a hot plate ($\approx$100 °C) and dry the residue at 105 °C for 30 min. The weight of the residue should not exceed 0.001 g.

WATER. (Page 56, Method 2). Use 1.0 mL (1.6 g) of the sample.

ULTRAVIOLET SPECTROPHOTOMETRY. Use the procedure on page 73 to determine the absorbance.

Periodic Acid
para-Periodic Acid

H_5IO_6 **Formula Wt 227.96**

CAS Number 10450–60–9

REQUIREMENTS

Assay . . . . 99.0–101.0% H_5IO_6

MAXIMUM ALLOWABLE

Insoluble matter. 0.01%
Residue after ignition 0.01%
Other halogens (as Cl) 0.01%
Sulfate (SO_4) 0.01%
Heavy metals (as Pb) 0.005%
Iron (Fe). 0.003%

TESTS

ASSAY. (By titration of oxidizing power). Weigh accurately 0.10–0.13 g, transfer to an iodine flask, and dissolve in about 50 mL of water. Add 1 mL of sulfuric acid and 3 g of potassium iodide, swirl, and let stand in the dark for 10 min. Titrate the liberated iodine with 0.1 N sodium thiosulfate, adding starch indicator near the end of the titration. One milliliter of 0.1 N thiosulfate corresponds to 0.02849 g of H_5IO_6.

INSOLUBLE MATTER. (Page 15). Use 10.0 g dissolved in 100 mL of water.

RESIDUE AFTER IGNITION. (Page 16). Use 10.0-g sample.

OTHER HALOGENS. Dissolve 1.0 g in 20 mL of water with 1 mL of phosphoric acid. Add 1 mL of hydrogen peroxide and boil until the iodine is expelled. Wash down the sides with water, add 1 mL of hydrogen peroxide, and boil. If iodine color forms, repeat the treatment until no more iodine is evolved. Boil an additional 10 min and dilute to 100 mL. Dilute 10 mL to 23 mL, and add 1 mL of nitric acid and 1 mL of silver nitrate reagent solution. The turbidity should not exceed that produced by 0.01 mg of chloride ion in an equal volume containing the quantities of all reagents used in the test.

Sample Solution A and Blank Solution B for the Determination of Sulfate, Heavy Metals, and Iron

Sample Solution A. Dissolve 10.0 g in 20 mL of water. Add 10 mg of sodium carbonate and 20 mL of hydrochloric acid, and evaporate to dryness on a hot plate ($\approx$100 °C). Repeat the evaporation twice, using 10 mL of hydrochloric acid. Dissolve the residue in 20 mL of water and dilute to 100 mL (1 mL = 0.10 g).

Blank Solution B. Use the quantity of hydrochloric acid added to sample solution A and 10 mg of sodium carbonate. Evaporate to dryness, dissolve in 20 mL of water, and dilute to 100 mL.

SULFATE. (Page 33, Method 1). Use 5 mL of sample solution A (0.5-g sample).

HEAVY METALS. (Page 28, Method 1). Dilute 4 mL of sample solution A (0.4-g sample) to 25 mL. Use 4 mL of blank solution B to prepare the standard.

IRON. (Page 30, Method 2). Use 3.3 mL of sample solution A (0.33-g sample). For the standard, use 3.3 mL of blank solution B.

Petroleum Ether
Ligroin

CAS Number 8032–32–4

Suitable for use in extraction–concentration analysis or general use. Product labeling shall designate the one or both of these uses for which suitability is represented on the basis of meeting the relevant requirements and tests. The extraction–concentration suitability requirements include only the general use requirement for color.

REQUIREMENTS

General Use

Color (APHA). . . . . ≤10
Boiling range . . . . 35–60 °C
Residue after evaporation . . . . ≤0.001%
Acidity. . . . . Passes test

Specific Use

Extraction–Concentration Suitability
GC–FID . . . . Passes test
GC–ECD . . . . Passes test

TESTS

COLOR (APHA). (Page 19).

BOILING RANGE. Distill 100 mL by the method described on page 18. The corrected initial boiling point should not be below 35 °C, and the corrected dry point should not be above 60 °C. The barometric pressure correction for petroleum ether is +0.039 °C per mm under 760 mm and –0.039 °C per mm above 760 mm.

RESIDUE AFTER EVAPORATION. (Page 16). Evaporate 100 g (150 mL) to dryness in a tared dish at 70–80 °C and dry the residue at 105 °C for 30 min.

ACIDITY. Thoroughly shake 10 mL with 5 mL of water for 2 min and allow to separate. The aqueous layer should not turn blue litmus paper red in 15 s.

EXTRACTION–CONCENTRATION SUITABILITY. Analyze the sample by using the general procedure cited on page 72.

1,10-Phenanthroline Monohydrate

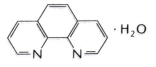

$C_{12}H_8N_2 \cdot H_2O$ **Formula Wt 198.22**

CAS Number 5144–89–8

REQUIREMENTS

Suitability as redox indicator . . . . Passes test
Suitability for determining iron . . . . Passes test

TESTS

Sample Solution A. Dissolve 0.10 g in water and dilute with water to 100 mL.

SUITABILITY AS REDOX INDICATOR. To 2.5 mL of sample solution A add 0.1 mL of 0.1 N ferrous ammonium sulfate solution and dilute to 50 mL with 12 N sulfuric acid. A distinct red color should be produced. Add 0.1 mL of 0.1 N potassium dichromate solution. The red color should be discharged.

SUITABILITY FOR DETERMINING IRON. To 20 mL of water add 0.001 mg of iron and 1 mL of hydroxylamine hydrochloride reagent solution. For the control, add 1 mL of hydroxylamine hydrochloride reagent solution to 20 mL of water. To each solution add 2.5 mL of sample solution A and allow to stand for 1 h. The color in the solution containing the added iron should be definitely pink compared with the control. The control containing no added iron should not show any turbidity or color.

Phenol

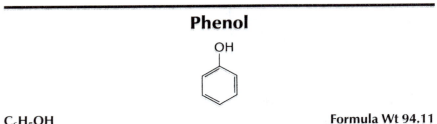

C_6H_5OH **Formula Wt 94.11**

CAS Number 108–95–2

NOTE. Phenol that conforms to this specification may contain a stabilizer. If a stabilizer is present, its presence should be stated on the label.

REQUIREMENTS

Assay . . . .≥99.0% C$_6$H$_5$OH
Freezing point . . . . Not below 40.5 °C (dry basis)
Clarity of solution . . . . Passes test

MAXIMUM ALLOWABLE

Residue after evaporation 0.05%
Water. 0.5%

TESTS

ASSAY. (By iodometry after bromination of hydroxyl groups). Weigh accurately about 1.5 g and dissolve in sufficient water to make exactly 1000 mL. Transfer 25 mL into a 500-mL glass-stoppered volumetric flask, add 30 mL of 0.1 N bromine and 5 mL of hydrochloric acid, and immediately stopper the flask. Shake frequently during one-half hour and allow to stand for 15 min. Then quickly add 10 mL of 10% potassium iodide, being careful that no bromine escapes, and stopper immediately. Shake, add 1 mL of Chloroform, and titrate the liberated iodine (which represents the excess of bromine) with 0.1 N sodium thiosulfate, using 3 mL of starch indicator solution. One milliliter of 0.1 N bromine corresponds to 0.001568 g of C$_6$H$_5$OH.

FREEZING POINT. Dry the sample by adding 15 g of molecular sieve zeolite having 4A-size pores to 100 g of molten sample in a 400-mL conical flask. (The zeolite may be in the form of a powder or of cylindrical granules about 3 mm in diameter.) Stopper the flask loosely and maintain the mixture at 55–60 °C for 20 min with frequent stirring. Quickly transfer a portion of the dried sample to a jacketed test tube, filling the tube to a depth of 8 cm. Insert a suitable thermometer, graduated in 0.1 °C units, and stir until the temperature falls to 10 °C above the expected freezing point. Immerse the tube in a water bath that is adjusted to about 4 °C below the expected freezing point and clamp in a vertical position. Insert a ring-type stirrer around the thermometer that is clamped to place the bottom of the bulb about 0.5 cm above the bottom of the test tube. Stir the sample continuously until crystallization occurs and the temperature does not change more than 0.05 °C in 2 min or, if supercooling occurs, the temperature rises rapidly to maximum accompanied by rapid crystallization. Record the constant temperature or the maximum temperature as the freezing point.

CLARITY OF SOLUTION. Dissolve 5.0 g in 100 mL of water. The solution should be free of turbidity and insoluble residue.

RESIDUE AFTER EVAPORATION. (Page 16). Evaporate 10.0 g to dryness in a tared dish on a hot plate (≈100 °C) and dry the residue at 105 °C for 30 min.

WATER. (Page 55, Method 1). Use 25.0 g of molten sample.

Phenolphthalein

3,3-Bis(4-hydroxyphenyl)-1(3*H*)-isobenzofuranone
3,3-Bis(*p*-hydroxyphenyl)phthalide

$C_{20}H_{14}O_4$ Formula Wt 318.33

CAS Number 77–09–8

REQUIREMENTS

Clarity of alcohol solution . . . . Passes test
Visual transition interval . . . . From pH 8.0 (colorless) to pH 10 (red)

TESTS

CLARITY OF ALCOHOL SOLUTION. Dissolve 1.0 g in 100 mL of ethyl alcohol. Not more than a faint trace of turbidity or insoluble matter should remain. Reserve the solution for the test for visual transition interval.

VISUAL TRANSITION INTERVAL. Dissolve 1.0 g of potassium chloride in 100 mL of water. Adjust the pH of the solution to 8.0 (using a pH meter as described on page 43) with 0.01 N acid or base. Add 0.15 mL of the 1% solution reserved from the test for clarity of alcohol solution. The solution should be colorless. Titrate the solution with 0.01 N sodium hydroxide until the pH is 8.2 (using the pH meter). The solution should have a pale pink color. Continue the titration until the pH is 8.6 (using the pH meter). The solution should have a definite pink color. Not more than 0.20 mL of 0.01 N sodium hydroxide should be consumed. Each additional 0.20 mL of 0.01 N sodium hydroxide should increase the amount of red color, until at pH 10 the solution should be very red.

Phenol Red

Phenolsulfonphthalein
4,4′-(3*H*-2,1-Benzoxathiol-3-ylidene)bisphenol S,S-Dioxide

$C_{19}H_{14}O_5S$ Formula Wt 354.38
$C_{19}H_{13}O_5SNa$ Formula Wt 376.36

CAS Number 143–74–8 (sultone), 34487–61–1 (sodium salt)

NOTE. This standard applies both to the free acid form and to the salt form of this indicator.

REQUIREMENTS

Clarity of solution. . . . . Passes test
Visual transition interval . . . . . From pH 6.8 (yellow) to pH 8.2 (red)

TESTS

CLARITY OF SOLUTION. If the indicator is the acid form, dissolve 0.1 g in 100 mL of alcohol. If the indicator is a salt form, dissolve 0.1 g in 100 mL of water. Not more than a faint trace of turbidity or insoluble matter should remain. Reserve the solution for the test for visual transition interval.

VISUAL TRANSITION INTERVAL. Dissolve 1.0 g of potassium chloride in 100 mL of water. Adjust the pH of the solution to 6.8 (using a pH meter as described on page 43) with 0.01 N acid or alkali. Add 0.15 mL of the 0.1% solution reserved from the test for clarity of solution. The color of the solution should be yellow, with not more than a faint trace of green color. Titrate the solution with 0.01 N sodium hydroxide to a pH of 7.0 (using a pH meter). The color of the solution should be orange. Continue the titration to pH 8.2. The color of the solution should be red. Not more than 0.20 mL of 0.01 N sodium hydroxide should be consumed in the entire titration.

Phosphomolybdic Acid
Molybdophosphoric Acid

CAS Number 51429–74–4

REQUIREMENTS

MAXIMUM ALLOWABLE

Insoluble matter. 0.01%
Chloride (Cl) 0.02%
Sulfate (SO_4) 0.025%
Ammonium (NH_4) 0.01%
Calcium (Ca) 0.02%
Heavy metals (as Pb) 0.005%
Iron (Fe). 0.005%

TESTS

INSOLUBLE MATTER. (Page 15). Use 10 g dissolved in 250 mL of water.

CHLORIDE. Dissolve 1.0 g in 50 mL of water, add 1 mL of nitric acid, filter through a chloride-free filter, and divide into two equal portions. For the control, add 0.5 mL of silver nitrate reagent solution to one-half of the solution, let stand for 10 min, filter, and add 0.1 mg of chloride ion. For the sample add 0.5 mL of silver nitrate reagent solution to the remaining half. Any turbidity in the solution of the sample should not exceed that in the control.

SULFATE. (Page 33, Method 1).

AMMONIUM. (Page 25). Dilute the distillate from a 1.0-g sample with water to 100 mL; 20 mL of the dilution shows not more than 0.02 mg of ammonium (NH_4).

CALCIUM. (By flame AAS, page 39).

Sample Stock Solution. Dissolve 1.0 g of sample in 50 mL of water in a 100-mL volumetric flask and dilute to the mark with water (1 mL = 0.01 g).

Element	Wavelength (nm)	Sample Wt (g)	Standard Added (mg)	Flame Type*	Background Correction
Ca	422.7	0.10	0.02; 0.04	N/A	No

*N/A is nitrous oxide/acetylene.

HEAVY METALS. Dissolve 3.0 g in 30 mL of water, filter, neutralize the filtrate to litmus with sodium hydroxide reagent solution, and add 5 mL of 1 N sodium hydroxide in excess. Heat the solution to boiling to destroy any blue color. For the standard, boil a solution containing 5 mL of 1 N sodium hydroxide and 0.15 mg of lead in 30 mL of water. Cool the solutions to about 10 °C and to each add 4.5 mL of 1 N hydrochloric acid and 10 mL of 10% (*W/V*) sodium sulfide reagent solution. Any color in the solution of the sample should not exceed that in the standard.

IRON. Dissolve 5.0 g in 30 mL of water, make alkaline with ammonium hydroxide, and add 0.25 to 0.30 mL of bromine water. Boil for a few minutes, filter, wash the precipitate several times, and discard the filtrate and washings. Pass 25 mL of dilute hydrochloric acid (1 + 1) through the filter to dissolve any precipitate, wash thoroughly, and dilute with water to 50 mL. Dilute 20 mL of this solution with water to 45 mL and add 2 mL of ammonium thiocyanate reagent solution. Any red color should not exceed that produced by 0.1 mg of ferric ion (Fe^{3+}) in an equal volume of solution containing 10 mL of dilute hydrochloric acid (1 + 1) and 2 mL of ammonium thiocyanate reagent solution.

Phosphoric Acid
Orthophosphoric Acid
Phosphoric(V) Acid

H_3PO_4 **Formula Wt 98.00**

CAS Number 7664–38–2

REQUIREMENTS

Assay . . . . ≥85.0% H_3PO_4

MAXIMUM ALLOWABLE

Color (APHA). 10
Insoluble matter. 0.001%
Chloride (Cl) . . . . 3 ppm
Nitrate (NO$_3$). . . . . 5 ppm
Sulfate (SO$_4$) 0.003%
Volatile acids (as CH$_3$COOH) 0.001%
Antimony (Sb) 0.002%
Calcium 0.002%
Magnesium 0.002%
Potassium (K). 0.005%
Sodium (Na) 0.025%
Arsenic (As) . . . . 1 ppm
Heavy metals (as Pb) 0.001%
Iron (Fe). 0.003%
Manganese (Mn) . . . . 0.5 ppm
Reducing substances . . . . Passes test

TESTS

ASSAY. (By acid–base titrimetry). Weigh accurately about 2.0 g and dilute with 120 mL of water. Titrate potentiometrically with carbonate-free 1 N sodium hydroxide to the end point at a pH near 9, using the first-derivative method (page 52). One milliliter of 1 N sodium hydroxide corresponds to 0.04900 g of H$_3$PO$_4$.

COLOR (APHA). (Page 19).

INSOLUBLE MATTER. (page 15). Dilute 20.0 g (12 mL) to 200 mL with water.

CHLORIDE. (Page 27). Dilute 3.4 g (2.0 mL) with water to 25 mL and add 0.5 mL of nitric acid.

NITRATE.

Sample Solution A. Add 4.0 g (2.4 mL) to 2 mL of water, dilute to 50 mL with brucine sulfate reagent solution, and mix.

Control Solution B. Add 4.0 g (2.4 mL) to 2 mL of the standard nitrate solution containing 0.01 mg of nitrate ion (NO$_3$) per mL, dilute to 50 mL with brucine sulfate reagent solution, and mix.

Continue with the procedure described on page 30, starting with the preparation of blank solution C.

SULFATE. (Page 33, Method 1). Use 1.7 g (1.0 mL), and neutralize with ammonium hydroxide. Allow 30 min for the turbidity to form.

VOLATILE ACIDS. Dilute 64.0 g (37.5 mL) with 75 mL of freshly boiled and cooled water in a distilling flask provided with a spray trap and distill off 50 mL. To the distillate, add 0.15 mL of phenolphthalein indicator solution and titrate with 0.01 N sodium hydroxide. Not more than 1.0 mL of the 0.01 N sodium hydroxide should be required to produce a pink color.

ANTIMONY, CALCIUM, MAGNESIUM, POTASSIUM, AND SODIUM. (By flame AAS, page 39).

Sample Stock Solution. Dilute 6.0 mL (10 g) of sample with 20 mL of water in a 100-mL volumetric flask and dilute to the mark with water (1 mL = 0.1 g). For antimony use a 10.0-g sample and standard additions.

Element	Wavelength (nm)	Sample Wt (g)	Standard Added (mg)	Flame Type*	Background Correction
Sb	217.6	10.0	0.20; 0.40	A/A	Yes
Ca	422.7	1.0	0.02; 0.04	N/A	No
Mg	285.2	1.0	0.01; 0.02	A/A	Yes
K	766.5	1.0	0.04; 0.08	A/A	No
Na	589.0	0.10	0.02; 0.04	A/A	No

*A/A is air/acetylene.

ARSENIC. (Page 25). To an arsine generator flask, add 10.0 g (6 mL) of sample, 30 mL of water, and 10 mL of ferric ammonium sulfate solution. Then add 2% potassium permanganate reagent solution dropwise to a persistent pink color. Add 1.5 g of sodium chloride and mix well until dissolution is complete. Heat nearly to boiling, remove from the heat, and add 1 mL of 40% stannous chloride dihydrate reagent solution. Dilute to 60 mL and cool to about 25 °C. For the standard, to a second arsine generator flask, add 0.01 mg of arsenic and 20 mL of dilute sulfuric acid (1 + 4), and treat as described for the sample. Assemble the test apparatus (page 26), and continue according to the procedure on page 25, using 10.0 g of No. 20-mesh granulated zinc. After 30 min, transfer the two silver diethyldithiocarbamate-absorbing solutions to 1-cm cells, and read the absorbance at 540 nm against the stock silver diethyldithiocarbamate solution. The absorbance for the sample should not exceed that for the standard.

Ferric Ammonium Sulfate Solution. Mix 84.0 g of ferric ammonium sulfate dodecahydrate, $FeNH_4(SO_4)_2 \cdot 12H_2O$, 2 mL of sulfuric acid, and 1 g of sodium chloride, NaCl. Dissolve and dilute with water to 1 L.

HEAVY METALS. (Page 28, Method 1). Dilute 4.0 g (2.4 mL) with water to 32 mL. Use 24 mL to prepare the sample solution, and use the remaining 8.0 mL to prepare the control solution. Adjust the pH to 4.5.

IRON. (Page 30, Method 2). Dilute 6.7 g (4.0 mL) to 100 mL. To 5.0 mL of the solution add 8 mL of 0.5 N sodium acetate.

MANGANESE. Add 25.0 g (15 mL) to 100 mL of dilute sulfuric acid (1 + 9). For the control, add 3 mL (5 g) of sample and 0.01 mg of manganese to 100 mL of dilute sulfuric acid (1 + 9). To each solution add 20 mL of nitric acid, heat to boiling, and boil gently for 5 min. Cool slightly, add 0.25 g of potassium periodate, and again boil for 5 min. Any red color in the solution of the sample should not exceed that in the control.

REDUCING SUBSTANCES. Add 5.0 mL of 0.1 N bromine solution to each of two 500-mL "iodine determination" flasks. To each flask add 120 mL of cold, dilute sulfuric acid (1 + 59) and to one of the flasks add rapidly 17 g (10 mL) of the sample. Stopper the flasks, shake gently, seal with water, and allow to stand at room temperature for 10 min. Cool in an ice bath, add 10 mL of potassium iodide reagent solution, and reseal with water. Shake gently to dissolve all fumes and place in the ice bath for about 5 min. Wash the stoppers and inside walls of the flasks with water and titrate each with 0.01 N sodium thiosulfate, adding 3 mL of starch indicator solution near the end of the titration. The difference between the two titrations should not exceed 1.0 mL of 0.01 N sodium thiosulfate.

Phosphoric Acid, Meta-
Vitreous Sodium Acid Metaphosphate

CAS Number 37267–86–0

NOTE. This reagent contains as a stabilizer a somewhat greater proportion of sodium metaphosphate than that corresponding to the empirical formula $NaH(PO_3)_2$.

REQUIREMENTS

Assay (HPO_3). 33.5–36.5%
Stabilizer ($NaPO_3$) 57.0–63.0%

MAXIMUM ALLOWABLE

Chloride (Cl) 0.001%
Nitrate (NO_3). 0.001%
Sulfate (SO_4) 0.005%
Arsenic (As) . . . . 1 ppm
Heavy metals (as Pb) 0.005%
Iron (Fe). 0.005%
Substances reducing permanganate (as H_3PO_3). 0.02%

TESTS

ASSAY. (By alkalimetry). Weigh accurately 5.0 g, dissolve in 400 mL of water, and titrate with 1 N sodium hydroxide solution to a pH of 4.4 (using a pH meter). One milliliter of 1 N sodium hydroxide corresponds to 0.07998 g of metaphosphoric acid (HPO_3).

STABILIZER. Weigh accurately 20.0 g, dissolve in water, and dilute with water to 1 L. To 50.0 mL of this solution in a 200-mL volumetric flask add 50 mL of water and 50 mL of a 10% solution of lead acetate, dilute with water to 200 mL, mix thoroughly, and decant through a filter. Treat 100 mL of the clear filtrate with hydrogen sulfide sufficiently to precipitate the excess of lead, filter, and wash with about 20 mL of water. To the filtrate add 2 mL of sulfuric acid, evaporate to dryness in a tared dish, and ignite to constant weight at 800 ± 25 °C. Calculate the weight of $NaPO_3$ by multiplying the weight of sodium sulfate by the factor 1.4356.

> **Sample Solution A for the Determination of Chloride, Arsenic, Heavy Metals, Iron, and Substances Reducing Permanganate.** Dissolve 20 g in 150 mL of water, cool, and dilute with water to 200 mL (1 mL = 0.1 g).

CHLORIDE. (Page 27). Dilute 10 mL of sample solution A (1-g sample) with water to 25 mL, and add 0.5 mL of nitric acid.

NITRATE. (Page 30). For sample solution A use 1.0 g in 4 mL of water. For control solution B use 1.0 g and 1 mL of standard nitrate solution containing 0.01 mg of nitrate ion (NO_3) per mL.

SULFATE. Dissolve 8.0 g in 180 mL of water, heat the solution to boiling, cool, filter, and heat again to boiling. Add 10 mL of barium chloride reagent solution, digest in a covered beaker on a hot plate (≈100 °C) for 2 h, and allow to stand overnight. If a precipitate is formed, filter, wash thoroughly, and ignite. The weight of the precipitate should not be more than 0.001 g greater than the weight obtained in a complete blank test, including filtration of a solution of 2 mL of hydrochloric acid in 200 mL of water.

ARSENIC. (Page 25). Dilute 30 mL of sample solution A (3-g sample) with water to 35 mL. For the standard, use 0.003 mg of arsenic.

HEAVY METALS. To 50 mL of sample solution A (5-g sample) add 0.15 mL of phenolphthalein indicator solution and neutralize with ammonium hydroxide. Adjust the pH of this solution to 4.5 (using a pH meter) with 1 N sulfuric acid and dilute with water to 250 mL. For the sample use 40 mL of the solution. For the control, add 0.02 mg of lead to 20 mL of the solution and dilute with water to 40 mL. Add 10 mL of freshly prepared hydrogen sulfide water to each and mix. Any color in the solution of the sample should not exceed that in the control.

IRON. To 2.0 mL of sample solution A (0.2-g sample) add 5 mL of water and 3 mL of hydrochloric acid, boil gently for 10 min, and cool. Add 3.5 mL of 0.5 N sodium acetate, 6 mL of hydroxylamine hydrochloride reagent solution, and 4 mL of 1,10-phenanthroline reagent solution. Adjust the pH to between 4 and 6 with ammonium hydroxide. Any red color produced within 1 h should not exceed that produced by 0.01 mg of iron in an equal volume of solution containing the quantities of reagents used in the test.

SUBSTANCES REDUCING PERMANGANATE. To 50 mL of sample solution A (5-g sample), add 10 mL of 10% sulfuric acid reagent solution, heat the solution to boiling, and titrate with 0.1 N potassium permanganate solution to a pink color that remains for 5 min. Not more than 0.25 mL of permanganate should be required.

Phosphoric Acid, Ultratrace
(suitable for use in ultratrace elemental analysis)

H_3PO_4 Formula Wt 98.00

CAS number 7664–38–2

NOTE. Reagent must be packaged in a preleached Teflon bottle and used in a clean environment to maintain purity.

REQUIREMENTS

Appearance . . . . Clear
Assay . . . . ≥85.0% H_3PO_4

MAXIMUM ALLOWABLE

Chloride (Cl) . . . . 3 ppm
Nitrate (NO_3) . . . . 5 ppm
Sulfate (SO_4) 0.003%
Aluminum (Al) . . . . 50 ppb
Barium (Ba) . . . . 5 ppb
Boron (B) . . . . 5 ppb
Cadmium (Cd) . . . . 5 ppb
Calcium (Ca) . . . . 50 ppb
Chromium (Cr) . . . . 5 ppb
Cobalt (Co) . . . . 10 ppb
Copper (Cu) . . . . 5 ppb
Iron (Fe) . . . . 20 ppb
Lead (Pb) . . . . 10 ppb

Lithium (Li). . . . . 5 ppb
Magnesium (Mg) . . . . 10 ppb
Manganese (Mn) . . . . 5 ppb
Molybdenum (Mo). . . . . 5 ppb
Mercury (Hg) . . . . 5 ppb
Potassium (K). . . . . 5 ppb
Sodium (Na). . . . . 50 ppb
Strontium (Sr) . . . . 5 ppb
Tin (Sn) . . . . 10 ppb
Titanium (Ti). . . . . 5 ppb
Vanadium (V) . . . . 5 ppb
Zinc (Zn) . . . . 50 ppb
Zirconium (Zr) . . . . 5 ppb

TESTS

ASSAY. (By acid–base titrimetry). Weigh accurately about 2.0 g, and dilute with 120 mL of water. Titrate potentiometrically with carbonate-free 1 N sodium hydroxide to the end point at a pH near 9, using the first-derivative method (page 52). One milliliter of 1 N sodium hydroxide corresponds to 0.04900 g of H_3PO_4.

$$\% \ H_3PO_4 = (mL \times N \ NaOH \times 4.90) \ / \ Sample \ wt \ (g)$$

CHLORIDE. (Page 27). Dilute 3.0 g sample with water to 25 mL and add 0.5 mL of nitric acid.

NITRATE. (Page 30). For sample solution A, use 2.0 g in 4 mL of water. For control solution B, use 1.0 g and 1 mL of standard nitrate containing 0.01 mg of nitrate ion (NO_3) per mL.

SULFATE. Use test for phosphoric acid, page 467.

TRACE METALS. Determine the aluminum, barium, cadmium, calcium, chromium, cobalt, copper, iron, lead, lithium, magnesium, manganese, molybdenum, nickel, potassium, sodium, strontium, tin, titanium, vanadium, zinc, and zirconium by the ICP–OES method described on page 42.

Determine the mercury by the CVAAS procedure described on page 40.

Phosphorus Pentoxide
Phosphorus(V) Oxide

P_2O_5 **Formula Wt 141.94**

CAS Number 1314–56–3

REQUIREMENTS

Assay. . . . . ≥98.0% P_2O_5

MAXIMUM ALLOWABLE

Insoluble matter. 0.02%
Phosphorus trioxide (P_2O_3) . . . . Passes test
Ammonium (NH_4) 0.01%
Heavy metals (as Pb) 0.01%

TESTS

NOTE. When making a solution, the phosphorus pentoxide must be added carefully and in small portions to water that has been cooled. Quiet dissolution may be achieved by allowing the sample to remain overnight in the uncovered weighing bottle placed in a covered beaker containing the necessary water. The pentoxide will absorb enough water from the air to dissolve without sputtering.

ASSAY. (By acid–base titrimetry). Weigh accurately 1.0 g, carefully dissolve in 100 mL of water, and evaporate to about 25 mL. Cool, dilute to 120 mL, add 0.5 mL of thymolphthalein indicator solution, and titrate with 1 N sodium hydroxide to the first appearance of a permanent blue color. Correct for an indicator blank, if any. One milliliter of 1 N sodium hydroxide corresponds to 0.03549 g of P_2O_5.

INSOLUBLE MATTER. Carefully dissolve 5.0 g in 40 mL of water, warming if necessary. Filter through a tared filtering crucible and set aside the filtrate for sample solution A. Wash the residue thoroughly and dry at 105 °C.

Sample Solution A for the Determination of Phosphorus Trioxide, Ammonium, and Heavy Metals. Dilute the filtrate, without washings, obtained in the preceding test with water to 100 mL (1 mL = 0.05 g).

PHOSPHORUS TRIOXIDE. To 60 mL of sample solution A (3-g sample) add 0.20 mL of 0.1 N potassium permanganate solution. Heat to boiling and allow to digest on a hot plate (≈100 °C) for 10 min. The pink color should not be entirely discharged. (Limit about 0.02%)

AMMONIUM. Dilute 2.0 mL of sample solution A (0.1-g sample) with water to 40 mL, and add 10 mL of freshly boiled 10% sodium hydroxide reagent solution and 2 mL of Nessler reagent. Any color should not exceed that produced by 0.01 mg of ammonium ion (NH_4) in an equal volume of solution containing the quantities of reagents used in the test.

HEAVY METALS. (Page 28, Method 1). Dilute 8.0 mL of sample solution A (0.4-g sample) with water to about 30 mL, boil for 5 min, cool, and dilute with water to 32 mL. Use 24 mL to prepare the sample solution, and use the remaining 8.0 mL to prepare the control solution.

Phthalic Acid

1,2-Benzenedicarboxylic Acid

$C_6H_4(COOH)_2$ **Formula Wt 166.13**

CAS Number 88–99–3

REQUIREMENTS

Assay . . . . $\geq$99.5% $C_6H_4(COOH)_2$

MAXIMUM ALLOWABLE

Insoluble matter. 0.05%
Residue after ignition (as SO_4) 0.02%
Chloride (Cl) 0.001%
Nitrate (NO_3) 0.005%
Sulfate (SO_4) 0.005%
Heavy metals (as Pb) 0.001%
Iron (Fe). 0.001%
Water (H_2O). 0.5%

TESTS

ASSAY. (By acid–base titrimetry). Weigh accurately 2.8 g, transfer to a conical flask, cool, and add 50.0 mL of standard 1 N sodium hydroxide. Add 25 mL of water, and boil on a hot plate until dissolved. Add 0.15 mL of phenolphthalein indicator solution, and titrate the excess sodium hydroxide with standard 1 N hydrochloric acid. Perform a blank determination, and make any necessary correction. One milliliter of 1 N sodium hydroxide consumed corresponds to 0.08306 g of $C_6H_4(COOH)_2$.

INSOLUBLE MATTER. (Page 15). Use 10.0 g dissolved in 250 mL of 10% sodium carbonate (Na_2CO_3) solution.

RESIDUE AFTER IGNITION. (Page 16). Use 4.0 g. Retain the residue to prepare sample solution B.

> *Sample Solution A for the Determination of Chloride, Nitrate, and Sulfate.* Dissolve 10.0 g in 100 mL of boiling water, and allow to cool to room temperature. The supernatant liquid after decantation from the crystallized phthalic acid is sample solution A (1 mL = 0.10 g).

CHLORIDE. (Page 27). Dilute 10 mL of sample solution A (1.0 g sample) to 20 mL with water. Filter if necessary through a chloride-free filter. For the standard, use 0.01 mg of chloride ion (Cl) and dilute to 20 mL with water. To each add 1 mL of nitric acid and 1 mL of silver nitrate reagent solution. Mix, allow to stand for 5 min protected from sunlight, and compare. The solutions can best be viewed visually against a black background. Any turbidity in the solution of the sample should not exceed that in the standard.

NITRATE. (Page 30). Use 10 mL of sample solution A (1-g sample) and 0.05 mg of nitrate ion (NO_3). Follow the general procedure for nitrate by the brucine sulfate method.

SULFATE. (Page 33, Method 3). Add 10 mg of sodium carbonate to 10 mL of sample solution A (1-g sample), and evaporate to dryness on a hot plate ($\approx$100 °C). Redissolve in a minimum volume of water and add 1 mL of dilute hydrochloric acid (1 + 19). If necessary, filter through a small filter, and wash with two 2-mL portions of water. Dilute to 10 mL and add 1 mL of barium chloride reagent solution. Any turbidity should not exceed that produced by 0.05 mg of sulfate ion (SO_4) in an equal volume of solution containing the quantities of reagents used in the test. Compare 30 min after adding the barium chloride to the sample and standard solutions.

> *Sample Solution B for the Determination of Heavy Metals and Iron.* To the residue after ignition add 10 mL of water and dilute to 20 mL (1 mL = 0.2 g).

HEAVY METALS. (Page 28). Use 10 mL of sample solution B (2.0-g sample).

IRON. (Page 30, Method 2). Use 5 mL of sample solution B (1.0-g sample).

WATER. (Page 55, Method 1). Use 5.0 g.

Phthalic Anhydride
1,3-Isobenzofurandione

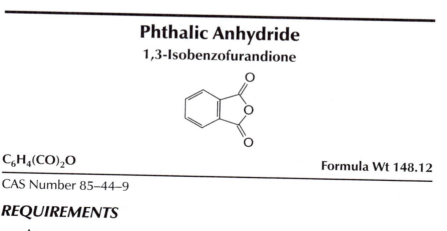

$C_6H_4(CO)_2O$ Formula Wt 148.12

CAS Number 85–44–9

REQUIREMENTS

Appearance . . . . .White flaky crystals
Assay. . . . . .99.0–100.2% $C_8H_4O_3$
Melting point. . . . . Not more than 3 °C range, including 131 °C

MAXIMUM ALLOWABLE

Residue after ignition 0.01%
Chloride (Cl) 0.002%
Sulfate (SO$_4$) 0.003%
Heavy metals (as Pb) . . . . 5 ppm
Iron (Fe) . . . . 5 ppm

TESTS

ASSAY. (By acid–base titrimetry). Accurately weigh about 0.3 g of sample into a 150-mL beaker containing 15.0 mL of 0.5 N morpholine in methanol and approximately 80 mL of methanol. Stir at least 30 min and titrate potentiometrically with 0.5 N hydrochloric acid. Similarly titrate a blank containing 15.0 mL of morpholine and 80 mL of methanol.

$$\% \text{ Phthalic anhydride } = \frac{(V_{sample} - V_{blank}) \times 148.12 \times 100}{\text{Sample wt (mg)}}$$

MELTING POINT. (Page 22).

RESIDUE AFTER IGNITION. (Page 16). Gently ignite 10.0 g in a tared crucible or dish. Finally, ignite at 800 ± 25 °C for 15 min. (Reserve this residue for the iron test.)

CHLORIDE. Mix 0.50 g of sample with 0.5 g of sodium carbonate and add 10–15 mL of water. Evaporate on a hot plate ($\approx$100 °C) and ignite until the mass is thoroughly charred, avoiding an excessively high temperature. When the sample is completely ignited, cool, and extract the fusion with 10 mL of water and 2 mL of nitric acid. Filter through chloride-free filter paper, wash with water to a volume of 20 mL, and add 1 mL of silver nitrate reagent solution. Any turbidity should not exceed that produced by 0.01 mg of chloride ion (Cl) in an equal volume of solution containing 0.5 g of sodium carbonate, 2 mL of nitric acid, and 1 mL of silver nitrate reagent solution.

SULFATE. Mix 3.0 g of sample with 1.0 g of sodium carbonate in a platinum crucible, add 20 mL of water, and evaporate to dryness on a hot plate ($\approx$100 °C). Ignite completely, taking care to protect the fusion from the flame because of the presence of sulfur compounds in the natural gas. Cool, dissolve the residue in 20 mL of water, add 2 mL of 30% hydrogen peroxide, and boil for 5 min. Add 3 mL of hydrochloric acid and evaporate to dryness on the hot plate. Dissolve the residue in 10 mL of water, filter, and wash with 10 mL of water. To the filtrate add 1 mL of 1 N hydrochloric acid and 2 mL of barium chloride reagent solution. Any turbidity should not be greater than that of a standard prepared as follows: evaporate 1 g of sodium carbonate, 2 mL of 30% hydrogen peroxide, and 3 mL of hydrochloric acid to dryness on a hot plate ($\approx$100 °C). Dissolve the residue and 0.09 mg of sulfate ion (SO$_4$) in sufficient water to make 20 mL, and add 1 mL of 1

N hydrochloric acid and 2 mL of barium chloride reagent solution. Compare 10 min after adding the barium chloride to the sample and standard solutions.

HEAVY METALS. (Page 28, Method 2). Use 4.0 g of sample.

IRON. (Page 30, Method 1). To the residue after ignition add 3 mL of hydrochloric acid, cover with a watch glass, and digest on a hot plate (≈100 °C) for 15–20 min. Remove the cover and evaporate to dryness. Dissolve the residue in 5 mL of hydrochloric acid, filter if necessary, and dilute with water to 100 mL. Use 20 mL of this solution.

Picric Acid
2,4,6-Trinitrophenol

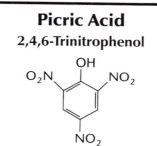

(NO$_2$)$_3$C$_6$H$_2$OH   **Formula Wt 229.11**

CAS Number 88–89–1

CAUTION. Trinitrophenol explodes when heated rapidly or when subjected to percussion. For safety in transportation, trinitrophenol is usually mixed with a minimum of 30% of water.

REQUIREMENTS

Water (H$_2$O) . . . . ≥30%
Melting point (dried) . . . . .121–123 °C

MAXIMUM ALLOWABLE

Insoluble and resinous matter 0.01%
Insoluble in toluene 0.1%
Sulfate (SO$_4$) 0.01%

TESTS

WATER. Dissolve 1.0 g, accurately weighed, in 200 mL of water. Titrate with 0.1 N sodium hydroxide, using 0.15 mL of phenolphthalein indicator solution. One milliliter of 0.1 N sodium hydroxide corresponds to 0.02291 g of (NO$_2$)$_3$C$_6$H$_2$OH. Subtract the percent of picric acid from 100% to determine the water content.

MELTING POINT (DRIED). Carefully transfer a small amount of dried sample (prepared in the test for insoluble in toluene) to a capillary tube, and run the melting point, using an oil immersion bath.

INSOLUBLE AND RESINOUS MATTER. Dissolve a sample equivalent to 10 g on the dry basis in 500 mL of water, add 1 mL of sulfuric acid, and digest on a hot plate (≈100 °C) for 1 h. Filter through a tared filtering crucible, wash thoroughly, dry at 105 °C, and weigh. No resinous material should be apparent in the crucible.

INSOLUBLE IN TOLUENE. Dry 5–10 g at 70 °C and dissolve 4.0 g of the dried sample in 100 mL of warm toluene. (NOTE. Add water to the remaining dried sample before discarding it.) Filter off any insoluble material using a tared filtering crucible, wash thoroughly with warm toluene, and dry at 105 °C.

SULFATE. (Page 33, Method 2). Use 25 mL of nitric acid. Allow 30 min for the turbidity to form.

Potassium Acetate

CH_3COOK Formula Wt 98.14

CAS Number 127–08–2

REQUIREMENTS

Assay . . . .≥99.0% CH_3COOK
pH of 5% solution . . . . 6.5–9.0 at 25 °C

MAXIMUM ALLOWABLE

Insoluble matter. 0.005%
Chloride (Cl) 0.003%
Phosphate (PO_4) 0.001%
Sulfate (SO_4) 0.002%
Heavy metals (as Pb) . . . . 5 ppm
Iron (Fe) . . . . 5 ppm
Calcium (Ca) 0.005%
Magnesium (Mg) 0.002%
Sodium (Na). 0.03%

TESTS

ASSAY. (By nonaqueous acid–base titrimetry). Weigh, to the nearest 0.1 mg, 0.4 g of sample, and dissolve in 5 mL of acetic anhydride and 25 mL of acetic acid. Add 0.15 mL of crystal violet indicator solution, and titrate in a closed system with 0.1 N perchloric acid in glacial acetic acid to a green end point. Perform a blank determination and make any necessary correction. One milliliter of 0.1 N perchloric acid corresponds to 0.00981 g of CH_3COOK.

pH OF A 5% SOLUTION. (Page 44). The pH should be 6.5–9.0 at 25 °C.

INSOLUBLE MATTER. (Page 15). Use 20 g dissolved in 200 mL of water.

CHLORIDE. (Page 27). Use 0.33 g.

PHOSPHATE. (Page 32, Method 2). Dissolve 1.0 g in 10 mL of sulfuric acid and evaporate to near dryness on a hot plate ($\approx$100 °C). Gently heat with a burner until the salt is dry. Add 10 mL of sulfuric acid and repeat the evaporation and drying. Dissolve in 80 mL of water, add 0.5 g of ammonium molybdate, and adjust the pH to 1.8 (using a pH meter) with dilute hydrochloric acid (1 + 9). Heat to boiling, cool, add 10 mL of hydrochloric acid, and proceed as described. Carry along a standard containing 0.01 mg of phosphate ion (PO_4) treated in the same manner as the sample after the addition of 80 mL of water.

SULFATE. (Page 33, Method 1).

HEAVY METALS. (Page 28, Method 1). Dissolve 6.0 g in about 10 mL of water, add 15 mL of 10% hydrochloric acid, and dilute with water to 30 mL. Use 25 mL to prepare the sample solution, and use the remaining 5.0 mL to prepare the control solution.

IRON. (Page 30, Method 1). Use 2.0 g.

CALCIUM, MAGNESIUM, AND SODIUM. (By flame AAS, page 39).

> *Sample Stock Solution.* Dissolve 5.0 g of sample in 80 mL of water. Transfer to a 100-mL volumetric flask and dilute to the mark with water (1 mL = 0.05 g).

Element	Wavelength (nm)	Sample Wt (g)	Standard Added (mg)	Flame Type*	Background Correction
Ca	422.7	0.50	0.025; 0.005	N/A	No
Mg	285.2	0.50	0.01; 0.02	A/A	Yes
Na	589.0	0.03	0.01; 0.02	A/A	No

*A/A is air/acetylene; N/A is nitrous oxide/acetylene.

Potassium Antimony Tartrate Trihydrate
Antimony Potassium Tartrate Trihydrate

$K_2(C_4H_2O_6Sb)_2 \cdot 3H_2O$ Formula Wt 667.87

CAS Number 11071–15–1

REQUIREMENTS

Assay . . . . 99.0–103.0% $K_2(C_4H_2O_6Sb)_2 \cdot 3H_2O$

MAXIMUM ALLOWABLE

Titrable acid or base . . . . 0.020 meq/g
Loss on drying 2.7%
Arsenic (As) 0.015%

TESTS

ASSAY. (By iodimetric titration of antimony). Weigh accurately 0.5 g of sample, transfer to a 250-mL beaker, and dissolve in 50 mL of water. Add 5.0 g of potassium sodium tartrate tetrahydrate, $KOCOCHOHCHOHCOONa \cdot 4H_2O$, 2 g of sodium borate decahydrate, $Na_2B_4O_7 \cdot 10H_2O$, and 3 mL of starch indicator solution. Titrate immediately with 0.1 N iodine to the production of a persistent blue color. One milliliter of 0.1 N iodine corresponds to 0.01670 g of $K_2(C_4H_2O_6Sb)_2 \cdot 3H_2O$.

TITRABLE ACID OR BASE. Dissolve 1.0 g in 50 mL of carbon dioxide-free water, and titrate to a pH of 4.5 using 0.01 N hydrochloric acid or 0.01 N sodium hydroxide as required. Not more than 2.0 mL of either titrant should be required.

LOSS ON DRYING. Weigh accurately 1.0 g, and dry at 105 °C to constant weight.

ARSENIC. Dissolve 0.1 g in 5 mL of hydrochloric acid, and add 10 mL of a freshly prepared solution of 20 g of stannous chloride dihydrate, $SnCl_2 \cdot 2H_2O$, in 30 mL of hydrochloric acid. Mix, transfer to a color-comparison tube, and allow to stand for 30 min. Viewed downward over a white surface, the color of the solution is not darker than that of a standard containing 0.015 mg of arsenic.

Potassium Bicarbonate
Potassium Hydrogen Carbonate

$KHCO_3$　　　　　　　　　　　　　　　　**Formula Wt 100.12**

CAS Number 298–14–6

REQUIREMENTS

Assay (dried basis) . . . . 99.7–100.5% $KHCO_3$

MAXIMUM ALLOWABLE

Insoluble matter. 0.01%
Chloride (Cl) 0.001%
Phosphate (PO_4) . . . . 5 ppm
Sulfur compounds (as SO_4) 0.003%
Ammonium (NH_4) . . . . 5 ppm

Heavy metals (as Pb) . . . . 5 ppm
Iron (Fe). . . . . 5 ppm
Calcium (Ca) 0.002%
Magnesium (Mg) 0.001%
Sodium (Na) 0.03%

TESTS

ASSAY. (By acid–base titrimetry). Weigh accurately 3.0 g, previously dried over sulfuric acid for 24 h, dissolve it in 50 mL of water, add 0.15 mL of methyl orange indicator solution, and titrate with 1 N hydrochloric acid. The potassium bicarbonate content calculated from the total alkalinity, as determined by the titration, should not be less than 99.7 nor more than 100.5% of the weight taken. One milliliter of 1 N hydrochloric acid corresponds to 0.1001 g of $KHCO_3$.

INSOLUBLE MATTER. (Page 15). Dissolve 10 g in 100 mL of hot water, heat to boiling, and boil gently for 2–3 min.

CHLORIDE. (Page 27). Use 1.0 g.

PHOSPHATE. (Page 32, Method 1). Dissolve 4.0 g in 20 mL of water, add 5 mL of hydrochloric acid, and evaporate to dryness on a hot plate ($\approx$100 °C). Dissolve the residue in 25 mL of approximately 0.5 N sulfuric acid, and proceed as described.

SULFUR COMPOUNDS. (Page 33, Method 4). Allow 30 min for the turbidity to form.

AMMONIUM. Dissolve 2.0 g in 40 mL of water, and add 10 mL of 10% sodium hydroxide reagent solution and 2 mL of Nessler reagent solution. Any color should not exceed that produced by 0.01 mg of ammonium ion (NH_4) in an equal volume of solution containing the quantities of reagents used in the test.

HEAVY METALS. (Page 28, Method 1). To 5.0 g in a 150-mL beaker add 10 mL of water, mix, and add cautiously 10 mL of hydrochloric acid. Evaporate to dryness on a hot plate ($\approx$100 °C), dissolve the residue in about 20 mL of water, and dilute with water to 25 mL. For the control, For the control, add 0.02 mg of lead ion (Pb) to 1.0 g of sample, and treat exactly as the 5.0 g of sample.

IRON. (Page 30, Method 1). Dissolve 2.0 g in 20 mL of water, cautiously add 5 mL of hydrochloric acid, and dilute with water to 50 mL. Use the solution without further acidification. In the standard use only 3 mL of hydrochloric acid.

CALCIUM, MAGNESIUM, AND SODIUM. (By flame AAS, page 39).

Sample Stock Solution. Cautiously dissolve 5.0 g of sample in 30 mL of dilute hydrochloric acid (1 + 2). Cool to room temperature, transfer to a 100-mL volumetric flask, and dilute to the mark with water (1 mL = 0.05 g).

Element	Wavelength (nm)	Sample Wt (g)	Standard Added (mg)	Flame Type*	Background Correction
Ca	422.7	1.0	0.02; 0.04	N/A	No
Mg	285.2	1.0	0.01; 0.02	A/A	Yes
Na	589.0	0.05	0.015; 0.03	A/A	No

*A/A is air/acetylene; N/A is nitrous oxide/acetylene.

Potassium Bromate

KBrO₃   **Formula Wt 167.00**

CAS Number 7758–01–2

REQUIREMENTS

Assay (dried basis) . . . . ≥99.8% $KBrO_3$
pH of a 5% solution . . . . 5.0–9.0 at 25 °C

MAXIMUM ALLOWABLE

Insoluble matter. 0.005%
Bromide (Br) . . . . Passes test
Nitrogen compounds (as N). 0.001%
Sulfate (SO_4) 0.005%
Heavy metals (as Pb) . . . . 5 ppm
Iron (Fe). 0.002%
Sodium (Na). 0.01%

TESTS

ASSAY. (By oxidation–reduction titration of bromate). Dry a powdered sample to constant weight at 150 °C. Weigh accurately 1.0 g, dissolve in water, and transfer to a 250-mL volumetric flask. Dilute to volume with water and mix thoroughly. Place a 25.0-mL aliquot of this solution in a glass-stoppered conical flask and add 3 g of potassium iodide and 3 mL of hydrochloric acid. Stopper, swirl, and allow to stand for 5 min. Titrate the liberated iodine with 0.1 N sodium thiosulfate, adding 3 mL of starch indicator solution near the end of the titration. One milliliter of 0.1 N sodium thiosulfate corresponds to 0.002783 g of $KBrO_3$.

pH OF A 5% SOLUTION. (Page 44). The pH should be 5.0–9.0 at 25 °C.

INSOLUBLE MATTER. (Page 15). Use 20 g dissolved in 150 mL of hot water.

BROMIDE. Dissolve 4.0 g in 80 mL of water and divide the solution into two equal portions. Add 0.15 mL of 1 N sulfuric acid to one portion. At the end of 2 min this portion should show no more yellow color than the portion to which no acid was added. (Limit about 0.05%)

NITROGEN COMPOUNDS. (Page 31). Use 1.0 g. For the standard, use 0.01 mg of nitrogen (N).

SULFATE. (Page 33, Method 2). Use 6 mL of dilute hydrochloric acid (1 + 1). Do two evaporations.

HEAVY METALS. (Page 28, Method 1). Dissolve 5.0 g in 20 mL of dilute hydrochloric acid (1 + 1). Evaporate the solution to dryness on a hot plate ($\approx$100 °C). Add 10 mL more of dilute hydrochloric acid (1 + 1), and again evaporate to dryness. Dissolve in about 20 mL of water and dilute with water to 25 mL. For the control, add 0.02 mg of lead to 1.0 g of sample, and treat exactly as the 5.0 g of sample.

IRON. (Page 30, Method 1). Dissolve 1.0 g in 20 mL of dilute hydrochloric acid (1 + 1), and evaporate on a hot plate ($\approx$100 °C) to dryness. Add 10 mL of additional dilute hydrochloric acid (1 + 1), and again evaporate to dryness. Dissolve the residue in water, add 4 mL of hydrochloric acid, and dilute with water to 100 mL. Use 50 mL of the solution. For the control, use 0.02 mg of iron (Fe) carried through the same procedure as the 1.0 g of sample.

SODIUM. (By flame AAS, page 39).

> *Sample Stock Solution.* Dissolve 1.0 g of sample in 25 mL of hydrochloric acid (1 + 3), and digest in a covered beaker on a hot plate ($\approx$100 °C) until the reaction ceases. Uncover, evaporate to dryness, add 10 mL of hydrochloric acid (1 + 3), and dilute with hydrochloric acid (1 + 99) to the mark in a 100-mL volumetric flask (1 mL = 0.01 g).

Element	Wavelength (nm)	Sample Wt (g)	Standard Added (mg)	Flame Type*	Background Correction
Na	589.0	0.10	0.01; 0.02	A/A	No

*A/A is air/acetylene.

Potassium Bromide

KBr   **Formula Wt 119.00**

CAS Number 7758–02–3

REQUIREMENTS

Assay. . . . .$\geq$99.0% KBr
pH of a 5% solution . . . . 5.0–8.8 at 25 °C

MAXIMUM ALLOWABLE

Insoluble matter. 0.005%
Bromate (BrO_3) 0.001%

Iodate (IO_3) 0.001%
Chloride (Cl) 0.2%
Iodide (I) 0.001%
Nitrogen compounds (as N) 0.005%
Sulfate (SO_4) 0.005%
Barium (Ba) 0.002%
Heavy metals (as Pb) . . . . 5 ppm
Iron (Fe) . . . . 5 ppm
Calcium (Ca) 0.002%
Magnesium (Mg) 0.001%
Sodium (Na) 0.02%

TESTS

ASSAY. (By argentimetric titration of bromide content). Weigh accurately 0.5 g of sample and dissolve in 50 mL of water. Add 50.0 mL of 0.1 N silver nitrate and 10 mL of dilute nitric acid, and titrate the excess silver nitrate with 0.1 N ammonium thiocyanate, using ferric ammonium sulfate as indicator.

$$\% \text{ KBr} = \frac{[(\text{mL} \times \text{N AgNO}_3) - (\text{mL} \times \text{N NH}_4\text{SCN})] \times 11.90}{\text{Sample wt (g)}}$$

pH OF A 5% SOLUTION. (Page 44). The pH should be 5.0–8.8 at 25 °C.

INSOLUBLE MATTER. (Page 15). Use 20 g dissolved in 150 mL of water.

BROMATE AND IODATE. (By differential pulse polarography, page 50). Use 10.0 g of sample in 25 mL of solution. For the standard, add 0.10 mg of bromate (BrO_3) and 0.10 mg of iodate (IO_3).

CHLORIDE. Dissolve 0.50 g in 15 mL of dilute nitric acid (1 + 2) in a small flask. Add 3 mL of 30% hydrogen peroxide and digest on a hot plate ($\approx$100 °C) until the solution is colorless. Wash down the sides of the flask with a little water, digest for an additional 15 min, cool, and dilute with water to 200 mL. Dilute 2.0 mL to 20 mL and add 1 mL of nitric acid and 1 mL of silver nitrate reagent solution. Any turbidity should not exceed that produced by 0.01 mg of chloride ion (Cl) in an equal volume of solution containing the quantities of reagents used in the test.

IODIDE. Dissolve 2.5 g in 50 mL of water in a 100-mL volumetric flask. For the control, dissolve 0.5 g in a similar flask and add 0.02 mg of iodide ion (I). Treat each solution as follows: Add 2 mL of bromine water, mix, and allow to stand for 5 min. Add 2 mL of sodium formate solution (25 g per 100 mL) and mix. Add 1 mL of 1 N sodium hydroxide or enough to bring the pH to 9 or greater, dilute to volume with water, and mix. Using differential pulse polarography, as described on page 50, starting with transfer of the solution to the polarographic cell, determine any iodate present. The wave for the sample, after correction for any iodate originally present, should not exceed that for the control (similarly corrected).

NITROGEN COMPOUNDS. (Page 31). Use 1.0 g. For the standard, use 0.05 mg of nitrogen (N).

SULFATE. (Page 33, Method 1).

BARIUM. For the sample dissolve 6.0 g in 15 mL of water. For the control, dissolve 1.0 g in 15 mL of water and add 0.1 mg of barium. To each solution add 5 mL of acetic acid, 5 mL of 30% hydrogen peroxide, and 1 mL of hydrochloric acid. Digest in a covered beaker on a hot plate (≈100 °C) until reaction ceases, uncover, and evaporate to dryness. Dissolve the residues in 15 mL of water, filter if necessary, and dilute with water to 23 mL. Add 2 mL of 10% potassium dichromate reagent solution and add ammonium hydroxide until the orange color is just dissipated and the yellow color persists. Add 25 mL of methanol, stir vigorously, and allow to stand for 10 min. The turbidity in the solution of the sample should not exceed that in the control.

HEAVY METALS. (Page 28, Method 1). Dissolve 6.0 g in about 20 mL of water, and dilute with water to 30 mL. Use 25 mL to prepare the sample solution, and use the remaining 5.0 mL to prepare the control solution.

IRON. (Page 30, Method 1). Use 2.0 g. In both sample and control solutions use 4 mL of hydrochloric acid.

CALCIUM, MAGNESIUM, AND SODIUM. (By flame AAS, page 39).

Sample Stock Solution. Dissolve 5.0 g of sample in 80 mL of water. Transfer to a 100-ml volumetric flask and dilute to the mark with water (1 mL = 0.05 g).

Element	Wavelength (nm)	Sample Wt (g)	Standard Added (mg)	Flame Type*	Background Correction
Ca	422.7	1.0	0.02; 0.04	N/A	No
Mg	285.2	1.0	0.01; 0.02	A/A	Yes
Na	589.0	0.05	0.01; 0.02	A/A	No

*A/A is air/acetylene; N/A is nitrous oxide/acetylene.

Potassium Carbonate
Potassium Carbonate, Anhydrous

K_2CO_3 Formula Wt 138.21

CAS Number 584–08–7

REQUIREMENTS

Assay. . . . . ≥99.0% K_2CO_3

MAXIMUM ALLOWABLE

Insoluble matter. 0.01%
Chloride (as Cl) 0.003%
Nitrogen compounds (as N). 0.001%
Phosphate (PO$_4$) 0.001%
Silica (SiO$_2$) 0.005%
Sulfur compounds (as SO$_4$) 0.004%
Ammonium hydroxide precipitate 0.01%
Heavy metals (as Pb) . . . . 5 ppm
Iron (Fe). . . . . 5 ppm
Calcium 0.005%
Magnesium 0.002%
Sodium (Na). 0.02%

TESTS

ASSAY. (By acid–base titrimetry). Weigh, to the nearest 0.1 mg, 3 g of sample, transfer to a 125-mL glass-stoppered flask, and dissolve with 50 mL of water. Add 0.15 mL of methyl orange indicator solution and titrate with 1 N hydrochloric acid. One milliliter of 1 N hydrochloric acid corresponds to 0.0691 g of K_2CO_3.

INSOLUBLE MATTER. (Page 15). Use 10 g dissolved in 100 mL of water.

CHLORIDE. Dissolve 1.0 g in about 40 mL of water plus 6 mL of nitric acid. Filter through a chloride-free filter, wash, and dilute the filtrate with water to 60 mL. To 20 mL add 1 mL of silver nitrate reagent solution. Any turbidity should not exceed that produced by 0.01 mg of chloride ion (Cl) in an equal volume of solution containing the quantities of reagents used in the test.

NITROGEN COMPOUNDS. (Page 31). Use 1.0 g. For the standard, use 0.01 mg of nitrogen (N).

PHOSPHATE. (Page 32, Method 2). Dissolve 1.0 g in 50 mL of water in a platinum dish and digest on a hot plate ($\approx$100 °C) for 30 min. Cool, neutralize with dilute sulfuric acid (1 + 19) to a pH of about 4, and dilute with water to about 75 mL. Add 0.5 g of ammonium molybdate to the solution, and when it is dissolved, adjust the pH to 1.8 (using a pH meter) with dilute hydrochloric acid (1 + 9). Heat to boiling, cool, add 10 mL of hydrochloric acid, and dilute with water to 100 mL. Proceed as described. Carry along a standard containing 0.01 mg of phosphate ion (PO$_4$) and 0.05 mg of silica (SiO$_2$) in about 75 mL of water treated as the 75 mL of sample solution. Reserve the aqueous phase for the determination of silica.

SILICA. Add 10 mL of hydrochloric acid to the solutions reserved from the determination of phosphate and transfer to separatory funnels. Add 40 mL of butyl alcohol, shake vigorously, and allow to separate. Draw off and discard the aqueous phase. Wash the butyl alcohol three times with 20-mL portions of dilute hydrochloric acid (1 + 99), discarding the washings each time. Dilute each butyl alcohol solution to 50 mL with butyl alcohol, take 10 mL from each, and dilute each to 50 mL with butyl alcohol. Add 0.5 mL of a freshly prepared 2% solution of stannous chloride dihydrate in hydrochloric acid. The blue color in the extract from the sample should not exceed that in the control. If the butyl alcohol extracts are turbid, wash them with 10 mL of dilute hydrochloric acid (1 + 99).

SULFUR COMPOUNDS. (Page 33, Method 4). Cautiously acidify with hydrochloric acid. Allow 30 min for the turbidity to form.

AMMONIUM HYDROXIDE PRECIPITATE. Dissolve 10 g in 100 mL of water. Cautiously add 12 mL of sulfuric acid to 12 mL of water, cool, add the mixture to the solution of the sample, and evaporate to dense fumes. Cool, dissolve the residue in 130 mL of hot water, and add ammonium hydroxide until the solution is just alkaline to methyl red. Heat to boiling and filter, reserving the filtrate and washings for the test for calcium and magnesium. Continue to wash the precipitate, but reject the washings, and finally ignite. The residue includes some, but not all, of the silica in the sample.

HEAVY METALS. (Page 28, Method 1). To 5.0 g in a 150-mL beaker add 10 mL of water, mix, and cautiously add 10 mL of hydrochloric acid. Evaporate the solution to dryness on a hot plate ($\approx$100 °C), dissolve the residue in about 20 mL of water, and dilute with water to 25 mL. For the control, add 0.02 mg of lead to 1.0 g of sample and treat exactly as the 5.0 g of sample.

IRON. (Page 30, Method 1). Dissolve 2.0 g in 20 mL of dilute hydrochloric acid (1 + 1), and evaporate to dryness on a hot plate ($\approx$100 °C). Dissolve the residue in 5 mL of dilute hydrochloric acid (1 + 1), and again evaporate to dryness. Dissolve the residue in 50 mL of dilute hydrochloric acid (1 + 24), and use the solution without further acidification. Use the residue from evaporation of 15 mL of hydrochloric acid to prepare the standard.

CALCIUM, MAGNESIUM, AND SODIUM. (By flame AAS, page 39).

Sample Stock Solution. Dissolve 5.0 g of sample with water in a 100-mL volumetric flask and dilute to the mark with water (1 mL = 0.05 g).

Element	Wavelength (nm)	Sample Wt (g)	Standard Added (mg)	Flame Type*	Background Correction
Ca	422.7	0.50	0.025; 0.05	N/A	No
Mg	285.2	0.50	0.01; 0.02	A/A	Yes
Na	589.0	0.05	0.01; 0.02	A/A	No

*A/A is air/acetylene; N/A is nitrous oxide/acetylene.

Potassium Carbonate Sesquihydrate

$K_2CO_3 \cdot 1.5H_2O$ **Formula Wt 165.23**

CAS Number 6381–79–9

REQUIREMENTS

Assay . . . . 98.5–101.0% $K_2CO_3 \cdot 1.5H_2O$
Loss on heating at 285 °C. 14.0–16.5%

MAXIMUM ALLOWABLE

Insoluble matter.	0.01%
Chloride (as Cl)	0.003%
Nitrogen compounds (as N).	0.001%
Phosphate (PO_4)	0.001%
Silica (SiO_2)	0.005%
Sulfur compounds (as SO_4)	0.004%
Ammonium hydroxide precipitate	0.01%
Heavy metals (as Pb)	5 ppm
Iron (Fe)	5 ppm
Calcium (Ca)	0.005%
Magnesium (Mg)	0.002%
Sodium (Na).	0.02%

TESTS

ASSAY. (By acidimetric titration of carbonate). Weigh accurately about 3.3 g and dissolve in 50 mL of water. Add 0.15 mL of methyl orange indicator solution, and titrate with 1 N hydrochloric acid to the change from yellow to orange. One milliliter of 1 N hydrochloric acid corresponds to 0.08262 g of $K_2CO_3 \cdot 1.5H_2O$.

LOSS ON HEATING AT 285 °C. Weigh accurately about 2.0 g and heat to constant weight at 270–300 °C.

INSOLUBLE MATTER. (Page 15). Use 10.0 g dissolved in 100 mL of water.

CHLORIDE. Dissolve 1.0 g in about 40 mL of water plus 6 mL of nitric acid. Filter through a chloride-free filter, wash, and dilute the filtrate with water to 60 mL. To 20 mL add 1 mL of silver nitrate reagent solution. Any turbidity should not exceed that produced by 0.01 mg of chloride ion (Cl) in an equal volume of solution containing the quantities of reagents used in the test.

NITROGEN COMPOUNDS. (Page 31). Use 1.0 g. For the standard, use 0.01 mg of nitrogen (N).

PHOSPHATE. (Page 32, Method 2). Dissolve 1.0 g in 50 mL of water in a platinum dish and digest on a hot plate ($\approx$100 °C) for 30 min. Cool, neutralize with dilute sulfuric acid (1 + 19) to a pH of about 4, and dilute with water to about 75 mL. Add 0.5 g of ammonium molybdate to the solution, and when it is dissolved, adjust the pH to 1.8 (using a pH meter) with dilute hydrochloric acid (1 + 9). Heat the solution to boiling, cool to room temperature, add 10 mL of hydrochloric acid, and dilute with water to 100 mL. Proceed as described. Carry along a standard containing 0.01 mg of phosphate ion (PO_4) and 0.05 mg of silica (SiO_2) in about 75 mL of water treated as the 75 mL sample solution. Reserve the aqueous phase for the determination of silica.

SILICA. Add 10 mL of hydrochloric acid to the solutions reserved from the determination of phosphate and transfer to separatory funnels. Add 40 mL of butyl alcohol, shake vigorously, and allow to separate. Draw off and discard the aqueous phase. Wash the butyl alcohol three times with 20-mL portions of dilute hydrochloric acid (1 + 99), discarding the washings each time. Dilute the butyl alcohol solutions to 50 mL with butyl alcohol, take 10 mL of each, and dilute each to 50 mL with butyl alcohol. Add 0.2 mL of a freshly prepared 2% solution of stannous chloride dihydrate in hydrochloric acid. The blue color in the extract from the sample should not exceed that in the control. If the butyl alcohol extracts are turbid, wash them with 10 mL of dilute hydrochloric acid (1 + 99).

SULFUR COMPOUNDS. (Page 33, Method 4). Cautiously acidify with hydrochloric acid. Allow 30 min for the turbidity to form.

AMMONIUM HYDROXIDE PRECIPITATE. Dissolve 10.0 g in 100 mL of water. Cautiously add 12 mL of sulfuric acid to 12 mL of water, cool, add the mixture to the solution of the sample, and evaporate to strong fuming. Cool, dissolve the residue in 130 mL of hot water, and add ammonium hydroxide until the solution is just alkaline to methyl red. Heat to boiling and filter, reserving the filtrate without the washings for the test for calcium and magnesium precipitate. Wash with hot water, rejecting the washings, and ignite. The residue includes some, but not all, of the silica in the sample.

HEAVY METALS. (Page 28, Method 1). To 5.0 g in a 150-mL beaker add 10 mL of water, mix, and add cautiously 10 mL of hydrochloric acid. Evaporate the solution to dryness on a hot plate ($\approx$100 °C), dissolve the residue in about 20 mL of water, and dilute with water to 25 mL. For the control, add 0.02 mg of lead to 1.0 g of sample and treat exactly as the 5.0 g of sample.

IRON. (Page 30, Method 1). Dissolve 2.0 g in 20 mL of dilute hydrochloric acid (1 + 1), and evaporate to dryness on a hot plate ($\approx$100 °C). Dissolve the residue in 5 mL of dilute hydrochloric acid (1 + 1), and again evaporate to dryness. Dissolve the residue in 50 mL of dilute hydrochloric acid (1 + 24), and use the solution without further acidification. Use the residue from evaporation of 15 mL of hydrochloric acid to prepare the standard.

CALCIUM, MAGNESIUM, AND SODIUM. (By flame AAS, page 39).

Sample Stock Solution. Dissolve 5.0 g of sample with water in a 100-mL volumetric flask and dilute to the mark with water (1 mL = 0.05 g).

Element	Wavelength (nm)	Sample Wt (g)	Standard Added (mg)	Flame Type*	Background Correction
Ca	422.7	0.50	0.025; 0.05	N/A	No
Mg	285.2	0.50	0.01; 0.02	A/A	Yes
Na	589.0	0.05	0.01; 0.02	A/A	No

*A/A is air/acetylene; N/A is nitrous oxide/acetylene.

Potassium Chlorate

KClO$_3$ **Formula Wt 122.55**

CAS Number 3811–04–9

DANGER! STRONG OXIDIZER. Causes irritation.

REQUIREMENTS

Assay . . . . ≥99.0% KClO$_3$

MAXIMUM ALLOWABLE

Insoluble matter. 0.005%
Bromate (BrO$_3$) 0.015%
Chloride (Cl) 0.001%
Nitrogen compounds (as N). 0.001%
Sulfate (SO$_4$) . . . . Passes test
Heavy metals (as Pb) . . . . 5 ppm
Iron (Fe) . . . . 3 ppm
Calcium (Ca) 0.002%
Magnesium (Mg) 0.002%
Sodium (Na). 0.01%

TESTS

ASSAY. (By oxidation–reduction titration of chlorate). Weigh, to the nearest 0.1 mg, 0.05 g of sample, and dissolve it with 10 mL of water in a 250-mL glass-stoppered flask. Add 50.0 mL of 0.1 N acid ferrous sulfate, cover the flask to prevent exposure to the air, and boil for 10 min. Cool, add 10 mL of 10% manganous sulfate solution and 5 mL of phosphoric acid, and titrate the excess ferrous sulfate with 0.1 N potassium permanganate. Run a blank in the same manner.

$$\% \text{ KClO}_3 = [(V_{blank} - V_{sample}) \times \text{N KMnO}_4 \times 2.043] / \text{Sample wt (g)}$$

where V_{blank} and V_{sample} are mL of $KMnO_4$ in titration of blank and sample solutions, respectively.

> **Acid Ferrous Sulfate Solution, 0.1 N.** Dissolve 27.8 g of ferrous sulfate crystals in about 500 mL of water containing 50 mL of sulfuric acid, and dilute to 1 L.

INSOLUBLE MATTER. (Page 15). Use 20 g dissolved in 250 mL of water.

BROMATE. (By differential pulse polarography, page 50). Use 1.0 g of sample and 0.5 g of calcium chloride dihydrate, $CaCl_2 \cdot 2H_2O$, in 25 mL of solution. For the standard, add 0.15 mg of bromate (BrO_3). Also prepare a reagent blank.

CHLORIDE. Dissolve 2.0 g in 40 mL of warm water, and filter if necessary through a chloride-free filter. Add 0.25 mL of nitric acid, free from lower oxides of nitrogen, and 1 mL of silver nitrate reagent solution. Any turbidity should not exceed that produced by 0.02 mg of chloride ion (Cl) in an equal volume of solution containing the quantities of reagents used in the test.

NITROGEN COMPOUNDS. (Page 31). Use 1.0 g. For the standard, use 0.01 mg of nitrogen (N).

SULFATE. Dissolve 5.0 g in 150 mL of water, filter if necessary, and add 1 mL of 10% hydrochloric acid reagent solution and 5 mL of barium chloride reagent solution. No precipitate should be produced on standing overnight. (Limit about 0.002%)

HEAVY METALS. (Page 28, Method 1). Dissolve 5.0 g in 20 mL of dilute hydrochloric acid (1 + 1). Evaporate the solution to dryness on a hot plate ($\approx$100 °C), add 5 mL more of dilute hydrochloric acid (1 + 1), and again evaporate to dryness. Dissolve the residue in about 20 mL of water and dilute with water to 25 mL. For the control, add 0.02 mg of lead to 1.0 g of sample and treat exactly as the 5.0 g of sample.

IRON. (Page 30). Dissolve 3.3 g in 20 mL of dilute hydrochloric acid (1 + 1) and evaporate to dryness on a hot plate ($\approx$100 °C). Add 5 mL of dilute hydrochloric acid (1 + 1) and again evaporate to dryness. Dissolve in 50 mL of dilute hydrochloric acid (1 + 25), and use the solution without further acidification. In preparing the control use the residue from evaporation of 15 mL of hydrochloric acid.

CALCIUM, MAGNESIUM, AND SODIUM. (By flame AAS, page 39).

> **Sample Stock Solution.** Dissolve 5.0 g in 25 mL of dilute hydrochloric acid (1 + 3), and digest in a covered beaker on a hot plate ($\approx$100 °C) until the reaction ceases. Uncover the beaker and evaporate to dryness. Add 10 mL of

dilute hydrochloric acid (1 + 3) and again evaporate to dryness. Dissolve in dilute hydrochloric acid (1 + 99) and dilute to 100 mL with the hydrochloric acid (1 + 99) in a volumetric flask (1 mL = 0.05 g).

Element	Wavelength (nm)	Sample Wt (g)	Standard Added (mg)	Flame Type*	Background Correction
Ca	422.7	1.0	0.02; 0.04	N/A	No
Mg	285.2	0.20	0.004; 0.008	A/A	Yes
Na	589.0	0.20	0.01; 0.02	A/A	No

*A/A is air/acetylene; N/A is nitrous oxide/acetylene.

Potassium Chloride

KCl   **Formula Wt 74.55**

CAS Number 7447–40–7

REQUIREMENTS

Assay . . . . 99.0–100.5% KCl
pH of a 5% solution . . . . 5.4–8.6 at 25 °C

MAXIMUM ALLOWABLE

Insoluble matter. 0.005%
Iodide (I) 0.002%
Bromide (Br) 0.01%
Chlorate and nitrate (as NO_3). 0.003%
Nitrogen compounds (as N). 0.001%
Phosphate (PO_4) . . . . 5 ppm
Sulfate (SO_4) 0.001%
Barium (Ba) . . . . Passes test
Heavy metals (as Pb) . . . . 5 ppm
Iron (Fe) . . . . : . . . . 3 ppm
Calcium (Ca) 0.002%
Magnesium (Mg) 0.001%
Sodium (Na). 0.005%

TESTS

ASSAY. (By argentimetric titration of chloride content). Weigh, to the nearest 0.1 mg, about 0.25 g of sample, and dissolve with 50 mL of water in a 250-mL glass-stoppered flask. Add, while agitating, exactly 50.0 mL of 0.1 N silver nitrate reagent solution, then add 3 mL of nitric acid and 10 mL of benzyl alcohol, and shake vigorously. Add 2 mL of ferric ammonium sulfate reagent solution (as indicator), and titrate the excess silver nitrate with 0.1 N ammonium thiocyanate.

$$\% \, KCl = \frac{[(50 \, mL \times N \, AgNO_3) - (mL \times N \, NH_4SCN)] \times 7.455}{Sample \, wt \, (g)}$$

pH OF A 5% SOLUTION. (Page 44). The pH should be 5.4–8.6 at 25 °C.

INSOLUBLE MATTER. (Page 15). Use 20 g dissolved in 150 mL of water.

IODIDE. Dissolve 11 g in 50 mL of water. Prepare a control by dissolving 1 g of the sample, 0.2 mg of iodide ion (I), and 1 mg of bromide ion (Br) in 50 mL of water. To each solution, in a separatory funnel, add 2 mL of hydrochloric acid and 5 mL of ferric chloride reagent solution. Allow to stand for 5 min. Add 10 mL of chloroform, shake for 1 min, allow the phases to separate, and draw off the chloroform layer. Reserve the water solution for the test for bromide. Any violet color in the chloroform extract from the solution of the sample should not exceed that in the extract from the control.

BROMIDE. Treat both the solution of the sample and the control obtained in the test for iodide as follows: wash twice by shaking with 10-mL portions of chloroform. Each time allow to separate, then draw off and discard the chloroform. To each of the water solutions add 10 mL of water, 65 mL of cold dilute sulfuric acid (1 + 1), and 15 mL of a solution of chromic acid prepared by dissolving 10 g in 100 mL of dilute sulfuric acid (1 + 3). Allow to stand for 5 min. Add 10 mL of chloroform, shake for 1 min, allow to settle, and draw off. (Half a 7-cm piece of filter paper rolled and placed in the stem of the separatory funnel will absorb any of the aqueous solution that may pass the stopcock, and thus assure a clear extract.) Any yellow-brown color in the chloroform extract from the solution of the sample should not exceed that in the extract from the control.

CHLORATE AND NITRATE.

Sample Solution A. Dissolve 0.50 g in 3 mL of water by heating in a boiling-water bath. Dilute to 50 mL with brucine sulfate reagent solution.

Control Solution B. Dissolve 0.50 g in 1.5 mL of water and 1.5 mL of the standard nitrate solution containing 0.01 mg of nitrate ion (NO_3) per mL by heating in a boiling-water bath. Dilute to 50 mL with brucine sulfate reagent solution.

Blank Solution C. Use 50 mL of brucine sulfate reagent solution.

Heat the three solutions in a preheated (boiling) water bath for 10 min. Cool rapidly in an ice bath to room temperature. Set a spectrophotometer at 410 nm and, using 1-cm cells, adjust the instrument to read 0 absorbance with blank solution C in the light path, then determine the absorbance of sample solution A. Adjust the instrument to read 0 absorbance with sample solution A in the light path and determine the absorbance of control solution B. The absorbance of sample solution A should not exceed that of control solution B.

NITROGEN COMPOUNDS. (Page 31). Use 1.0 g. For the standard, use 0.01 mg of nitrogen (N).

PHOSPHATE. (Page 32, Method 1). Dissolve 4.0 g in 25 mL of approximately 0.5 N sulfuric acid, and continue as described.

SULFATE. Dissolve 25.0 g in 200 mL of water, add 2 mL of hydrochloric acid, heat to boiling, add 10 mL of barium chloride reagent solution, digest in a covered beaker on a hot plate ($\approx$100 °C) for 2 h, and allow to stand overnight. If a precipitate is formed, filter, wash thoroughly, and ignite. Correct for the weight obtained in a complete blank test.

BARIUM. Dissolve 4.0 g in 20 mL of water, filter if necessary, and divide into two portions. To one portion add 2 mL of 10% sulfuric acid reagent solution and to the other 2 mL of water. The solutions should be equally clear at the end of 2 h. (Limit about 0.001%)

HEAVY METALS. (Page 28, Method 1). Dissolve 6.0 g in about 20 mL of water and dilute with water to 30 mL. Use 25 mL to prepare the sample solution, and use the remaining 5.0 mL to prepare the control solution.

IRON. (Page 30, Method 1). Use 3.3 g.

CALCIUM, MAGNESIUM, AND SODIUM. (By flame AAS, page 39).

Sample Stock Solution. Dissolve 5.0 g of sample in water and dilute to the mark with water in a 100-mL volumetric flask (1 mL = 0.05 g).

Element	Wavelength (nm)	Sample Wt (g)	Standard Added (mg)	Flame Type*	Background Correction
Ca	422.7	1.0	0.02; 0.04	N/A	No
Mg	285.2	1.0	0.01; 0.02	A/A	Yes
Na	589.0	0.10	0.005; 0.01	A/A	No

*A/A is air/acetylene; N/A is nitrous oxide/acetylene.

Potassium Chromate

K₂CrO₄   **Formula Wt 194.19**

CAS Number 7789–00–6

REQUIREMENTS

Assay . . . . ≥99.0% K₂CrO₄
pH of a 5% solution . . . . 8.6–9.8 at 25 °C

MAXIMUM ALLOWABLE

Insoluble matter 0.005%
Chloride (Cl) 0.005%

Sulfate (SO$_4$) 0.03%
Calcium (Ca) 0.005%
Sodium (Na) 0.02%

TESTS

ASSAY. (By iodometric oxidation–reduction titration). Weigh, to the nearest 0.1 mg, 0.25 g of sample, and dissolve in 200 mL of carbon dioxide-free water in a 500-mL glass-stoppered iodine flask. Add 3 g of potassium iodide and 7 mL of hydrochloric acid, and allow to stand in the dark for 10 min. Titrate the liberated iodine with 0.1 N sodium thiosulfate, adding 5 mL of starch indicator near the end point. One milliliter of 0.1 N sodium thiosulfate corresponds to 0.006473 grams K$_2$CrO$_4$.

$$\% \ K_2CrO_4 = (mL \times N \ Na_2S_2O_3 \times 6.473) \ / \ Sample \ wt \ (g)$$

pH OF A 5% SOLUTION. (Page 44). The pH should be 8.6–9.8 at 25 °C.

INSOLUBLE MATTER. (Page 15). Use 20 g dissolved in 150 mL of water.

CHLORIDE. Dissolve 0.20 g in 10 mL of water, filter if necessary through a small chloride-free filter, and add 1 mL of ammonium hydroxide, 1 mL of silver nitrate reagent solution, and 2 mL of nitric acid. Any turbidity should not exceed that produced by 0.01 mg of chloride ion (Cl) in an equal volume of solution containing the quantities of reagents used in the test. The comparison is best made by the general method for chloride in colored solutions, page 27.

SULFATE. Dissolve 10.0 g of sample in 250 mL of water. Add 2.5 g of barium chloride solution made in 15 mL of water, and add 5 mL of hydrochloric acid. Digest on low heat for 2 h and allow to stand at room temperature for 12 h. If any precipitate is formed, filter, wash thoroughly, and ignite in a platinum crucible. Fuse the ignited residue with 1 g of anhydrous sodium carbonate, extract the fused mass with water, and filter off the insoluble residue. Add 5 mL of hydrochloric acid to the filtrate, dilute to about 150 mL with water, heat the solution to boiling, and add 10 mL of 95% alcohol. Digest on low heat until reduction of chromate is complete as indicated by the change to a clear green or colorless solution. Neutralize the solution with ammonium hydroxide, filter off any chromic hydroxide precipitate, and dilute to 200 mL in a volumetric flask. Use 10.0 mL of this solution for the sample (0.5 g). Dilute to 30 mL with water. For the control, take 0.15 mg of sulfate ion (SO$_4$) in 30 mL of water. To each, add 1 mL of (1 + 19) hydrochloric acid and 1 mL of barium chloride reagent solution. Compare after 10 min. Sample turbidity should not exceed that of the control solution.

CALCIUM AND SODIUM. (By flame AAS, page 39).

Sample Stock Solution. Dissolve 5.0 g in water and dilute to 100 mL in a volumetric flask with water (1 mL = 0.05 g).

Element	Wavelength (nm)	Sample Wt (g)	Standard Added (mg)	Flame Type*	Background Correction
Ca	422.7	0.50	0.025; 0.05	N/A	No
Na	589.0	0.05	0.01; 0.02	A/A	No

*A/A is air/acetylene; N/A is nitrous oxide/acetylene.

Potassium Cyanide

KCN   **Formula Wt 65.12**

CAS Number 151–50–8

REQUIREMENTS

Assay . . . . ≥96.0% KCN

MAXIMUM ALLOWABLE

Chloride (Cl) 0.5%
Phosphate (PO_4) 0.005%
Sulfate (SO_4) 0.04%
Sulfide (S). 0.003%
Thiocyanate (SCN) . . . . Passes test
Iron, total (as Fe) 0.03%
Lead (Pb) . . . . 2 ppm
Sodium (Na). 0.5%

TESTS

ASSAY. (By argentimetric titration of cyanide content). Weigh accurately 0.5 g and dissolve in 30 mL of water. Add 0.2 mL of 10% potassium iodide reagent solution and 1 mL of ammonium hydroxide and titrate with 0.1 N silver nitrate to a slight yellowish permanent turbidity. One milliliter of 0.1 N silver nitrate corresponds to 0.01302 g of KCN.

> *Sample Solution A for the Determination of Chloride, Phosphate, Sulfate, Sulfide, Thiocyanate, and Iron.* Dissolve 10 g in water and dilute with water to 200 mL. Filter, if necessary, under a hood into a dry flask (1 mL = 0.05 g).

CHLORIDE. (Page 27). Dilute 1.0 mL of sample solution A with water to 50 mL. To 2.0 mL (0.002-g sample) of this solution add 2 mL of 30% hydrogen peroxide and allow to stand in a covered beaker until reaction ceases, then digest in the covered beaker on a hot plate (≈100 °C) for 20–30 min. Cool, and dilute with water to 25 mL.

PHOSPHATE. (Page 32, Method 1). To 20 mL of sample solution A add 2 mL of hydrochloric acid, and evaporate to dryness in a well-ventilated hood. Add 5 mL of dilute hydrochloric acid (1 + 1) and evaporate to dryness again. Dissolve in 50 mL of approximately 0.5 N sulfuric acid. To 20 mL of the solution add 5 mL of approximately 0.5 N sulfuric acid, and continue as described.

SULFATE. (Page 33, Method 2). Use 2.5 mL of sample solution A plus 1 mL of dilute hydrochloric acid (1 + 1) in a hood.

SULFIDE. To 20 mL of sample solution A (measured in a graduated cylinder) (1-g sample) add 0.15 mL of alkaline lead solution (made by adding 10% sodium hydroxide reagent solution to a 10% lead acetate reagent solution until the precipitate is redissolved). The color should not be darker than is produced by 0.03 mg of sulfide ion (S) in an equal volume of solution when treated with 0.15 mL of the alkaline lead solution.

THIOCYANATE. To 20 mL of sample solution A (measured in a graduated cylinder) (1-g sample) add 4 mL of hydrochloric acid and 0.20 mL of ferric chloride reagent solution. At the end of 5 min the solution should show no reddish tint when compared with 20 mL of water to which have been added the quantities of hydrochloric acid and ferric chloride used in the test. (Limit about 0.02%)

IRON, TOTAL. (Page 30, Method 1). Transfer 10 mL of sample solution A (0.5-g sample) to a platinum evaporating dish, add 3 mL of hydrochloric acid, and evaporate to dryness in a well-ventilated hood. Heat the residue at 650 °C for 30 min. Cool, add 2 mL of hydrochloric acid and 10 mL of water, and digest on a hot plate ($\approx$100 °C) until dissolution is complete. Dilute with water to 30 mL, and use 2.0 mL of this solution.

LEAD. Dissolve 1.2 g in 10 mL of water in a separatory funnel. Add 5 mL of ammonium citrate reagent solution, 2 mL of hydroxylamine hydrochloride reagent solution for the dithizone test, and 0.10 mL of phenol red indicator solution, and make the solution alkaline if necessary by the addition of ammonium hydroxide. Add 5 mL of standard dithizone solution in chloroform, shake gently but well for 1 min, and allow the layers to separate. The intensity of the red color of the chloroform layer should be no greater than that of a control made with 0.002 mg of lead and 0.2 g of the sample treated exactly like the solution of 1.2 g of sample in 10 mL of water.

SODIUM. (By flame AAS, page 39).

Sample Stock Solution. Dissolve 0.5 g of sample in a 100-mL volumetric flask and dilute to the mark with water (1 mL = 0.005 g).

Element	Wavelength (nm)	Sample Wt (g)	Standard Added (mg)	Flame Type*	Background Correction
Na	589.0	0.01	0.05; 0.10	A/A	No

*A/A is air/acetylene.

Potassium Dichromate

K$_2$Cr$_2$O$_7$ **Formula Wt 294.18**

CAS Number 7778–50–9

REQUIREMENTS

Assay . . . .≥99.0% K$_2$Cr$_2$O$_7$

MAXIMUM ALLOWABLE

Insoluble matter. 0.005%
Loss on drying 0.05%
Chloride (Cl) 0.001%
Sulfate (SO$_4$) 0.005%
Calcium (Ca) 0.003%
Iron (Fe) 0.001%
Sodium (Na). 0.02%

TESTS

ASSAY. (By iodometric oxidation–reduction titration). Weigh, to the nearest 0.1 mg, 0.2 g of sample, transfer to a 500-mL glass-stoppered iodine flask, and dissolve with 125 mL of water. Add 5 g of potassium iodide and 5 mL of hydrochloric acid (1 + 1) with constant swirling. Carefully wash down the sides of the flask with water so that a layer of water is formed on top of the solution, stopper the flask, and allow to stand in the dark for 10 min. Add about 160 mL of water and titrate with 0.1 N sodium thiosulfate, adding 5 mL of starch indicator solution near the end point. One milliliter of 0.1 N sodium thiosulfate corresponds to 0.004903 grams K$_2$Cr$_2$O$_7$.

$$\% \ K_2Cr_2O_7 = (mL \times N \ Na_2S_2O_3 \times 4.903) \ / \ \text{Sample wt (g)}$$

INSOLUBLE MATTER. (page 15). Dissolve 20 g in 200 mL of water

LOSS ON DRYING. Weigh accurately about 2 g and dry to constant weight at 105 °C.

CHLORIDE. Dissolve 1.0 g in 10 mL of water, filter if necessary through a small chloride-free filter, and add 1 mL of ammonium hydroxide and 1 mL of silver nitrate reagent solution. Prepare a standard containing 0.01 mg of chloride ion (Cl) in 10 mL of water, and add 1 mL of ammonium hydroxide and 1 mL of silver nitrate reagent solution. Add 2 mL of nitric acid to each. The comparison is best made by the general method for chloride in colored solutions, page 27.

SULFATE. Dissolve 10 g in 250 mL of water, filter if necessary, and heat to boiling. Add 25 mL of a solution containing 1 g of barium chloride and 2 mL of hydrochloric acid per 100 mL of solution. Digest in a covered beaker on a hot

plate (≈100 °C) for 2 h and allow to stand overnight. If a precipitate is formed, filter, wash thoroughly, and ignite. Fuse the residue with 1 g of sodium carbonate.
Extract the fused mass with water and filter off the insoluble residue. Add 5 mL of
hydrochloric acid to the filtrate, dilute with water to about 200 mL, heat to boiling, and add 10 mL of alcohol. Digest in a covered beaker on the hot plate (≈100
°C) until reduction of chromate is complete, as indicated by the change to a clear
green or colorless solution. Neutralize the solution with ammonium hydroxide
and add 2 mL of hydrochloric acid. Heat to boiling and add 10 mL of barium
chloride reagent solution. Digest in a covered beaker on a hot plate (≈100 °C) for
2 h and allow to stand overnight. Filter, wash thoroughly, and ignite. Correct for
the weight obtained in a complete blank test. If the original precipitate of barium
sulfate weighs less than the requirement allows, the fusion with sodium carbonate
is not necessary.

CALCIUM, IRON, AND SODIUM. (By flame AAS, page 39).

Sample Stock Solution for Calcium and Sodium. Dissolve 5.0 g of sample in 20 mL of water, transfer the solution to a 100-mL volumetric flask, and
dilute to the mark with water (1 mL = 0.05 g).

Sample Stock Solution for Iron. Dissolve 10.0 g of sample in 20 mL of
water, transfer the solution to a 100-mL volumetric flask, and dilute to the
mark with water (1 mL = 0.10 g).

Element	Wavelength (nm)	Sample Wt (g)	Standard Added (mg)	Flame Type*	Background Correction
Ca	422.7	0.50	0.015; 0.03	N/A	No
Fe	248.3	2.50	0.05; 0.10	A/A	Yes
Na	589.0	0.05	0.01; 0.02	A/A	No

*A/A is air/acetylene; N/A is nitrous oxide/acetylene.

Potassium Ferricyanide
Potassium Hexacyanoferrate(III)

$K_3Fe(CN)_6$ **Formula Wt 329.25**

CAS Number 13746–66–2

REQUIREMENTS

Assay. . . . . ≥99.0% $K_3Fe(CN)_6$

MAXIMUM ALLOWABLE

Insoluble matter. 0.005%
Chloride (Cl) 0.01%
Sulfate (SO_4) 0.01%
Ferro compounds . . . . . . .0.05% as ferrocyanide radical $[Fe(CN)_6]^{4-}$

TESTS

ASSAY. (By iodometric titration). Weigh accurately 1.4 g and dissolve in 50 mL of water in a glass-stoppered conical flask. Add 3 g of potassium iodide and 1 drop of glacial acetic acid. Then add 3.0 g of zinc sulfate heptahydrate, $ZnSO_4 \cdot 7H_2O$, in 20 mL of water. Stopper, swirl, and allow to stand in the dark for 30 min. Titrate the liberated iodine with 0.1 N sodium thiosulfate, adding 3 mL of starch indicator solution near the end of the titration. Correct for a blank. One milliliter of 0.1 N sodium thiosulfate corresponds to 0.03292 g of $K_3Fe(CN)_6$.

INSOLUBLE MATTER. Dissolve 20 g in 200 mL of water at room temperature, filter promptly through a tared filtering crucible, wash thoroughly, and dry at 105 °C.

CHLORIDE. Dissolve 2.0 g in 175 mL of water in a cylinder, add 2.5 g of cupric sulfate crystals dissolved in 25 mL of water, mix thoroughly, and allow the precipitate to settle. Filter the supernatant liquid, if necessary, through a chloride-free filter. To 10 mL of the clear solution add 1 mL of nitric acid, 10 mL of water, and 1 mL of silver nitrate reagent solution. Any turbidity should not exceed that produced by 0.01 mg of chloride ion (Cl) in an equal volume of solution containing the quantities of nitric acid and silver nitrate used in the test and enough cupric sulfate to match the color in the test solution.

SULFATE. Dissolve 5.0 g in 100 mL of water without heating, filter promptly, and to 10 mL (0.5 g) of the filtrate add 0.2 mL of glacial acetic acid and 5 mL of barium chloride reagent solution. Any turbidity should not exceed that produced by a standard containing 0.05 mg of sulfate ion (SO_4) in an equal volume of solution containing 0.2 mL of glacial acetic acid and 5 mL of barium chloride reagent solution. Compare 10 min after adding the barium chloride to the sample and standard solutions.

FERRO COMPOUNDS. Mix 400 mL of water with 10 mL of 25% sulfuric acid reagent solution and add 0.1 N potassium permanganate until the pink color persists for 1 min. Dissolve 4.0 g of the sample in the solution, add 0.10 mL of the 0.1 N potassium permanganate solution, and stir. The solution should retain a pink tint in comparison with a blank prepared with the same quantities of potassium ferricyanide, water, and sulfuric acid.

Potassium Ferrocyanide Trihydrate
Potassium Hexacyanoferrate(II) Trihydrate

$K_4Fe(CN)_6 \cdot 3H_2O$ **Formula Wt 422.39**

CAS Number 14459–95–1

REQUIREMENTS

Assay . . . . 98.5–102.0% $K_4Fe(CN)_6 \cdot 3H_2O$

MAXIMUM ALLOWABLE

Insoluble matter 0.005%
Chloride (Cl) 0.01%
Sulfate (SO_4) . . . . Passes test

TESTS

ASSAY. (By titration of reducing power). Weigh accurately about 1.6 g and dissolve in 250 mL of water. Add 30 mL of 25% sulfuric acid, and titrate with 0.1 N potassium permanganate to a change from yellow to orange-yellow. One milliliter of 0.1 N potassium permanganate corresponds to 0.04224 g of $K_4Fe(CN)_6 \cdot 3H_2O$.

INSOLUBLE MATTER. (Page 15). Use 20 g dissolved in 200 mL of water.

CHLORIDE. Dissolve 2.0 g in 175 mL of water, filter if necessary through a chloride-free filter, add 2.5 g of cupric sulfate crystals dissolved in 25 mL of water, mix thoroughly, and allow the precipitate to settle in a cylinder. To 10 mL of the clear supernatant liquid add 1 mL of nitric acid, 10 mL of water, and 1 mL of silver nitrate reagent solution. Any turbidity should not exceed that produced by 0.01 mg of chloride ion (Cl) in an equal volume of solution containing the quantities of nitric acid and silver nitrate used in the test and enough cupric sulfate to match the color in the test solution.

SULFATE. Dissolve 5.0 g in 100 mL of water without heating, filter promptly, and to 10 mL (0.5 g) of the filtrate add 0.2 mL of glacial acetic acid and 5 mL of barium chloride reagent solution. Any turbidity should not exceed that produced by a standard containing 0.05 mg of sulfate ion (SO_4) in an equal volume of solution containing 0.2 mL of glacial acetic acid and 5 mL of barium chloride reagent solution. Compare 10 min after adding the barium chloride to the sample and standard solutions. (Limit about 0.01%)

Potassium Fluoride

KF

Formula Wt 58.10

CAS Number 7789–23–3

REQUIREMENTS

Assay . . . . ≥99.0% KF

MAXIMUM ALLOWABLE

Chloride (Cl) 0.005%

Titrable acid . . . . 0.03 meq/g
Titrable base. . . . . 0.01 meq/g
Potassium fluosilicate (K_2SiF_6) 0.1%
Sulfate (SO_4) 0.005%
Heavy metals (as Pb) 0.001%
Iron (Fe). 0.001%
Sodium (Na). 0.2%

TESTS

ASSAY. (By indirect acid–base titrimetry). Weigh accurately 0.2 g of sample and dissolve in 50 mL of water in a polyethylene vessel. Pass the solution through the cation column at a rate of about 5 mL/min and collect the eluate in a 500-mL polyethylene titration flask. Then wash the resin in the column with water at a rate of about 10 mL/min, and collect in the same titration flask. Add 0.15 mL of phenolphthalein indicator solution to the flask and titrate with 0.1 N sodium hydroxide. Continue the titration as the washing proceeds until 50 mL of eluate requires no further titration. One milliliter of 0.1 N sodium hydroxide corresponds to 0.005810 g of KF.

CHLORIDE. (Page 27). Dissolve 1.0 g plus 0.2 g of boric acid in 80 mL of water, filter if necessary through a chloride-free filter, and dilute to 100 mL with water. Use 20 mL (0.2-g sample) of this solution.

TITRABLE ACID. Dissolve 5.0 g in 40 mL of carbon dioxide-free water in a platinum dish, add 10 mL of a freshly-prepared saturated solution of potassium nitrate, cool the solution to 0 °C, and add 0.15 mL of phenolphthalein indicator solution. If a pink color is produced, omit the titration for titrable acid and reserve the solution for titrable base. If no pink color is produced, titrate with 0.1 N sodium hydroxide until the pink color persists for 15 s while the temperature is near 0 °C. Not more than 1.50 mL of the 0.1 N sodium hydroxide should be required. Reserve the solution for the test for potassium fluosilicate.

TITRABLE BASE. If a pink color was produced in the solution prepared for the test for titrable acid, add 0.1 N hydrochloric acid, stirring the liquid only gently, until the pink color is discharged. Not more than 0.50 mL of the 0.1 N hydrochloric acid should be required. Reserve the solution for the test for potassium fluosilicate.

POTASSIUM FLUOSILICATE. Heat the solution reserved from the test for titrable acid or titrable base to boiling and titrate while hot with 0.01 N sodium hydroxide to a permanent pink color. Not more than 9.0 mL of 0.01 N sodium hydroxide should be required.

SULFATE. (Page 33, Method 2). Evaporate 1.0 g to dryness with 2 mL of hydrochloric acid in a platinum dish. Repeat the evaporation four additional times.

HEAVY METALS. (Page 28, Method 1). Use 2.0 g. Adjust the pH, using pH indicator paper.

IRON. (Page 30, Method 1). To 1.0 g in a platinum dish add 10 mL of hydrochloric acid, and evaporate to dryness. Dissolve the residue in 10 mL of hydrochloric acid, and repeat the evaporation to dryness. Warm the residue with a few drops of hydrochloric acid, and dilute with water to 40 mL. In preparing the standard, use the residue from the evaporation of 20 mL of hydrochloric acid.

SODIUM. (By flame AAS, page 39).

Sample Stock Solution. Dissolve 0.10 g in 100 mL of water, using suitable plastic volumetric flasks (1 mL = 0.001 g).

Element	Wavelength (nm)	Sample Wt (g)	Standard Added (mg)	Flame Type*	Background Correction
Na	589.0	0.005	0.005; 0.01	A/A	No

*A/A is air/acetylene.

Potassium Hydrogen Diiodate
Potassium Bi-iodate

KH(IO$_3$)$_2$ **Formula Wt 389.91**

CAS Number 13455–24–8

REQUIREMENTS

Assay (acid–base titrimetry) . . . . 99.95–100.05% KH(IO$_3$)$_2$
Assay (iodometric) . . . . 99.9–100.1% KH(IO$_3$)$_2$

MAXIMUM ALLOWABLE

Insoluble matter 0.01%
Chloride (Cl) 0.02%
Sulfate (SO$_4$) 0.005%
Heavy metals (as Pb) 0.001%
Iron (Fe) 0.001%

TESTS

NOTE. Both assay procedures represent common uses of this reagent. The sample should be thoroughly crushed and dried for 1 h at 105 °C before use in either assay procedure.

ASSAY. (By acid–base titrimetery). Weigh in duplicate (to the nearest 0.1 mg) 1.980 ± 0.001 g of dried sample. Dissolve in 150 mL of carbon-dioxide-free water. Add 0.610 g of NIST standard tris(hydroxymethyl)aminomethane (TRIS), stir, and titrate potentiometrically with a solution containing 1 mg of NIST standard TRIS per mL to a pH of 5.00. Calculate the assay value of the sample as follows:

$$\% \ KH(IO_3)_2 = \frac{[(V \times C) + A] \times F \times 38.991}{Sample \ wt \ (g) \times 121.14}$$

where

$V =$ volume (mL) of TRIS titrant solution
$C =$ concentration (mg/mL) TRIS titrant solution
$A =$ weight (mg) of solid TRIS added
$F =$ assay value of NIST standard TRIS (%100).

ASSAY. (By iodometric titration of oxidizing power). Weigh accurately about 0.13 g of dried sample and dissolve 100 mL of water in a glass-stoppered conical flask. Add 3.0 g of potassium iodide and 1 mL of sulfuric acid. Stopper, swirl, and let stand for 30 min. Titrate the liberated iodine with 0.1 N sodium thiosulfate and add 3 mL of starch indicator solution near the end of the titration. Correct for a blank. One milliliter of 0.1 N sodium thiosulfate corresponds to 0.003249 g of $KH(IO_3)_2$.

INSOLUBLE MATTER. (Page 15). Use 10.0 g dissolved in 200 mL of water.

CHLORIDE. Dissolve 1.0 g in 100 mL of water in a conical flask, filter if necessary through a chloride-free filter, and add 6 mL of hydrogen peroxide and 1 mL of phosphoric acid. Heat to boiling and boil gently until all the iodine is expelled and the solution is colorless. Cool, wash down the sides of the flask, and add 0.5 mL of hydrogen peroxide. If an iodine color develops, boil until the solution is colorless and then for an additional 10 min. If no color develops, boil for 10 min. Dilute with water to 100 mL and dilute a 10-mL portion to 25 mL. Prepare a standard containing 0.02 mg of chloride ion and 0.6 mL of hydrogen peroxide in 23 mL water. Add 1 mL of nitric acid and 1 mL of silver nitrate reagent solution to each. Any turbidity in the sample solution should not exceed that in the standard.

SULFATE. (Page 33, Method 2). Evaporate 1.0 g to dryness with 3 mL of hydrochloric acid and repeat the evaporation three additional times. Allow 30 min for the turbidity to form.

HEAVY METALS. Treat 10.0 g with 10 mL of hydrochloric acid and evaporate to dryness. Repeat the evaporation two additional times. Dissolve the residue in water and add 1 mL of 1 N acetic acid and 5 mL of hydrogen sulfide water. For the standard, use 0.1 mg of lead treated exactly as the sample. Any brown color produced in the sample solution within 5 min should not exceed that in the standard.

IRON. (Page 30, Method 2). Treat 1.0 g with 1 mL of 50% sulfuric acid and 5 mL of hydroxylamine hydrochloride reagent solution. Evaporate to expel the iodine.

Dissolve the residue in 10 mL of water and add 1 mL of hydroxylamine hydrochloride reagent solution.

Potassium Hydrogen Phthalate, Acidimetric Standard
Potassium Acid Phthalate
Potassium Biphthalate
Phthalic Acid, Monopotassium Salt

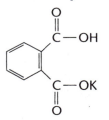

HOCOC$_6$H$_4$COOK   **Formula Wt 204.22**

CAS Number 877–24–7

NOTE. This reagent is satisfactory for use as a pH standard. For use as an acidimetric standard, this material should be lightly crushed and dried for 2 h at 120 °C to remove any absorbed moisture.

REQUIREMENTS

Assay (dried basis). . . . . 99.95–100.05% C$_8$H$_5$O$_4$K
pH of a 0.05 M solution. . . . . 4.00–4.02 at 25.0 ± 0.2 °C

MAXIMUM ALLOWABLE

Insoluble matter. 0.005%
Chlorine compounds (as Cl) 0.003%
Sulfur compounds (as S) 0.002%
Heavy metals (as Pb) . . . . 5 ppm
Iron (Fe). . . . . 5 ppm
Sodium (Na) 0.005%

TESTS

ASSAY. (By acid–base titrimetry). This is a comparative procedure in which the potassium hydrogen phthalate to be assayed is compared with NIST SRM Potassium Hydrogen Phthalate. Accurately weighed portions of anhydrous sodium carbonate, dissolved in water, are allowed to react with accurately weighed portions of the two potassium hydrogen phthalate samples of such size as to provide a small excess of the latter. The carbon dioxide is removed by boiling and the excess potassium hydrogen phthalate is determined by titration with carbonate-free sodium hydroxide using a potentiometric end point.

> *CAUTION.* Due to the small tolerance in assay limits for this acidimetric standard, extreme care must be observed in the weighing, transferring, and titrating operations in the following procedure. Strict adherence to the specified sample weights and final titration volumes is absolutely necessary. It is recommended that the titrations be run at least in duplicate on both the sample and the SRM. Duplicate values of M, the number of milliequivalents of total alkali required per gram of potassium hydrogen phthalate, obtained on the sample and on the SRM, should agree within 2 parts in 5000 to be acceptable for averaging.

Procedure. Place duplicate 1.300 ± 0.001-g portions of a uniform sample of anhydrous sodium carbonate, Na_2CO_3, in clean, dry weighing bottles. Dry at 285 °C for 3 h, stopper the bottles, and cool in a desiccator for at least 2 h. Open the weighing bottles momentarily to equalize the pressure, weigh each bottle and its contents to the nearest 0.1 mg, and transfer the entire contents of each bottle to a clean, dry 50-mL beaker. Store the beaker in a desiccator until ready for use, weigh each bottle again, and determine by difference, to the nearest 0.1 mg, the weight of each portion of sodium carbonate.

Transfer the portions of sodium carbonate to 500-mL titration flasks by pouring them through powder funnels. Rinse the beakers, funnels, and sides of the flasks thoroughly with several small portions of water. In the same manner add to one of these flasks 5.100 ± 0.003 g of the potassium hydrogen phthalate to be assayed and to the other flask the same weight of the NIST SRM potassium hydrogen phthalate—both of which have been previously crushed (not ground) in an agate or mullite mortar to approximately 100-mesh fineness, dried at 120 °C for 2 h, and cooled in a desiccator for at least 2 h. Dilute the contents of each titration flask with water to about 90 mL, swirl to dissolve the samples, stopper, and bubble carbon dioxide-free air through the solutions at a moderately fast rate during all subsequent operations. Boil the solutions gently for 20 min to complete the removal of carbon dioxide, taking care to prevent loss of solutions by spraying. Cool the solutions to room temperature in an ice bath, and dilute to 75–80 mL with carbon dioxide-free water. Titrate each solution with 0.02 N carbonate-free sodium hydroxide using a combination electrode (glass–calomel or glass–silver, silver chloride) for the measurement of E in millivolts or pH. The end point of the titration is determined by the second-derivative method (page 52).

Calculations. Calculate the number of milliequivalents of total alkali (sodium carbonate plus sodium hydroxide) required to neutralize exactly 1 g of the sample of potassium hydrogen phthalate and of the NIST SRM by the following equations:

$$M_u = \frac{\dfrac{B}{0.052994} + (C \times D)}{W_u}$$

$$M_s = \frac{\dfrac{B}{0.052994} + (C \times D)}{W_s \times F}$$

where

M_u = number of milliequivalents of total alkali required per gram of potassium hydrogen phthalate being assayed

M_s = number of milliequivalents of total alkali required per gram of NIST standard potassium hydrogen phthalate

B = weight, in grams, of sodium carbonate

C = volume, in mL, of NaOH standard solution

D = normality of NaOH standard solution

W_u = weight, in grams, of potassium hydrogen phthalate being assayed

W_s = weight, in grams, of NIST SRM potassium hydrogen phthalate

F = assay value of NIST SRM in percent divided by 100

Calculate the assay value of the sample of potassium hydrogen phthalate as follows:

$$A = (M_u \times 100) / M_s$$

where A is percent of potassium hydrogen phthalate in the sample assayed and M_u and M_s are defined as before.

pH OF A 0.05 M SOLUTION. Dissolve 1.021 g in 100 g of carbon dioxide- and ammonia-free water (or 1.012 g in a volume of 100 mL). Standardize the pH meter and electrodes at pH 4.00 at 25.0 ± 0.2 °C with 0.05 M NIST SRM Potassium Hydrogen Phthalate. Determine the pH of the solution by the method described on page 46. The pH should be 4.00–4.02 at 25.0 ± 0.2 °C.

INSOLUBLE MATTER. (Page 15). Use 20 g dissolved in 200 mL of water.

CHLORINE COMPOUNDS. Mix 1.0 g with 0.5 g of sodium carbonate, moisten the mixture, and ignite until thoroughly charred, avoiding an unduly high temperature. Treat the residue with 30 mL of water, cautiously add 3 mL of nitric acid, and filter through a chloride-free filter. Wash the residue with a few milliliters of hot water, dilute the filtrate with water to 50 mLs, and to 25 mL of the filtrate add 1 mL of silver nitrate reagent solution. Any turbidity should not exceed that produced by 0.015 mg of chloride ion (Cl) in an equal volume of solution containing the quantities of reagents used in the test.

SULFUR COMPOUNDS. Mix 2.0 g with 1 g of sodium carbonate, add in small portions 15 mL of water, evaporate, and thoroughly ignite, taking care to protect the sample from the flame because of the presence of sulfur compounds in the natural gas. Treat the residue with 20 mL of water and 2 mL of 30% hydrogen peroxide and heat on a hot plate ($\approx$100 °C) for 15 min. Add 5 mL of hydrochloric acid and evaporate to dryness on the hot plate. Dissolve the residue in 10 mL of water, filter, wash with several portions of water, and dilute the filtrate with water to 25 mL. Add to this solution 0.5 mL of 1 N hydrochloric acid and 2 mL of barium chloride reagent solution. Any turbidity should not exceed that in a standard prepared as follows: treat 1 g of sodium carbonate with 2 mL of 30% hydrogen peroxide and 5 mL of hydrochloric acid and evaporate to dryness on a hot plate ($\approx$100 °C). Dissolve the residue and 0.12 mg of sulfate ion (SO_4) in sufficient water to make 25 mL and add 0.5 mL of 1 N hydrochloric acid and 2 mL of barium chloride reagent solution.

HEAVY METALS. (Page 28, Method 1). Dissolve 6.6 g in warm water, and dilute with water to 50 mL. Use 40 mL to prepare the sample solution, and use the remaining 10 mL to prepare the control solution.

IRON. (Page 30, Method 2). For the test, use 2.0 g, dissolve in 30 mL of water, and dilute to 50 mL.

SODIUM. (By flame AAS, page 39).

Sample Stock Solution. Dissolve 1.0 g of sample in water and dilute to 100 mL in a volumetric flask with water (1 mL = 0.01 g).

Element	Wavelength (nm)	Sample Wt (g)	Standard Added (mg)	Flame Type*	Background Correction
Na	589.0	0.20	0.01; 0.02	A/A	No

*A/A is air/acetylene.

Potassium Hydrogen Sulfate, Fused
Potassium Pyrosulfate

CAS Number 7790–62–7

NOTE. This product is a mixture of potassium pyrosulfate, $K_2S_2O_7$, and potassium hydrogen sulfate, $KHSO_4$.

REQUIREMENTS

Acidity (as H_2SO_4) 37.5–38.6%

MAXIMUM ALLOWABLE

Water (H_2O). 2.5%
Insoluble matter. 0.01%
Ammonium hydroxide precipitate 0.01%

Chloride (Cl) 0.002%
Phosphate (PO$_4$) 0.001%
Heavy metals (as Pb) 0.001%
Iron (Fe). 0.002%
Calcium (Ca) 0.002%
Magnesium (Mg) 0.001%
Sodium (Na) 0.01%

TESTS

ACIDITY. (By acid–base titrimetry). Weigh accurately 4.0 g, dissolve in 50 mL of water, and titrate with 1 N sodium hydroxide, using methyl orange as indicator. One milliliter of 1 N alkali corresponds to 0.04904 g of H$_2$SO$_4$.

WATER. Weigh accurately about 5 g, dissolve in water, and dilute with water to 250 mL. Evaporate 25.0 mL of the solution to dryness in a tared platinum container and ignite to remove all sulfuric acid and leave K$_2$SO$_4$. Calculate the percentage of K$_2$SO$_4$ (B). Calculate the percentage of SO$_3$ (A) from the percentage of H$_2$SO$_4$ obtained in the test for acidity. One percent of H$_2$SO$_4$ is equivalent to 0.8163% of SO$_3$. Calculate the percentage of H$_2$O (C) as the difference between 100% and the sum of the percentages of SO$_3$ and K$_2$SO$_4$.

$$C = 100 - (A + B)$$

INSOLUBLE MATTER. (Page 15). Dissolve 20 g in 200 mL of water.

AMMONIUM HYDROXIDE PRECIPITATE. Dissolve 20 g in 200 mL of water, filter, add ammonium hydroxide until the solution is alkaline to methyl red, boil for 1 min, and digest in a covered beaker on a hot plate (≈100 °C) for 1 h. Filter through a tared filtering crucible, saving the filtrate separate from the washings for the test for calcium and magnesium precipitate. Wash thoroughly and dry at 105 °C.

CHLORIDE. (Page 27). Use 0.50 g.

PHOSPHATE. Dissolve 2.0 g in 15 mL of dilute ammonium hydroxide (1 + 2), and evaporate to dryness. Dissolve the residue in 25 mL of approximately 0.5 N hydrochloric acid, add 1 mL of ammonium molybdate reagent solution and 1 mL of 4-(methylamino)phenol sulfate reagent solution, and allow to stand for 2 h at room temperature. Any blue color should not exceed that produced by 0.02 mg of phosphate ion (PO$_4$) in an equal volume of solution containing the quantities of reagents used in the test, including the residue from evaporation of 5 mL of ammonium hydroxide.

HEAVY METALS. (Page 28, Method 1). Dissolve 4.0 g in about 20 mL of water, and dilute with water to 32 mL. Use 24 mL to prepare the sample solution, and use the remaining 8.0 mL to prepare the control solution.

IRON. (Page 30, Method 1). Dissolve 5.0 g in 80 mL of dilute hydrochloric acid (1 + 3), and boil gently for 10 min. Cool, dilute with water to 100 mL, and use 10 mL of the solution.

CALCIUM, MAGNESIUM, AND SODIUM. (By flame AAS, page 39).

Sample Stock Solution. Dissolve 5.0 g of sample with water in a 100-mL volumetric flask and dilute to the mark with water (1 mL = 0.05 g).

Element	Wavelength (nm)	Sample Wt (g)	Standard Added (mg)	Flame Type*	Background Correction
Ca	422.7	1.0	0.02; 0.04	N/A	No
Mg	285.2	1.0	0.01; 0.02	A/A	Yes
Na	589.0	0.10	0.01; 0.02	A/A	No

*A/A is air/acetylene; N/A is nitrous oxide/acetylene.

Potassium Hydroxide

KOH   **Formula Wt 56.11**

CAS Number 1310–58–3

NOTE. Reagent potassium hydroxide usually contains 10–15% water.

REQUIREMENTS

Assay and potassium carbonate . . . . .≥85% KOH and ≤2.0% K_2CO_3

MAXIMUM ALLOWABLE

Chloride (Cl) 0.01%
Nitrogen compounds (as N). 0.001%
Phosphate (PO_4) . . . . 5 ppm
Sulfate (SO_4) 0.003%
Ammonium hydroxide precipitate 0.02%
Heavy metals (as Ag) 0.001%
Iron (Fe) 0.001%
Nickel (Ni) 0.001%
Sodium (Na). 0.05%

TESTS

NOTE. Special care must be taken in sampling to obtain a representative sample and to avoid absorption of water and carbon dioxide by the sample taken.

ASSAY AND POTASSIUM CARBONATE. (By acid–base titrimetry). Weigh rapidly 35–40 g, to within 0.1 g, dissolve, cool, and dilute to 1 L, using carbon dioxide-free water throughout. Dilute 50.0 mL of the well-mixed solution to 200 mL with carbon dioxide-free water, add 5 mL of barium chloride reagent solution, shake, and allow to stand for a few minutes. Titrate with 1 N hydrochloric acid to the phenolphthalein end point. One milliliter of 1 N hydrochloric acid corresponds to 0.05611 g of KOH. Continue the titration with 1 N hydrochloric acid to the methyl orange end point to determine carbonate. One milliliter of 1 N hydrochloric acid corresponds to 0.06911 g of K_2CO_3.

Sample Solution A for the Determination of Chloride, Nitrogen Compounds, Phosphate, Sulfate, Heavy Metals, Iron, and Nickel.
Dissolve 50.0 g in carbon dioxide- and ammonia-free water, cool, and dilute with the water to 500 mL in a volumetric flask (1 mL = 0.10 g).

CHLORIDE. (Page 27). Use 1.0 mL of sample solution A (0.1-g sample).

NITROGEN COMPOUNDS. (Page 31). Use 20 mL of sample solution A (2-g sample). For the control, use 10 mL of sample solution A and 0.01 mg of nitrogen (N).

PHOSPHATE. (Page 32, Method 1). To 40 mL of sample solution A (4-g sample) add 10 mL of hydrochloric acid, and evaporate to dryness on a hot plate ($\approx$100 °C). Dissolve the residue in 25 mL of approximately 0.5 N sulfuric acid, and continue as described. For the standard, include the residue from evaporation of 5 mL of hydrochloric acid.

SULFATE. (Page 33, Method 1). Neutralize 16.7 mL of sample solution A, and evaporate to 10 mL. Allow 30 min for the turbidity to form.

AMMONIUM HYDROXIDE PRECIPITATE. Weigh about 10 g and dissolve in about 100 mL of water. Cautiously add 12 mL of sulfuric acid to 12 mL of water, cool, add the mixture to the solution of the sample, and evaporate to dense fumes. Cool, dissolve the residue in 130 mL of hot water, and add ammonium hydroxide until the solution is just alkaline to methyl red. Heat to boiling, filter, wash with hot water, and ignite. The residue includes some, but not all, of the silica in the sample.

HEAVY METALS. To 60 mL of sample solution A (6-g sample) cautiously add 15 mL of nitric acid. For the control, add 0.05 mg of silver to 10 mL of sample solution A and cautiously add 15 mL of nitric acid. Evaporate both solutions to dryness over a low flame or on an electric hot plate. Dissolve each residue in about 20 mL of water, filter if necessary through a chloride-free filter, and dilute with water to 25 mL. Adjust the pH of the control and sample solutions to between 3 and 4 (using a pH meter) with 1 N acetic acid or ammonium hydroxide (10% NH_3), dilute with water to 40 mL, and mix. Add 10 mL of freshly prepared hydrogen sulfide water to each and mix. Any yellow-brown color in the solution of the sample should not exceed that in the control.

IRON. (Page 30, Method 1). Neutralize 10 mL of sample solution A (1-g sample) with hydrochloric acid, using phenolphthalein indicator, add 2 mL in excess, and dilute with water to 50 mL. Use this solution. For the standard, use the residue obtained by evaporating the quantity of acid used to neutralize the sample solution.

NICKEL. Dilute 20 mL of sample solution A (2-g sample) to 50 mL with water and neutralize with hydrochloric acid. Dilute with water to 85 mL and adjust the pH to 8 with ammonium hydroxide. Add 5 mL of bromine water, 5 mL of 1% dimethylglyoxime solution in alcohol, and 5 mL of 10% sodium hydroxide reagent solution. Any red color should not exceed that produced by 0.02 mg of nickel in an equal volume of solution containing the quantities of reagents used in the test.

SODIUM. (By flame AAS, page 39).

> ***Sample Stock Solution.*** Dissolve 1.0 g of sample in 50 mL of water, add 5 mL of nitric acid, transfer to a 100-mL volumetric flask, and dilute to the mark with water (1 mL = 0.01 g).

Element	Wavelength (nm)	Sample Wt (g)	Standard Added (mg)	Flame Type*	Background Correction
Na	589.0	0.05	0.02; 0.04	A/A	No

*A/A is air/acetylene.

Potassium Iodate

KIO_3 Formula Wt 214.00

CAS Number 7758–05–6

REQUIREMENTS

Assay . . . . 99.4–100.4% KIO_3
pH of a 5% solution . . . . 5.0–8.0 at 25 °C

MAXIMUM ALLOWABLE

Insoluble matter. 0.005%
Chloride and bromide (as Cl) 0.01%
Iodide (I) 0.001%
Nitrogen compounds (as N). 0.005%
Sulfate (SO_4) 0.005%
Heavy metals (as Pb) . . . . .5 ppm
Iron (Fe). 0.001%
Sodium (Na). 0.005%

TESTS

ASSAY. (By iodometric titration of oxidizing power). Weigh accurately 0.14 g and dissolve in 50 mL of water in a glass-stoppered conical flask. Add 3 g of potassium iodide and 1 mL of sulfuric acid. Stopper, swirl, and allow to stand for 3 min. Titrate the liberated iodine with 0.1 N sodium thiosulfate, adding 3 mL of starch indicator solution near the end of the titration. Correct for a blank. One milliliter of 0.1 N sodium thiosulfate corresponds to 0.003567 g of KIO_3.

pH OF A 5% SOLUTION. (Page 44). The pH should be 5.0–8.0 at 25 °C.

INSOLUBLE MATTER. (Page 15). Use 20 g dissolved in 250 mL of water.

CHLORIDE AND BROMIDE. Dissolve 1.0 g in 100 mL of water in a flask, filter if necessary through a chloride-free filter, and add 6 mL of hydrogen peroxide and 1 mL of phosphoric acid. Heat to boiling and boil gently until all the iodine is expelled and the solution is colorless. Cool, wash down the sides of the flask, and add 0.5 mL of hydrogen peroxide. If an iodine color develops, boil until the solution is colorless and for 10 min longer. If no color develops, boil for 10 min. Dilute with water to 100 mL, take 10 mL, and dilute with water to 23 mL. Prepare a standard containing 0.01 mg of chloride and 0.6 mL of hydrogen peroxide in 23 mL of water. Add 1 mL of nitric acid and 1 mL of silver nitrate reagent solution to each. Any turbidity in the solution of the sample should not exceed that in the standard.

IODIDE. Dissolve 11 g in 160 mL of water, and add 1 g of citric acid and 5 mL of chloroform. Prepare a control solution containing 1 g of the sample, 1 g of citric acid, and 0.1 mg of iodide in 160 mL of water; add 5 mL of chloroform. Shake vigorously and allow the chloroform to separate. Any pink color in the chloroform layer from the sample should not exceed that in the chloroform layer from the control.

NITROGEN COMPOUNDS. (Page 31). Dilute the distillate of a 1.0-g sample to 50 mL. Use 10 mL of the distillate. For the standard, use 0.01 mg of nitrogen (N).

SULFATE. (Page 33, Method 2). Evaporate 1.0 g to dryness with 3 mL of hydrochloric acid. Repeat the evaporation two more times. Allow 30 min for the turbidity to form.

HEAVY METALS. (Page 28, Method 1). Dissolve 7.0 g in 50 mL of water, heat to boiling, add 1.5 mL of dilute formic acid (1 + 9), and continue heating until the initial reaction ceases. Add dropwise 50 mL of dilute formic acid (1 + 9), and continue boiling until the solution appears to be free of iodine. Wash down the sides of the container, and boil again if the iodine color reappears. Repeat the washing down and boiling until the solution remains colorless. Evaporate to dryness on a hot plate (≈100 °C). Dissolve the residue in about 20 mL of water, and dilute with water to 35 mL. Use 25 mL to prepare the sample solution, use 5.0 mL to prepare the control solution, and reserve the remaining 5.0 mL for the test for iron.

IRON. (Page 30, Method 2). Use 5.0 mL of the solution reserved from the test for heavy metals.

SODIUM. (By flame AAS, page 39).

Sample Stock Solution. Dissolve 1.0 g of sample in water, add 5 mL of nitric acid, transfer to a 100-mL volumetric flask, and dilute to the mark with water (1 mL = 0.01 g).

Element	Wavelength (nm)	Sample Wt (g)	Standard Added (mg)	Flame Type*	Background Correction
Na	589.0	0.20	0.005; 0.01	A/A	No

*A/A is air/acetylene.

Potassium Iodide

KI   **Formula Wt 166.00**

CAS Number 7681–11–0

REQUIREMENTS

Assay . . . . ≥99.0% KI
pH of a 5% solution . . . . 6.0–9.2 at 25 °C

MAXIMUM ALLOWABLE

Insoluble matter. 0.005%
Loss on drying at 150 °C. 0.2%
Chloride and bromide (as Cl) 0.01%
Iodate (IO_3) . . . . 3 ppm
Nitrogen compounds (as N). 0.001%
Phosphate (PO_4) 0.001%
Sulfate (SO_4) 0.005%
Barium (Ba) 0.002%
Heavy metals (as Pb) . . . . 5 ppm
Iron (Fe). . . . . 3 ppm
Calcium (Ca) 0.002%
Magnesium (Mg) 0.001%
Sodium (Na). 0.005%

TESTS

ASSAY. (By titration of reductive power). Weigh, to the nearest 0.1 mg, about 0.5 g of sample, and dissolve with 20 mL of water in a 200-mL glass-stoppered titration flask. Add 30 mL of hydrochloric acid and 5 mL of chloroform, cool if necessary, and titrate with 0.05 M potassium iodate solution until the iodine color disappears from the aqueous layer. Stopper, shake vigorously for 30 s, and continue the titration, shaking vigorously after each addition of the iodate, until the iodine color in the chloroform is discharged.

$$\% \text{ KI} = (\text{mL} \times \text{M KIO}_3 \times 33.20) / \text{Sample wt (g)}$$

pH OF A 5% SOLUTION. (Page 44). The pH should be 6.0–9.2 at 25 °C.

INSOLUBLE MATTER. (Page 15). Use 20 g dissolved in 150 mL of water. Retain the filtrate, without the washings, for the test for calcium, magnesium, and R_2O_3 precipitate.

LOSS ON DRYING. Weigh accurately about 2.0 g of a sample in which the large crystals have been crushed, but which has not been ground for very long in a mortar. Dry in a weighed dish for 6 h at 150 °C.

CHLORIDE AND BROMIDE. Dissolve 1.0 g in 100 mL of water in a distilling flask. Add 1 mL of hydrogen peroxide and 1 mL of phosphoric acid, heat to boiling, and boil gently until all the iodine is expelled and the solution is colorless. Cool, wash down the sides of the flask, and add 0.5 mL of hydrogen peroxide. If an iodine color develops, boil until the solution is colorless and then for an additional 10 min. If no color develops, boil for 10 min, filter if necessary through a chloride-free filter, and dilute with water to 100 mL. Dilute 10 mL with water to 23 mL and add 1 mL of nitric acid and 1 mL of silver nitrate reagent solution. Any turbidity should not exceed that produced by 0.01 mg of chloride ion (Cl) in an equal volume of solution containing the quantities of nitric acid and silver nitrate used in the test.

IODATE. (By differential pulse polarography, page 50). Use 10.0 g of sample in 25 mL of solution. For the standard, add 0.03 mg of iodate (IO_3).

NITROGEN COMPOUNDS. (Page 31). Use 1.0 g. For the standard, use 0.01 mg of nitrogen (N).

PHOSPHATE. (Page 32, Method 1). Dissolve 2.0 g in water, and add 10 mL of nitric acid and 5 mL of hydrochloric acid. Evaporate to dryness on a hot plate ($\approx$100 °C). Dissolve the residue in 25 mL of approximately 0.5 N sulfuric acid and continue as described. For the standard, include the residue from evaporation of the quantities of acids used with the sample.

SULFATE. (Page 33, Method 1). Allow 30 min for the turbidity to form.

BARIUM. Dissolve 3.0 g of the sample in 10 mL of dilute hydrochloric acid (1 + 1) and add 5 mL of nitric acid. For the control, dissolve 0.5 g of sample and 0.05 mg of barium in 10 mL of dilute hydrochloric acid (1 + 1) and add 5 mL of nitric acid. Evaporate both solutions to dryness. Dissolve the residues in 10 mL of dilute hydrochloric acid (1 + 1), add 5 mL of nitric acid, and again evaporate to dryness. Dissolve each residue in water and dilute with water to 23 mL. To each solution add 2 mL of potassium dichromate reagent solution and 10% ammonium hydroxide reagent solution until the orange color is just dissipated and the yellow color persists. To each solution add with constant stirring 25 mL of methanol. Any turbidity in the solution of the sample should not exceed that in the control.

HEAVY METALS. (Page 28, Method 1). Dissolve 6.0 g in water, and dilute with water to 30 mL. Use 25 mL to prepare the sample solution, and use the remaining 5.0 mL to prepare the control solution.

IRON. (Page 30, Method 2). Dissolve 3.3 g in 50 mL of water, and add 0.1 mL of ammonium hydroxide.

CALCIUM, MAGNESIUM, AND SODIUM. (By flame AAS, page 39).

Sample Stock Solution. Dissolve 5.0 g of sample in water in a 100-mL volumetric flask, and dilute to the mark with water (1 mL = 0.05 g).

Element	Wavelength (nm)	Sample Wt (g)	Standard Added (mg)	Flame Type*	Background Correction
Ca	422.7	1.0	0.02; 0.04	N/A	No
Mg	285.2	1.0	0.01; 0.02	A/A	Yes
Na	589.0	0.20	0.005; 0.01	A/A	No

*A/A is air/acetylene; N/A is nitrous oxide/acetylene.

Potassium Nitrate

KNO_3 Formula Wt 101.10

CAS Number 7757–79–1

REQUIREMENTS

Assay . . . .≥99.0% KNO_3
pH of a 5% solution . . . . 4.5–8.5 at 25 °C

MAXIMUM ALLOWABLE

Insoluble matter. 0.005%
Chloride (Cl) 0.002%
Iodate (IO_3) . . . . .5 ppm
Nitrite (NO_2) 0.001%
Phosphate (PO_4) . . . . .5 ppm
Sulfate (SO_4) 0.003%
Heavy metals (as Pb) . . . . .5 ppm
Iron (Fe). . . . . .3 ppm

Calcium (Ca) 0.005%
Magnesium (Mg) 0.002%
Sodium (Na) 0.005%

TESTS

ASSAY. (By acid–base titrimetry). Weigh, to the nearest 0.1 mg, 0.4 g of sample and dissolve in 100 mL of water. Pass the solution through a cation-exchange column (see page 56) with water at a rate of about 5 mL/min and collect the eluate in a 500-mL titration flask. Then wash the resin in a column at a rate of about 10 mL/min and collect in the same titration flask. Add 0.15 mL of phenolphthalein to the flask and titrate with 0.1 N sodium hydroxide. Continue the titration as the washing proceeds until 50 mL of the eluate requires no further titration. One milliliter on 0.1 N sodium hydroxide corresponds to 0.01011 g KNO_3.

pH OF A 5% SOLUTION. (Page 44). The pH should be 4.5–8.5 at 25 °C.

INSOLUBLE MATTER. (Page 15). Use 20 g dissolved in 150 mL of water.

CHLORIDE. (Page 27). Use 0.50 g.

IODATE. (By differential pulse polarography, page 50). Use 10.0 g of sample in 25 mL of solution. For the standard, add 0.05 mg of iodate (IO_3).

NITRITE. (page 51). Use 1.0 g of sample and 0.01 mg of nitrite.

PHOSPHATE. (Page 32, Method 1). Dissolve 4.0 g in 25 mL of approximately 0.5 N sulfuric acid and continue as directed.

SULFATE. (Page 33, Method 2). Evaporate 3.0 g using 7 mL of dilute hydrochloric acid (1 + 1). Repeat the evaporation, and heat the last residue for 1 h at 120 °C.

HEAVY METALS. (Page 28, Method 1). Dissolve 6.0 g in about 20 mL of water, and dilute with water to 30 mL. Use 25 mL to prepare the sample solution, and use the remaining 5.0 mL to prepare the control solution.

IRON. (Page 30, Method 1). Use 3.3 g.

CALCIUM, MAGNESIUM, AND SODIUM. (By flame AAS, page 39).

> **Sample Stock Solution.** Dissolve 5.0 g of sample in water in a 100-mL volumetric flask and dilute to the mark with water (1 mL = 0.05 g).

Element	Wavelength (nm)	Sample Wt (g)	Standard Added (mg)	Flame Type*	Background Correction
Ca	422.7	0.50	0.025; 0.05	N/A	No
Mg	285.2	0.50	0.01; 0.02	A/A	Yes
Na	589.0	0.20	0.005; 0.01	A/A	No

*A/A is air/acetylene; N/A is nitrous oxide/acetylene.

Potassium Nitrite

KNO$_2$ **Formula Wt 85.10**

CAS Number 7758–09–0

REQUIREMENTS

Assay . . . . ≥96.0% KNO$_2$

MAXIMUM ALLOWABLE

Insoluble matter. 0.01%
Chloride (Cl) 0.03%
Sulfate (SO$_4$) 0.01%
Heavy metals (as Pb) 0.001%
Iron (Fe) 0.001%
Calcium (Ca) 0.005%
Magnesium 0.002%
Sodium (Na). 0.5%

TESTS

ASSAY. (By oxidation–reduction titrimetry). Weigh accurately 6–7 g, dissolve in water, and dilute with water to 1 L in a volumetric flask. Add 5 mL of sulfuric acid to 300 mL of water and, while the solution is still warm, add 0.1 N potassium permanganate until a faint pink color that persists for 2 min is produced. Then add 40.0 mL of 0.1 N potassium permanganate and mix gently. Add, slowly and with constant agitation, 25.0 mL of the sample solution, holding the tip of the pipet well below the surface of the liquid. Add 15.0 mL of 0.1 N ferrous ammonium sulfate, allow the solution to stand for 5 min, and titrate the excess of ferrous ammonium sulfate with 0.1 N potassium permanganate. Each milliliter of 0.1 N potassium permanganate consumed by the potassium nitrite corresponds to 0.004255 g of KNO$_2$.

INSOLUBLE MATTER. (Page 15). Use 20 g dissolved in 200 mL of water.

CHLORIDE. Dissolve 3.3 g in 20 mL of water, filter if necessary through a chloride-free filter, and dilute with water to 100 mL. Dilute 1.0 mL of this solution with 15 mL of water and slowly add 1 mL of acetic acid. Heat to boiling, boil gently for 5 min, cool, and dilute with water to 20 mL. Add 1 mL of nitric acid and 1 mL of silver nitrate reagent solution. Any turbidity should not exceed that produced by 0.01 mg of chloride ion (Cl) in an equal volume of solution containing the quantities of reagents used in the test.

Sample Solution A for the Determination of Sulfate, Heavy Metals, and Iron. Dissolve 10.0 g in water and dilute with water to 100 mL (1 mL = 0.1 g).

SULFATE. (Page 33, Method 2). Use 5.0 mL of sample solution A (0.5-g sample) plus 1 mL of hydrochloric acid.

HEAVY METALS. (Page 28, Method 1). To 30 mL of sample solution A (3-g sample) add 5 mL of hydrochloric acid, and evaporate to dryness on a hot plate ($\approx$100 °C). Add 5 mL of hydrochloric acid and again evaporate to dryness. Dissolve in about 20 mL of water, and dilute with water to 25 mL. For the control, add 0.02 mg of lead to 10 mL of sample solution A (1-g sample), add 5 mL of hydrochloric acid, and treat exactly as the 30 mL of sample solution A.

IRON. (Page 30, Method 1). To 10 mL of sample solution A (1-g sample) add 5 mL of hydrochloric acid, and evaporate to dryness on a hot plate ($\approx$100 °C). Dissolve the residue in 15–20 mL of water, add 2 mL of hydrochloric acid, filter if necessary, and dilute with water to 50 mL. Use the solution without further acidification.

CALCIUM, MAGNESIUM, AND SODIUM. (By flame AAS, page 39).

Sample Stock Solution A. Dissolve 0.40 g of sample in water in a 100-mL volumetric flask, and dilute to the mark with water (1 mL = 0.004 g).

Sample Stock Solution B. Dissolve 5.0 g of sample in 80 mL of water. Transfer to a 100-mL volumetric flask and dilute to the mark with water (1 mL = 0.05 g).

Element	Wavelength (nm)	Sample Wt (g)	Standard Added (mg)	Flame Type*	Background Correction
Ca	422.7	0.50	0.025; 0.05	N/A	No
Mg	285.2	0.50	0.01; 0.02	A/A	Yes
Na	589.0	0.004	0.01; 0.02	A/A	No

*A/A is air/acetylene; N/A is nitrous oxide/acetylene.

Potassium Oxalate Monohydrate
Oxalic Acid, Potassium Salt, Monohydrate

$(COOK)_2 \cdot H_2O$ Formula Wt 184.23

CAS Number 6487–48–5

REQUIREMENTS

Assay . . . . 98.5–101.0% $K_2C_2O_4 \cdot H_2O$
Substances darkened by hot sulfuric acid . . . . Passes test
Neutrality . . . . Passes test

MAXIMUM ALLOWABLE

Insoluble matter. 0.01%
Chloride (Cl) 0.002%
Sulfate (SO_4) 0.01%
Ammonium (NH_4) 0.002%
Heavy metals (as Pb) 0.002%
Iron (Fe) 0.001%
Sodium (Na). 0.02%

TESTS

ASSAY. (By oxidation–reduction titrimetry). Weigh accurately 1.8 g, and dissolve in water in a 250-mL volumetric flask. Dilute to volume and mix. To a 50-mL aliquot add 50 mL of water and 5 mL of sulfuric acid. Titrate slowly with 0.1 N potassium permanganate until about 25 mL has been added. Heat the solution to about 70 °C, and complete the titration to a permanent faint pink color. One milliliter of 0.1 N potassium permanganate consumed corresponds to 0.009212 g of $K_2C_2O_4 \cdot H_2O$.

SUBSTANCES DARKENED BY HOT SULFURIC ACID. Heat 1.0 g in a recently ignited test tube with 10 mL of sulfuric acid until the appearance of dense fumes. The acid when cooled should have no more color than a mixture of the following composition: 0.2 mL of cobalt chloride solution (5.95 g of $CoCl_2 \cdot 6H_2O$ and 2.5 mL of hydrochloric acid in 100 mL), 0.3 mL of ferric chloride solution (4.50 g of $FeCl_3 \cdot 6H_2O$ and 2.5 mL of hydrochloric acid in 100 mL), 0.3 mL of cupric sulfate solution (6.24 g of $CuSO_4 \cdot 5H_2O$ and 2.5 mL of hydrochloric acid in 100 mL), and 9.2 mL of water.

NEUTRALITY. Dissolve 2.0 g in 200 mL of water, and add 10.0 mL of 0.01 N oxalic acid and 0.20 mL of a 1% solution of phenolphthalein. Boil the solution in a flask for 10 min, passing through it a stream of carbon dioxide-free air. Cool the solution rapidly to room temperature while keeping the flow of carbon dioxide-free air passing through it and titrate with 0.01 N sodium hydroxide. Not less than 9.2 nor more than 10.5 mL of 0.01 N sodium hydroxide should be required to match the pink color produced in a buffer solution containing 0.20 mL of a 1% solution of phenolphthalein. The buffer solution contains: 3.1 g of boric acid (H_3BO_3), 3.8 g of potassium chloride, and 5.90 mL of 1 N sodium hydroxide in 1 L.

INSOLUBLE MATTER. (Page 15). Use 10 g dissolved in 100 mL of water.

CHLORIDE. (Page 27). Dissolve 2.0 g in water plus 10 mL of nitric acid, filter if necessary through a chloride-free filter, and dilute with water to 100 mL. To 25 mL of the solution add 1 mL of silver nitrate reagent solution. Any turbidity should not exceed that produced by 0.01 mg of chloride ion (Cl) in an equal volume of solution containing the quantities of reagents used in the test.

SULFATE. (Page 33, Method 4).

AMMONIUM. Dissolve 1.0 g in 50 mL of ammonia-free water and add 2 mL of Nessler reagent. Any color should not exceed that produced by 0.02 mg of ammonium ion (NH_4) in an equal volume of solution containing 2 mL of Nessler reagent.

HEAVY METALS. (Page 28, Method 1). Mix 2.0 g with 2 mL of water, add 2 mL of nitric acid and 2 mL of 30% hydrogen peroxide, and digest in a covered beaker on a hot plate ($\approx$100 °C) until reaction ceases. Remove the cover and evaporate to dryness. Dissolve the residue in 4 mL of dilute nitric acid (1 + 1) and again evaporate to dryness. Dissolve the residue in about 20 mL of water, dilute with water to 32 mL, and use 24 mL to prepare the sample solution. Use the remaining 8.0 mL, plus the residue from evaporation of 1 mL of 30% hydrogen peroxide and 3 mL of nitric acid to dryness, to prepare the control solution.

IRON. (Page 30, Method 1). Mix 2.0 g with 2 mL of water, add 2 mL of nitric acid and 2 mL of 30% hydrogen peroxide, and digest in a covered beaker on a hot plate ($\approx$100 °C) until reaction ceases. Remove the cover and evaporate to dryness. Add 5 mL of hydrochloric acid, and again evaporate to dryness. Dissolve the residue in a few milliliters of water, add 4 mL of hydrochloric acid, and dilute with water to 100 mL. Use 50 mL without further acidification. For the control, use the residue from evaporation to dryness of 1 mL of nitric acid, 1 mL of 30% hydrogen peroxide, and 2.5 mL of hydrochloric acid.

SODIUM. (By flame AAS, page 39).

Sample Stock Solution. Dissolve 1.0 g of sample in water, and dilute to 100 mL in a volumetric flask with water (1 mL = 0.01 g).

Element	Wavelength (nm)	Sample Wt (g)	Standard Added (mg)	Flame Type*	Background Correction
Na	589.0	0.05	0.01; 0.02	A/A	No

*A/A is air/acetylene.

Potassium Perchlorate

$KClO_4$ **Formula Wt 138.55**

CAS Number 7778–74–7

REQUIREMENTS

Assay. . . . . 99.0–100.5% $KClO_4$

MAXIMUM ALLOWABLE

Insoluble matter. 0.005%
Chloride (Cl) 0.003%
Sulfate (SO$_4$) 0.001%
Calcium (Ca) 0.005%
Sodium (Na). 0.02%
Heavy metals (as Pb) . . . . 5 ppm
Iron (Fe). . . . . 5 ppm

TESTS

ASSAY. (By argentimetric titration after decomposition to chloride). Weigh accurately 0.25–0.30 g into a platinum crucible or dish. Add 3 g of sodium carbonate, Na$_2$CO$_3$, and heat, gently at first, then to fusion until a clear melt is obtained. Leach with minimal dilute nitric acid, and boil gently to expel carbon dioxide. Cool to room temperature, add 10 mL of 30% ammonium acetate solution, and titrate with 0.1 N silver nitrate to the dichlorofluorescein end point. One milliliter of 0.1 N silver nitrate corresponds to 0.01385 g of KClO$_4$.

INSOLUBLE MATTER. (Page 15). Use 50 g dissolved in 600 mL of hot water.

CHLORIDE. (Page 27). Dissolve 0.33 g in 20 mL of hot water, filter, and cool.

SULFATE. Dissolve 40 g in 400 mL of hot water, add 4 mL of hydrochloric acid, and heat the solution to boiling. Add 5 mL of barium chloride reagent solution, digest in a covered beaker on a hot plate ($\approx$100 °C) for 2 h, and allow to stand overnight. Warm to dissolve any potassium perchlorate that may have crystallized. If a precipitate remains, filter, wash thoroughly, and ignite. The weight of the precipitate should not be more than 0.001 g greater than the weight obtained in a complete blank test.

CALCIUM AND SODIUM. (By flame AAS, page 39).

Sample Stock Solution. Dissolve 5.0 g of sample in water and dilute to 100 mL in a volumetric flask with water (1 mL = 0.05 g).

Element	Wavelength (nm)	Sample Wt (g)	Standard Added (mg)	Flame Type*	Background Correction
Ca	422.7	0.50	0.025; 0.05	N/A	No
Na	589.0	0.05	0.01; 0.02	A/A	No

*A/A is air/acetylene; N/A is nitrous oxide/acetylene.

HEAVY METALS. (Page 28, Method 1). Dissolve 3.0 g in hot water, dilute with water to 60 mL, and use 50 mL to prepare the sample solution. For the control, add 0.01 mg of lead to the remaining 10 mL, and dilute with water to 50 mL.

IRON. (Page 30, Method 1). Use 2.0 g.

Potassium Periodate

Potassium Metaperiodate
(suitable for determination of manganese)

KIO_4 Formula Wt 230.00

CAS Number 7790–21–8

REQUIREMENTS

Assay (dried basis). . . . . 99.8–100.3% KIO_4

MAXIMUM ALLOWABLE
Other halogens (as Cl). 0.01%
Manganese (Mn) . . . . 1 ppm

TESTS

ASSAY (DRIED BASIS). (By titration of oxidative capacity). Weigh accurately 1 g, previously dried over magnesium perchlorate or phosphorus pentoxide for 6 h, dissolve in water, and dilute with water to 500 mL in a volumetric flask. To 50.0 mL in a glass-stoppered flask add 10 g of potassium iodide and 10 mL of a cooled solution of dilute sulfuric acid (1 + 5). Allow to stand for 5 min, add 100 mL of cold water, and titrate the liberated iodine with 0.1 N sodium thiosulfate, adding starch indicator solution near the end point. Carry out a complete blank test and make any necessary correction. One milliliter of 0.1 N sodium thiosulfate corresponds to 0.002875 g of KIO_4.

OTHER HALOGENS. Dissolve 0.50 g in a mixture of 10 mL of water and 16 mL of sulfurous acid and boil for 3 min. Cool, add 5 mL of ammonium hydroxide and 20 mL of 2.5% silver nitrate solution, and filter. Dilute the filtrate with water to 100 mL, and to 20 mL of this solution add 1.5 mL of nitric acid. Any turbidity should not exceed that produced by 0.01 mg of chloride ion (Cl) in an equal volume of solution containing 0.5 mL of ammonium hydroxide, 1.5 mL of nitric acid, and 1 mL of silver nitrate reagent solution.

MANGANESE. Add 5.5 g to 85 mL of 10% sulfuric acid reagent solution. For the control, add 0.5 g of the sample and 0.005 mg of manganese ion (Mn) to 85 mL of 10% sulfuric acid reagent solution. Add 10 mL of nitric acid and 5 mL of phosphoric acid to each, boil gently for 10 min, and cool. Any pink color in the solution of the sample should not exceed that in the control.

Potassium Permanganate

KMnO$_4$ **Formula Wt 158.03**

CAS Number 7722–64–7

REQUIREMENTS

Assay . . . . ≥99.0% KMnO$_4$

MAXIMUM ALLOWABLE

Insoluble matter. 0.2%
Chloride and chlorate (as Cl) 0.005%
Nitrogen compounds (as N). 0.005%
Sulfate (SO$_4$) 0.02%

TESTS

ASSAY. (By titration of oxidizing power). Weigh, to the nearest 0.1 mg, about 1.5 g of sample. Transfer to a 500-mL volumetric flask, dissolve with water, then dilute to the mark with water, and mix thoroughly. Weigh accurately about 0.25 g of NIST or primary standard sodium oxalate (dried at 105 °C), and dissolve in a 500-mL glass-stoppered flask. Add 15 mL of 25% sulfuric acid, then add from a buret 30 mL of the potassium permanganate solution, heat the solutions to about 70 °C, and complete the titration with the permanganate solution.

$$\% \, KMnO_4 \; = \; \frac{g \, Na_2C_2O_4 \, \times \, 23585}{g \, sample \, \times \, mL \, KMnO_4}$$

INSOLUBLE MATTER. (Page 15). Dissolve 2.0 g in 150 mL of warm water on a hot plate (≈100 °C). Filter at once through a tared filtering crucible, wash thoroughly, and dry at 105 °C.

CHLORIDE AND CHLORATE. Dissolve 1.0 g in 75 mL of water and add 5 mL of nitric acid and 3 mL of 30% hydrogen peroxide. After reduction is complete, filter if necessary through a chloride-free filter, and dilute with water to 100 mL. To 20 mL of this solution add 1 mL of silver nitrate reagent solution. Any turbidity should not exceed that produced by 0.01 mg of chloride ion (Cl) in an equal volume of solution containing the quantities of reagents used in the test.

NITROGEN COMPOUNDS. (Page 31). Dilute the distillate of a 1.0-g sample, treated with 40 mL of 10% sodium hydroxide solution, to 100 mL and use 20 mL. For the standard, use 0.01 mg of nitrogen (N).

SULFATE. Dissolve 5.0 g in 75 mL of water, and add 10 mL of alcohol and 10 mL of hydrochloric acid. Heat until colorless, adding more alcohol and a little hydrochloric acid if necessary. Evaporate to dryness. Dissolve the residue in 100 mL of dilute hydrochloric acid (1 + 99), filter, and heat to boiling. Add 5 mL of barium chloride reagent solution, digest in a covered beaker on a hot plate (≈100 °C) for 2 h, and allow to stand overnight. Filter, wash thoroughly, and ignite. The weight of the precipitate should not be more than 0.0024 g greater than the weight obtained in a complete blank test.

Potassium Permanganate, Low in Mercury

KMnO₄

$KMnO_4$

Formula Wt 158.03

CAS Number 7722–64–7

REQUIREMENTS

Assay. . . . .≥99.0% $KMnO_4$

MAXIMUM ALLOWABLE

Insoluble matter. 0.2%
Chloride and chlorate (as Cl) 0.005%
Nitrogen compounds (as N). 0.005%
Sulfate (SO_4) 0.02%
Mercury (Hg). . . . . 0.05 ppm

TESTS

Assay and other tests except mercury are the same as for potassium permanganate, page 524.

MERCURY. For the sample place 1.0 g in a 125-mL conical flask and add 10 mL of water. For the standard, place 10 mL of water in a similar flask and add 0.05 µg of mercury (0.5 mL of a solution prepared freshly by diluting 1.0 mL of a standard mercury solution, page 95, with water to 500 mL). Add 2 mL of 4% potassium permanganate solution and 2 mL of 25% sulfuric acid to each flask. Heat on a hot plate (≈100 °C) for 30 min, then cool to room temperature. Transfer the contents of the sample flask to the aeration vessel (page 39), making the total volume about 50 mL. Rinse the flask with 4 mL of 10% hydroxylamine hydrochloride solution and add this to the aeration vessel also. Complete the reduction of manganese dioxide and mercury with 2 mL of 10% stannous chloride solution. Treat the standard solution in the same way. Obtain the peak height for the sample and standard solutions by CVAAS (page 40). The peak height for the sample should not exceed that for the standard.

Potassium Peroxydisulfate
Potassium Persulfate

$K_2S_2O_8$ **Formula Wt 270.32**

CAS Number 7727–21–1

REQUIREMENTS

Assay . . . .≥99.0% $K_2S_2O_8$

MAXIMUM ALLOWABLE

Insoluble matter. 0.005%
Chlorine compounds (as Cl). 0.001%
Nitrogen compounds (as N) 0.001%
Heavy metals (as Pb) 0.001%
Iron (Fe) . . . . 5 ppm
Manganese (Mn) . . . . 2 ppm

TESTS

ASSAY. (By titration of oxidizing power). Weigh accurately 0.5 g and add it to 25.0 mL of freshly prepared standard ferrous sulfate solution (about 0.2 N) in a glass-stoppered flask. Stopper the flask, allow it to stand for 1 h with frequent shaking, add 5 mL of phosphoric acid, and titrate the excess ferrous sulfate with 0.1 N potassium permanganate. Add 5 mL of phosphoric acid to 25.0 mL of the standard ferrous sulfate solution and titrate with 0.1 N potassium permanganate. The difference in the volume of permanganate consumed in the two titrations is equivalent to the potassium peroxydisulfate. One milliliter of 0.1 N potassium permanganate corresponds to 0.01352 g of $K_2S_2O_8$.

> **Standard Ferrous Sulfate Solution.** Dissolve 7 g of clear crystals of ferrous sulfate heptahydrate, $FeSO_4 \cdot 7H_2O$, in 90 mL of freshly boiled and cooled distilled water and add sulfuric acid to make 100 mL. Standardize the solution with 0.1 N potassium permanganate. The solution must be freshly prepared.

INSOLUBLE MATTER. (Page 15). Use 20 g dissolved in 200 mL of water.

CHLORINE COMPOUNDS. Mix 1.0 g with 1 g of anhydrous sodium carbonate and fuse in a porcelain dish in a muffle furnace. Cool, dissolve the residue in 20 mL of water, neutralize with nitric acid, and add an excess of 1 mL of the acid. Filter if necessary through a chloride-free filter and add 1 mL of silver nitrate reagent solution. Any turbidity should not exceed that in a standard containing 0.01 mg of chloride ion (Cl) and 1 g of sodium carbonate fused and acidified as in the test.

NITROGEN COMPOUNDS. (Page 31). Use 1.0 g. For the standard, use 0.01 mg of nitrogen (N).

HEAVY METALS. (Page 28, Method 1). Dissolve 2.0 g in 20 mL of hot water, add 5 mL of hydrochloric acid, and evaporate to about 5 mL. Cool, and dilute with water to 25 mL. For the standard, use 0.02 mg of lead in 5 mL of hydrochloric acid, add 10 mL of water and 0.25 mL of sulfuric acid, evaporate to fumes of sulfur trioxide, cool, and dilute with water to 25 mL.

IRON. (Page 30, Method 1). To 2.0 g add 20 mL of hot water and 5 mL of hydrochloric acid, and evaporate to dryness. Moisten the residue with 2 mL of hydrochloric acid, dilute with water to 50 mL, and use the solution without further acidification.

MANGANESE. To 5.0 g in a 150-mL beaker add 5 mL of sulfuric acid and 5 mL of phosphoric acid. Heat to dissolve the sample and continue heating to fumes of sulfur trioxide. Cool, dilute with water to 45 mL, add 10 mL of nitric acid, and boil for 5 min. Cool, add about 0.25 g of potassium periodate, and boil for 5 min. Cool and compare with a standard prepared by treating 0.01 mg of manganese in the same manner, using the same quantities of reagents (with no sample), except that the standard solution need not be fumed before adding the nitric acid. Any color in the solution of the sample should not exceed that in the standard.

Potassium Phosphate, Dibasic
Dipotassium Hydrogen Phosphate

K_2HPO_4 Formula Wt 174.18

CAS Number 7758–11–4

REQUIREMENTS

Assay . . . . $\geq$98.0% K_2HPO_4
pH of a 5% solution . . . . 8.5–9.6 at 25 °C

MAXIMUM ALLOWABLE

Insoluble matter. 0.01%
Loss on drying at 105 °C 1.0%
Chloride (Cl) 0.003%
Nitrogen compounds (as N). 0.001%
Sulfate (SO_4) 0.005%
Heavy metals (as Pb) . . . . 5 ppm
Iron (Fe). 0.001%
Sodium (Na) 0.05%

TESTS

ASSAY. (By acid–base titrimetry). Weigh accurately 8.0 g of sample, add 50 mL of water and exactly 50.0 mL of 1 N hydrochloric acid, and stir until the sample is completely dissolved. Titrate the excess acid, constantly stirring, with 1 N sodium hydroxide to the inflection point occurring at pH 4, as measured with a pH meter and glass electrode system. Calculate A, the volume of 1 N hydrochloric acid consumed by the sample. Continue the titration with 1 N sodium hydroxide to the inflection point occurring at about pH 8.8. Calculate B, the volume of 1 N sodium hydroxide required in the titration between the two inflection points. One milliliter of 1 N sodium hydroxide corresponds to 0.1742 g of K_2HPO_4. If A is equal to or less than B, calculate the K_2HPO_4 percentage from titration A. If A is greater than B, calculate the percentage from $2B - A$.

If A is equal to or less than B:

$$\% \ K_2HPO_4 = (A \times 0.1742 \times 100) \ / \ \text{Sample wt (g)}$$

If (A) is greater than (B):

$$\% \ K_2HPO_4 = [(2B - A) \times 0.1742 \times 100] \ / \ \text{Sample wt (g)}$$

pH OF A 5% SOLUTION. (Page 44). The pH should be 8.5–9.6 at 25 °C.

INSOLUBLE MATTER. (Page 15). Use 10 g dissolved in 100 mL of water.

LOSS ON DRYING AT 105 °C. Weigh accurately about 2 g. Dry to constant weight at 105 °C.

CHLORIDE. (Page 27). Use 0.33 g.

NITROGEN COMPOUNDS. (Page 31). Use 1.0 g. For the standard, use 0.01 mg of nitrogen (N).

SULFATE. Dissolve 1.0 g in 15 mL of water, and add 4.0 mL of 10% hydrochloric acid (solution approximately pH 2). Filter through washed filter paper, and add 10 mL of water through the same filter. For the control, add 0.05 mg of sulfate ion (SO_4) in 20 mL of water, and add 1.0 mL of (1 + 19) hydrochloric acid. Dilute both sample and control solutions to 35 mL with water, add 5.0 mL of 30% barium chloride solution to each, and mix. Observe turbidity after 10 min. Sample solution turbidity should not exceed that of the control solution.

HEAVY METALS. (Page 28, Method 1). Dissolve 6.0 g in about 20 mL of water, add 8 mL of 4 N hydrochloric acid, and dilute with water to 30 mL. Use 25 mL to prepare the sample solution, and use the remaining 5.0 mL to prepare the control solution.

IRON. (Page 30, Method 2). Use 1.0 g.

SODIUM. (By flame AAS, page 39).

> **Sample Stock Solution.** Dissolve 2.0 g of sample in a 100-mL volumetric flask and dilute to the mark with water (1 mL = 0.02 g).

Element	Wavelength (nm)	Sample Wt (g)	Standard Added (mg)	Flame Type*	Background Correction
Na	589.0	0.02	0.01; 0.02	A/A	No

*A/A is air/acetylene.

Potassium Phosphate, Monobasic
Potassium Dihydrogen Phosphate

KH_2PO_4 **Formula Wt 136.09**

CAS Number 7778–77–0

REQUIREMENTS

Assay. . . . . $\geq$99.0% KH_2PO_4
pH of a 5% solution . . . . 4.1–4.5 at 25 °C

MAXIMUM ALLOWABLE

Insoluble matter. 0.01%
Loss on drying at 105 °C 0.2%
Chloride (Cl) 0.001%
Nitrogen compounds (as N). 0.001%
Sulfate (SO_4) 0.003%
Heavy metals (as Pb) 0.001%
Iron (Fe). 0.002%
Sodium (Na) 0.005%

TESTS

ASSAY. (By acid–base titrimetry). Weigh accurately 5.0 g of sample and dissolve in 50 mL of water. Titrate with 1 N sodium hydroxide to the inflection point at about pH 8.8 as measured with a pH meter and glass electrode system. One milliliter of 1 N sodium hydroxide corresponds to 0.1361 g of KH_2PO_4.

pH OF A 5% SOLUTION. (Page 44). The pH should be 4.1–4.5 at 25 °C.

INSOLUBLE MATTER. (Page 15). Dissolve 10 g in 100 mL of water.

LOSS ON DRYING AT 105 °C. Weigh accurately about 2 g, and dry for 2 h at 105 °C.

CHLORIDE. (Page 27). Use 1.0 g of sample and 2 mL of nitric acid.

NITROGEN COMPOUNDS. (Page 31). Use 1.0 g treated with 20 mL of freshly boiled 10% sodium hydroxide. For the standard, use 0.01 mg of nitrogen (N).

SULFATE. Dissolve 2.0 g in 15 mL of water, and add 2.5 mL of 10% hydrochloric acid (solution approximately pH 2). Filter through washed filter paper, and add 10 mL of water through the same filter. For the control, add 0.06 mg of sulfate ion (SO_4) in 20 mL of water, and add 1.0 mL of (1 + 19) hydrochloric acid. Dilute both sample and control solutions to 35 mL with water, add 5.0 mL of 30% barium chloride solution to each, and mix. Observe turbidity after 10 min. Sample solution turbidity should not exceed that of the control solution.

HEAVY METALS. (Page 28, Method 1). Dissolve 4.0 g in about 20 mL of water, and dilute with water to 32 mL. Use 24 mL to prepare the sample solution, and use the remaining 8.0 mL to prepare the control solution.

IRON. (Page 30, Method 2). Dissolve 1.0 g in water, dilute with water to 20 mL, and use 10 mL.

SODIUM. (By flame AAS, page 39).

> **Sample Stock Solution.** Dissolve 1.0 g of sample in water and dilute to 100 mL in a volumetric flask with water (1 mL = 0.01 g).

Element	Wavelength (nm)	Sample Wt (g)	Standard Added (mg)	Flame Type*	Background Correction
Na	589.0	0.20	0.01; 0.02	A/A	No

*A/A is air/acetylene.

Potassium Phosphate, Tribasic
Tripotassium Orthophosphate

K_3PO_4 **Formula Wt 212.27**

$K_3PO_4 \cdot 7H_2O$ **Formula Wt 338.38**

CAS Number 7778–53–2 anhydrous; 22763–02–6 heptahydrate

> *NOTE.* This reagent must be labeled to indicate whether it is the anhydrous or heptahydrate form.

REQUIREMENTS

Assay (as-is basis) . . . . ≥98% K_3PO_4 or $K_3PO_4 \cdot 7H_2O$

MAXIMUM ALLOWABLE

Dibasic potassium phosphate (K_2HPO_4) 1%
Excess alkali (as KOH) 1%
Insoluble matter. 0.01%

Chloride (Cl) 0.005%
Nitrogen compounds (as N). 0.002%
Heavy Metals (as Pb) 0.002%
Iron (Fe). 0.001%
Sulfate (SO_4) 0.005%
Sodium (Na) 0.1%

TESTS

ASSAY. (By acid-base titrimetry) Dissolve 3.0 g for the anhydrous salt and 5.0 g for the heptahydrate salt, accurately weighed, in 50 mL of water. Add exactly 100.0 mL of 1 N hydrochloric acid and heat to boiling to expel carbon dioxide. Cool and titrate with 1 N sodium hydroxide to the first inflection point at about pH 4.1 as measured with a standardized pH meter and a glass electrode system. Calculate A, the volume of 1 N hydrochloric acid consumed by the sample. Protect the solution from absorbing carbon dioxide from the air and continue the titration with 1 N sodium hydroxide to the second inflection point at about pH 8.8. Calculate B, the volume of 1 N sodium hydroxide required in the titration between the two inflection points.

If A is equal to or greater than $2B$, then

$$\% \ K_3PO_4 = \frac{B \times F}{\text{Sample wt (g)}} \times 100$$

If A is less than $2B$, then

$$\% \ K_3PO_4 = \frac{(A - B) \times F}{\text{Sample wt (g)}} \times 100$$

where F is 0.2123 for anhydrous and 0.3384 for heptahydrate.

DIBASIC POTASSIUM PHOSPHATE AND EXCESS ALKALI. Calculate the amount of dibasic potassium phosphate as K_2HPO_4 or the amount of excess alkali as KOH from the titration values obtained in the assay.

If A is less than $2B$, then

$$\% \ K_2HPO_4 = \frac{(2B - A) \times 0.1742}{\text{Sample wt (g)}} \times 100$$

If A is greater than $2B$, then

$$\% \ KOH = \frac{(A - 2B) \times 0.0561}{\text{Sample wt (g)}} \times 100$$

INSOLUBLE MATTER. (Page 15). Dissolve 10.0 g in 100 mL of water, add 0.10 mL of methyl red indicator solution, add hydrochloric acid until the solution is slightly acid to the indicator, and continue as described.

CHLORIDE. (Page 27). Dissolve 0.20 g of sample in 10 mL of water, add 3 mL of nitric acid, and dilute to 20 mL.

NITROGEN COMPOUNDS. (Page 31). Use 0.50 g treated with 20 mL of freshly boiled 10% sodium hydroxide. For the standard, use 0.01 mg of nitrogen (N).

HEAVY METALS. (Page 28, Method 1). Dissolve 2.0 g in about 20 mL of water, and dilute with water to 32 mL. Use 24 mL to prepare the sample solution, and use the remaining 8.0 mL to prepare the control solution.

IRON. (Page 30, Method 2). Use 1.0 g.

SULFATE. Dissolve 1.0 g in 15 mL of water, and add 4.0 mL of 10% hydrochloric acid (solution pH 2). Filter through washed filter paper, and add 10 mL of water through the same filter. For the control, add 0.05 mg of sulfate ion (SO_4).in 20 mL of water, and then add 1.0 mL of (1 + 19) hydrochloric acid. Dilute both to 35 mL with water, add 5.0 mL of 30% barium chloride solution, and mix. Observe turbidity after 10 min. The turbidity of the sample solution should not exceed that of the control.

SODIUM. (By flame AAS, page 39).

Sodium Stock Solution. Dissolve 1.0 g of sample in water and dilute to 100 mL in a volumetric flask with water (1 mL = 0.01 g).

Element	Wavelength (nm)	Sample Wt (g)	Standard Added (mg)	Flame Type*	Background Correction
Na	589.0	0.02	0.01;0.02	A/A	No

*A/A is air/acetylene; N/A is nitrous oxide/acetylene.

Potassium Sodium Tartrate Tetrahydrate
2,3-Dihydroxybutanedioic Acid, Potassium Sodium Salt, Tetrahydrate

$$NaO-\overset{O}{\overset{\|}{C}}-\overset{OH}{\overset{|}{CH}}-\overset{OH}{\overset{|}{CH}}-\overset{O}{\overset{\|}{C}}-OK \cdot 4H_2O$$

$KNaC_4H_4O_6 \cdot 4H_2O$ **Formula Wt 282.22**

CAS Number 6381–59–5

REQUIREMENTS

Assay . . . . 99.0–102.0% $KNaC_4H_4O_6 \cdot 4H_2O$
pH of a 5% solution . . . . 6.0–8.5 at 25 °C

MAXIMUM ALLOWABLE

Insoluble matter. 0.005%
Chloride (Cl) 0.001%
Phosphate (PO$_4$) 0.002%
Sulfate (SO$_4$) 0.005%
Ammonium (NH$_4$) 0.002%
Calcium (Ca) 0.005%
Heavy metals (as Pb) . . . . 5 ppm
Iron (Fe). 0.001%

TESTS

ASSAY. (Total alkalinity by nonaqueous titration). Weigh accurately 0.5 g and mix with 50 mL of glacial acetic acid, 30 mL of 96% formic acid, and 45 mL of acetic anhydride. Heat and stir until dissolution is complete, and titrate with 0.1 N perchloric acid in glacial acetic acid to a green end point with crystal violet indicator. Correct for a reagent blank. One milliliter of 0.1 N perchloric acid corresponds to 0.01411 g of KNaC$_4$H$_4$O$_6 \cdot$ 4H$_2$O.

pH OF A 5% SOLUTION. (Page 44). The pH should be 6.0–8.5 at 25 °C.

INSOLUBLE MATTER. (Page 15). Use 20 g in 200 mL of water.

CHLORIDE. (Page 27). Use 1.0 g.

PHOSPHATE. (Page 32, Method 1). Ignite 1.0 g in a platinum dish. Dissolve the residue in 5 mL of water, add 5 mL of nitric acid, and evaporate to dryness. Dissolve the residue in 25 mL of approximately 0.5 N sulfuric acid and continue as described.

SULFATE. To 5.0 g of sample, add 4 mL of hydrochloric acid and 8 mL of 30% hydrogen peroxide. Digest in a covered beaker on a hot plate (≈100 °C) until reaction ceases. Carefully rinse down cover glass and sides of the beaker with about 5 mL of water. Add 4 mL of hydrochloric acid and 3 mL of 30% hydrogen peroxide, and evaporate to dryness. Add 10 mL of water, and evaporate to dryness. Dissolve the residue in 5 mL of 10% hydrochloric acid, and dilute to 25 mL with water. (5 ml = 1 g). For the sample, take 15 mL and, for the control, take 5 mL of the sample solution. Add 0.10 mg of sulfate ion (SO$_4$) to each. Dilute both solutions to 25 mL, add 1 mL of barium chloride solution to each, and compare after 10 min. Sample solution turbidity should not exceed that of the control solution.

AMMONIUM. Dissolve 1.0 g in 50 mL of water. To 25 mL of the solution add 2 mL of 10% sodium hydroxide reagent solution, dilute with water to 50 mL, and add 2 mL of Nessler reagent. Any color should not exceed that produced by 0.01 mg of ammonium ion (NH$_4$) in an equal volume of solution containing the quantities of reagents used in the test.

CALCIUM. (By flame AAS, page 39).

> **Sample Stock Solution.** Dissolve and dilute 4.0 g to 100 mL with water in a volumetric flask (1 mL = 0.04 g).

Element	Wavelength (nm)	Sample Wt (g)	Standard Added (mg)	Flame Type*	Background Correction
Ca	422.7	0.40	0.02; 0.04	N/A	No

*N/A is nitrous oxide/acetylene.

HEAVY METALS. (Page 28, Method 1). Dissolve 8.0 g in about 20 mL of water, and dilute with water to 32 mL. Use 24 mL to prepare the sample solution, and use the remaining 8.0 mL to prepare the control solution.

IRON. (Page 30, Method 1). Use 1.0 g.

Potassium Sulfate

K_2SO_4 **Formula Wt 174.26**

CAS Number 7778–80–5

REQUIREMENTS

Assay . . . . ≥99.0% K_2SO_4
pH of a 5% solution . . . . 5.5–8.5 at 25 °C

MAXIMUM ALLOWABLE

Insoluble matter. 0.01%
Chloride (Cl) 0.001%
Nitrogen compounds (as N). . . . . 5 ppm
Heavy metals (as Pb) . . . . 5 ppm
Iron (Fe). . . . . 5 ppm
Calcium (Ca) 0.01%
Magnesium (Mg) 0.005%
Sodium (Na). 0.02%

TESTS

ASSAY. (By acid–base titrimetry). Weigh, to the nearest 0.1 mg, 0.4 g of sample and dissolve in 50 mL of water. Pass the solution through the cation-exchange column (see page 56) at a rate of about 5 mL/min, and collect the eluate in a 500-mL titration flask. Then wash the resin in the column with water at a rate of about 10 mL/min, and collect in the same titration flask. Add 0.15 mL of phenolpthalein indicator solution to the flask and titrate with 0.1 N sodium hydroxide. Continue the titration as the washing proceeds until 50 mL requires no further titration.

$$\% \ K_2SO_4 = (mL \times N \ NaOH \times 8.713) \ / \ Sample \ wt \ (g)$$

pH OF A 5% SOLUTION. (Page 44). The pH should be 5.5–8.5 at 25 °C.

INSOLUBLE MATTER. (Page 15). Use 10 g dissolved in 150 mL of water.

CHLORIDE. (Page 27). Use 1.0 g.

NITROGEN COMPOUNDS. (Page 31). Use 2.0 g. For the standard, use 0.01 mg of nitrogen (N).

HEAVY METALS. (Page 28, Method 1). Dissolve 6.0 g in water, and dilute with water to 54 mL. Use 45 mL to prepare the sample solution, and use the remaining 9.0 mL to prepare the control solution.

IRON. (Page 30, Method 1). Use 2.0 g.

CALCIUM, MAGNESIUM, AND SODIUM. (By flame AAS, page 39).

Sample Stock Solution. Dissolve 2.0 g of sample with water in a 100-mL volumetric flask, and dilute to the mark with water (1 mL = 0.02 g).

Element	Wavelength (nm)	Sample Wt (g)	Standard Added (mg)	Flame Type*	Background Correction
Ca	422.7	0.20	0.02; 0.04	N/A	No
Mg	285.2	0.20	0.01; 0.02	A/A	Yes
Na	589.0	0.10	0.01; 0.02	A/A	No

*A/A is air/acetylene; N/A is nitrous oxide/acetylene.

Potassium Thiocyanate

KSCN

Formula Wt 97.18

CAS Number 333–20–0

REQUIREMENTS

Appearance . . . . Colorless or white crystals
Assay . . . . ≥98.5% KSCN
pH of a 5% solution . . . . 5.3–8.7 at 25 °C

MAXIMUM ALLOWABLE

Insoluble in water 0.005%
Chloride (Cl) 0.005%

Sulfate (SO_4) 0.005%
Ammonium (NH_4) 0.003%
Heavy metals (as Pb) . . . . 5 ppm
Iron (Fe) . . . . 2 ppm
Sodium (Na) 0.005%
Iodine-consuming substances . . . . Passes test

TESTS

CAUTION. Cyanide gases are given off during ignition and with addition of acids. Use a well-ventilated hood.

ASSAY. (By argentimetric titration of thiocyanate content). Weigh, to the nearest 0.1 mg, 7 g of sample, dissolve with 100 mL of water in a 1000-mL volumetric flask, dilute to the mark with water, and mix thoroughly. Transfer a 50.0-mL aliquot to a 250-mL glass-stoppered iodine-type flask, add 5 mL of 1 N nitric acid, then add with agitation exactly 50.0 mL of 0.1 N silver nitrate solution, and shake vigorously. Add 2 mL of ferric ammonium sulfate indicator solution, and chill in an ice bath to approximately 10 °C or lower. Titrate the excess silver nitrate with 0.1 N ammonium thiocyanate solution. Near the end point, shake after the addition of each drop.

$$\% \text{ KCNS} = \frac{[(50.0 \times N \text{ AgNO}_3) - (\text{mL} \times N \text{ NH}_4\text{SCN})] \times 9.718}{\text{Sample wt (g)} / 20}$$

pH OF A 5% SOLUTION. (Page 44). The pH should be 5.3–8.7 at 25 °C.

INSOLUBLE IN WATER. Dissolve 20 g in 200 mL of water, heat to boiling, and digest in a covered beaker on a hot plate ($\approx$100 °C) for 1 h. Filter through a tared filtering crucible, wash thoroughly, and dry at 105 °C.

CHLORIDE. Dissolve 1.0 g in 20 mL of water, filter if necessary through a chloride-free filter, and add 10 mL of 25% sulfuric acid reagent solution and 7 mL of 30% hydrogen peroxide. Evaporate to 20 mL by boiling in a well-ventilated hood, add 15–20 mL of water, and evaporate again. Repeat until all the cyanide has been volatilized, cool, and dilute with water to 100 mL. To 20 mL of this solution add 1 mL of nitric acid and 1 mL of silver nitrate reagent solution. Any turbidity should not exceed that produced by 0.01 mg of chloride ion (Cl) in an equal volume of solution containing the quantities of reagents used in the test.

SULFATE. (Page 33, Method 1). Allow 30 min for the turbidity to form.

AMMONIUM. (Page 25). Dilute the distillate from a 1.0-g sample to 100 mL; 50 mL of the dilution should not show more than 0.015 mg of ammonium (NH_4).

HEAVY METALS. (Page 28, Method 1). Dissolve 6.0 g in about 20 mL of water, and dilute with water to 30 mL. Use 25 mL to prepare the sample solution, and use the remaining 5.0 mL to prepare the control solution.

IRON. (Page 30, Method 2). Dissolve 10 g in 60 mL of water. To 30 mL of the solution add 0.1 mL of hydrochloric acid, and continue as described.

SODIUM. (By flame AAS, page 39).

Sample Stock Solution. Dissolve 1.0 g of sample with water in a 100-mL volumetric flask, and dilute to the mark with water (1 mL = 0.01 g).

Element	Wavelength (nm)	Sample Wt (g)	Standard Added (mg)	Flame Type*	Background Correction
Na	589.0	0.10	0.005; 0.01	A/A	No

*A/A is air/acetylene.

IODINE-CONSUMING SUBSTANCES. Dissolve 5.0 g in 50 mL of water and add 1.7 mL of 10% sulfuric acid reagent solution. Add 1 g of potassium iodide and 1 mL of starch indicator solution and titrate with 0.1 N iodine solution. Not more than 1.0 mL of 0.1 N iodine solution should be required.

1,2-Propanediol
Propylene Glycol

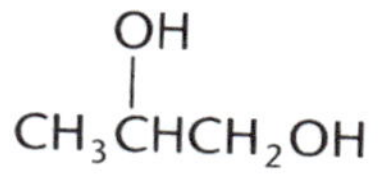

$$CH_3CHCH_2OH$$
with OH on the second carbon

CH$_3$CHOHCH$_2$OH   **Formula Wt 76.10**

CAS Number 57–55–6

REQUIREMENTS

Assay. . . . .≥99.5% CH$_3$CHOHCH$_2$OH

MAXIMUM ALLOWABLE

Color (APHA). 10
Residue after ignition. 0.005%
Titrable acid. . . . . 0.0005 meq/g
Chloride (Cl) . . . . 1 ppm
Water. 0.2%

TESTS

ASSAY. Analyze the sample by gas chromatography using the general parameters cited on page 70. The following specific conditions are also required.

Column: Type III, polyethylene glycol

Measure the area under all peaks and calculate the 1,2-propanediol content in area percent. Correct for water content.

COLOR (APHA). (Page 19).

RESIDUE AFTER IGNITION. (Page 16). Use 50 g (48.5 mL).

TITRABLE ACID. Place 60 g (58 mL) in a 250-mL conical flask. Add 0.15 mL of phenol red indicator, and titrate with 0.01 N sodium hydroxide to a pink color. Not more than 3.0 mL should be required.

CHLORIDE. (Page 27). Use 10 g (9.7 mL).

WATER. (Page 55, Method 1). Use 20 mL (19.4 g).

Propionic Acid
Propanoic Acid

Formula Wt 74.08

$C_3H_6O_2$

CAS Number 79–09–4

REQUIREMENTS

Assay . . . . ≥99.5% CH_3CH_2COOH

MAXIMUM ALLOWABLE

Color (APHA) 20
Residue after evaporation. 0.01%
Readily oxidizable substances (as HCOOH). 0.10%
Heavy metals (as Pb) 0.001%
Carbonyl compounds [formaldehyde, acetone, or
  acetaldehyde plus propionaldehyde (as the latter)] . 0.002%
Water (H_2O). 0.15%

TESTS

ASSAY. (By acid–base titrimetry). Weigh accurately 3.0 g and dissolve in 100 mL of carbon dioxide-free water. Add 0.15 mL of phenolphthalein indicator solution and titrate with 1 N sodium hydroxide to the first appearance of a faint pink end point that persists for at least 30 s. One milliliter of 1 N sodium hydroxide corresponds to 0.07408 g of CH_3CH_2COOH.

$$\% \ CH_3CH_2COOH = (mL \times N \ NaOH \times 74.08) / Sample \ wt \ (g)$$

COLOR (APHA). (Page 19).

RESIDUE AFTER EVAPORATION. (Page 16). Evaporate 20 g (20 mL) to dryness in a tared dish on a hot plate (≈100 °C) and dry the residue at 105 °C for 30 min.

READILY OXIDIZABLE SUBSTANCES. Dissolve 7.5 g of sodium hydroxide in 50 mL of water, cool, add 3 mL of bromine, stir until dissolved, and dilute to 1000 mL with water. Transfer 25.0 mL of this solution to a glass-stoppered conical flask containing 100 mL of water, and add 10 mL of a 1 in 5 solution of sodium acetate trihydrate and 10.0 mL of the sample. Allow to stand for 15 min, then add 1.0 g of potassium iodide dissolved in 5 mL of water, mix, and add 10 mL of hydrochloric acid. Titrate with 0.1 N sodium thiosulfate just to the disappearance of the brown color. Perform a blank determination. One milliliter of 0.1 N sodium thiosulfate corresponds to 0.004603 g of formic acid (HCOOH). The difference between the blank and sample titrations should not exceed 2.2 mL.

HEAVY METALS. (Page 28, Method 1). Add about 25 mg of sodium carbonate to 3.0 g of the sample in a platinum dish and evaporate to dryness on a hot plate ($\approx$100 °C). Dissolve the residue in 1 mL of dilute hydrochloric acid (1 + 19) and dilute with water to 30 mL. Use 25 mL to prepare the sample solution, and use the remaining 5 mL to prepare the control solution.

CARBONYL COMPOUNDS. (By differential pulse polarography, page 50). Use 1.0-g (1.0-mL) samples diluted with 4 mL of water and neutralized to a pH between 6.5 and 6.8 with 10% sodium hydroxide. For the standards use 0.02 mg each of formaldehyde (HCHO), propionaldehyde (CH_3CH_2CHO), and acetone (CH_3COCH_3). The peak potentials for acetaldehyde (CH_3CHO) and propionaldehyde overlap at about −1.15 to −1.25V.

WATER. (Page 55, Method 1). Use 10.0 mL (9.9 g) of the sample.

---

n-Propyl Alcohol
1-Propanol

$CH_3(CH_2)_2OH$

Formula Wt 60.10

CAS Number 71–23–8

REQUIREMENTS

Assay (GC). . . . . ≥99.5% $CH_3(CH_2)_2OH$
Solubility in water . . . . Passes test

MAXIMUM ALLOWABLE

Color (APHA). 10
Residue after evaporation 0.001%
Titrable acid. . . . . 0.0004 meq/g
Carbonyl compounds (as C_2H_5CHO). 0.03%
Ethyl alcohol (CH_3CH_2OH) 0.01%
Methanol (CH_3OH) 0.01%

Isopropyl alcohol (CH$_3$CHOHCH$_3$) 0.05%

Water (H$_2$O) 0.2%

TESTS

ASSAY, ETHYL ALCOHOL, METHANOL, AND ISOPROPYL ALCOHOL.
Analyze the sample by gas chromatography using the general parameters cited on page 70. The following specific conditions are also required.

   Column: Type I, methyl silicone

   Detector: Flame ionization

Measure the area under all peaks and calculate the area percent for methanol, ethyl alcohol, isopropyl alcohol, and *n*-propyl alcohol. Correct for water content.

SOLUBILITY IN WATER. Mix 10 mL with 40 mL of water and allow to stand for 1 h. The solution should be as clear as an equal volume of water.

COLOR (APHA). (Page 19).

RESIDUE AFTER EVAPORATION. (Page 16). Evaporate 100 g (124 mL) to dryness in a tared dish on a hot plate (≈100 °C) and dry the residue at 105 °C for 30 min.

TITRABLE ACID. Add 10 mL of sample to 30 mL of carbon dioxide-free water in a glass-stoppered flask and add 0.5 mL of phenolphthalein indicator solution. Add 0.01 N sodium hydroxide until a slight pink color persists after shaking for 30 s. Add 25 g (31 mL) of the sample, mix well, and titrate with 0.01 N sodium hydroxide until the pink color is restored. Not more than 1.0 mL of the sodium hydroxide solution should be required.

CARBONYL COMPOUNDS. (By differential pulse polarography, page 50). Use 1.0 g (1.25 mL). For the standard, use 0.30 mg of propionaldehyde (C$_2$H$_5$CHO).

WATER. (Page 55, Method 1). Use 10.0 mL (8.0 g) of the sample.

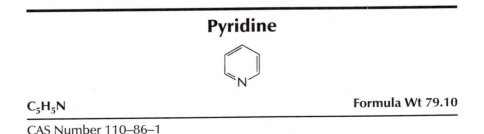

Pyridine

C$_5$H$_5$N   **Formula Wt 79.10**

CAS Number 110–86–1

REQUIREMENTS

Assay . . . . ≥99.0% C_5H_5N
Solubility in water . . . . Passes test

MAXIMUM ALLOWABLE

Residue after evaporation 0.002%
Water (H_2O) 0.1%
Chloride (Cl) 0.001%
Sulfate (SO_4) 0.001%
Ammonia (NH_3) 0.002%
Copper (Cu) . . . . 5 ppm
Reducing substances . . . . Passes test

TESTS

ASSAY. Analyze the sample by gas chromatography using the general parameters cited on page 70. The following specific conditions are also required.

Column: Type I, methyl silicone

Measure the area under all peaks and calculate the pyridine content in area percent. Correct for water content.

SOLUBILITY IN WATER. Dilute 10 mL with 90 mL of water. The solution should show no turbidity in 30 min.

RESIDUE AFTER EVAPORATION. (Page 16). Evaporate 50 g (51 mL) to dryness on a hot plate (≈100 °C) and dry the residue at 105 °C for 30 min. Save residue after weighing for copper test.

WATER. (Page 55, Method 1). Use 20 mL (19.6 g) of the sample.

CHLORIDE. (Page 27). Dilute 1.0 g (1.0 mL) with water to 25 mL.

SULFATE. (Page 33, Method 3).

AMMONIA. Add 2 g (2.0 mL) to 10 mL of carbon dioxide-free water. Add 0.10 mL of phenolphthalein indicator solution. If a pink color develops it should be discharged by not more than 0.24 mL of 0.01 N hydrochloric acid.

COPPER. (By flame AAS, page 39). Add 5 mL of nitric acid to the residue from residue after evaporation and digest on a hot plate (≈100 °C) for about 5 min. Cool and add 25 mL of water. Transfer to a 50-mL volumetric flask, dilute to the mark with water, and mix (1 mL = 1.0 g).

Element	Wavelength (nm)	Sample Wt (g)	Standard Added (mg)	Flame Type*	Background Correction
Cu	324.8	5.0	0.025; 0.05	A/A	Yes

*A/A is air/acetylene.

REDUCING SUBSTANCES. To 5 g (5.0 mL) add 0.50 mL of 0.1 N potassium permanganate solution. The pink color should not be entirely discharged in 30 min. If a brown color interferes, the solution should be centrifuged or filtered through a sintered glass filter. If a pink color persists, the sample passes.

Pyrogallol
1,2,3-Trihydroxybenzene
1,2,3-Benzenetriol

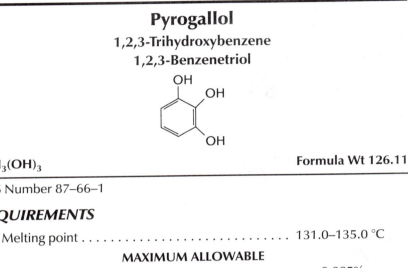

$C_6H_3(OH)_3$ **Formula Wt 126.11**

CAS Number 87–66–1

REQUIREMENTS

Melting point . . . . 131.0–135.0 °C

MAXIMUM ALLOWABLE

Residue after ignition 0.005%
Chloride (Cl) 0.001%
Sulfate (SO_4) 0.005%
Heavy metals (as Pb) . . . . 5 ppm
Iron (Fe) 0.001%

TESTS

MELTING POINT. (Page 22).

RESIDUE AFTER IGNITION. (Page 16). Ignite 20 g, and use 1 mL of sulfuric acid. Retain the residue.

CHLORIDE. (Page 27). Dissolve 1.0 g in 20 mL of water, and add 3 mL of 10% nitric acid.

SULFATE. (Page 33, Method 1).

HEAVY METALS. (Page 28, Method 2). Use 4.0 g.

Sample Solution A for the Determination of Iron. To the residue retained from the test for residue after ignition, add 3 mL of hydrochloric acid and 1 mL of nitric acid, and evaporate to dryness on a hot plate (≈100 °C). Take up the residue in 10 mL of water and 1 mL of hydrochloric acid (1 + 19), filter if necessary, and wash to a volume of 40 mL (1 mL = 0.5 g).

IRON. (Page 30, Method 1). Dilute 20 mL of sample solution A (10-g sample) with water to 100 mL, and use 10 mL.

8-Quinolinol
8-Hydroxyquinoline
Oxine

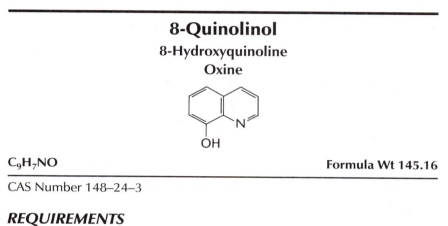

C_9H_7NO   **Formula Wt 145.16**

CAS Number 148–24–3

REQUIREMENTS

Melting point. . . . . 72.5–74.0 °C
Suitability for magnesium determination . . . . Passes test

MAXIMUM ALLOWABLE

Insoluble in alcohol 0.05%
Residue after ignition. 0.05%
Sulfate (SO_4) 0.02%

TESTS

MELTING POINT. (Page 22).

SUITABILITY FOR MAGNESIUM DETERMINATION. Dissolve 2.5 g in 5 mL of glacial acetic acid, warming if necessary, and dilute with water to 100 mL. Add 3.5 mL of this solution to 50 mL of a solution containing 6 mg of magnesium ion (Mg). Heat to 80 °C, add with stirring 2 mL of ammonium hydroxide, allow to stand for 10 min, and filter. To the filtrate, which should be alkaline and yellow in color, add a solution containing 3 mg of magnesium ion (Mg) and heat to 80 °C. The characteristic yellow magnesium quinolate precipitate should form.

INSOLUBLE IN ALCOHOL. Dissolve 3.0 g in 40 mL of alcohol, filter through a tared filtering crucible, wash with 95% alcohol, and dry at 105 °C.

RESIDUE AFTER IGNITION. (Page 16). Ignite 2.0 g. Use 1 mL of sulfuric acid.

SULFATE. Dissolve 2.0 g in 15 mL of methanol and 5 mL of hydrochloric acid (1 + 3), and add 0.20 mg of sulfate ion (SO_4). Dilute the sample and control solutions to 40 mL with water, and add 1 mL of barium chloride reagent solution to each. After 10 min, turbidity of sample solution should not exceed that of the control solution.

Reagent Alcohol
Alcohol, Reagent

NOTE. This material is a denatured form of ethyl alcohol, approved for sale under U.S. Regulations governing the stated description or equivalent, consisting of about 5 volumes of isopropyl alcohol and about 95 volumes of formula 3-A specially denatured alcohol (which consists of about 5 volumes of methanol and about 100 volumes of ethyl alcohol).

REQUIREMENTS

Assay . . . . Within 94.0–96.0% (v/v) methanol and ethyl alcohol
and 4.0–6.0% (v/v) isopropyl alcohol

MAXIMUM ALLOWABLE

Water (H_2O). 0.5%
Color (APHA) 10
Residue after evaporation. 0.001%

TESTS

ASSAY. Analyze the sample by gas chromatography using the general parameters cited on page 70. The following specific conditions are also required.

Column: Type I, methyl silicone

Measure the area under all peaks and calculate the ethyl alcohol plus methanol and isopropyl alcohol content, correcting for response factors. Correct for water content.

WATER. (Page 55, Method 1). Use 10 mL (7.4 g) of the sample.

COLOR (APHA). (Page 19).

RESIDUE AFTER EVAPORATION. (Page 16). Evaporate 100 g (124 mL) to dryness in a tared dish on a hot plate ($\approx$100 °C) and dry the residue at 105 °C for 30 min.

Reinecke Salt
Ammonium Tetra(thiocyanato)diamminechromate(III) Monohydrate
Ammonium Diamminetetrakis(thiocyanato-*N*)chromate(1–)

$NH_4[Cr(SCN)_4(NH_3)_2] \cdot H_2O$

Formula Wt 354.45

CAS Number 13573–16–5

CAUTION. Aqueous solutions of Reinecke salt decompose slowly with the evolution of hydrogen cyanide gas, HCN. Decomposition is rapid above 65 °C.

REQUIREMENTS

Assay . . . . ≥93.0% $NH_4[Cr(SCN)_4(NH_3)_2] \cdot H_2O$
Insoluble in dilute hydrochloric acid . . . . ≤0.05%
Sensitivity . . . . Passes test

TESTS

ASSAY. (By gravimetric determination as mercury salt). Weigh accurately 0.5 g and dissolve in 100 mL of water containing 0.5 mL of hydrochloric acid. Add a hot solution containing 0.5 g of mercuric acetate dissolved in a mixture of 140 mL of water and 10 mL of hydrochloric acid. Digest on a hot plate (≈100 °C) for 5 min, filter through a tared porcelain filtering crucible, and wash several times with hot water. The precipitate can be dried at 105 °C and weighed as $Hg[Cr(SCN)_4(NH_3)_2]_2$:

% Reinecke salt = (Wt of precipitate × 85.67) / Sample wt (g)

Alternatively, the precipitate can be ignited in a hood and heated to constant weight at 800 °C as Cr_2O_3:

% Reinecke salt = (Wt of Cr_2O_3 × 466.3) / Sample wt (g)

INSOLUBLE IN DILUTE HYDROCHLORIC ACID. Dissolve 2.0 g in 200 mL of dilute hydrochloric acid (1 + 99), heat to boiling, and digest in a covered beaker on a hot plate (≈100 °C) for 1 h. Filter through a tared filtering crucible, wash thoroughly with water, and dry at 105 °C.

SENSITIVITY. Dissolve 50 mg in 10 mL of water. Add 0.2 mL of this solution to 1 mL of a solution of 10 mg of choline chloride in 20 mL of water, and shake gently. A distinct red precipitate should form within 10 s.

Salicylic Acid

2-Hydroxybenzoic Acid

C$_7$H$_6$O$_3$ **Formula Wt 138.12**

CAS Number 69–72–7

REQUIREMENTS

Assay . . . . ≥99.0% C$_7$H$_6$O$_3$
Melting point . . . . 158.0–161.0 °C

MAXIMUM ALLOWABLE

Residue after ignition 0.01%
Chloride (Cl) 0.001%
Sulfate (SO$_4$) 0.003%
Heavy metals (as Pb) . . . . 5 ppm
Iron (Fe) . . . . 2 ppm
Substances darkened by sulfuric acid . . . . Passes test

TESTS

ASSAY. Analyze the sample by liquid chromatography using the general method described on page 71. The parameters cited have given satisfactory results.

Weigh 0.5 g of sample in a vial and dissolve in 5 mL of acetonitrile plus 5 mL of mobile phase. Inject duplicate 10-µL aliquots into the chromatograph, and compare the peak areas with those obtained from standards under the same conditions.

Mobile Phase: Acetonitrile–0.01 M phosphoric acid (25/75) at 2 mL/min. Prepare 1 M phosphoric acid by diluting 7 mL of the concentrated (85%) acid to 100 mL with water. Mix 10 mL of this solution with 740 mL of water and 250 mL of acetonitrile. De-gas under vacuum or with a sonic bath.

Column: Decyl (C-8 monomeric phase), 250 × 4.6 mm i.d., 5 µm, 9% C loading, end-capped.

Column Temperature: Ambient

Flow Rate: 2 mL/min

Sample Size: 10 µL

Detector: Ultraviolet at 280 nm

Approximate Retention Times (min): *p*-hydroxybenzoic acid, 1.3; *p*-hydroxy-isophthalic acid, 1.8; phenol, 3.0; salicylic acid, 4.1. Actual times may vary, depending upon the age or condition of the column.

MELTING POINT. (Page 22).

RESIDUE AFTER IGNITION. (Page 16). Ignite 20 g. Use 1 mL of sulfuric acid.
Retain the residue to prepare sample solution A for the determination of heavy metals and iron.

CHLORIDE. (Page 27). Dissolve 1.0 g in 20 mL of 95% alcohol.

SULFATE. Mix 4.0 g with 1 g of sodium carbonate, add 30 mL of hot water in
small portions, and evaporate to dryness. Ignite gently, taking care to protect the mixture from the flame because of the presence of sulfur compounds in the natural gas. To the residue add 15 mL of water and 1 mL of 30% hydrogen peroxide, boil for 5 min, and add 2 mL of hydrochloric acid. Evaporate to dryness on a hot plate ($\approx$100 °C), cool, and add 10 mL of water. Filter, wash the filter with two 5-mL portions of water, and dilute the combined filtrate and washings with water to 25 mL. For the standard, evaporate 1 mL of 30% hydrogen peroxide, 2 mL of hydrochloric acid, and 1 g of sodium carbonate to dryness on the hot plate. Take up the residue in 10 mL of water, add 0.12 mg of sulfate ion (SO_4), and dilute with water to 25 mL. To both sample and standard solutions add 0.5 mL of 1 N hydrochloric acid and 2 mL of barium chloride reagent solution. Any turbidity in the solution of the sample should not exceed that in the standard. Compare 10 min after adding the barium chloride to the sample and standard solutions.

Sample Solution A for the Determination of Heavy Metals and Iron.
To the residue retained from the test for residue after ignition, add 2 mL of hydrochloric acid and 0.5 mL of nitric acid, and evaporate to dryness on a hot plate ($\approx$100 °C). Warm the residue with 1 mL of 1 N hydrochloric acid, add 30 mL of hot water, cool, and dilute with water to 40 mL (1 mL = 0.5 g).

HEAVY METALS. (Page 28, Method 2). Use 8.0 mL of sample solution A (4-g sample).

IRON. (Page 30, Method 1). Use 10 mL of sample solution A (5-g sample).

SUBSTANCES DARKENED BY SULFURIC ACID. Dissolve 0.5 g in 10 mL of
sulfuric acid. The solution should have not more than a pale yellow color.

Silica Gel Desiccant

CAS Number 7631–86–9

REQUIREMENTS

Suitability for moisture absorption . . . . Passes test

TESTS

SUITABILITY FOR MOISTURE ABSORPTION*. Weigh accurately about 5 g in a tared weighing dish. Place the dish for 24 h in a desiccator in which the atmosphere possesses a relative humidity of 80%, maintained by equilibrium with sulfuric acid having a specific gravity of 1.19 (27% H_2SO_4). The increase in weight should not be less than 27%.

Silver Diethyldithiocarbamate

$(C_2H_5)_2NCS_2Ag$ **Formula Wt 256.14**

CAS Number 1470–61–7

NOTE. To enhance stability, storage below 8 °C is recommended.

REQUIREMENTS

Solubility in pyridine . . . . Passes test
Suitability for determination of arsenic. . . . . Passes test

TESTS

SOLUBILITY IN PYRIDINE. Transfer 1 g of sample to a 200-mL volumetric flask and dilute to the mark with freshly distilled pyridine. The solution should be clear, bright yellow, and dissolution should be complete. (Save this solution for the suitability for determination of arsenic test.)

SUITABILITY FOR DETERMINATION OF ARSENIC. Transfer 0.002 mg of arsenic (As) to an arsine generator flask and dilute with water to 35 mL. (For a description of the apparatus used in this test, see the section on arsenic, page 25.) Add 20 mL of dilute sulfuric acid (1 + 4), 2 mL of 16.5% potassium iodide reagent solution, and 0.5 mL of 40% stannous chloride dihydrate reagent solution in concentrated hydrochloric acid. Mix and allow to stand for 30 min at room

---

*Adapted from Rosin, J., *Reagent Chemicals and Standards,* Litton Educational Publishing Inc.: 1967.

temperature. Pack the scrubber tube loosely with two pledgets of lead acetate cotton. Place 3.0 mL of sample solution, retained from the solubility in pyridine test, in the absorber tube. Quickly add 3.0 g of granulated (No. 20 mesh) zinc to the arsine generator flask and immediately connect the flask to the scrubber–absorber assembly. Allow the evolution of hydrogen to proceed at room temperature for 45 min, swirling the solution every 10 min. Disconnect the absorber tube and determine the absorbance of the solution in a 1.00-cm cell at 525 nm, against a fresh portion of the sample solution in a similar matched cell set at zero absorbance as the reference liquid. The absorbance should not be less than 0.10.

Silver Nitrate

AgNO$_3$ **Formula Wt 169.87**

CAS Number 7761–88–8

REQUIREMENTS

Assay. . . . . $\geq$99.0% AgNO$_3$
Clarity of solution . . . . Passes test

MAXIMUM ALLOWABLE

Chloride (Cl) . . . . 5 ppm
Free acid . . . . Passes test
Substances not precipitated by hydrochloric acid 0.01%
Sulfate (SO$_4$) 0.002%
Copper (Cu). . . . . 2 ppm
Iron (Fe). . . . . 2 ppm
Lead (Pb) 0.001%

TESTS

ASSAY. (By argentometric titrimetry). Weigh, to the nearest 0.1 mg, about 0.65 g of sample in a 250-mL Erlenmeyer flask. Dissolve in 100 mL of water and add 5 mL of nitric acid. Add 2 mL of ferric ammonium sulfate indicator solution and titrate with 0.1 N potassium thiocyanate to the first appearance of a reddish-brown end point. One milliliter of 0.1 N potassium thiocyanate corresponds to 0.01699 g of AgNO$_3$.

CLARITY OF SOLUTION. Dissolve 20 g in 100 mL of water in a 250-mL conical flask. The solution should be clear, and no significant amount of insoluble matter, such as minute fibers and/or other particles of any shape or color, should be observed. Reserve the solution for the test for substances not precipitated by hydrochloric acid.

CHLORIDE. Dissolve 2.0 g in 40 mL of water, add ammonium hydroxide drop-wise until the precipitate first formed is redissolved, and dilute with water to 50 mL. Transfer the solution to a 100-mL platinum dish that is made the cathode. Insert a rotating anode and electrolyze for 1 h, starting with a current of 1 ampere. Decant the solution and evaporate to approximately 25 mL. Neutralize the solution to the phenolphthalein end point with nitric acid, and add 1 mL of nitric acid in excess and 1 mL of silver nitrate reagent solution. Any turbidity should not exceed that produced by 0.01 mg of chloride ion (Cl) in an equal volume of solution containing 1 mL of nitric acid and 1 mL of silver nitrate reagent solution.

FREE ACID. Dissolve 5.0 g in 50 mL of water, add 0.25 mL of bromcresol green indicator solution (0.04%), and mix well. The solution should be colored blue, not green or yellow.

SUBSTANCES NOT PRECIPITATED BY HYDROCHLORIC ACID. Dilute the solution obtained in the test for clarity of solution to about 600 mL. Heat to boiling and add hydrochloric acid to precipitate the silver completely (about 11 mL). Allow to stand overnight and filter. Evaporate the filtrate to dryness, and add 0.15 mL of hydrochloric acid and 10 mL of water. Heat, filter, and wash with about 10 mL of water. Evaporate the filtrate to dryness in a tared dish or crucible and dry at 105 °C. Correct for the weight obtained in a complete blank test. Reserve the residues for the preparation of sample solution A and blank solution B.

> ***Sample Solution A and Blank Solution B for the Determination of Sulfate, Copper, Iron, and Lead.*** To each of the residues remaining from the test for substances not precipitated by hydrochloric acid add 3 mL of dilute hydrochloric acid (1 + 1), cover with a watch glass, and digest on a hot plate ($\approx$100 °C) for 15–20 min. Cool and dilute with water to 100 mL. One is sample solution A (1 mL = 0.2 g); the other is blank solution B.

SULFATE. (Page 33, Method 1). Use 12.5 mL of sample solution A. Use 12.5 mL of blank solution B to prepare the standard solution.

COPPER. Add a slight excess of ammonium hydroxide to 25 mL of sample solution A (5-g sample). For the standard, add 0.01 mg of copper ion (Cu) to 25 mL of blank solution B and make slightly alkaline with ammonium hydroxide. Add 10 mL of 0.1% solution of sodium diethyldithiocarbamate to each. Any yellow color in the solution of the sample should not exceed that in the standard. Estimate the copper content to aid in the test for lead.

IRON. (Page 30, Method 1). Use 25 mL of sample solution A (5-g sample). Add the standard to 25 mL of blank solution B.

LEAD. Dilute 25 mL of sample solution A (5-g sample) to 35 mL. For the standard, add 0.05 mg of lead ion (Pb) and the amount of copper estimated to be present in the test for copper to 25 mL of blank solution B, and dilute with water to 35 mL. Adjust the pH of the standard and sample solutions to between 3 and 4 (using a pH meter) with 1 N acetic acid or ammonium hydroxide (10% NH_3), dilute with water to 40 mL, and mix. Add 10 mL of freshly prepared hydrogen sulfide water to each and mix. Any color in the solution of the sample should not exceed that in the standard.

Silver Sulfate

Ag_2SO_4 **Formula Wt 311.80**

CAS Number 10294–26–5

REQUIREMENTS

Assay. . . . .≥98.0% Ag_2SO_4

MAXIMUM ALLOWABLE

Insoluble matter and silver chloride. 0.02%
Nitrate (NO_3). 0.001%
Substances not precipitated by hydrochloric acid 0.03%
Iron (Fe). 0.001%

TESTS

ASSAY. (By argentometric titrimetry). Weigh, to the nearest 0.1 mg, about 0.5 g of sample in a 250-mL Erlenmeyer flask. Dissolve in 50 mL of water and 5 mL of nitric acid. Add 2 mL of ferric ammonium sulfate indicator solution and titrate with 0.1 N potassium thiocyanate to the first appearance of a reddish-brown end point. One milliliter of 0.1 N potassium thiocyanate corresponds to 0.01559 g of Ag_2SO_4.

INSOLUBLE MATTER AND SILVER CHLORIDE. Add 5.0 g of the powdered salt to 500 mL of boiling water and boil gently until the silver sulfate is dissolved. If any insoluble matter remains, filter while hot through a tared filtering crucible (retain the filtrate for the test for substances not precipitated by hydrochloric acid), wash thoroughly with hot water (discard the washings), and dry at 105 °C.

NITRATE. To 0.50 g of the powdered salt add 2 mL of phenoldisulfonic acid reagent solution and heat on a hot plate (≈100 °C) for 15 min. Cool, dilute with 20 mL of water, filter, and make alkaline with ammonium hydroxide. Any yellow color should not exceed that produced when a solution containing 0.005 mg of nitrate ion (NO_3) is evaporated to dryness and the residue is treated in the same manner as the sample.

SUBSTANCES NOT PRECIPITATED BY HYDROCHLORIC ACID. Heat to boiling the filtrate obtained in the test for insoluble matter and silver chloride, add 5 mL of hydrochloric acid, and allow to stand overnight. Dilute with water to 500 mL and filter. Evaporate 250 mL of the filtrate to dryness, and add 0.15 mL of hydrochloric acid and 10 mL of water. Heat and filter. Add 0.10 mL of sulfuric acid to the resulting filtrate, evaporate to dryness in a tared dish, and ignite at 800 ± 25 °C for 15 min. Correct for the weight obtained in a complete blank test. Retain the residue for the test for iron.

IRON. (Page 30, Method 1). To the residue obtained in the preceding test add 3 mL of dilute hydrochloric acid (1 + 1), cover with a watch glass, and digest on a hot plate (≈100 °C) for 15–20 min. Remove the watch glass and evaporate to dryness. Dissolve the residue in 35 mL of dilute hydrochloric acid (1 + 6), filter if necessary, and dilute with water to 50 mL. Use 20 mL of the solution without further acidification.

Soda Lime

CAS Number 8006–28–8

> *NOTE.* Soda lime is a mixture of variable proportions of sodium hydroxide with calcium oxide or hydroxide. This reagent may or may not include an indicator.

REQUIREMENTS

Carbon dioxide absorption capacity . . . . ≥19.0%
MAXIMUM ALLOWABLE
Loss on drying at 200 °C. 7%
Fines 1%

TESTS

CARBON DIOXIDE ABSORPTION CAPACITY. Fill the lower transverse section of a U-shaped drying tube of about 15-mm internal diameter and 15-cm height with loosely packed glass wool. Place in one arm of the tube about 5 g of anhydrous calcium chloride, and accurately weigh the tube and its contents. Into the other arm of the tube place 9.5–10.5 g of soda lime, and again weigh accurately. Insert stoppers in the open arms of the U-tube, and connect the side tube of the arm filled with soda lime to a calcium chloride drying tube, which in turn is connected to a suitable source of carbon dioxide. Pass the carbon dioxide though the U-tube at a rate of 75 mL per minute for 30 min, accurately timed. Disconnect the U-tube, cool to room temperature, remove the stoppers, and weigh.

LOSS ON DRYING AT 200 °C. Weigh accurately about 10 g and dry at 200 °C for 18 h.

FINES. Place 100 g on a clean U.S. No. 100 standard sieve nested in a receiving pan. Cover the sieve and shake on a mechanical shaker for 5 min. The weight of the fine material in the receiving pan should not exceed 1.0 g.

Sodium

Na

Atomic Wt 22.99

CAS Number 7440–23–5

REQUIREMENTS

MAXIMUM ALLOWABLE

Chloride (Cl) 0.002%
Nitrogen (N) 0.003%
Phosphate (PO$_4$) . . . . 5 ppm
Sulfate (SO$_4$) 0.002%
Heavy metals (as Pb) . . . . 5 ppm
Iron (Fe). 0.001%

TESTS

Sample Solution A for All Tests. If the metal contains any adhering oil or other foreign material, shave off a thin layer and use only bright clean metal for the sample. Weigh 20 g and cut it into small pieces. Cool about 100 mL of water in a beaker in an ice bath. Add the small pieces of sodium one at a time to the ice-cold water. Keep the solution cool and do not add another piece until the preceding one has completely reacted and dissolved. If desired, a magnetic stirrer may be used to aid dissolution. After all the sample has been dissolved, cool, and dilute with water to 500 mL (1 mL = 0.04 g).

CHLORIDE. (Page 27). Neutralize 12.5 mL of sample solution A (0.5-g sample) with nitric acid.

NITROGEN. (Page 30). Use 12.5 mL of sample solution A (0.5-g sample). For the standard, use 0.015 mg of nitrogen (N).

PHOSPHATE. (Page 32, Method 1). To 100 mL of sample solution A (4-g sample), add 15 mL of hydrochloric acid and evaporate to about 30 mL. Add 30 mL of hydrochloric acid, filter through a filtering crucible, and wash the precipitated sodium chloride twice with 5-mL portions of hydrochloric acid. Evaporate the filtrate and washings to dryness on a hot plate ($\approx$100 °C). Dissolve the residue in 25 mL of approximately 0.5 N sulfuric acid, and continue as described.

SULFATE. (Page 33, Method 1). Neutralize 62.5 mL of sample solution A (2.5-g sample) with hydrochloric acid and evaporate to about 20 mL. Add 1 mL of dilute hydrochloric acid (1 + 19).

HEAVY METALS. (Page 28, Method 1). To 125 mL of sample solution A add 25 mL of hydrochloric acid, and evaporate to dryness on a hot plate (≈100 °C). Dissolve the residue in about 20 mL of water, and dilute with water to 25 mL. For the control, add 0.02 mg of lead to 25 mL of sample solution A, and treat exactly as the 125 mL of sample solution A.

IRON. (Page 30, Method 1). Neutralize 25 mL of sample solution A (1-g sample) with hydrochloric acid, add 2 mL of hydrochloric acid in excess, and dilute with water to 50 mL. Use the solution without further acidification.

Sodium Acetate
Sodium Acetate, Anhydrous

CH_3COONa Formula Wt 82.03

CAS Number 127–09–3

REQUIREMENTS

Assay . . . . ≥99.0% $C_2H_3O_2Na$
pH of a 5% solution . . . . 7.0–9.2 at 25 °C

MAXIMUM ALLOWABLE

Insoluble matter. 0.01%
Loss on drying at 120 °C. 1.0%
Chloride (Cl) 0.002%
Phosphate (PO_4) 0.001%
Sulfate (SO_4) 0.003%
Calcium (Ca) 0.005%
Magnesium (Mg) 0.002%
Heavy metals (as Pb) 0.001%
Iron (Fe) 0.001%

TESTS

ASSAY. (Total alkalinity by nonaqueous titration). Weigh accurately 0.3 g of sample into a 125-mL flask and dissolve in 50 mL of acetic acid and 5 mL of acetic anhydride. Use a second flask containing 50 mL of acetic acid and 5 mL of acetic anyhdride as a blank. Allow to stand for 15 min. Add 0.10 mL of crystal violet indicator solution to both flasks, and titrate with 0.1 N perchloric acid in acetic acid until the solution color changes from violet to emerald green.

$$\% \ CH_3COONa = \frac{(V_s - V_b) \times N \ HClO_4 \times 8.203}{\text{Sample wt (g)}}$$

where V_s, V_b = mL of perchloric acid to titrate sample and blank, respectively.

pH OF A 5% SOLUTION. (Page 44). The pH should be 7.0–9.2 at 25 °C.

INSOLUBLE MATTER. (Page 15). Use 20 g dissolved in 150 mL of water.

LOSS ON DRYING AT 120 °C. Weigh accurately about 2 g in a tared weighing bottle, and dry at 120 °C to constant weight.

CHLORIDE. (Page 27). Use 0.50 g.

PHOSPHATE. (Page 32, Method 2). Dissolve 2.0 g in 10 mL of nitric acid and evaporate to dryness on a hot plate (≈100 °C). Add 10 mL of nitric acid and repeat the evaporation. Dissolve in 80 mL of water, add 0.5 g of ammonium molybdate, and adjust the pH to 1.8 (using a pH meter) with dilute hydrochloric acid (1 + 9). Heat to boiling, cool, add 10 mL of hydrochloric acid, and continue as described. Carry along a standard containing 0.02 mg of phosphate treated in the same manner as the sample after the addition of 80 mL of water.

SULFATE. (Page 33, Method 1). Allow 30 min for the turbidity to form.

CALCIUM AND MAGNESIUM. (By flame AAS, page 39).

> **Sample Stock Solution.** Dissolve 5.0 g in 80 mL of water and transfer to a 100-mL volumetric flask. Dilute to the mark with water (1 mL = 0.05 g).

Element	Wavelength (nm)	Sample Wt (g)	Standard Added (mg)	Flame Type*	Background Correction
Ca	422.7	0.50	0.025; 0.05	N/A	No
Mg	285.2	0.50	0.01; 0.02	A/A	Yes

*A/A is air/acetylene; N/A is nitrous oxide/acetylene.

HEAVY METALS. (Page 28, Method 1). Dissolve 6.0 g in about 10 mL of water, add 15 mL of dilute hydrochloric acid (10%), and dilute with water to 60 mL. Use 25 mL to prepare the sample solution, and use 5.0 mL of the remaining solution to prepare the control solution.

IRON. (Page 30, Method 1). Dissolve 1.0 g in 50 mL of dilute hydrochloric acid (1 + 24), and use the solution without further acidification.

Sodium Acetate Trihydrate

$NaC_2H_3O_2 \cdot 3H_2O$　　　　　　　　　　　　**Formula Wt 136.08**

CAS Number 6131–90–4

REQUIREMENTS

Assay . . . . 99.0–101% $NaC_2H_3O_2 \cdot 3H_2O$
pH of a 5% solution . . . . 7.5–9.2 at 25 °C
Substances reducing permanganate . . . . Passes test

MAXIMUM ALLOWABLE

Insoluble matter. 0.005%
Chloride (Cl) 0.001%
Phosphate (PO$_4$) . . . . 5 ppm
Sulfate (SO$_4$) 0.002%
Heavy metals (as Pb) . . . . 5 ppm
Iron (Fe). . . . . 5 ppm
Calcium (Ca) 0.005%
Magnesium (Mg) 0.002%
Potassium (K). 0.005%

TESTS

ASSAY. (Total alkalinity by nonaqueous titration). Weigh accurately 0.5 g of sample into a 125-mL flask, and dissolve in 50 mL of acetic acid and 5 mL of acetic anhydride. Use a second flask containing 50 mL of acetic acid and 5 mL of acetic anhydride as a blank. Add 0.10 mL of crystal violet indicator solution to both flasks, and titrate with 0.1 N perchloric acid in acetic acid until the solution color changes from violet to emerald green.

$$\% \ CH_3COONa \cdot 3H_2O = \frac{(V_s - V_b) \times N \ HClO4 \times 13.61}{Sample \ wt \ (g)}$$

where V_s, V_b = mL of perchloric acid to titrate sample and blank, respectively

pH OF A 5% SOLUTION. (Page 44). The pH should be 7.5–9.2 at 25 °C.

SUBSTANCES REDUCING PERMANGANATE. Dissolve 5.0 g in 50 mL of water, and add 5 mL of 10% sulfuric acid reagent solution and 0.10 mL of 0.1 N potassium permanganate. The pink color should persist for at least 1 h.

INSOLUBLE MATTER. (Page 15). Use 20 g dissolved in 150 mL of water.

CHLORIDE. (Page 27). Use 1.0 g.

PHOSPHATE. Dissolve 2.0 g in 10 mL of nitric acid and evaporate to dryness on a hot plate ($\approx$100 °C). Add 10 mL of sulfuric acid and repeat the evaporation. Dissolve in 80 mL of water, add 0.5 g of ammonium molybdate tetrahydrate, $(NH_4)_6Mo_7O_24 \cdot 4H_2O$, and adjust the pH to 1.8 (using a pH meter) with dilute hydrochloric acid (1 + 9). Heat to boiling, cool, and add 10 mL of hydrochloric acid, and continue as described. Carry along a standard containing 0.01 mg of phosphate ion (PO$_4$) treated in the same manner as the sample after the addition of 80 mL of water.

SULFATE. (Page 33, Method 1). Allow 30 min for the turbidity to form.

HEAVY METALS. (Page 28, Method 1). Dissolve 6.0 g in about 10 mL of water, add 15 mL of dilute hydrochloric acid (10%), and dilute with water to 30 mL. Use

25 mL to prepare the sample solution, and use 5.0 mL of the remaining solution to prepare the control solution.

IRON. (Page 30, Method 1). Dissolve 2.0 g in 50 mL of dilute hydrochloric acid (1 + 24), and use the solution without further acidification.

CALCIUM, MAGNESIUM, AND POTASSIUM. (By flame AAS, page 39).

Sample Stock Solution. Dissolve 5.0 g of sample in water in a 100-mL volumetric flask and dilute to the mark with water (1 mL = 0.05 g).

Element	Wavelength (nm)	Sample Wt (g)	Standard Added (mg)	Flame Type*	Background Correction
Ca	422.7	0.50	0.025; 0.05	N/A	No
Mg	285.2	0.50	0.01; 0.02	A/A	Yes
K	766.5	0.50	0.025; 0.05	A/A	No

*A/A is air/acetylene; N/A is nitrous oxide/acetylene.

Sodium Arsenate Heptahydrate
Disodium Hydrogen Arsenate Heptahydrate

$Na_2HAsO_4 \cdot 7H_2O$ **Formula Wt 312.01**

CAS Number 10048–95–0

REQUIREMENTS

Assay . . . . .98.0–102.0% $Na_2HAsO_4 \cdot 7H_2O$

MAXIMUM ALLOWABLE

Insoluble matter. 0.005%
Arsenite (As_2O_3) 0.01%
Chloride (Cl) 0.001%
Nitrate (NO_3). 0.005%
Sulfate (SO_4) 0.01%
Heavy metals (as Pb) 0.002%
Iron (Fe). 0.001%

TESTS

ASSAY. (By titration of oxidizing power of arsenate). Weigh accurately 0.55 g and dissolve in 50 mL of water in a glass-stoppered conical flask. Heat to 80 °C and add 10 mL of hydrochloric acid and 3 g of potassium iodide. Stopper the flask, swirl, and maintain at 80 °C for 15 min. Cool to room temperature and titrate the liberated iodine with 0.1 N sodium thiosulfate, adding 3 mL of starch indicator solution near the end of the titration. One milliliter of 0.1 N sodium thiosulfate corresponds to 0.01560 g of $Na_2HAsO_4 \cdot 7H_2O$.

INSOLUBLE MATTER. (Page 15). Use 20 g dissolved in 200 mL of water.

ARSENITE. Dissolve 10 g in 75 mL of water, make the solution just acidic to litmus with 10% sulfuric acid, and add 2 g of sodium bicarbonate. When dissolution is complete, add starch indicator solution, and titrate with 0.02 N iodine to a blue end point. Not more than 1.0 mL of 0.02 N iodine should be required.

CHLORIDE. (Page 27). Use 1.0 g of sample and 5 mL of nitric acid.

NITRATE.

Sample Solution A. Dissolve 0.20 g in 3 mL of water by heating on a hot plate (≈100 °C). Dilute to 50 mL with brucine sulfate reagent solution.

Control Solution B. Dissolve 0.20 g in 2 mL of water and 1 mL of the standard nitrate solution containing 0.01 mg of nitrate ion (NO_3) per milliliter by heating on a hot plate (≈100 °C). Dilute to 50 mL with brucine sulfate reagent solution.

Continue with the procedure described on page 30, starting with the preparation of blank solution C.

SULFATE. Dissolve 10 g in 10 mL of water, add 5 mL of hydrochloric acid, and heat the solution to boiling. Add 5 mL of barium chloride reagent solution, digest in a covered beaker on a hot plate (≈100 °C) for 2 h, and allow to stand overnight. If any precipitate is formed, filter, wash thoroughly, and ignite. Correct for the weight obtained on a complete blank test.

HEAVY METALS. (Page 28, Method 1). Dissolve 2.5 g in 20 mL of water in a small dish. Add 2 g of potassium iodide, 10 mL of hydrobromic acid, and 0.10 mL of sulfuric acid. Evaporate to dryness on a hot plate (≈100 °C), wash down the sides of the dish with a few milliliters of water, add 5 mL of hydrochloric acid, and again evaporate to dryness on a hot plate (≈100 °C). Dissolve the residue in a few milliliters of water, neutralize to litmus with ammonium hydroxide (10% NH_3), and dilute with water to 50 mL. Use 30 mL to prepare the sample solution, and use 10 mL of the remaining solution to prepare the control solution.

IRON. (Page 30, Method 1). Use 1.0 g of sample and 5 mL of hydrochloric acid.

Sodium Bicarbonate
Sodium Hydrogen Carbonate

$NaHCO_3$ Formula Wt 84.01

CAS Number 144–55–8

REQUIREMENTS

Assay (dried basis) . . . . 99.7–100.3% NaHCO$_3$

MAXIMUM ALLOWABLE

Insoluble matter. 0.015%
Chloride (Cl) 0.003%
Phosphate (PO$_4$) 0.001%
Sulfur compounds (as SO$_4$) 0.003%
Ammonium (NH$_4$) . . . . 5 ppm
Heavy metals (as Pb) . . . . 5 ppm
Iron (Fe). 0.001%
Calcium (Ca) 0.02%
Magnesium (Mg) 0.005%
Potassium (K). 0.005%

TESTS

ASSAY. (By acid–base titrimetry). Weigh accurately 3 g, previously dried over indicating type silica gel desiccant for 24 h, dissolve it in 50 mL of water, add methyl orange indicator solution, and titrate with 1 N hydrochloric acid. The sodium bicarbonate content calculated from the total alkalinity, as determined by the titration, should not be less than 99.7 nor more than 100.3% of the weight taken. One milliliter of 1 N hydrochloric acid corresponds to 0.08401 g of NaHCO$_3$.

INSOLUBLE MATTER. (Page 15). Use 10.0 g dissolved in 100 mL of hot water.

CHLORIDE. (Page 27). Use 0.33 g, and neutralize with nitric acid.

PHOSPHATE. (Page 32, Method 1). Dissolve 2.0 g in 15 mL of dilute hydrochloric acid (1 + 2), and evaporate to dryness on a hot plate ($\approx$100 °C). Dissolve the residue in 25 mL of approximately 0.5 N sulfuric acid, and continue as described.

SULFUR COMPOUNDS. Dissolve 2.0 g of sample in 20 mL of water, evaporate to 5 mL, add 1 mL of bromine water, and evaporate to dryness. Cool. Dissolve in 10 mL of 10% hydrochloric acid and evaporate to dryness. Cool. Add 5 mL of 10% hydrochloric acid and evaporate to dryness. Cool. Dissolve with 10 mL of water and evaporate to dryness. Cool. Dissolve in 10 mL of water, adjust to approximately pH 2 with 10% hydrochloric acid or (1 + 3) ammonium hydroxide. Filter through a washed filter paper, wash with two 2-mL portions of water, and dilute with water to 30 mL.

For the control, take 0.06 mg of sulfate ion (SO$_4$) standard in 15 mL of water. Add 1 mL of (1 + 19) hydrochloric acid, dilute with water to 30 mL, and add 1 mL of barium chloride solution. Compare the sample and control solutions after 30 min. Sample solution turbidity should not exceed that of the control solution.

AMMONIUM. Dissolve 2.0 g in 40 mL of ammonia-free water, and add 10 mL of 10% sodium hydroxide reagent solution and 2 mL of Nessler reagent. Any color should not exceed that produced by 0.01 mg of ammonium ion (NH_4) in an equal volume of solution containing the quantities of reagents used in the test.

HEAVY METALS. (Page 28, Method 1). To 5.0 g in a 150-mL beaker add 10 mL of water, mix, and cautiously add 10 mL of hydrochloric acid. Evaporate to dryness on a hot plate ($\approx$100 °C), dissolve residue in about 20 mL of water, and dilute with water to 25 mL. For the control, add 0.02 mg of lead ion (Pb) to 1.0 g of sample, and treat exactly as the 5.0 g of sample.

IRON. (Page 30, Method 1). Dissolve 1.0 g in 30 mL of dilute hydrochloric acid (1 + 9), dilute with water to 50 mL, and use the solution without further acidification.

CALCIUM, MAGNESIUM, AND POTASSIUM. (By flame AAS, page 39).

Sample Stock Solution. Dissolve 4.0 g of sample with water in a 100-mL volumetric flask, and dilute to the mark with water (1 mL = 0.04 g).

Element	Wavelength (nm)	Sample Wt (g)	Standard Added (mg)	Flame Type*	Background Correction
Ca	422.7	0.20	0.04; 0.08	N/A	No
Mg	285.2	0.20	0.01; 0.02	A/A	Yes
K	766.5	0.20	0.01; 0.02	A/A	No

*A/A is air/acetylene; N/A is nitrous oxide/acetylene.

Sodium Bismuthate

NaBiO$_3$ **Formula Wt 279.97**

CAS Number 12232–99–4

REQUIREMENTS

Assay . . . . $\geq$80.0% NaBiO$_3$
Oxidizing efficiency . . . . $\geq$99.6%

MAXIMUM ALLOWABLE

Chloride (Cl) 0.002%
Manganese (Mn) . . . . 5 ppm

TESTS

ASSAY. (By titration of oxidizing power). Weigh accurately 0.7 g, place in a flask, add 25.0 mL of ferrous sulfate solution, and stopper the flask. Transfer 25.0 mL of the ferrous sulfate solution to another flask and stopper the flask. Allow each flask to stand for 30 min, shaking frequently, and titrate the ferrous sulfate in each with 0.1 N potassium permanganate. The difference in the volume of permanganate consumed in the two titrations is equivalent to the sodium bismuthate. One milliliter of 0.1 N potassium permanganate corresponds to 0.01400 g of $NaBiO_3$.

> *Ferrous Sulfate Solution.* Dissolve 7 g of clear crystals of ferrous sulfate, $FeSO_4 \cdot 7H_2O$, in 90 mL of freshly boiled and cooled water, and add sulfuric acid to make 100 mL. The solution must be freshly prepared.

CHLORIDE. Add 1.0 g to 25 mL of water, heat to boiling, and keep at the boiling temperature for 10 min. Dilute with water to 50 mL and filter through a chloride-free filter. To 25 mL of the filtrate add 0.15 mL of 30% hydrogen peroxide to clear the solution, and then add 1 mL of nitric acid and 1 mL of silver nitrate reagent solution. Any turbidity should not exceed that produced by 0.01 mg of chloride ion (Cl) in an equal volume of solution containing the quantities of reagents used in the test.

MANGANESE. Dissolve 2.0 g in 35 mL of dilute nitric acid (5 + 2), heat to boiling, and boil gently for 5 min. Prepare a standard containing 0.01 mg of manganese ion (Mn) in 35 mL of dilute nitric acid (5 + 2). To each add 5 mL of sulfuric acid, 5 mL of phosphoric acid, and 0.5 mL of sulfurous acid. Boil gently to expel oxides of nitrogen, cool the solutions to 15 °C, and add 0.5 g of sodium bismuthate to each. Allow to stand for 5 min with occasional stirring, dilute each with 25 mL of water, and filter through a filter other than paper. Any pink color in the solution of the sample should not exceed that in the standard.

> *Manganese Metal for Use as Oxidimetric Standard.* Assay a selected lot of commercial high-purity electrolytic manganese, previously screened through a number 10 and retained on a number 20 screen, by determining the concentration of impurities. Metals at levels below 0.03% can be evaluated with sufficient accuracy by the spectrographic semiquantitative method. Metals at higher concentrations are determined by suitable quantitative methods; carbon and sulfur can be determined by classical combustion methods. Oxygen, hydrogen, and nitrogen are determined by vacuum fusion analysis employing a 25-g iron bath containing 2–3 g of tin. The sample is placed in a tin capsule and dropped into the bath which is held at 1500–1550 °C. Up to three samples can be analyzed before discarding the bath. The assay of the manganese metal for use as an oxidimetric standard should not be less than 99.8% Mn. (Because of the tendency of the metal to react with oxygen, it must be stored in a tightly sealed container after the assay has been performed.)

OXIDIZING EFFICIENCY. To 0.20 g of oxidimetric standard manganese metal in a 1-L conical flask, add 15 mL of dilute nitric acid (1 + 3), and heat cautiously until the manganese is dissolved. Add 8 mL of 70% perchloric acid and boil gently until the acid fumes strongly and manganese dioxide begins to separate. Cool, add 5 mL of water and 25 mL of dilute nitric acid (1 + 3), and boil for several minutes to expel free chlorine. Add sufficient sulfurous acid or sodium nitrite solution to just dissolve the manganese dioxide. Boil the solution to expel completely the oxides of nitrogen. Cool to room temperature, add 225 mL of colorless, dilute nitric acid (2 + 5) and sufficient water to bring the total volume to 250 mL, and cool to 10–15 °C.

Add 7 g of sodium bismuthate (weighed to the nearest 10 mg) to the flask, agitate briskly for 1 min, dilute with 250 mL of cold water (10–15 °C), and filter immediately through a fine-porosity fritted glass filter (pretreat the frit in hot nitric acid and then wash it free of acid with hot water). The filter can be washed free of manganese more readily if not allowed to run dry during the filtering and washing. Wash the filter with cold, freshly boiled dilute nitric acid (3 + 97) until the washings are entirely colorless, and immediately treat the filtrate and washings as directed in the next paragraph.

Add 8.5 g of ferrous ammonium sulfate heptahydrate (weighed to the nearest mg) to the filtered solution of permanganic acid. Stir briskly. As soon as reduction is complete and all the salt is dissolved, add 0.01 M 1,10-phenanthroline indicator solution, and titrate the excess of ferrous ion with 0.1 N potassium permanganate to a clear green color that persists for at least 30 s. Standardize the 0.1 N potassium permanganate against oxidimetric standard sodium oxalate from NIST.

Determine the manganese equivalent of the ferrous ammonium sulfate heptahydrate by titrating 1.75 g of the salt with the 0.1 N potassium permanganate in 500 mL of cold, dilute nitric acid that has been pretreated with 2 g of sodium bismuthate under the conditions described.

Calculate the oxidizing efficiency of the sodium bismuthate as follows:

$$\text{Oxidizing efficiency} = [(A - B) \times 0.00110 \times 100] / C$$

where

$A =$ milliliters of exactly 0.1 N potassium permanganate equivalent to the ferrous sulfate added

$B =$ milliliters of exactly 0.1 N potassium permanganate required to titrate the excess ferrous ions

$C =$ grams of manganese used, taking into account the assay of the metal

Sodium Bisulfite

CAS Number 7631–90–5 ($NaHSO_3$); 7681–57–4 ($Na_2S_2O_5$)

NOTE. This reagent is usually a mixture of sodium bisulfite, $NaHSO_3$, and sodium metabisulfite, $Na_2S_2O_5$.

REQUIREMENTS

Assay. . . . . ≥58.5% SO_2

MAXIMUM ALLOWABLE

Insoluble matter. 0.005%
Chloride (Cl) 0.02%
Heavy metals (as Pb) 0.001%
Iron (Fe). 0.002%

TESTS

ASSAY. (Titration of reducing power). Weigh accurately 0.47 g and add to a mixture of 100.0 mL of 0.1 N iodine and 5 mL of 10% hydrochloric acid solution. Swirl gently until the sample is dissolved completely. Titrate the excess of iodine with 0.1 N sodium thiosulfate, adding 3 mL of starch indicator solution near the end of the titration. One milliliter of 0.1 N iodine consumed corresponds to 0.003203 g of SO_2.

INSOLUBLE MATTER. (Page 15). Use 20 g dissolved in 200 mL of water.

CHLORIDE. Dissolve 0.50 g in 100 mL of water and transfer 10 mL of the well-mixed solution to a platinum dish. Add 10% sodium hydroxide solution until the solution is slightly alkaline to litmus, making note of the volume of sodium hydroxide added. Prepare a standard containing 0.01 mg of chloride ion (Cl) in 10 mL of water, and add the same volume of 10% sodium hydroxide solution as was added to the sample solution. To each solution add, dropwise, 2 mL of 30% hydrogen peroxide and allow to stand at room temperature for 10 min. Evaporate the solutions to dryness on a hot plate (≈100 °C), dissolve the residues in 10 mL of water, and add 1 mL of nitric acid and 1 mL of silver nitrate reagent solution to each. Any turbidity in the solution of the sample should not exceed that of the standard.

HEAVY METALS. (Page 28, Method 1). Dissolve 3.0 g in a solution of 15 mL of water and 8 mL of hydrochloric acid. Evaporate to dryness on a hot plate (≈100 °C), dissolve the residue in about 20 mL of water, and dilute with water to 25 mL. For the control, add 0.02 mg of lead ion (Pb) to 1.0 g of sample, and treat exactly as the 3.0 g of sample.

IRON. (Page 30, Method 1). Dissolve 1.0 g in 10 mL of water, add 2 mL of hydrochloric acid, and evaporate to dryness on a hot plate (≈100 °C). Dissolve the residue in a mixture of 5 mL of water and 2 mL of hydrochloric acid, and again evaporate to dryness. Dissolve the residue in 4 mL of hydrochloric acid, dilute with water to 100 mL, and use 50 mL of the solution without further acidification.

Sodium Borate Decahydrate
Borax, Sodium Tetraborate Decahydrate

$Na_2B_4O_7 \cdot 10H_2O$ **Formula Wt 381.37**

CAS Number 1303–96–4

REQUIREMENTS

Assay . . . . 99.5–105.0% $Na_2B_4O_7 \cdot 10H_2O$
pH of a 0.01 M solution . . . . 9.15–9.20 at 25 °C

MAXIMUM ALLOWABLE

Insoluble matter. 0.005%
Chloride (Cl) 0.001%
Phosphate (PO_4) 0.001%
Sulfate (SO_4) 0.005%
Calcium (Ca) 0.005%
Heavy metals (as Pb) 0.001%
Iron (Fe) . . . . 5 ppm

TESTS

ASSAY. (By acid–base titrimetry). Weigh accurately 1 g, and dissolve in 50 mL of water. Make slightly acid to methyl red indicator with 1 N hydrochloric acid, cover with a watch glass, and boil gently for 2 min to expel any carbon dioxide. Cool, and adjust to the methyl red end point (pinkish yellow) with carbonate-free 0.5 N sodium hydroxide. Add phenolphthalein indicator and 8 g of mannitol, then titrate with 0.5 N sodium hydroxide through the development of a yellow color to a permanent pink end point. One milliliter of 0.5 N sodium hydroxide corresponds to 0.09534 g of $Na_2B_4O_7 \cdot 10H_2O$. The pH adjustment may be done potentiometrically. The end points are pH 5.4 and 8.5.

pH OF A 0.01 M SOLUTION. Dissolve 0.381 g in 100 g of carbon dioxide- and ammonia-free water (or 0.380 g in a volume of 100 mL). Standardize the pH meter and electrode at pH 9.18 at 25 °C with 0.01 M NIST SRM Sodium Tetraborate Decahydrate prepared from NIST SRM. Determine the pH by the method described on page 46. The pH should be 9.15–9.20 at 25 °C.

INSOLUBLE MATTER. (Page 15). Use 20 g dissolved in 300 mL of water.

CHLORIDE. (Page 27). Use 1.0 g.

PHOSPHATE. (Page 32, Method 1). Dissolve 2.0 g in 10 mL of warm water, add 2 mL of hydrochloric acid, and evaporate to dryness on a hot plate ($\approx$100 °C). Dissolve in 25 mL of approximately 0.5 N sulfuric acid, filter to remove any boric acid, and use the clear filtrate. Continue as described.

SULFATE. Dissolve 8.0 g in 120 mL of warm water plus 6 mL of hydrochloric acid. Filter and wash with 30 mL of water. Heat to boiling, add 5 mL of barium chloride reagent solution, digest in a covered beaker on a hot plate ($\approx$100 °C) for 2 h, and allow to stand overnight. Heat to dissolve any boric acid that may be crystallized. If a precipitate is formed, filter, wash thoroughly, and ignite. Correct for the weight obtained in a complete blank test.

CALCIUM. (By flame AAS, page 39).

Sample Stock Solution. Dissolve 10.0 g of sample with water in a 200-mL volumetric flask, add 10 mL of hydrochloric acid, and dilute to the mark with water (1 mL = 0.05 g).

Element	Wavelength (nm)	Sample Wt (g)	Standard Added (mg)	Flame Type*	Background Correction
Ca	422.7	1.0	0.05; 0.10	N/A	No

*N/A is nitrous oxide/acetylene.

HEAVY METALS. (Page 28, Method 1). Dissolve 4.0 g in 40 mL of hot water, add 5 mL of glacial acetic acid, and dilute with water to 48 mL. Use 36 mL to prepare the sample solution, and use the remaining 12 mL to prepare the control solution.

IRON. (Page 30, Method 1). Use 2.0 g of sample and 3 mL of hydrochloric acid.

Sodium Bromide

NaBr

Formula Wt 102.89

CAS Number 7647–15–6

REQUIREMENTS

Assay (corrected). . . . .≥99.0% NaBr
pH of a 5% solution . . . . 5.0–8.8 at 25 °C

MAXIMUM ALLOWABLE

Insoluble matter. 0.005%
Bromate (BrO_3) 0.001%
Chloride (Cl) 0.2%
Nitrogen compounds (as N). . . . . 5 ppm
Sulfate (SO_4) 0.002%
Barium (Ba) 0.002%
Heavy metals (as Pb) . . . . 5 ppm
Iron (Fe). . . . . 5 ppm
Calcium (Ca) 0.002%
Magnesium (Mg) 0.001%
Potassium (K). 0.1%

TESTS

ASSAY. (By argentimetric titration of bromide content). Weigh, to the nearest 0.1 mg, 0.4 g of sample. Transfer to a 250-mL titration flask and dissolve in 25 mL of water. Add slowly, while agitating, 50.0 mL of 0.1 N silver nitrate, then add 3 mL of nitric acid and 10 mL of benzyl alcohol, and shake vigorously. Add 2 mL of ferric ammonium nitrate and titrate the excess silver nitrate with 0.1 N ammonium thiocyanate.

$$\% \text{NaBr (uncorrected)} = \frac{[(\text{mL} \times \text{N AgNO}_3) - (\text{mL} \times \text{N NH}_4\text{SCN})] \times 10.29}{\text{Sample wt (g)}}$$

$$\% \text{NaBr (corrected)} = [\% \text{NaBr (uncorrected)}] - (2.90 \times \% \text{Cl})$$

pH OF A 5% SOLUTION. (Page 44). The pH should be 5.0–8.8 at 25 °C.

INSOLUBLE MATTER. (Page 15). Use 20 g dissolved in 150 mL of water.

BROMATE. (By differential pulse polarography, page 50). Use 10.0 g of sample in 25 mL of solution. For the standard, add 0.10 mg of bromate (BrO_3).

CHLORIDE. Dissolve 0.50 g in 15 mL of dilute nitric acid (1 + 2) in a small flask. Add 3 mL of 30% hydrogen peroxide and digest on a hot plate ($\approx$100 °C) until the solution is colorless. Wash down the sides of the flask with a little water, digest for an additional 15 min, cool, and dilute with water to 200 mL. Dilute 2.0 mL with water to 20 mL, and add 1 mL of nitric acid and 1 mL of silver nitrate reagent solution. Any turbidity should not exceed that produced by 0.01 mg of chloride ion (Cl) in an equal volume of solution containing the quantities of reagents used in the test.

NITROGEN COMPOUNDS. (Page 31). Use 1.0 g. For the standard, use 0.005 mg of nitrogen (N).

SULFATE. (Page 33, Method 1).

BARIUM. For the sample dissolve 6.0 g in 15 mL of water. For the control, dissolve 1.0 g in 15 mL of water and add 0.1 mg of barium ion (Ba). To each solution add 5 mL of acetic acid, 5 mL of 30% hydrogen peroxide, and 1 mL of hydrochloric acid. Digest in a covered beaker on a hot plate ($\approx$100 °C) until reaction ceases, uncover, and evaporate to dryness. Dissolve the residues in 15 mL of water, filter if necessary, and dilute with water to 23 mL. Add 2 mL of 10% potassium dichromate reagent solution and add ammonium hydroxide until the orange color is just dissipated and the yellow color persists. Add 25 mL of methanol, stir vigorously, and allow to stand for 10 min. Any turbidity in the solution of the sample should not exceed that in the control.

HEAVY METALS. Dissolve 6.0 g in about 20 mL of water and dilute with water to 30 mL. For the control, add 0.02 mg of lead ion (Pb) to 5.0 mL of the solution and dilute with water to 25 mL. For the sample use the remaining 25-mL portion. Adjust the pH of the control and sample solutions to between 3 and 4 (using a pH meter) with 1 N acetic acid or ammonium hydroxide (10% NH_3), dilute with water to 40 mL, and mix. Add 10 mL of freshly prepared hydrogen sulfide water to each and mix. Any color in the solution of the sample should not exceed that in the control.

IRON. Dissolve 2.0 g in 40 mL of water plus 2 mL of hydrochloric acid, and dilute with water to 50 mL. Add 30–50 mg of ammonium peroxydisulfate crystals and 3 mL of ammonium thiocyanate reagent solution. Any red color should not exceed that produced by 0.01 mg of iron (Fe) in an equal volume of solution containing the quantities of reagents used in the test.

CALCIUM, MAGNESIUM, AND POTASSIUM. (By flame AAS, page 39).

Sample Stock Solution. Dissolve 5.0 g of sample with water in a 100-mL volumetric flask, and dilute to the mark with water (1 mL = 0.05 g).

Element	Wavelength (nm)	Sample Wt (g)	Standard Added (mg)	Flame Type*	Background Correction
Ca	422.7	1.0	0.02; 0.04	N/A	No
Mg	285.2	1.0	0.01; 0.02	A/A	Yes
K	766.5	0.05	0.025; 0.05	A/A	No

*A/A is air/acetylene; N/A is nitrous oxide/acetylene.

Sodium Carbonate
Sodium Carbonate, Anhydrous

Na_2CO_3 **Formula Wt 105.99**

CAS Number 497–19–8

REQUIREMENTS

Assay (dried basis). . . . . ≥99.5% Na_2CO_3

MAXIMUM ALLOWABLE

Insoluble matter. 0.01%
Loss on heating at 285 °C. 1.0%
Chloride (Cl) 0.001%
Nitrogen compounds (as N). 0.001%
Phosphate (PO_4) 0.001%

Silica (SiO_4) 0.005%
Sulfur compounds (as SO_4) 0.003%
Ammonium hydroxide precipitate 0.01%
Heavy metals (as Pb) . . . . .5 ppm
Iron (Fe). . . . . .5 ppm
Calcium (Ca) 0.03%
Magnesium (Mg) 0.005%
Potassium (K). 0.005%

TESTS

ASSAY. (By acid–base titrimetry of carbonate). Weigh, to the nearest 0.1 mg, 2 g of the dried sample from the test for loss on heating at 285 °C. Transfer to a 125-mL glass-stoppered flask, dissolve with 50 mL of water, add 0.10 mL of methyl orange indicator solution, and titrate with 1 N hydrochloric acid.

$$\% \; Na_2CO_3 = (mL \times N \; HCl \times 5.300) \; / \; Sample \; wt \; (g)$$

INSOLUBLE MATTER. (Page 15). Use 10 g dissolved in 100 mL of water.

LOSS ON HEATING AT 285 °C. Crush sample and accurately weigh 10 g in a low-form weighing bottle. Heat to constant weight at 270–300 °C. Retain this sample for assay determination.

CHLORIDE. (Page 27). Use 1.0 g of sample and 2 mL of nitric acid.

NITROGEN COMPOUNDS. (Page 31). Use 1.0 g. For the standard, use 0.01 mg of nitrogen (N).

PHOSPHATE. (Page 32, Method 2). Dissolve 1.0 g in 50 mL of water in a platinum dish and digest on a hot plate (≈100 °C) for 30 min. Cool, neutralize with dilute sulfuric acid (1 + 19) to a pH of about 4, and dilute with water to about 75 mL. Add 0.5 g of ammonium molybdate and adjust the pH to 1.8 (using a pH meter) with dilute hydrochloric acid (1 + 9). Heat to boiling, cool, add 10 mL of hydrochloric acid, and dilute with water to 100 mL. Continue as described, and concurrently prepare a standard containing 0.01 mg of phosphate ion (PO_4) and 0.05 mg of silica (SiO_2) in about 75 mL of water treated as the 75 mL of sample solution. Reserve the aqueous phase for the determination of silica.

SILICA. Add 10 mL of hydrochloric acid to the solutions reserved from the determination of phosphate and transfer to separatory funnels. Add 40 mL of butyl alcohol, shake vigorously, and allow to separate. Draw off and discard the aqueous phase. Wash the butyl alcohol three times with 20-mL portions of dilute hydrochloric acid (1 + 99), discarding the washings each time. Dilute each butyl alcohol solution with butyl alcohol to 50 mL, take 10 mL from each, and dilute each to 50 mL with butyl alcohol. Add 0.5 mL of a freshly prepared 2% solution of stannous chloride dihydrate in hydrochloric acid. The blue color in the extract from the sample should not exceed that in the standard. If the butyl alcohol extracts are turbid, wash with 10 mL of dilute hydrochloric acid (1 + 99).

SULFUR COMPOUNDS. Dissolve 2.0 g of sample in 20 mL of water, evaporate to 5 mL, add 1 mL of bromine water, and evaporate to dryness. Cool. Dissolve in 10 mL of 10% hydrochloric acid and evaporate to dryness. Cool. Add 5 mL of 10% hydrochloric acid and evaporate to dryness. Cool. Dissolve with 10 mL of water and evaporate to dryness. Cool. Dissolve in 10 mL of water, adjust the pH to approximately pH 2 with 10% hydrochloric acid or ammonium hydroxide (1 + 3). Filter through a washed, fine-porosity filter paper, wash with two 2-mL portions of water, and dilute with water to 20 mL.

To prepare 0.06 mg of sulfate ion (SO_4) standard, add 1 mL of (1 + 19) hydrochloric acid and dilute with water to 20 mL.

To sample and standard solutions, add 1 mL of barium chloride reagent solution and allow to stand 30 min. Any turbidity in the sample solution must not exceed that of the standard.

AMMONIUM HYDROXIDE PRECIPITATE. Dissolve 10 g in 100 mL of water and filter if necessary. Cautiously add 30 mL of cooled dilute sulfuric acid (1 + 1) and evaporate until dense fumes of sulfur trioxide appear. Cool, dissolve the residue in 130 mL of hot water, and add ammonium hydroxide until the solution is just alkaline to methyl red. Heat to boiling, boil gently for 5 min, and filter, reserving the filtrate for the test for calcium and magnesium precipitate. Wash the precipitate with hot water, rejecting the washings, and ignite. The residue includes some, but not all, of the silica in the sample.

HEAVY METALS. (Page 28, Method 1). To 5.0 g in a 150-mL beaker add 10 mL of water, mix, and cautiously add 10 mL of hydrochloric acid. Evaporate to dryness on a hot plate ($\approx$100 °C), dissolve the residue in about 20 mL of water, and dilute with water to 25 mL. For the control, add 0.02 mg of lead ion (Pb) to 1.0 g of sample, and treat exactly as the 5.0 g of sample.

IRON. (Page 30, Method 1). Dissolve 2.0 g in 50 mL of dilute hydrochloric acid (1 + 9). Use the solution without further acidification.

CALCIUM, MAGNESIUM, AND POTASSIUM. (By flame AAS, page 39).

Sample Stock Solution. Dissolve 5.0 g of sample with water in a 100-mL volumetric flask and dilute to the mark with water (1 mL = 0.05 g).

Element	Wavelength (nm)	Sample Wt (g)	Standard Added (mg)	Flame Type*	Background Correction
Ca	422.7	0.10	0.03; 0.06	N/A	No
Mg	285.2	0.10	0.005; 0.01	A/A	Yes
K	766.5	0.20	0.01; 0.02	A/A	No

*A/A is air/acetylene; N/A is nitrous oxide/acetylene.

Sodium Carbonate, Alkalimetric Standard

Na_2CO_3 Formula Wt 105.99

CAS Number 497–19–8

NOTE. For use as an alkalimetric standard this reagent should be heated at 285 °C for 2 h.

REQUIREMENTS

Assay (dried basis) . . . . .99.95–100.05% Na_2CO_3

MAXIMUM ALLOWABLE

Insoluble matter. 0.01%
Loss on heating at 285 °C. 1.0%
Chloride (Cl) 0.001%
Nitrogen compounds (as N). 0.001%
Phosphate (PO_4) 0.001%
Silica (SiO_4) 0.005%
Sulfur compounds (as SO_4) 0.003%
Ammonium hydroxide precipitate 0.01%
Heavy metals (as Pb) . . . . .5 ppm
Iron (Fe). . . . . .5 ppm
Potassium (K). 0.005%
Calcium (Ca) 0.02%
Magnesium (Mg) 0.004%

TESTS

Except for assay and calcium and magnesium, other tests are the same as for sodium carbonate, page 567.

ASSAY. (By acid–base titrimetry of carbonate). Transfer a quantity of NIST SRM Potassium Hydrogen Phthalate to an agate or mullite mortar and grind to approximately 100-mesh fineness. Place 5.100 ± 0.003 g of the 100-mesh material in a weighing bottle, dry at 120 °C for 2 h, and cool in a desiccator for at least 2 h. Weigh the bottle and contents accurately, transfer the contents to the titration flask, and weigh the empty bottle to obtain the exact weight of the potassium hydrogen phthalate. Place 1.300 g of the alkalimetric standard sodium carbonate in a clean, dry weighing bottle. Heat at 285 °C for 2 h, cool in a desiccator for at least 2 h, and weigh the bottle and contents accurately. Transfer the contents to a clean, dry 50-mL beaker and weigh the empty bottle to obtain the exact weight of the sodium carbonate.

Transfer the sodium carbonate to the titration flask by pouring it through a powder funnel. Rinse the beaker and funnel thoroughly with carbon dioxide-free water, using several small portions of the water to rinse the funnel and sides of the flask. Dilute the solution with carbon dioxide-free water to 90 mL, swirl to dissolve the sample, stopper the flask, and bubble carbon dioxide-free air through the solution during all subsequent operations. Boil the solution for 15 min, cool to room temperature in an ice bath, and dilute with carbon dioxide-free water to 75–80 mL. Titrate with 0.02 N sodium hydroxide, using a combination electrode (glass–calomel or glass–silver, silver chloride) for the measurement of E in millivolts or pH. The end point of the titration is determined by the second derivative method (page 52). Calculate the % Na_2CO_3 from the following formula:

$$A = 25.949F / B$$

where

A = % Na_2CO_3 in the sample
B = weight, in grams, of the Na_2CO_3 sample
F = calculation factor

The calculation factor, F, is obtained as follows:

$$F = (C \times D) - 0.20422EG$$

where

C = weight, in grams, of the potassium hydrogen phthalate
D = assay value of the potassium hydrogen phthalate
E = normality of the sodium hydroxide solution, as determined against NIST standard potassium hydrogen phthalate, using the method in the NIST assay certificate
G = volume, in milliliters, of sodium hydroxide consumed

CALCIUM AND MAGNESIUM. (By flame AAS, page 39).

Sample Stock Solution. Dissolve 5.0 g of sample with water in a 100-mL volumetric flask and dilute to the mark with water (1 mL = 0.05 g).

Element	Wavelength (nm)	Sample Wt (g)	Standard Added (mg)	Flame Type*	Background Correction
Ca	422.7	0.10	0.02; 0.04	N/A	No
Mg	285.2	0.10	0.004; 0.008	A/A	Yes

*A/A is air/acetylene; N/A is nitrous oxide/acetylene.

Sodium Carbonate Monohydrate

$Na_2CO_3 \cdot H_2O$ **Formula Wt 124.00**

CAS Number 5968–11–6

REQUIREMENTS

Assay. . . . . ≥99.5% $Na_2CO_3 \cdot H_2O$
Loss on drying at 150 °C 13.0–15.0%

MAXIMUM ALLOWABLE

Insoluble matter.	0.01%
Chloride (Cl)	0.001%
Nitrogen compounds (as N).	0.001%
Phosphate (PO$_4$)	5 ppm
Silica (SiO$_4$)	0.005%
Sulfur compounds (as SO$_4$)	0.004%
Ammonium hydroxide precipitate	0.01%
Heavy metals (as Pb)	5 ppm
Iron (Fe).	5 ppm
Calcium (Ca)	0.03%
Magnesium (Mg)	0.005%
Potassium (K).	0.005%

TESTS

ASSAY. (By acid–base titrimetry of carbonate). Weigh, to the nearest 0.1 mg, 2.5 g of sample. Transfer to a 125-mL glass-stoppered flask, dissolve with 50 mL of water, add 0.10 mL of methyl orange indicator solution, and titrate with 1 N hydrochloric acid.

$$\% \, Na_2CO_3 \cdot H_2O = (mL \times N \, HCl \times 6.201) \, / \, Sample \, wt \, (g)$$

LOSS ON DRYING AT 150 °C. Weigh accurately about 1 g and dry at 150 °C to constant weight.

INSOLUBLE MATTER. (Page 15). Use 10 g dissolved in 100 mL of water.

CHLORIDE. (Page 27). Use 1.0 g of sample and 2 mL of nitric acid.

NITROGEN COMPOUNDS. (Page 31). Use 1.0 g. For the standard, use 0.01 mg of nitrogen (N).

PHOSPHATE. (Page 32, Method 2). Dissolve 2.0 g in 50 mL of water in a platinum dish and digest on a hot plate ($\approx$100 °C) for 30 min. Cool, neutralize with dilute sulfuric acid (1 + 19) to a pH of about 4, and dilute with water to about 75 mL. Add 0.5 g of ammonium molybdate and adjust the pH to 1.8 (using a pH meter) with dilute hydrochloric acid (1 + 9). Heat to boiling, cool, add 10 mL of hydrochloric acid, and dilute with water to 100 mL. Continue as described, and concurrently prepare a standard containing 0.01 mg of phosphate ion (PO$_4$) and 0.10 mg of silica (SiO$_2$) in about 75 mL of water treated as the 75 mL of sample solution. Reserve the aqueous phase for the determination of silica.

SILICA. Add 10 mL of hydrochloric acid to the solutions reserved from the determination of phosphate and transfer to separatory funnels. Add 40 mL of butyl alcohol, shake vigorously, and allow to separate. Draw off and discard the aqueous phase. Wash the butyl alcohol three times with 20-mL portions of dilute hydrochloric acid (1 + 99), discarding the washings each time. Dilute each butyl alcohol solution with butyl alcohol to 50 mL, take 10 mL from each, and dilute each to 50 mL with butyl alcohol. Add 0.5 mL of a freshly prepared 2% solution of stannous chloride dihydrate in hydrochloric acid. The blue color in the extract from the sample should not exceed that in the standard. If the butyl alcohol extracts are turbid, wash with 10 mL of dilute hydrochloric acid (1 + 99).

SULFUR COMPOUNDS. Dissolve 2.0 g of sample in 20 mL of water, evaporate to 5 mL, add 1 mL of bromine water, and evaporate to dryness. Cool. Dissolve in 10 mL of 10% hydrochloric acid and evaporate to dryness. Cool. Add 5 mL of 10% hydrochloric acid and evaporate to dryness. Cool. Dissolve with 10 mL of water and evaporate to dryness. Cool. Dissolve in 10 mL of water and adjust the pH to approximately pH 2 with 10% hydrochloric acid or ammonium hydroxide (1 + 3). Filter through a washed, fine-porosity filter paper, wash with two 2-mL portions of water, and dilute with water to 20 mL.

To prepare 0.06 mg of sulfate ion (SO_4) standard, add 1 mL of (1 + 19) hydrochloric acid and dilute with water to 20 mL.

To sample and standard solutions, add 1 mL of barium chloride reagent solution and allow to stand 30 min. Any turbidity in the sample solution must not exceed that of the standard solution.

AMMONIUM HYDROXIDE PRECIPITATE. Dissolve 10 g in 100 mL of water and filter if necessary. Cautiously add 30 mL of cooled dilute sulfuric acid (1 + 1). Evaporate until dense fumes of sulfur trioxide appear. Cool, dissolve the residue in 130 mL of hot water, and add ammonium hydroxide until the solution is just alkaline to methyl red. Heat to boiling, boil gently for 5 min, and filter, reserving the filtrate for the test for calcium and magnesium precipitate. Wash the precipitate with hot water, rejecting the washings, and ignite. The residue includes some, but not all, of the silica in the sample.

HEAVY METALS. (Page 28, Method 1). To 5.0 g in a 150-mL beaker add 10 mL of water, mix, and cautiously add 10 mL of hydrochloric acid. Evaporate to dryness on a hot plate ($\approx$100 °C), dissolve the residue in about 20 mL of water, and dilute with water to 25 mL. For the control, add 0.02 mg of lead ion (Pb) to 1.0 g of sample, and treat exactly as the 5.0 g of sample.

IRON. (Page 30, Method 1). Dissolve 2.0 g in 50 mL of dilute hydrochloric acid (1 + 9). Use the solution without further acidification.

CALCIUM, MAGNESIUM, AND POTASSIUM. (By flame AAS, page 39).

Sample Stock Solution. Dissolve 5.0 g of sample with water in a 100-mL volumetric flask, and dilute to the mark with water (1 mL = 0.05 g).

Element	Wavelength (nm)	Sample Wt (g)	Standard Added (mg)	Flame Type*	Background Correction
Ca	422.7	0.10	0.03; 0.06	N/A	No
Mg	285.2	0.10	0.005; 0.01	A/A	Yes
K	766.5	0.20	0.01; 0.02	A/A	No

*A/A is air/acetylene; N/A is nitrous oxide/acetylene.

Sodium Chlorate

NaClO$_3$ **Formula Wt 106.44**

CAS Number 7775–09–9

REQUIREMENTS

Assay . . . . ≥99.0% NaClO$_3$

MAXIMUM ALLOWABLE

Insoluble matter. 0.005%
Bromate (BrO$_3$) 0.015%
Chloride (Cl) 0.005%
Nitrogen compounds (as N). 0.001%
Sulfate (SO$_4$) 0.001%
Heavy metals (as Pb) 0.001%
Iron (Fe) . . . . 5 ppm
Calcium (Ca) 0.002%
Magnesium (Mg) 0.002%
Potassium (K). 0.01%

TESTS

ASSAY. (By argentimetric titration of chloride after reduction). Weigh accurately 0.5 g of sample, transfer to a glass-stoppered conical flask, and dissolve in 100 mL of water. Add 25 mL of fresh sulfurous acid (5% SO$_2$) and 5 mL of 25% nitric acid and heat for 30 min to remove the excess SO$_2$. Cool, add 50.0 mL of 0.1 N silver nitrate solution and 10 mL of toluene, and shake vigorously. Titrate the excess silver nitrate with 0.1 N ammonium thiocyanate, using ferric ammonium sulfate indicator solution. One milliliter of 0.1 N silver nitrate corresponds to 0.01064 g of NaClO$_3$.

INSOLUBLE MATTER. (Page 15). Use 20 g dissolved in 250 mL of water.

BROMATE. (By differential pulse polarography, page 50). Use 1.0 g of sample and 0.5 g of calcium chloride dihydrate, CaCl$_2$ · 2H$_2$O, in 25 mL of solution. For the standard, add 0.15 mg of bromate ion (BrO$_3$). Also prepare a reagent blank.

CHLORIDE. (Page 27). Use 0.2 g. Use only nitric acid that is free from lower oxides of nitrogen in the test.

NITROGEN COMPOUNDS. (Page 31). Use 1.0 g. For the standard, use 0.01 mg of nitrogen (N).

SULFATE. (Page 33, Method 1). Use 5.0 g.

HEAVY METALS. (Page 28, Method 1). Dissolve 2.5 g in 20 mL of dilute hydrochloric acid (1 + 1). Evaporate the solution to dryness on a hot plate (≈100 °C), add 5 mL more of dilute hydrochloric acid (1 + 1), and again evaporate to dryness. Dissolve the residue in about 20 mL of water and dilute with water to 25 mL. For the standard–control solution, add 0.02 mg of lead ion (Pb) to 0.5 g of sample and treat exactly as the 2.5-g sample.

IRON. (Page 30, Method 1). Dissolve 2.0 g in 20 mL of dilute hydrochloric acid (1 + 1) and evaporate to dryness on a hot plate (≈100 °C). Add 5 mL of dilute hydrochloric acid (1 + 1) and again evaporate to dryness. Dissolve the residue in 50 mL of dilute hydrochloric acid (1 + 25) and use the solution without further acidification. In preparing the standard, use the residue from evaporation of 15 mL of hydrochloric acid.

CALCIUM, MAGNESIUM, AND POTASSIUM. (By flame AAS, page 39).

Sample Stock Solution. Dissolve 5.0 g in 25 mL of dilute hydrochloric acid (1 + 3) and digest in a covered beaker on a hot plate (≈100 °C) until the reaction ceases. Uncover the beaker and evaporate to dryness. Add 10 mL of dilute hydrochloric acid (1 + 3) and again evaporate to dryness. Dissolve in dilute hydrochloric acid (1 + 99) and dilute to 100 mL with dilute hydrochloric acid (1 + 99) (1 mL = 0.05 g).

Element	Wavelength (nm)	Sample Wt (g)	Standard Added (mg)	Flame Type*	Background Correction
Ca	422.7	1.0	0.02; 0.04	N/A	No
Mg	285.2	0.20	0.004; 0.008	A/A	Yes
K	766.5	0.20	0.01; 0.02	A/A	No

*A/A is air/acetylene; N/A is nitrous oxide/acetylene.

Sodium Chloride

NaCl   **Formula Wt 58.44**

CAS Number 7647–14–5

REQUIREMENTS

Assay. . . . . ≥99.0% NaCl
pH of a 5% solution . . . . 5.0–9.0 at 25 °C

MAXIMUM ALLOWABLE

Insoluble matter. 0.005%
Iodide (I) 0.002%
Bromide (Br) 0.01%

Chlorate and nitrate (as NO_3). 0.003%
Nitrogen compounds (as N). 0.001%
Phosphate (PO_4) . . . . 5 ppm
Sulfate (SO_4) 0.004%
Barium (Ba) . . . . Passes test
Heavy metals (as Pb) . . . . 5 ppm
Iron (Fe). . . . . 2 ppm
Calcium (Ca) 0.002%
Magnesium (Mg) 0.001%
Potassium (K). 0.005%

TESTS

ASSAY. (By argentimetric titration of chloride content). Weigh accurately about 0.25 g and dissolve with 50 mL of water in a 250-mL glass-stoppered flask. Add, while agitating, exactly 50.0 mL of 0.1 N silver nitrate solution, then add 3 mL of nitric acid and 10 mL of benzyl alcohol. Shake vigorously, add 2 mL of ferric ammonium sulfate reagent solution (as indicator), and titrate the excess silver nitrate with 0.1 N ammonium thiocyanate solution from a 10-mL buret.

$$\% \text{ NaCl} = \frac{[(\text{mL} \times \text{N AgNO}_3) - (\text{mL} \times \text{N NH}_4\text{SCN})] \times 5.844}{\text{Sample wt (g)}}$$

pH OF A 5% SOLUTION. (Page 44). The pH should be 5.0–9.0 at 25 °C.

INSOLUBLE MATTER. (Page 15). Use 20 g dissolved in 200 mL of water.

IODIDE. Dissolve 11 g in 50 mL of water. Prepare a control by dissolving 1 g of the sample, 0.2 mg of iodide ion (I), and 1.0 mg of bromide ion (Br) in 50 mL of water. To each solution, in a separatory funnel, add 2 mL of hydrochloric acid and 5 mL of ferric chloride reagent solution. Allow to stand for 5 min. Add 10 mL of carbon tetrachloride, shake for 1 min, allow the carbon tetrachloride to settle, and draw it off. Reserve the water solution for the test for bromide. Any violet color in the carbon tetrachloride extract from the solution of the sample should not exceed that in the extract from the control.

BROMIDE. Treat both the solution of the sample and the control, obtained in the test for iodide, as follows: wash twice by shaking with 10-mL portions of carbon tetrachloride. Each time allow to settle, then draw off and discard the carbon tetrachloride. To each of the water solutions add 10 mL of water, 65 mL of cool, dilute sulfuric acid (1 + 1), and 15 mL of a solution of chromic acid prepared by dissolving 10 g of chromium trioxide (CrO_3) in 100 mL of dilute sulfuric acid (1 + 3). Allow to stand for 5 min. Add 10 mL of carbon tetrachloride, shake for 1 min, allow to settle, and draw off. (Half of a 7-cm piece of filter paper rolled and placed in the stem of the separatory funnel will absorb any of the aqueous solution that may pass the stopcock and thus assure a clear extract.) Any yellow-brown color in the carbon tetrachloride extract from the solution of the sample should not exceed that in the extract from the control.

CHLORATE AND NITRATE.

Sample Solution A. Dissolve 0.50 g in 3 mL of water by heating in a boiling-water bath. Dilute to 50 mL with brucine sulfate reagent solution.

Control Solution B. Dissolve 0.50 g in 1.5 mL of water and 1.5 mL of the standard nitrate solution containing 0.01 mg of nitrate ion (NO_3) per mL by heating in a boiling-water bath. Dilute to 50 mL with brucine sulfate reagent solution.

Continue with the procedure described on page 30, starting with the preparation of blank solution C.

NITROGEN COMPOUNDS. (Page 31). Use 1.0 g. For the standard, use 0.01 mg of nitrogen (N).

PHOSPHATE. (Page 32, Method 1). Dissolve 4.0 g in 25 mL of approximately 0.5 N sulfuric acid and continue as described.

SULFATE. Dissolve 3.0 g of the sample in 20 mL of water, and add 1.0 mL of 10% hydrochloric acid. For the control, take 1.0 g of the sample, dissolve in 20 mL of water, and add 0.08 mg of sulfate ion (SO_4) and 1.0 mL of 10% hydrochloric acid. Dilute both solutions to 35 mL and add 1 mL of barium chloride solution to each. Sample solution turbidity after 10 min should not exceed that of the control solution.

BARIUM. Dissolve 4.0 g in 20 mL of water, filter if necessary, and divide into two equal portions. To one portion add 2 mL of 10% sulfuric acid reagent solution and to the other 2 mL of water. The solutions should be equally clear at the end of 2 h.

HEAVY METALS. (Page 28, Method 1). Dissolve 6.0 g in about 20 mL of water, and dilute with water to 30 mL. Use 25 mL to prepare the sample solution and use the remaining 5.0 mL to prepare the control solution.

IRON. (Page 30, Method 1). Use 5.0 g.

CALCIUM, MAGNESIUM, AND POTASSIUM. (By flame AAS, page 39).

Sample Stock Solution. Dissolve 10.0 g of sample in 75 mL of water, transfer to a 100-mL volumetric flask, and dilute to the mark with water (1 mL = 0.10 g).

Element	Wavelength (nm)	Sample Wt (g)	Standard Added (mg)	Flame Type*	Background Correction
Ca	422.7	1.0	0.02; 0.04	N/A	No
Mg	285.2	1.0	0.01; 0.02	A/A	Yes
K	766.5	1.0	0.025; 0.05	A/A	No

*A/A is air/acetylene; N/A is nitrous oxide/acetylene.

Sodium Citrate Dihydrate

2-Hydroxy-1,2,3-propanetricarboxylic Acid, Trisodium Salt, Dihydrate

$Na_3C_6H_5O_7 \cdot 2H_2O$ Formula Wt 294.10

CAS Number 6132–04–3

REQUIREMENTS

Assay . . . . ≥99.0% $Na_3C_6H_5O_7 \cdot 2H_2O$
pH of a 5% solution . . . . 7.0–9.0 at 25 °C

MAXIMUM ALLOWABLE

Insoluble matter 0.005%
Chloride (Cl) 0.003%
Sulfate (SO_4) 0.005%
Ammonia (NH_3) 0.003%
Calcium (Ca) 0.005%
Heavy metals (as Pb) . . . . 5 ppm
Iron (Fe) . . . . 5 ppm

TESTS

ASSAY. (Total alkalinity by nonaqueous titration). Weigh accurately 0.35 g and transfer to a 250-mL beaker. Dissolve in 100 mL of glacial acetic acid, add 5 mL of acetic anyhdride, stir until dissolution is complete, and titrate with 0.1 N perchloric acid in glacial acetic acid, determining the end point potentiometrically. Correct for a reagent blank. One milliliter of 0.1 N perchloric acid corresponds to 0.009803 g of $Na_3C_6H_5O_7 \cdot 2H_2O$.

pH OF A 5% SOLUTION. (Page 44). The pH should be 7.0–9.0 at 25 °C.

INSOLUBLE MATTER. (Page 15). Use 20.0 g dissolved in 200 mL of water.

CHLORIDE. (Page 27). Use 0.33 g.

SULFATE. Drive off any moisture by heating a sample in a dish on a hot plate. Then carefully ignite 1 g in an electric muffle furnace until nearly free of carbon. Boil the residue with 10 mL of water and 0.5 mL of 30% hydrogen peroxide for 5 min. Add 5 mg of sodium carbonate and 1 mL of hydrochloric acid and evaporate on a hot plate (≈100 °C) to dryness. Dissolve the residue in 4 mL of hot water, to which has been added 1 mL of dilute hydrochloric acid (1 + 19), filter through a small filter, wash with 2-mL portions of water, and dilute the filtrate to 10 mL. Add 1 mL of barium chloride reagent solution and mix well. Any turbidity produced should not be greater than that in a standard prepared as described below and to which the barium chloride solution is added at the same time as it is added to the sample solution. To 10 mL of water, add 5 mg of sodium carbonate, 1 mL of hydrochloric acid, 0.5 mL of 30% hydrogen peroxide, and 0.05 mg of sulfate (SO_4) and evaporate on a hot plate ($\approx$100 °C) to dryness. Dissolve the residue in 9 mL of water and add 1 mL of dilute hydrochloric acid (1 + 19). If necessary, adjust with water to the same volume as the sample solution, add 1 mL of barium chloride reagent solution, and mix well.

AMMONIA. Dissolve 1 g in 50 mL of ammonia-free water and add 2 mL of Nessler reagent. Any color should not exceed that produced by 0.03 mg of ammonia (NH_3) in an equal volume of solution containing 2 mL of Nessler reagent.

CALCIUM. (By flame AAS, page 39).

> **Sample Stock Solution.** Dissolve 10.0 g of sample with water in a 100-mL volumetric flask, add 5 mL of hydrochloric acid, and dilute to the mark with water (1 mL = 0.10 g).

Element	Wavelength (nm)	Sample Wt (g)	Standard Added (mg)	Flame Type*	Background Correction
Ca	422.7	1.0	0.05; 0.10	N/A	No

*N/A is nitrous oxide/acetylene.

HEAVY METALS. (Page 28, Method 1). Dissolve 6.0 g in water, add 7.5 mL of dilute hydrochloric acid (1 + 1), and dilute with water to 42 mL. Use 35 mL to prepare the sample solution and use the remaining 7.0 mL to prepare the control solution.

IRON. (Page 30, Method 1). Use 2.0 g.

Sodium Cobaltinitrite
Trisodium Hexakis(nitrito-*N*)cobaltate(3–)
Sodium Hexanitritocobaltate(III)
(for determination of potassium)

$Na_3Co(NO_2)_6$ **Formula Wt 403.94**

CAS Number 14649–73–1

REQUIREMENTS

Insoluble matter. . . . . ≤0.02%
Suitability for determination of potassium . . . . Passes test

TESTS

INSOLUBLE MATTER. (Page 15). Use 5.0 g dissolved in 25 mL of dilute acetic acid (1 + 25), and allow to stand in a covered beaker overnight.

SUITABILITY FOR DETERMINATION OF POTASSIUM. Dissolve 1.583 g of potassium chloride in water and dilute with water to 500 mL. To 10.0 mL of this solution (20 mg of K_2O) in a 50-mL beaker, add 2 mL of 1 N nitric acid and 8 mL of water. Dissolve 5 g of the sodium cobaltinitrite in water, dilute with water to 25 mL, and filter. Cool both solutions to approximately 20 °C and add 10 mL of the cobaltinitrite solution to the potassium chloride solution. Allow to stand for 2 h and filter through a sintered-glass crucible that has been washed with alcohol, dried at 105 °C, and weighed. Wash the precipitate with cobaltinitrite wash solution, finally wash with 5–10 mL of alcohol, and dry at 105 °C. The weight of the potassium sodium cobaltinitrite precipitate $[K_2NaCo(NO_2)_6 \cdot H_2O]$ should be between 0.0945 and 0.0985 g.

> *Cobaltinitrite Wash Solution.* (0.01 N nitric acid saturated with potassium sodium cobaltinitrite.) Transfer 75–100 mg of potassium sodium cobaltinitrite to a 125-mL glass-stoppered flask. Add 50 mL of 0.01 N nitric acid, shake on a mechanical shaker for 1 h, and filter through a fine-porosity sintered-glass funnel. The filtrate is cobaltinitrite wash solution.

> *NOTE.* If no potassium sodium cobaltinitrite is available, a sufficient amount may be prepared by following the suitability for determination of potassium test, but substituting 0.01 N nitric acid in place of the cobaltinitrite wash solution. When it is possible to follow the suitability procedure as written, the analytical precipitate of potassium sodium cobaltinitrite so obtained may be used to prepare the wash solution.

Sodium Cyanide

NaCN	Formula Wt 49.01

CAS Number 143–33–9

REQUIREMENTS

Assay . . . . ≥95.0% NaCN

MAXIMUM ALLOWABLE

Chloride (Cl) 0.15%
Phosphate (PO_4) 0.02%
Sulfate (SO_4) 0.05%
Sulfide (S) 0.005%
Thiocyanate (SCN) 0.02%
Iron, total (as Fe) 0.005%
Lead (Pb) . . . . 5 ppm

TESTS

ASSAY. (By argentimetric titration of cyanide content). Weigh accurately about 0.4 g and dissolve in 30 mL of water. Add 0.20 mL of 10% potassium iodide reagent solution and 1 mL of ammonium hydroxide and titrate with 0.1 N silver nitrate until a slight permanent yellowish turbidity forms. One milliliter of 0.1 N silver nitrate corresponds to 0.009802 g of NaCN.

Sample Solution A for the Determination of Chloride, Sulfate, Sulfide, Thiocyanate, and Iron. Dissolve 10.0 g in water, filter if necessary through a chloride-free filter, and dilute with water to 200 mL (1 mL = 0.05 g).

CHLORIDE. Dilute 1.0 mL of sample solution A (0.05-g sample) with water to 50 mL. To 10 mL of this solution add 5 mL of 30% hydrogen peroxide and cover the beaker until the reaction ceases. Digest in the covered beaker on a hot plate ($\approx$100 °C) for about one-half hour, cool, and neutralize with nitric acid. Dilute with water to 25 mL and add 1 mL of nitric acid and 1 mL of silver nitrate reagent solution. The turbidity should not exceed that produced by 0.015 mg of chloride ion (Cl) in an equal volume of solution containing the quantities of reagents used in the test.

PHOSPHATE. (Page 32, Method 1). Dissolve 1.0 g in 5 mL of water, add 2 mL of hydrochloric acid, and evaporate to dryness in a well-ventilated hood. Add 5 mL of dilute hydrochloric acid (1 + 1) and evaporate to dryness again. Dissolve in 50 mL of approximately 0.5 N sulfuric acid. To 5.0 mL of the solution add 20 mL of approximately 0.5 N sulfuric acid and continue as described.

SULFATE. (Page 33, Method 2). Use 2 mL of sample solution A plus 1 mL of hydrochloric acid in a well-ventilated hood.

SULFIDE. To 10 mL of sample solution A (measured in a graduated cylinder) (0.5-g sample) add 10 mL of water and 0.15 mL of alkaline lead solution (made by adding 10% sodium hydroxide reagent solution to a 10% lead acetate reagent solution until the precipitate first formed is redissolved). The color should not be darker than that produced by 0.025 mg of sulfide ion (S) in an equal volume of solution when treated with 0.15 mL of the alkaline lead solution.

THIOCYANATE. To 20 mL of sample solution A (measured in a graduated cylinder) (1-g sample) add 4 mL of hydrochloric acid and 0.20 mL of ferric chloride reagent solution. At the end of 5 min the solution should show no reddish tint when compared with 20 mL of water to which have been added the quantities of hydrochloric acid and ferric chloride used in the test.

IRON, TOTAL. (Page 30, Method 1). Transfer 4.0 mL of sample solution A (0.2-g sample) to a platinum dish, add 3 mL of hydrochloric acid, and evaporate to dryness in a well-ventilated hood. Heat the residue at 650 °C for 30 min, cool, and add 2 mL of hydrochloric acid and 10 mL of water. Cover with a watch glass, digest on a hot plate ($\approx$100 °C) until dissolution is complete, and use the solution without further acidification.

LEAD. Dissolve 0.50 g in 10 mL of water in a separatory funnel. Add 5 mL of ammonium citrate reagent solution (lead-free), 2 mL of hydroxylamine hydrochloride reagent solution for the dithizone test, and 0.10 mL of phenol red indicator solution, and make the solution alkaline if necessary by adding ammonium hydroxide. Add 5 mL of dithizone standard solution in chloroform, shake gently but well for 1 min, and allow the layers to separate. The intensity of the red color of the chloroform layer should be no greater than that of a control made with 0.002 mg of lead and 0.1 g of the sample dissolved in 10 mL of water and treated exactly like the solution of 0.50 g of sample in 10 mL of water.

Sodium Dichromate Dihydrate

$Na_2Cr_2O_7 \cdot 2H_2O$ Formula Wt 298.00

CAS Number 7789–12–0

REQUIREMENTS

Assay . . . . 99.5–100.5% $Na_2Cr_2O_7 \cdot 2H_2O$

MAXIMUM ALLOWABLE

Insoluble matter.	0.005%
Ammonium hydroxide precipitate	0.005%
Chloride (Cl)	0.005%
Sulfate (SO_4)	0.01%
Calcium (Ca)	0.003%
Aluminum (Al)	0.002%

TESTS

ASSAY. (By titration of oxidizing power). Weigh accurately about 0.2 g and dissolve in 200 mL of water. Add 7 mL of hydrochloric acid and 3 g of potassium iodide, mix, and allow to stand in the dark for 10 min. Titrate the liberated iodine with 0.1 N sodium thiosulfate, using starch indicator near the end point, to a greenish blue color. One milliliter of 0.1 N sodium thiosulfate corresponds to 0.004967 g of $Na_2Cr_2O_7 \cdot 2H_2O$.

INSOLUBLE MATTER. (Page 15). Dissolve 20.0 g in 200 mL of water.

AMMONIUM HYDROXIDE PRECIPITATE. Dissolve 20.0 g in 200 mL of water, add 2 mL of ammonium hydroxide, heat to boiling, and digest in a covered beaker on a hot plate ($\approx$100 °C) for 1 h. Filter, wash thoroughly, and ignite.

CHLORIDE. Dissolve 0.2 g in 10 mL of water, filter if necessary through a chloride-free filter, and add 1 mL of ammonium hydroxide and 1 mL of silver nitrate reagent solution. Prepare a standard containing 0.01 mg of chloride ion (Cl) in 10 mL of water and add 1 mL of ammonium hydroxide and 1 mL of silver nitrate reagent solution. Add 2 mL of nitric acid to each. The comparison is best made by the general method for chloride in colored solutions, page 27.

SULFATE. Dissolve 10 g in 250 mL of water, filter if necessary, and heat to boiling. Add 25 mL of a solution containing 1 g of barium chloride and 2 mL of hydrochloric acid in 100 mL of solution. Digest in a covered beaker on a hot plate ($\approx$100 °C) for 2 h and allow to stand overnight. If a precipitate forms, filter, wash thoroughly, and ignite. Fuse the residue with 1 g of sodium carbonate. Extract the fused mass with water and filter off the insoluble residue. Add 5 mL of hydrochloric acid to the filtrate, dilute with water to about 200 mL, heat to boiling, and add 10 mL of ethyl alcohol. Digest in a covered beaker on the hot plate until the reduction of chromate is complete, as indicated by the change to a clear green or colorless solution. Neutralize the solution with ammonium hydroxide and add 2 mL of hydrochloric acid. Heat to boiling, add 10 mL of barium chloride reagent solution, digest in a covered beaker on a hot plate ($\approx$100 °C) for 2 h, and allow to stand overnight. Filter, wash thoroughly, and ignite. Correct for the weight obtained in a complete blank test. If the original precipitate of barium sulfate weighs less than the requirement permits, the fusion with sodium carbonate is not necessary.

CALCIUM. (By flame AAS, page 39).

Sample Stock Solution. Dissolve 5.0 g in 50 mL of water, transfer to a 100-mL volumetric flask, and dilute to the mark with water (1 mL = 0.05 g).

Element	Wavelength (nm)	Sample Wt (g)	Standard Added (mg)	Flame Type*	Background Correction
Ca	422.7	1.0	0.03; 0.06	N/A	No

*N/A is nitrous oxide/acetylene.

ALUMINUM. Dissolve 20.0 g in 140 mL of water, filter, and add 5 mL of glacial acetic acid to the filtrate. Make alkaline with ammonium hydroxide. Digest for 2 h on a hot plate ($\approx$100 °C), filter, wash, ignite, and weigh.

Sodium Diethyldithiocarbamate Trihydrate

$(CH_3CH_2)_2NCSSNa \cdot 3H_2O$ **Formula Wt 225.31**

CAS Number 20624–25–3

REQUIREMENTS

Solubility in water . . . . Passes test

Sodium (as Na_2SO_4). 30.5%–32.5%

Sensitivity to copper. . . . . Passes test

TESTS

SOLUBILITY IN WATER. Dissolve 1 g in 50 mL of water. There should be no undissolved residue and the solution should be substantially clear. Retain the solution for the test for sensitivity.

> *NOTE.* Because of inherent instability, this reagent may be expected to react with oxygen of the atmosphere to form insoluble oxidation products. After storage for some time the reagent may fail to meet the requirement for solubility in water.

SODIUM. Weigh accurately 2 g in a tared dish or crucible. Moisten the sample with concentrated sulfuric acid and ignite cautiously to char the material. Finally, ignite to constant weight at 800 ± 25 °C.

SENSITIVITY TO COPPER.

> *Solution A.* Dilute 5 mL of the solution retained from the test for solubility in water to 100 mL with water.

For the sample, prepare a solution containing 0.002 mg of copper ion (Cu) in 100 mL of dilute ammonium hydroxide (1 + 99). For the control, use 100 mL of dilute ammonium hydroxide (1 + 99). Transfer the solutions to separatory funnels and add to each 10 mL of solution A and 10 mL of isopentyl alcohol. Shake for about 1 min and allow to separate (about 30 min). The isopentyl alcohol separated from the solution containing the 0.002 mg of copper should show a distinct yellow color when compared with the isopentyl alcohol from the control.

Sodium Dodecyl Sulfate

SDS; Sodium Lauryl Sulfate

$CH_3(CH_2)_{11}SO_4Na$ **Formula Wt 288.38**

CAS Number 151–21–3

REQUIREMENTS

Assay (acidimetric, on dried basis) . . . .≥99.0% $C_{12}H_{25}NaO_4S$

Fatty alcohols (as $C_{12}H_{25}OH$) . . . .≥96.0%

MAXIMUM ALLOWABLE

Absorbance (3% w/v) . . . . 0.1 AU

Loss on drying at 105 °C 1.0%

Titrable base . . . . 0.06 meq/g
Heavy metals (as Pb) 0.002%
Sodium chloride and sodium sulfate . . . . 8.0% total
Unsulfated alcohols 4.0%

TESTS

ASSAY. Accurately weigh 5.0 g, transfer to a 500-mL round-bottom flask equipped with a ground-glass joint. Add 25.0 mL of 1 N sulfuric acid, and attach a condenser. Using a heating mantle and a magnetic stirrer, heat cautiously and gently at about 100 °C until foaming ceases (about 4 h). Boil under reflux for at least 3–4 h or more after foaming ceases. Cool and wash the condenser with 30 mL of ethanol and then 30 mL of water. Titrate with 1 N sodium hydroxide by using 0.15 mL phenolphthalein as the indicator. Perform a complete blank determination. One milliliter of 1 N sodium hydroxide corresponds to 0.2884 g $C_{12}H_{25}NaO_4S$. Correct the assay value for loss on drying.

ABSORBANCE. Determine the absorbance of 3.0% w/v aqueous solution in 10-mm silica or quartz cuvettes over the range 220 to 350 nm against water as the reference. The absorbance does not exceed 0.1.

LOSS ON DRYING AT 105 °C. Weigh accurately about 2 g, and dry to constant weight.

TITRABLE BASE. Dissolve 1.0 g in 100 mL of water, add 0.15 mL phenol red indicator, and titrate with 0.1 N hydrochloric acid. Not more than 0.60 mL is required for neutralization.

HEAVY METALS. (Page 28, Method 2). Use 1.0 g.

SODIUM CHLORIDE. Accurately weigh about 5 g, and dissolve in about 50 mL of water. Neutralize the solution with 10% nitric acid using litmus paper as the indicator, add 2 mL of potassium chromate reagent solution, and titrate with 0.1 N silver nitrate. Each mL of 0.1 N silver nitrate is equivalent to 5.844 mg of NaCl.

SODIUM SULFATE.

Lead nitrate solution. Dissolve 3.31 g of lead nitrate in water to make 100 mL.

Transfer about 1 g of sodium dodecyl sulfate, accurately weighed, to a 250-mL beaker, add 35 mL of water, and warm to dissolve. To the warm solution, add 2.0 mL of 1 N nitric acid, mix, and add 50 mL of alcohol. Heat the solution to boiling, and slowly add 20 mL of lead nitrate solution with stirring. Cover the beaker, simmer for 5 min, and let settle. If the supernatant liquid is hazy, let stand for 10 min, heat to boiling, and decant as much liquid as possible through a 9-cm filter paper (Whatman No. 41 or equivalent). Wash four times by decantation, each time using 50 mL of 50% alcohol, and bring the mixture to a boil. Finally, transfer the filter paper to the original beaker, and immediately add 30 mL of water, 20.0 mL of 0.05 M EDTA, and 2 mL of ammoniacal buffer (pH 10). Warm to dissolve the precipitate, add 0.3 mL of eriochrome black T, and titrate with 0.05 M zinc sulfate. Each milliliter of 0.05 M EDTA is equivalent to 7.102 mg of Na_2SO_4.

UNSULFATED ALCOHOLS. Dissolve about 10 g, accurately weighed, in 100 mL of water, and add 100 mL of alcohol. Transfer the solution to a separator, and extract with three 50-mL portions of solvent hexane. If an emulsion forms, sodium chloride may be added to promote separation of the two layers. Wash the combined solvent hexane extracts with three 50-mL portions of water, and dry with anhydrous sodium sulfate. Filter the solvent hexane extract into a tared beaker, evaporate on a steam bath until the hexane is completely volatized, dry the residue at 105 °C for 30 min, cool, and weigh. The weight of the residue is not more than 4.0% of the weight of the sodium dodecyl sulfate taken.

FATTY ALCOHOLS.

Sample Preparation. Using a 150-mL glass-stoppered conical flask, warm 25 mL of distilled water on a hot plate ($\approx$100 °C). Add approximately 2 g of sample and dissolve. Add 5 mL of hydrochloric acid, connect a cold finger condenser, and let reflux for 2 h. When hydrolysis is complete and the molten fatty alcohol has separated as a completely clear liquid, cool the flask to 0 °C by using an ice bath. (*NOTE*: This should be done with minimum disturbance to the separated fatty alcohol. It is desirable to solidify the fatty alcohol in one piece without increasing the surface area or allowing occluded water to be introduced.) Prepare a funnel with a suitable filter paper, and transfer the fatty alcohol and solution to the funnel. Wash the solid fatty alcohol with cold water. The fatty alcohol may be broken up and transferred to the filter. However, this should only be done after the first washing. Using a dry filter paper, blot the fatty alcohol dry.

Chromatographic Conditions

Column: Type I, methyl silicone

Column Temperature: 80 °C to 230 °C, programmed at 10 °C/min, 5 min hold at 230 °C

Injector Temperature: 220 °C

Detector Temperature: 250 °C

Sample Size: 0.2 μL

Carrier Gas: Helium at 6 mL/min

Detector: Flame ionization

Standard Solution. Accurately weigh 0.10 g of pure dodecyl alcohol, and transfer to a 50-mL volumetric flask. Dissolve and dilute to volume with hexanes. Mix well and stopper.

Sample Solution. Accurately weigh 0.10 g of the dried fatty alcohol from the sample preparation, and transfer to a 50 mL-volumetric flask. Dissolve and dilute to volume with hexanes. Mix well and stopper.

Consecutively inject the standard and sample solutions into a prepared gas chromatograph. Record the chromatogram for the duration of the temperature programmed run. Electronically integrate the area under the peaks of interest. Determine the amount of dodecyl alcohol (C_{12}) in the sample by a conventional area normalization calculation. Correct for assay content of dodecyl alcohol.

Sodium Fluoride

NaF   **Formula Wt 41.99**

CAS Number 7681–49–4

REQUIREMENTS

Assay. . . . . ≥99% NaF

MAXIMUM ALLOWABLE

Insoluble matter. 0.02%
Loss on drying at 150 °C 0.3%
Chloride (Cl) 0.005%
Titrable acid. . . . . 0.03 meq/g
Titrable base . . . . 0.01 meq/g
Sodium fluosilicate (Na_2SiF_6) 0.1%
Sulfate (SO_4) 0.03%
Sulfite (SO_2). 0.005%
Heavy metals (as Pb) 0.003%
Iron (Fe). 0.003%
Potassium (K). 0.02%

TESTS

ASSAY. (By indirect acid–base titrimetry). Weigh, to the nearest 0.1 mg, 1.6 g of sample into a 100-mL volumetric flask. Dissolve and dilute to the mark with water. Dilute 10.0 mL of the diluted sample with water to 50 mL. Pass the solution through a cation-exchange column (see page 56) at a rate of about 5 mL/min, and collect the eluate in a 500-mL titration flask. Then wash the resin in the column with water at a rate of about 10 mL/min and collect in the titration flask. Add 0.15 mL of phenolphthalein to the flask, and titrate with 0.1 N sodium hydroxide. Continue the titration, as the washing proceeds, until 50 mL of eluate requires no further titration.

$$\% \text{ NaF} = (\text{mL NaOH} \times \text{N NaOH} \times 4.119) / \text{Sample wt (g)}$$

NOTE. All water used in the assay must be carbon dioxide- and ammonia-free.

INSOLUBLE MATTER. (Page 15). Use 3.5 g dissolved in 100 mL of warm water in a platinum dish.

LOSS ON DRYING AT 150 °C. Weigh accurately about 1 g in a tared platinum dish or crucible and dry to constant weight at 150 °C.

CHLORIDE. (Page 27). Dissolve 1.0 g plus 1 g of boric acid in 80 mL of water, filter if necessary through a chloride-free filter, and dilute with water to 100 mL. Use 20 mL of this solution.

TITRABLE ACID. Dissolve 2.0 g in 50 mL of water (free from carbon dioxide) in a platinum dish, add 10 mL of a saturated solution of potassium nitrate, cool the solution to 0 °C, and add 0.15 mL of phenolphthalein indicator solution. If a pink color is produced, omit the titration for free acid and reserve the solution for the test for titrable base. If no pink color is produced, titrate with 0.01 N sodium hydroxide until the pink color persists for 15 s while the temperature of the solution is near 0 °C. Not more than 6.0 mL of the 0.01 N sodium hydroxide should be required. Reserve the solution for the test for sodium fluosilicate.

TITRABLE BASE. If a pink color was produced in the solution prepared in the test for titrable acid, add 0.01 N hydrochloric acid, stirring the liquid only gently, until the pink color is discharged. Not more than 2.0 mL of the acid should be required. Reserve the solution for the test for sodium fluosilicate.

SODIUM FLUOSILICATE. Heat the solution reserved from one of the preceding tests to boiling and titrate while hot with 0.01 N sodium hydroxide until a permanent pink color is obtained. Not more than 4.3 mL of the sodium hydroxide should be required.

SULFATE. (Page 33, Method 2). Evaporate 0.17 g with 2 mL of hydrochloric acid in a platinum or Teflon dish. Repeat the evaporation four more times.

SULFITE. Dissolve 8.0 g in 150 mL of water, and add 2 mL of hydrochloric acid and 0.25 mL of starch indicator solution. Titrate immediately with 0.01 N iodine. Not more than 1.0 mL of the 0.01 N iodine should be required to produce a blue color.

HEAVY METALS. (Page 28, Method 1). Dissolve 1.0 g in about 25 mL of water, and dilute with water to 30 mL. Use 25 mL to prepare the sample solution, and use the remaining 5.0 mL to prepare the control solution. Use pH paper to adjust the pH.

IRON. (Page 30, Method 1). To 1.0 g in a platinum dish add 10 mL of hydrochloric acid, and evaporate to dryness. Dissolve the residue in 10 mL of hydrochloric acid, and repeat the evaporation to dryness. Warm the residue with a few drops of hydrochloric acid, dilute with water to 30 mL, and use 10 mL. In preparing the standard use the residue from evaporation of 7 mL of hydrochloric acid.

POTASSIUM. (By flame AAS, page 39).

Sample Stock Solution. Dissolve 1.0 g of sample with water in a 100-mL volumetric flask, and dilute to the mark with water (1 mL = 0.01 g).

Element	Wavelength (nm)	Sample Wt (g)	Standard Added (mg)	Flame Type*	Background Correction
K	766.5	0.10	0.02; 0.04	A/A	No

*A/A is air/acetylene.

Sodium Formate

HCOONa   **Formula Wt 68.01**

CAS Number 141–53–7

REQUIREMENTS

Assay . . . .≥99.0% HCOONa

MAXIMUM ALLOWABLE

Insoluble matter. 0.005%
Chloride (Cl) 0.001%
Sulfate (SO_4) 0.001%
Calcium (Ca) 0.005%
Iron (Fe). . . . . 5 ppm
Heavy metals (as Pb) . . . . 5 ppm

TESTS

ASSAY. (By titration of reducing power). Weigh accurately about 0.88 g, transfer to a 250-mL volumetric flask (or 0.70 g to a 200-mL flask), dissolve with 50 mL of water, dilute with water to volume, and mix thoroughly. To 25.0 mL of this solution, in a 250-mL glass-stoppered flask, add 3 mL of 10% sodium hydroxide reagent solution and 50.0 mL of 0.1 N potassium permanganate. Heat on a hot plate (≈100 °C) for 20 min, then cool completely to room temperature. Add 4 g of potassium iodide crystals, mix, and add 5 mL of 25% sulfuric acid. Titrate the liberated iodine, representing the excess permanganate, with 0.1 N sodium thiosulfate, using 5 mL of starch indicator solution near the end point. Run a blank on 25 mL of water with the same quantities of permanganate and other reagents, in the same manner as the sample. One milliliter of 0.1 N potassium permanganate consumed corresponds to 0.003401 g of HCOONa.

INSOLUBLE MATTER. (Page 15). Use 20 g dissolved in 200 mL of water.

CHLORIDE. (Page 27). Use 1.0 g.

SULFATE. (Page 33, Method 1). Use 5.0 g of sample. Allow 30 min for the turbidity to form.

CALCIUM AND IRON. (By flame AAS, page 39).

> **Sample Stock Solution.** Ignite 20.0 g of sample in a platinum crucible at 600 °C. Dissolve the residue in 10 mL of 20% hydrochloric acid, dilute with water to 50 mL, and filter. Dilute to the mark with water in a 100-mL volumetric flask (1 mL = 0.20 g).

Element	Wavelength (nm)	Sample Wt (g)	Standard Added (mg)	Flame Type*	Background Correction
Ca	422.7	0.40	0.02; 0.04	N/A	No
Fe	248.3	4.0	0.02; 0.04	A/A	Yes

*A/A is air/acetylene; N/A is nitrous oxide/acetylene.

HEAVY METALS. (Page 28, Method 1). Substitute glacial acetic acid for 1 N acetic acid to adjust pH in Method 1. Dissolve 6.0 g in water, and dilute with water to 30 mL. Use 25 mL to prepare the sample solution, and use the remaining 5.0 mL to prepare the control solution.

Sodium Hydrogen Sulfate, Fused

CAS Number 7681-38-1

> *NOTE.* This product is usually a mixture of sodium pyrosulfate, $Na_2S_2O_7$, and sodium hydrogen sulfate, $NaHSO_4$.

REQUIREMENTS

Acidity (as H_2SO_4) 39.0–42.0%

MAXIMUM ALLOWABLE

Insoluble matter. 0.01%
Ammonium hydroxide precipitate 0.01%
Chloride (Cl) 0.001%
Phosphate (PO_4) 0.001%

Heavy metals (as Pb) . . . . 5 ppm
Iron (Fe) 0.002%
Calcium (Ca) 0.002%
Magnesium (Mg) 0.001%

TESTS

ACIDITY. Weigh accurately 4.0 g, dissolve in 50 mL of water, and titrate with 1 N sodium hydroxide, using methyl orange as indicator. One milliliter of 1 N sodium hydroxide corresponds to 0.04904 g of H_2SO_4.

INSOLUBLE MATTER. (Page 15). Dissolve 20.0 g in 200 mL of water.

AMMONIUM HYDROXIDE PRECIPITATE. Dissolve 20.0 g in 200 mL of water, filter, add ammonium hydroxide until the solution is alkaline to methyl red, boil for 1 min, and digest in a covered beaker on a hot plate ($\approx$100 °C) for 1 h. Filter through a tared filtering crucible. Wash thoroughly and dry at 105 °C.

CHLORIDE. (Page 27). Use 1.0 g.

PHOSPHATE. Dissolve 2.0 g in 15 mL of dilute ammonium hydroxide (1 + 2) and evaporate to dryness. Dissolve the residue in 25 mL of 0.5 N hydrochloric acid, add 1 mL of ammonium molybdate–nitric acid reagent solution and 1 mL of 4-(methylamino)phenol sulfate reagent solution, and allow to stand for 2 h at room temperature. Any blue color should not exceed that produced by 0.02 mg of phosphate ion (PO_4) in an equal volume of solution containing the quantities of reagents used in the test, including the residue from evaporation of 5 mL of ammonium hydroxide.

HEAVY METALS. (Page 28, Method 1). Dissolve 6.0 g in about 25 mL of water, and dilute with water to 30 mL. Use 25 mL to prepare the sample solution, and use the remaining 5.0 mL to prepare the control solution.

IRON. (Page 30, Method 1). Dissolve 5.0 g in 80 mL of dilute hydrochloric acid (1 + 3), and boil gently for 10 min. Cool, dilute with water to 100 mL, and use 10 mL.

CALCIUM AND MAGNESIUM. (By flame AAS, page 39).

Sample Stock Solution. Dissolve 10.0 g of sample in a 100-mL volumetric flask and dilute to mark with water (1 mL = 0.10 g).

Element	Wavelength (nm)	Sample Wt (g)	Standard Added (mg)	Flame Type*	Background Correction
Ca	422.7	1.0	0.02; 0.04	N/A	No
Mg	285.2	1.0	0.01; 0.02	A/A	Yes

*A/A is air/acetylene; N/A is nitrous oxide/acetylene.

Sodium Hydroxide

NaOH   **Formula Wt 40.00**

CAS Number 1310–73–2

REQUIREMENTS

Assay . . . . ≥97.0% NaOH
Sodium Carbonate . . . . ≤1.0% Na_2CO_3

MAXIMUM ALLOWABLE

Chloride (Cl) 0.005%
Nitrogen compounds (as N). 0.001%
Phosphate (PO_4) 0.001%
Sulfate (SO_4) 0.003%
Ammonium hydroxide precipitate 0.02%
Heavy metals (as Ag) 0.002%
Iron (Fe) 0.001%
Mercury (Hg) . . . . 0.1 ppm
Nickel (Ni) 0.001%
Potassium (K). 0.02%

TESTS

GENERAL. Special care must be taken in sampling to obtain a representative sample and to avoid absorption of water and carbon dioxide by the sample taken.

ASSAY AND SODIUM CARBONATE. (By acid–base titrimetry). Weigh accurately 35–40 g, dissolve and dilute to 1 L, using carbon dioxide-free water. Dilute 50.0 mL of this solution with carbon dioxide-free water to 200 mL. Add 5 mL of barium chloride reagent solution, stopper, and allow to stand for 5 min. Add 0.15 mL of phenolphthalein and titrate with 1 N hydrochloric acid to the end point to determine the hydroxide. One milliliter of 1 N hydrochloric acid corresponds to 0.04000 g of NaOH. Add 0.15 mL of methyl orange and continue the titration with 1 N hydrochloric acid to the end point to determine the carbonate. One milliliter of 1 N hydrochloric acid corresponds to 0.05300 g of Na_2CO_3.

> ***Sample Solution A for the Determination of Chloride, Nitrogen Compounds, Phosphate, Heavy Metals, Iron, and Nickel.*** Dissolve 50.0 g in carbon dioxide- and ammonia-free water, cool, and dilute with the water to 500 mL (1 mL = 0.10 g).

CHLORIDE. (Page 27). Use 2.0 mL of sample solution A (0.2-g sample) diluted with water to 20 mL.

NITROGEN COMPOUNDS. (Page 31). Use 20 mL of sample solution A. For the control, use 10 mL of sample solution A and 0.01 mg of nitrogen diluted to 20 mL.

PHOSPHATE. (Page 32, Method 1). To 20 mL of sample solution A (2-g sample) add 5 mL of hydrochloric acid, and evaporate to dryness on a hot plate ($\approx$100 °C). Dissolve the residue in 25 mL of approximately 0.5 N sulfuric acid and continue as described.

SULFATE. Dissolve 3.0 g of sample in 75 mL of water in a 400-mL beaker. Cautiously neutralize with (1 + 1) hydrochloric acid to litmus paper. Add 1 mL of excess hydrochloric acid. Evaporate to dryness. Add 20 mL of water and re-evaporate to dryness. Dissolve the residue in 20 mL of water and filter through a small, washed filter paper. Add two 3-mL portions of water through filter. For the control, take 0.10 mg of sulfate ion (SO_4) in 25 mL of water. To both sample and control, add 1 mL of (1 + 19) hydrochloric acid and dilute each to 35 mL. Add 1 mL of barium chloride solution to each. Turbidity in the sample solution after 10 min should not exceed that of the control solution.

AMMONIUM HYDROXIDE PRECIPITATE. Weigh 10.0 g and dissolve in about 100 mL of water. Cautiously add 15 mL of sulfuric acid to 15 mL of water, cool, add the mixture to the solution of the sample, and evaporate to strong fuming. Cool, dissolve the residue in 130 mL of hot water, and add ammonium hydroxide until the solution is just alkaline to methyl red. Heat to boiling, filter, wash with hot water, and ignite. The residue includes some, but not all, of the silica in the sample.

HEAVY METALS. To 30 mL of sample solution A (3.0-g sample) cautiously add 10 mL of nitric acid. For the control, add 0.05 mg of silver to 5.0 mL of sample solution A and cautiously add 10 mL of nitric acid. Evaporate both solutions to dryness over a low flame or on an electric hot plate. Dissolve each residue in about 20 mL of water, filter if necessary through a chloride-free filter, and dilute with water to 25 mL. Adjust the pH of the control and sample solutions to between 3 and 4 (using a pH meter) with 1 N acetic acid or ammonium hydroxide (10% NH_3), dilute with water to 40 mL, and mix. Add 10 mL of freshly prepared hydrogen sulfide water to each and mix. Any yellowish brown color in the solution of the sample should not exceed that in the control.

IRON. (Page 30, Method 1). Neutralize 10 mL of sample solution A (1.0-g sample) with hydrochloric acid, using 0.15 mL of phenolphthalein indicator, add 2 mL in excess, dilute with water to 50 mL, and use this solution. In preparing the standard use the residue from evaporation of the quantity of acid used to neutralize the sample solution.

MERCURY. To each of two 250-mL conical flasks add 20 mL of water and 1 mL of 4% potassium permanganate solution. For the sample, add to one flask 5.5 g of the sodium hydroxide. For the control, add to the other flask 0.5 g of sample and 0.5 µg of mercury (1.0 mL of a solution prepared freshly by diluting 1.0 mL of standard mercury solution, page 95, with water to 100 mL). To each flask add slowly, with constant swirling, 18 mL of hydrochloric acid. Heat to boiling, allow to cool, and dilute with water to 100 mL. Determine mercury in 10-mL aliquots by the CVAAS method (page 40), using 1 mL of 10% hydroxylamine hydrochloride solution and 2 mL of 10% stannous chloride solution for the reduction. The peak obtained for the sample should not exceed that obtained for the control.

NICKEL. Dilute 20 mL of sample solution A (2.0-g sample) to 50 mL with water and neutralize with hydrochloric acid. Dilute with water to 85 mL and adjust the pH to 8 with ammonium hydroxide. Add 5 mL of bromine water, 5 mL of 1% dimethylglyoxime solution in alcohol, and 5 mL of 10% sodium hydroxide reagent solution. Any red color should not exceed that produced by 0.02 mg of nickel in an equal volume of solution containing the quantities of reagents used in the test.

POTASSIUM. (By flame AAS, page 39).

Sample Stock Solution. Dissolve 2.0 g of sample in 50 mL of water in a 100-mL volumetric flask. Neutralize with nitric acid, add 2 mL in excess, and dilute to the mark with water (1 mL = 0.02 g).

Element	Wavelength (nm)	Sample Wt (g)	Standard Added (mg)	Flame Type*	Background Correction
K	766.5	0.10	0.02; 0.04	A/A	No

*A/A is air/acetylene.

Sodium Iodide

NaI   **Formula Wt 149.89**

CAS Number 7681–82–5

REQUIREMENTS

Assay . . . . ≥99.5% NaI
pH of a 5% solution . . . . 6.0–9.0 at 25 °C

MAXIMUM ALLOWABLE

Insoluble matter. 0.01%
Chloride and bromide (as Cl) 0.01%
Iodate (IO_3) . . . . 3 ppm
Nitrogen compounds (as N). 0.002%
Phosphate (PO_4) 0.001%

Sulfate (SO_4) 0.005%
Barium (Ba) 0.002%
Heavy metals (as Pb) . . . . 5 ppm
Iron (Fe) . . . . 5 ppm
Calcium (Ca) 0.002%
Magnesium (Mg) 0.001%
Potassium (K) 0.01%

TESTS

ASSAY. (By titration of iodide content). Weigh accurately 0.5 g and dissolve in 20 mL of water in a glass-stoppered conical flask. Add 30 mL of hydrochloric acid and 5 mL of chloroform, cool if necessary, and titrate with standard 0.05 M potassium iodate until the iodine color disappears from the aqueous layer. Shake vigorously for 30 s and continue the titration, shaking vigorously after each addition, until the iodine color in the chloroform is discharged. One milliliter of 0.05 M potassium iodate corresponds to 0.01499 g of NaI.

pH OF A 5% SOLUTION. (Page 44). The pH should be 6.0–9.0 at 25 °C.

INSOLUBLE MATTER. (Page 15). Use 20 g dissolved in 200 mL of water.

CHLORIDE AND BROMIDE. Dissolve 1.0 g in 100 mL of water in a distillation flask. Add 1 mL of hydrogen peroxide (30%) and 1 mL of phosphoric acid, heat to boiling, and boil gently until all the iodine is expelled and the solution is colorless. Cool, wash down the sides of the flask with water, and add 0.5 mL of hydrogen peroxide. If an iodine color develops, boil until the solution is colorless and for 10 min longer. If no color develops, boil for 10 min, filter if necessary through a chloride-free filter, and dilute with water to 100 mL. Dilute 10 mL with water to 23 mL and add 1 mL of nitric acid and 1 mL of silver nitrate reagent solution. Any turbidity should not exceed that produced by 0.01 mg of chloride ion (Cl) in an equal volume of solution containing the quantities of nitric acid and silver nitrate used in the test.

IODATE. (By differential pulse polarography, page 50). Use 10.0 g of sample in 25 mL of solution. For the standard, add 0.03 mg of iodate (IO_3).

NITROGEN COMPOUNDS. (Page 31). Use 1.0 g. For the standard, use 0.02 mg of nitrogen (N).

PHOSPHATE. (Page 32, Method 1). Dissolve 2.0 g in water, and add 10 mL of nitric acid and 5 mL of hydrochloric acid. Evaporate to dryness on a hot plate ($\approx$100 °C). Dissolve the residue in 25 mL of approximately 0.5 N sulfuric acid and continue as described. For the standard, include the residue from evaporation of the quantities of acids used with the sample.

SULFATE. (Page 33, Method 1). Use 1.0 g. Any yellow iodine color that develops can be discharged by the addition of a few milligrams of ascorbic acid.

BARIUM. Dissolve 3.0 g in 10 mL of dilute hydrochloric acid (1 + 1) and add 5 mL of nitric acid. For the control, dissolve 0.5 g of sample and 0.05 mg of barium in 10 mL of dilute hydrochloric acid (1 + 1) and add 5 mL of nitric acid. Evaporate both solutions to dryness. Dissolve the residues in 10 mL of dilute hydrochloric acid (1 + 1), add 5 mL of nitric acid, and again evaporate to dryness. Dissolve each residue in water and dilute with water to 23 mL. To each solution add 2 mL of potassium dichromate reagent solution and 10% ammonium hydroxide reagent solution until the orange color is just dissipated and the yellow color persists. To each solution add with constant stirring 25 mL of methanol. Any turbidity in the solution of the sample should not exceed that in the control.

HEAVY METALS. (Page 28, Method 1). Dissolve 6.0 g in water, and dilute with water to 30 mL. Use 25 mL to prepare the sample solution, and use the remaining 5.0 mL to prepare the control solution.

IRON. (Page 30, Method 2). Use 2.0 g.

CALCIUM, MAGNESIUM, AND POTASSIUM. (By flame AAS, page 39).

> *Sample Stock Solution.* Dissolve 50 g in about 70 mL of water and dilute to 100 mL with water (1 mL = 0.50 g).

Element	Wavelength (nm)	Sample Wt (g)	Standard Added (mg)	Flame Type*	Background Correction
Ca	422.7	1.00	0.02; 0.04	N/A	No
Mg	285.2	1.00	0.01; 0.02	A/A	Yes
K	766.5	0.20	0.01; 0.02	A/A	No

*A/A is air/acetylene; N/A is nitrous oxide/acetylene.

Sodium Metabisulfite
Disodium Disulfite

$Na_2S_2O_5$ **Formula Wt 190.11**

CAS Number 7681–57–4

REQUIREMENTS

Assay . . . .≥97.0% $Na_2S_2O_5$

MAXIMUM ALLOWABLE

Insoluble matter. 0.005%
Chloride (Cl) 0.05%
Thiosulfate (S_2O_3) 0.05%
Heavy metals (as Pb) 0.001%
Iron (Fe). 0.002%

TESTS

ASSAY. (By titration of reductive capacity). Weigh accurately 0.45 g and add to a mixture of 100.0 mL of 0.1 N iodine and 5 mL of 10% hydrochloric acid solution. Swirl gently until the sample is dissolved completely. Titrate the excess of iodine with 0.1 N sodium thiosulfate, adding 3 mL of starch indicator solution near the end of the titration. One milliliter of 0.1 N iodine corresponds to 0.004753 g of $Na_2S_2O_5$.

INSOLUBLE MATTER. (Page 15). Use 20 g dissolved in 200 mL of water.

CHLORIDE. Dissolve 1.0 g in 10 mL of water, filter if necessary through a small chloride-free filter, and add 6 mL of 30% hydrogen peroxide. Add 1 N sodium hydroxide until the solution is slightly alkaline to phenolphthalein and dilute with water to 100 mL. Dilute 2.0 mL of this solution with water to 20 mL, and add 1 mL of nitric acid and 1 mL of silver nitrate reagent solution. Any turbidity should not exceed that produced by 0.01 mg of chloride ion (Cl) in an equal volume of solution containing the quantities of reagents used in the test.

THIOSULFATE. Dissolve 2.2 g in 10 mL of 10% hydrochloric acid solution in a 50-mL beaker. Boil gently for 5 min, cool, and transfer to a small test tube. Any turbidity should not be greater than that produced by a standard of 1.0 mL of 0.01 N sodium thiosulfate solution carried through the entire procedure.

HEAVY METALS. (Page 28, Method 1). Dissolve 3.0 g in 25 mL of dilute hydrochloric acid (2 + 3), and evaporate to dryness on a hot plate ($\approx$100 °C). Add 5 mL of hydrochloric acid, and again evaporate to dryness. Dissolve the residue in 20 mL of water, and dilute with water to 25 mL. For the control, add 0.02 mg of lead to 1.0 g of sample, and treat exactly as the 3.0 g of sample.

IRON. (Page 30, Method 1). Dissolve 0.50 g in 14 mL of dilute hydrochloric acid (2 + 5), and evaporate on a hot plate ($\approx$100 °C) to dryness. Dissolve the residue in 7 mL of dilute hydrochloric acid (2 + 5), and again evaporate to dryness. Dissolve the residue in 2 mL of hydrochloric acid, dilute with water to 50 mL, and use the solution without further acidification.

Sodium Methoxide, 0.5 M Methanolic Solution
Sodium Methylate

CH_3ONa Formula Wt 54.02

CAS Number 124–41–4

REQUIREMENTS

Assay . . . . 0.48–0.52 mole of CH_3ONa
per liter of solution (25.9–28.1 g/L)
Clarity of solution. . . . . Passes test

TESTS

ASSAY. (By acid–base titrimetry). Pipet 5.0 mL of the sample into a conical flask, and slowly add 100 mL of water, with stirring. Add 0.10 mL of phenolphthalein indicator solution and titrate with 0.1 N hydrochloric acid to a colorless end point. One milliliter of 0.1 N hydrochloric acid corresponds to 0.005402 g of CH_3ONa.

CLARITY OF SOLUTION. Place 50 mL in a 50-mL color comparison tube. The sample has not more than a trace of turbidity or insoluble matter.

Sodium Molybdate Dihydrate

$Na_2MoO_4 \cdot 2H_2O$ Formula Wt 241.95

CAS Number 10102–40–6

REQUIREMENTS

Assay . . . . 99.5–103.0% $Na_2MoO_4 \cdot 2H_2O$
pH of 5% solution . . . . 7.0–10.5 at 25 °C

MAXIMUM ALLOWABLE

Insoluble matter. 0.005%
Chloride (Cl) 0.005%
Phosphate (PO_4) . . . . 5 ppm
Sulfate (SO_4) 0.015%
Ammonium (NH_4) 0.001%
Heavy metals (as Pb) . . . . 5 ppm
Iron (Fe) 0.001%

TESTS

ASSAY. (By oxidative titration after reduction of Mo^{VI}). Weigh accurately 0.3 g and dissolve in 10 mL of water in a 150-mL beaker. Activate the zinc amalgam of a Jones reductor by passing 100 mL of 1 N sulfuric acid through the column. Discard this acid, and place 25 mL of ferric ammonium sulfate solution in the receiver under the column. To the sample solution add 100 mL of 1 N sulfuric acid, and pass this mixture through the reductor, followed by 100 mL of 1 N sulfuric acid and then 100 mL of water (using these to rinse the beaker). Add 5 mL of phosphoric acid to the solution in the receiver, and titrate with 0.1 N potassium permanganate. Run a blank and make any necessary correction. One milliliter of 0.1 N potassium permanganate corresponds to 0.008066 g of $Na_2MoO_4 \cdot 2H_2O$.

> **Ferric Ammonium Sulfate Solution.** Dissolve 50 g of ferric ammonium sulfate dodecahydrate, $FeNH_4(SO_4)_2 \cdot 12H_2O$, in 500 mL of solution containing 25 mL of sulfuric acid.

pH OF A 5% SOLUTION. (Page 44). The pH should be 7.0–10.5 at 25 °C.

INSOLUBLE MATTER. (Page 15). Use 20 g dissolved in 200 mL of water.

CHLORIDE. Dissolve 1.0 g in 50 mL of water and mix. Pipet 10 mL of the solution into a color-comparison tube, add 2 mL of nitric acid, dilute with water to 25 mL, and add 1 mL of silver nitrate reagent solution. Any turbidity should not exceed that produced by 0.01 mg of chloride ion (Cl) in an equal volume of solution containing the quantities of reagents used in the test.

PHOSPHATE. (Page 32, Method 3).

SULFATE. (Page 33, Method 2). Dissolve in a mixture of 5 mL of water and 5 mL of nitric acid. Evaporate to dryness on a hot plate. Digest the residue with a mixture of 1 mL of hydrochloric acid and 10 mL of water. Dilute to 20 mL and filter through two pre-washed fine-porosity papers. Add barium chloride to the clear filtrate and use Method 1.

AMMONIUM. Dissolve 1.0 g in 20 mL of water, add 10 mL of 10% sodium hydroxide solution, and dilute with water to 50 mL. For the standard, add 0.01 mg of ammonium ion (NH_4) to 20 mL of water, add 10 mL of 10% sodium hydroxide solution, and dilute with water to 50 mL. To each solution add 2 mL of Nessler reagent. Any color produced in the sample solution should not exceed that in the standard.

HEAVY METALS. Dissolve 6.0 g in 35 mL of 10% sodium hydroxide solution, add 5 mL of ammonium hydroxide, and dilute with water to 42 mL. For the control, add 0.02 mg of lead to 7.0 mL of the solution and dilute with water to 40 mL. For the sample dilute the remaining 35-mL portion with water to 40 mL. Add 10 mL of freshly prepared hydrogen sulfide water to each and mix. Any color in the solution of the sample should not exceed that in the control.

IRON. (By flame AAS, page 39).

> **Sample Stock Solution.** Dissolve 10.0 g in water and dilute with water to 100 mL (1 mL = 0.10 g).

Element	Wavelength (nm)	Sample Wt (g)	Standard Added (mg)	Flame Type*	Background Correction
Fe	248.3	2.50	0.05; 0.10	A/A	Yes

*A/A is air/acetylene.

Sodium Nitrate

$NaNO_3$ **Formula Wt 84.99**

CAS Number 7631–99–4

REQUIREMENTS

Assay . . . .≥99.0% $NaNO_3$
pH of a 5% solution . . . . 5.5–8.3 at 25 °C

MAXIMUM ALLOWABLE

Insoluble matter. 0.005%
Chloride. 0.001%
Iodate (IO_3) . . . . 5 ppm
Nitrite (NO_2) 0.001%
Phosphate (PO_4) . . . . 5 ppm
Sulfate (SO_4) 0.003%
Calcium (Ca) 0.005%
Magnesium (Mg) 0.002%
Heavy metals (as Pb) . . . . 5 ppm
Iron (Fe). . . . . 3 ppm

TESTS

ASSAY. (By indirect acid–base titrimetry). Weigh, to the nearest 0.1 mg, 0.3 g of sample and dissolve in 100 mL of water. Pass the solution through a cation-exchange column (*see* page 56) at a rate of about 5 mL/min, and collect the eluate in a 500-mL titration flask. Then wash the resin in the column with water at a rate of about 10 mL/min, and collect in the same titration flask. Add 0.15 mL of phenolphthalein to the flask and titrate with 0.1 N sodium hydroxide. Continue the titration, as the washing proceeds, until 50 mL of eluate requires no further titration. One milliliter of 0.1 N sodium hydroxide corresponds to 0.008499 g $NaNO_3$.

NOTE. All water used in the assay must be carbon dioxide- and ammonia-free.

pH OF A 5% SOLUTION. (Page 44). The pH should be 5.5–8.3 at 25 °C.

INSOLUBLE MATTER. (Page 15). Use 20 g dissolved in 100 mL of water.

CHLORIDE. (Page 27). Use 1.0 g.

IODATE. (By differential pulse polarography, page 50). Use 10.0 g of sample and dissolve in 25 mL. For the standard, add 0.05 mg of iodate (IO_3).

NITRITE. (page 51). Use 1.0 g of sample and 0.01 mg of nitrite.

PHOSPHATE. (Page 33, Method 1). Dissolve 4.0 g in 25 mL of approximately 0.5 N sulfuric acid and continue as described.

SULFATE. (Page 33, Method 2). Evaporate 3.0 g using 7 mL of dilute hydrochloric acid (1 + 1). Repeat the evaporation. Allow 30 min for the turbidity to form.

CALCIUM AND MAGNESIUM. (By flame AAS, page 39).

Sample Stock Solution. Dissolve 5.0 g of sample in water in a 100-mL volumetric flask, and dilute to the mark with water (1 mL = 0.05 g).

Element	Wavelength (nm)	Sample Wt (g)	Standard Added (mg)	Flame Type*	Background Correction
Ca	422.7	0.50	0.025; 0.05	N/A	No
Mg	285.2	0.50	0.01; 0.02	A/A	Yes

*A/A is air/acetylene; N/A is nitrous oxide/acetylene.

HEAVY METALS. (Page 28, Method 1). Dissolve 6.0 g in about 20 mL of water, and dilute with water to 30 mL. Use 25 mL to prepare the sample solution, and use the remaining 5.0 mL to prepare the control solution.

IRON. (Page 30, Method 1). Use 3.3 g.

Sodium Nitrite

$NaNO_2$ **Formula Wt 69.00**

CAS Number 7632–00–0

REQUIREMENTS

Assay . . . . ≥97.0% $NaNO_2$

MAXIMUM ALLOWABLE

Insoluble matter. 0.01%
Chloride (Cl) 0.005%
Sulfate (SO_4) 0.01%
Calcium (Ca) 0.01%
Potassium (K). 0.005%
Heavy metals (as Pb) 0.001%
Iron (Fe). 0.001%

TESTS

Sample Solution A for the Assay and the Determination of Chloride, Sulfate, Heavy Metals, and Iron. Weigh accurately 10.0 g, dissolve in water, and dilute with water to 100 mL in a volumetric flask (1 mL = 0.1 g).

ASSAY. (By indirect titration of reductive capacity). Dilute a 10.0-mL aliquot of sample solution A to 100 mL in a volumetric flask. Add 5 mL of sulfuric acid to 300 mL of water, and while the solution is still warm, add 0.1 N potassium permanganate until a faint pink color that persists for 2 min is produced. Then add 40.0 mL of 0.1 N potassium permanganate and mix gently. Add, slowly and with constant agitation, 10.0 mL of the diluted sample solution, holding the tip of the pipet well below the surface of the liquid. Add 15.0 mL of 0.1 N ferrous ammonium sulfate, allow the solution to stand for 5 min, and titrate the excess of ferrous ammonium sulfate with 0.1 N potassium permanganate. Each milliliter of 0.1 N potassium permanganate consumed by the sodium nitrite corresponds to 0.003450 g of $NaNO_2$.

INSOLUBLE MATTER. (Page 15). Use 10 g dissolved in 100 mL of water.

CHLORIDE. (Page 27). To 2.0 mL of sample solution A (0.2-g sample) add 10 mL of water, slowly add 1 mL of glacial acetic acid, and boil gently for 5 min.

SULFATE. (Page 33, Method 2). Use 5.0 mL of sample solution A plus 1 mL of hydrochloric acid.

CALCIUM AND POTASSIUM. (By flame AAS, page 39).

Sample Stock Solution. Dissolve 10.0 g of sample with water in a 100-mL volumetric flask, and dilute to the mark with water (1 mL = 0.10 g).

Element	Wavelength (nm)	Sample Wt (g)	Standard Added (mg)	Flame Type*	Background Correction
Ca	422.7	0.50	0.05; 0.10	N/A	No
K	766.5	0.50	0.025; 0.05	A/A	No

*A/A is air/acetylene; N/A is nitrous oxide/acetylene.

HEAVY METALS. (Page 28, Method 1). To 30 mL of sample solution A (3-g sample) add 5 mL of hydrochloric acid and evaporate to dryness on a steam bath. Add

5 mL of hydrochloric acid and again evaporate to dryness. Dissolve in about 20 mL of water and dilute with water to 25 mL. For the control, add 0.02 mg of lead to 10 mL of sample solution A (1-g sample), add 5 mL of hydrochloric acid, and treat exactly as the 30 mL of sample solution A.

IRON. (Page 30, Method 1). To 10 mL of sample solution A (1-g sample) add 5 mL of hydrochloric acid, and evaporate on a hot plate ($\approx$100 °C) to dryness. Dissolve the residue in 2 mL of hydrochloric acid plus about 20 mL of water, filter if necessary, and dilute with water to 50 mL. Use the solution without further acidification.

Sodium Nitroferricyanide Dihydrate
Disodium Pentakis(cyano-C)nitrosylferrate(2–) Dihydrate
Sodium Pentacyanonitrosoferrate(III) Dihydrate

$Na_2Fe(CN)_5NO \cdot 2H_2O$ **Formula Wt 297.95**

CAS Number 13755–38–9

REQUIREMENTS

Assay . . . . 99.0–102.0% $Na_2Fe(CN)_5NO \cdot 2H_2O$

MAXIMUM ALLOWABLE

Insoluble matter. 0.01%
Chloride (Cl) 0.02%
Sulfate (SO_4) . . . . Passes test

TESTS

ASSAY. (By argentimetric titration of ferricyanide content). Weigh accurately 1.5 g, and transfer to a 100-mL volumetric flask with about 80 mL of water. Dilute to the mark and mix after complete dissolution. To a 25-mL aliquot of this solution in a 250-mL conical flask, add 50 mL of water and 1 mL of 10% potassium chromate solution. Titrate with 0.1 N silver nitrate solution to a reddish brown color. One milliliter of 0.1 N silver nitrate corresponds to 0.01490 g of $Na_2Fe(CN)_5NO \cdot 2H_2O$.

INSOLUBLE MATTER. Dissolve 10 g in 50 mL of water at room temperature, filter promptly through a tared filtering crucible, wash thoroughly, and dry at 105 °C.

CHLORIDE. Dissolve 1.0 g in 175 mL of water, add 1.25 g of cupric sulfate crystals dissolved in 25 mL of water, mix thoroughly, and allow to stand until the precipitate has settled. Filter through a chloride-free filter, rejecting the first 50 mL of the filtrate. Dilute 10 mL of the filtrate with water to 20 mL, and add 1 mL of nitric acid and 1 mL of silver nitrate reagent solution. Any turbidity should not exceed that produced by 0.01 mg of chloride ion (Cl) in an equal volume of solution containing the quantities of reagents used in the test, with enough cupric sulfate added to match the color of the test.

SULFATE. Dissolve 5.0 g in 100 mL of water without heating, filter, and to the filtrate add 0.25 mL of glacial acetic acid and 5 mL of barium chloride reagent solution. Stir and pour into a Nessler tube for observation. No turbidity should be produced in 10 min. (Limit about 0.01%)

Sodium Oxalate
Ethanedioic Acid, Disodium Salt

$(COONa)_2$ Formula Wt 134.00

CAS Number 62–76–0

REQUIREMENTS

Assay . . . . $\geq$99.5% $(COONa)_2$

MAXIMUM ALLOWABLE

Insoluble matter. 0.005%
Loss on drying at 105 °C. 0.01%
Neutrality. . . . . Passes test
Chloride (Cl) 0.002%
Sulfate (SO_4) 0.002%
Ammonium (NH_4) 0.002%
Heavy metals (as Pb) 0.002%
Iron (Fe) 0.001%
Potassium (K). 0.005%
Substances darkened by hot sulfuric acid. . . . . Passes test

TESTS

ASSAY. (By indirect titration of reducing power of oxalate against permanganate). Weigh, to the nearest 0.1 mg, 0.25 g of the dried sample, and dissolve with 100 mL of water in a 250-mL glass-stoppered flask. Add 3 mL of sulfuric acid and then slowly add 20 mL of 0.1 N potassium permanganate solution. Heat the solution to 70 °C and complete the titration with the permanganate until a pale pink color persists for 30 s. One millliliter of 0.1 N potassium permanganate corresponds to 0.0067 grams $Na_2C_2O_4$.

$$\% \ Na_2C_2O_4 = (mL \times N \ KMnO_4 \times 6.70) \ / \ Sample \ wt \ (g)$$

INSOLUBLE MATTER. (Page 15). Use 20 g dissolved in 500 mL of water.

LOSS ON DRYING AT 105 °C. Weigh accurately 10 g in a tared dish or crucible and dry to constant weight at 105 °C.

NEUTRALITY. Dissolve 2.0 g in 200 mL of water, and add 10.0 mL of 0.01 N oxalic acid and 0.15 mL of phenolphthalein solution. Boil the solution in a flask for 10 min, passing through it a stream of carbon dioxide-free air. Cool the solution rapidly to room temperature while keeping the flow of carbon dioxide-free air passing through it. Titrate with 0.01 N sodium hydroxide. Not less than 9.2 nor more than 10.5 mL of 0.01 N sodium hydroxide should be required to match the pink color of a buffer solution containing 0.15 mL of phenolphthalein solution. The buffer solution contains in 1 L 3.1 g of boric acid (H_3BO_3), 3.8 g of potassium chloride (KCl), and 5.90 mL of 1 N sodium hydroxide.

CHLORIDE. Dissolve 2.0 g in water plus 10 mL of nitric acid, filter if necessary through a chloride-free filter, and dilute with water to 100 mL. To 25 mL of the solution add 1 mL of silver nitrate reagent solution. Any turbidity should not exceed that produced by 0.01 mg of chloride ion (Cl) in an equal volume of solution containing the quantities of reagents used in the test.

SULFATE. Mix 20.0 g with 50 mL of water, add 25 mL of nitric acid and 25 mL of 30% hydrogen peroxide, and digest in a covered beaker on a hot plate ($\approx$100 °C) until reaction ceases. Remove the cover and evaporate to dryness. Add 80 mL of dilute hydrochloric acid (1 + 1) and again evaporate to dryness. Repeat the evaporation with dilute hydrochloric acid. Dissolve the residue in 200 mL of water, add 2 mL of hydrochloric acid, and filter if necessary. Heat the filtrate to boiling, add 10 mL of barium chloride reagent solution, digest in a covered beaker on a hot plate ($\approx$100 °C) for 2 h, and allow to stand overnight. If a precipitate is formed, filter, wash thoroughly, and ignite. Correct for the weight obtained in a complete blank test.

AMMONIUM. Dissolve 1.0 g in 50 mL of ammonia-free water and add 2 mL of Nessler reagent. Any color should not exceed that produced by 0.02 mg of ammonium ion (NH_4) in an equal volume of solution containing 2 mL of Nessler reagent.

HEAVY METALS. (Page 28, Method 1). Mix 2.0 g with 2 mL of water, add 2 mL of nitric acid and 2 mL of 30% hydrogen peroxide, and digest in a covered beaker on a hot plate ($\approx$100 °C) until reaction ceases. Remove the cover and evaporate to dryness. Dissolve the residue in 2 mL of nitric acid plus 2 mL of water and again evaporate to dryness. Dissolve the residue in about 20 mL of water and dilute with water to 32 mL. Use 24 mL to prepare the sample solution. Use the remaining 8.0 mL, plus the residue from evaporating 1 mL of 30% hydrogen peroxide and 2 mL of nitric acid to dryness, to prepare the control solution.

IRON. (Page 30, Method 1). Mix 2.0 g with 2 mL of water, add 2 mL of nitric acid and 2 mL of 30% hydrogen peroxide, and digest in a covered beaker on a hot plate ($\approx$100 °C) until reaction ceases. Remove the cover and evaporate to dryness. Add 5 mL of hydrochloric acid and again evaporate to dryness. Dissolve the residue in a few milliliters of water, add 4 mL of hydrochloric acid, and dilute with water to 100 mL. Use 50 mL of the solution without further acidification. Use the residue from evaporation of 1 mL of nitric acid, 1 mL of 30% hydrogen peroxide, and 2 mL of hydrochloric acid to prepare the standard solution.

POTASSIUM. (By flame AAS, page 39).

> *Sample Stock Solution.* Dissolve 5.0 g of sample in 5 mL of nitric acid in a 200-mL volumetric flask, and dilute to the mark with water (1 mL = 0.025 g).

Element	Wavelength (nm)	Sample Wt (g)	Standard Added (mg)	Flame Type*	Background Correction
K	766.5	0.50	0.025; 0.05	A/A	No

*A/A is air/acetylene.

SUBSTANCES DARKENED BY HOT SULFURIC ACID. Heat 1.0 g in a recently ignited test tube with 10 mL of sulfuric acid until the appearance of dense fumes. The acid when cooled should have no more color than a mixture of the following composition: 0.2 mL of cobalt chloride solution (5.95 g of $CoCl_2 \cdot 6H_2O$ and 2.5 mL of hydrochloric acid in 100 mL), 0.3 mL of ferric chloride solution (4.50 g of $FeCl_3 \cdot 6H_2O$ and 2.5 mL of hydrochloric acid in 100 mL), 0.3 mL of cupric sulfate solution (6.24 g of $CuSO_4 \cdot 5H_2O$ and 2.5 mL of hydrochloric acid in 100 mL), and 9.2 mL of water.

Sodium Perchlorate

$NaClO_4$ **Formula Wt 122.44**
$NaClO_4 \cdot H_2O$ **Formula Wt 140.46**

CAS Number 7601–89–0 (anhydrous); 7791–07–3 (monohydrate)

> *NOTE.* This specification includes the anhydrous form and the monohydrate.

REQUIREMENTS

Assay . . . . 98.0–102.0% $NaClO_4$ (anhydrous);
                                        85.0–90.0% $NaClO_4$ (monohydrate)
pH of a 5% solution . . . . 6.0–8.0 at 25 °C

MAXIMUM ALLOWABLE

Insoluble matter. 0.005%
Chloride (Cl) 0.003%
Sulfate (SO_4) 0.002%
Calcium (Ca) 0.02%
Potassium (K). 0.05%
Heavy metals (as Pb) . . . . 5 ppm
Iron (Fe). . . . . 5 ppm

TESTS

ASSAY. (By argentimetric titration of chloride content after perchlorate reduction to chloride). Weigh accurately about 0.25–0.30 g of sample into a platinum crucible or dish. Add 3 g of sodium carbonate, Na_2CO_3, and heat, gently at first, then to fusion until a clear melt is obtained. Leach with minimal dilute nitric acid and boil gently to expel carbon dioxide. Cool to room temperature, add 10 mL of 30% ammonium acetate solution, and titrate with 0.1 N silver nitrate to the dichlorofluorescein end point. One milliliter of 0.1 N silver nitrate corresponds to 0.01225 g of $NaClO_4$ or 0.01405 g of $NaClO_4 \cdot H_2O$.

pH OF A 5% SOLUTION. (Page 44). The pH should be 6.0–8.0 at 25 °C.

INSOLUBLE MATTER. (Page 15). Use 5.0 g dissolved in 50 mL of water.

CHLORIDE. (Page 27). Dissolve 0.33 g in 20 mL of water.

SULFATE. (Page 33, Method 1). Dissolve 2.5 g in 5 mL of water.

CALCIUM AND POTASSIUM. (By flame AAS, page 39).

> **Sample Stock Solution.** Dissolve 1.0 g of sample with water in a 100-mL volumetric flask, add 10 mL of hydrochloric acid, and dilute to the mark with water (1 mL = 0.01 g).

Element	Wavelength (nm)	Sample Wt (g)	Standard Added (mg)	Flame Type*	Background Correction
Ca	422.7	0.10	0.02; 0.04	N/A	No
K	766.5	0.10	0.05; 0.10	A/A	No

*A/A is air/acetylene; N/A is nitrous oxide/acetylene.

HEAVY METALS. (Page 28, Method 1). Dissolve 3.0 g in water, dilute with water to 60 mL, and use 50 mL to prepare the sample solution. For the control, add 0.01 mg of lead to the remaining 10 mL and dilute with water to 50 mL.

IRON. (Page 30, Method 1). Use 2.0 g.

Sodium Periodate
Sodium Metaperiodate
Sodium Tetraoxoiodate(VII)
(suitable for determination of manganese)

NaIO$_4$ Formula Wt 213.89

CAS Number 7790–28–5

REQUIREMENTS

Assay (dried basis) . . . . 99.8–100.3% NaIO$_4$

MAXIMUM ALLOWABLE

Other halogens (as Cl) 0.02%

Manganese (Mn) . . . . 3 ppm

TESTS

ASSAY. (By indirect titration of reductive capacity). Dry a sample over phosphorus pentoxide or magnesium perchlorate for 6 h. Weigh accurately 1 g, dissolve in water, and transfer to a 500-mL volumetric flask. Dilute to volume with water and mix thoroughly. Place a 50.0-mL aliquot of this solution in a glass-stoppered flask and add 10 g of potassium iodide and 10 mL of a cooled solution of sulfuric acid (1 + 5). Stopper, swirl, allow to stand for 5 min, and add 100 mL of cold water. Titrate the liberated iodine with 0.1 N sodium thiosulfate, adding 3 mL of starch indicator solution near the end of the titration. Correct for a complete blank. One milliliter of 0.1 N sodium thiosulfate corresponds to 0.002674 g of NaIO$_4$.

OTHER HALOGENS. Dissolve 0.50 g in 25 mL of a dilute solution of sulfurous acid (3 + 2). Boil for 3 min, cool, and add 5 mL of ammonium hydroxide and 20 mL of 2.5% silver nitrate solution. Allow the precipitate to coagulate, filter through a chloride-free filter, and dilute the filtrate with water to 200 mL. To 20 mL of this solution add 1.5 mL of nitric acid. Any turbidity should not exceed that produced by 0.01 mg of chloride ion (Cl) in an equal volume of solution containing 0.5 mL of ammonium hydroxide, 1.5 mL of nitric acid, and 1 mL of silver nitrate reagent solution.

MANGANESE. Dissolve 2.5 g in 40 mL of 10% sulfuric acid reagent solution. For the control, add 0.5 g of the sample and 0.006 mg of manganese to 40 mL of 10% sulfuric acid reagent solution. Add 5 mL of nitric acid and 5 mL of phosphoric acid to each, boil gently for 10 min, and cool. Any pink color in the solution of the sample should not exceed that in the control.

Sodium Peroxide

Na_2O_2 | **Formula Wt 77.98**

CAS Number 1313–60–6

CAUTION. This reagent should be stored in airtight containers in a cool place. Avoid contact with organic materials; fire or explosion may result.

NOTE. When dissolving sodium peroxide, the material should be added slowly and in small portions to well-cooled water; when neutralizing, the acid must be added cautiously in small portions, and the solution kept cool.

REQUIREMENTS

Assay. . . . . ≥93.0% Na_2O_2

MAXIMUM ALLOWABLE

Chloride (Cl) 0.002%
Nitrogen compounds (as N). 0.003%
Phosphate (PO_4) . . . . 5 ppm
Sulfate (SO_4) 0.001%
Heavy metals (as Pb) 0.002%
Iron (Fe). 0.005%

TESTS

ASSAY. (By titration of reducing power against permanganate). Weigh accurately 0.7 g and add slowly to 400 mL of dilute sulfuric acid (1 + 99) that has been cooled to 10 °C. Dilute to volume in a 500-mL volumetric flask with dilute sulfuric acid (1 + 99), and titrate 100.0 mL with 0.1 N potassium permanganate. One milliliter of 0.1 N potassium permanganate corresponds to 0.003899 g of Na_2O_2.

CHLORIDE. Add 1.0 g in small portions to 35 mL of water, cool, and slowly add 4 mL of nitric acid. Filter if necessary through a chloride-free filter and dilute the filtrate with water to 40 mL. To 20 mL of the filtrate add 1 mL of silver nitrate reagent solution. Any turbidity should not exceed that produced by 0.01 mg of chloride ion (Cl) in an equal volume of solution containing the quantities of reagents used in the test.

NITROGEN COMPOUNDS. (Page 31). Dissolve 1.0 g in 20 mL of ammonia-free water cooled to 10 °C, add acetic acid until neutral plus an excess of 0.15 mL, and boil down to a volume of 10 mL. For the standard, use 0.03 mg of nitrogen (N).

PHOSPHATE. (Page 32, Method 1). Dissolve 4.0 g in 50 mL of water, cautiously add 10 mL of nitric acid, and evaporate to dryness on a hot plate ($\approx$100 °C). Dissolve the residue in 25 mL of approximately 0.5 N sulfuric acid, and continue as described.

SULFATE. Completely dissolve 20 g in 300 mL of cold water (10 °C), neutralize with hydrochloric acid, and note the volume used. Add an excess of 2 mL of the acid and evaporate to a volume of about 200 mL. Filter if necessary, heat the filtrate to boiling, add 10 mL of barium chloride reagent solution, digest in a covered beaker on a hot plate ($\approx$100 °C) for 2 h, and allow to stand overnight. If a precipitate is formed, filter, wash thoroughly, and ignite. Correct for the weight obtained in a complete blank test. For the blank add about 10 mg of sodium carbonate and 1 mL of bromine water to the volume of hydrochloric acid used to neutralize the sodium peroxide, and evaporate to dryness. Dissolve in 2 mL of hydrochloric acid, add 200 mL of water, boil, and filter if necessary. Add 10 mL of barium chloride reagent solution and carry along with the sample.

HEAVY METALS. (Page 28, Method 1). Add 2.0 g cautiously to 15 mL of dilute hydrochloric acid (1 + 2) cooled to 10 °C. Evaporate the solution to dryness on a hot plate ($\approx$100 °C), add 15 mL of dilute hydrochloric acid, and repeat the evaporation to dryness. Dissolve the residue in 20 mL of water, and dilute with water to 25 mL. For the control, add 0.02 mg of lead to 15 mL of dilute hydrochloric acid, cool the acid to 10 °C, and cautiously add 1.0 g of the sample, continuing with the evaporations exactly as for the 2.0 g of sample.

IRON. (Page 30, Method 1). Add 10 g, cautiously, to 75 mL of a cooled solution of dilute hydrochloric acid (1 + 2). Evaporate on a hot plate ($\approx$100 °C) to dryness, add 10 mL of hydrochloric acid, and again evaporate to dryness. Dissolve the residue in 80 mL of dilute hydrochloric acid (1 + 40), dilute with water to 100 mL, and use 2.0 mL of the solution.

Sodium Phosphate, Dibasic
Disodium Hydrogen Phosphate
(suitable for buffer solutions)

Na_2HPO_4 Formula Wt 141.96

CAS Number 7558–79–4

REQUIREMENTS

Assay . . . . $\geq$99.0% Na_2HPO_4
pH of a 5% solution . . . . 8.7–9.3 at 25 °C

MAXIMUM ALLOWABLE

Insoluble matter. 0.01%
Loss on drying at 105 °C 0.2%
Chloride (Cl) 0.002%
Nitrogen compounds (as N). 0.002%
Sulfate (SO$_4$) 0.005%
Heavy metals (as Pb) 0.001%
Iron (Fe). 0.002%

TESTS

ASSAY. (By indirect acid–base titrimetry). Weigh accurately 6.0 g and dissolve in 50 mL of water. Add exactly 50.0 mL of 1 N hydrochloric acid, and stir until the sample is completely dissolved. Titrate the excess acid, constantly stirring, with 1 N sodium hydroxide to the inflection point occurring at about pH 4 as measured with a standardized pH meter and glass electrode system. Calculate A, the volume of 1 N hydrochloric acid consumed by the sample. Continue the titration with 1 N sodium hydroxide to the inflection point occurring at about pH 8.8. Calculate B, the volume of 1 N sodium hydroxide required in the titration between the two inflection points. One milliliter of 1 N sodium hydroxide corresponds to 0.1420 g of Na_2HPO_4.

If A is equal to or less than B, then

$$\% \ Na_2HPO_4 = \frac{A \times 0.1420}{\text{Sample wt (g)}} \times 100$$

If A is greater than B, then

$$\% \ Na_2HPO_4 = \frac{(2B - A) \times 0.1420}{\text{Sample wt (g)}} \times 100$$

pH OF A 5% SOLUTION. (Page 44). Use the sample retained from the test for loss on drying at 105 °C. The pH should be 8.7–9.3.

INSOLUBLE MATTER. (Page 15). Use 10.0 g dissolved in 100 mL of water. Do not use a glass filtering crucible.

LOSS ON DRYING AT 105 °C. Weigh accurately 6.0 g and dry at 105 °C to constant weight. Save the material for the determination of pH.

CHLORIDE. (Page 27). Use 0.50 g of sample and 1.5 mL of nitric acid.

NITROGEN COMPOUNDS. (Page 31). Use 0.50 g. For the standard, use 0.01 mg of nitrogen (N).

SULFATE. Dissolve 1.0 g in 15 mL of water and add 4.0 mL of 10% hydrochloric acid (solution pH approximately 2). Filter through washed filter paper and add 10 mL of water through the same filter. For control, add 0.05 mg of sulfate ion (SO$_4$)

in 20 mL of water and add 1.0 mL of (1 + 19) hydrochloric acid. Dilute both to 35 mL with water, add 5.0 mL of 30% barium chloride solution, and mix. Observe turbidity after 10 min. Sample solution turbidity should not exceed that of the control solution.

HEAVY METALS. (Page 28, Method 1). Dissolve 4.0 g in about 20 mL of water, add 14.5 mL of 2 N hydrochloric acid, and dilute with water to 40 mL. Use 30 mL to prepare the sample solution, and use the remaining 10 mL to prepare the control solution.

IRON. (Page 30, Method 2). Dissolve 1.0 g in water, dilute with water to 20 mL, and use 10 mL of the solution.

Sodium Phosphate, Dibasic, Heptahydrate
Disodium Hydrogen Phosphate Heptahydrate

$Na_2HPO_4 \cdot 7H_2O$ Formula Wt 268.07

CAS Number 7782–85–6

REQUIREMENTS

Assay . . . . 98.0–102.0% $Na_2HPO_4 \cdot 7H_2O$
pH of a 5% solution . . . . 8.7–9.3 at 25 °C

MAXIMUM ALLOWABLE

Insoluble matter. 0.005%
Chloride (Cl) 0.001%
Nitrogen compounds (as N). 0.001%
Sulfate (SO_4) 0.005%
Heavy metals (as Pb) 0.001%
Iron (Fe) 0.001%

TESTS

ASSAY. (By indirect acid–base titrimetry). Weigh accurately 10.0 g and dissolve in 50 mL of water. Add exactly 50.0 mL of 1 N hydrochloric acid and stir until the sample is completely dissolved. Titrate the excess acid, constantly stirring, with 1 N sodium hydroxide to the inflection point occurring at about pH 4 as measured with a standardized pH meter and glass electrode system. Calculate *A*, the volume of 1 N hydrochloric acid consumed by the sample. Continue the titration with 1 N sodium hydroxide to the inflection point occurring at about pH 8.8. Calculate *B*, the volume of 1 N sodium hydroxide required in the titration between the two inflection points. One milliliter of 1 N sodium hydroxide corresponds to 0.2681 g of $Na_2HPO_4 \cdot 7H_2O$.

If A is equal to or less than B, then

$$\% \ Na_2HPO_4 \cdot 7H_2O = \frac{A \times 0.2681}{Sample \ wt \ (g)} \times 100$$

If A is greater than B, then

$$\% \ Na_2HPO_4 \cdot 7H_2O = \frac{(2B - A) \times 0.2681}{Sample \ wt \ (g)} \times 100$$

pH OF A 5% SOLUTION. (Page 44). The pH should be 8.7–9.3 at 25 °C.

INSOLUBLE MATTER. (Page 15). Use 20 g dissolved in 200 mL of water.

CHLORIDE. (Page 27). Use 1.0 g of sample and 3 mL of nitric acid.

NITROGEN COMPOUNDS. (Page 31). Use 1.0 g. For the standard, use 0.01 mg of nitrogen (N).

SULFATE. Dissolve 1.0 g in 15 mL of water and add 2.5 mL of 10% hydrochloric acid (solution approximately pH 2). Filter through washed filter paper and add 10 mL of water through the same filter. For the control, add 0.05 mg of sulfate ion (SO_4) in 20 mL of water and add 1.0 mL of (1 + 19) hydrochloric acid. Dilute both to 35 mL with water, add 5.0 mL of 30% barium chloride solution, and mix. Observe turbidity after 10 min. Sample solution turbidity should not exceed that of the control solution.

HEAVY METALS. (Page 28, Method 1). Dissolve 4.0 g in about 20 mL of water, add 8 mL of 2 N hydrochloric acid, and dilute with water to 32 mL. Use 24 mL to prepare the sample solution, and use the remaining 8.0 mL to prepare the control solution.

IRON. (Page 30, Method 2). Use 1.0 g.

Sodium Phosphate, Monobasic, Monohydrate
Sodium Dihydrogen Phosphate Monohydrate

$NaH_2PO_4 \cdot H_2O$ Formula Wt 137.99

CAS Number 10049–21–5

REQUIREMENTS

Assay. . . . . 98.0–102.0% $NaH_2PO_4 \cdot H_2O$
pH of a 5% solution . . . . 4.1–4.5 at 25 °C

MAXIMUM ALLOWABLE

Insoluble matter. 0.01%
Chloride (Cl) . . . . 5 ppm
Nitrogen compounds (as N). 0.001%
Sulfate (SO_4) 0.003%
Calcium (Ca) 0.005%
Potassium (K). 0.01%
Heavy metals (as Pb) 0.001%
Iron (Fe). 0.001%

TESTS

ASSAY. (By acid–base titrimetry). Weigh accurately 5.0 g and dissolve in 50 mL of water. Titrate with 1 N sodium hydroxide to the inflection point at about pH 8.8 as measured with a standardized glass electrode system. One milliliter of 1 N sodium hydroxide corresponds to 0.1380 g of $NaH_2PO_4 \cdot H_2O$.

pH OF A 5% SOLUTION. (Page 44). The pH should be 4.1–4.5 at 25 °C.

INSOLUBLE MATTER. (Page 15). Dissolve 10.0 g in 100 mL of water.

CHLORIDE. (Page 27). Use 2.0 g of sample and 2 mL of nitric acid.

NITROGEN COMPOUNDS. (Page 31). Use 1.0 g. For the standard, use 0.01 mg of nitrogen (N).

SULFATE. Dissolve 1.0 g in 15 mL of water and add 2.5 mL of 10% hydrochloric acid (solution pH approximately 2). Filter through washed filter paper and add 10 mL of water through the same filter. For control, add 0.03 mg of sulfate ion (SO_4) in 20 mL of water and add 1.0 mL of (1 + 19) hydrochloric acid. Dilute both to 35 mL with water, add 5.0 mL of 30% barium chloride solution, and mix. Observe turbidity after 10 min. Sample solution turbidity should not exceed that of the control solution.

CALCIUM AND POTASSIUM. (By flame AAS, page 39).

Sample Stock Solution. Dissolve 10.0 g of sample in 70 mL of water and transfer to a 100 mL volumetric flask. Add 5 mL of nitric acid and dilute to the mark with water (1 mL = 0.10 g).

Element	Wavelength (nm)	Sample Wt (g)	Standard Added (mg)	Flame Type*	Background Correction
Ca	422.7	0.40	0.02; 0.04	N/A	No
K	766.5	0.40	0.02; 0.04	N/A	No

* N/A is nitrous oxide/acetylene.

HEAVY METALS. (Page 28, Method 1). Dissolve 4.0 g in about 20 mL of water, and dilute with water to 32 mL. Use 24 mL to prepare the sample solution, and use the remaining 8.0 mL to prepare the control solution.

IRON. (Page 30, Method 2). Use 1.0 g.

Sodium Phosphate, Tribasic, Dodecahydrate

$Na_3PO_4 \cdot 12H_2O$ **Formula Wt 380.12**

CAS Number 10101–89–0

> *NOTE.* This salt, when stored under ordinary conditions, may lose some water of crystallization. Any loss in water will result in an assay of more than 100% $Na_3PO_4 \cdot 12H_2O$ but does not affect the determination of the relative amount of free alkali present.

REQUIREMENTS

Assay . . . . 98.0–102.0% $Na_3PO_4 \cdot 12H_2O$

MAXIMUM ALLOWABLE

Excess alkali (as NaOH) 2.5%
Insoluble matter. 0.01%
Chloride (Cl) 0.001%
Nitrogen compounds (as N). 0.001%
Sulfate (SO_4) 0.01%
Heavy metals (as Pb) 0.001%
Iron (Fe). 0.001%

TESTS

ASSAY. (By acid–base titrimetry). Weigh accurately 13–14 g and dissolve in 50 mL of water. Add exactly 100.0 mL of 1 N hydrochloric acid and pass a stream of carbon dioxide-free air or nitrogen, in fine bubbles, through the solution for 30 min to expel carbon dioxide. The beaker must be covered with a perforated cover to prevent loss of any of the solution by spraying. Wash down the cover and sides of the beaker. Titrate the solution with 1 N sodium hydroxide to the inflection point occurring at about pH 4 as measured with a standardized pH meter and glass electrode system. Calculate *A*, the milliliters of 1 N hydrochloric acid consumed in this titration. Protect the solution from absorbing carbon dioxide from the air, and continue the titration with 1 N sodium hydroxide to the inflection point occurring at about pH 8.8. Calculate *B*, the volume of 1 N sodium hydroxide required in the titration between the two inflection points. One milliliter of 1 N sodium hydroxide corresponds to 0.3801 g of $Na_3PO_4 \cdot 12H_2O$.

If *A* is equal to or greater than 2*B*, then

$$\% \, Na_3PO_4 \cdot 12H_2O = \frac{B \times 0.3801}{\text{Sample wt (g)}} \times 100$$

If A is less than $2B$, then

$$\% \, Na_3PO_4 \cdot 12H_2O = \frac{(A-B) \times 0.3801}{Sample \, wt \, (g)} \times 100$$

EXCESS ALKALI. Calculate the amount of excess alkali as NaOH from the titration values obtained in the test for assay. If A is equal to or less than $2B$, there is no excess alkali present when the salt is dissolved. If A is greater than $2B$, then

$$\% \, NaOH = \frac{(A-2B) \times 0.040}{Sample \, wt \, (g)} \times 100$$

INSOLUBLE MATTER. (Page 15). Dissolve 10.0 g in 100 mL of water, add 0.10 mL of methyl red indicator solution, add hydrochloric acid until the solution is slightly acid to the methyl red indicator, and continue as described.

CHLORIDE. (Page 27). Use 1.0 g of sample and 3 mL of nitric acid.

NITROGEN COMPOUNDS. (Page 31). Use 1.0 g. For the standard, use 0.01 mg of nitrogen (N).

SULFATE. Dissolve 1.0 g in 15 mL of water and add 4.0 mL of 10% hydrochloric acid (solution approximately pH 2). Filter through washed filter paper and add 10 mL of water through the same filter. For the control, add 0.10 mg of sulfate ion (SO_4) in 20 mL of water and add 1.0 mL of (1 + 19) hydrochloric acid. Dilute both to 35 mL with water, add 5.0 mL of 30% barium chloride solution, and mix. Observe turbidity after 10 min. Sample solution turbidity should not exceed that of the control solution.

HEAVY METALS. (Page 28, Method 1). Dissolve 4.0 g in about 20 mL of water, add 6 mL of 2 N hydrochloric acid, and dilute with water to 32 mL. Use 24 mL to prepare the sample solution, and use the remaining 8.0 mL to prepare the control solution.

IRON. (Page 30, Method 2). Dissolve 1.0 g in 10 mL of water, and add 3 mL of 1 N hydrochloric acid. Omit the addition of acid to the standard.

Sodium Pyrophosphate Decahydrate
Diphosphoric Acid, Tetrasodium Salt, Decahydrate

$Na_4P_2O_7 \cdot 10H_2O$ **Formula Wt 446.06**

CAS Number 13472–36–1

REQUIREMENTS

Assay . . . . 99.0–103.0% $Na_4P_2O_7 \cdot 10H_2O$
pH of a 5% solution . . . . 9.5–10.5 at 25 °C

MAXIMUM ALLOWABLE

Insoluble matter. 0.01%
Chloride (Cl) 0.002%
Sulfate (SO_4) 0.005%
Nitrogen compounds (as N). 0.001%
Heavy metals (as Pb) 0.001%
Iron (Fe). 0.001%

TESTS

ASSAY. (By acid–base titrimetry). Weigh accurately 8.0 g and dissolve in 100 mL of water. Titrate with 1 N hydrochloric acid to a green color with 0.15 mL of bromphenol blue indicator. One milliliter of 1 N hydrochloric acid corresponds to 0.2230 g of $Na_4P_2O_7 \cdot 10H_2O$.

pH OF A 5% SOLUTION. (Page 44). The pH should be from 9.5–10.5 at 25 °C.

INSOLUBLE MATTER. (Page 15). Use 10.0 g dissolved in 150 mL of water.

CHLORIDE. (Page 27). Use 0.50 g of sample and 3 mL of nitric acid.

SULFATE. Dissolve 1.0 g in 15 mL of water and add 5.0 mL of 10% hydrochloric acid (solution pH approximately 2). Filter through washed filter paper and add 10 mL of water through the same filter. For control, add 0.05 mg of sulfate ion (SO_4) in 20 mL of water and add 1.0 mL of (1 + 19) hydrochloric acid. Dilute both to 35 mL with water, add 5.0 mL of 30% barium chloride solution, and mix. Observe turbidity after 10 min. Sample solution turbidity should not exceed that of the control solution.

NITROGEN COMPOUNDS. (Page 31). Use 1.0 g. For the standard, use 0.01 mg of nitrogen (N).

HEAVY METALS. (Page 28, Method 1). Dissolve 3.0 g in about 20 mL of water, warming if necessary, add 8 mL of 2 N hydrochloric acid, and dilute with water to 42 mL. Use 35 mL to prepare the sample solution, and use the remaining 7.0 mL to prepare the control solution.

IRON. (Page 30, Method 2). Use 1.0 g.

Sodium Sulfate, Anhydrous

Na_2SO_4 **Formula Wt 142.04**

CAS Number 7757–82–6

NOTE: Sodium sulfate anhydrous is suitable for use in extraction–concentration analysis or for general use. Product labeling shall designate the uses for which suitability is represented on the basis of meeting the relevant requirements and tests. The extraction–concentration suitability requirements include all of the requirements for general use.

REQUIREMENTS

General Use
Assay . . . . $\geq$99.0% Na_2SO_4
pH of a 5% solution . . . . 5.2–9.2 at 25 °C

MAXIMUM ALLOWABLE
Insoluble matter. 0.01%
Loss on ignition 0.5%
Chloride (Cl) 0.001%
Nitrogen compounds (as N). . . . . 5 ppm
Phosphate (PO_4) 0.001%
Heavy metals (as Pb) . . . . 5 ppm
Iron (Fe). 0.001%
Calcium (Ca) 0.01%
Magnesium (Mg) 0.005%
Potassium (K). 0.01%

Specific Use
Extraction–concentration suitability . . . . Passes test

TESTS

ASSAY. (By indirect acid–base titrimetry after ion exchange). Weigh accurately 0.3 g of sample and dissolve in about 50 mL of water. Pass the sample solution through the column at a rate of about 5 mL/min, collecting the eluate in a 500 mL titration flask. Wash the column with water at a rate of about 10 mL/min, collecting the washings in the same flask. Add 0.15 mL of phenolphthalein indicator solution, and titrate the eluate with 0.1 N sodium hydroxide. Continue to wash the column and to titrate the eluate until the passage of 50 mL of water requires no further titrant. One milliliter of 0.1 N sodium hydroxide corresponds to 0.007102 g of Na_2SO_4.

NOTE. All water used in the assay must be free of carbon dioxide and ammonia.

pH OF A 5% SOLUTION. (Page 44). The pH should be 5.2–9.2 at 25 °C.

INSOLUBLE MATTER. (Page 15). Use 10.0 g dissolved in 100 mL of water.

LOSS ON IGNITION. Weigh accurately about 2 g in a tared dish or crucible and ignite at 800 ± 25 °C for 15 min.

CHLORIDE. (Page 27). Use 1.0 g.

NITROGEN COMPOUNDS. (Page 31). Use 2.0 g. For the standard, use 0.01 mg of nitrogen (N).

PHOSPHATE. (Page 32, Method 1). Dissolve 2.0 g in 20 mL of 0.5 N sulfuric acid. For the standard, dilute 0.02 mg of phosphate ion (PO_4) to 20 mL with 0.5 N sulfuric acid.

HEAVY METALS. (Page 28, Method 1). Dissolve 6.0 g in water and dilute with water to 42 mL. Use 35 mL to prepare the sample solution and use the remaining 7 mL to prepare the control solution.

IRON. (Page 30, Method 1). Use 1.0 g of sample.

CALCIUM, MAGNESIUM, AND POTASSIUM. (By flame AAS, page 39).

> **Sample Stock Solution.** Dissolve 4.0 g in 50 mL of water and transfer to a 100-mL volumetric flask. Dilute to the mark with water (1 mL = 0.04 g).

Element	Wavelength (nm)	Sample Wt (g)	Standard Added (mg)	Flame Type*	Background Correction
Ca	422.7	0.40	0.02; 0.04	N/A	No
Mg	285.2	0.40	0.01; 0.02	A/A	Yes
K	766.5	0.10	0.01; 0.02	A/A	No

*A/A is air/acetylene; N/A is nitrous oxide/acetylene.

EXTRACTION–CONCENTRATION SUITABILITY. (Page 71). Prepare a method blank to check the glassware and dichloromethane before beginning the sample analysis. Follow the sample procedure, below, except omit the sample.

Dry about 55 g of sample in a shallow tray at 400 °C for 1 h, and allow to cool in a desiccator. Transfer 50 ± 0.1 g of the dried sample, accurately weighed, to a 250-mL conical flask, add 60 mL of extraction–concentration grade dichloromethane, and allow to stand for 2–3 min. Swirl every 15 s to mix. Allow to stand until any fines that may be present have settled, and then decant the dichloromethane into a Kuderna–Danish flask equipped with a 3-ball Snyder column and a 10-mL concentrator tip. Extract the sample with two additional 60-mL portions of dichloromethane, as described above, combine the extracts, and concentrate the contents of the flasks to 1 mL. Perform solvent exchange using 15 mL of extraction–concentration grade hexanes in the Kuderna–Danish flask, and concentrate to 1 mL. Place a micro Snyder column on the 10-mL tip, perform two additional solvent exchanges, using 5 mL of hexanes each time, and concentrate to 1.0 mL. Inject 5.0 µL of the concentrate from the method blank into a gas chromatograph equipped with an electron capture detector (ECD) and a flame ionization detector (FID), as described on page 70. Dilute each standard specified in the ECD and FID suitability tests with an equal volume of hexanes, giving concentrations of 0.5 µg/L heptachlor epoxide and 0.5 mg/L of 2-octanol, respectively, and inject 1.0 µL of each into the chromatograph. Finally, inject 5.0 µL of the 180:1 concentrate from the sample into the chromatograph. Calculate the concentration of contaminants in the sample by comparing the sample peak heights with the height of the heptachlor epoxide and 2-octanol peaks from the injection of the standards, correcting for any contribution of the solvents as determined by the method blank. No peak in the extract of the sample shall exceed the height of the standard when measured as described on page 73.

Sodium Sulfate, Decahydrate

$Na_2SO_4 \cdot 10H_2O$ **Formula Wt 322.19**

CAS Number 7727–73–3

REQUIREMENTS

General Use

Assay . . . . ≥99.0% $Na_2SO_4 \cdot 10H_2O$
pH of a 5% solution . . . . 5.2–9.2 at 25 °C

MAXIMUM ALLOWABLE

Insoluble matter. 0.01%
Chloride (Cl) . . . . 5 ppm
Nitrogen compounds (as N). . . . . 3 ppm
Phosphate (PO_4) . . . . 5 ppm
Heavy metals (as Pb) . . . . 3 ppm
Iron (Fe) . . . . 5 ppm
Calcium (Ca) 0.005%
Magnesium (Mg) 0.003%
Potassium (K). 0.005%

TESTS

ASSAY. (By indirect acid–base titrimetry after ion exchange). Weigh accurately 0.7 g of sample and dissolve in about 50 mL of water. Pass the sample solution through the column at a rate of about 5 mL/min, collecting the eluate in a 500-mL titration flask. Wash the column with water at a rate of about 10 mL/min, collecting the washings in the same flask. Add 0.15 mL of phenolphthalein indicator solution, and titrate the eluate with 0.1 N sodium hydroxide. Continue to wash the column and to titrate the eluate until the passage of 50 mL of water requires no further titrant. One milliliter of 0.1 N sodium hydroxide corresponds to 0.01611 g of $Na_2SO_4 \cdot 10H_2O$.

pH OF A 5% SOLUTION. (Page 44). The pH should be 5.2–9.2 at 25 °C.

INSOLUBLE MATTER. (Page 15). Use 10.0 g dissolved in 100 mL of water.

CHLORIDE. (Page 27). Use 2.0 g.

NITROGEN COMPOUNDS. (Page 31). Use 5.0 g. For the standard, use 0.015 mg of nitrogen (N).

PHOSPHATE. (Page 32, Method 1). Dissolve 4.0 g in 20 mL of 0.5 N sulfuric acid. For the standard, dilute 0.02 mg of phosphate ion (PO_4) to 20 mL with 0.5 N sulfuric acid.

HEAVY METALS. (Page 28, Method 1). Dissolve 8.0 g in water and dilute with water to 48 mL. Use 42 mL to prepare the sample solution and use the remaining 6 mL to prepare the control solution.

IRON. (Page 30, Method 1). Use 2.0 g.

CALCIUM, MAGNESIUM, AND POTASSIUM. (By flame AAS, page 39).

> *Sample Stock Solution.* Dissolve 4.0 g in 50 mL of water in a 100-mL volumetric flask. Dilute to the mark with water (1 mL = 0.04 g).

Element	Wavelength (nm)	Sample Wt (g)	Standard Added (mg)	Flame Type*	Background Correction
Ca	422.7	0.40	0.02; 0.04	N/A	No
Mg	285.2	0.40	0.01; 0.02	A/A	Yes
K	766.5	0.40	0.01; 0.02	A/A	No

*A/A is air/acetylene; N/A is nitrous oxide/acetylene.

Sodium Sulfide, Nonahydrate

$Na_2S \cdot 9H_2O$ **Formula Wt 240.18**

CAS Number 1313–84–4

> *NOTE:* To enhance stability, storage below 10 °C is recommended.

REQUIREMENTS

Appearance . . . . Crystals, colorless or with
only a slight yellow color

Assay . . . . ≥98.0% Na$_2$S · 9H$_2$O

MAXIMUM ALLOWABLE

Ammonium (NH$_4$) 0.005%
Sulfite and thiosulfate (as SO$_4$) 0.1%
Iron . . . . Passes test

TESTS

ASSAY. (By titration of reductive capacity). Weigh accurately 2.5 g and in a 250-mL volumetric flask dilute with water through which nitrogen gas has been freshly bubbled to expel oxygen. Dilute to volume with similar water and displace air from the headspace with nitrogen. Stopper and mix thoroughly. Place a 50.0-mL aliquot of this solution into a mixture of 50.0 mL of 0.1 N iodine and 25 mL of 0.1 N hydrochloric acid in 400 mL of water. Titrate the excess of iodine with 0.1 N sodium thiosulfate, adding 3 mL of starch indicator solution near the end of the titration. One milliliter of 0.1 N iodine consumed corresponds to 0.01201 g of Na$_2$S · 9H$_2$O.

AMMONIUM. (Page 25). Dissolve 1.0 g in 80 mL of water, add 20 mL of lead acetate reagent solution, and allow to stand until the precipitate has settled. Decant 50 mL of the clear supernatant liquid, and use the 50 mL after distillation. For the standard, use 0.025 mg of ammonium ion (NH$_4$) treated exactly as the sample.

SULFITE AND THIOSULFATE. (*NOTE.* The water used in this test must be free from dissolved oxygen.) Dissolve 3.0 g in 200 mL of water and add 100 mL of 5% zinc sulfate solution. Mix thoroughly; allow to stand for 30 min. Filter and titrate 100 mL of the filtrate with 0.01 N iodine, using starch indicator. Not more than 3.0 mL of the iodine solution should be required.

IRON. Dissolve 5.0 g in 100 mL of water. The solution should be clear and colorless.

Sodium Sulfite
Sodium Sulfite, Anhydrous

Na$_2$SO$_3$ Formula Wt 126.04

CAS Number 7757–83–7

REQUIREMENTS

Assay . . . . ≥98.0% Na$_2$SO$_3$

MAXIMUM ALLOWABLE

Insoluble matter. 0.005%

Free acid . . . . Passes test

Titrable free base . . . . 0.03 meq/g

Chloride (Cl) 0.02%

Heavy metals (as Pb) 0.001%

Iron (Fe). 0.001%

TESTS

ASSAY. (By titration of reductive capacity). Place 0.5 g, accurately weighed, into 100 mL of 0.1 N iodine. Mix, allow to stand for 5 min, and add 1 mL of hydrochloric acid. Titrate the excess of iodine with 0.1 N sodium thiosulfate, adding 3 mL of starch indicator solution near the end of the titration. One milliliter of 0.1 N iodine consumed corresponds 0.006302 g of Na_2SO_3.

INSOLUBLE MATTER. (Page 15). Use 20.0 g dissolved in 200 mL of water.

FREE ACID. Dissolve 1.0 g in 10 mL of water and add 0.10 mL of phenolphthalein indicator solution. A pink color should be produced.

TITRABLE FREE BASE. Dissolve 1.0 g in 10 mL of water and add 3 mL of 30% hydrogen peroxide that has been previously neutralized to 0.15 mL of methyl red. Shake well, allow to stand for 5 min, and titrate with 0.01 N hydrochloric acid. Not more than 3.0 mL of the acid should be required to neutralize the solution.

CHLORIDE. Dissolve 1.0 g in 10 mL of water, filter if necessary through a small chloride-free filter, add 3 mL of 30% hydrogen peroxide, and dilute with water to 100 mL. Dilute 5 mL of this solution with water to 20 mL, and add 1 mL of nitric acid and 1 mL of silver nitrate reagent solution. Any turbidity should not exceed that produced by 0.01 mg of chloride ion (Cl) in an equal volume of solution containing the quantities of nitric acid and silver nitrate used in the test.

HEAVY METALS. (Page 28, Method 1). Dissolve 3.0 g in 20 mL of dilute hydrochloric acid (1 + 1), and evaporate the solution to dryness on a hot plate ($\approx$100 °C). Add 10 mL more of dilute hydrochloric acid (1 + 1), and again evaporate to dryness. Dissolve the residue in about 20 mL of water, and dilute with water to 25 mL. For the control, add 0.02 mg of lead to 1.0 g of sample, and treat exactly as the 3.0 g of sample.

IRON. (Page 30, Method 1). Dissolve 1.0 g in 20 mL of dilute hydrochloric acid (1 + 9), and evaporate on a hot plate ($\approx$100 °C) to dryness. Dissolve the residue in 5 mL of dilute hydrochloric acid (1 + 1), and again evaporate to dryness. Dissolve the residue in 4 mL of dilute hydrochloric acid (1 + 1), dilute with water to 50 mL, and use the solution without further acidification.

Sodium Tartrate Dihydrate
2,3-Dihydroxybutanedioic Acid, Disodium Salt, Dihydrate

$$\text{NaO} - \overset{\overset{\displaystyle O}{\|}}{C} - \overset{\overset{\displaystyle OH}{|}}{CH} - \overset{\overset{\displaystyle OH}{|}}{CH} - \overset{\overset{\displaystyle O}{\|}}{C} - \text{ONa} \cdot 2H_2O$$

$(CHOHCOONa)_2 \cdot 2H_2O$ **Formula Wt 230.08**

CAS Number 6106–24–7

NOTE. This reagent is suitable for standardization of Karl Fischer reagent as used for the determination of trace amounts of water (less than 1%).

REQUIREMENTS

Assay . . . . .99.0–101.0% $Na_2C_4H_4O_6 \cdot 2H_2O$
Loss on drying at 150 °C. 15.61%–15.71%
pH of a 5% solution . . . . 7.0–9.0 at 25 °C

MAXIMUM ALLOWABLE

Insoluble matter. 0.005%
Chloride (Cl) . . . . 5 ppm
Phosphate (PO₄) . . . . 5 ppm
Sulfate (SO₄) 0.005%
Ammonium (NH₄) 0.003%
Calcium (Ca) 0.01%
Heavy metals (as Pb) . . . . 5 ppm
Iron (Fe) 0.001%

TESTS

ASSAY. (Total alkalinity by nonaqueous titration). Weigh accurately 0.45 g, and dissolve in 100 mL of glacial acetic acid. Titrate with 0.1 N perchloric acid in glacial acetic acid to a green end point with crystal violet indicator. One milliliter of 0.1 N perchloric acid corresponds to 0.01151 g of $Na_2C_4H_4O_6 \cdot 2H_2O$.

LOSS ON DRYING AT 150 °C. Weigh accurately 3.0 g into a low-form weighing bottle and dry at 150 °C to constant weight (minimum 4 hours).

pH OF A 5% SOLUTION. (Page 44). The pH should be 7.0–9.0 at 25 °C.

INSOLUBLE MATTER. (Page 15). Use 20 g dissolved in 200 mL of water.

CHLORIDE. (Page 27). Use 2.0 g.

PHOSPHATE. (Page 32, Method 1). Ignite 4.0 g in a platinum dish. Dissolve the residue in 5 mL of water, add 5 mL of nitric acid, and evaporate to dryness. Dissolve the residue in 25 mL of approximately 0.5 N sulfuric acid and continue as described.

SULFATE. Drive off any moisture on a hot plate and then carefully ignite 1.0 g in an electric muffle furnace until it is nearly free of carbon. Boil the residue with 10 mL of water and 0.5 mL of 30% hydrogen peroxide for 5 min. Add 5 mg of sodium carbonate and 1 mL of hydrochloric acid and evaporate on a hot plate ($\approx$100 °C) to dryness. Dissolve the residue in 4 mL of hot water to which has been added 1 mL of dilute hydrochloric acid (1 + 19), filter through a small filter, wash with two 2-mL portions of water, and dilute the filtrate to 10 mL. Add 1 mL of barium chloride reagent solution and mix well. Any turbidity produced should not be greater than that in a standard prepared as described and to which the barium chloride solution is added at the same time it is added to the sample solution. To 10 mL of water, add 5 mg of sodium carbonate, 1 mL of hydrochloric acid, 0.5 mL of 30% hydrogen peroxide, and 0.05 mg of sulfate ion (SO_4) and evaporate on a hot plate ($\approx$100 °C) to dryness. Dissolve the residue in 9 mL of water and add 1 mL of dilute hydrochloric acid (1 + 19). If necessary, adjust with water to the same volume as the sample solution, add 1 mL of barium chloride reagent solution, and mix well.

AMMONIUM. Dissolve 1.0 g in 60 mL of water. To 20 mL add 1 mL of 10% sodium hydroxide reagent solution, dilute with water to 50 mL, and add 2 mL of Nessler reagent. Any color should not exceed that produced by 0.01 mg of ammonium ion (NH_4) in an equal volume of solution containing the quantities of reagents used in the test.

CALCIUM. (By flame AAS, page 39).

Sample Stock Solution. Dissolve 10.0 g of sample with water in a 100-mL volumetric flask, add 5 mL of hydrochloric acid, and dilute to the mark with water (1 mL = 0.1 g).

Element	Wavelength (nm)	Sample Wt (g)	Standard Added (mg)	Flame Type*	Background Correction
Ca	422.7	0.50	0.05; 0.10	N/A	No

*N/A is nitrous oxide/acetylene.

HEAVY METALS. (Page 28, Method 1). Dissolve 6.0 g in about 20 mL of water, and dilute with water to 30 mL. Use 25 mL to prepare the sample solution, and use the remaining 5.0 mL to prepare the control solution.

IRON. (Page 30, Method 1). Use 1.0 g.

Sodium Tetraphenylborate
Sodium Tetraphenylboron

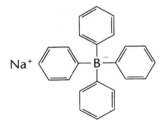

NaB(C$_6$H$_5$)$_4$ **Formula Wt 342.22**

CAS Number 143–66–8

REQUIREMENTS

Assay . . . . ≥99.5% NaB(C$_6$H$_5$)$_4$

MAXIMUM ALLOWABLE

Loss on drying at 105 °C. 0.5%

Clarity of solution. . . . . Passes test

TESTS

ASSAY. (By gravimetry). Weigh accurately about 0.5 g, dissolve in 100 mL of water, and add 1 mL of acetic acid. Add, with constant stirring, 25 mL of 5% potassium hydrogen phthalate solution, and allow to stand for 2 h. Filter through a tared, fine-porosity filtering crucible, wash with three 5-mL portions of saturated potassium tetraphenylborate solution, and dry at 105 °C for 1 h. The weight of the potassium tetraphenylborate multiplied by 0.9551 corresponds to the weight of the sodium tetraphenylborate.

LOSS ON DRYING AT 105 °C. Weigh accurately about 1.5 g and dry to constant weight at 105 °C.

CLARITY OF SOLUTION. Weigh 1.5 g, add 250 mL of water and 0.75 g of hydrated aluminum oxide, stir for 5 min, and filter through a filter paper. Filter the first 25-mL portion again. The filtrate should be clear.

Sodium Thiocyanate

NaSCN   **Formula Wt 81.07**

CAS Number 540–72–7

REQUIREMENTS

Assay. . . . . ≥98.0% NaSCN

MAXIMUM ALLOWABLE

Insoluble matter. 0.005%
Carbonate (as Na_2CO_3) 0.2%
Chloride (Cl) 0.01%
Sulfate (SO_4) 0.01%
Sulfide (S) 0.001%
Ammonium (NH_4) 0.002%
Heavy metals (as Pb) . . . . 5 ppm
Iron (Fe). . . . . 2 ppm

TESTS

ASSAY. (By indirect argentometric titration). Weigh, to the nearest 0.1 mg, 6.0 g of sample, dissolve with 100 mL of water in a 1000-mL volumetric flask, dilute to volume with water, and mix well. To 50.0 mL in a 250-mL glass-stoppered flask, add while agitating, exactly 50.0 mL of 0.1 N silver nitrate solution, then add 3 mL of nitric acid and 2 mL of ferric ammonium sulfate reagent solution (as indicator), and titrate the excess silver nitrate with 0.1 N ammonium thiocyanate solution.

$$\% \text{ NaSCN} = \frac{[(50.0 \times \text{N AgNO}_3) - (\text{mL} \times \text{N NH}_4\text{SCN})] \times 20 \times 8.107}{\text{Sample wt (g)}}$$

INSOLUBLE MATTER. (Page 15). Use 20.0 g dissolved in 150 mL of water.

CARBONATE. Dissolve 10.0 g in 100 mL of carbon dioxide-free water, add 0.10 mL of methyl orange indicator, and titrate with 0.1 N sulfuric acid. Not more than 3.8 mL of the acid should be required.

CHLORIDE. Dissolve 0.50 g in 20 mL of water, filter if necessary through a chloride-free filter, and add 10 mL of 25% sulfuric acid solution and 7 mL of 30% hydrogen peroxide. Evaporate to 20 mL by boiling in a well-ventilated hood, add 15–20 mL of water, and evaporate again. Repeat until all of the cyanide has been volatilized, cool, and dilute with water to 100 mL. To 20 mL of this solution add 1 mL of nitric acid and 1 mL of silver nitrate reagent solution. Any turbidity should not exceed that produced by 0.01 mg of chloride ion (Cl) in an equal volume of solution containing the quantities of reagents used in the test.

SULFATE. (Page 33, Method 1).

SULFIDE. Add 5.0 g to 80 mL of solution prepared by mixing 10 mL of ammonium hydroxide and 10 mL of silver nitrate reagent solution in 60 mL of water. Heat on a hot plate (≈100 °C) with occasional shaking for 20 min and transfer to a color comparison tube. Any color should not exceed that produced by 0.05 mg of sulfide ion (S) in an equal volume of solution containing the quantities of reagents used in the test.

AMMONIUM. (Page 25). Dilute the distillate from a 1.0-g sample to 100 mL, and use 50 mL of the dilution. For the standard, use 0.01 mg of ammonium ion (NH_4).

HEAVY METALS. (Page 28, Method 1). Dissolve 6.0 g in about 20 mL of water and dilute with water to 30 mL. Use 25 mL to prepare the sample solution, and use the remaining 5.0 mL to prepare the control solution.

IRON. Dissolve 10.0 g in 60 mL of water. To 30 mL of the solution add 0.1 mL of hydrochloric acid, 6 mL of hydroxylamine hydrochloride reagent solution, and 4 mL of 1,10-phenanthroline reagent solution, and add ammonium hydroxide to bring the pH to approximately 5. Any red color should not exceed that produced by 0.01 mg of iron in an equal volume of solution containing the quantities of reagents used in the test. Compare 1 h after adding the reagents to the sample and standard solutions.

Sodium Thiosulfate Pentahydrate

$Na_2S_2O_3 \cdot 5H_2O$ **Formula Wt 248.19**

CAS Number 10102–17–7

REQUIREMENTS

Assay . . . . 99.5–101.0% $Na_2S_2O_3 \cdot 5H_2O$
pH of a 5% solution . . . . 6.0–8.4 at 25 °C

MAXIMUM ALLOWABLE

Insoluble matter. 0.005%
Nitrogen compounds (as N). 0.002%
Sulfate and sulfite (as SO_4). 0.1%
Sulfide (S). . . . . Passes test

TESTS

ASSAY. (By titration of reductive capacity). Weigh accurately 1.0 g and dissolve in 30 mL of water. In the pH of a 5% solution test, if a value greater than 7.5 was obtained, add 0.05 mL of 1 N acetic acid. Titrate with 0.1 N iodine to a blue color with starch indicator. One milliliter of 0.1 N iodine corresponds to 0.02482 g of $Na_2S_2O_3 \cdot 5H_2O$.

pH OF A 5% SOLUTION. (Page 44). The pH should be 6.0–8.4 at 25 °C.

INSOLUBLE MATTER. (Page 15). Use 10.0 g dissolved in 100 mL of water.

NITROGEN COMPOUNDS. (Page 31). Dissolve 1.5 g in 10 mL of water in a flask, and add 10 mL of 25% sulfuric acid reagent solution. Connect the flask to a water-cooled reflux condenser, heat to boiling, and reflux for about 5 min. Cool and rinse down the condenser with a small amount of water. Filter, using a filter that has been washed free of ammonia, into a flask, and wash the residue with 10–15 mL of water. Add to the flask 30 mL of freshly boiled 10% sodium hydroxide reagent solution and distil. To the distillate add 2 mL of 10% sodium hydroxide reagent solution, dilute with water to 100 mL, and use 50 mL of the dilution. For the control, use 0.01 mg of nitrogen (N) and 0.25 g of sample.

SULFATE AND SULFITE. Dissolve 1.0 g in 50 mL of water and add approximately 0.1 N iodine until the liquid has a faint yellow color. Dilute with water to a volume of 100 mL and mix thoroughly. To 5.0 mL of the solution add 5 mL of water, 1 mL of 1 N hydrochloric acid, and 1 mL of barium chloride reagent solution. Any turbidity should not exceed that produced by 0.05 mg of sulfate ion (SO_4) in an equal volume of solution containing the quantities of reagents used in the test. Compare 10 min after adding the barium chloride to the sample and standard solutions.

SULFIDE. Dissolve 1.0 g in 10 mL of water and add 0.5 mL of alkaline lead solution (made by adding sufficient 10% sodium hydroxide reagent solution to 10% lead acetate reagent solution to redissolve the precipitate that is first formed). No dark color should be produced in 1 min. (Limit about 1 ppm)

Sodium Tungstate Dihydrate

$Na_2WO_4 \cdot 2H_2O$ **Formula Wt 329.86**

CAS Number 10213–10–2

REQUIREMENTS

Assay . . . . 99.0–101.0% $Na_2WO_4 \cdot 2H_2O$

MAXIMUM ALLOWABLE

Insoluble matter. 0.01%
Titrable free base . . . . 0.02 meq/g
Chloride (Cl) 0.005%
Molybdenum (Mo) 0.001%
Nitrogen compounds (as N). 0.001%
Sulfate (SO_4) 0.01%
Heavy metals and iron (as Pb) 0.001%

TESTS

ASSAY. (By gravimetry). Weigh accurately 1.0 g and dissolve in 50 mL of water in a 150-mL beaker. Add 2 mL of 30% hydrogen peroxide and 45 mL of nitric acid. Heat on a steam bath for 90 min, cool, and filter through a fine filter paper. Wash with 50 mL of hot 1% nitric acid, dry in a tared crucible, burn off the paper at a low temperature, and ignite at 825 °C to constant weight. The weight of the WO_3 multiplied by 1.4228 corresponds to the weight of the of $Na_2WO_4 \cdot 2H_2O$.

INSOLUBLE MATTER. (Page 15). Use 10.0 g dissolved in 100 mL of water.

TITRABLE FREE BASE. Dissolve 2.0 g in 50 mL of cold water and add 0.10 mL of thymol blue indicator solution. A blue color should be produced, which is changed to yellow by the addition of not more than 4.0 mL of 0.01 N hydrochloric acid.

CHLORIDE. Dissolve 1.0 g in 20 mL of water, filter if necessary through a chloride-free filter, add 3 mL of phosphoric acid, mix well, and dilute with water to 50 mL. Dilute 10 mL of the solution with water to 25 mL and add 1 mL of nitric acid and 1 mL of silver nitrate reagent solution. Any turbidity should not exceed that produced by 0.01 mg of chloride ion (Cl) in an equal volume of solution containing the quantities of reagents used in the test.

MOLYBDENUM.

> **Sample Solution A.** Dissolve 0.50 g in 15 mL of ammonium citrate solution (300 g of dibasic ammonium citrate dissolved in 500 mL of water and filtered).

> **Standard Solution B.** Add a standard molybdenum solution containing 5 µg of molybdenum to 15 mL of the ammonium citrate solution.

> **Blank Solution C.** Use 15 mL of the ammonium citrate solution.

Dilute the three solutions with water to 18 mL, then to each add the following reagents in order while stirring: 30 mL of dilute hydrochloric acid (1 + 1), 0.05 mL of 0.1 M cupric chloride in dilute hydrochloric acid (1 + 99), 3 mL of ammonium thiocyanate reagent solution (filtered if necessary), 3 mL of potassium iodide solution (50 g in 100 mL of water), and 2 mL of 1% sodium sulfite solution. Set a spectrophotometer at 460 nm and, using 1-cm cells, adjust the instrument to read 0 absorbance with blank solution C in the light path, then determine the absorbance of sample solution A and standard solution B. The absorbance of sample solution A should not exceed that of standard solution B. Compare within 30 min after development of the colors.

NITROGEN COMPOUNDS. (Page 31). Use 1.0 g. For the standard, use 0.01 mg of nitrogen (N).

SULFATE. Dissolve 2.0 g in 100 mL of water and add slowly with stirring 5 mL of hydrochloric acid. Evaporate to dryness and heat for 20 min at 110 °C. Add 30 mL of water, 2.5 mL of hydrochloric acid, and 2 mL of cinchonine solution [made by dissolving 5 g of cinchonine in 50 mL of dilute hydrochloric acid (1 + 3)] and heat just below the boiling point for 30 min. Dilute with water to 30 mL and allow to stand until cool. Filter, and to 15 mL of the filtrate add ammonium hydroxide drop by drop until a slight permanent precipitate forms. Add just enough hydrochloric acid to redissolve the precipitate and add 2 mL of barium chloride reagent solution. Any turbidity should not exceed that produced in a standard made as follows: to 1 mL of the cinchonine solution add 0.1 mg of sulfate ion (SO_4), dilute with water to 15 mL, and add 2 mL of barium chloride reagent solution. Compare 10 min after adding the barium chloride to the sample and standard solutions.

HEAVY METALS AND IRON. Dissolve 2.0 g in 90 mL of water plus 2 mL of ammonium hydroxide, and dilute with water to 100 mL. To 10 mL of this solution in a separatory funnel add 10 mL of dithizone extraction solution in chloroform. Shake well for 20 s and discard the aqueous layer. Wash the chloroform layer with successive 10-mL portions of 0.5% ammonium hydroxide until the ammonium hydroxide solution remains nearly colorless. The pink color remaining in the chloroform should not exceed that produced by 0.002 mg of lead in 10 mL of dilute ammonium hydroxide (1 + 99) treated exactly as the 10-mL portion of the solution of the sample.

Stannous Chloride Dihydrate
Tin(II) Chloride Dihydrate

$SnCl_2 \cdot 2H_2O$ Formula Wt 225.65

CAS Number 10025–69–1

REQUIREMENTS

Assay . . . . .98.0–103.0% $SnCl_2 \cdot 2H_2O$

MAXIMUM ALLOWABLE

Solubility in hydrochloric acid . . . . Passes test
Sulfate (SO_4) . . . . Passes test
Calcium (Ca) 0.005%
Iron (Fe). 0.003%
Lead (Pb) 0.01%
Potassium (K). 0.005%
Sodium (Na) 0.01%

TESTS

ASSAY. (By titration of reductive capacity of Sn^{II}). In a weighing bottle, weigh accurately 2.0 g. Transfer to a 250-mL volumetric flask, first with three 5-mL portions of hydrochloric acid and then with water through which nitrogen gas has been freshly bubbled to expel oxygen. Dilute to volume with similar water. Displace air from the headspace with nitrogen. Stopper and mix thoroughly. Place a 50.0-mL aliquot of this solution in a 500-mL conical flask filled with nitrogen gas. Add 5 g of potassium sodium tartrate tetrahydrate and 60 mL of a cold saturated solution of sodium bicarbonate. Titrate promptly with 0.1 N iodine to a blue color with starch indicator. One milliliter of 0.1 N iodine corresponds to 0.01128 g of $SnCl_2 \cdot 2H_2O$.

SOLUBILITY IN HYDROCHLORIC ACID. Dissolve 5.0 g in 5 mL of hydrochloric acid, heat to 40 °C if necessary, and dilute with 5 mL of water. The salt should dissolve completely.

SULFATE. Dissolve 5.0 g in 5 mL of hydrochloric acid, dilute the solution with water to 50 mL, filter if necessary, and heat to boiling. Add 5 mL of barium chloride reagent solution, digest in a covered beaker on a hot plate ($\approx$100 °C) for 2 h, and allow to stand overnight. No precipitate should be formed. (Limit about 0.003%)

CALCIUM, IRON, LEAD, POTASSIUM, AND SODIUM. (By flame AAS, page 39).

Sample Stock Solution. Dissolve 20.0 g of sample in a mixture of 50 mL of water and 20 mL of hydrochloric acid. Heat to 50 °C if necessary. Cool to room temperature, transfer to a 100-mL volumetric flask, and dilute to the mark with water (1 mL = 0.20 g).

Element	Wavelength (nm)	Sample Wt (g)	Standard Added (mg)	Flame Type*	Background Correction
Ca	422.7	1.0	0.05; 0.10	N/A	No
Fe	248.3	4.0	0.12; 0.24	A/A	Yes
Pb	217.0	1.0	0.10; 0.20	A/A	Yes
K	766.5	0.20	0.01; 0.02	A/A	No
Na	589.0	0.20	0.01; 0.02	A/A	No

*A/A is air/acetylene; N/A is nitrous oxide/acetylene.

Starch, Soluble
(for iodometry)

CAS Number 9005–84–9

REQUIREMENTS

Solubility . . . . Passes test
pH of a 2% solution . . . . 5.0–7.0 at 25 °C
Residue after ignition . . . . ≤0.4%
Sensitivity . . . . Passes test

TESTS

SOLUBILITY. Prepare a paste of 2.0 g of the sample with a little cold water and add to it with stirring 100 mL of boiling water. The solution should be no more than opalescent, and on cooling it should remain liquid and not increase in opalescence.

pH OF A 2% SOLUTION. (Page 46). Use the solution obtained in the test for solubility. The pH should be 5.0–7.0 at 25 °C.

RESIDUE AFTER IGNITION. (Page 16). Ignite 1.0 g and moisten the char with 1 mL of sulfuric acid.

SENSITIVITY. Prepare a paste of 1.0 g of the sample with a little cold water and add it with stirring to 200 mL of boiling water. Cool and add 5 mL of this solution to 100 mL of water containing 50 mg of potassium iodide, and add 0.05 mL of 0.1 N iodine solution. A deep blue color should be produced, which will be discharged by 0.05 mL of 0.1 N sodium thiosulfate solution.

Strontium Chloride Hexahydrate

$SrCl_2 \cdot 6H_2O$ **Formula Wt 266.62**

CAS Number 10025–70–4

REQUIREMENTS

Assay . . . . 99.0–103.0% $SrCl_2 \cdot 6H_2O$
pH of a 5% solution . . . . 5.0–7.0 at 25 °C

MAXIMUM ALLOWABLE

Insoluble matter. 0.005%
Sulfate (SO_4) 0.001%
Barium (Ba) 0.05%
Calcium (Ca) 0.05%
Magnesium (Mg) . . . . 2 ppm
Heavy metals (as Pb) . . . . 5 ppm
Iron (Fe) . . . . 5 ppm

TESTS

ASSAY. (By complexometric titration of strontium). Weigh accurately about 1.0 g of sample, transfer to a 250-mL beaker, and dissolve in 50 mL of water. Add 5 mL of diethylamine and 35 mg of methylthymol blue indicator mixture. Titrate immediately with 0.1 M EDTA until the blue color turns to colorless or gray. One milliliter of 0.1 N EDTA corresponds to 0.02666 g of $SrCl_2 \cdot 6H_2O$.

pH OF A 5% SOLUTION. (Page 44). The pH should be 5.0–7.0 at 25 °C.

INSOLUBLE MATTER. (Page 15). Use 20 g dissolved in 150 mL of water.

SULFATE. (Page 33, Method 1). Use a 5.0-g sample. Allow 30 min for the turbidity to form.

BARIUM, CALCIUM, AND MAGNESIUM. (By flame AAS, page 39).

Sample Stock Solution. Dissolve 10.0 g of sample with water in a 100-mL volumetric flask, add 5 mL of nitric acid, and dilute to the mark with water (1 mL = 0.10 g).

Element	Wavelength (nm)	Sample Wt (g)	Standard Added (mg)	Flame Type*	Background Correction
Ba	553.6	0.20	0.10; 0.20	N/A	No
Ca	422.7	0.10	0.05; 0.10	N/A	No
Mg	285.2	2.50	0.005; 0.01	A/A	No

*A/A is air/acetylene; N/A is nitrous oxide/acetylene.

HEAVY METALS. (Page 28, Method 1). Dissolve 6.0 g in about 20 mL of water, and dilute with water to 30 mL. Use 25 mL to prepare the sample solution, and use the remaining 5.0 mL to prepare the control solution.

IRON. Dissolve 2.0 g in 25 mL of water, add 2 mL of hydrochloric acid and 0.10 mL of 0.1 N potassium permanganate, and allow to stand for 5 min. Add 3 mL of ammonium thiocyanate reagent solution. Any red color should not exceed that produced by 0.01 mg of iron in an equal volume of solution containing the quantities of reagents used in the test.

Strontium Nitrate

$Sr(NO_3)_2$ **Formula Wt 211.63**

CAS Number 10042–76–9

REQUIREMENTS

Assay . . . . ≥99.0% $Sr(NO_3)_2$

pH of a 5% solution . . . . 5.0–7.0 at 25 °C

MAXIMUM ALLOWABLE

Insoluble matter. 0.01%
Loss on drying at 105 °C 0.1%
Chloride (Cl) 0.002%
Sulfate (SO_4) 0.005%
Barium (Ba). 0.05%
Calcium (Ca) 0.05%
Magnesium (Mg) 0.10%
Sodium (Na) 0.10%
Heavy metals (as Pb) . . . . 5 ppm
Iron (Fe). . . . . 5 ppm

TESTS

ASSAY. (By complexometric titration of strontium). Weigh accurately 0.8 g, transfer to a 250-mL beaker, and dissolve in 50 mL of water. Add 5 mL of diethylamine and 25 mg of methylthymol blue indicator mixture. Titrate immediately with 0.1 M EDTA until the blue color just turns to colorless or gray. One milliliter of 0.1 M EDTA corresponds to 0.02116 g of $Sr(NO_3)_2$.

pH OF A 5% SOLUTION. (Page 44). The pH should be 5.0–7.0 at 25 °C.

INSOLUBLE MATTER. (Page 15). Use 10.0 g dissolved in 100 mL of water.

LOSS ON DRYING AT 105 °C. Weigh accurately 2.0 g and dry at 105 °C for 4 h.

CHLORIDE. (Page 27). Use 0.50 g.

SULFATE. (Page 33, Method 2). Use 6.0 mL of dilute hydrochloric acid (1 + 1), and do two evaporations. Allow 30 min for the turbidity to form.

BARIUM, CALCIUM, MAGNESIUM, AND SODIUM. (By flame AAS, page 39).

> **Sample Stock Solution.** Dissolve 1.0 g of sample in 50 mL of water in a 100-mL volumetric flask, add 5 mL of nitric acid, and dilute to the mark with water (1 mL = 0.01 g).

Element	Wavelength (nm)	Sample Wt (g)	Standard Added (mg)	Flame Type*	Background Correction
Ba	553.6	0.20	0.10; 0.20	N/A	No
Ca	422.7	0.05	0.02; 0.04	N/A	No
Mg	285.2	0.02	0.01; 0.02	A/A	Yes
Na	589.0	0.02	0.01; 0.02	A/A	No

*A/A is air/acetylene; N/A is nitrous oxide/acetylene.

HEAVY METALS. (Page 28, Method 1). Dissolve 6.0 g in about 20 mL of water, and dilute with water to 30 mL. Use 25 mL to prepare the sample solution, and use the remaining 5.0 mL to prepare the control solution.

IRON. Dissolve 2.0 g in 10 mL of dilute hydrochloric acid (1 + 1) and evaporate to dryness. Repeat the evaporation. Dissolve the residue in 20 mL of water and add 2 mL of hydrochloric acid. Add 0.10 mL of 0.1 N potassium permanganate, dilute with water to 50 mL, and allow to stand for 5 min. Add 3 mL of ammonium thiocyanate reagent solution. Any red color should not exceed that produced by 0.01 mg of iron treated exactly as the sample.

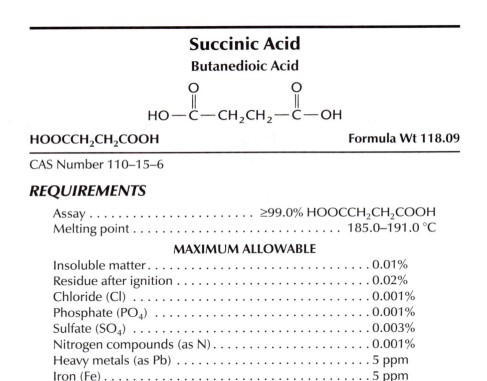

Succinic Acid
Butanedioic Acid

HOOCCH$_2$CH$_2$COOH   **Formula Wt 118.09**

CAS Number 110–15–6

REQUIREMENTS

Assay . . . . ≥99.0% HOOCCH$_2$CH$_2$COOH
Melting point . . . . 185.0–191.0 °C

MAXIMUM ALLOWABLE

Insoluble matter. 0.01%
Residue after ignition 0.02%
Chloride (Cl) 0.001%
Phosphate (PO$_4$) 0.001%
Sulfate (SO$_4$) 0.003%
Nitrogen compounds (as N). 0.001%
Heavy metals (as Pb) . . . . 5 ppm
Iron (Fe). . . . . 5 ppm

TESTS

ASSAY. (By acid–base titrimetry). Weigh accurately 0.3 g, dissolve in 25 mL of water in a conical flask, add 0.15 mL of phenolphthalein indicator solution, and titrate with 0.1 N sodium hydroxide. One milliliter of 0.1 N sodium hydroxide corresponds to 0.005905 g of HOOCCH$_2$CH$_2$COOH.

MELTING POINT. (Page 22).

INSOLUBLE MATTER. (Page 15). Use 10.0 g dissolved in 150 mL of water.

RESIDUE AFTER IGNITION. (Page 16). Ignite 10.0 g, omitting the use of sulfuric acid.

CHLORIDE. (Page 27). Use 1.0 g.

PHOSPHATE. (Page 33, Method 1). Dissolve 1.0 g of sample in 40 mL of water. Add 5 mL of 5 N sulfuric acid and continue as directed, using equal amount of reagents and volume for the control solution.

SULFATE. Dissolve 1.0 g of sample in 20 mL of water and filter through small, fine-porosity filter paper. Add two 2-mL portions of water through filter and collect in sample. For control, use 0.03 mg of sulfate ion (SO_4) in same volume of water. To sample and control solutions, add 5 mL of alcohol, 1 mL of dilute hydrochloric acid (1 + 19), and 1 mL of barium chloride reagent solution. After 30 min, sample turbidity should not exceed that of control.

NITROGEN COMPOUNDS. (Page 31). Use 1.0 g. For the standard, use 0.01 mg of nitrogen (N).

HEAVY METALS. To 30 mL of ammonium hydroxide (1 + 9), add 25 mL of water. While stirring, add 6.0 g of sample, dissolve, and dilute to 60 mL with water. For sample, take 50 mL of sample solution. For control, take 10 mL of sample solution, add 0.02 mg of lead, and dilute to 50 mL. Add 10 mL of freshly prepared hydrogen sulfide water to each and mix. Any brown color produced in the test solution within 5 min should not be darker than that produced in the control.

IRON. (Page 30, Method 1). Use 2.0 g.

Sucrose

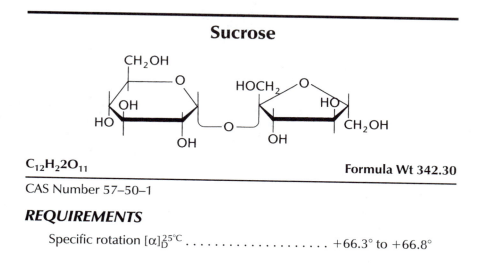

$C_{12}H_{22}O_{11}$

Formula Wt 342.30

CAS Number 57–50–1

REQUIREMENTS

Specific rotation $[\alpha]_D^{25°C}$ . . . . +66.3° to +66.8°

MAXIMUM ALLOWABLE

Insoluble matter. 0.005%
Loss on drying at 105 °C. 0.03%
Residue after ignition 0.01%
Titrable acid . . . . 0.0008 meq/g
Chloride (Cl) 0.005%
Sulfate and sulfite (as SO_4). 0.005%
Heavy metals (as Pb) . . . . 5 ppm
Iron (Fe). . . . . 5 ppm
Invert sugar 0.05%

TESTS

SPECIFIC ROTATION. (Page 23). Weigh accurately 26.0 g and dissolve in 90 mL of water in a 100-mL volumetric flask. Dilute with water to volume at 25 °C. Observe the optical rotation in a polarimeter at 25 °C using sodium light, and calculate the specific rotation.

INSOLUBLE MATTER. (Page 15). Use 40.0 g dissolved in 150 mL of water. Reserve the filtrate, without the washings, for preparation of sample solution A.

LOSS ON DRYING AT 105 °C. Weigh accurately 4.9–5.1 g in a tared dish or crucible and dry at 105 °C for 2 h.

RESIDUE AFTER IGNITION. (Page 16). Ignite 10.0 g and moisten the char with 1 mL of sulfuric acid.

TITRABLE ACID. To 100 mL of carbon dioxide-free water add 0.10 mL of phenolphthalein indicator solution and 0.01 N sodium hydroxide until a pink color is produced. Dissolve 10.0 g of the sample in this solution and titrate with 0.01 N sodium hydroxide to the same end point. Not more than 0.83 mL should be required.

> **Sample Solution A for the Determination of Chloride, Sulfate and Sulfite, Heavy Metals, and Iron.** Transfer the filtrate from the test for insoluble matter to a 200-mL volumetric flask, and dilute to volume with water.

CHLORIDE. Dilute 10 mL of sample solution A with water to 50 mL. Dilute 5 mL of this solution with water to 25 mL and add 1 mL of nitric acid and 1 mL of silver nitrate reagent solution. Any turbidity should not exceed that produced by 0.01 mg of chloride ion (Cl) in an equal volume of solution containing the quantities of reagents used in the test.

SULFATE AND SULFITE. (Page 33, Method 2). Use 5.0 mL of sample solution A, add 1 mL of bromine water, and boil. Omit evaporation to dryness, and follow Method 1 for the remainder of the test, starting with adding 1 mL of dilute hydrochloric acid (1 + 19).

HEAVY METALS. (Page 28, Method 1). Use 25.0 mL of sample solution A to prepare the sample solution, and use 5.0 mL of sample solution A to prepare the control solution.

IRON. (Page 30, Method 1). Use 10.0 mL of sample solution A (2-g sample).

INVERT SUGAR. Prepare a reagent solution containing in 1 L 150 g of potassium bicarbonate, $KHCO_3$, 100 g of potassium carbonate, K_2CO_3, and 6.928 g of cupric sulfate pentahydrate, $CuSO_4 \cdot 5H_2O$. Transfer 50 mL of this reagent to a 400-mL beaker, cover, heat to boiling, and allow to boil for 1 min. Dissolve 10.0 g of the sample in water, dilute with water to 50 mL, and add this solution to the solution in the 400-mL beaker. Heat to boiling and boil for 5 min. At the end of this period stop the reaction by adding 100 mL of cold, recently boiled water. Filter through a tared filtering crucible, wash thoroughly, and dry at 105 °C. The weight of the precipitate should not exceed 0.0277 g.

Sulfamic Acid

NH_2SO_3H **Formula Wt 97.09**

CAS Number 5329–14–6

REQUIREMENTS

Assay (dried basis) . . . . 99.3–100.3% NH_2SO_3H
MAXIMUM ALLOWABLE
Insoluble matter. 0.01%
Residue after ignition 0.01%
Chloride (Cl) 0.001%
Sulfate (SO_4) 0.05%
Heavy metals (as Pb) 0.001%
Iron (Fe) . . . . 5 ppm

TESTS

ASSAY. (By acid–base titrimetry). Weigh accurately 0.4 g, previously dried over sulfuric acid for 2 h, and dissolve in about 30 mL of water. Add 0.15 mL of phenolphthalein indicator solution and titrate with 0.1 N sodium hydroxide. One milliliter of 0.1 N sodium hydroxide corresponds to 0.009709 g of NH_2SO_3H.

INSOLUBLE MATTER. (Page 15). Use 10.0 g dissolved in 200 mL of water.

RESIDUE AFTER IGNITION. (Page 16). Ignite 10.0 g without addition of sulfuric acid.

CHLORIDE. (Page 27). Use 1.0 g.

SULFATE. (Page 33, Method 1). Dissolve 1.0 g of sample in 100 mL of water and use 10 mL (0.1 g) of this solution for the test.

HEAVY METALS. (Page 28, Method 1). Dissolve 4.0 g in 30 mL of water, neutralize to litmus with ammonium hydroxide, and dilute with water to 40 mL. Use 30 mL to prepare the sample solution and use the remaining 10 mL to prepare the control solution.

IRON. (Page 30, Method 1). Use 2.0 g.

Sulfanilic Acid
4-Aminobenzenesulfonic Acid

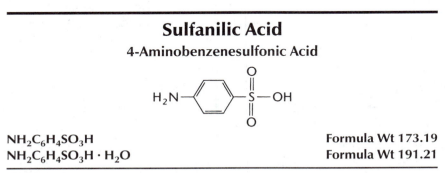

$NH_2C_6H_4SO_3H$ **Formula Wt 173.19**
$NH_2C_6H_4SO_3H \cdot H_2O$ **Formula Wt 191.21**

CAS Number 121–57–3

> *NOTE.* This reagent is available in both the anhydrous and monohydrate forms. The identity should be indicated on the label.

REQUIREMENTS

Assay . . . . 98.0–102.0% of the form offered

MAXIMUM ALLOWABLE

Residue after ignition 0.01%
Insoluble in sodium carbonate solution 0.02%
Chloride (Cl) 0.002%
Nitrite (NO$_2$) . . . . 0.5 ppm
Sulfate (SO$_4$) 0.01%

TESTS

ASSAY. (By acid–base titrimetry). Weigh accurately 0.7 g, and dissolve by warming gently in 50 mL of water. Cool and titrate with 0.1 N sodium hydroxide, using 0.15 mL of phenolphthalein indicator. One milliliter of 0.1 N sodium hydroxide corresponds to 0.01912 g of $NH_2C_6H_4SO_3H \cdot H_2O$ and to 0.01732 g of $NH_2C_6H_4SO_3H$.

RESIDUE AFTER IGNITION. Gently ignite 10.0 g in a tared crucible or dish until charred. Slowly raise the temperature until all carbon is removed, and finally heat at 800 ± 25 °C for 15 min.

INSOLUBLE IN SODIUM CARBONATE SOLUTION. Dissolve 5.0 g in 50 mL of a clear 5% solution of sodium carbonate and allow to stand in a covered beaker for 1 h. If an insoluble residue remains, filter through a tared porous porcelain or a platinum filtering crucible, wash with cold water, and dry at 105 °C.

CHLORIDE. Boil 5.0 g with 100 mL of water until dissolved. Cool, dilute with water to 100 mL, mix well, and filter through a chloride-free filter. Dilute 10 mL of the filtrate with water to 20 mL, and add 1 mL of nitric acid and 1 mL of silver nitrate reagent solution. Any turbidity should not exceed that produced by 0.01 mg of chloride ion (Cl) in an equal volume of solution containing the quantities of reagents used in the test. Save the remaining filtrate for the test for sulfate.

NITRITE. Dissolve 0.70 g in 100 mL of water, warming if necessary but keeping the temperature below 30 °C. For the control, dissolve 0.20 g in about 75 mL of water, add 0.00025 mg of nitrite ion (NO_2), and dilute with water to 100 mL. To each solution add 5 mL of sulfanilic-1-naphthylamine reagent solution and allow to stand for 10 min. Any pink color produced in the solution of the sample should not exceed that in the control.

> *Sulfanilic-1-Naphthylamine Solution.* Dissolve 0.5 g of sulfanilic acid in 150 mL of 36% acetic acid. Dissolve 0.1 g of *N*-(1-naphthyl)ethylenediamine dihydrochloride in 150 mL of 36% acetic acid. Mix the two solutions. If a pink color develops on standing, it may be discharged with a little zinc dust.

SULFATE. Cool about 30 mL of the solution reserved from the test for chloride to about 0 °C and filter. To 10 mL of the filtrate add 1 mL of dilute hydrochloric acid (1 + 19) and 1 mL of barium chloride reagent solution. Any turbidity should not exceed that produced by 0.05 mg of sulfate ion (SO_4) in an equal volume of solution containing the quantities of reagents used in the test. Compare 10 min after adding the barium chloride to the sample and standard solutions.

5-Sulfosalicylic Acid Dihydrate
2-Hydroxy-5-sulfobenzoic Acid Dihydrate

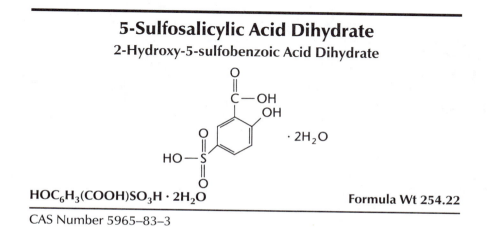

$$HOC_6H_3(COOH)SO_3H \cdot 2H_2O$$

Formula Wt 254.22

CAS Number 5965–83–3

REQUIREMENTS

Assay . . . . .99.0–101.0% HOC$_6$H$_3$(COOH)SO$_3$H · 2H$_2$O

MAXIMUM ALLOWABLE

Insoluble matter. 0.02%
Residue after ignition 0.1%
Chloride (Cl) 0.001%
Salicylic acid (HOC$_6$H$_4$COOH) 0.04%
Sulfate (SO$_4$) 0.02%
Heavy metals (as Pb) 0.002%
Iron (Fe) 0.001%

TESTS

ASSAY. (By acid–base titrimetry). Weigh accurately 5.0 g, dissolve in 50 mL of water, add 0.15 mL of phenolphthalein indicator solution, and titrate with 1 N sodium hydroxide. One milliliter of 1 N sodium hydroxide corresponds to 0.1271 g of HOC$_6$H$_3$(COOH)SO$_3$H · 2H$_2$O.

INSOLUBLE MATTER. (Page 15). Use 5.0 g dissolved in 50 mL of water.

RESIDUE AFTER IGNITION. (Page 16). Ignite 1.0 g in a dish, other than platinum, and moisten the char with 1 mL of sulfuric acid.

CHLORIDE. (Page 27). Use 1.0 g.

SALICYLIC ACID. Dissolve 5.0 g in 15 mL of cold water, add 10 mL of dilute hydrochloric acid (1 + 1), and extract with 50 mL of benzene in a 250-mL separatory funnel. Shake for 1.5 min, allow the layers to separate, and discard the aqueous layer. Add about 2 g of anhydrous sodium sulfate to the benzene layer, shake for 1 min, and allow the sodium sulfate to settle. Pour some of the clear benzene extract into a beaker, transfer 10 mL to a test tube, and add 10 mL of ferric ammonium sulfate solution. Shake the tube vigorously for 15 s and centrifuge at about 2500 rpm (about 1700 rcf) for 3 min. The color in the clear lower aqueous layer should not exceed that produced by 2.0 mg of salicyclic acid treated in exactly the same manner as the sample. In cases of borderline results or apparent nonconformity, aspirate off the benzene layer and determine the absorbance of the aqueous layer from the sample and the standard in 1-cm cells at 540 nm, using the ferric ammonium sulfate solution as the reference liquid in the spectrophotometer. Calculate the percent of salicylic acid as follows:

$$\% \text{ Salicylic acid} = \frac{C \times \dfrac{A_u}{A_s}}{W \times 10}$$

where

$C =$ weight, in milligrams, of salicylic acid in standard
$A_u =$ absorbance of sample solution
$A_s =$ absorbance of standard solution
$W =$ weight, in grams, of sample

Ferric Ammonium Sulfate Solution. To 200 mL of water in a beaker add 10% sulfuric acid until the pH is about 3, dissolve 200 mg of ferric ammonium sulfate dodecahydrate, $FeNH_4(SO_4)_2 \cdot 12H_2O$, in the solution, and adjust the pH to 2.45 (using a pH meter) with 10% sulfuric acid.

SULFATE. (Page 33, Method 1). Neutralize with dilute ammonium hydroxide $(1 + 9)$ to phenolphthalein (0.15 mL of indicator).

Sample Solution A for the Determination of Heavy Metals and Iron. Gently ignite 4.0 g in a crucible or dish, other than platinum, until charred. Cool, moisten the char with 1 mL of sulfuric acid, and ignite again slowly until all the carbon and excess sulfuric acid have been volatilized. Add 3 mL of hydrochloric acid and 0.5 mL of nitric acid, cover, and digest on a hot plate ($\approx$100 °C) until dissolved. Remove the cover, evaporate to dryness on a hot plate ($\approx$100 °C), dissolve the residue in 4 mL of 1 N acetic acid, and dilute with water to 100 mL (1 mL = 0.04 g).

HEAVY METALS. (Page 28, Method 1). Use 25 mL of sample solution A (1.0 g) to prepare the sample solution.

IRON. (Page 30, Method 1). Use 25 mL of sample solution A (1.0 g).

Sulfuric Acid

H_2SO_4   **Formula Wt 98.08**

CAS Number 7664–93–9

REQUIREMENTS

Appearance . . . . Free from suspended or insoluble matter
Assay. . . . . .95.0–98.0% H_2SO_4

MAXIMUM ALLOWABLE

Color (APHA). 10
Residue after ignition. . . . . 5 ppm
Chloride (Cl) . . . . 0.2 ppm
Nitrate (NO_3). . . . . 0.5 ppm
Ammonium (NH_4). . . . . 2 ppm
Substances reducing permanganate (as SO_2). . . . . 2 ppm
Arsenic (As) . . . . 0.01 ppm

Heavy metals (as Pb) . . . . 1 ppm
Iron (Fe) . . . . 0.2 ppm
Mercury (Hg) . . . . 5 ppb

TESTS

APPEARANCE. Mix the material in the original container, pour 10 mL into a test tube (20 × 150 mm), and compare with distilled water in a similar tube. The liquids should be equally clear and free from suspended matter.

ASSAY. (By acid–base titrimetry). Tare a small glass-stoppered flask, add about 1 mL of the sample, and weigh accurately. Cautiously add 30 mL of water, cool, add 0.15 mL of methyl orange indicator solution, and titrate with 1 N sodium hydroxide. One milliliter of 1 N sodium hydroxide corresponds to 0.04904 g of H_2SO_4.

COLOR (APHA). (Page 19).

RESIDUE AFTER IGNITION. (Page 16). Evaporate 200.0 g (110 mL) to dryness in a tared platinum dish and ignite at 800 ± 25 °C for 15 min.

CHLORIDE. Place 40 mL of water in each of three beakers. To one, carefully add 50 g (27.2 mL) of sample, and to the others, carefully add 50 g (27.2 mL) of chloride-free sulfuric acid. To the second beaker, add 0.01 mg of chloride (Cl). Cool to room temperature, mix, and to each beaker add 1 mL of nitric acid and 1 mL of silver nitrate reagent solution. Mix well, and if necessary, make the volume of each solution identical by adding water. Transfer equal portions of each solution to Nessler tubes. After 10 min, any turbidity of the sample solution should not exceed that of the standard. The blank should be free of turbidity. The comparison can be aided by use of a nephelometer.

NITRATE.

    Sample Solution A. Cautiously add 40.0 g (22 mL) to 2.0 mL of water, dilute to 50 mL with brucine sulfate reagent solution, and mix.

    Control Solution B. Cautiously add 40.0 g (22 mL) to 2.0 mL of the standard nitrate solution containing 0.01 mg of nitrate ion (NO_3) per milliliter, dilute to 50 mL with brucine sulfate reagent solution, and mix.

Continue with the procedure described on page 30, starting with the preparation of blank solution C.

AMMONIUM. (By differential pulse polarography, page 49). Carefully add 2.0 g (1.1 mL) to 4 mL of water in a 25-mL volumetric flask in an ice bath. Use 9.5 mL of ammonia-free 6 N sodium hydroxide for neutralization and 0.004 mg of ammonium (NH_4) for the standard.

SUBSTANCES REDUCING PERMANGANATE. Cool 50 mL of water in a 150-mL beaker in an ice bath. Slowly add 25 mL (46 g) of sample while stirring and keeping the solution cool. Remove from the bath and add 0.30 mL of 0.01 N potassium permanganate. The solution should remain pink for at least 5 min.

ARSENIC. (Page 25). To 300.0 g (165 mL) add 3 mL of nitric acid, and evaporate on a hot plate ($\approx$100 °C) to about 10 mL. Cool, cautiously dilute with about 20 mL of water, and evaporate just to dense fumes of sulfur trioxide. Cool, and cautiously wash the solution into a generator flask with the aid of 50 mL of water. Omit the addition of dilute sulfuric acid. For the standard, use 0.003 mg of arsenic (As).

HEAVY METALS. (Page 28, Method 1). Add 20.0 g (11 mL) to about 10 mg of sodium carbonate dissolved in a small quantity of water. Heat over a low flame until nearly dry, then add 1 mL of nitric acid. Evaporate to dryness, add about 20 mL of water, and dilute with water to 25 mL.

IRON. (Page 30, Method 1). Add 50.0 g (27 mL) to about 10 mg of sodium carbonate dissolved in a small quantity of water. Evaporate to dryness by heating on an electric hot plate. Cool, add 5 mL of dilute hydrochloric acid (1 + 1), cover with a watch glass, and digest on a hot plate ($\approx$100 °C) for 15 min. Cool, and dilute with water to 25 mL.

MERCURY. To each of two 125-mL conical flasks, add 20 mL of water and 5 mL of 4% potassium permanganate solution. For the sample, add to one flask, slowly and with cooling, 11.0 g (6.0 mL) of the sulfuric acid. For the control, add to the second flask 1.0 g (0.5 mL) of the sulfuric acid and 0.05 μg of mercury (Hg) (0.5 mL of a solution freshly prepared by diluting 1 mL of standard mercury solution, page 95, with water to 500 mL). Place both flasks on a hot plate ($\approx$100 °C) for 15 min, then cool to room temperature. Determine the mercury in each by CVAAS (page 40), using 1 mL of 10% hydroxylamine hydrochloride solution and 2 mL of 10% stannous chloride solution for the reduction. Any mercury found in the sample should not exceed that in the control.

Sulfuric Acid, Fuming

CAS Number 8014–95–7

> *NOTE.* This specification applies to fuming sulfuric acid with nominal contents of 15, 20, or 30% free SO_3.

REQUIREMENTS

Appearance . . . . Colorless to very light brown color
Assay (free SO_3) . . . . 12.0–17.0%, 18.0–24.0%, or 26.0–29.5%

MAXIMUM ALLOWABLE

Residue after ignition 0.002%

Nitrate (NO_3) . . . . .1 ppm
Ammonium (NH_4) . . . . .3 ppm
Arsenic (As) . . . . .0.03 ppm
Iron (Fe) . . . . .2 ppm

TESTS

ASSAY. (By acid–base titrimetry). Weigh accurately 4 g in a tared Dely weighing tube. Carefully transfer to a casserole containing 100 mL of carbon dioxide-free water by placing the tip of the tube beneath the surface and flushing the tube with carbon dioxide-free water. Cool, add 0.15 mL of phenolphthalein indicator solution, and titrate with 1 N sodium hydroxide.

$$\% \ H_2SO_4 = (mL \times N \ NaOH \times 4.904) \ / \ Sample \ wt \ (g)$$

where % free SO_3 = 4.445 (% H_2SO_4 – 100).

NOTE. For accurate results, a weight buret should be used. The Dely weighing tube and its use are described in standard reference books, such as *Standard Methods of Analysis*, 6th ed., Vol. IIA, F. J. Welcher, Ed., Van Nostrand, 1963, pp. 537–538.

RESIDUE AFTER IGNITION. (Page 16). Evaporate 50.0 g (27 mL) to dryness in a tared platinum dish and ignite at 800 ± 25 °C for 15 min.

NITRATE.

Sample Solution A. Cautiously add 10.0 g (5 mL) to 1.0 mL of water, dilute to 50 mL with brucine sulfate reagent solution, and mix.

Control Solution B. Cautiously add 10.0 g (5 mL) to 1.0 mL of the standard nitrate solution containing 0.01 mg of nitrate ion (NO_3) per mL, dilute to 50 mL with brucine sulfate reagent solution, and mix.

Continue with the procedure described on page 30, starting with the preparation of blank solution C.

AMMONIUM. (By differential pulse polarography, page 49). Carefully add 2.0 g (1.1 mL) to 4 mL of water in a 25-mL volumetric flask in an ice bath. Use approximately 10 mL of ammonia-free 6 N sodium hydroxide for neutralization and 0.006 mg of ammonium (NH_4) for the standard. The volume of buffer solution may be lowered to 4 mL.

ARSENIC. (Page 25). To 100.0 g (52 mL) add 3 mL of nitric acid, and evaporate on a hot plate (≈100 °C) to about 10 mL. Cool, cautiously dilute with about 20 mL of water, and evaporate to about 5 mL. Cool, cautiously dilute again with about 20 mL of water, and evaporate just to dense fumes of sulfur trioxide. Cool, and cautiously wash the solution into a generator flask with the aid of 50 mL of water. Omit the addition of 20 mL of dilute sulfuric acid (1 + 4). For the standard, use 0.003 mg of arsenic (As).

IRON. (Page 30, Method 1). Cautiously add 5.0 g (2.6 mL) to about 10 mg of sodium carbonate, and evaporate to dryness by heating on a hot plate ($\approx$100 °C). Cool, add 5 mL of dilute hydrochloric acid (1 + 1), cover with a watch glass, and digest on a hot plate ($\approx$100 °C) for 15 min. Cool, and dilute with water to 25 mL.

Sulfuric Acid, Ultratrace
(suitable for use in ultratrace elemental analysis)

H_2SO_4 Formula Wt 98.08

CAS number 7664–93–9

> *NOTE.* Reagent must be packaged in a preleached Teflon bottle and used in a clean environment to maintain purity.

REQUIREMENTS

Appearance . . . . Clear
Assay . . . . 85–98% H_2SO_4

MAXIMUM ALLOWABLE

Chloride (Cl) . . . . 0.2 ppm
Nitrate (NO_3) . . . . 0.5 ppm
Aluminum (Al) . . . . 1 ppb
Barium (Ba) . . . . 1 ppb
Boron (B) . . . . 5 ppb
Cadmium (Cd) . . . . 1 ppb
Calcium (Ca) . . . . 1 ppb
Chromium (Cr) . . . . 1 ppb
Cobalt (Co) . . . . 1 ppb
Copper (Cu) . . . . 1 ppb
Iron (Fe) . . . . 5 ppb
Lead (Pb) . . . . 1 ppb
Lithium (Li) . . . . 1 ppb
Magnesium (Mg) . . . . 1 ppb
Manganese (Mn) . . . . 1 ppb
Molybdenum (Mo) . . . . 1 ppb
Mercury (Hg) . . . . 1 ppb
Potassium (K) . . . . 1 ppb
Selenium . . . . 100 ppb
Silicon (Si) . . . . 5 ppb
Sodium (Na) . . . . 5 ppb
Strontium (Sr) . . . . 1 ppb
Tin (Sn) . . . . 1 ppb
Titanium (Ti) . . . . 1 ppb

Vanadium (V) . . . . 1 ppb
Zinc (Zn) . . . . 1 ppb
Zirconium (Zr) . . . . 1 ppb

TESTS

ASSAY. (By acid–base titrimetry). Tare a small, glass-stoppered flask, add about 1 mL of the sample, and weigh accurately. Cautiously add 30 mL of water, cool, add 0.15 mL of methyl orange indicator solution, and titrate with 1 N sodium hydroxide. One milliliter of 1 N sodium hydroxide corresponds to 0.04904 g of H_2SO_4.

$$\% \ H_2SO_4 = (mL \times N \ NaOH \times 4.904) \ / \ Sample \ wt \ (g)$$

CHLORIDE. (Page 27).

NITRATE. (Page 30).

TRACE METALS. Determine the aluminum, barium, cadmium, calcium, chromium, cobalt, copper, iron, lead, lithium, magnesium, manganese, molybdenum, nickel, potassium, sodium, strontium, tin, titanium, vanadium, zinc, and zirconium by the ICP–OES method described on page 42.

Determine the mercury by the CVAAS procedure described on page 40.

Determine the selenium by the HGAAS method described on page 41. To a set of 3 100-mL volumetric flasks, transfer 10 g (5.4 mL) of sample. To 2 flasks, add the specified amount of selenium standard from the following table. Dilute each to 100 mL with (1 + 1) hydrochloric acid. Mix. Prepare fresh working standard before use.

Element	Wavelength (nm)	Sample Wt (g)	Standard Addition (μg)
Se	196.0	10.0	1.0; 2.0 (= 100 and 200 ppb)

Sulfurous Acid
(a solution of SO₂ in water)

CAS Number 7782–99–2

REQUIREMENTS

Assay . . . . ≥6.0% SO_2

MAXIMUM ALLOWABLE

Residue after ignition 0.005%
Chloride (Cl) . . . . 5 ppm
Heavy metals (as Pb) . . . . 2 ppm
Iron (Fe) . . . . 5 ppm

TESTS

ASSAY. (By iodometric titration). Tare a glass-stoppered conical flask containing 50.0 mL of 0.1 N iodine. Quickly introduce about 2 mL of the sample, stopper, and weigh again. Titrate the excess iodine with 0.1 N sodium thiosulfate, adding 3 mL of starch indicator solution near the end of the titration. One milliliter of 0.1 N iodine consumed corresponds to 0.003203 g of SO_2.

$$\% \ SO_2 = \frac{[(mL \ I_2 \times N \ I_2) - (mL \ Na_2S_2O_3 \times N \ Na_2S_2O_3)] \times 0.03203 \times 100}{\text{Sample wt (g)}}$$

RESIDUE AFTER IGNITION. Evaporate 20.0 g (20 mL) to dryness on a hot plate ($\approx$100 °C) in a tared crucible or dish and ignite at 800 $\pm$ 25 °C for 15 min.

CHLORIDE. Digest 10.0 g (10 mL) with 2 mL of nitric acid on a hot plate ($\approx$100 °C) for 1 h. Cool and dilute with water to 100 mL. To 20 mL of the solution add 1 mL of nitric acid and 1 mL of silver nitrate reagent solution. Any turbidity should not exceed that produced by 0.01 mg of chloride ion (Cl) in an equal volume of solution containing the quantities of reagents used in the test.

HEAVY METALS. (Page 28, Method 1). To 10.0 g (10 mL) add 10 mL of water, boil to expel the sulfur dioxide, and dilute with water to 25 mL.

IRON. (Page 30, Method 1). To 2.0 g (2.0 mL) add about 10 mg of sodium carbonate, and evaporate to dryness. Dissolve the residue with 0.5 mL of hydrochloric acid, add 0.5 mL of nitric acid, and evaporate again to dryness. Dissolve the residue in 2 mL of hydrochloric acid, dilute with water to 50 mL, and use the solution without further acidification.

Tannic Acid

CAS Number 1401–55–4

> *NOTE.* For analytical purposes, tannic acid should be the hydrolyzable type, such as that isolated from nutgalls, sumac, or seed pods of Tara.

REQUIREMENTS

Identification . . . . Passes test
MAXIMUM ALLOWABLE
Loss on drying at 105 °C 12.0%
Residue after ignition 0.5%
Heavy metals (as Pb) 0.003%
Zinc (Zn) 0.005%
Sugars, dextrin. . . . . Passes test

TESTS

IDENTIFICATION. To 2 mL of a 10% aqueous solution of tannic acid add 0.1 mL of a 5% aqueous lead nitrate solution. A precipitate forms immediately, but completely dissolves on continued swirling to yield a clear solution.

LOSS ON DRYING AT 105 °C. Weigh accurately 1.0 g and dry to constant weight at 105 °C.

RESIDUE AFTER IGNITION. (Page 16). Ignite 5.0 g and moisten the char with 1 mL of sulfuric acid. Retain the residue.

HEAVY METALS. (Page 28, Method 1). Transfer 0.67 g into a 150-mL beaker and cautiously add 15 mL of nitric acid and 5 mL of 70% perchloric acid. Evaporate the mixture to dryness on a hot plate (≈100 °C) in a suitable hood, cool, add 2 mL of hydrochloric acid, and wash down the sides of the beaker with water. Carefully evaporate the solution to dryness on the hot plate (≈100 °C), rotating the beaker to avoid spattering. Repeat the addition of 2 mL of hydrochloric acid, wash down the sides of the beaker with water, and again evaporate to dryness on a hot plate (≈100 °C). Cool the residue, take up in 1 mL of hydrochloric acid and 10 mL of water, and dilute with water to 25 mL.

ZINC. Dissolve the residue from the residue after ignition test in 2 mL of glacial acetic acid, dilute with 8 mL of water, and filter if necessary. Add 0.5 g of sodium acetate and 5 mL of hydrogen sulfide water and mix. Any white turbidity should not exceed that produced by 0.25 mg of zinc treated exactly as the sample residue.

SUGARS, DEXTRIN. Dissolve 2.0 g in 10 mL of water and add 20 mL of alcohol. The mixture should be clear and should remain clear after standing for 1 h. Add 0.5 mL of ether. No turbidity should be produced.

Tartaric Acid
2,3-Dihydroxybutanedioic Acid

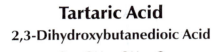

HOOC(CHOH)₂COOH   **Formula Wt 150.09**

CAS Number 87–69–4

REQUIREMENTS

Assay . . . . ≥99.0%

MAXIMUM ALLOWABLE

Insoluble matter. 0.005%
Residue after ignition 0.02%
Chloride (Cl) 0.001%
Oxalate (C_2O_4) . . . . Passes test
Phosphate (PO_4) 0.001%
Sulfur compounds (as SO_4) 0.002%
Heavy metals (as Pb) . . . . 5 ppm
Iron (Fe). . . . . 5 ppm

TESTS

ASSAY. (By acid–base titrimetry). Weigh accurately a sample of approximately 3.0 g and transfer to a 250 mL conical flask with about 75 mL of carbon-dioxide-free water. Add 0.15 mL of phenolphthalein indicator and titrate with 1 N sodium hydroxide to a faint-pink endpoint that persists for at least 30 s. One milliliter of 1 N sodium hydroxide corresponds to 0.07504 g of $H_2C_4H_4O_6$.

INSOLUBLE MATTER. (Page 15). Use 20.0 g dissolved in 200 mL of water.

RESIDUE AFTER IGNITION. (Page 16). Ignite 5.0 g, and moisten the char with 2 mL of sulfuric acid.

CHLORIDE. (Page 27). Use 1.0 g.

OXALATE. Dissolve 5.0 g in 30 mL of water and divide into two equal portions. Neutralize one portion with ammonium hydroxide, using litmus paper as indicator. Add the other portion and dilute with water to 40 mL. Shake well, cool, and allow to stand at 15 °C for 15 min. Filter, and to 20 mL of the filtrate add an equal volume of a saturated, filtered solution of calcium sulfate. No turbidity or precipitate should appear in 2 h. (Limit about 0.1%)

> ***Sample Solution A for the Determination of Phosphate, Sulfur Compounds, and Heavy Metals.*** To 10.0 g add about 10 mg of sodium carbonate, 5 mL of nitric acid, and 5 mL of 30% hydrogen peroxide. Digest in a covered beaker on a hot plate ($\approx$100 °C) until the reaction ceases. Wash down the cover glass and the sides of the beaker and evaporate to dryness. Repeat the treatment with nitric acid and peroxide and again evaporate to dryness. Dissolve the residue in about 30 mL of water, filter if necessary, and dilute with water to 50 mL (1 mL = 0.2 g).

PHOSPHATE. (Page 32, Method 1). Dilute 10 mL of sample solution A (2-g sample) with water to 20 mL, add 25 mL of 0.5 N sulfuric acid, and continue as described.

SULFUR COMPOUNDS. (Page 33, Method 1). Use 4.2 mL of sample solution A.

HEAVY METALS. (Page 28, Method 1). Dilute 20 mL of sample solution A (4-g sample) with water to 25 mL.

IRON. (Page 30, Method 1). Use 2.0 g.

Tetrabromophenolphthalein Ethyl Ester

$C_{22}H_{14}Br_4O_4$ **Formula Wt 661.97**

CAS Number 1176–74–5

REQUIREMENTS

Appearance . . . . Yellow to red powder
Assay . . . . ≥95.0%
Melting point . . . . 208.0–213.0 °C (within an
                                           interval of no more than 3.0°)
Solubility . . . . Passes test

TESTS

ASSAY. (By spectrophotometry). Weigh accurately 50.0 mg, transfer to a 100-mL volumetric flask, dissolve in about 75 mL of a 50/50 mixture of alcohol and buffer solution, and dilute to volume with the mixture. Pipet 2 mL of this solution into a 100-mL volumetric flask and dilute to volume with buffer solution. Determine the absorbance of this solution in 1.00-cm cells at the wavelength of maximum absorption, approximately 593 nm, using the buffer solution as the blank. Determine the molar absorptivity by dividing the absorbance by 1.5×10^{-5} (that is, the molar concentration of the tetrabromophenolphthalein ethyl ester) and calculate the assay value as follows:

$$\% \ C_{22}H_{14}Br_4O_4 = (\text{Molar absorptivity at the maximum} / 76{,}922) \times 100$$

Buffer Solution (pH = 7.00). Add 291 mL of 0.1 M sodium hydroxide to 500 mL of 0.1 M monobasic potassium phosphate, KH_2PO_4, and dilute with water to 1 L.

NOTE. Because the solutions are light-sensitive, protect from direct light.

MELTING POINT. (Page 22).

SOLUBILITY. Dissolve 0.1 g in 100 mL of toluene. The solution should be clear and dissolution should be complete.

Tetrabutylammonium Bromide
TBAB

$(CH_3CH_2CH_2CH_2)_4NBr$ **Formula Wt 322.37**

CAS Number 1643–19–2

REQUIREMENTS

General Use

Appearance . . . . White crystals
Assay . . . . $\geq 98.0\%$ $(CH_3CH_2CH_2CH_2)_4NBr$

MAXIMUM ALLOWABLE

Tributylamine (Bu_3N) 0.5%
Tributylamine hydrobromide $(Bu_3N \cdot HBr)$ 0.5%

Specific Use

Polarographic test . . . . Passes test

TESTS

ASSAY. (By argentimetric titration of bromide content). Weigh accurately about 1.0 g and dissolve in a 250-mL conical flask with 50 mL of water. Add 0.05 g of sodium bicarbonate and 0.2 mL of eosin Y solution. Titrate with 0.1 N silver nitrate to the first color change. One milliliter of 0.1 N silver nitrate corresponds to 0.03224 g of tetrabutylammonium bromide.

TRIBUTYLAMINE AND TRIBUTYLAMINE HYDROBROMIDE. (By potentiometric titration, page 51).

Sample Preparation. Accurately weigh and dissolve 20.0 g of sample in 90 mL of isopropyl alcohol in a 400-mL beaker.

Measure the pH of the sample solution on a standardized pH meter. If the pH is less than 5, then no tributylamine is present. Potentiometrically titrate with 1 N sodium hydroxide to determine the tributylamine hydrobromide content. The first inflection point, A mL, will be approximately pH 5. A second inflection point, B mL, will occur at approximately pH 10.5.

$$\% \ Bu_3N \cdot HBr = [(B - A) \times N \ NaOH \times 26.6265] / Sample \ wt \ (g)$$

If the pH is greater than 5, then potentiometrically titrate with 1 N hydrochloric acid to determine the tributylamine content. The inflection point, C mL, will be approximately pH 5. In the calculations, D mL is the total mL of 1 N hydrochloric acid added. Titrate the same sample with 1.0 N sodium hydroxide to an inflection point approximately pH 10.5 at E mL to determine the tributylamine hydrobromide content.

$$\% \ Bu_3N = (C \times N \ HCl \times 18.5353) / Sample \ wt \ (g)$$

$\% \ Bu_3N \cdot HBr$
$$= \{[(E \times N \ NaOH) - (D \times N \ HCl)] \times N \ NaOH \times 26.6265\} / Sample \ wt \ (g)$$

POLAROGRAPHIC TEST. (Page 46).

Cathode: Dropping Hg electrode

Anode: Pt wire

Reference: Similar saturated calomel electrode: make up with 1.0 M tetrabutylammonium chloride.

Prepare a 1.0 M solution (32.24 g in 100 mL of water) and add 0.2 mL of 1.0 M tetrabutylammonium hydroxide. Deoxygenate with nitrogen until minimum background current is achieved (approx. 5 min). By using a drop time of 1 s, scan at 0.01 V/s from -1.2 to -2.6 V vs. electrode with a sensitivity of 0.5 µA full scale with modulation amplitude of 50 mV. The material is acceptable if no impurity current greater than 0.05 µA is observed.

Tetrabutylammonium Hydroxide, 1.0 M Aqueous Solution
TBAH

$(CH_3CH_2CH_2CH_2)_4 \cdot NOH$ **Formula Wt 259.47**

CAS Number 2052–49–5

REQUIREMENTS

General Use
Assay . . . . 1.0 ± 0.02 M as $(CH_3CH_2CH_2CH_2)_4 \cdot NOH$

MAXIMUM ALLOWABLE

Color (APHA) 30
Carbonate (CO_3) 0.1%
Tributylamine (Bu_3N) 0.2%
Halide (as Cl) 0.08%

Potassium (K). . . . . 2 ppm
Sodium (Na) . . . . 2 ppm

Specific Use

Ultraviolet Spectrophotometry

Wavelength (nm)	Absorbance (AU)
280	0.04
254	0.07
245	0.10
240	0.15
Polarographic test	Passes test

TESTS

ASSAY. (By acid–base titrimetry). Pipet 25.0 mL into a 250-mL conical flask. Add 100 mL of water and 0.15 mL of phenolphthalein indicator solution, and titrate with 1 N hydrochloric acid until the solution becomes completely colorless.

$$\text{Molarity } (CH_3CH_2CH_2CH_2)_4 \cdot NOH = (\text{mL HCl} \times \text{N HCl}) / 25.0 \text{ mL}$$

COLOR (APHA). (Page 19).

CARBONATE AND TRIBUTYLAMINE. Determine carbonate and tributylamine content by a two-step potentiometric titration. Refer to page 51 for a general procedure for potentiometric titrations.

Part A. Pipet 25.0 mL of sample into a 400-mL beaker and dilute with water to 200 mL. Standardize a pH meter with a glass electrode and reference electrode, usually a saturated calomel electrode (SCE), or a combination pH electrode. Use a magnetic stir plate and magnetic stir bar or equivalent agitation throughout the titration. Add dilute nitric acid (1 + 9) to sample just short of the phenolphthalein endpoint (or pH 9.0). Titrate with standardized hydrochloric acid potentiometrically to the first endpoint. Save the sample for the tributylamine determination in Part B.

Part B. Deoxygenate the sample from Part A with nitrogen gas for at least 5 min. Dropwise, add 1 N sodium hydroxide to a pH of 11 to 12. Titrate with standardized 0.1 N hydrochloric acid.

$$\text{ppm } CO_3 \text{ (part A)} = (\text{mL of 1.0 N HCl} \times \text{N HCl} \times 60{,}000) / 25.0 \text{ mL}$$

$$\text{ppm } CO_3 \text{ (part B)} = (\text{mL of 0.1 N HCl} \times \text{N HCl} \times 60{,}000) / 25.0 \text{ mL}$$

$$\text{ppm Carbonate} = \text{ppm } CO_3 \text{ (Part A)} - \text{ppm } CO_3 \text{ (Part B)}$$

$$\text{ppm Tributylamine} = [\text{ppm } CO_3 \text{ (Part B)} \times 185] / 60$$

HALIDE (as Cl). (By potentiometric titration, page 51). Pipet 25.0 mL of TBAH into a 400-mL beaker and dilute to 200 mL. Add 0.05 mL of phenolphthalein to the solution and agitate with a magnetic stirrer and stir bar. Add nitric acid dropwise until the sample is acidic. Titrate with standardized 0.1 N silver nitrate using a silver electrode and a suitable millivolt meter.

$$\% \text{ Halide} = (\text{mL of } 0.1 \text{ N AgNO}_3 \times \text{N AgNO}_3 \times 3.5453) / 25.0 \text{ g}$$

POTASSIUM AND SODIUM. (By flame AAS, page 39).

Sample Stock Solution. Evaporate 50.0 g to dryness in a platinum crucible on a hot plate ($\approx$100 °C), and take up the residue in 10 mL of 5% nitric acid. Transfer to a 100-mL volumetric flask and dilute to the mark with 5% nitric acid (1 mL = 0.50 g).

Element	Wavelength (nm)	Sample Wt (g)	Standard Added (mg)	Flame Type*	Background Correction
K	766.5	10.0	0.02; 0.04	A/A	No
Na	589.0	10.0	0.01; 0.02	A/A	No

* A/A is air/acetylene.

ULTRAVIOLET SPECTROPHOTOMETRY. Use procedure on page 73 to determine absorbance.

POLAROGRAPHIC TEST. (Page 46).

Cathode: Dropping Hg electrode

Anode: Pt wire

Reference: Saturated calomel electrode: make up with 1.0 M tetrabutylammonium chloride

Warm 1.0 M TBAH (25.95 g in 100 mL) to 30–35 °C. Deoxygenate with nitrogen until minimum background current is achieved (approximately 5 min). By using a drop time of 1 s, scan at 0.01 V/s from –1.2 to –2.6 V vs. SCE with a sensitivity of 0.5 µA full scale with modulation amplitude of 50 mV. The material is acceptable if no impurity current greater than 0.05 µA is observed.

Tetrahydrofuran

C_4H_8O **Formula Wt 72.11**

CAS Number 109–99–9

NOTE: Generally, a stabilizer is present to retard peroxide formation.

REQUIREMENTS

Assay . . . . $\geq$99.0% C_4H_8O

MAXIMUM ALLOWABLE

Color (APHA) 20
Peroxide (as H_2O_2) 0.015%
Residue after evaporation 0.03%
Water 0.05%

TESTS

ASSAY. Analyze the sample by gas chromatography using the general parameters cited on page 70. The following specific conditions are also required.

Column: Type I, methyl silicone

Measure the area under all peaks and calculate the tetrahydrofuran content in area percent. Correct for water content.

COLOR (APHA). (Page 19).

PEROXIDE (AS H_2O_2). To 44.0 g (50 mL) in a conical flask, add 10 mL of 10% potassium iodide solution and 2 mL of dilute sulfuric acid. Titrate the liberated iodine with 0.1 N sodium thiosulfate. Not more than 3.9 mL should be required.

CAUTION. If peroxide is present, do not perform the test for residue after evaporation.

RESIDUE AFTER EVAPORATION. Evaporate 20.0 g (17.8 mL) to dryness in a tared dish on a hot plate ($\approx$100 °C), and dry the residue at 105 °C for 30 min.

WATER. (Page 56, Method 2). Use 100 µL (88 mg).

Tetramethylammonium Bromide
TMAB

$(CH_3)_4NBr$ **Formula Wt 154.05**

CAS Number 64–20–0

REQUIREMENTS

General Use

Appearance . . . . White crystals
Assay . . . . $\geq$98.0% as $(CH_3)_4NBr$

MAXIMUM ALLOWABLE

Trimethylamine (Me_3N) 0.5%
Trimethylamine hydrobromide ($Me_3N \cdot HBr$) 0.5%

Specific Use

Polarographic test . . . . Passes test

TESTS

ASSAY. (By argentimetric titration of bromide content). Weigh accurately about 0.5 g and dissolve in a 250-mL conical flask with 50 mL of water. Add 0.05 g of sodium bicarbonate and 0.2 mL of eosin Y. Titrate with 0.1 N silver nitrate to the first color change. One milliliter of 0.1 N silver nitrate corresponds to 0.01540 g of tetramethylammonium bromide.

TRIMETHYLAMINE AND TRIMETHYLAMINE HYDROBROMIDE. (By potentiometric titration, page 51).

Sample Preparation. Accurately weigh and dissolve 20 of g sample in 90 mL of isopropyl alcohol in a 400-mL beaker.

Measure the pH of the sample solution on a standardized pH meter. If the pH is less than 5, then no trimethylamine is present. Potentiometrically titrate with 1 N sodium hydroxide to determine the trimethylamine hydrobromide content. The first inflection point, A mL will be approximately pH 5. A second inflection point, B mL, will occur at approximately pH 10.5.

$$\% \ Me_3N \cdot HBr = [(B - A) \times N \ NaOH \times 14.0023] / \text{Sample wt (g)}$$

If the pH is greater than 5, then potentiometrically titrate with 1 N hydrochloric acid to determine the trimethylamine content. The inflection point, C mL, will be approximately pH 5. In the calculations, D mL is the total milliliters of 1 N hydrochloric acid added. Titrate the same sample with 1 N sodium hydroxide to an inflection point at approximately pH 10.5 at E mL.

$$\% \ Me_3N = (C \times N \ HCl \times 8.9111) / \text{Sample wt (g)}$$

$$\% \ Me_3N \cdot HBr = [(E - D) \times N \ NaOH \times 14.0023] / \text{Sample wt (g)}$$

POLAROGRAPHIC TEST. (Page 46).

Cathode: Dropping Hg electrode

Anode: Pt wire

Reference: Similar saturated calomel electrode: make up with 1.0 M tetrabutylammonium chloride

Prepare a 1.0 M solution (15.4 g in 100 mL of water) and add 0.2 mL of 1.0 M tetrabutylammonium hydroxide. Deoxygenate with nitrogen until minimum background current is achieved (approximately 5 min). By using a drop time of 1 s, scan at 0.01 V/s from −1.2 to −2.6 V vs. SCE with a sensitivity of 0.5 μA full scale with modulation amplitude of 50 mV. The material is acceptable if no impurity current greater than 0.05 μA is observed.

Tetramethylammonium Hydroxide, 1.0 M Aqueous Solution

TMAH

N,N,N-Trimethylmethanaminium Hydroxide

(CH$_3$)$_4$NOH   **Formula Wt 91.16**

CAS Number 75–59–2

REQUIREMENTS

General Use

Assay . . . . .1.0 ± 0.02 M as (CH$_3$)$_4$NOH

MAXIMUM ALLOWABLE

Color (APHA). 10
Chloride (Cl) 0.03%
Residue after ignition. . . . . 5 ppm
Potassium (K). . . . . 2 ppm
Sodium (Na) . . . . 2 ppm

Specific Use

Polarographic test . . . . Passes test

TESTS

ASSAY. (By acid–base titrimetry). Pipet 25.0 mL into a 250-mL conical flask. Add 100 mL of water and 0.15 mL of phenolphthalein indicator solution, and titrate with 1 N hydrochloric acid until the solution becomes completely colorless.

$$M \ (CH_3)_4NOH = (mL \ HCl \times N \ HCl) \ / \ 25.0 \ mL$$

COLOR (APHA). (Page 19).

CHLORIDE. (Page 27). Dilute 1.0 mL to 10.0 mL with water in a volumetric flask. Mix well. Pipet 1.0 mL into 15 mL of dilute nitric acid (1 + 15). Filter, if necessary, through a small chloride-free filter and add 1 mL of silver nitrate reagent solution. Any turbidity in the sample solution should not exceed that produced by 0.03 mg of chloride ion (Cl–) in an equal volume of solution containing the quantities of reagents used in the test.

RESIDUE AFTER IGNITION. (Page 16). Evaporate 200.0 g in a tared platinum dish to dryness in a hood and ignite.

POTASSIUM AND SODIUM. (By flame AAS, page 39).

Sample Stock Solution. Evaporate 50.0 g to dryness in a platinum crucible and take up the residue in 5 mL of 5% nitric acid solution. Transfer to a 100-mL volumetric flask and dilute to mark with 5% nitric acid solution (1 mL = 0.50 g).

Element	Wavelength (nm)	Sample Wt (g)	Standard Added (mg)	Flame Type*	Background Correction
K	766.5	10.0	0.02; 0.04	A/A	No
Na	589.0	10.0	0.01; 0.02	A/A*	No

* A/A is air/acetylene.

POLAROGRAPHIC TEST. (Page 46).

Cathode: Dropping Hg electrode

Anode: Pt wire

Reference: Similar saturated calomel electrode: make up with 1.0 M tetrabutylammonium chloride

Deoxygenate the TMAH with nitrogen until minimum background current is achieved (approximately 5 min). By using a drop time of 1 s, scan at 0.01 V/s from −1.2 to −2.6 V vs. SCE with a sensitivity of 0.5 µA full scale with modulation amplitude of 50 mV. The material is acceptable if no impurity current greater than 0.05 µA is observed.

Tetramethylsilane

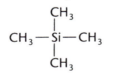

(CH$_3$)$_4$Si   **Formula Wt 88.22**

CAS Number 75–76–3

REQUIREMENTS

Residue after evaporation. . . . .≤0.05%
Suitability for proton NMR position reference . . . . Passes test

TESTS

RESIDUE AFTER EVAPORATION. (Page 16). Evaporate 2.0 g (3.1 mL) to dryness in a tared dish and dry the residue at 105 °C for 30 min.

SUITABILITY FOR PROTON NMR POSITION REFERENCE. An undiluted sample should display no impurity peak in its proton NMR reference spectrum more intense than 20% of the ^{13}C satellite line at ±59.1 Hz from tetramethylsilane.

Thioacetamide

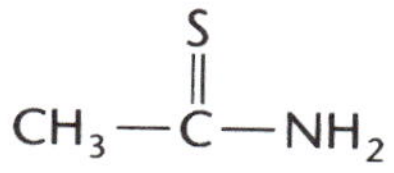

$$CH_3 - \overset{\overset{\displaystyle S}{\|}}{C} - NH_2$$

CH₃CSNH₂   **Formula Wt 75.13**

CAS Number 62–55–5

NOTE. This reagent, when stored under ordinary conditions, may decompose slightly and fail the test for clarity of a 2% solution.

REQUIREMENTS

Assay. . . . .≥99.0% CH_3CSNH_2
Melting point. . . . . .111–114 °C
Clarity of a 2% solution . . . . Passes test

MAXIMUM ALLOWABLE
Residue after ignition. 0.05%

TESTS

ASSAY. (By argentimetric titration). Weigh accurately 1.5 g, dissolve in water, and dilute with water to 500 mL in a volumetric flask. To 50.0 mL of this solution add 1 mL of ammonium hydroxide and 50.0 mL of 0.1 N silver nitrate. Allow to stand for 20 min, carefully filter through a filtering crucible or a sintered glass funnel that has been cleaned with dilute nitric acid, and wash the funnel and flask well with water. To the clear solution add 5 mL of nitric acid and 2 mL of ferric ammonium sulfate indicator solution, and titrate with 0.1 N potassium thiocyanate. One milliliter of 0.1 N silver nitrate corresponds to 0.003757 g of CH_3CSNH_2.

MELTING POINT. (Page 22).

CLARITY OF A 2% SOLUTION. Dissolve 2.0 g in 100 mL of water. The solution should be clear and colorless.

RESIDUE AFTER IGNITION. (Page 16). Ignite 2.0 g, and moisten the char with 1 mL of sulfuric acid.

Thiourea

$$H_2N-\overset{\overset{\displaystyle S}{\|}}{C}-NH_2$$

NH₂CSNH₂   **Formula Wt 76.12**

CAS Number 62–56–6

REQUIREMENTS

Assay (dried basis) . . . . ≥99.0% NH₂CSNH₂
Melting point . . . . 174–177 °C
Sensitivity to bismuth. . . . . Passes test
Solubility in water . . . . Passes test

MAXIMUM ALLOWABLE

Residue after ignition 0.1%
Loss on drying 0.5%

TESTS

ASSAY. (By argentimetric titration). Take the sample from the loss on drying test. Weigh accurately 1.0 g, dissolve in water, and dilute to 250 mL in a volumetric flask. To 20.0 mL of this solution in a glass-stoppered flask, add 25.0 mL of 0.1 N silver nitrate and 10 mL of ammonium hydroxide (10% NH₃). Stopper the flask, shake vigorously for 2 min, heat to boiling, and cool. To the cooled solution add 60 mL of dilute nitric acid, shake vigorously, and filter through a filtering crucible or a sintered glass funnel that has been cleaned with dilute nitric acid, and wash the funnel and flask well with water. To the filtrate plus washings add 2 mL of ferric ammonium sulfate indicator solution, and titrate with 0.1 N potassium thiocyanate. One milliliter of 0.1 N silver nitrate corresponds to 0.003806 g of NH₂CSNH₂.

MELTING POINT. (Page 22).

SENSITIVITY TO BISMUTH. Dissolve 0.10 g in 10 mL of water. Add 1.0 mL of this solution to 10 mL of bismuth standard solution. A distinct yellow color is produced immediately.

SOLUBILITY IN WATER. Dissolve 1 g in 20 mL of water (50 °C). The solution should be clear and colorless.

RESIDUE AFTER IGNITION. (Page 16). Ignite 10.0 g.

LOSS ON DRYING. Weigh accurately about 1.5 g and dry to constant weight at 105 °C.

Thorium Nitrate Tetrahydrate

$Th(NO_3)_4 \cdot 4H_2O$ **Formula Wt 552.12**

CAS Number 13470–07–0

REQUIREMENTS

Assay . . . . .98.0–102.0% $Th(NO_3)_4 \cdot 4H_2O$

MAXIMUM ALLOWABLE

Insoluble matter. 0.01%
Chloride (Cl) 0.002%
Sulfate (SO_4) 0.01%
Substances not precipitated by ammonium hydroxide. . 0.2%
Heavy metals (as Pb) 0.002%
Iron (Fe). 0.002%
Rare earth elements (as La) 0.2%
Titanium (Ti) 0.01%

TESTS

ASSAY. (By gravimetry). Weigh accurately 1.0 g, and dissolve in 100 mL of water in a beaker. Add 2 mL of 10% sulfuric acid, heat to boiling, and add 20 mL of a hot 10% solution of oxalic acid. Cool, filter, wash with a little water, and ignite the precipitate to constant weight. Multiply the oxide weight by 209.0, and divide by the weight of the sample to obtain % $Th(NO_3)_4 \cdot 4H_2O$.

INSOLUBLE MATTER. (Page 15). Use 10.0 g dissolved in 100 mL of water. Reserve the filtrate and washings for the preparation of sample solution A.

> ***Sample Solution A for the Determination of Chloride, Sulfate, Substances Not Precipitated by Ammonium Hydroxide, Heavy Metals, Iron, Rare Earth Elements, and Titanium.*** Transfer the filtrate and washings reserved from the test for insoluble matter to a 200-mL volumetric flask and dilute with water to 200 mL (1 mL = 0.05 g).

CHLORIDE. (Page 27). Dilute 10 mL (0.5-g sample) of sample solution A with water to 20 mL.

SULFATE. Dilute 10 mL (0.5-g sample) of sample solution A with water to 20 mL. Add 5 mL of 2 N ammonium acetate and transfer the solution to a separatory funnel. Extract with successive 10-mL portions of 0.2 M *N*-phenylbenzohydroxamic acid in Chloroform until the color of the Chloroform layer remains unchanged. Wash the aqueous layer twice with 5-mL portions of Chloroform. Transfer the aqueous solution to a beaker and evaporate on a hot plate ($\approx$100 °C) to 2–3 mL. Add 2 mL of nitric acid and 2 mL of hydrochloric acid, cover, and digest on the hot plate ($\approx$100 °C) until any reaction ceases. Uncover, wash down the sides of the beaker, and evaporate to dryness. Dissolve the residue in 4 mL of water plus 1 mL of dilute hydrochloric acid (1 + 19). Filter if necessary through a small filter, wash with two 2-mL portions of water, and dilute with water to 10 mL. Add 1 mL of barium chloride reagent solution. Any turbidity should not exceed that produced by 0.05 mg of sulfate ion (SO_4) in an equal volume of solution that has been treated exactly like the sample. Compare 10 min after adding the barium chloride to the sample and standard solutions.

SUBSTANCES NOT PRECIPITATED BY AMMONIUM HYDROXIDE. Dilute 40 mL (2-g sample) of sample solution A with water to 100 mL. Add 0.10 mL of methyl red indicator solution and add ammonium hydroxide until the solution is alkaline to methyl red. Heat the solution to boiling and digest on a hot plate ($\approx$100 °C) for 15 min. Dilute the solution with water to 150 mL, mix thoroughly, and filter. Evaporate 75 mL (1-g sample) to dryness in a tared dish on a hot plate ($\approx$100 °C). Moisten the residue with 0.10 mL of sulfuric acid. Heat gently to volatilize the excess salts and acid, and finally ignite at 800 ± 25 °C for 15 min.

HEAVY METALS. (Page 28, Method 1). Use 30 mL of sample solution A (1.5-g sample) to prepare the sample solution, and use 10 mL of sample solution A to prepare the control solution.

IRON. (Page 30, Method 1). Use 10 mL of sample solution A (0.5-g sample).

RARE EARTH ELEMENTS. Dilute 5.0 mL (0.25-g sample) of sample solution A to 90 mL with water. Use a pH meter to adjust to pH 4.5 with sodium acetate–acetic acid buffer solution and dilute with water to 100 mL. To 10 mL in a separatory funnel add 10 mL of N-phenylbenzohydroxamic acid reagent solution, shake vigorously, allow the layers to separate, and draw off and discard the Chloroform layer. Repeat the extraction to 5-mL portions of N-phenylbenzohydroxamic acid reagent solution two more times, drawing off and discarding the Chloroform layer each time. Wash the aqueous layer with 5 mL of Chloroform and draw off and discard the Chloroform. Add 2 mL of alizarin red S solution and dilute with water to 50 mL. Adjust the pH of the solution to 4.68 with 0.5 N sodium acetate. Any red color should not exceed that produced by 0.05 mg of lanthanum that has been treated exactly like the 10 mL of the sample.

> *Sodium Acetate–Acetic Acid Buffer.* Dissolve 54.4 g of sodium acetate in water, add 23 mL of acetic acid, and dilute with water to 200 mL.

> *N-Phenylbenzohydroxamic Acid Reagent Solution.* Dissolve 4.26 g of N-phenylbenzohydroxamic acid in Chloroform and dilute to 100 mL with Chloroform.

> *Alizarin Red S Solution.* Dissolve 0.10 g of 3,4-dihydroxy-2-anthraquinonesulfonic acid, sodium salt, in water and dilute with water to 100 mL.

TITANIUM. Dilute 10 mL (0.5-g sample) of sample solution A with water to 20 mL and add 0.5 mL of 30% hydrogen peroxide. Any yellow color should not exceed that produced by 0.05 mg of titanium (Ti) in an equal volume of solution containing the quantities of reagents used in the test.

Thymol Blue

Thymolsulfonphthalein

4,4′-(3H-2,1-Benzoxathiol-3-ylidene)bis-[5-methyl-2-(1-methylethyl)phenol] S,S-Dioxide

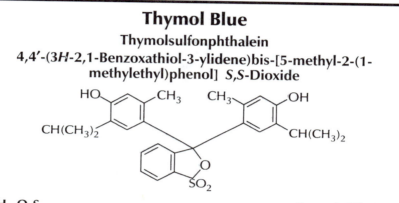

$C_{27}H_{30}O_5S$ **Formula Wt 466.60**

CAS Number 76–61–9

NOTE. This specification applies to both the free acid form and the salt form of this indicator.

REQUIREMENTS

Clarity of solution . . . . Passes test
Visual transition interval (acid range) . . . . From pH 1.2 (red)
                                                               to pH 2.8 (yellow)
Visual transition interval (alkaline range) . . . . From pH 8.0 (yellow)
                                                               to pH 9.2 (blue)

TESTS

CLARITY OF SOLUTION. If the indicator is the acid form, dissolve 0.1 g in 100 mL of alcohol. If the indicator is a salt form, dissolve 0.1 g in 100 mL of water. Not more than a faint trace of turbidity or insoluble matter should remain. Reserve the solution for the tests for visual transition interval.

VISUAL TRANSITION INTERVAL (ACID RANGE). Dissolve 1 g of potassium chloride in 100 mL of water. Adjust the pH of the solution to 1.20 (using a pH meter as described on page 43) with 1.0 N hydrochloric acid. Add 0.1–0.3 mL of the 0.1% solution reserved from the test for clarity of solution. The color of the solution should be pink. Titrate the solution with 1.0 N sodium hydroxide until the pH is 2.2 (using the pH meter). The color of the solution should be orange.

Continue the titration until the pH is 2.8. The color of the solution should be yellow. Not more than 8.6 mL of the 1.0 N sodium hydroxide should be consumed in the titration.

VISUAL TRANSITION INTERVAL (ALKALINE RANGE). Dissolve 1 g of potassium chloride in 100 mL of water. Adjust the pH of the solution to 8.0 (using a pH meter as described on page 43) by adding 0.01 N hydrochloric acid or sodium hydroxide. Add 0.1–0.3 mL of the 0.1% solution reserved from the test for clarity of solution. The color of the solution should be yellow. Titrate the solution with 0.01 N sodium hydroxide until the pH is 8.4 (using the pH meter). The color of the solution should be green. Continue the titration until the pH is 9.2. The color of the solution should be blue. Not more than 0.50 mL of the 0.01 N sodium hydroxide should be consumed in the titration.

Thymolphthalein
5',5"-Diisopropyl-2',2"-dimethylphenolphthalein
3,3-Bis[4-hydroxy-2-methyl-5-(1-methylethyl)phenyl]-1(3H)-isobenzofuranone

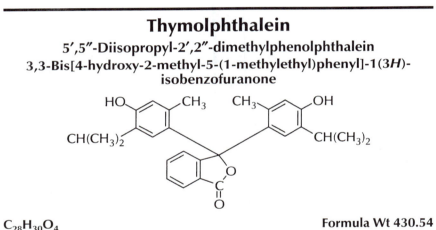

$C_{28}H_{30}O_4$ **Formula Wt 430.54**

CAS Number 125–20–2

REQUIREMENTS

Clarity of solution. . . . . Passes test
Visual transition interval. . . . . From pH 8.8 (colorless)
to pH 10.5 (blue)

TESTS

CLARITY OF SOLUTION. Dissolve 0.1 g in 100 mL of alcohol. Not more than a faint trace of turbidity or insoluble matter should remain. Reserve the solution for the test for visual transition interval.

VISUAL TRANSITION INTERVAL. Dissolve 1 g of potassium chloride in 100 mL of water. Adjust the pH of the solution to 8.80 (using a pH meter as described on page 43) with 0.01 N sodium hydroxide. Add 0.5–1.0 mL of the 0.1% solution reserved from the test for clarity of solution. The solution should be colorless. Titrate the solution with 0.01 N sodium hydroxide to pH 9.4 (using the pH meter). The solution should have a pale grayish-blue color. Continue the titration to pH 9.9. The solution should have a blue color. Not more than 2.0 mL of 0.01 N sodium hydroxide should be consumed in the titration. Each additional 0.05 mL of 0.01 N sodium hydroxide should increase the amount of blue color until at pH 10.5 the color should be an intense blue.

Tin

Sn   **Atomic Wt 118.71**

CAS Number 7440–31–5

REQUIREMENTS

Assay . . . . ≥99.5% Sn

MAXIMUM ALLOWABLE

Antimony (Sb) 0.02%
Copper (Cu). 0.005%
Iron (Fe) 0.01%
Lead (Pb) 0.005%
Arsenic (As) . . . . 1 ppm

TESTS

ASSAY. (By complexometric titration). Weigh, to the nearest 0.1 mg, 0.25 g of sample, and digest on a hot plate ($\approx$100 °C) in a covered beaker with 5 mL of hydrochloric acid until the sample is dissolved. Add 30.0 mL of standardized 0.1 M EDTA and heat the solution almost to boiling. Cool and adjust the pH to 5.5, using a pH meter, with a saturated hexamethylenetetramine solution. Cool, dilute with water to about 250 mL, and add a few milligrams of xylenol orange indicator. Titrate with 0.1 M lead solution to a change from yellow to reddish purple.

$$\% \text{ Sn} = \frac{[(30.0 \times \text{M EDTA}) - (\text{mL} \times \text{M Pb})] \times 11.87}{\text{Sample wt (g)}}$$

ANTIMONY, COPPER, IRON, AND LEAD. (By flame AAS, page 39).

Sample Stock Solution. Dissolve 10.0 g of sample in 30 mL of hydrochloric acid and 3 mL of nitric acid by allowing to stand in a covered beaker until dissolution is complete. Transfer to a 100-mL volumetric flask and dilute to the mark with water (1 mL = 0.10 g).

Element	Wavelength (nm)	Sample Wt (g)	Standard Added (mg)	Flame Type*	Background Correction
Sb	217.6	1.0	0.20; 0.40	A/A	Yes
Cu	324.8	0.50	0.025; 0.05	A/A	Yes
Fe	248.3	0.50	0.05; 0.10	A/A	Yes
Pb	217.0	1.0	0.05; 0.10	A/A	Yes

*A/A is air/acetylene.

ARSENIC. Dissolve 1.0 g in a mixture of 5 mL of water and 10 mL of nitric acid in a generator flask in a hood. Add 10 mL of dilute sulfuric acid (1 + 1), evaporate on a hot plate (≈100 °C) to about 5 mL, and heat just to fumes of sulfur trioxide. (At this point the metastannic acid should go into solution.) Cool, cautiously wash down the flask with 5–10 mL of water, and again heat just to fumes of sulfur trioxide. Cool, cautiously wash down the flask with 5–10 mL of water, and repeat the fuming. Cool, dilute with water to 55 mL, and proceed as described in the general method for arsenic on page 25, omitting the addition of the dilute sulfuric acid and using 2 mL of hydrobromic acid (48%) instead of the potassium iodide solution specified under procedure. Swirl the contents of the flask occasionally to break up the zinc mass. When evolution of gas stops, open the generator flask, and add 1-g portions of granulated zinc until the solution is clear and colorless. Any red color in the silver diethyldithiocarbamate solution of the sample should not exceed that in a standard containing 0.001 mg of arsenic (As).

o-Tolidine Dihydrochloride
3,3'-Dimethyl-[1,1'-biphenyl]-4,4'-diamine Dihydrochloride

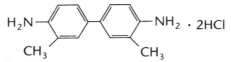

[-C₆H₃(CH₃)-4-NH₂]₂ · 2HCl $\quad\quad\quad$ **Formula Wt 285.22**

CAS Number 612–82–8

REQUIREMENTS

Sensitivity to chlorine . . . . Passes test
Solubility . . . . Passes test

MAXIMUM ALLOWABLE

Residue after ignition 0.1%

TESTS

SENSITIVITY TO CHLORINE. Dissolve 0.7 g in 50 mL of water. Add this solution, with constant stirring, to 50 mL of dilute hydrochloric acid (3 + 7). Add 1 mL of this solution to 20 mL of water containing 5 mL of chlorine standard. A distinct yellow color is produced that, when measured in a 1-cm cell against the chlorine standard at 430 nm, has an absorbance of not less than 0.42. (Absorbance readings should be made within 10 min.) (Limit 0.1 ppm free chlorine)

Chlorine Standard. Pipet 1.0 mL of a 2.5% sodium hypochlorite solution into a 1-L volumetric flask, and dilute with carbon dioxide-free water to volume. (Standardize the 2.5% sodium hypochlorite solution, and in the dilution correct for any departure from 2.5%.)

SOLUBILITY. Dissolve 0.50 g in 50 mL of water. The resulting solution is complete and colorless.

RESIDUE AFTER IGNITION. Place 1.0 g in a tared porcelain crucible, add 0.5 mL of sulfuric acid, and slowly ignite in a hood. Heat in a furnace at 800 ± 25 °C to constant weight.

Toluene
Methylbenzene

$C_6H_5CH_3$

Formula Wt 92.14

CAS Number 108–88–3

Suitable for use in Ultraviolet Spectrophotometry or general use. Product labeling shall designate the one or both of these uses for which suitability is represented on the basis of meeting the relevant requirements and tests. The Ultraviolet Spectrophotometry requirements include all of the requirements for general use.

REQUIREMENTS

General Use

Assay . . . . $\geq$99.5% $C_6H_5CH_3$

MAXIMUM ALLOWABLE

Color (APHA) 10
Residue after evaporation 0.001%
Substances darkened by sulfuric acid. . . . . Passes test
Sulfur compounds (as S) 0.003%
Water (H_2O) 0.03%

Specific Use

Ultraviolet Spectrophotometry

Wavelength (nm)	Absorbance (AU)
350–400	0.01
335	0.02
310	0.05
300	0.10
293	0.20
288	0.50
286	1.00

TESTS

ASSAY. Analyze the sample by gas chromatography using the general parameters cited on page 70. The following specific conditions are also required.

Column: Type I, methyl silicone

Measure the area under all peaks and calculate the toluene content in area percent. Correct for water content.

COLOR (APHA). (Page 19).

RESIDUE AFTER EVAPORATION. (Page 16). Evaporate 100.0 g (115 mL) to dryness in a tared dish on a hot plate ($\approx$100 °C) and dry the residue at 105 °C for 30 min.

SUBSTANCES DARKENED BY SULFURIC ACID. Shake 15 mL with 5 mL of sulfuric acid for 15–20 s and allow to stand for 15 min. The toluene layer should be colorless and the color of the acid should not exceed that of a color standard composed of 2 volumes of water plus 1 volume of a color standard containing 5 g of $CoCl_2 \cdot 6H_2O$, 40 g of $FeCl_3 \cdot 6H_2O$, and 20 mL of hydrochloric acid in a liter.

SULFUR COMPOUNDS. Place 30 mL of approximately 0.5 N potassium hydroxide in methanol in a conical flask, add 5.0 g (6.0 mL) of the sample, and boil the mixture gently for 30 min under a reflux condenser, avoiding the use of a rubber stopper or connection. Detach the condenser, dilute with 50 mL of water, and heat on a hot plate ($\approx$100 °C) until the toluene and methanol are evaporated. Add 50 mL of bromine water and heat for 15 min longer. Transfer the solution to a beaker, neutralize with dilute hydrochloric acid (1 + 3), add an excess of 1 mL of the acid, and evaporate to about 50 mL. Filter if necessary, heat the filtrate to boiling, add 5 mL of barium chloride reagent solution, digest in a covered beaker on a hot plate ($\approx$100 °C) for 2 h, and allow to stand overnight. If a precipitate is formed, filter, wash thoroughly, and ignite. Correct for the weight obtained in a complete blank test.

WATER. (Page 56, Method 2). Use 100 µL (87 mg) of the sample.

ULTRAVIOLET SPECTROPHOTOMETRY. Use the procedure on page 73 to determine the absorbance.

---

para-Toluenesulfonic Acid Monohydrate
4-Methylbenzenesulfonic Acid Monohydrate

$CH_3C_6H_4SO_3H \cdot H_2O$ **Formula Wt 190.22**

CAS Number 6192–52–5

REQUIREMENTS

Assay . . . .≥98.5% $CH_3C_6H_4SO_3H \cdot H_2O$
Water (H_2O) 9.5–11.5%

MAXIMUM ALLOWABLE

Clarity of solution . . . . Passes test
Residue after ignition 0.1%
Sulfate (SO_4) 0.3%
Heavy metals (as Pb) 0.001%
Iron (Fe) 0.01%
Sodium (Na) 0.002%

TESTS

ASSAY. (By acid–base titrimetry). Weigh accurately about 0.8 g, dissolve in 100 mL of water, add 0.10 mL of phenolphthalein indicator solution, and titrate with 0.1 N sodium hydroxide to a pink end point. One milliliter of 0.1 N sodium hydroxide corresponds to 0.01902 g of $CH_3C_6H_4SO_3H \cdot H_2O$.

WATER. (Page 55, Method 1). Use 0.5 g.

CLARITY OF SOLUTION. Dissolve 10 g in 50 mL of water. The solution should be clear and complete.

RESIDUE AFTER IGNITION. (Page 16). Use 5.0 g. Retain the residue to prepare sample solution A.

SULFATE. Dissolve 0.70 g in 180 mL of water, add 10 mL of hydrochloric acid, and heat to boiling. Add 10 mL of barium chloride reagent solution, digest in a covered beaker on a hot plate (≈100 °C) for 2 h, and allow to stand overnight. If a precipitate is formed, filter, wash thoroughly, and ignite. The weight of the precipitate should not be more than 0.005 g greater than the weight obtained in a complete blank test.

Sample Solution A for the Determination of Heavy Metals. Dissolve the residue from the test for residue after ignition with a mixture of 2 mL of nitric acid and 4 mL of hydrochloric acid, and evaporate to dryness on a hot plate (≈100 °C). Take up the residue with 1 mL of dilute hydrochloric acid (1 + 1) and 8 mL of hot water. Cool and dilute to 50 mL with water.

HEAVY METALS. (Page 28, Method 1). Use 20 mL of sample solution A (2.0 g).

IRON AND SODIUM. (By flame AAS, page 39).

Sample Stock Solution. Dissolve 10.0 g in water and dilute with water to 100 mL (1 mL = 0.10 g).

Element	Wavelength (nm)	Sample Wt (g)	Standard Added (mg)	Flame Type*	Background Correction
Fe	248.3	1.0	0.05; 0.10	A/A	Yes
Na	589.0	0.50	0.005; 0.01	A/A	No

*A/A is air/acetylene.

Trichloroacetic Acid

$$Cl_3C-\overset{\displaystyle O}{\overset{\displaystyle \|}{C}}-OH$$

CCl₃COOH   **Formula Wt 163.39**

CAS Number 76–03–9

REQUIREMENTS

Assay . . . . ≥99.0% CCl₃COOH

MAXIMUM ALLOWABLE

Residue after ignition 0.03%
Chloride (Cl) 0.001%
Nitrate (NO₃) 0.002%
Phosphate (PO₄) . . . . 5 ppm
Sulfate (SO₄) 0.02%
Heavy metals (as Pb) 0.002%
Iron (Fe) 0.001%
Substances darkened by sulfuric acid . . . . Passes test

TESTS

ASSAY. (By acid–base titrimetry). Weigh accurately 5.0 g, previously dried for 24 h over an efficient desiccant, and dissolve in 20 mL of water in a 500-mL reflux flask. Add 0.15 mL of methyl red indicator solution and titrate with 1 N sodium hydroxide to a distinct yellow color, then add 50.0 mL of 1 N sulfuric acid. Add sufficient dioxane to make the final solution at least 55% dioxane. Add silicon carbide boiling chips and boil gently under complete reflux for 1 h. Cool, add 0.15 mL of methyl red indicator solution, and titrate the excess of sulfuric acid with 1 N sodium hydroxide, the end point being the distinct transition from a light orange color to a definite yellow color. Correct for the acid present in the dioxane by running a complete blank, containing 2.5 mL of Chloroform.

$$\% \ CCl_3COOH = (V \times 0.1634 \times 100) / \text{Sample wt (g)}$$

where V is the volume, in milliliters, of 1 N sulfuric acid consumed by the sample. This quantity is the volume of 1 N sodium hydroxide consumed by the blank minus the volume of 1 N sodium hydroxide consumed by the sample.

RESIDUE AFTER IGNITION. (Page 16). Ignite 5.0 g.

CHLORIDE. (Page 27). Use 1.0 g.

NITRATE. (Page 30). For sample solution A use 0.25 g. For control solution B use 0.25 g and 0.5 mL of standard nitrate solution.

> *Sample Solution A for the Determination of Phosphate, Sulfate, Heavy Metals, and Iron.* Evaporate 5.0 g to dryness in a beaker on a hot plate ($\approx$100 °C). Dissolve in about 10 mL of water, filter if necessary, and dilute with water to 100 mL (1 mL = 0.05 g).

PHOSPHATE. (Page 32, Method 2). Dilute 40 mL of sample solution A (2.0-g sample) with water to about 80 mL. Add 0.5 g of ammonium molybdate and adjust the pH to 1.8 (using a pH meter) with dilute hydrochloric acid (1 + 9). Heat to boiling, cool to room temperature, dilute with water to 90 mL, and add 10 mL of hydrochloric acid. Continue as described, and carry along a standard containing 0.01 mg of phosphate ion (PO_4) treated in the same manner as the 40 mL of sample solution A.

SULFATE. (Page 33, Method 1). Use 5.0 mL of sample solution A (0.25-g sample).

HEAVY METALS. (Page 28, Method 1). Use 20 mL of sample solution A (1-g sample) with water to 25 mL.

IRON. (Page 30, Method 1). Use 20 mL of sample solution A (1-g sample).

SUBSTANCES DARKENED BY SULFURIC ACID. Dissolve 1.0 g in 10 mL of sulfuric acid and digest in a covered beaker on a hot plate ($\approx$100 °C) for 15 min. Any color should not exceed that produced when 10 mL of sulfuric acid is added to 1 mL of water containing 0.1 mg of sucrose and heated in the same way as the sample.

1,1,1-Trichloroethane
Methyl Chloroform

CH_3CCl_3 Formula Wt 133.41

CAS Number 71–55–6

NOTE. This solvent usually contains additives to retard decomposition or to inhibit the corrosion of metals, particularly aluminum,
with which it might come into contact. The nature and concentrations of such compounds should be stated on the label. Typical stabilizers include dioxane and alcohols such as 2-butanol.

REQUIREMENTS

Assay . . . . 98.5–101.0% (CH_3CCl_3 + active additives)

MAXIMUM ALLOWABLE

Color (APHA) 10
Water (H_2O). 0.10%
Residue after evaporation. 0.001%
Titrable acid . . . . 0.00030 meq/g

TESTS

ASSAY. Analyze the sample by gas chromatography using the general parameters
cited on page 70. The following specific conditions are also required.

Column: 6 m × 3.2 mm stainless steel, packed with 15% SP 1000 on Supelcoport

Column Temperature: 70 °C for 8 min, then programmed at 8 °C/min to
200 °C

Injection Port Temperature: 200 °C

Detector Temperature: 250 °C

Carrier Gas: Nitrogen at 18 mL/min

Sample Size: 1 µL

Detector: Flame ionization

Range: 100

Approximate Retention Times (min): Vinylidene chloride, 4.8; butylene
oxide, 7.9; 1,1-dichloroethane, 8.9; carbon tetrachloride, 9.0; 1,1,1-trichloro-
ethane, 9.8; 2-propanol, 10.5; dichloromethane, 10.8; 1,1,2-trichloroethyl-
ene, 13.5; Chloroform, 14.2; *n*-propanol, 14.7; perchloroethylene, 15.2; 1,2-
dichloroethane, 16.0; dioxane, 16.2; 2-methyl-1-propanol, 17.0; *n*-butanol,
17.8; 1,1,2-trichloroethane, 21.3. Carbon tetrachloride and 1,1-dichloro-
ethane are not resolved from 1,1,1-trichloroethane.

Measure the area under all peaks and calculate the 1,1,1-trichloroethane content in area percent plus intentional additives. Correct for water content.

COLOR (APHA). (Page 19).

WATER. (Page 55, Method 1). Use 13.3 g (10 mL) of the sample.

RESIDUE AFTER EVAPORATION. Evaporate 100.0 g (75 mL) to dryness in a tared dish on a hot plate ($\approx$100 °C), and dry the residue at 105 °C for 30 min.

TITRABLE ACID. Measure 100 g (75 mL) into a dry conical 250-mL flask. Add 0.2 mL of bromthymol blue indicator, and titrate with 0.02 N sodium hydroxide in methanol to a greenish-blue color. Not more than 1.50 mL should be required.

Trichloroethylene
Trichloroethene

$$Cl_2C=CHCl$$

$CHCl:CCl_2$　　　　　　　　　　　　　　　　　　　　　**Formula Wt 131.39**

CAS Number 79–01–6

NOTE. This material usually contains a stabilizer. If a stabilizer is present, the amount and type should be stated on the label.

REQUIREMENTS

Assay . . . . $\geq$99.5% $CHCl:CCl_2$

MAXIMUM ALLOWABLE

Color (APHA) 10
Residue after evaporation 0.001%
Titrable acid . . . . 0.0001 meq/g
Titrable base . . . . 0.0003 meq/g
Water (H_2O) 0.02%
Heavy metals (as Pb) . . . . 1 ppm
Free halogens . . . . Passes test

TESTS

ASSAY. Analyze the sample by gas chromatography using the general parameters cited on page 70. The following specific conditions are also required.

Column: Type I, methyl silicone

Measure the area under all peaks and calculate the trichloroethylene content in area percent. Correct for water content and added stabilizer.

COLOR (APHA). (Page 19).

RESIDUE AFTER EVAPORATION. (Page 16). Evaporate 100.0 g (69 mL) to dryness in a tared dish on a hot plate ($\approx$100 °C) in a well-ventilated hood and dry the residue at 105 °C for 30 min.

TITRABLE ACID AND TITRABLE BASE. To 25 mL of water and 0.10 mL of phenolphthalein indicator solution in a 250-mL glass-stoppered flask, add 0.01 N sodium hydroxide until a slight pink color appears. Add 36.0 g (25 mL) of sample and shake for 30 s. If the pink color persists, titrate with 0.01 N hydrochloric acid, shaking repeatedly, until the pink color just disappears. Not more than 0.90 mL of 0.01 N hydrochloric acid should be required (titrable base). If the pink color is discharged when the sample is added, titrate with 0.01 N sodium hydroxide until the pink color is restored. Not more than 0.50 mL of 0.01 N sodium hydroxide should be required (titrable acid).

WATER. (Page 56, Method 2). Use 100 µL (140 mg) of the sample.

HEAVY METALS. Evaporate 10.0 g (7 mL) to dryness in a glass evaporating dish on a hot plate ($\approx$100 °C) inside a well-ventilated hood. For the standard, evaporate a solution containing 0.01 mg of lead to dryness. Cool, add 2 mL of hydrochloric acid to each, and slowly evaporate to dryness on a hot plate ($\approx$100 °C). Moisten the residues with 0.05 mL of hydrochloric acid, add 10 mL of hot water, and digest for 2 min. Filter if necessary through a small filter, wash the evaporating dish and the filter with about 10 mL of water, and dilute each with water to 25 mL. Adjust the pH of the standard and sample solutions to between 3 and 4 (using a pH meter) with 1 N acetic acid or ammonium hydroxide (10% NH_3), dilute with water to 40 mL, and mix. Add 10 mL of freshly prepared hydrogen sulfide water to each and mix. Any color in the solution of the sample should not exceed that in the standard.

FREE HALOGENS. Shake 10 mL for 2 min with 10 mL of water to which 0.10 mL of 10% potassium iodide reagent solution has been added and allow to separate. The lower layer should not show a violet tint.

1,1,2-Trichlorotrifluoroethane

$Cl_2FCCClF_2$ **Formula Wt 187.40**

CAS Number 76–13–1

Suitable for use in extraction–concentration analysis or ultraviolet spectrophotometry or for general use. Product labeling shall designate the one or more of these uses for which suitability is presented on the basis of meeting the relevant requirements and tests. The ultraviolet spectrophotometry requirements include all the requirements for general use. The extraction–concentration suitability requirements include only the general use requirement for color.

REQUIREMENTS

General Use

Assay (GC). . . . .≥99.8% $C_2Cl_3F_3$

MAXIMUM ALLOWABLE

Color (APHA). 10
Titrable acid. . . . . 0.0003 meq/g
Residue after evaporation. . . . . 5 ppm
Water. 0.01%

Specific Use

Ultraviolet Spectrophotometry

Wavelength (nm)	Absorbance (AU)
300	0.01
254	0.02
240	0.20
231	1.0

Extraction–Concentration Suitability

Absorbance . . . . Passes test
GC–FID . . . . Passes test
GC–ECD . . . . Passes test

TESTS

ASSAY. Analyze the sample by gas chromatography using the general parameters cited on page 70. The following specific conditions are also required.

Column: Type 1, methyl silicone

Measure the area under all the peaks and calculate the 1,1,2-trichloro-1,2,2-trifluoroethane content in area percent. Correct for water content.

COLOR (APHA). (Page 19).

TITRABLE ACID. To 25 mL of water in a glass-stoppered flask, add 0.2 mL of phenolphthalein indicator solution and add 0.01 N sodium hydroxide until a slight pink color appears. Add 33.0 g (21 mL) of sample and shake for 30 s. Titrate with 0.01 N sodium hydroxide until the pink color is restored. Not more than 1.0 mL should be required.

RESIDUE AFTER EVAPORATION. (Page 16). Evaporate 129 mL (200.0 g) to dryness in a tared dish on a low-setting hot plate and dry the residue at 105 °C for 30 min. The weight of the residue should not exceed 0.001 g.

WATER. (Page 56, Method 2). Use 1.0 mL (1.6 g) of the sample.

ULTRAVIOLET SPECTROPHOTOMETRY. Use the procedure on page 73 to determine the absorbance.

EXTRACTION–CONCENTRATION SUITABILITY. Analyze the sample by using the general procedure cited on page 72. In the sample preparation step, exchange the sample with hexanes, page 324, suitable for extraction–concentration.

2,2,4-Trimethylpentane

Isooctane

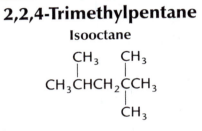

(CH₃)₃CCH₂CH(CH₃)₂   **Formula Wt 114.23**

CAS Number 540–84–1

Suitable for use in extraction–concentration analysis or ultraviolet spectrophotometry or for general use. Product labeling shall designate the one or more of these uses for which suitability is represented based on meeting the relevant requirements and tests. The ultraviolet spectrophotometry suitability requirements include all of the suitability requirements for general use. The extraction–concentration requirements include only the general use requirement for color.

REQUIREMENTS

General Use

Assay . . . . ≥99.0% (CH₃)₃CCH₂CH(CH₃)₂

MAXIMUM ALLOWABLE

Color (APHA) 10
Residue after evaporation. 0.001%
Water-soluble titrable acid . . . . 0.0003 meq/g
Sulfur compounds (as S) 0.005%

Specific Use

Ultraviolet Spectrophotometry

Wavelength (nm)	Absorbance (AU)
250–400	0.01
240	0.04
230	0.10
220	0.20
210	1.0

Extraction–Concentration Suitability

Absorbance	Passes test
GC-FID	Passes test
GC-ECD	Passes test

TESTS

ASSAY. Analyze the sample by gas chromatography using the general parameters cited on page 70. The following specific conditions are also required.

Column: Type I, methyl silicone

Measure the area under all peaks and calculate the 2,2,4-trimethylpentane content in area percent. Correct for water content.

COLOR (APHA). (Page 19).

RESIDUE AFTER EVAPORATION. (Page 16). Evaporate 100.0 g (145 mL) to dryness in a tared dish on a hot plate ($\approx$100 °C) and dry the residue at 105 °C for 30 min.

WATER-SOLUBLE TITRABLE ACID. To 30.0 g (44 mL) in a separatory funnel, add 50 mL of water, and shake vigorously for 2 min. Allow the layers to separate, draw off the aqueous layer, and add 0.15 mL of phenolphthalein indicator solution to the aqueous layer. Not more than 1.0 mL of 0.01 N sodium hydroxide should be required to produce a pink color.

SULFUR COMPOUNDS. To 30 mL of 0.5 N potassium hydroxide in methanol in a conical flask, add 5.5 g (8 mL) of the sample, and boil the mixture gently for 30 min under a reflux condenser, avoiding the use of a rubber stopper or connection. Detach the condenser, dilute with 50 mL of water, and heat on a hot plate ($\approx$100 °C) until the 2,2,4-trimethylpentane and methanol are evaporated. Add 50 mL of bromine water, heat for 15 min longer, and transfer to a beaker. Neutralize with dilute hydrochloric acid (1 + 3), add an excess of 1 mL of the acid, evaporate to about 50 mL, and filter if necessary. Heat the filtrate to boiling, add 5 mL of barium chloride reagent solution, digest in a covered beaker on a hot plate ($\approx$100 °C) for 2 h, and allow to stand overnight. If a precipitate is formed, filter, wash thoroughly, and ignite. Correct for the weight obtained in a complete blank test.

ULTRAVIOLET SPECTROPHOTOMETRY. Use the procedure on page 73 to determine the absorbance.

EXTRACTION–CONCENTRATION SUITABILITY. Analyze the sample using the general procedure cited on page 72.

Tris(hydroxymethyl)aminomethane
2-Amino-2-(hydroxymethyl)-1,3-propanediol
"Tris"

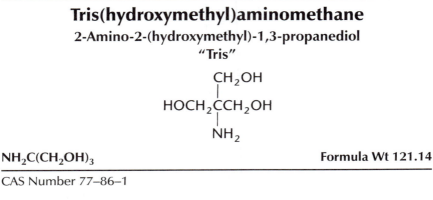

$NH_2C(CH_2OH)_3$	**Formula Wt 121.14**

CAS Number 77–86–1

REQUIREMENTS

Assay (dry basis) .99.8–100.1% $C_4H_{11}NO_3$

MAXIMUM ALLOWABLE

Absorbance . . . . Passes test
Water (H_2O). 2%
Insoluble matter. 0.005%
Heavy metals (as Pb) . . . . 5 ppm
Iron (Fe) . . . . 5 ppm

TESTS

ASSAY. (By acid–base titrimetry). Weigh accurately 0.5 g of sample previously dried at 105 °C for 3 hours. Transfer to a 200-mL beaker, dissolve in 50 mL of carbon dioxide- and ammonia-free water, and titrate the solution with 0.1 N hydrochloric acid to a pH of 4.7, using a suitable pH meter. One milliliter of 0.1 N hydrochloric acid corresponds to 0.012114 g of $C_4H_{11}NO_3$.

ABSORBANCE. Determine the absorbance of a 40% solution of the sample in water in a 1.00-cm cell at 290 nm against water in a similar matched cell set at zero absorbance as the reference liquid. The absorbance should not exceed 0.2 absorbance units.

WATER. (Page 55, Method 1). Use 1.0 g of the sample.

INSOLUBLE MATTER. Dissolve 20 g in 200 mL of water. Filter through a tared filtering crucible, wash thoroughly, and dry at 105 °C.

HEAVY METALS. (Page 28, Method 2). Dissolve 6.0 g in about 20 mL of water, and dilute with water to 30 mL. Use 25 mL to prepare the sample solution, and use the remaining 5.0 mL to prepare the control solution.

IRON. (Page 30, Method 1). Use a 2.0-g sample.

Uranyl Acetate Dihydrate

$UO_2(CH_3COO)_2 \cdot 2H_2O$ **Formula Wt 424.15**

CAS Number 6159–44–0

> *NOTE.* The formula weight of this reagent is likely to deviate from the value cited since the natural distribution of uranium isotopes is often altered in current sources of uranium compounds.

REQUIREMENTS

Assay . . . . 98.0–102.0% $UO_2(CH_3COO)_2 \cdot 2H_2O$

MAXIMUM ALLOWABLE

Insoluble matter. 0.01%
Chloride (Cl) 0.003%
Sulfate (SO_4) 0.01%
Alkalies and alkaline earths (as sulfates). 0.05%
Heavy metals (as Pb) 0.002%
Iron (Fe). 0.001%
Substances reducing permanganate (as U^{IV}) 0.06%

TESTS

ASSAY. (Total uranium by oxidimetry). Weigh accurately 0.6 g, dissolve in 100 mL of 1 N sulfuric acid, and add enough 0.1 N potassium permanganate to give a distinct pink color. Pass 100 mL of 1 N sulfuric acid through a Jones reductor (page 86), followed by 100 mL of water, and discard. Pass the sample through the reductor, followed in succession by 100 mL each of 1 N sulfuric acid and water, and collect the effluent in a titration flask. Bubble air through the solution for 5–10 min, then titrate with 0.1 N potassium permanganate to a pink color. Run a reagent blank and subtract it from the sample titration. One milliliter of 0.1 N potassium permanganate corresponds to 0.02121 g of $UO_2(CH_3COO)_2 \cdot 2H_2O$.

INSOLUBLE MATTER. Dissolve 10.0 g in 185 mL of water plus 5 mL of glacial acetic acid at room temperature. Filter through a tared filtering crucible, wash thoroughly with water, and dry at 105 °C.

Sample Solution A for the Determination of Chloride, Sulfate, Alkalies and Alkaline Earths, Heavy Metals, and Iron. Dissolve 15 g in 275 mL of water plus 5 mL of glacial acetic acid, add 10 mL of 30% hydrogen peroxide, and heat to coagulate the precipitate. Decant through a fritted-glass filter, without washing, and dilute with water to 300 mL (1 mL = 0.05 g).

CHLORIDE. (Page 27). Dilute 6.7 mL of sample solution A (0.33-g sample) with water to 20 mL.

SULFATE. (Page 33, Method 1). Use 10 ml of sample solution A (0.5-g sample).

ALKALIES AND ALKALINE EARTHS. To 40 mL of sample solution A (2-g sample) add 0.10 mL of sulfuric acid, evaporate to dryness, and gently ignite at about 450 °C. Digest the residue with 25 mL of hot water, filter, evaporate the filtrate to dryness in a tared dish, and ignite at 800 ± 25 °C for 15 min.

HEAVY METALS. (Page 28, Method 1). Evaporate 20 mL of sample solution A (1-g sample) to dryness, dissolve the residue in about 20 mL of water, and dilute with water to 25 mL.

IRON. (Page 30, Method 1). Evaporate 20 mL of sample solution A (1-g sample) to dryness, dissolve the residue in 2 mL of hydrochloric acid, and dilute with water to 50 mL. Use the solution without further acidification.

SUBSTANCES REDUCING PERMANGANATE. Prepare two solutions, each containing 3.0 g in 200 mL of dilute sulfuric acid (1 + 99). Titrate one solution with 0.1 N potassium permanganate. Not more than 0.20 mL of permanganate should be required to cause a color change that can be observed by comparison with the solution not titrated. The 0.20 mL includes the 0.05 mL allowed to produce the color change in the absence of uranous compounds.

Uranyl Nitrate Hexahydrate

$UO_2(NO_3)_2 \cdot 6H_2O$ **Formula Wt 502.13**

CAS Number 13520–83–7

> *NOTE.* The formula weight of this reagent is likely to deviate from the value cited, since the natural distribution of uranium isotopes is often altered in current sources for uranium compounds.

REQUIREMENTS

Assay . . . . 98.0–102.0% $UO_2(NO_3)_2 \cdot 6H_2O$

MAXIMUM ALLOWABLE

Insoluble matter. 0.005%
Chloride (Cl) 0.002%
Sulfate (SO_4) 0.005%
Alkalies and alkaline earths (as sulfates). 0.1%
Heavy metals (as Pb) 0.002%
Iron (Fe). 0.002%
Substances reducing permanganate (as U^{IV}) 0.06%

TESTS

ASSAY. (By gravimetry for uranium). Weigh accurately 0.5 g and dissolve in 100 mL of water in a 250-mL beaker. Heat the solution to boiling and add ammonium hydroxide, free from carbonate, until precipitation is complete. Filter through a tared Gooch crucible, wash with 1% ammonium nitrate solution, and ignite gently with free access of air, to constant weight. Multiply the oxide weight by 178.9 and divide by the sample weight to obtain %$UO_2(NO_3)_2 \cdot 6H_2O$.

INSOLUBLE MATTER. (Page 15). Use 20 g dissolved in 200 mL of water.

Sample Solution A for the Determination of Chloride, Sulfate, Alkalies and Alkaline Earths, Heavy Metals, and Iron. Dissolve 10.0 g in 200 mL of water, add 7 g of ammonium acetate and 10 mL of 30% hydrogen peroxide, and heat to coagulate the precipitate. Decant through a fritted-glass filter, without washing, and dilute with water to 250 mL in a volumetric flask (1 mL = 0.025 g).

CHLORIDE. (Page 27). Dilute 10 mL of sample solution A (0.5-g sample) with water to 20 mL.

SULFATE. (Page 33, Method 1). Evaporate 20 mL of sample solution A (1-g sample) to 10 mL.

ALKALIES AND ALKALINE EARTHS. To 20 mL of sample solution A (1-g sample) add 0.10 mL of sulfuric acid, evaporate to dryness in a tared evaporating dish, and ignite at 800 ± 25 °C for 15 min.

HEAVY METALS. (Page 28, Method 1). Evaporate 20 mL of sample solution A (1-g sample) to dryness, dissolve the residue in about 20 mL of water, and dilute with water to 25 mL.

IRON. (Page 30, Method 1). Use 10 mL of sample solution A (0.5-g sample).

SUBSTANCES REDUCING PERMANGANATE. Prepare two solutions, each containing 3.0 g in 200 mL of dilute sulfuric acid (1 + 99). Titrate one solution with 0.1 N potassium permanganate. Not more than 0.20 mL of permanganate should be required to cause a color change that can be observed by comparison with the solution not titrated. The 0.20 mL includes 0.05 mL allowed to produce the color change in the absence of uranous compounds.

Urea

$$H_2N-\overset{\overset{\displaystyle O}{\|}}{C}-NH_2$$

NH$_2$CONH$_2$ **Formula Wt 60.06**

CAS Number 57–13–6

REQUIREMENTS

Assay 99.0–100.5%
Melting point . . . . 132–135 °C

MAXIMUM ALLOWABLE

Insoluble matter. 0.01%
Residue after ignition 0.01%
Chloride (Cl) . . . . .5 ppm
Sulfate (SO$_4$) 0.001%
Heavy metals (as Pb) 0.001%
Iron (Fe). 0.001%

TESTS

ASSAY. (By nonaqueous titration). Accurately weigh about 0.06 g of sample into a 150-mL beaker. Add 75 mL of acetonitrile and about 2 mL of 0.1 N perchloric acid in glacial acetic acid from a buret. When the sample is completely dissolved, finish the titration potentiometrically using a glass electrode set (with 10% lithium perchlorate in acetonitrile as the filling solution in a fast flow reference electrode). Each mL of the 0.1 N perchloric acid titer is equivalent to 0.006006 g of urea.

MELTING POINT. (Page 22).

INSOLUBLE MATTER. (Page 15). Use 10.0 g dissolved in 100 mL of water.

RESIDUE AFTER IGNITION. (Page 16). Ignite 10.0 g.

CHLORIDE. (Page 27). Use 2.0 g.

SULFATE. Dissolve 4.0 g and about 10 mg of sodium carbonate in 10 mL of hydrochloric acid. Add 10 mL of nitric acid and digest in a covered beaker on a hot plate (≈100 °C) until reaction ceases. Add 5 mL of hydrochloric acid and 10 mL of nitric acid and again digest in a covered beaker on a hot plate (≈100 °C) until reaction ceases. Remove the cover and evaporate to dryness. Dissolve the residue in 4 mL of water plus 1 mL of dilute hydrochloric acid (1 + 19) and filter through a small filter. Wash with two 2-mL portions of water, dilute with water to 10 mL, and add 1 mL of barium chloride reagent solution. Any turbidity should not exceed that produced when a solution containing 0.04 mg of sulfate ion (SO_4) is treated exactly like the 4.0-g sample. Compare 10 min after adding the barium chloride to the sample and standard solutions.

HEAVY METALS. (Page 28, Method 1). Dissolve 2.0 g in about 20 mL of water, and dilute with water to 25 mL.

IRON. (Page 30, Method 1). Use 1.0 g.

Variamine Blue B
N-[p-Methoxyphenyl]-p-phenylenediamine Hydrochloride

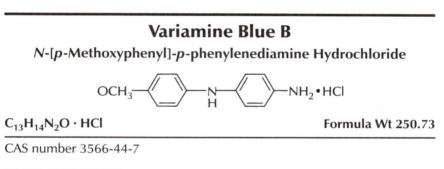

$C_{13}H_{14}N_2O \cdot HCl$ **Formula Wt 250.73**

CAS number 3566-44-7

REQUIREMENTS

Suitability as a complexometric ion indicator . . . . Passes test

TESTS

SUITABILITY AS A COMPLEXOMETRIC ION INDICATOR. Weigh 0.2 g of sample and suspend in 20 mL of water. Stir for five minutes. Mix 200 mL of water, 10 mL of 0.1 M EDTA, and 2 mL of 20% hydrochloric acid. Heat the solution to 40–50 °C. Adjust the pH to 2.5 ± 0.2 with 10% sodium acetate solution using a pH meter. Add 0.2 mL of the sample solution and titrate with ferric chloride hexahydrate solution. The color change is from bright yellow to burgundy. Add back gradually 0.10 ± 0.05 mL of EDTA in 0.05 mL increments. The color should return to bright yellow.

> ***Ferric Chloride Hexahydrate Solution.*** Dissolve 2.5 g ferric chloride hexahydrate in 100 mL of dilute hydrochloric acid (1+99).

Water, Reagent
Distilled Water, Deionized Water

H₂O   **Formula Wt 18.02**

CAS Number 7732–18–5

NOTE. The requirements for water are for freshly prepared material and not intended for bottled or stored product.

REQUIREMENTS

General Use

MAXIMUM ALLOWABLE

Specific conductance at 25 °C . . . . 2.0×10^{-6} ohm^{-1} cm^{-1}
Silicate (as SiO_2). . . . . 0.01 ppm
Heavy metals (as Pb) . . . . 0.01 ppm
Substances reducing permanganate . . . . Passes test
Chloride (Cl) . . . . 0.4 ppm
Nitrate (NO_4) . . . . 0.4 ppm
Phosphate (PO_4) . . . . 1.0 ppm
Sulfate (SO_4) . . . . 1.0 ppm

Specific Use

Ultraviolet Spectrophotometry

Wavelength (nm)	Absorbance (AU)
300	0.005
254	0.005
210	0.010
200	0.010

Liquid Chromatography Suitability

Absorbance . . . . Passes test
Gradient elution. . . . . Passes test

TESTS

SPECIFIC CONDUCTANCE. Fill a conductance cell that has been properly standardized. Take precautions to prevent the water from absorbing carbon dioxide, ammonia, hydrogen chloride, or any other gases commonly present in a laboratory. Immerse in a constant-temperature bath at 25 °C, equilibrate, and measure the resistance with a suitable conductance bridge. The calculated specific conductance should not exceed 2.0×10^{-6} ohm^{-1} cm^{-1}.

SILICATE. Evaporate 1500 mL to 80 mL in a platinum dish on a hot plate (≈100 °C) (in portions, if necessary). Add 10 mL of silica-free ammonium hydroxide,* cool, and add 5 mL of a 10% solution of ammonium molybdate. Adjust the pH to between 1.7 and 1.9 (using a pH meter) with hydrochloric acid or silica-free ammonium hydroxide. Heat the solution just to boiling, cool to room temperature, and dilute with water to 90 mL. Transfer the solution to a separatory funnel, add 10 mL of hydrochloric acid and 35 mL of ethyl ether, and shake vigorously for a few minutes. Allow the two layers to separate, draw off the aqueous layer into another separatory funnel, and add 10 mL of hydrochloric acid and 50 mL of butyl alcohol. Shake vigorously, allow to separate, draw off the aqueous phase, and discard. Wash the butyl alcohol phase three times with 20 mL of dilute hydrochloric acid (1 + 99), discarding the aqueous solution each time. Add 0.5 mL of freshly prepared 2% stannous chloride dihydrate solution in hydrochloric acid to the butyl alcohol. Any blue color should not exceed that produced by 0.015 mg of silica (SiO_2) when treated exactly like the sample. Because the blue color fades on standing, compare immediately after the reduction with stannous chloride. Treatment again with stannous chloride will restore a fading blue color to its original intensity.

Silicate Solution in Ammonium Hydroxide. (0.01 mg of SiO_2 in 1 mL). To prepare a stock solution of silica dissolve 0.946 g of assayed sodium silicate, $Na_2SiO_3 \cdot 9H_2O$, in 9 mL of silica-free ammonium hydroxide and dilute to 100 mL (99.7 g at 25 °C) with water. Dilute 5.0 mL of this solution plus 5 mL of silica-free ammonium hydroxide to 1 L (997.1 g at 25 °C) with water. If volumetric polyolefin ware is unavailable, dilute the silica solution by weight.

HEAVY METALS. (Page 28, Method 1). Add 10 mg of sodium chloride and 1 mL of 1 N hydrochloric acid to 2 L, and evaporate to dryness on a hot plate (≈100 °C). Dissolve the residue in about 20 mL of water and dilute with water to 25 mL. For the control, evaporate 1 mL of 1 N hydrochloric acid to dryness on a hot plate (≈100 °C), cool, add 0.02 mg of lead ion (Pb), and dilute with water to 25 mL.

SUBSTANCES REDUCING PERMANGANATE. To 500 mL add 1 mL of sulfuric acid and 0.30 mL of 0.01 N potassium permanganate and allow to stand for 1 h at room temperature. The pink color should not be entirely discharged.

CHLORIDE, NITRATE, PHOSPHATE, AND SULFATE. To each of three 100-mL beakers, add 2.0 mg of sodium carbonate. To two beakers, add 50 mL of the sample. To one of these beakers add 0.02 mg each of chloride and nitrate ions and

*Ammonium Hydroxide (Silica-Free). Aqueous ammonia in contact with glass, even momentarily, will dissolve sufficient silica to make the reagent useless for this test. Because ammonium hydroxide is ordinarily supplied in glass containers by commercial sources, silica-free ammonium hydroxide must be prepared in the laboratory. The most convenient method is the saturation of water with ammonia gas from a cylinder of compressed anhydrous ammonia. Plastic tubing and bottles—for example, polyethylene—must be used throughout.

0.05 mg each of phosphate and sulfate ions. To the third beaker (blank) add 20 mL of 0.06 micromho cm^{-1} water. Evaporate each beaker on a steam bath or low-temperature hot plate to about 20 mL. Cool. Dilute to 25 mL. Analyze 100-µL aliquots by ion chromatography and measure the various peaks. The differences between the sample and the blank should not be greater than the corresponding differences between the standard and the sample.

ULTRAVIOLET SPECTROPHOTOMETRY. Use the procedure on page 73 to determine the absorbance.

LIQUID CHROMATOGRAPHY SUITABILITY. Analyze the sample by using the procedure on page 71.

Water, Ultratrace
(suitable for use in ultratrace elemental analysis)

H$_2$O   **Formula Wt 18.02**

CAS number 7732–18–5

> *NOTE.* Reagent must be packaged in a pre-cleaned Teflon or polyethylene bottle and used in a clean environment to maintain purity.

REQUIREMENTS

Appearance . . . . Clear

MAXIMUM ALLOWABLE
Aluminum (Al) . . . . 0.5 ppb
Barium (Ba) . . . . 0.5 ppb
Boron (B) . . . . 0.5 ppb
Cadmium (Cd) . . . . 0.5 ppb
Calcium (Ca) . . . . 0.5 ppb
Chromium (Cr) . . . . 0.5 ppb
Cobalt (Co) . . . . 0.5 ppb
Copper (Cu) . . . . 0.5 ppb
Iron (Fe) . . . . 0.5 ppb
Lead (Pb) . . . . 0.5 ppb
Lithium (Li) . . . . 0.5 ppb
Magnesium (Mg) . . . . 0.5 ppb
Manganese (Mn) . . . . 0.5 ppb
Molybdenum (Mo) . . . . 0.5 ppb
Mercury (Hg) . . . . 0.5 ppb
Potassium (K) . . . . 0.5 ppb
Selenium . . . . 5 ppb

Silicon (Si) . . . . 0.5 ppb
Sodium (Na) . . . . 0.5 ppb
Strontium (Sr) . . . . 0.5 ppb
Tin (Sn) . . . . 0.5 ppb
Titanium (Ti) . . . . 0.5 ppb
Vanadium (V). . . . . 0.5 ppb
Zinc (Zn) . . . . 0.5 ppb
Zirconium (Zr). . . . . 0.5 ppb

TESTS

TRACE METALS. Determine the aluminum, barium, boron, cadmium, calcium, chromium, cobalt, copper, iron, lead, lithium, magnesium, manganese, molybdenum, nickel, potassium, sodium, strontium, tin, titanium, vanadium, zinc, and zirconium by the ICP–OES method described on page 42.

Determine the mercury by the CVAAS procedure described on page 40.

Determine the selenium by the HGAAS method described on page 41. To a set of 3 100-mL volumetric flasks, transfer 50 g (50 mL) of sample. To 2 flasks, add the specified amount of selenium standard from the following table. Dilute each to 100 mL with (1 + 1) hydrochloric acid. Mix. Prepare fresh working standard before use.

Element	Wavelength (nm)	Sample Wt (g)	Standard Addition (µg)
Se	196.0	50.0	0.25; 0.50(= 5 and 10 ppb)

Xylenes
Dimethylbenzenes

$C_6H_4(CH_3)_2$ **Formula Wt 106.17**

CAS Number 1330–20–7

NOTE. This reagent is generally a mixture of the ortho, meta, and para isomers and may contain some ethylbenzene.

REQUIREMENTS

Assay . . . . ≥98.5% xylene isomers plus ethylbenzene
(ethylbenzene not to exceed 25%)

MAXIMUM ALLOWABLE

Color (APHA) 10
Residue after evaporation 0.002%
Substances darkened by sulfuric acid . . . . Passes test
Sulfur compounds (as S) 0.003%
Water (H_2O) 0.05%

TESTS

ASSAY. Analyze the sample by gas chromatography using the general parameters cited on page 70. The following specific conditions are also required.

Column: Type I, methyl silicone

Measure the area under all peaks and calculate the area percent of each peak. The total percent of xylenes is the sum of the percents of ortho, meta, and para isomers and of ethylbenzene. Correct for water content.

COLOR (APHA). (Page 19).

RESIDUE AFTER EVAPORATION. (Page 16). Evaporate 100 g (115 mL) to dryness in a tared dish on a hot plate (≈100 °C) and dry the residue at 105 °C for 30 min.

SUBSTANCES DARKENED BY SULFURIC ACID. Shake 15 mL with 5 mL of sulfuric acid for 15–20 s and allow to stand for 15 min. The xylene layer should be colorless, and the color of the acid should not exceed that of a color standard composed of 1 volume of water and 3 volumes of a color standard containing 5 g of $CoCl_2 \cdot 6H_2O$, 40 g of $FeCl_3 \cdot 6H_2O$, and 20 mL of hydrochloric acid in a liter.

SULFUR COMPOUNDS. Place 30 mL of approximately 0.5 N potassium hydroxide in methanol in a conical flask, add 6.0 mL of the sample, and boil the mixture gently for 30 min under a reflux condenser, avoiding the use of a rubber stopper or connection. Detach the condenser, dilute with 50 mL of water, and heat on a hot plate (≈100 °C) until the xylenes and methanol have evaporated. Add 50 mL of bromine water and heat for 15 min longer. Transfer the solution to a beaker, neutralize with dilute hydrochloric acid (1 + 3), add an excess of 1 mL of the acid, and concentrate to about 50 mL. Filter if necessary, heat the filtrate to boiling, add 5 mL of barium chloride reagent solution, digest in a covered beaker on a hot plate (≈100 °C) for 2 h, and allow to stand overnight. If a precipitate is formed, filter, wash thoroughly, and ignite. Correct for the weight obtained in a complete blank test.

WATER. (Page 56, Method 2). Use 100 μL (87 mg) of the sample.

Xylenol Orange

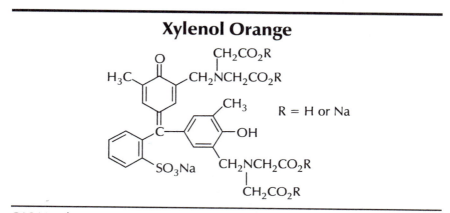

R = H or Na

CAS Number 1611–35–4 (acid); 3618–43–7 (salt)

NOTE. This standard applies to both the free acid form and the salt form of this reagent.

REQUIREMENTS

Clarity of solution . . . . Passes test
Suitability for zinc titration . . . . Passes test

TESTS

CLARITY OF SOLUTION. If the indicator is the acid form, dissolve 0.1 g in 100 mL of alcohol. If the indicator is a salt form, dissolve 0.1 g in 100 mL of water. Not more than a faint trace of turbidity or insoluble matter should remain. Reserve the solution for the test for suitability for zinc titration.

SUITABILITY FOR ZINC TITRATION. Dissolve 2.5 g of hexamethylenetetramine in 75 mL of water. Add 3 mL of 10% nitric acid and 1 mL of 0.1% sample solution reserved from the test for clarity of solution. The color should be yellow. Add 0.03 mL of 0.1 M zinc chloride solution, which should change the color to a magenta purple. Addition of 0.05 mL of 0.1 M EDTA solution should restore the yellow color.

Zinc

Zn   **Atomic Wt 65.39**

CAS Number 7440–66–6

REQUIREMENTS

Particle size . . . . 20–30 mesh
Assay . . . . ≥99.8% Zn
Suitability for determination of arsenic (As) . . . . Passes test

MAXIMUM ALLOWABLE

Iron (Fe) 0.01%
Lead (Pb) 0.01%

TESTS

ASSAY. (By complexometric titration for zinc). Weigh, to the nearest 0.1 mg, 0.2–0.25 g of sample, and dissolve with 10 mL of (1 + 1) nitric acid. When completely dissolved, dilute with water to about 300 mL and, with the aid of magnetic stirring, neutralize with 10% sodium hydroxide to methyl red. Add a few milligrams of ascorbic acid, 15 mL of pH 10 buffer solution, and about 50 mg of Eriochrome Black T indicator mixture. Titrate immediately with 0.1 M EDTA to a blue end point. One milliliter of 0.1 M EDTA corresponds to 0.006539 g of Zn.

SUITABILITY FOR DETERMINATION OF ARSENIC. Note that the apparatus used for this test is described in the general method for arsenic on page 25. Add 13 g of the sample to a generator flask containing a mixture of 35 mL of water, 10 mL of hydrochloric acid, 2 mL of potassium iodide solution (16.5% KI in water), and 0.5 mL of stannous chloride solution (40% $SnCl_2 \cdot 2H_2O$ in concentrated HCl). Immediately connect the scrubber–absorber assembly to the generator flask. When the gas evolution stops, disconnect the flask, and add successive 5-mL portions of hydrochloric acid. Immediately connect the scrubber–absorber assembly to the flask after each addition. After four or five portions of acid have been added, decant the spent acid and add a mixture of 35 mL of water, 10 mL of hydrochloric acid, 2 mL of potassium iodide solution (16.5% KI in water), and 0.5 mL of stannous chloride solution (40% $SnCl_2 \cdot 2H_2O$ in concentrated HCl). Immediately reconnect the scrubber–absorber assembly to the flask. Continue the addition of 5-mL portions of hydrochloric acid and the decantations, if necessary, until all of the zinc sample is dissolved. Disconnect the tubing from the generator flask, and transfer the silver diethyldithiocarbamate solution to a suitable color-comparison tube. Any red color in the silver diethyldithiocarbamate solution of the sample should not exceed that in a control prepared with 3 g of the zinc sample and containing 0.001 mg of arsenic (As).

IRON AND LEAD. (By flame AAS, page 39).

Sample Stock Solution. Dissolve 10.0 g of sample in 75 mL of (1 + 1) nitric acid. When dissolution is nearly complete, add 5 mL of nitric acid, and heat to boiling, or until any residue from the zinc is dissolved. Cool, transfer with water to a 100-mL volumetric flask, and dilute to the mark with water (1 mL = 0.1 g).

Element	Wavelength (nm)	Sample Wt (g)	Standard Added (mg)	Flame Type*	Background Correction
Fe	248.3	1.0	0.05; 0.1	A/A	Yes
Pb	217.0	1.0	0.05; 0.1	A/A	Yes

*A/A is air/acetylene.

Zinc Acetate Dihydrate

$(CH_3COO)_2Zn \cdot 2H_2O$ **Formula Wt 219.51**

CAS Number 5970–45–6

REQUIREMENTS

Assay . . . . 98.0–101.0% $(CH_3COO)_2Zn \cdot 2H_2O$
pH of a 5% solution . . . . 6.0–7.0 at 25 °C

MAXIMUM ALLOWABLE

Insoluble matter. 0.005%
Chloride (Cl) . . . . 5 ppm
Sulfate (SO_4) 0.005%
Calcium (Ca) 0.005%
Magnesium (Mg) 0.005%
Potassium (K). 0.01%
Sodium (Na) 0.05%
Iron (Fe). . . . . 5 ppm
Lead (Pb) 0.002%

TESTS

ASSAY. (By complexometric titration for zinc). Weigh accurately 0.8 g, transfer to a 500-mL beaker, and dilute to about 300 mL with water. Add 15 mL of saturated aqueous hexamethylenetetramine solution and 50 mg of xylenol orange indicator mixture. Titrate with standard 0.1 M EDTA to a color change of purple-red to lemon-yellow. One milliliter of 0.1 M EDTA corresponds to 0.02195 g of $(CH_3COO)_2Zn \cdot 2H_2O$.

pH OF A 5% SOLUTION. (Page 44). The pH should be 6.0–7.0 at 25 °C.

INSOLUBLE MATTER. (Page 15). Use 20 g dissolved in a mixture of 2 mL of glacial acetic acid and 200 mL of water.

CHLORIDE. (Page 27). Use 2.0 g.

SULFATE. Dissolve 20 g in 200 mL of water, add 1 mL of hydrochloric acid, filter, heat to boiling, and add 10 mL of barium chloride reagent solution. Digest in a covered beaker on a hot plate ($\approx$100 °C) for 2 h and allow to stand overnight. If a precipitate forms, filter, wash with water containing about 0.2% hydrochloric acid, wash finally with plain water, and ignite. Correct for a complete blank determination. One gram of $BaSO_4$ so obtained corresponds to 0.41 g of sulfate (SO_4).

CALCIUM, MAGNESIUM, POTASSIUM, AND SODIUM. (By flame AAS, page 39).

> *Sample Stock Solution.* Dissolve 10.0 g in water in a 100-mL volumetric flask, add 0.5 mL of hydrochloric acid, and dilute with water to the mark (1 mL = 0.10 g).

Element	Wavelength (nm)	Sample Wt (g)	Standard Added (mg)	Flame Type*	Background Correction
Ca	422.7	0.80	0.02; 0.04	N/A	No
Mg	285.2	0.20	0.005; 0.01	A/A	Yes
K	766.5	0.20	0.01; 0.02	A/A	No
Na	589.0	0.04	0.01; 0.02	A/A	No

*A/A = Air Acetylene; N/A is nitrous oxide/acetylene.

IRON. (Page 30, Method 1). Use 2.0 g.

LEAD. (By differential pulse polarography, page 51). Use 5.0 g of sample and 0.25 mL of hydrochloric acid. For the standard, add 0.10 mg of lead ion (Pb).

Zinc Chloride

$ZnCl_2$ **Formula Wt 136.30**

CAS Number 7646–85–7

REQUIREMENTS

Assay . . . . $\geq$97.0% $ZnCl_2$

MAXIMUM ALLOWABLE

Oxychloride . . . . Passes test

Insoluble matter. 0.005%
Nitrate (NO_3). 0.003%
Sulfate (SO_4) 0.01%
Ammonium (NH_4) 0.005%
Calcium (Ca) 0.06%
Iron (Fe). 0.001%
Lead (Pb) 0.005%
Magnesium (Mg) 0.01%
Potassium (K). 0.02%
Sodium (Na) 0.05%

TESTS

ASSAY. (By argentometric titration). Using precautions to avoid absorption of moisture, weigh accurately 0.3 g, dissolve in about 100 mL of water in a 200-mL volumetric flask, and add 5 mL of nitric acid and 50.0 mL of 0.1 N silver nitrate. Shake vigorously, dilute to volume, mix well, and filter through a dry paper into a dry flask or beaker, rejecting the first 20 mL of the filtrate. To 100 mL of the filtrate subsequently collected, add 2 mL of ferric ammonium sulfate indicator solution, and titrate the excess of silver nitrate with 0.1 N ammonium thiocyanate. One milliliter of 0.1 N silver nitrate consumed corresponds to 0.006815 g of $ZnCl_2$.

OXYCHLORIDE. Dissolve 20 g in 200 mL of water. On the addition of 6 mL of 1 N hydrochloric acid, any flocculent precipitate should entirely dissolve. Retain the solution for the test for insoluble matter.

INSOLUBLE MATTER. To the solution obtained in the test for oxychloride add 6 mL of 1 N hydrochloric acid. If insoluble matter is present, filter through a tared filtering crucible, wash thoroughly with water containing about 0.2% hydrochloric acid, and dry at 105 °C.

NITRATE. (Page 30). For sample solution A use 0.50 g. For control solution B use 0.50 g and 1.5 mL of standard nitrate solution.

SULFATE. (Page 33, Method 1). Allow 30 min for the turbidity to form.

AMMONIUM. Dissolve 1.0 g in water and dilute with water to 50 mL. Pour 10 mL of the solution into 10 mL of freshly boiled 10% sodium hydroxide reagent solution. Dilute with water to 50 mL and add 2 mL of Nessler reagent. Any color should not exceed that produced by 0.01 mg of ammonium ion (NH_4) in an equal volume of solution containing the quantities of reagents used in the test.

CALCIUM, IRON, LEAD, MAGNESIUM, POTASSIUM, AND SODIUM. (By flame AAS, page 39).

Sample Stock Solution. Dissolve 10.0 g in water in a 100-mL volumetric flask, add 0.5 mL of hydrochloric acid, and dilute with water to the mark (1 mL = 0.10 g).

Element	Wavelength (nm)	Sample Wt (g)	Standard Added (mg)	Flame Type*	Background Correction
Ca	422.7	0.10	0.03; 0.06	N/A	No
Fe	248.3	2.0	0.02; 0.04	A/A	Yes
Pb	217.0	2.0	0.05; 0.10	A/A	Yes
Mg	285.2	0.10	0.005; 0.01	A/A	Yes
K	766.5	0.10	0.01; 0.02	A/A	No
Na	589.0	0.02	0.005; 0.01	A/A	No

*A/A is air/acetylene; N/A is nitrous oxide/acetylene.

Zinc Oxide

ZnO   **Formula Wt 81.39**

CAS Number 1314–13–2

REQUIREMENTS

Assay . . . . ≥99.0% ZnO

MAXIMUM ALLOWABLE

Insoluble in dilute sulfuric acid. 0.01%
Alkalinity . . . . Passes test
Chloride (Cl) 0.001%
Nitrate (NO_3) 0.003%
Sulfur compounds (as SO_4) 0.01%
Calcium (Ca) 0.005%
Iron (Fe) 0.001%
Lead (Pb) 0.005%
Magnesium (Mg) 0.005%
Manganese (Mn) . . . . 5 ppm
Potassium (K). 0.01%
Sodium (Na). 0.05%

TESTS

ASSAY. (By complexometric titration of zinc). Weigh accurately 0.25 g and transfer to a 500-mL beaker. Dissolve in 15 mL of hydrochloric acid (1 + 1) and dilute to about 300 mL with water. Add 0.15 mL of methyl red indicator solution and neutralize (yellow color) by dropwise addition of 10% sodium hydroxide solution. Add 15 mL of saturated aqueous hexamethylenetetramine solution and 50 mg of xylenol orange indicator mixture. Titrate with standard 0.1 M EDTA to a color change of purple-red to lemon-yellow. One milliliter of 0.1 M EDTA corresponds to 0.008139 g of ZnO.

INSOLUBLE IN DILUTE SULFURIC ACID. Dissolve 10 g by heating in a covered beaker on a steam bath with 160 mL of dilute sulfuric acid (1 + 15) for 1 h. Filter through a tared filtering crucible, wash thoroughly, and dry at 105 °C.

ALKALINITY. Suspend 2 g in 20 mL of water, boil for 1 min, and filter. Add 0.10 mL of phenolphthalein indicator solution to 10 mL of the filtrate. No red color should be produced.

CHLORIDE. (Page 27). Suspend 1.0 g in 20 mL of water, and dissolve by adding 3 mL of nitric acid.

NITRATE. (Page 30). For sample solution A, dissolve 0.5 g of sample in 5 mL of hydrochloric acid (1 + 1). For control solution B, use 5 mL of hydrochloric acid (1 + 1) and 1.5 mL of standard nitrate solution.

SULFUR COMPOUNDS. Suspend 5.0 g in 50 mL of water and add about 50 mg of sodium carbonate and 1 mL of bromine water. Boil for 5 min, cautiously add hydrochloric acid in small portions until the zinc oxide is dissolved, then add 1 mL more of the acid. Filter, wash, and dilute the filtrate with water to about 150 mL. Heat the filtrate to boiling, add 5 mL of barium chloride reagent solution, digest in a covered beaker on a hot plate ($\approx$100 °C) for 2 h, and allow to stand overnight. If a precipitate is formed, filter, wash two or three times with dilute hydrochloric acid (1 + 99), complete the washing with water alone, and ignite. Correct for the weight obtained in a complete blank test.

CALCIUM, IRON, LEAD, MAGNESIUM, MANGANESE, POTASSIUM, AND SODIUM. (By flame AAS, page 39).

Sample Stock Solution. Dissolve 10.0 g in 10 mL of hydrochloric acid in a 100-mL volumetric flask, and dilute with water to volume (1 mL = 0.10 g).

Element	Wavelength (nm)	Sample Wt (g)	Standard Added (mg)	Flame Type*	Background Correction
Ca	422.7	0.80	0.02; 0.04	N/A	No
Fe	248.3	2.0	0.02; 0.04	A/A	Yes
Pb	217.0	2.0	0.05; 0.10	A/A	Yes
Mg	285.2	0.20	0.005; 0.01	A/A	Yes
Mn	279.5	2.0	0.01; 0.02	A/A	Yes
K	766.5	0.20	0.01; 0.02	A/A	No
Na	589.0	0.04	0.01; 0.02	A/A	No

*A/A is air/acetylene; N/A is nitrous oxide/acetylene.

Zinc Sulfate Heptahydrate

$ZnSO_4 \cdot 7H_2O$ **Formula Wt 287.56**

CAS Number 7446–20–0

REQUIREMENTS

Assay . . . . .99.0–103.0% $ZnSO_4 \cdot 7H_2O$
pH of a 5% solution . . . . 4.4–6.0 at 25 °C

MAXIMUM ALLOWABLE

Insoluble matter. 0.01%
Chloride (Cl) . . . . 5 ppm
Nitrate (NO_3) 0.002%
Ammonium (NH_4) 0.001%
Calcium (Ca) 0.005%
Iron (Fe). 0.001%
Lead (Pb) 0.003%
Magnesium (Mg) 0.005%
Manganese (Mn) . . . . 3 ppm
Potassium (K). 0.01%
Sodium (Na). 0.05%

TESTS

ASSAY. (By complexometric determination of zinc). Weigh accurately 1.0 g, transfer to a 500-mL beaker, and dissolve in about 300 mL of water. Add 15 mL of saturated aqueous hexamethylenetetramine solution and 50 mg of xylenol orange indicator mixture. Titrate with standard 0.1 M EDTA to a color change of purple-red to lemon-yellow. One milliliter of 0.1 M EDTA corresponds to 0.02876 g of $ZnSO_4 \cdot 7H_2O$.

pH OF A 5% SOLUTION. (Page 44). The pH should be 4.4–6.0 at 25 °C.

INSOLUBLE MATTER. (Page 15). Use 10.0 g dissolved in 100 mL of water.

CHLORIDE. (Page 27). Use 2.0 g.

NITRATE. (Page 30). For sample solution A, use 1.0 g. For control solution B, use 1.0 g and 2 mL of standard nitrate solution.

AMMONIUM. Dissolve 1.0 g in 25 mL of water, pour into 25 mL of freshly boiled and cooled 10% sodium hydroxide reagent solution, and add 2 mL of Nessler reagent. Any color should not exceed that produced by 0.01 mg of ammonium ion (NH_4) in an equal volume of solution containing the quantities of reagents used in the test.

CALCIUM, IRON, LEAD, MAGNESIUM, MANGANESE, POTASSIUM, AND SODIUM. (By flame AAS, page 39).

Sample Stock Solution. Dissolve 10.0 g of sample in water in a 100-mL volumetric flask, add 0.5 mL of hydrochloric acid, and dilute with water to the mark (1 mL = 0.10 g).

Element	Wavelength (nm)	Sample Wt (g)	Standard Added (mg)	Flame Type*	Background Correction
Ca	422.7	0.80	0.02; 0.04	N/A	No
Fe	248.3	2.0	0.02; 0.04	A/A	Yes
Pb	217.0	2.0	0.06; 0.12	A/A	Yes
Mg	285.2	0.20	0.005; 0.01	A/A	Yes
Mn	279.5	2.0	0.01; 0.02	A/A	Yes
K	766.5	0.20	0.01; 0.02	A/A	No
Na	589.0	0.04	0.01; 0.02	A/A	No

*A/A is air/acetylene; N/A is nitrous oxide/acetylene.

Specifications for Standard-Grade Reference Materials

Requirements and Tests

As a result of needs from the analytical testing community, the Committee on Analytical Reagents has developed specifications and analytical protocols for standard-grade reference materials. The following section includes specifications for a new grade of chemicals denoted as "ACS Standard-Grade Reference Materials."

Materials designated as standard grade are suitable for preparation of analytical standards used for a variety of applications, including instrument calibration, quality control, analyte identification, and method performance. These compounds are also suitable for other general applications requiring high-purity materials.

The initial list of compounds included in this volume is limited to selected materials of interest in environmental analysis. Future supplements and editions will include additional reference materials that cover a broader range of applications.

This section of the book is formatted differently than the previous section, "Specifications for Reagent Chemicals." Instead of an entry for each compound, the materials are grouped by classes. Each class of compounds has requirements and tests for the entire class. Following the requirements and tests are tables of compound-specific information, including physical properties, analytical data, and references to analytical methods.

The classes of materials included in this section are: chlorinated phenols, halogenated hydrocarbons, monocyclic aromatic hydrocarbons, organochlorine pesticides, polychlorinated biphenyl congeners, polycyclic aromatic hydrocarbons, and straight-chain hydrocarbons.

Chlorinated Phenols

REQUIREMENTS

Identity by
Infrared spectroscopy . . . . Passes test
Gas chromatography/mass spectrometry . . . . Passes test

Assay by
Gas chromatography . . . . ≥98.0%
Acid–base titrimetry 98–102%
Thin layer chromatography . . . . Passes test

TESTS

IDENTITY BY INFRARED SPECTROSCOPY. Analyze the sample using the general procedure cited on page 79. Identifying absorbances for the compound listed in Table XII must be present. The spectrum is compared to the NIST standardized library of compounds, if available. Absorbances in the test spectrum must match those found in the NIST spectrum.

IDENTITY BY GAS CHROMATOGRAPHY/MASS SPECTROMETRY. Analyze the sample by gas chromatography/mass spectrometry using method GCMS-CP1 described below.

Method GCMS-CP1

Ionization Mode: Electron ionization/70eV

Column: 100% Dimethylpolysiloxane, 30 M × 0.25 mm i.d., 0.25 μm film thickness

Temperature Program: 80 °C for 12 min, then 5 °C/min to 250 °C

Carrier Gas: Helium at 0.8 mL/min

Injector: Split/splitless (20:1 split)

Injector Temperature: 250 °C

Scan Range: 45–500 amu

Sample Size: 1 μL of a 400 μg/mL solution

Identifying ions for the compound listed in Table XII must be present. The spectrum is compared to the NIST standardized library of compounds, if available. Ions in the test spectrum must match those found in the NIST spectrum.

ASSAY BY GAS CHROMATOGRAPHY. Analyze the sample by gas chromatography using method GC-CP1 described below. Alternate method GC-CP2 should be used for specific compounds as noted in Table XII.

Table XII. Chlorinated Phenols Compound Data

Name	CAS No.	Chemical Formula	Formula Weight	IR (cm^{-1})			MS (m/z)		
				IR-1	IR-2	IR-3	MS-1	MS-2	MS-3
2-Chlorophenol	95–57–8	C_6H_5ClO	128.56	3584	1486	740	128	64	130
3-Chlorophenol[a]	108–43–0	C_6H_5ClO	128.56	3649	1600	885	128	65	130
4-Chlorophenol	106–48–9	C_6H_5ClO	128.56	3650	1495	825	128	65	130
2,3-Dichlorophenol	576–24–9	$C_6H_4Cl_2O$	163.00	3575	1450	770	162	164	126
2,4-Dichlorophenol	120–83–2	$C_6H_4Cl_2O$	163.00	3584	1480	720	162	164	63
2,5-Dichlorophenol	583–78–8	$C_6H_4Cl_2O$	163.00	3579	1478	906	162	164	63
2,6-Dichlorophenol[a]	87–65–0	$C_6H_4Cl_2O$	163.00	3575	1458	770	162	164	63
3,4-Dichlorophenol	95–77–2	$C_6H_4Cl_2O$	163.00	3450	1470	650	162	164	99
3,5-Dichlorophenol	591–35–5	$C_6H_4Cl_2O$	163.00	3640	1590	828	162	164	63
2,3,4,5,6-Pentachlorophenol	87–86–5	C_6HCl_5O	266.34	3557	1383	770	266	264	268
2,3,4,5-Tetrachlorophenol	4901–51–3	$C_6H_2Cl_4O$	231.89	3520	1410	725	232	230	234
2,3,4,6-Tetrachlorophenol	58–90–2	$C_6H_2Cl_4O$	231.89	3570	1445	750	131	232	230
2,3,5,6-Tetrachlorophenol	935–95–5	$C_6H_2Cl_4O$	231.89	3563	1404	700	232	230	234
2,3,4-Trichlorophenol	15950–66–0	$C_6H_3Cl_3O$	197.45	3575	1453	783	196	198	97
2,3,5-Trichlorophenol	933–78–8	$C_6H_3Cl_3O$	197.45	3571	1585	840	196	198	97
2,3,6-Trichlorophenol	933–75–5	$C_6H_3Cl_3O$	197.45	3567	1449	800	196	198	200
2,4,5-Trichlorophenol	95–95–4	$C_6H_3Cl_3O$	197.45	3584	1458	870	196	198	97
2,4,6-Trichlorophenol	88–06–2	$C_6H_3Cl_3O$	197.45	3575	1470	730	196	198	97
3,4,5-Trichlorophenol	609–19–8	$C_6H_3Cl_3O$	197.45	3653	1420	816	196	198	133

[a] Use GC–CP2.

Method GC-CP1

Column: 5% Diphenyl–95% dimethylpolysiloxane, 30 M × 0.53 mm i.d., 1.5 μm film thickness

Detector: Thermal conductivity

Detector Temperature: 300 °C

Injector: Split/splitless (10:1 split)

Injector Temperature: 250 °C

Sample Size: 2 μL of a 1000 μg/mL solution

Carrier Gas: Helium at 3.0 mL/min

Temp. Program: 80 °C for 12 min, then 5 °C/min to 240 °C

Method GC-CP2

Column: 100% Dimethylpolysiloxane, 30 M × 0.25 mm i.d., 0.25 μm film thickness

Detector: Thermal conductivity

Detector Temperature: 300 °C

Injector: Split/splitless (10:1 split)

Injector Temperature: 250 °C

Sample Size: 2 μL of a 1000 μg/mL solution

Carrier Gas: Helium at 1.0 mL/min

Temperature Program: 80 °C for 12 min, then 5 °C/min to 240 °C

Measure the area under each peak (excluding the solvent peak) and calculate the analyte content in area percent.

ASSAY BY ACID–BASE TITRIMETRY. Follow the general procedure cited on page 51. The following specific conditions are also required.

Weigh enough sample so that a minimum of 25 mL of titrant will be used. Dissolve the sample in 2-propanol. Titrate the sample using a standardized solution of KOH in methanol.

ASSAY BY THIN LAYER CHROMATOGRAPHY. Analyze the sample by thin layer chromatography using the general procedure described on page 73. The following specific conditions are also required.

Stationary Phase: Silica

Mobile Phase: 1:1 (v/v) Ethyl acetate:hexane

Detection Methods: UV at 254 nm and iodine

Compound must exhibit a single spot.

Halogenated Hydrocarbons

REQUIREMENTS

Identity by

Gas chromatography/mass spectrometry. . . . . Passes test

Assay by

Gas chromatography (using two methods) . . . . ≥98.0%

TESTS

IDENTITY BY GAS CHROMATOGRAPHY/MASS SPECTROMETRY. Analyze the sample by gas chromatography/mass spectrometry using method GCMS-HHC1 described below.

Method GCMS-HHC1

Ionization Mode: Electron ionization/70eV

Column: 5% Diphenyl–95% dimethylpolysiloxane, 30 M × 0.25 mm i.d., 0.25 µm film thickness

Temperature Program: 35 °C for 10 min, then 4 °C/min to 200 °C, hold 10 min

Carrier Gas: Helium at 0.8 mL/min

Injector: Split/splitless (100:1 split)

Injector Temperature: 200 °C

Scan Range: 35–400 amu

Sample Size: 1 µL of a 100 µg/mL solution

Identifying ions for the compound listed in Table XIII must be present. The spectrum is compared to the NIST standardized library of compounds, if available. Ions in the test spectrum must match those found in the NIST spectrum.

ASSAY BY GAS CHROMATOGRAPHY. Analyze the sample by gas chromatography using method GCMS-HHC1 described above and method GC-HHC1 described below.

Method GC-HHC1

Column: 3% Diphenyl–3% cyanopropyl–94% dimethylpolysiloxane, 30 M × 0.53 mm i.d., 3.0 µm film thickness

Detector: Thermal conductivity

Detector Temperature: 300 °C

Table XIII. Halogenated Hydrocarbons Compound Data

Name	CAS No.	Chemical Formula	Formula Weight	MS (m/z) MS-1	MS-2	MS-3
Allyl chloride	107–05–1	C_3H_5Cl	76.53	41	76	78
1-Bromo-2-chloroethane	107–04–0	C_2H_4BrCl	143.42	63	142	65
2-Bromo-1-chloropropane	3017–95–6	C_3H_6BrCl	157.44	77	156	79
Bromochloromethane	74–97–5	CH_2BrCl	129.39	49	128	130
Bromodichloromethane	75–27–4	$CHBrCl_2$	163.83	83	127	85
Bromoethane	74–96–4	C_2H_5Br	108.97	108	110	79
Bromoform	75–25–2	$CHBr_3$	252.75	173	250	171
Carbon tetrachloride	56–23–5	CCl_4	153.82	117	119	82
3-Chloro-2-methylpropene	563–47–3	C_3H_7Cl	90.55	55	90	75
1-Chlorobutane	109–69–3	C_4H_8Cl	92.57	56	63	41
Chlorodibromomethane	124–48–1	$CHBr_2Cl$	208.29	127	206	129
Chloroform	67–66–3	$CHCl_3$	119.38	83	117	85
1-Chlorohexane	544–10–5	$C_6H_{13}Cl$	120.62	91	93	69
1,2-Dibromo-3-chloropropane	96–12–8	$C_3H_5ClBr_2$	236.34	157	155	75
1,2-Dibromoethane	106–93–4	$C_2H_4Br_2$	187.87	107	186	109
Dibromofluoromethane	1868–53–7	$CHBr_2F$	191.84	111	190	113
Dibromomethane	74–95–3	CH_2Br_2	173.85	174	172	93
cis-1,4-Dichloro-2-butene	1476–11–5	$C_4H_6Cl_2$	125.00	53	88	75
trans-1,4-Dichloro-2-butene	110–57–6	$C_4H_6Cl_2$	125.00	75	124	53
1,3-Dichlorobutane	1190–22–3	$C_4H_8Cl_2$	127.01	55	90	63
1,4-Dichlorobutane	110–56–5	$C_4H_8Cl_2$	127.01	55	90	62
1,1-Dichloroethane	75–34–3	$C_2H_4Cl_2$	98.96	63	98	83
1,2-Dichloroethane	107–06–2	$C_2H_4Cl_2$	98.96	62	98	64
1,1-Dichloroethene	75–35–4	$C_2H_2Cl_2$	96.94	61	96	63
cis-1,2-Dichloroethene	156–59–2	$C_2H_2Cl_2$	96.94	61	96	63
trans-1,2-Dichloroethene	156–60–5	$C_2H_2Cl_2$	96.94	61	96	98
1,2-Dichloropropane	78–87–5	$C_3H_6Cl_2$	112.99	63	112	76
1,3-Dichloropropane	142–28–9	$C_3H_6Cl_2$	112.99	76	112	78
1,1-Dichloropropene	563–58–6	$C_3H_4Cl_2$	110.97	75	109	77
2,2-Dichloropropane	590–20–7	$C_3H_6Cl_2$	112.99	77	97	41
Diiodomethane	75–11–6	CH_2I_2	267.84	268	141	127
Hexachlorobutadiene	87–68–3	C_4Cl_6	260.76	225	258	190
Hexachlorocyclopentadiene	77–47–4	C_5Cl_6	272.77	237	270	235
Hexachloroethane	67–72–1	C_2Cl_6	236.74	117	201	119
Hexachloropropene	1888–71–7	C_3Cl_6	248.75	213	246	141
Iodomethane	74–88–4	CH_3I	141.94	142	127	141
Methylene chloride	75–09–2	CH_2Cl_2	84.93	49	84	86
Pentachloroethane	76–01–7	C_2HCl_5	202.29	117	167	119
Tetrachloroethene	127–18–4	C_2Cl_4	165.83	166	164	131
1,1,1,2-Tetrachloroethane	630–20–6	$C_2H_2Cl_4$	167.85	131	117	95
1,1,2,2-Tetrachloroethane	79–34–5	$C_2H_2Cl_4$	167.85	83	166	95
1,1,1-Trichloroethane	71–55–6	$C_2H_3Cl_3$	133.41	97	99	61
1,1,2-Trichloroethane	79–00–5	$C_2H_3Cl_3$	133.41	97	132	83
Trichloroethene	79–01–6	C_2HCl_3	131.39	95	130	132
1,2,3-Trichloropropane	96–18–4	$C_3H_5Cl_3$	147.43	75	110	61

Injector: Split/splitless (10:1 split)

Injector Temperature: 200 °C

Sample Size: 2 µL of a 1000 µg/mL solution

Carrier Gas: Helium at 3.0 mL/min

Temperature Program: 35 °C for 10 min, then 4 °C/min to 200 °C, hold 9 min

Measure the area under all peaks (excluding the solvent peak) and calculate the analyte content in area percent for each analysis.

Monocyclic Aromatic Hydrocarbons

REQUIREMENTS

Identity by

Gas chromatography/mass spectrometry. . . . . Passes test
Boiling point (liquids). . . . . Passes test
Melting point (solids). . . . . Passes test
Relative retention index (for *m*-, *o*-, and *p*-xylene only) . Passes test

Assay by

Gas chromatography (using two methods) . . . . ≥98.0%

TESTS

IDENTITY BY GAS CHROMATOGRAPHY/MASS SPECTROMETRY. Analyze the sample by gas chromatography/mass spectrometry using method GCMS-MCA1 described below.

Method GCMS-MCA1

Ionization Mode: Electron ionization/70eV

Column: 5% Diphenyl–95% dimethylpolysiloxane, 30 M × 0.25 mm i.d., 0.25 µm film thickness

Temperature Program: 35 °C for 5 min, then 5 °C/min to 300 °C, hold 10 min

Carrier Gas: Helium 1.0 mL/min

Injector: Split/splitless (100:1 split)

Injector Temperature: 240 °C

Scan Range: 35-550 amu

Sample Size: 2 µL of a 1000 µg/mL solution

Table XIV. Monocyclic Aromatic Hydrocarbons Compound Data

Name	CAS No.	Chemical Formula	Formula Weight	MS (m/z)			MP (°C)	BP (°C)
				MS-1	MS-2	MS-3		
Benzene	71–43–2	C_6H_6	78.11	78	52			77–82
Benzal chloride	98–87–3	$C_7H_6Cl_2$	161.03	125	160	89		202–207
Benzyl chloride	100–44–7	C_7H_7Cl	126.59	91	126	65		176–181
n-Butylbenzene	104–51–8	$C_{10}H_{14}$	134.22	91	134	65		180–185
sec-Butylbenzene	135–98–8	$C_{10}H_{14}$	134.22	105	134	91		170–175
tert-Butylbenzene	98–06–6	$C_{10}H_{14}$	134.22	119	91	134		166–171
Bromobenzene	108–86–1	C_6H_5Br	157.02	77	156	51		153–158
Chlorobenzene	108–90–7	C_6H_5Cl	112.56	112	77	51		129–134
2-Chlorotoluene	95–49–8	C_7H_7Cl	126.59	91	126	63		156–161
4-Chlorotoluene	106–43–4	C_7H_7Cl	126.59	91	126	63		159–164
1,2-Dichlorobenzene	95–50–1	$C_6H_4Cl_2$	147.00	146	111	75		178–183
1,3-Dichlorobenzene	541–73–1	$C_6H_4Cl_2$	147.00	146	111	75		170–175
1,4-Dichlorobenzene	106–46–7	$C_6H_4Cl_2$	147.00	146	111	75	50–55	
Ethylbenzene	100–41–4	C_8H_{10}	106.17	91	106	51		134–139
Hexachlorobenzene	118–74–1	C_6Cl_6	284.78	284	142	249	227–232	
Isopropylbenzene	98–82–8	C_9H_{12}	120.20	105	120	77		150–155
4-Isopropyltoluene	99–87–6	$C_{10}H_{14}$	134.22	119	134	91		174–179
Pentachlorobenzene	608–93–5	C_6HCl_5	250.34	250	215	108	83–88	

Compound	CAS	Formula	MW					
n-Propylbenzene	103-65-1	C_9H_{12}	120.20	91	120	38		156–161
Styrene	100-42-5	C_8H_8	104.15	104	91	78		142–147
1,2,3,4-Tetrachlorobenzene	634-66-2	$C_6H_2Cl_4$	215.89	216	179	109		251–256
1,2,4,5-Tetrachlorobenzene	95-94-3	$C_6H_2Cl_4$	215.89	216	179	108	137–142	
1,2,3,5-Tetrachlorobenzene	634-90-2	$C_6H_2Cl_4$	215.89	216	179	108	52–57	
Toluene	108-88-3	C_7H_8	92.14	91	65	51		108–113
1,2,3-Trichlorobenzene	87-61-6	$C_6H_3Cl_3$	181.45	180	145	109		211–216
1,2,4-Trichlorobenzene	120-82-1	$C_6H_3Cl_3$	181.45	180	145	109	51–56	
1,3,5-Trichlorobenzene	108-70-3	$C_6H_3Cl_3$	181.45	180	145	109	61–66	
a,a,a-Trichlorotoluene	98-07-7	$C_7H_5Cl_3$	195.48	159	194	123		218–223
1,2,3-Trimethylbenzene	526-73-8	C_9H_{12}	120.20	105	120	77		173–178
1,2,4-Trimethylbenzene	95-63-6	C_9H_{12}	120.20	105	120	77		166–171
1,3,5-Trimethylbenzene	108-67-8	C_9H_{12}	120.20	105	120	77		162–167
m-Xylene	108-38-3	C_8H_{10}	106.17	91	106	51		136–141
o-Xylene	95-47-6	C_8H_{10}	106.17	91	106	51		141–146
p-Xylene	106-42-3	C_8H_{10}	106.17	91	106	51		135–140

Identifying ions for the compound listed in Table XIV must be present. The spectrum is compared to the NIST standardized library of compounds, if available. Ions in the test spectrum must match those found in the NIST spectrum.

IDENTITY BY BOILING POINT. Analyze the liquid sample using the procedure cited on page 18. Assayed values must be within the ranges listed in Table XIV.

IDENTITY BY MELTING POINT. Analyze the solid sample using the procedure cited on page 22. Assayed values must be within the ranges listed in Table XIV.

IDENTITY BY RELATIVE RETENTION INDEX. Prepare a solution by adding equal amounts of the sample and toluene in a suitable solvent for a final concentration of 500 µg/mL. Analyze the sample by gas chromatography using method GC-MCA1 described below. Calculate the relative retention time of the sample versus toluene by dividing the retention time of the sample by the retention time of toluene. Assayed valued must be within 1% of 1.63 for *m*-xylene, 1.90 for *o*-xylene, and 1.59 for *p*-xylene.

ASSAY BY GAS CHROMATOGRAPHY. Analyze the sample by gas chromatography using method GC-MCA1 and method GC-MCA2 described below.

Method GC-MCA1

Column: Polyethylene glycol, 30 M × 0.53 mm i.d., 1.0 µm film thickness

Detector: Flame ionization

Detector Temperature: 200 °C

Injector: Splitless

Injector Temperature: 200 °C

Sample Size: 2 µL of a 1000 µg/mL solution

Carrier Gas: Helium at 5 mL/min

Temperature Program: 35 °C for 10 min, then 2.5 °C/min to 200 °C

Method GC-MCA2

Column: 6% Cyanopropylphenyl–94% dimethylpolysiloxane, 75 M × 0.53 mm i.d., 3.0 µm film thickness

Detector: Flame ionization

Detector Temperature: 200 °C

Injector: Splitless

Injector Temperature: 200 °C

Sample Size: 2 µL of a 1000 µg/mL solution

Carrier Gas: Helium at 10 mL/min

Temperature Program: 70 °C, then 5 °C/min to 210 °C, hold 10 min

Measure the area under all peaks (excluding the solvent peak) and calculate the analyte content in area percent for each analysis.

Organochlorine Pesticides

REQUIREMENTS

Identity by

Infrared spectroscopy . . . . Passes test
Gas chromatography/mass spectrometry. . . . . Passes test

Assay by

Gas chromatography . . . . ≥98.0%
Thin-layer chromatography . . . . Passes test

Specific Use

GC-ECD suitability. . . . . Passes test

TESTS

IDENTITY BY INFRARED SPECTROSCOPY. Analyze the sample using the general procedure cited on page 79. Identifying absorbances for the compound listed in Table XV must be present. The spectrum is compared to the NIST standardized library of compounds, if available. Absorbances in the test spectrum must match those found in the NIST spectrum.

IDENTITY BY GAS CHROMATOGRAPHY/MASS SPECTROMETRY. Analyze the sample by gas chromatography/mass spectrometry using method GCMS-OCP1 described below.

Method GCMS-OCP1

Ionization Mode: Electron ionization/70eV

Column: 5% Diphenyl–95% dimethylpolysiloxane, 30 M × 0.25 mm i.d., 0.25 µm film thickness

Temperature Program: 45 °C for 1 min, then 10 °C/min to 300 °C, hold 10 min

Carrier Gas: Helium at 1.0 mL/min

Injector: Split/splitless (100:1 split)

Table XV. Organochlorine Pesticides Compound Data

Name	CAS No.	Chemical Formula	Formula Weight	IR (cm⁻¹)			MS (m/z)		
				IR-1	IR-2	IR-3	MS-1	MS-2	MS-3
Aldrin	309–00–2	$C_{12}H_8Cl_6$	364.91	2987	1597	693	66	362	263
a-BHC	319–84–6	$C_6H_6Cl_6$	290.83	2963	1340	793	183	219	181
b-BHC	319–85–7	$C_6H_6Cl_6$	290.83	2952	1309	754	109	253	181
d-BHC	319–86–8	$C_6H_6Cl_6$	290.83	2932	1236	774	109	219	183
g-BHC	58–89–9	$C_6H_6Cl_6$	290.83	2948	1344	689	181	288	219
Butachlor	23184–66–9	$C_{17}H_{26}ClNO_2$	311.85	2971	1703	1074	176	311	57
cis-Chlordane	5103–71–9	$C_{10}H_6Cl_8$	409.78	2963	1604	534	373	406	272
trans-Chlordane	5103–74–2	$C_{10}H_6Cl_8$	409.78	2964	1603	560	373	406	272
Chlordecone	143–50–0	$C_{10}Cl_{10}O$	490.63	1814	1043	646	272	274	237
Chlorobenzilate	510–15–6	$C_{16}H_{14}Cl_2O_3$	325.20	3495	1719	1485	251	139	111
Chloroneb	2675–77–6	$C_8H_8Cl_2O_2$	207.05	2950	1490	775	191	206	141
Chlorthal	1861–32–1	$C_{10}H_6Cl_4O_4$	331.96	2963	1762	1240	301	330	299
2,4'-DDD	53–19–0	$C_{14}H_{10}Cl_4$	320.04	3075	1490	770	235	318	165
4,4'-DDD	72–54–8	$C_{14}H_{10}Cl_4$	320.04	3090	1494	765	235	318	165
2,4'-DDE	3424–82–6	$C_{14}H_8Cl_4$	318.02	3075	1590	862	246	316	248
4,4'-DDE	72–55–9	$C_{14}H_8Cl_4$	318.02	3090	1490	858	246	316	248
2,4'-DDT	789–02–6	$C_{14}H_9Cl_5$	354.48	3075	1490	777	335	352	165
4,4'-DDT	50–29–3	$C_{14}H_9Cl_5$	354.48	3094	1494	774	235	352	165
Dibutylchlorendate	1770–80–5	$C_{17}H_{20}Cl_6O_4$	501.06	2960	1745	845	501	499	427
Dicamba	1918–00–9	$C_8H_6Cl_2O_3$	221.04	3565	1776	1167	173	220	175

Name	CAS	Formula	MW						
Diclorofop methyl	51338–27–3	$C_{16}H_{14}Cl_2O_4$	341.20	3000	1750	1219	253	340	281
Dicofol	115–32–2	$C_{14}H_9Cl_5O$	370.50	3532	805	770	139	251	111
Dieldrin	60–57–1	$C_{12}H_8Cl_6O$	380.91	2986	1599	850	79	378	263
Endosulfan I	959–98–8	$C_9H_6Cl_6O_3S$	406.93	2937	1605	752	195	404	339
Endosulfan II	33213–65–9	$C_9H_6Cl_6O_3S$	406.92	2959	1205	673	195	404	339
Endosulfan sulfate	1031–07–8	$C_9H_6Cl_6O_4S$	422.93	1437	1201	808	272	420	387
Endrin	72–20–8	$C_{12}H_8Cl_6O$	380.91	2979	1600	855	81	378	263
Endrin aldehyde	7421–93–4	$C_{12}H_8Cl_6O$	380.91	2940	1741	835	67	345	250
Endrin ketone	53494–70–5	$C_{12}H_7Cl_6O$	380.91	2963	1748	754	317	378	250
Heptachlor	76–44–8	$C_{10}H_5Cl_7$	373.32	2959	1602	789	100	370	272
Heptachlor epoxide	1024–57–3	$C_{10}H_5Cl_7$	389.32	3044	1602	858	81	386	353
Hexachlorophene	70–30–4	$C_{13}H_6Cl_6O_2$	406.90	3502	1438	1281	196	404	209
Isodrin	465–73–6	$C_{12}H_8Cl_6$	364.91	2963	1603	1468	193	362	263
Mecoprop	7085–19–0	$C_{10}H_{11}ClO_3$	214.65	3575	1803	1490	142	214	107
Metolachlor	51218–45–2	$C_{15}H_{22}ClNO_2$	283.79	2982	1691	1113	162	238	146
Mirex	2385–85–5	$C_{10}Cl_{12}$	545.54	1148	1059	650	272	274	237
4,4'-Methoxychlor	72–43–5	$C_{16}H_{15}Cl_3O_2$	345.65	2938	1511	1252	227	344	228
cis-Nonachlor	5103–73–1	$C_{10}H_5Cl_9$	444.23	2980	1599	1267	409	440	237
trans-Nonachlor	39765–80–5	$C_{10}H_5Cl_9$	444.23	2954	1599	1257	409	440	272
Pentachloroanisole	1825–21–4	$C_7H_3Cl_5O$	280.36	2948	1375	1032	280	278	265
Propachlor	1918–16–7	$C_{11}H_{14}ClNO_4$	211.69	2982	1691	704	120	211	176

Injector Temperature: 250 °C

Scan Range: 33–525 amu

Sample Size: 1 μL of a 500 μg/mL solution

Identifying ions for the compound listed in Table XV must be present. The spectrum is compared to the NIST standardized library of compounds, if available. Ions in the test spectrum must match those found in the NIST spectrum.

ASSAY BY GAS CHROMATOGRAPHY. Analyze the sample by gas chromatography using method GC-OCP1 described below.

Method GC-OCP1

Column: 5% Diphenyl–95% dimethylpolysiloxane, 30 M × 0.53 mm i.d., 1.0 μm film thickness

Detector: Flame ionization

Detector Temperature: 200 °C

Injector: Splitless

Injector Temperature: 200 °C

Sample Size: 2 μL of a 1000 μg/mL solution

Carrier Gas: Helium at 3.0 mL/min

Temperature Program: 70 °C, then 5 °C/min to 210 °C, hold 10 min

Measure the area under all peaks (excluding the solvent peak) and calculate the analyte content in area percent.

ASSAY BY THIN LAYER CHROMATOGRAPHY. Analyze the sample by thin layer chromatography using the general procedure described on page 73. The following specific conditions are also required.

Stationary Phase: Silica

Mobile Phase: Hexane

Detection Method: UV at 254 nm and iodine

Compound must exhibit a single spot.

GC-ECD SUITABILITY. Prepare a 200 ng/mL solution of the sample in an ECD nonresponsive solvent. Analyze the sample using method GC-OCP2 described below.

Method GC-OCP2

Column: 5% Diphenyl–95% dimethylpolysiloxane, 30 M × 0.53 mm i.d., 1.5 μm film thickness

Detector: Electron capture

Detector Temperature: 325 °C

Injector: Splitless

Injector Temperature: 250 °C

Sample Size: 1 µL

Carrier Gas: Nitrogen at 3.5 mL/min

Temperature Program: 45 °C for 1 min, then 10 °C/min to 250 °C, hold 10 min

Measure the area under all peaks (excluding the solvent peak) and calculate the analyte and total impurities content in area percent. The sum of all impurity peaks should not exceed 5%, with no single peak greater than 3%.

Polychlorinated Biphenyl Congeners

REQUIREMENTS

Identity by

Infrared spectroscopy . . . . Passes test
Gas chromatography/mass spectrometry. . . . . Passes test

Assay by

Gas chromatography (using two methods) . . . . ≥98.0%
Melting point (for solids) . . . . Passes test

TESTS

IDENTITY BY INFRARED SPECTROSCOPY. Analyze the sample using the general procedure cited on page 79. Identifying absorbances for the compound listed in Table XVI must be present. The spectrum is compared to the NIST standardized library of compounds, if available. Absorbances in the test spectrum must match those found in the NIST spectrum.

IDENTITY BY GAS CHROMATOGRAPHY/MASS SPECTROMETRY. Analyze the sample by gas chromatography/mass spectrometry using method GCMS-PCB1 described below.

Method GCMS-PCB1

Ionization Mode: Electron ionization/70eV

Table XVI. Polychlorinated Biphenyl Cogeners Compound Data

BZ No.	Name	CAS No.	Chemical Formula	Formula Weight	IR (cm^{-1})					MS (m/z)			MP
					IR-1	IR-2	IR-3	IR-4	IR-5	MS-1	MS-2	MS-3	
1	2-Chlorobiphenyl	2051-60-7	$C_{12}H_9Cl$	188.65	750	699	1471	1039	771	188	152	190	28–33
2	3-Chlorobiphenyl	2051-61-8	$C_{12}H_9Cl$	188.65	756	698	1477	1597	1570	188	152	190	—
3	4-Chlorobiphenyl	2051-62-9	$C_{12}H_9Cl$	188.65	1482	760	1097	834	1009	188	152	190	74–79
4	2,2'-Dichlorobiphenyl	13029-08-8	$C_{12}H_8Cl_2$	223.10	752	762	1467	1462	1064	222	152	224	58–63
5	2,3-Dichlorobiphenyl	16605-91-7	$C_{12}H_8Cl_2$	223.10	759	699	1453	1410	1044	222	152	224	—
6	2,3'-Dichlorobiphenyl	25569-80-6	$C_{12}H_8Cl_2$	223.10	756	1463	1467	1040	825	222	224	152	—
7	2,4-Dichlorobiphenyl	33284-50-3	$C_{12}H_8Cl_2$	223.10	1469	817	700	1107	825	222	224	152	—
8	2,4'-Dichlorobiphenyl	34883-43-7	$C_{12}H_8Cl_2$	223.10	757	1471	830	1098	1008	222	224	152	42–47
9	2,5-Dichlorobiphenyl	34883-39-1	$C_{12}H_8Cl_2$	223.10	1463	1101	700	766	815	222	224	152	—
10	2,6-Dichlorobiphenyl	33146-45-1	$C_{12}H_8Cl_2$	223.10	1428	698	781	761	1441	222	224	152	33–38
11	3,3'-Dichlorobiphenyl	2050-67-1	$C_{12}H_8Cl_2$	223.10	722	1597	780	1468	1566	222	224	152	27–32
12	3,4-Dichlorobiphenyl	2974-92-7	$C_{12}H_8Cl_2$	223.10	1467	761	1033	1138	1139	222	224	152	46–51
13	3,4'-Dichlorobiphenyl	2974-90-5	$C_{12}H_8Cl_2$	223.10	1098	803	785	1475	832	222	224	152	—
14	3,5-Dichlorobiphenyl	34883-41-5	$C_{12}H_8Cl_2$	223.10	1561	761	806	1596	697	222	224	152	30–35
15	4,4'-Dichlorobiphenyl	2050-68-2	$C_{12}H_8Cl_2$	223.10	1097	815	1477	1488	1007	222	224	152	148–153
16	2,2',3-Trichlorobiphenyl	38444-78-9	$C_{12}H_7Cl_3$	257.55	756	1411	787	810	1035	256	186	258	—
17	2,2',4-Trichlorobiphenyl	37680-66-3	$C_{12}H_7Cl_3$	257.55	1463	821	758	1590	1105	256	258	186	—
18	2,2',5-Trichlorobiphenyl	37680-65-2	$C_{12}H_7Cl_3$	257.55	758	1463	1100	1019	816	256	258	186	42–47
19	2,2',6-Trichlorobiphenyl	38444-73-4	$C_{12}H_7Cl_3$	257.55	754	1431	779	794	1444	256	258	186	88–93
20	2,3,3'-Trichlorobiphenyl	38444-84-7	$C_{12}H_7Cl_3$	257.55	781	1451	1398	1561	702	256	258	186	42–47
21	2,3,4-Trichlorobiphenyl	55702-46-0	$C_{12}H_7Cl_3$	257.55	1444	1450	700	1364	837	256	258	186	100–105
22	2,3,4'-Trichlorobiphenyl	38444-85-8	$C_{12}H_7Cl_3$	257.55	1094	1452	804	785	1494	256	258	186	70–75
23	2,3,5-Trichlorobiphenyl	55720-44-0	$C_{12}H_7Cl_3$	257.55	1414	699	766	1554	1121	256	258	186	37–42
24	2,3,6-Trichlorobiphenyl	55702-45-9	$C_{12}H_7Cl_3$	257.55	1433	1181	698	1385	763	256	258	186	54–59
25	2,3',4-Trichlorobiphenyl	55712-37-3	$C_{12}H_7Cl_3$	257.55	1463	1109	836	1604	787	256	258	186	34–39

No.	Name	CAS	Formula	MW									Range
26	2,3',5-Trichlorobiphenyl	38444-81-4	$C_{12}H_7Cl_3$	257.55	1458	1103	1032	1464	696	256	258	186	40–45
27	2,3',6-Trichlorobiphenyl	38444-76-7	$C_{12}H_7Cl_3$	257.55	786	1434	1444	695	1561	256	258	186	—
28	2,4,4'-Trichlorobiphenyl	7012-37-5	$C_{12}H_7Cl_3$	257.55	1469	1098	816	738	1006	256	258	186	56–61
29	2,4,5-Trichlorobiphenyl	15862-07-4	$C_{12}H_7Cl_3$	257.55	1458	699	1092	1048	1447	256	258	186	75–80
30	2,4,6-Trichlorobiphenyl	35693-92-6	$C_{12}H_7Cl_3$	257.55	833	1544	1581	698	1371	222	258	186	60–65
31	2,4',5-Trichlorobiphenyl	16606-02-3	$C_{12}H_7Cl_3$	257.55	1460	1098	1499	1027	833	256	258	186	63–68
32	2,4',6-Trichlorobiphenyl	38444-77-4	$C_{12}H_7Cl_3$	257.55	787	1433	780	1440	1114	256	258	186	53–58
33	2',3,4-Trichlorobiphenyl	38444-86-9	$C_{12}H_7Cl_3$	257.55	1461	758	1136	1035	1041	256	258	186	59–64
34	2',3,5-Trichlorobiphenyl	37680-68-5	$C_{12}H_7Cl_3$	257.55	757	1562	1590	809	1045	256	258	186	55–60
35	3,3',4-Trichlorobiphenyl	37680-69-6	$C_{12}H_7Cl_3$	257.55	1465	737	786	1138	1553	256	258	186	64–69
36	3,3',5-Trichlorobiphenyl	38444-87-0	$C_{12}H_7Cl_3$	257.55	1561	1592	807	786	712	256	258	186	76–81
37	3,4,4'-Trichlorobiphenyl	38444-90-5	$C_{12}H_7Cl_3$	257.55	1466	1098	815	1015	1376	256	258	186	86–91
38	3,4,5-Trichlorobiphenyl	53555-66-1	$C_{12}H_7Cl_3$	257.55	1434	1546	761	1377	813	256	258	186	69–74
39	3,4',5-Trichlorobiphenyl	38444-88-1	$C_{12}H_7Cl_3$	257.55	1097	1498	1559	804	827	256	258	186	85–90
40	2,2',3,3'-Tetrachlorobiphenyl	38444-93-8	$C_{12}H_6Cl_4$	291.99	1441	782	1407	1420	1041	290	292	220	121–126
41	2,2',3,4-Tetrachlorobiphenyl	52663-59-9	$C_{12}H_6Cl_4$	291.99	1444	1365	757	1436	1177	290	292	220	46–51
42	2,2',3,4'-Tetrachlorobiphenyl	36559-22-5	$C_{12}H_6Cl_4$	291.99	837	1444	787	1452	1417	290	292	220	66–71
43	2,2',3,5-Tetrachlorobiphenyl	70362-46-8	$C_{12}H_6Cl_4$	291.99	758	1412	1388	1123	1557	290	292	220	43–48
44	2,2',3,5'-Tetrachlorobiphenyl	41464-39-5	$C_{12}H_6Cl_4$	291.99	1450	1033	1102	788	755	290	292	220	46–51
45	2,2',3,6-Tetrachlorobiphenyl	70362-45-7	$C_{12}H_6Cl_4$	291.99	1434	757	1386	1177	1440	290	292	220	78–83
46	2,2',3,6'-Tetrachlorobiphenyl	41464-47-5	$C_{12}H_6Cl_4$	291.99	779	1435	1412	1563	811	290	292	220	123–128
47	2,2',4,4'-Tetrachlorobiphenyl	2437-79-8	$C_{12}H_6Cl_4$	291.99	1465	791	1107	1588	817	290	292	294	44–49
48	2,2',4,5-Tetrachlorobiphenyl	70362-47-9	$C_{12}H_6Cl_4$	291.99	1456	1075	760	1347	735	290	292	220	84–89
49	2,2',4,5'-Tetrachlorobiphenyl	41464-40-8	$C_{12}H_6Cl_4$	291.99	1459	1100	1106	843	1018	290	292	220	64–69
50	2,2',4,6-Tetrachlorobiphenyl	62796-65-0	$C_{12}H_6Cl_4$	291.99	757	1546	1429	1582	1373	290	292	220	45–50
51	2,2',4,6'-Tetrachlorobiphenyl	68194-04-7	$C_{12}H_6Cl_4$	291.99	1435	790	820	1107	780	290	292	220	—
52	2,2',5,5'-Tetrachlorobiphenyl	35693-99-3	$C_{12}H_6Cl_4$	291.99	1463	1102	817	1456	1026	290	292	220	84–89
53	2,2',5,6'-Tetrachlorobiphenyl	41464-41-9	$C_{12}H_6Cl_4$	291.99	1434	1444	793	1099	780	290	292	220	99–104

Continued on next page

Table XVI. Continued

BZ No.	Name	CAS No.	Chemical Formula	Formula Weight	IR (cm^{-1})					MS (m/z)			MP
					IR-1	IR-2	IR-3	IR-4	IR-5	MS-1	MS-2	MS-3	
54	2,2',6,6'-Tetrachlorobiphenyl	15968-05-5	$C_{12}H_6Cl_4$	291.99	798	1434	777	1564	1084	290	292	294	195–200
55	2,3,3',4-Tetrachlorobiphenyl	74338-24-2	$C_{12}H_6Cl_4$	291.99	1446	1359	789	743	1178	290	292	294	84–89
56	2,3,3',4'-Tetrachlorobiphenyl	41464-43-1	$C_{12}H_6Cl_4$	291.99	1449	1036	787	1136	752	290	292	294	96–101
57	2,3,3',5-Tetrachlorobiphenyl	70424-67-8	$C_{12}H_6Cl_4$	291.99	1555	1414	1576	1125	697	290	292	294	85–90
58	2,3,3',5'-Tetrachlorobiphenyl	41464-49-7	$C_{12}H_6Cl_4$	291.99	785	1560	1387	1570	1585	290	292	294	123–128
59	2,3,3',6-Tetrachlorobiphenyl	74472-33-6	$C_{12}H_6Cl_4$	291.99	1434	1442	786	709	812	290	292	294	—
60	2,3,4,4'-Tetrachlorobiphenyl	33025-41-1	$C_{12}H_6Cl_4$	291.99	1447	1093	816	836	1363	290	292	294	143–148
61	2,3,4,5-Tetrachlorobiphenyl	33284-53-6	$C_{12}H_6Cl_4$	291.99	1414	1350	708	766	824	290	292	294	91–96
62	2,3,4,6-Tetrachlorobiphenyl	54230-22-7	$C_{12}H_6Cl_4$	291.99	1422	1345	697	819	770	290	292	294	76–81
63	2,3,4',5-Tetrachlorobiphenyl	74472-34-7	$C_{12}H_6Cl_4$	291.99	1433	1098	1543	808	1373	290	292	294	159–164
64	2,3,4',6-Tetrachlorobiphenyl	52663-58-8	$C_{12}H_6Cl_4$	291.99	1437	787	1498	1094	1181	290	292	294	85–90
65	2,3,5,6-Tetrachlorobiphenyl	33284-54-7	$C_{12}H_6Cl_4$	291.99	700	1392	1062	680	1380	290	292	294	75–80
66	2,3',4,4'-Tetrachlorobiphenyl	32598-10-0	$C_{12}H_6Cl_4$	291.99	1463	817	1107	1036	777	290	292	294	122–127
67	2,3',4,5-Tetrachlorobiphenyl	73557-53-8	$C_{12}H_6Cl_4$	291.99	1459	1453	886	697	789	290	292	294	60–65
68	2,3',4,5'-Tetrachlorobiphenyl	73575-52-7	$C_{12}H_6Cl_4$	291.99	847	805	1482	1569	1597	290	292	294	90–95
69	2,3',4,6-Tetrachlorobiphenyl	60233-24-1	$C_{12}H_6Cl_4$	291.99	845	1446	1580	1372	787	290	292	294	50–55
70	2,3',4',5-Tetrachlorobiphenyl	32598-11-1	$C_{12}H_6Cl_4$	291.99	1460	1033	1139	1105	824	290	292	294	102–107
71	2,3',4',6-Tetrachlorobiphenyl	41464-46-0	$C_{12}H_6Cl_4$	291.99	789	1439	1435	781	1136	290	292	294	34–39
72	2,3',5,5'-Tetrachlorobiphenyl	41464-42-0	$C_{12}H_6Cl_4$	291.99	1558	808	1487	1104	1569	290	292	294	104–109
73	2,3',5',6-Tetrachlorobiphenyl	74338-23-1	$C_{12}H_6Cl_4$	291.99	789	1413	1568	1600	811	290	292	294	68–73
74	2,4,4',5-Tetrachlorobiphenyl	32690-93-0	$C_{12}H_6Cl_4$	291.99	1458	1099	1088	1016	830	290	292	294	125–130
75	2,4,4',6-Tetrachlorobiphenyl	32598-12-2	$C_{12}H_6Cl_4$	291.99	1099	1583	1447	1544	1430	290	292	294	60–65
76	2',3,4,5-Tetrachlorobiphenyl	70362-48-0	$C_{12}H_6Cl_4$	291.99	757	1430	1544	1378	812	290	292	294	132–137
77	3,3',4,4'-Tetrachlorobiphenyl	32598-13-3	$C_{12}H_6Cl_4$	291.99	1464	1137	816	1033	1363	290	292	294	178–183
78	3,3',4,5-Tetrachlorobiphenyl	70362-49-1	$C_{12}H_6Cl_4$	291.99	1546	1436	811	1369	786	290	292	294	116–121

No.	Name	CAS	Formula	MW									
79	3,3',4,5'-Tetrachlorobiphenyl	41464-48-6	$C_{12}H_6Cl_4$	291.99	806	1590	1552	1481	1434	290	292	294	119–124
80	3,3',5,5'-Tetrachlorobiphenyl	33284-52-5	$C_{12}H_6Cl_4$	291.99	1559	1589	805	1129	1378	290	292	294	169–174
81	3,4,4',5-Tetrachlorobiphenyl	70362-50-4	$C_{12}H_6Cl_4$	291.99	1495	1121	830	1416	1040	290	292	294	86–91
82	2,2',3,3',4-Pentachlorobiphenyl	52663-62-4	$C_{12}H_5Cl_5$	326.43	1439	1415	791	753	1179	324	326	328	116–121
83	2,2',3,3',5-Pentachlorobiphenyl	60145-20-2	$C_{12}H_5Cl_5$	326.43	1413	787	1386	1041	1558	324	326	328	81–86
84	2,2',3,3',6-Pentachlorobiphenyl	52663-60-2	$C_{12}H_5Cl_5$	326.43	1434	1044	1178	812	785	324	326	328	107–112
85	2,2',3,4,4'-Pentachlorobiphenyl	65510-45-4	$C_{12}H_5Cl_5$	326.43	1440	794	1108	1179	817	324	326	328	46–51
86	2,2',3,4,5-Pentachlorobiphenyl	55312-69-1	$C_{12}H_5Cl_5$	326.43	1409	1350	759	736	1178	324	326	328	83–88
87	2,2',3,4,5'-Pentachlorobiphenyl	38380-02-8	$C_{12}H_5Cl_5$	326.43	1444	819	1101	1363	1179	324	326	328	108–113
88	2,2',3,4,6-Pentachlorobiphenyl	55215-17-3	$C_{12}H_5Cl_5$	326.43	1346	1424	743	761	1570	324	326	328	64–69
89	2,2',3,4,6'-Pentachlorobiphenyl	73575-57-2	$C_{12}H_5Cl_5$	326.43	1430	791	1435	1368	778	324	326	328	83–88
90	2,2',3,4',5-Pentachlorobiphenyl	68194-07-1	$C_{12}H_5Cl_5$	326.43	850	1416	819	1109	1389	324	326	328	49–54
91	2,2',3,4',6-Pentachlorobiphenyl	68194-05-8	$C_{12}H_5Cl_5$	326.43	1438	846	809	1107	1178	324	326	328	60–65
92	2,2',3,5,5'-Pentachlorobiphenyl	52663-61-3	$C_{12}H_5Cl_5$	326.43	1475	1101	1036	832	1390	324	326	328	58–63
93	2,2',3,5,6-Pentachlorobiphenyl	73575-56-1	$C_{12}H_5Cl_5$	326.43	1395	741	678	1044	1166	324	326	328	94–99
94	2,2',3,5,6'-Pentachlorobiphenyl	73575-55-0	$C_{12}H_5Cl_5$	326.43	1408	794	794	1385	780	324	326	328	80–85
95	2,2',3,5',6-Pentachlorobiphenyl	38379-99-6	$C_{12}H_5Cl_5$	326.43	816	1033	1180	1436	1098	324	326	328	91–96
96	2,2',3,6,6'-Pentachlorobiphenyl	73575-54-9	$C_{12}H_5Cl_5$	326.43	1429	796	1437	1179	812	324	326	328	117–122
97	2,2',3',4,5-Pentachlorobiphenyl	41464-51-1	$C_{12}H_5Cl_5$	326.43	1445	1417	789	1060	889	324	326	328	76–81
98	2,2',3',4,6-Pentachlorobiphenyl	60233-25-2	$C_{12}H_5Cl_5$	326.43	848	1431	1413	1547	785	324	326	328	92–97
99	2,2',4,4',5-Pentachlorobiphenyl	38380-01-1	$C_{12}H_5Cl_5$	326.43	1456	1450	801	1098	1107	324	326	328	57–62
100	2,2',4,4',6-Pentachlorobiphenyl	39485-83-1	$C_{12}H_5Cl_5$	326.43	800	1432	1580	1098	1547	324	326	328	—
101	2,2',4,5,5'-Pentachlorobiphenyl	37680-73-2	$C_{12}H_5Cl_5$	326.43	1457	1101	1075	1144	1480	324	326	328	76–81
102	2,2',4,5,6'-Pentachlorobiphenyl	68194-06-9	$C_{12}H_5Cl_5$	326.43	1438	793	781	1096	1477	324	326	328	69–74
103	2,2',4,5',6-Pentachlorobiphenyl	60145-21-3	$C_{12}H_5Cl_5$	326.43	1439	851	1097	818	1371	324	326	328	67–72
104	2,2',4,6,6'-Pentachlorobiphenyl	56558-16-8	$C_{12}H_5Cl_5$	326.43	1436	1416	836	1579	795	324	326	328	85–90
105	2,3,3',4,4'-Pentachlorobiphenyl	32598-14-4	$C_{12}H_5Cl_5$	326.43	1443	1136	1355	787	1036	324	326	328	115–120
106	2,3,3',4,5-Pentachlorobiphenyl	70424-69-0	$C_{12}H_5Cl_5$	326.43	1412	1348	1400	698	789	324	326	328	83–88

Continued on next page

Table XVI. Continued

BZ No.	Name	CAS No.	Chemical Formula	Formula Weight	IR-1	IR-2	IR-3	IR-4	IR-5	MS-1	MS-2	MS-3	MP
107	2,3,3',4',5-Pentachlorobiphenyl	70424-68-9	$C_{12}H_5Cl_5$	326.43	1480	1138	821	1415	1036	324	326	328	94–99
108	2,3,3',4,5'-Pentachlorobiphenyl	70362-41-3	$C_{12}H_5Cl_5$	326.43	806	1567	1427	1357	1594	324	326	328	118–123
109	2,3,3',4,6-Pentachlorobiphenyl	74472-35-8	$C_{12}H_5Cl_5$	326.43	704	1343	1430	1415	824	324	326	328	68–73
110	2,3,3',4',6-Pentachlorobiphenyl	38380-03-9	$C_{12}H_5Cl_5$	326.43	1435	1177	1037	813	1135	324	326	328	81–86
111	2,3,3',5,5'-Pentachlorobiphenyl	39635-32-0	$C_{12}H_5Cl_5$	326.43	1555	1414	1377	1568	809	324	326	328	105–110
112	2,3,3',5,6-Pentachlorobiphenyl	74472-36-9	$C_{12}H_5Cl_5$	326.43	719	1390	1063	1165	708	324	326	328	89–94
113	2,3,3',5',6-Pentachlorobiphenyl	68194-10-5	$C_{12}H_5Cl_5$	326.43	1568	814	1178	1377	701	324	326	328	54–59
114	2,3,4,4',5-Pentachlorobiphenyl	74472-37-0	$C_{12}H_5Cl_5$	326.43	1413	1496	1346	738	1018	324	326	328	96–101
115	2,3,4,4',6-Pentachlorobiphenyl	74472-38-1	$C_{12}H_5Cl_5$	326.43	1342	1419	747	1097	821	324	326	328	63–68
116	2,3,4,5,6-Pentachlorobiphenyl	18259-05-7	$C_{12}H_5Cl_5$	326.43	701	1351	1329	1386	1372	324	326	328	121–126
117	2,3,4',5,6-Pentachlorobiphenyl	68194-11-6	$C_{12}H_5Cl_5$	326.43	1495	1059	1385	677	1095	324	326	328	166–171
118	2,3',4,4',5-Pentachlorobiphenyl	31508-00-6	$C_{12}H_5Cl_5$	326.43	1455	1094	1142	1051	824	324	326	328	109–114
119	2,3',4,4',6-Pentachlorobiphenyl	56558-17-9	$C_{12}H_5Cl_5$	326.43	1431	853	1447	1579	1544	324	326	328	73–78
120	2,3',4,5,5'-Pentachlorobiphenyl	68194-12-7	$C_{12}H_5Cl_5$	326.43	1470	1565	808	1430	1569	324	326	328	129–134
121	2,3',4,5',6-Pentachlorobiphenyl	56558-18-0	$C_{12}H_5Cl_5$	326.43	854	1596	1564	801	1372	324	326	328	91–96
122	2',3,3',4,5-Pentachlorobiphenyl	76842-07-4	$C_{12}H_5Cl_5$	326.43	1430	785	1544	1164	817	324	326	328	115–120
123	2',3,4,4',5-Pentachlorobiphenyl	65510-44-3	$C_{12}H_5Cl_5$	326.43	1431	810	1365	1483	1110	324	326	328	132–137
124	2',3,4,5,5'-Pentachlorobiphenyl	70424-70-3	$C_{12}H_5Cl_5$	326.43	1433	1544	817	814	1102	324	326	328	114–119
125	2',3,4,5,6'-Pentachlorobiphenyl	74472-39-2	$C_{12}H_5Cl_5$	326.43	1419	790	1434	817	1547	324	326	328	124–129
126	3,3',4,4',5-Pentachlorobiphenyl	57465-28-8	$C_{12}H_5Cl_5$	326.43	1432	811	1033	1144	1359	324	326	328	158–163
127	3,3',4,5,5'-Pentachlorobiphenyl	39635-33-1	$C_{12}H_5Cl_5$	326.43	1545	808	1420	1586	1416	324	326	328	150–155
128	2,2',3,3',4,4'-Hexachlorobiphenyl	38380-07-3	$C_{12}H_4Cl_6$	360.88	1434	1184	796	1356	817	358	360	362	148–153
129	2,2',3,3',4,5-Hexachlorobiphenyl	52215-18-4	$C_{12}H_4Cl_6$	360.88	1410	1348	1420	788	735	358	360	362	100–105
130	2,2',3,3',4,5'-Hexachlorobiphenyl	52663-66-8	$C_{12}H_4Cl_6$	360.88	1413	818	1390	1460	1361	358	360	362	112–117
131	2,2',3,3',4,6-Hexachlorobiphenyl	61798-70-7	$C_{12}H_4Cl_6$	360.88	1412	1345	1425	740	788	358	360	362	133–138

132	2,2',3,3',4,6'-Hexachlorobiphenyl	38380-05-1	C$_{12}$H$_4$Cl$_6$	360.88	1431	1181	811	1365	875	358	360	362	115–120
133	2,2',3,3',5,5'-Hexachlorobiphenyl	35694-04-3	C$_{12}$H$_4$Cl$_6$	360.88	1557	1125	835	1378	869	358	360	362	125–130
134	2,2',3,3',5,6-Hexachlorobiphenyl	52704-70-8	C$_{12}$H$_4$Cl$_6$	360.88	735	1397	1071	1049	1167	358	360	362	130–135
135	2,2',3,3',5,6'-Hexachlorobiphenyl	52744-13-5	C$_{12}$H$_4$Cl$_6$	360.88	1043	1417	1181	812	1369	358	360	362	100–105
136	2,2',3,3',6,6'-Hexachlorobiphenyl	38411-22-2	C$_{12}$H$_4$Cl$_6$	360.88	1179	1430	813	1049	1407	358	360	362	111–116
137	2,2',3,4,4',5-Hexachlorobiphenyl	35694-06-5	C$_{12}$H$_4$Cl$_6$	360.88	1413	1350	789	1106	821	358	360	362	79–84
138	2,2',3,4,4',5'-Hexachlorobiphenyl	35065-28-2	C$_{12}$H$_4$Cl$_6$	360.88	1440	805	1368	1054	1180	358	360	362	77–82
139	2,2',3,4,4',6-Hexachlorobiphenyl	56030-56-9	C$_{12}$H$_4$Cl$_6$	360.88	1424	801	1346	1108	819	358	360	362	76–81
140	2,2',3,4,4',6'-Hexachlorobiphenyl	59291-64-4	C$_{12}$H$_4$Cl$_6$	360.88	1426	804	859	1366	1576	358	360	362	66–71
141	2,2',3,4,5,5'-Hexachlorobiphenyl	52712-04-6	C$_{12}$H$_4$Cl$_6$	360.88	1421	1348	1099	1474	819	358	360	362	86–91
142	2,2',3,4,5,6-Hexachlorobiphenyl	41411-61-4	C$_{12}$H$_4$Cl$_6$	360.88	1353	742	1329	1387	690	358	360	362	132–137
143	2,2',3,4,5,6'-Hexachlorobiphenyl	68194-15-0	C$_{12}$H$_4$Cl$_6$	360.88	1404	1349	798	1436	1565	358	360	362	87–92
144	2,2',3,4,5',6-Hexachlorobiphenyl	68194-14-9	C$_{12}$H$_4$Cl$_6$	360.88	1344	1433	1428	1099	818	358	360	362	70–75
145	2,2',3,4,6,6'-Hexachlorobiphenyl	74472-40-5	C$_{12}$H$_4$Cl$_6$	360.88	1409	1346	780	796	1572	358	360	362	134–139
146	2,2',3,4',5,5'-Hexachlorobiphenyl	51908-16-8	C$_{12}$H$_4$Cl$_6$	360.88	1470	1416	1079	1146	1389	358	360	362	86–91
147	2,2',3,4',5,6-Hexachlorobiphenyl	68194-13-8	C$_{12}$H$_4$Cl$_6$	360.88	1400	848	1482	1386	1167	358	360	362	134–139
148	2,2',3,4',5,6'-Hexachlorobiphenyl	74472-41-6	C$_{12}$H$_4$Cl$_6$	360.88	1409	857	1588	1126	1371	358	360	362	79–84
149	2,2',3,4',5',6-Hexachlorobiphenyl	38380-04-0	C$_{12}$H$_4$Cl$_6$	360.88	1436	1475	1181	1049	1392	358	360	362	76–81
150	2,2',3,4',6,6'-Hexachlorobiphenyl	68194-08-1	C$_{12}$H$_4$Cl$_6$	360.88	854	1430	819	1549	1374	358	360	362	74–79
151	2,2',3,5,5',6-Hexachlorobiphenyl	52663-63-5	C$_{12}$H$_4$Cl$_6$	360.88	1393	1099	1409	1045	1087	358	360	362	96–101
152	2,2',3,5,6,6'-Hexachlorobiphenyl	68194-09-2	C$_{12}$H$_4$Cl$_6$	360.88	1397	782	675	1436	797	358	360	362	124–129
153	2,2',4,4',5,5'-Hexachlorobiphenyl	35065-27-1	C$_{12}$H$_4$Cl$_6$	360.88	1449	1455	1087	1148	1048	358	360	362	101–106
154	2,2',4,4',5,6'-Hexachlorobiphenyl	60145-22-4	C$_{12}$H$_4$Cl$_6$	360.88	1434	818	1105	1578	1083	358	360	362	66–71
155	2,2',4,4',6,6'-Hexachlorobiphenyl	33979-03-2	C$_{12}$H$_4$Cl$_6$	360.88	1577	817	1419	1550	859	358	360	362	109–114
156	2,3,3',4,4',5-Hexachlorobiphenyl	38380-08-4	C$_{12}$H$_4$Cl$_6$	360.88	1413	1136	1343	1036	773	358	360	362	128–133
157	2,3,3',4,4',5'-Hexachlorobiphenyl	69782-90-7	C$_{12}$H$_4$Cl$_6$	360.88	1426	810	1353	1167	789	358	360	362	159–164
158	2,3,3',4,4',6-Hexachlorobiphenyl	74472-42-7	C$_{12}$H$_4$Cl$_6$	360.88	1414	1343	820	1132	1037	358	360	362	108–113
159	2,3,3',4,5,5'-Hexachlorobiphenyl	39635-35-3	C$_{12}$H$_4$Cl$_6$	360.88	1404	1342	809	1569	1595	358	360	362	147–152

Continued on next page

Table XVI. Continued

BZ No.	Name	CAS No.	Chemical Formula	Formula Weight	IR (cm⁻¹)					MS (m/z)			MP
					IR-1	IR-2	IR-3	IR-4	IR-5	MS-1	MS-2	MS-3	
160	2,3,3',4,5,6-Hexachlorobiphenyl	41411-62-5	$C_{12}H_4Cl_6$	360.88	722	1328	1349	1381	1366	358	360	362	90-95
161	2,3,3',4,5',6-Hexachlorobiphenyl	74474-43-8	$C_{12}H_4Cl_6$	360.88	1409	1567	1341	810	1595	358	360	362	103-108
162	2,3,3',4,5,5'-Hexachlorobiphenyl	39635-34-2	$C_{12}H_4Cl_6$	360.88	816	1546	1413	1391	1367	358	360	362	141-146
163	2,3,3',4',5,6-Hexachlorobiphenyl	74472-44-9	$C_{12}H_4Cl_6$	360.88	1400	1383	729	1066	1479	358	360	362	119-124
164	2,3,3',4',5',6-Hexachlorobiphenyl	74472-45-0	$C_{12}H_4Cl_6$	360.88	1425	818	1180	808	1371	358	360	362	90-95
165	2,3,3',5,5',6-Hexachlorobiphenyl	74472-46-1	$C_{12}H_4Cl_6$	360.88	1078	724	1385	1569	812	358	360	362	149-154
166	2,3,4,4',5,6-Hexachlorobiphenyl	41411-63-6	$C_{12}H_4Cl_6$	360.88	1329	1379	1353	751	737	358	360	362	162-167
167	2,3',4,4',5,5'-Hexachlorobiphenyl	52663-72-6	$C_{12}H_4Cl_6$	360.88	1431	1066	1471	813	1553	358	360	362	123-128
168	2,3',4,4',5',6-Hexachlorobiphenyl	59291-65-5	$C_{12}H_4Cl_6$	360.88	1420	1552	1367	862	822	358	360	362	107-112
169	3,3',4,4',5,5'-Hexachlorobiphenyl	32774-16-6	$C_{12}H_4Cl_6$	360.88	1423	1538	808	1357	1529	358	360	362	206-211
170	2,2',3,3',4,4',5-Heptachlorobiphenyl	35065-30-6	$C_{12}H_3Cl_7$	395.32	1409	1181	794	1348	666	392	394	396	135-140
171	2,2',3,3',4,4',6-Heptachlorobiphenyl	52663-71-5	$C_{12}H_3Cl_7$	395.32	1419	1347	806	1177	818	392	394	396	115-120
172	2,2',3,3',4,5,5'-Heptachlorobiphenyl	52663-74-8	$C_{12}H_3Cl_7$	395.32	1346	1409	1382	841	1558	392	394	396	133-138
173	2,2',3,3',4,5,6-Heptachlorobiphenyl	68194-16-1	$C_{12}H_3Cl_7$	395.32	1352	737	1328	1384	1372	392	394	396	202-207
174	2,2',3,3',4,5,6'-Heptachlorobiphenyl	38411-25-5	$C_{12}H_3Cl_7$	395.32	1418	1182	814	1345	1380	392	394	396	122-127
175	2,2',3,3',4,5',6-Heptachlorobiphenyl	40186-70-7	$C_{12}H_3Cl_7$	395.32	1407	1343	1389	823	1124	392	394	396	119-124
176	2,2',3,3',4,6,6'-Heptachlorobiphenyl	52663-65-7	$C_{12}H_3Cl_7$	395.32	1423	1343	1180	898	813	392	394	396	100-105
177	2,2',3,3',4',5,6-Heptachlorobiphenyl	52663-70-4	$C_{12}H_3Cl_7$	395.32	1401	1378	1367	686	1166	392	394	396	150-155
178	2,2',3,3',5,5',6-Heptachlorobiphenyl	52663-67-9	$C_{12}H_3Cl_7$	395.32	1393	1358	1095	1048	1401	392	394	396	108-113
179	2,2',3,3',5,6,6'-Heptachlorobiphenyl	52663-64-6	$C_{12}H_3Cl_7$	395.32	1062	1399	813	1410	1348	392	394	396	127-132
180	2,2',3,4,4',5,5'-Heptachlorobiphenyl	35065-29-3	$C_{12}H_3Cl_7$	395.32	1413	1469	1353	847	1066	392	394	396	111-116
181	2,2',3,4,4',5,6-Heptachlorobiphenyl	74472-47-2	$C_{12}H_3Cl_7$	395.32	1330	1390	803	1352	1368	392	394	396	123-128
182	2,2',3,4,4',5,6'-Heptachlorobiphenyl	60145-23-5	$C_{12}H_3Cl_7$	395.32	1406	1349	800	1548	1587	392	394	396	106-111
183	2,2',3,4,4',5',6-Heptachlorobiphenyl	52663-69-1	$C_{12}H_3Cl_7$	395.32	1424	1133	1095	1350	1357	392	394	396	91-96
184	2,2',3,4,4',6,6'-Heptachlorobiphenyl	74472-48-3	$C_{12}H_3Cl_7$	395.32	1413	813	1346	1569	1373	392	394	396	113-118

185	2,2',3,4,5,5',6-Heptachlorobiphenyl	52712–05–7	C$_{12}$H$_3$Cl$_7$	395.32	1353	1330	1361	1408	1098	392	394	396	146–151
186	2,2',3,4,5,6,6'-Heptachlorobiphenyl	74472–49–4	C$_{12}$H$_3$Cl$_7$	395.32	1353	1363	782	1328	1437	392	394	396	193–198
187	2,2',3,4',5,5',6-Heptachlorobiphenyl	52663–68–0	C$_{12}$H$_3$Cl$_7$	395.32	1398	1468	1389	1167	908	392	394	396	102–107
188	2,2',3,4',5,6,6'-Heptachlorobiphenyl	74487–85–7	C$_{12}$H$_3$Cl$_7$	395.32	1400	859	1587	1369	672	392	394	396	132–137
189	2,3,3',4,4',5,5'-Heptachlorobiphenyl	39635–31–9	C$_{12}$H$_3$Cl$_7$	395.32	1406	817	1339	771	840	392	394	396	160–165
190	2,3,3',4,4',5,6-Heptachlorobiphenyl	41411–64–7	C$_{12}$H$_3$Cl$_7$	395.32	1329	731	1368	1350	1387	392	394	396	120–125
191	2,3,3',4,4',5',6-Heptachlorobiphenyl	74472–50–7	C$_{12}$H$_3$Cl$_7$	395.32	1415	1341	1405	817	1547	392	394	396	111–116
192	2,3,3',4,5,5',6-Heptachlorobiphenyl	74472–51–8	C$_{12}$H$_3$Cl$_7$	395.32	1365	1327	1350	728	1569	392	394	396	169–174
193	2,3,3',4',5,5',6-Heptachlorobiphenyl	69782–91–8	C$_{12}$H$_3$Cl$_7$	395.32	1373	1167	724	1401	819	392	394	396	137–142
194	2,2',3,3',4,4',5,5'-Octachlorobiphenyl	35694–08–7	C$_{12}$H$_2$Cl$_8$	429.77	1403	1349	1181	1342	852	426	430	428	153–158
195	2,2',3,3',4,4',5,6-Octachlorobiphenyl	52663–78–2	C$_{12}$H$_2$Cl$_8$	429.77	1372	1328	1350	806	1368	426	430	428	168–173
196	2,2',3,3',4,4',5,6'-Octachlorobiphenyl	42740–50–1	C$_{12}$H$_2$Cl$_8$	429.77	1402	1341	1350	802	1335	426	430	428	125–130
197	2,2',3,3',4,4',6,6'-Octachlorobiphenyl	33091–17–7	C$_{12}$H$_2$Cl$_8$	429.77	1403	814	1564	1356	1336	426	430	428	135–140
198	2,2',3,3',4,5,5',6-Octachlorobiphenyl	68194–17–2	C$_{12}$H$_2$Cl$_8$	429.77	1360	1353	1331	1401	749	426	430	428	193–198
199	2,2',3,3',4,5,6,6'-Octachlorobiphenyl	52663–73–7	C$_{12}$H$_2$Cl$_8$	429.77	1353	1329	1411	1181	1073	426	430	428	173–178
200	2,2',3,3',4,5',6,6'-Octachlorobiphenyl	40186–71–8	C$_{12}$H$_2$Cl$_8$	429.77	1403	1338	1368	685	1169	426	430	428	139–144
201	2,2',3,3',4',5,5',6-Octachlorobiphenyl	52663–75–9	C$_{12}$H$_2$Cl$_8$	429.77	1403	1338	1377	1169	745	426	430	428	155–160
202	2,2',3,3',5,5',6,6'-Octachlorobiphenyl	2136–99–4	C$_{12}$H$_2$Cl$_8$	429.77	1406	1331	1073	1169	760	426	430	428	155–160
203	2,2',3,4,4',5,5',6-Octachlorobiphenyl	52663–76–0	C$_{12}$H$_2$Cl$_8$	429.77	1356	1329	1333	1383	1393	426	430	428	109–114
204	2,2',3,4,4',5,6,6'-Octachlorobiphenyl	74472–52–9	C$_{12}$H$_2$Cl$_8$	429.77	1365	817	1353	1331	1550	426	430	428	174–179
205	2,3,3',4,4',5,5',6-Octachlorobiphenyl	74472–53–0	C$_{12}$H$_2$Cl$_8$	429.77	1370	1327	1383	1327	727	426	430	428	195–200
206	2,2',3,3',4,4',5,5',6-Nonachlorobiphenyl	40186–72–9	C$_{12}$HCl$_9$	464.23	1378	1354	1331	1341	805	460	464	462	200–205
207	2,2',3,3',4,4',5,6,6'-Nonachlorobiphenyl	52663–79–3	C$_{12}$HCl$_9$	464.23	1374	1338	1357	818	693	460	464	462	211–216
208	2,2',3,3',4,5,5',6,6'-Nonachlorobiphenyl	52663–77–1	C$_{12}$HCl$_9$	464.23	1343	1338	1409	1082	691	460	464	196	178–183
209	2,2',3,3',4,4',5,5',6,6'-Decachlorobiphenyl	2051–24–3	C$_{12}$Cl$_{10}$	498.66	1345	1328	829	696	760	494	214	213	316–321

Column: 5% Diphenyl–95% dimethylpolysiloxane, 30 M × 0.25 mm i.d., 0.25 µm film thickness

Temperature Program: 140 °C for 1 min, then 10 °C/min to 320 °C, then hold 5 min

Carrier Gas: Helium at 1 mL/min

Injector: Split/splitless (50:1 split)

Injector Temperature: 280 °C

Scan Range: 40-600 amu

Sample Size: 0.5 µL of a 2000 µg/mL solution

Identifying ions for the compound listed in Table XVI must be present. The spectrum is compared to the NIST standardized library of compounds, if available. Ions in the test spectrum must match those found in the NIST spectrum.

ASSAY BY GAS CHROMATOGRAPHY. Analyze the sample by gas chromatography using method GCMS-PCB1 described above and method GC-PCB1 described below.

Method GC-PCB1

Column: 50% Diphenyl–50% dimethylpolysiloxane, 25 M × 0.53 mm i.d., 1.0 µm film thickness

Detector: Flame ionization

Detector Temperature: 300 °C

Injector: Splitless

Injector Temperature: 280 °C

Sample Size: 1 µL of a 1000 µg/mL solution

Carrier Gas: Helium at 3 mL/min

Temperature Program: 100 °C for 1 min, then 10 °C/min to 300 °C, then hold 5 min

Measure the area under all peaks (excluding the solvent peak) and calculate the analyte content in area percent for each analysis.

ASSAY BY MELTING POINT. Analyze the sample using the general procedure listed on page 22. Assayed values must be within the temperature ranges stated in Table XVI. The melting point assay should have a melting range of no more than 3 °C.

Polycyclic Aromatic Hydrocarbons

REQUIREMENTS

Identity by

Gas chromatography/mass spectrometry. . . . . Passes test

Melting point. . . . . Passes test

Assay by

Gas chromatography . . . . ≥98.0%

Thin-layer chromatography . . . . Passes test

Specific Use

Liquid chromatography suitability . . . . Passes test

TESTS

IDENTITY BY GAS CHROMATOGRAPHY/MASS SPECTROMETRY. Analyze the sample by gas chromatography/mass spectrometry using method GCMS-PAH1 described below.

Method GCMS-PAH1

Ionization Mode: Electron ionization/70eV

Column: 5% Diphenyl–95% dimethylpolysiloxane, 30 M × 0.25 mm i.d., 0.25 μm film thickness

Temperature Program: 50 °C, then 10 °C/min to 310 °C, hold 10 min

Carrier Gas: Helium at 0.8 mL/min

Injector: Split/splitless (100:1 split)

Injector Temperature: 250 °C

Scan Range: 50–550 amu

Sample Size: 0.5 μL of a 1,000 μg/mL solution

Identifying ions for the compound listed in Table XVII must be present. The spectrum is compared to the NIST standardized library of compounds, if available. Ions in the test spectrum must match those found in the NIST spectrum.

IDENTITY BY MELTING POINT. Analyze the sample by using the general procedure cited on page 22. Assayed values must be within the ranges stated in Table XVII.

Table XVII. Polycyclic Aromatic Hydrocarbons Compound Data

Name	CAS No.	Chemical Formula	Formula Weight	MS (m/z)			MP (°C)
				MS-1	MS-2	MS-3	
Acenaphthene	83-32-9	$C_{12}H_{10}$	154.21	154	153	76	91–96
Acenaphthylene	208-96-8	$C_{12}H_8$	152.20	152	151	76	90–95
Anthracene	120-12-7	$C_{14}H_{10}$	178.23	178	176	89	214–219
Benz[a]anthracene	56-55-3	$C_{18}H_{12}$	228.29	228	226	114	158–163
Benzo[a]pyrene	50-32-8	$C_{20}H_{12}$	252.31	252	126	113	173–178
Benzo[b]fluoranthene	205-99-2	$C_{20}H_{12}$	252.31	252	126	113	166–171
Benzo[g,h,i]perylene	191-24-2	$C_{22}H_{12}$	276.34	276	277	138	277–282
Benzo[j]fluoranthene	205-82-3	$C_{20}H_{12}$	252.31	252	250	126	161–166
Benzo[k]fluoranthene	207-08-9	$C_{20}H_{12}$	252.31	252	126	133	215–220
Chrysene	218-01-9	$C_{18}H_{12}$	228.29	228	114	101	254–259
Dibenz[a,h]acridine	226-36-8	$C_{21}H_{13}N$	279.34	279	280	139	226–231
Dibenz[a,h]anthracene	53-70-3	$C_{22}H_{14}$	278.35	278	279	139	266–271
Dibenz[a,j]acridine	224-42-0	$C_{21}H_{13}N$	279.34	279	280	139	217–222
Dibenzo[a,e]pyrene	192-65-4	$C_{24}H_{14}$	302.37	302	303	151	242–247
Dibenzo[a,i]pyrene	189-55-9	$C_{24}H_{14}$	302.37	302	303	152	280–285
Fluoranthene	206-44-0	$C_{16}H_{10}$	202.26	202	101	88	108–113
Fluorene	86-73-7	$C_{13}H_{10}$	166.22	166	165	83	114–119
Indeno[1,2,3-c,d]pyrene	193-39-5	$C_{22}H_{12}$	276.34	276	274	138	158–163
3-Methylcholanthrene	56-49-5	$C_{20}H_{16}$	268.36	268	252	126	177–182
Naphthalene	91-20-3	$C_{10}H_8$	128.17	128	127	102	78–83
Phenanthrene	85-01-8	$C_{14}H_{10}$	178.23	178	176	152	97–102
Pyrene	129-00-0	$C_{16}H_{10}$	202.26	202	203	101	150–155

ASSAY BY GAS CHROMATOGRAPHY. Analyze the sample by gas chromatography using method GC-PAH1 described below.

Method GC-PAH1

Column: 5% Diphenyl–95% dimethylpolysiloxane, 30 M × 0.53 mm i.d., 1.5 µm film thickness

Detector: Flame ionization

Detector Temperature: 325 °C

Injector: Splitless

Injector Temperature: 250 °C

Sample Size: 1 µL of a 1000 µg/mL solution

Carrier Gas: Hydrogen at 3.5 mL/min

Temperature Program: 70 °C, then 10 °C/min to 310 °C, hold 10 min

Measure the area under all peaks (excluding the solvent peak) and calculate the analyte content in area percent.

ASSAY BY THIN LAYER CHROMATOGRAPHY. Analyze the sample by thin layer chromatography using the general procedure described on page 73. The following specific conditions are also required.

Stationary Phase: Silica

Mobile Phase: 9:1 (v/v) Hexane:dichloromethane

Detection Methods: UV at 254 nm and iodine

Compound must exhibit a single spot.

LIQUID CHROMATOGRAPHY SUITABILITY. Analyze the sample by liquid chromatography using method HPLC-PAH1 described below. The sum of all impurity peaks should not exceed 5% with no single peak greater than 3%.

Method HPLC-PAH1

Column: Octadecyl, 250 × 4.6 mm i.d., 5 µm

Mobile Phase: Acetonitrile:Water 80:20

Flow Rate: 2.0 mL/min

Detector: UV, 254 nm

Sensitivity: 0.05 AUFS

Sample Size: 1 µL of a 100 µg/mL solution in acetonitrile

Measure the area under all peaks (excluding the solvent peak) and calculate the analyte and total impurities content in area percent.

Straight-Chain Hydrocarbons

REQUIREMENTS

Identity by
Gas chromatography/mass spectrometry . . . . Passes test
Relative retention index . . . . Passes test

Assay by
Gas chromatography (using two methods) . . . . ≥98.0%

TESTS

IDENTITY BY GAS CHROMATOGRAPHY/MASS SPECTROMETRY. Analyze the sample by gas chromatography/mass spectrometry using method GCMS-SCHC1 described below.

Method GCMS-SCHC1

Ionization Mode: Electron ionization/70eV

Column: 5% Diphenyl–95% dimethylpolysiloxane, 30 M × 0.25 mm i.d., 0.25 µm film thickness

Temperature Program: 35 °C for 5 min, then 30 °C/min to 320 °C, then hold 5 min

Carrier Gas: Helium at 1 mL/min

Injector: Split/splitless (100:1 split)

Injector Temperature: 250 °C

Scan Range: 32–600 amu

Sample Size: 1 µL of a 1000 µg/mL solution

Identifying ions for the compound listed in Table XVIII must be present. The spectrum is compared to the NIST standardized library of compounds, if available. Ions in the test spectrum must match those found in the NIST spectrum.

IDENTITY BY RELATIVE RETENTION INDEX. Prepare a solution by adding equal amounts of the sample and *n*-decane in a suitable solvent for a final concentration of 500 µg/mL. Analyze the sample by gas chromatography using method GC-SCHC1 described below.

Method GC-SCHC1

Column: 5% Diphenyl–95% dimethylpolysiloxane, 30 M × 0.53 mm i.d., 1.5 µm film thickness

Table XVIII. Straight-Chain Hydrocarbons Compound Data

Name	CAS No.	Chemical Formula	Formula Weight	MS (m/z)			Relative Retention
				MS-1	MS-2	MS-3	
Decane	124–18–5	$C_{10}H_{22}$	142.28	43	142	57	1.000
Docosane	629–97–0	$C_{22}H_{46}$	310.61	43	57	71	1.578
Dodecane	112–40–3	$C_{12}H_{26}$	170.34	43	170	57	1.132
Dotriacontane	544–85–4	$C_{32}H_{66}$	450.88	43	57	71	1.989
Eicosane	112–95–8	$C_{20}H_{42}$	282.55	57	282	71	1.503
Heneicosane	629–94–7	$C_{21}H_{44}$	296.58	43	57	71	1.541
Hentriacontane	630–04–6	$C_{31}H_{64}$	436.85	43	57	71	1.927
Heptacosane	593–49–7	$C_{27}H_{56}$	380.74	43	57	71	1.752
Heptadecane	629–78–7	$C_{17}H_{36}$	240.47	57	240	71	1.382
Heptane	142–82–5	C_7H_{16}	100.20	43	100	71	0.528
Heptatriacontane	NA	$C_{37}H_{76}$	521.01	43	57	71	2.514
Hexacosane	630–01–3	$C_{26}H_{54}$	366.71	43	57	71	1.718
Hexadecane	544–76–3	$C_{16}H_{34}$	226.45	57	226	71	1.337
Hexane	110–54–3	C_6H_{14}	86.18	57	86	43	0.268
Hexatriacontane	630–06–8	$C_{36}H_{74}$	506.98	43	57	71	2.362
Nonacosane	630–03–5	$C_{29}H_{60}$	408.80	43	57	71	1.828
Nonadecane	629–92–5	$C_{19}H_{40}$	268.53	57	268	71	1.465
Nonane	111–84–2	C_9H_{20}	128.26	43	128	57	0.913
Octacosane	630–02–4	$C_{28}H_{58}$	394.77	43	57	71	1.788
Octadecane	593–45–3	$C_{18}H_{38}$	254.50	57	254	43	1.424
Octane	111–65–9	C_8H_{18}	114.23	43	114	85	0.791
Octatriacontane	7194–85–6	$C_{38}H_{78}$	535.04	43	57	71	2.666
Pentacosane	629–99–2	$C_{25}H_{52}$	352.69	43	57	71	1.684
Pentadecane	629–62–9	$C_{15}H_{32}$	212.42	57	212	71	1.291
Pentatriacontane	630–07–9	$C_{35}H_{72}$	492.96	43	57	71	2.243
Tetracontane	4181–95–7	$C_{40}H_{82}$	563.09	43	57	71	3.097
Tetracosane	646–31–1	$C_{24}H_{50}$	338.66	43	57	71	1.649
Tetradecane	629–59–4	$C_{14}H_{30}$	198.39	43	198	57	1.241
Tetratriacontane	14167–59–0	$C_{34}H_{70}$	478.93	43	57	71	9.88
Triacontane	638–68–6	$C_{30}H_{62}$	422.82	43	57	71	1.874
Tricosane	638–67–5	$C_{23}H_{48}$	324.63	43	57	71	1.614
Tridecane	629–50–5	$C_{13}H_{28}$	184.37	43	184	57	1.189
Tritriacontane	630–05–7	$C_{33}H_{68}$	464.90	43	57	71	2.060
Undecane	1120–21–4	$C_{11}H_{24}$	156.31	43	156	57	1.070

Detector: Flame ionization

Detector Temperature: 300 °C

Injector: Splitless

Injector Temperature: 250 °C

Sample Size: 1 µL

Carrier Gas: Helium at 5 mL/min

Temperature Program: 35 °C for 5 min, then 30 °C/min to 320 °C, then hold 10 min.

Calculate the relative retention time of the sample versus *n*-decane by dividing the retention time of the sample by the retention time of *n*-decane. Assayed values must be within 1% of the stated values in the Table XVIII.

ASSAY BY GAS CHROMATOGRAPHY. Analyze the sample by gas chromatography using method GC-SCHC1 described above and method GC-SCHC2 described below.

Method GC-SCHC2

Column: 100% Dimethylpolysiloxane, 25 M × 0.25 mm i.d., 0.25 µm film thickness

Detector: Flame ionization

Detector Temperature: 300 °C

Injector: Split/splitless (100:1 split)

Injector Temperature: 250 °C

Sample Size: 1 µL of a 1000 µg/mL solution

Carrier Gas: Helium at 5 mL/min

Temperature Program: 35 °C for 5 min, then 30 °C/min to 320 °C, then hold 10 min

Measure the area under all peaks (excluding the solvent peak) and calculate the analyte content in area percent for each analysis.

Index

Index